Jiří Komárek · Konstantinos Anagnostidis

# Cyanoprokaryota · 1. Teil / 1st Part: Chroococcales

# Süßwasserflora von Mitteleuropa

Begründet von/Founded by A. Pascher

Herausgegeben von/Edited by
H. Ettl · G. Gärtner · H. Heynig ·
D. Mollenhauer

Band/Volume 19/1:
Jiří Komárek
Konstantinos Anagnostidis
Cyanoprokaryota
1. Teil/1st Part: Chroococcales

**Süßwasserflora von Mitteleuropa**

| | |
|---|---|
| Bd./Vol. 1/1 | Chrysophyte and Haptophyte Algae, 2nd ed., Part 1 (Chrysophyceae s. str., Dictyochophyceae, Phaeothamniophyceae and Haptophyceae) |
| Bd./Vol. 1/2 | Chrysophyte and Haptophyte Algae, 2nd ed., Part 2 (Synurophyceae) (2007) |
| Bd./Vol. 2/1 | Bacillariophyceae (Naviculaceae) (1999) |
| Bd./Vol. 2/2 | Bacillariophyceae (Bacillariaceae, Epithemiaceae, Surirellaceae) (1997) |
| Bd./Vol. 2/3 | Bacillariophyceae (Centrales; Fragilariaceae, Eunotiaceae) (2004) |
| Bd./Vol. 2/4 | Bacillariophyceae (Achnanthaceae, Literaturverzeichnis) (2004) |
| Bd./Vol. 2/5 | Bacillariophyceae (English and French translation of the keys) (2000) |
| Bd./Vol. 3 | Xanthophyceae 1. Teil (1978) |
| Bd./Vol. 4 | Xanthophyceae 2. Teil (1980) |
| Bd./Vol. 5 | Cryptophyceae und Raphidophyeae |
| Bd./Vol. 6 | Dinophyceae (Dinoflagellida) (1990) |
| Bd./Vol. 7 | Phaeophyceae und Rhodophyceae |
| Bd./Vol. 8 | Euglenophyceae |
| Bd./Vol. 9 | Chlorophyta I (Phytomonadina) (1983) |
| Bd./Vol. 10 | Chlorophyta II (Tetrasporales, Chlorococcales, Gloeodendrales) (1988) |
| Bd./Vol. 11 | Chlorophyta III (Chlorophyceae p. p.: Chlorellales, Protosiphonales) |
| Bd./Vol. 12 | Chlorophyta IV (Chlorophyceae p. p.: Stichococcales, Microsporales; Codiolophyceae: Ulotrichales, Monostromatales) |
| Bd./Vol. 13 | Chlorophyta V (Chlorophyceae p. p.: Chaetophorales, Trentepohliales, Chlorosphaerales) |
| Bd./Vol. 14 | Chlorophyta VI (Oedogoniophyceae: Oedogoniales) (1985) |
| Bd./Vol. 15 | Chlorophyta VII (Bryopsidophyceae: Cladophorales, Sphaeropleales) |
| Bd./Vol. 16 | Chlorophyta VIII (Conjugatophyceae I: Zygnemales) (1984) |
| Bd./Vol. 17 | Chlorophyta IX (Conjugatophyceae II: Zygnematales: Mesotaeniaceae; Desmidiales) |
| Bd./Vol. 18 | Charales (Charophyceae) (1997) |
| Bd./Vol. 19/1 | Cyanoprokaryota I (Chroococcales) (1998) |
| Bd./Vol. 19/2 | Cyanoprokaryota II (Oscillatoriales) (2005) |
| Bd./Vol. 19/3 | Cyanoprokaryota III (Nostocales, Stigonematales) |
| Bd./Vol. 20 | Schizomycetes (1982 vergriffen/out of print) |
| Bd./Vol. 21/1 | Fungi (Lichens) |
| Bd./Vol. 21/2 | Fungi |
| Bd./Vol. 22 | Bryophyta |
| Bd./Vol. 23 | Pterido- und Anthophyta 1. Teil: Lycopodiaceae bis Orchidaceae (1980) |
| Bd./Vol. 24 | Pterido- und Anthophyta 2. Teil: Saururaceae bis Asteraceae (1981) |

# Cyanoprokaryota
# 1. Teil / 1st Part: Chroococcales

Jiří Komárek
Konstantinos Anagnostidis

643 Abbildungen / figures

Spektrum
AKADEMISCHER VERLAG

**Authors' adresses**
Professor Dr. Jiří Komárek
Institute of Botany CAS, Department of Algology, Dukelská 145,
37982 Třeboň, Czech Republik or
University of South Bohemia, Faculty of Biological Sciences, Branišovská 31,
37005 České Budějovice, Czech Republik
Prof. Dr. rer. nat. Konstantinos Anagnostidis †
formerly University of Athens, Institute of Systematic Botany, Section Ecology and Systematics, Panepistimiopolis, GR-157784 Athens, Greece



**Bibliographic information published by die Deutsche Bibliothek**
Die Deutsche Bibliothek lists this publication in the Deutsche Nationalbibliografie;
Detailed bibliographic data is available in the Internet at http://dnb.d-nb.de.


First published 1998, unaltered reprints 1999, 2000,
Spektrum Akademischer Verlag Heidelberg
unaltered reprint 2008

Spektrum Akademischer Verlag is an imprint of Springer

Springer is a part of Springer Science+Business Media

08 09 10 11 5 4 3 2



Editor: Dr. Ulrich G. Moltmann
Book production: Detlef Mädje
Cover Design: SpieszDesign, Neu-Ulm
Composed by: Typomedia Satztechnik GmbH, Ostfildern
Printed: Books on Demand GmbH, Norderstedt
Graphics and Photographs: by the authors if not stated otherwise in the figure captions

Printed in Germany

ISBN 978–3–7482–2111–1

*«Am Anfang war Dunkelheit und nichts, mit dem Licht kamen die Blaualgen . . .»*
Hanuš Ettl

# Introduction

usually overlooked by users of identification books. Please, **read this chapter!** It is not too long, and we believe, that comprises important information for you.
Cyanoprokaryotes (Cyanophyceae, Cyanobacteria) are probably the oldest organisms with oxyphototrophic (plant) metabolism. However, their cell structure and type of asexual reproduction prove their eubacterial origin. In spite of their long existence, from Precambrian onwards, they have not lost their vitality, and still are able to colonize all possible biotopes, including very extreme environments (thermal springs up to 80 °C, rocks in extremely hot or cold deserts in both the tropics and Antarctica, hypersaline lakes and marshes, biotopes in volcanic regions, and other harsh habitats). Their viability and diversity is probably connected with their adaptive capabilities. Their old origin, variation, stability of distinguishable types, diversification abilities, adaptation to various ecological situations, and structural and metabolic specificities, make Cyanoprokaryotes one of the most fascinating groups of organisms.
This book is a review of all the cyanoprokaryotic taxonomic units, registered and recognizable from European natural biotopes, including the marine coasts. The limitation of the flora purely to Central Europe is impractical so far as cyanoprokaryotes are concerned. The non-European species and species with vague taxonomic descriptions are listed under the appropriate genera.
Our determination keys and taxonomic descriptions should permit the identification of natural populations, and should be related to known culture material. Numerous strains were isolated, kept in strain collections and used as important model organisms for experiments of various kinds. Unfortunately, the majority of these strains are designated by arbitrary and incorrect names, which subsequently have appeared in important experimental studies (*«Anacystis nidulans»*, *«Agmenellum quadruplicatum»*, *«Anabaena variabilis»*, *«Fremyella diplosiphon»*, etc.). Only a few cyanoprokaryotes grow in culture in the typical form known from nature, the others usually need special procedures to induce their typical forms and life cycles. The identification of strains without previous knowledge of natural material needs to be studied with a special care.
The taxonomic name cannot be used arbitrarily. It is not simply a conventional formality, but the symbol, which represents a defined set of features and structural, biochemical, morphological and ecophysiological characters (stabilized in a genotype). If a deviation from features, characteristic for any higher taxon (and symbolized by a taxonomic name) is recognized, the name cannot be used. If the incorrect name is accepted, all the information about any given population/strain is *a priori* wrong.
This book was written for identification and standardization of known, morphologically and ecologically distinguishable European cyanoprokaryotes that have been described from natural populations. In the introductory part only data important for identification practice are therefore included. There is extensive literature on cyanoprokaryotes: the basic cytomorphology, ecophysiological properties, specificities and reproduction types have been well described and are so well known that we believe it is not necessary to describe them again. At the

end of this chapter we append a list of this basic literature, which comprises all the necessary detailed information.
We have selected accurate drawings of described species. Their correctness was the primary criterion for selection, but sometimes, we have included drawings of a given species by several authors, because the variation of the same specific taxa is better understood if illustrated by various drawing techniques. The figures of the majority of European species can be found in the literature. We also present illustrations of some non-European species when such figures are needed in order to understand the generic diacritical features or infrageneric diversity.

## Summarized important information and instructions for users of this book

- This flora contains a review of European species only, based mainly on knowledge of natural populations. Many of them occur elsewhere in the world (see «Occurrence»), however, the keys should be applied to specimens from outside Europe with great care. A few species only have a cosmopolitan distribution (and under restricted ecological conditions). Species outside Europe are mentioned in additional lists.
- Do not forget that the name of any taxon (species, genus, etc.) is not an arbitrary concept, but symbolizes and represents a set of distinct and defined cytomorphological, ultrastructural, biological, ecological and biochemical characters. If one of these characters is different from features described as diacritical for the superior taxon, the name can not be applied (monothetic principle of taxa).
- Identification of any species is correct only if your material corresponds fully (in all characteristics including ecology) with the description. The variability of species can be broader than described in this book, but you must be convinced of the taxonomic identity of your material.
- Do not try to designate all species you have found with a name from this book, or to identify every specimen in your samples. If the identity with a described taxon is not certain or if doubts arise, the designation «*in suspenso*» with «*cf.*» (*confertim* = close together), «*aff.*» (*affinis* = neighbouring, similar, allied to), «*sp.*» (*species*), or with «?» clearly is more correct than to use an incorrect name. This is important particularly for non-European samples (from tropical and circumpolar regions, where the diversity of biotopes and species is substantially wider than in the temperate zones).
- Many species occur in nature in different morphological stages (status), which are influenced by environmental conditions and/or seasonally dependent. These status can sometimes deviate morphologically from the typical generic and specific characters and assume the appearance of quite another taxon at the generic and specific levels. It does not mean that they are taxonomically identical.
- Species with complex life and/or vegetative cycles occur and the various stages can be distinctly different from one another. Some stages of the life cycles can correspond to generic features of another genus. It does not mean that they are taxonomically identical. Knowledge of the entire life cycle of the studied taxa is therefore sometimes needed.
- Large populations sometimes develop with distinctive and unique morphological features, though their real potential variation range can be much broader. On the other hand, populations developing under suboptimal conditions usually possess wide plasticity, forming numerous morphological deviations of no taxonomic significance. Avoid describing such morphological deviations as «new» taxonomic units (species, varieties, forms).

- Morphological and physiological variation (plasticity) can differ from one taxon to another. The limiting and diacritical features can have diverse values within different genera.
- Ecological data are an integral part of species characterisation; if you find similar morphological types with different ecologies (living in different environments), they probably belong to different genotypes. It is generally recognized that all species studied in detail from the viewpoint of ecological demands are only typical for a restricted range of habitats.
- Cytological and biochemical properties must coincide with morphological ones; they should be used as much as possible to characterize species and/or populations.
- In floristic papers, notes about complete correspondence of described taxa, or morphological deviations are necessary. Deviations in population features should be mentioned and depicted. The documentation of all described taxa is desirable (at least by micrographs).
- In experimental and ecological papers, it is necessary to document your model strain(s) or dominant species, with short morphological and biological descriptions and with good microphotos, illustrating your material and demonstrating the correctness of the identification(s).
- Many cyanoprokaryotes in culture produce atypical or anomalous stages, which are sometimes different from their characteristic morphology in nature. Therefore, the potential variation range under various conditions has to be studied. Material from the original locality should be compared to assure correctness of identification.
- The investigation of cyanoprokaryotes is still far from satisfactory. The review compiled in this book is a manual for the identification of all known species and should be used as a basis for further research.
- Good luck in research on Cyanoprokaryota!

## Basic literature

The following list comprises important monographs, regional floras and books, in which the morphology, ecophysiology, biochemistry, taxonomic reviews and previous identification keys of cyanoprokaryotes (Cyanophytes, Cyanobacteria) are summarized. The list is organized chronologically.

Bornet E. & Flahault C: Revision des Nostocacées hétérocystées. – Ann. Sci. Nat. 7 – Bot., Paris, **3**: 328–381, **4**: 343–373, **5**: 51–129, **7**: 171–262. [Latin, French commentary] (1886–1888).

Gomont M.: Monographie des Oscillariées (Nostocacées homocystées). – Ann. Sci. Nat. 7 – Bot., Paris, **15**:263–368, **16**: 91–264. [Latin, French commentary] (1892).

Tilden J.: Minnesota algae I. The Myxophyceae of North America and adjacent regions. – Minneapolis, 328 pp. [English] (1910).

Geitler L.: Cyanophyceae. – In: Süsswasserfl. Deutschlands, Österreichs u. d. Schweiz 12, Gustav Fischer-Verl., Jena, 481 pp. [German] (1925).

Frémy P.: Cyanophycées des côtes d'Europe. – Mém. Soc. Nat. Sci. Natur. Math. Cherbourg, Saint-Lô, 232 pp. + 66 tables. [French] (1929–1933).

Frémy P.: Les Myxophycées de l'Afrique équatoriale française. – Arch. Bot. **2**: 1–508. [French] (1930).

Geitler L.: Cyanophyceae. – In: Rabenhorst's Kryptog.-Fl. 14, Akad. Verlagsges., Leipzig, 1196 pp. [German] (1932).

Elenkin A.A.: Monographia algarum cyanophycearum aquidulcium et terrest-

rium in finibus URSS inventarum. [Sinezelenye vodorosli SSSR.] – Izd. AN SSSR, Moskva-Leningrad, **1**,**2**(1–2): 1–1908. [Russian, Latin diagnoses and descriptions] (1936, 1938,1949).
Huber-Pestalozzi G.: Das Phytoplankton des Süsswassers. Systematik und Biologie. I. – In: Die Binnengewässer, Stuttgart, **16**: 1–342. [German] (1938).
Geitler L.: Schizophyta (Klasse Schizophyceae). – In: Engler A. & Prantl K. ed., Natürl. Pflanzenfamilien **1b**, Duncker & Humblot, Berlin, 232 pp. [German] (1942).
Fritsch F.E.: The structure and reproduction of algae, II. – The Univ. Press, Cambridge, 939 pp. [English] (1945).
Kosinskaja E.K.: Opredelitel' morskich sinezelenych vodoroslej. [Identification key for marine blue-green algae.] – Izd. AN SSSR, Moskva-Leningrad, 272 pp. [Russian] (1948).
Smith G.M.: Division Cyanophyta. – In: The fresh-water algae of the United States, 2nd ed., McGraw-Hill Book Comp. Inc., New York – Toronto – London, p. 539–604. [English] (1950).
Hollerbach M.M., Kosinskaja E.K. & Poljanskij V.I.: Sinezelenye vodorosli. [Blue-green algae.] – In: Opred. presnovodn. vodorosl. SSSR 2, Izd. Sov. nauka, Moskva, 652 pp. [Russian] (1953).
Drouet F. & Daily W.A.: Revision of the coccoid Myxophyceae. – Butler Univ. Bot. Studies 12, Ann. Arbor-Michigan, 222 pp. [English, Latin diagnoses] (1956).
Komárek J.: Revision der planktischen Cyanophyceen der Tschechoslowakei. – In: Ettl H. & Komárek J., Algologische Studien, Academia, Praha, p. 10–206. [German] (1958).
Desikachary T.V.: Cyanophyta. – I.C.A.R. Monographs on Algae, New Delhi, 686 pp. [English] (1959).
Geitler L.: Schizophyceen. – Gebr. Borntraeger, Berlin, 131 pp. [German] (1960).
Anagnostidis K.: Untersuchungen über die Cyanophyceen einiger Thermen in Griechenland. – Thessaloniki, 322 pp. + 38 Tafeln. [Greek with German summary] (1961).
Starmach K.: Cyanophyta – sinice, Glaucophyta – glaukofity. – In: Flora slodkowodna Polski **2**, PAN, Warszawa, 807 pp. [Polish] (1966).
Drouet F.: Revision of the classification of the Oscillatoriaceae. – Acad. Nat. Sci. Philadelphia – Monogr. 15., Philadelphia, 370 pp. [English] (1968).
Kondrateva N.V.: Sin'o-zeleni vodorosti – cyanophyta. [Blue-green algae – Cyanophyta.] – In: Vizn. prisnov. vodorost. URSR 1/2, AN URSR, Kiev, 524 pp. [Ukrainian](1968).
Bourrelly P.: Les algues d'eau douce III. – N. Boubée & Cie., Paris, 512 pp. [French] (1970).
Desikachary T.V.: Taxonomy and biology of blue-green algae. – Univ. of Madras, 591 pp. [English] (1972).
Carr N.G. & Whitton B.A. ed.: The biology of blue-green algae. – Blackwell Sci. Publ., Oxford – London – Edinbourgh – Melbourne, 676 pp. [English] (1973).
Drouet F.: Revision of the Nostocaceae with cylindrical trichomes. – Hafner Press, New York – London, 292 pp. [English] (1973).
Fogg G.E., Stewart W.D.P., Fay P. & Walsby A.E.: The blue-green algae. – Academic Press, London – New York, 459 pp. [English] (1973).
Wolk C.P.: Physiological and cytological chemistry of blue-green algae. – Bacteriol. Rev. **37**(1): 32–101. [English] (1973).
Kondrateva N.V.: Morfogenez i osnovnye puti evoljucii gormogonievych vodoroslej. [Morphogenesis and the main evolutionary trends in hormogonal algae.] – Izd. Naukova dumka, Kiev, 302 pp. [Russian] (1975).

Drouet F.: Revision of the Nostocaceae with constricted trichomes. – In: Beih. Nova Hedwigia 57, J. Cramer, Vaduz 258 pp. [English] (1977).
Waterbury J. & Stanier R.: Pattern of growth and development in pleurocapsalean Cyanobacteria. – Microbiol. Rev. **42**: 2–44. [English] (1978).
Rippka R., Deruelles J., Waterbury J.B., Herdman M. & Stanier R.Y.: Generic assignments strain histories and properties of pure cultures of cyanobacteria. – J. Gener. Microbiol. **111**: 1–61. [English] (1978).
Humm H.J. & Wicks S.R.: Introduction and guide to the marine bluegreen algae. – A. Wiley & Sons, New York – Chichester – Brisbane – Toronto, 194 pp. [English] (1980).
Kondrateva N.V.: Vnutrividovaja morfologičeskaja izmenčivost' sinezelenych vodoroslej (Cyanophyta). [Infraspecific morphological variation in blue-green algae (Cyanophyta).] – Izd. Naukova dumka, Kiev, 284 pp. [Russian] (1980).
Hirose H. & Hirano M.: Class Cyanophyceae. – In: Hirose H. & Yamagishi T. ed., Illustrations of the Japanese fresh-water algae, 2nd ed., Uchidarokakuho Publ. Co., Tokyo, 1–151 pp. [Japanese] (1981).
Carr N.G. & Whitton B.A. ed.: The Biology of Cyanobacteria. – Bot. monographs 19, Blackwell Sci. Publ., Oxford – London – Edinburgh – Boston – Melbourne, 686 pp. [English] (1982).
Fay P.: The Blue-greens (Cyanophyta – Cyanobacteria). – Studies in Biology no. 160, E. Arnold, London, 88 pp. [English] (1983).
Kondrateva N.V., Kovalenko O.V. & Prichod'kova L.P.: Sin'o-zeleni vodorosti – Cyanophyta. [Blue-green algae – Cyanophyta.] – In: Vizn. prisnov. vodoroslej URSR 1/1, AN URSR, Kiev, 388 pp. [Ukrainian] (1984).
Bourrelly P.: Les algues d'eau douce III. – N. Boubée & Cie., Paris, 606 pp. [French] (1985).
Jensen T.E.: Cell inclusions in the cyanobacteria. – Arch. Hydrobiol./Algol. Studies, Stuttgart, 38/39: 33–73 [English] (1985).
Anagnostidis K. & Komárek J.: Modern approach to the classification system of cyanophytes 1. Introduction, 3. Oscillatoriales, 5. Stigonematales. – Arch. Hydrobiol./Algolog. Studies, Stuttgart, **38/39**: 291–302, **50/53**: 327–472, **59**: 1–73. [English] (1985, 1988, 1990).
Komárek J. & Anagnostidis K.: Modern approach to the classification system of cyanophytes 2. Chroococales, 4. Nostocales. – Arch. Hydrobiol./Algolog. Studies, Stuttgart, **43**: 157–226, **56**: 247–345. [English] (1986, 1989).
Fay P. & Van Baalen C. ed.: The Cyanobacteria. – Elsevier, Amsterdam – New York – Oxford, 543 pp. [English] (1987).
Castenholz R.W. & Waterbury J.B.: Group I. Cyanobacteria. – In: Staley J.T. & al. ed., Bergey's Manual of Systematic Bacteriology, **3**: 1710–1799. [English] (1989).
Chu H. & al. ed.: Flora algarum sinicarum aquae dulcis II. Chroococcophyceae. – Science Press, 161 pp. [Chinese] (1991).
Wilmotte A. & Golubić S.: Morphological and genetic criteria in the taxonomy of Cyanophyta/Cyanobacteria. – Arch. Hydrobiol./Algolog. Studies, Stuttgart, **64**: 1–24. [English] (1991).
Castenholz R.W.: Species usage, concept and evolution in the Cyanobacteria (Blue-green algae). – J. Phycol. **28**: 737–745. [English] (1992).
Wilmotte A.: Oxygenic phototrophic Bacteria. – In: Bergey's Manual of Determinative Bacteriology – **9**, Group 11, p. 377–425. [English] (1993).
Komárek J.: Current trends and species delimitation in the cyanoprokaryote taxonomy. – Arch. Hydrobiol./Algolog. Studies, Stuttgart, **75**: 11–29. [English] (1994).
Bryant D.A.: The molecular biology of cyanobacteria. – Kluwer Acad. Publ., 916 pp. [English] (1994).

Pennisi E.: Static evolution. – Science News, Washington, **145** (11): 168–169. [English] (1994).
Turner S.: Molecular Systematics of Oxygenic Photosynthetic Bacteria. – In: Bhattacharya D. ed., «The origin of the Algae and their Plastids», Springer-Verlag. [English] (1997).

# Characters of Chroococcales (important for identification)

The order *Chroococcales* embraces all the unicellular or colonial cyanoprokaryotes, which do not form true filaments with the direct physiological interference between the cells. However, a polarized thallus or cell differentiation within a colony can occur in more complex types.

## Cell morphology (form, polarity)

Spherical, oval, rod-shaped, irregular, rarely spindle-shaped cells occur in coccal unicellular and colonial cyanoprokaryotes. Their structure is principially uniform and has a similar inner arrangement of organelles (Table **1**). The form of cells is stable in all the genera within certain variational limits (influenced by environmental factors) and in connection with modification of the basic type of division (during life cycle and cell growth). This complex of cytomorphological characters is used as an intergeneric or interfamiliar feature.

Within all simple cyanoprokaryotes there exists a tendency towards polarity. The polarized cells are supposed to be evolutionarily derived from the non-polarized types and so are more advanced. The tendency to polarization of solitary cells, or the arrangement of cells into pseudofilamentous formations and formation of polarized colonies evidently expresses the evolutionary trends in simple cyanoprokaryotes. The heteropolar character of cells has clearly developed into numerous modifications, from simple unicellular types (usually attached to the substrate) to the complex thallus of *Hyellaceae*, in which both the morphological and functional diversification in different parts of thallus are recognizable. The differences between non-polarized and polarized types, or between simple colonial genera and genera with polarized thallus, are considered evolutionarily important and accepted for the definition of families. This criterion coincides well with other superior features such as cell division (reproduction process). E.g., all families with polarized thallus (*Entophysalidaceae, Hydrococcaceae, Chamaesiphonaceae, Hyellaceae*) also have a characteristic type of cell division. A negligible exception is found in the unicellular family *Dermocarpellaceae*, which is characterized by unicellular, sessile cells with a special and characteristic life cycle and reproductive process (cell division obligatorily by multiple fission); the genus *Stanieria* with sessile, but not polarized cells is therefore placed into this family together with genera that have clearly polarized cells (*Cyanocystis, Dermocarpella*). This genus has more or less spherical cells, which adhere to the substrate or to other cells within their aggregates by a random side of the daughter cell (baeocyte).

A certain degree of polarity also occurs in reproductive cells, particularly in cases in which small solitary cells liberated from the divided «mother» cell, or from the colony. In several simple types, no polarity occurs but in several species with polarized (elongate) vegetative and reproductive cells apical polarity is present, congruent with the original longer cell axis of the mother cell (*Stichosiphon*). On the other hand, the changed perpendicular (lateral) polarity of germinating exocytes is known in the genus *Cyanophanon* (Geitler 1957).

Under suboptimal conditions or near the lethal limits of several species appear aberrant, atypical cells (so called involuted cells), which have special characteristics in various species (irregular, filamentous). The formation and function of involution cells is not clear, but the special morphological type of these cells seems to be characteristic both for species and genera, and also for subfamilies (*Aphanothecoideae, Synechococcoideae*).

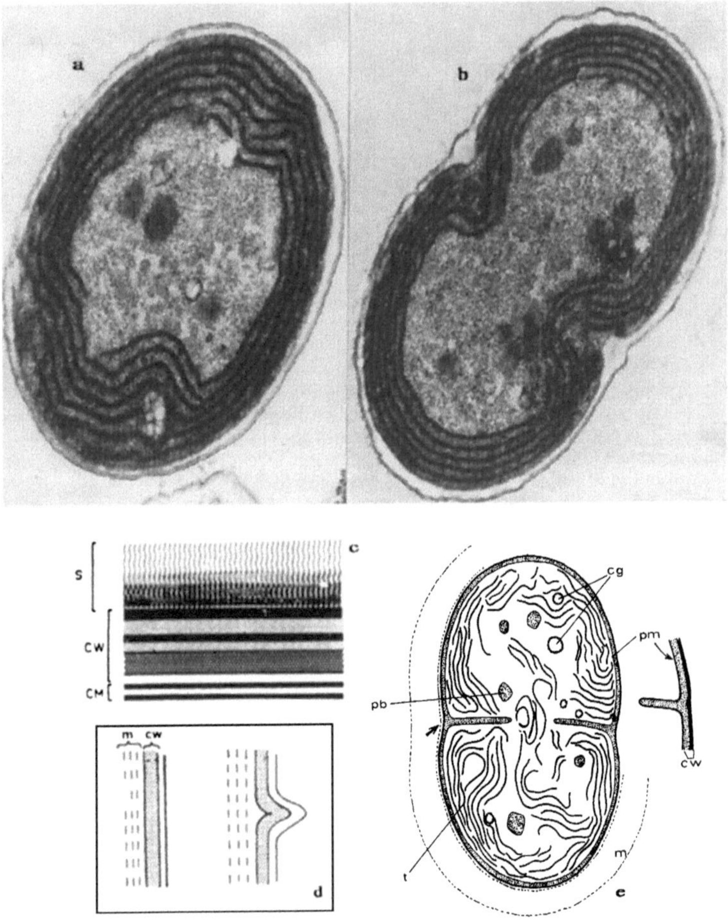


**Table 1.** Cell structure and cell division (binary fission): **a-b** - section through the simple cyanoprokaryotic cell (*Cyanobium diatomicola*) and initial state of perpendicular cell division (orig.), **c** - scheme of the cell wall (CM = cell membrane, CW = layered cell wall, S = mucilaginous envelope) (after Jost 1965), **d–e** - scheme of cell division (after Komárek & Anagnostidis 1986).

# Cell structure, mucilaginous envelopes

The cytology of cyanoprokaryotes is simple, and all cells possess the same basic type of cell structure. However, numerous modifications have been recognized, which are stable and characteristic of various strains, groups of strains and so amenable to taxonomic treatment. The Gram-negative cell wall is uniform in all cyanoprokaryotes, composed of three obligatory layers, and its chemical composition and properties are well known. The structure and composition of cell walls are important characteristics of the eubacterial origin of cyanoprokaryotes. However, sometimes additional layers occur facultatively. The crystalline S-layer, present in certain cyanoprokaryotic genera (and in several eubacteria), is of special interest. The function of this layer is not yet known, but various its types have been recognized. In several species of coccal genera with spherical cells (*Synechocystis, Aphanocapsa, Microcystis*) the structure of this layer is hexagonal, in baeocytic *Chroococcidiopsis* it is quadratic; in a few filamentous strains it is monoclonal. The occurrence of the S-layer (or ability to form it) certainly is connected with the genotypes of several cyanoprokaryote groups and probably should be used in the recognition of relationships between various genera and for classification. However, the necessary data are still insufficient. The review of Šmarda (1991) is the most current summary of all the knowledge of the S-layer.

The outer mucilaginous envelopes around cell walls are more variable, and are useful for taxonomic classification. They are not integral parts of the cell walls, and their form, structure, stratification, stability and precipitates therein are species- or genus-specific. The gelatinous envelopes can vary widely, but their character is usually specific for given taxa. The consistency of the mucilage, its delimitation by a distinct margin, lamellation, width and colouration (ability to produce sheath pigments) are characteristic for various species and genera. Thin, firm sheaths similar to those of filamentous types (pseudovaginae) often appear in polarized and morphologically more complicated families of coccal cyanoprokaryotes. In the simplest unicellular and sessile *Chamaesiphonaceae*, there are small adherent pads (gelatinous clusters) at the cell bases, which in other types develop into a cup-shaped pseudovagina. In the more complex genera *Chamaesiphon* and *Chamaecalyx* these pseudovaginae envelop the whole cells and the sheath apex opens only after cell division and exocyte or baeocyte formation). Thin and firm sheaths (considered as cell walls in the past) are also characteristic for the families *Dermocarpellaceae, Xenococcaceae* and *Hyellaceae.*

Thylakoid arrangement is the most important feature of the inner cell structures which is usable for taxonomic classification. In more complex and advanced cyanoprokaryotic orders and families there is a uniformity in this feature. The analogous pattern of thylakoid arrangements was found, e.g., in nostocalean and stigonematalean species (Komárek & Anagnostidis 1989, Anagnostidis & Komárek 1990). The type of thylakoid position in *Oscillatoriales* is characteristic of various families (Komárek & Cáslavská 1991). By contrast, in coccal types there is great diversity in this feature. In simple types differences between various genera occur which coincide with other characteristics (*Cyanobium, Cyanothece*, etc.). The number of thylakoids and the type of thylakoid arrangement are characteristic (and different) in various related genera (*Cyanobacterium – Synechococcus, Merismopedia – Microcrocis, Cyanophanon – Stichosiphon*, etc.). Some genera exist in which certain differences between different species (e.g., in *Chamaesiphon*) were found. It is not yet clear, whether these differences indicate taxonomic relationships.

Main pattern in thylakoid arrangements in coccal cyanoprokaryotes (Table **2**):
(**i**) The parietal position of one or several thylakoids, which are parallel to one

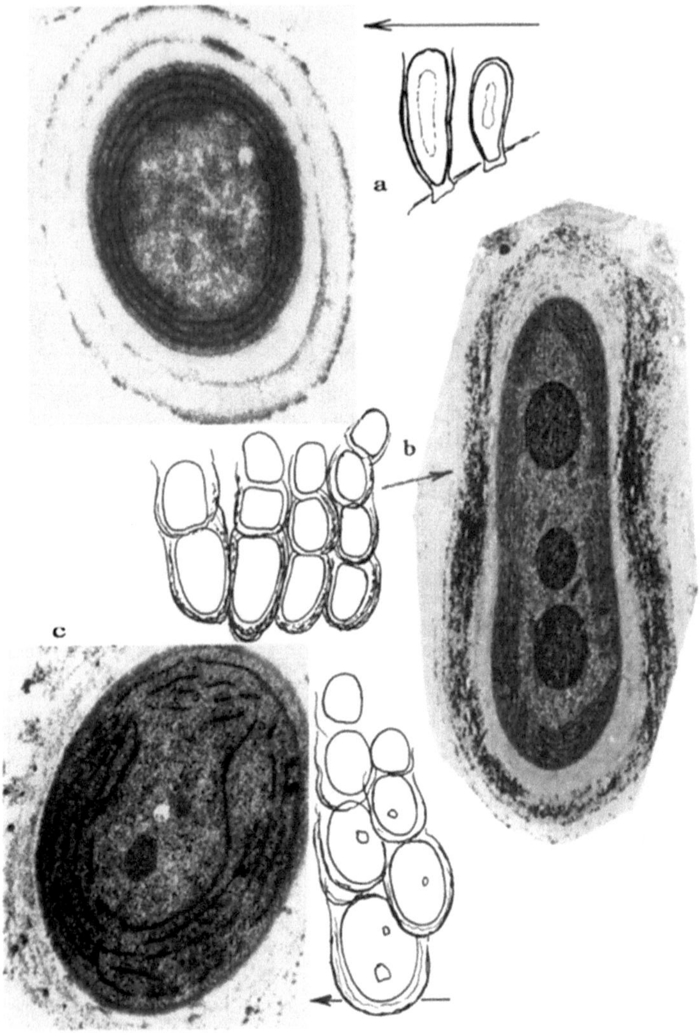


**Table 2.** Arrangement of thylakoids within cells of various chroococcal cyanoprokaryotes: **2A/a** - parietal thylakoids within cells of *Chamaesiphon* sp. (cross section!), **b** - parietally and densely arranged short thylakoids in *Chamaesiphon polonicus*, **c** - parietally, loosely arranged thylakoids in *Chamaesiphon polymorphus*.

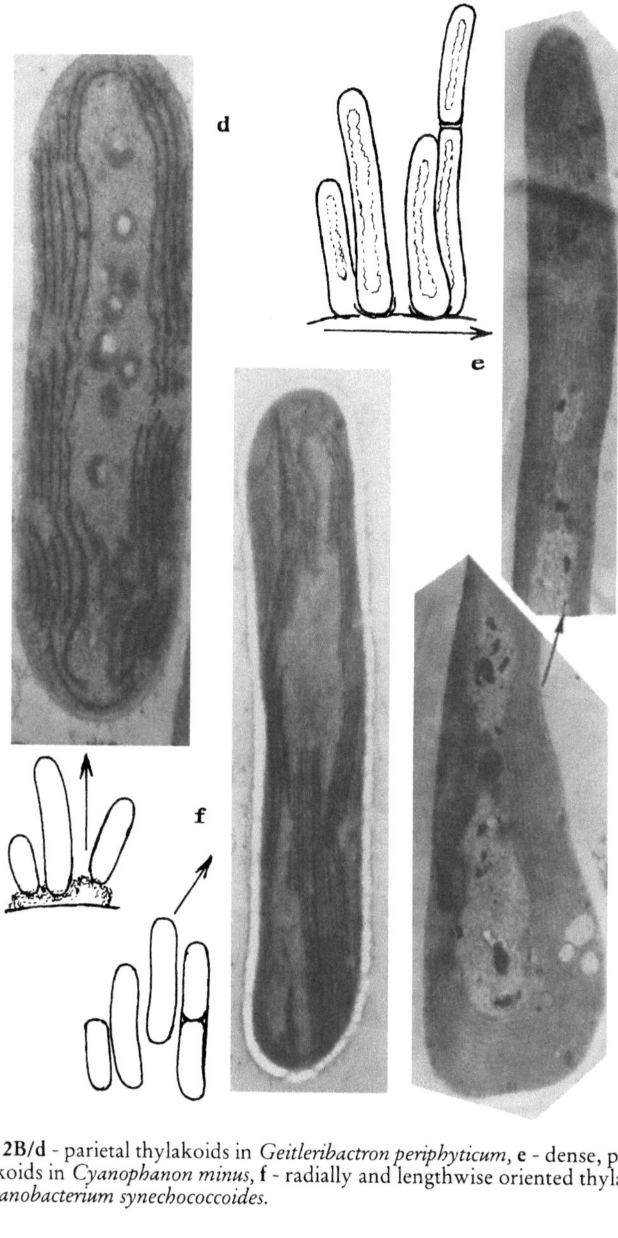


**Table 2B/d** - parietal thylakoids in *Geitleribactron periphyticum*, **e** - dense, parietal thylakoids in *Cyanophanon minus*, **f** - radially and lengthwise oriented thylakoids in *Cyanobacterium synechococcoides*.

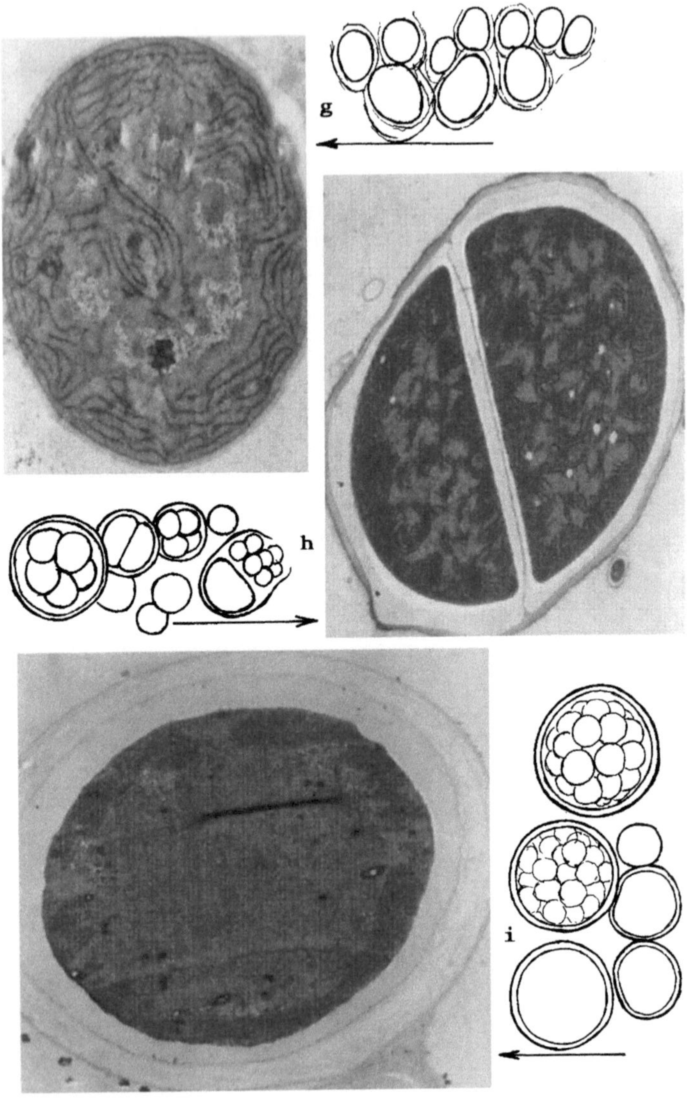


**Table 2C/g** - parietally and loosely arranged thylakoids, part of thylakoids in the central region of the cell, in *Chamaesiphon subglobosus*, **h** - slightly fasciculate thylakoids, regularly distributed over the whole cell volume in *Chroococcidiopsis thermalis*, **i** - fasciculate thylakoids localized near the cell wall in *Stanieria* sp. (marine strain 1416–1).

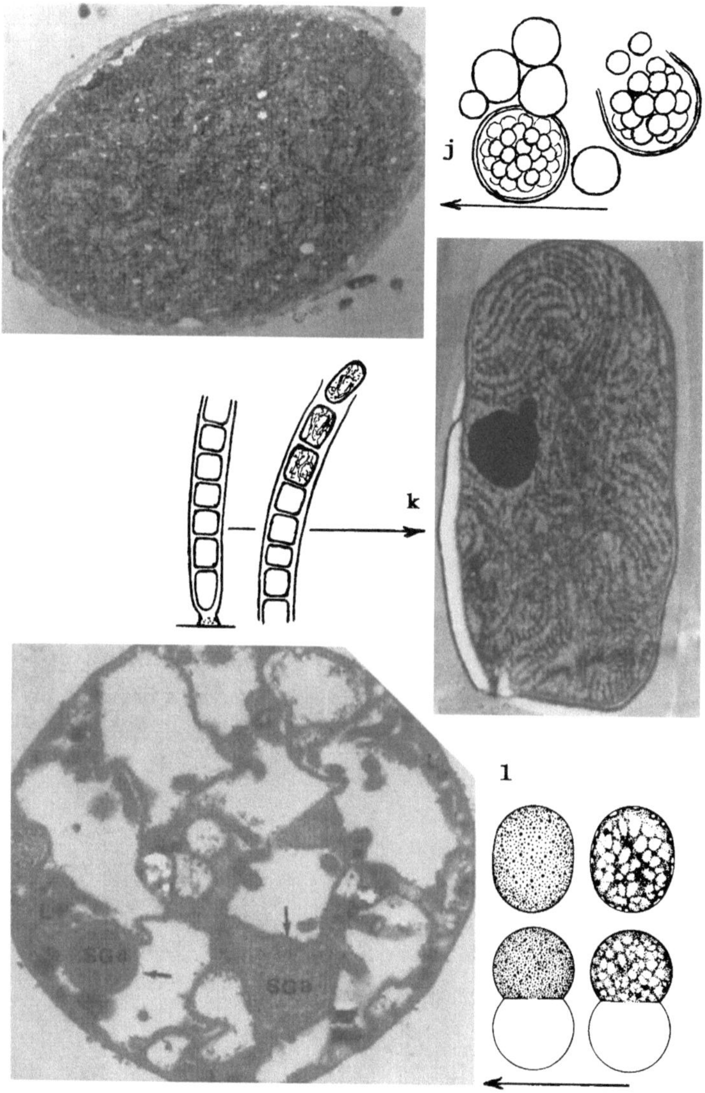


**Table 2D/j** - short and wavy thylakoids regularly distributed over the whole cell volume in *Stanieria cyanosphaera*, **k** - coiled thylakoids distributed over the whole cell volume in *Stichosiphon willei*, **l** - thylakoids with distinct intrathylakoidal spaces over the whole cell content in *Cyanothece halobia;* (2D/l after Roussomoustakaki & Anagnostidis 1991, other orig.).

another along the cell wall in long sections of the cell; in cross section they are arranged concentrically or triangularly. This position is usually recognizable as the chromatoplasma by light microscopy, i.e., the more intensely coloured peripheral part of the protoplast;

(**ii**) The similar peripheral arrangement with a few other thylakoids localized in the central part of cells;

(**iii**) More or less radial arrangement of thylakoids; the free fascicles of thylakoids are sometimes recognizable as a slight net-like keritomy in the cell protoplast (fine, lengthwise striation of the cell content);

(**iv**) small fascicles of thylakoids which are mainly situated on the periphery of a cell;

(**v**) thylakoids are more or less regularly distributed throughout the whole cell, in various densities; the same situation, but with short, wavy, and very dense, irregularly intricate thylakoids throughout the whole cell volume.

Widening of thylakoids that form intra-thylakoid spaces is a special phenomenon. Under a light microscope, they are similar to «keritomy» or indistinct «vacuolization». Keritomy of this type is not wholly dependent on ecological factors, but the radiation intensity plays a role in their development. Several species with this phenomenon occur in marine or hypersaline biotopes. However, typical intra-thylakoid spaces are found also in other ecological types, e.g., from oligotrophic or dystrophic waters. Formation of intrathylakoid spaces is probably irreversible, and its function is unknown. Recent results indicate that inside such thylakoids structures appear, the function and nature of which are still unclear. They occur regularly in several species and are particularly common in certain genera (*Cyanothece, Chroococcidiopsis,* and others) (Komárek & al. 1975, Roussomoustakaki & Anagnostidis 1991, Komárek & Albertano 1994, Šmarda & al. 1996). Their formation evidently represents an important biological, ecological and therefore taxonomic feature.

All characteristics connected with thylakoid arrangements are dependent to a certain degree on the environmental factors. Changes have also been noted during the life cycle and during the vegetation season. However, the principal pattern (position and number) of the thylakoid system is species specific within certain limits and can be used for taxonomic classification.

Cell inclusions are important ultrastructural features. The usual inclusions in centroplasma, such as polyphosphate granules, carboxysomes and cyanophycin granules have a more or less similar position within the cells of all cyanoprokaryotes. Their occurrence (quantity) is dependent upon environmental factors. It is not yet clear, whether their position in some species is stable and characteristic. Numerous other special inclusions were noted by other authors in cyanobacteria (Jensen 1985). However, the data available are still unsufficient to be applied in taxonomy.

The presence of gas vesicles and aerotopes (groups of gas vesicles aggregated parallely into clusters recognizable in a light microscope) is considered to be a valuable taxonomic characteristics. Their nature and function are well known and they represent an interesting ecological and cytological phenomena. They occur mainly in planktic types and represent an important taxonomic feature in spite of their reversibility and formation depending on environmental factors. The ability to form gas vesicles evidently is genotypically fixed and, together with their characteristic shape, size and position in cells, have been used as an interspecific (in the genera *Woronichinia, Chroococcus,* etc.) andintergeneric (*Microcystis*) character. Species exist, in which gas vesicles occur obligatorily during the whole vegetation season (with the exception of resting stages; e.g. in the g. *Microcystis*). By contrast, the facultative appearance of aerotopes during the vegetation period, under special environmental conditions, or in special phases of a life cycle, is also common (*Snowella*). Special localization of gas vesicles and aerotopes in cells (in the family *Pseudanabaenaceae*) is characteristic of several filamentous genera. Stable modifications in the size and structure of gas vesicles were also recognized in various ecologically diverse or ecologically identical cyanoprokaryote species (Walsby 1972, 1978, 1981, Walsby & al. 1983, Šmarda 1996, Walsby 1997), but the available data are still insufficient for evaluating this diversity for taxonomic purposes. In various filamentous cyanoprokaryotes aerotopes are only known in special reproductive stages (e.g., in hormogonia of several nostocalean species).

## Cell division, reproduction

In principle, cell division is similar for all cyanoprokaryotes. The plasmatic membrane followed by cell wall layers ingrow (after the genome replication) centripetally into the protoplast and divide the cell into two parts (binary fission). However, this process occurs in numerous stable modifications, which are used as the primary criterion for classification at level of family (Table **3**):

- Simple binary fission, dividing the mother cell into two equal parts. The daughter cells grow to approximately the original size and shape before the next division. This process is combined with a very important characteristics, which is probably very stable and is used for the delimitation of the three simplest families of the coccal cyanoprokaryotes: (i) The division plane is oriented in all succeeding generations at right angle to one and the same cell axis (this orientation is easily recognized, particularly in elongate cells; in this case the division plane is oriented perpendicularly to the longer axis of a cell; family *Synechococcaceae* subf. *Aphanothecoideae*); (ii) the division plane changes regularly in two perpendicular directions in succeeding generations (*Merismopediaceae*); (iii) the division plane changes regularly in three perpendicular directions (usually in types with spherical cells; in several genera cubic shaped colonies arise as a result of this type of division; *Microcystaceae*). The mechanisms leading to the manifestation of one or other of the above mentioned modifications are so far unclear; (Table **3/I**).
- Asymmetrical binary fission is a modification of the previous case (i). It differs from the described type by the occasional shifted plane of division, away from the middle towards one, arbitrary end of a cell. This division is typical for populations that have grown under suboptimal conditions (*Synechococcaceae*, subf. *Synechococcoideae*). A very similar type of cell division exists in simple *Chamaesiphonaceae* with polarized cells (attached by one end to the substrate). The division plane is shifted usually towards the upper (free) cell end in these species; the apical daughter cells (exocytes) separate from the basal parts and exhibit slight morphological differences from the «smaller» daughter cells of asymmetrically divided synechococcoid species.
- Irregular binary fission is characterized by a similar process as described in the previous categories, but the dividing planes are not regularly oriented to any cell axis (the cells divide along «various» planes in succeeding generations). The cells grow to approximately the original size, but not always into the original shape before the next division. Families characterized by this type of cell division usually contain genera with colonies that have irregularly organized cells of variable shape. In such colonies the dominant cell division sometimes occurs in one direction, which results in cells arranged in rows, but the ability to divide along various planes is always detectible.
- Exocyte division is the asymmetric binary fission of heteropolar, sessile cells, after which the smaller (apical) part separates from the free end of the attached cells (Table **3/II**). The exocytes can separate individually, or in rows (after division in rapid sequence), or three-dimensionally: the first division is perpendicular to

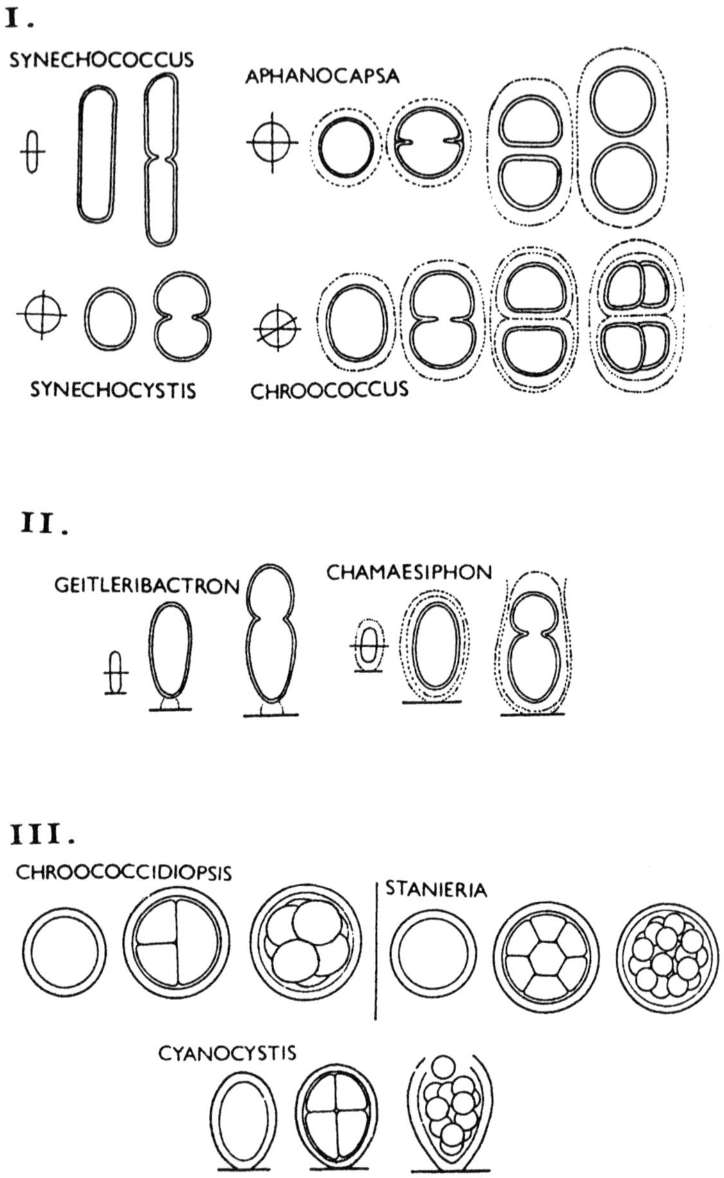


**Table 3.** Types of cell division in various types of chroococcal cyanoprokaryotes. (After Komárek & Anagnostidis 1986.)

the cell axis (horizontal to the substrate), and the upper part divides simultaneously or in a rapid sequence into several exocytes. In several genera the differentiated exocytes have the ability to divide again. The last cases of more or less simultaneous reproduction of exocytes is evidently dependent on the repeated genome replication before the division process. Exocytes separate from the mother cell and attach to a substrate, or remain sessile at the top of mother cells or at the margin of their pseudovaginae. Some exocytes possess a certain degree of polarity, which is congruent with the polarity of mother cells (*Stichosiphon*), or it is different from the mother cells (e.i., it is perpendicular to the original cell axis; *Cyanophanon*). All the above mentioned modifications are stable and so characteristic that they can be used as intergeneric diacritical features.

– Nanocyte and baeocyte division (multiple fission) is characterized by the simultaneous (*Stanieria*) or successive (in rapid sequence, in other *Dermocarpellaceae*) division of a mother cell into numerous small daughter cells (Table 3/III). Multiple fission is obligatory in the family *Dermocarpellaceae*, and occurs facultatively in genera of *Xenococcaceae* and *Hyellaceae* (combined with binary fission). Nanocyte (within mucous envelopes) and baeocyte (within firm sheaths) division is evidently the result of the cell cycle, in which the DNA replicates repeatedly before cell division. Stable and distinct modifications of this process can be used as intergeneric criteria within *Dermocarpellaceae*, but in genera with a more highly differentiated thallus (*Xenococcaceae, Hyellaceae*) transitions exist between various types of cell division, and different combinations of division types can be recognized only at the specific level.

Formation of baeocytes and/or nanocytes was interpreted in the past as endogenous differentiation of daughter cells, which release from the mother cell wall after its rupture or gelatinization (as in autosporine eucaryotic reproduction). With electron microscopy we can see that this cell division does not differ substantially from normal binary fission and the only difference is the sequence of divisions.

Reproduction in coccal cyanoprokaryotes is by solitary cells, or by gelatinous clusters containing several cells (parts of a gelatinous thallus), which are liberated from colonies or separate from mother cells. The liberated cells can be nanocytes, baeocytes, exocytes, young cells after binary fission or vegetative cells which have separated from mother colonies (*Siphononema*). Sometimes, the whole colony disintegrates into solitary cells, e.g., the genus *Woronichinia* cells are liberated from the colony by a special mode at the end of the vegetation cycle («expulsing» cells). Reproduction type (or combination of various reproductive procedures) is characteristic of different families; modifications of the various types usually coincide with other important features and can be used for delimitation of certain genera.

Solitary reproductive cells, separating from colonies, or exocytes, baeocytes and nanocytes have been described as immotile or motile (or facultatively motile). Motility of reproductive cells (immotile monocytes and nanocytes – motile planocytes) is an important characteristic at the intergeneric level (*Chroococcidiopsis – Myxosarcina*). This approach should be accepted (it is an important biological character), but it is sometimes difficult to detect motility both in the living and cultured material, because motility is facultative. In numerous genera and species, motility of reproductive cells is unknown. The wider application of this character in the taxonomic classification is therefore as yet impossible.

## Colonies (thallus structure)

A few genera (*Cyanobium, Cyanothece, Synechococcus, Synechocystis*) live as solitary cells, around which a thin layer of very fine, diffluent mucilage can develop.

In all other cases the cells aggregate into various kinds of colonies. The structure of the mucilage and resulting colony form are diagnostically characteristic of the genera. The lamellation or stratification of slime arise from the periodic excretion of mucilage (usually from cells of successive generations). The structure, consistency and delimitation of the margin of gelatinous envelopes (diffluent or clearly delimited) is used as an intergeneric feature as is the ability to form various kinds of gelatinous strands or stalks.
The colony form results from the characteristic slime production combined with the type of cell division and cell polarity. Colonies can be spherical, irregularly spherical, or form irregular mucilaginous clusters, microscopic or macroscopic, free-floating or attached to a substrate. All such characters and morphological types of colony are stable and usually have intergeneric (or suprageneric) values within certain variation limits.
The internal structure of colonies of more complicated families (*Entophysalidaceae, Chamaesiphonaceae, Xenococcaceae, Hyellaceae*) is interesting. Both polarity of cells and of colonies can occur, sometimes connected with morphological and functional differentiation of the thallus (the mechanisms leading to the thallus differentiation in chroococcal species are unclear). Particularly in *Hyellaceae*, the thallus diversifies distinctly morphologically into basal and apical pseudofilaments, epi- and endolithic (or endophytic) parts with cells dividing into baeocytes, apical cells of pseudofilaments, cells initiating pseudobranching, etc.

# Classification system

## Supraspecific criteria

The system of unicellular and colonial cyanoprokaryotes is based on the combination of various features which are known or considered to be constant and genetically stable. Families are classified according to the following criteria:
- Form of cells in relation to the cell cycle (type of cell division).
- Polarity of cells and colonies is classified as the diacritical feature between simple and more complex families.
- Type of cell division (modifications of binary and multiple fission) is constant to such a degree, that it can serve well for the delimitation of families. Two or several cell division types can occur in several families, and the special combination of various division types is accepted to the family definition (Table 4).

The unicellular or colonial mode of life is the main criterion for separation of genera within families. The form and structure of a colony, formation of various types of mucilaginous strands, layers, and the position of cells in a colony are particularly important.

## Species concept

At present, the species concept in cyanoprokaryotes is under intense discussion. (1) From the results of modern molecular genetics only a small number of different cyanoprokaryotic genotypes is assumed (comp. Castenholz 1992). The concept of a restricted number of cyanoprokaryote species was already used speculatively by Drouet in his monographs (Drouet & Daily 1956, Drouet 1968, 1973, 1977), but his results do not correspond with common knowledge of cyanoprokaryote diversity both in nature and cultures. The modern and precise scientific methods should yield a final image of the genotype diversity in this group, but this investigative process is still in its early stage. It is not yet clear, whether the presupposition of a restricted number of genotypes in cyanoprokaryotes is influenced mainly by the small number of strains investigated so far. From the entire spectrum of cyanoprokaryotes known from nature (and particularly including the populations from tropical and extreme biotopes which are far from those that are well known), only a very small percentage of taxonomic units are in culture.

(2) In contrast to the presupposition of a restricted number of genotypes, many small, morphologically stable deviations exist, which occur repeatedly in ecologically restricted biotopes, sometimes in geographically distant regions. Also the great number of stabilized cyanoprokaryotic types (strains) in culture collections indicate the existence of numerous stable morphotypes. That there is a problem connected with species delimitation in cyanoprokaryotes, supported by the comparative studies of strains, which indicate that virtually there are no wholly identical populations (and strains). From the list of 51 strains of small *Synechococcus/Cyanobium/Cyanobacterium* types, evaluated by Waterbury & Rippka (in Castenholz & Waterbury 1989, Bergey's Manual) follows a continual series of slightly deviating strains varying in G/C ratio, which are stable in culture (Table **5**). The authors divided this entire set of strains into seven clusters; according to the criteria of traditional cyanoprokaryote taxonomy, it can be hardly found more traditional species among them (and they should be almost identical with clusters defined by molecular biologists). It is worth noting that all these strains have retained their deviating characters intact even under changes in external conditions (despite the fact that they all belong possibly only to one genotype determined by the strict method of molecular genetics). Also in strains included together in one taxonomic cluster there exist recognizable differences in their morphology or ecophysiology, one from another (Table **5**). Similar strain variation was illustrated by Komárek (1972) in *Phormidium autumnale*, Stam (1978, 1980, Stam & Holeman 1975, 1979) in LPP-group B (= *Leptolyngbya*), Whitton (1987) in *Rivulariaceae*, Kato & al. (1991) in *Microcystis aeruginosa*, etc. These results indicate that there is an enormous number of stable populations (which, when cultured, produce stable strains); after detailed studies, some of them can be given separate taxonomic status. Only in some cases, strains should be grouped together in larger natural and delimited units, or to which can be ascribed the rank of species. The number of such small units, derived from cultured populations, always is larger than the present number of described species. (The relation of these numerous stable morphotypes with the proposed restricted number of genotypes is not clear yet.)

(3) Another (traditional) approach to the species delimitation (derived mainly from the study of natural populations) is based on morphologically defined units, that in different biotopes repeatedly occur and form morphologically and ecologically uniform and stable types, consisting of slightly different or, on the contrast, variable populations (morphotypes of the above mentioned stable populations). They occur often in widely separated areas far distant from one another, and appear over various time periods (occurring continually in the localities concerned, or periodically every year, or after several to many years). It is improbable, that all of them can arise from one or only a few genotypes that repeatedly produce (sometimes after years) morphologically identical types in similar biotopes. The occurrence of identical phenotypes in corresponding biotopes (distant from one another) and the stability of various morphological entities occurring together in the same ecological niche indicate the existence of several genotype sources.

This last characteristic (paragraph 3) is found representative of traditional species in their natural habitat. The problem to solve is, that particularly in newly studied

| Division pattern | Examples | Terms from literature | Accepted term | Fam. |
|---|---|---|---|---|
|  | Cyanothece<br>Aphano-thece | process:<br>cell division<br>binary fission<br><br>product:<br>daughter cells | daughter cells | Syn. |
| 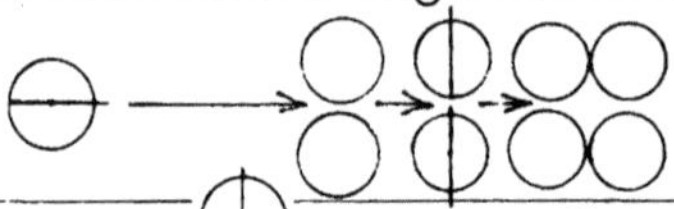 | Synecho-cystis<br>Aphano-capsa<br>Merismo-pedia | | | Meris. |
| 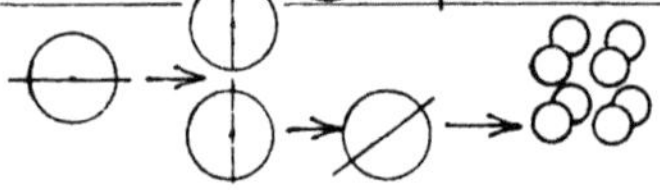 | Eucapsis | | | Micr. |
|  | Chro-ococcus | cells | daughter cells | Chrooc. |
| 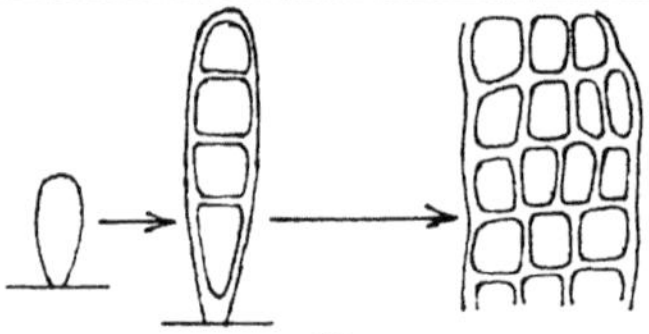 | Entophysalis | process:<br>cell division<br>binary fission<br><br>product:<br>daughter cells | daughter cells | Entophysalidaceae |
| 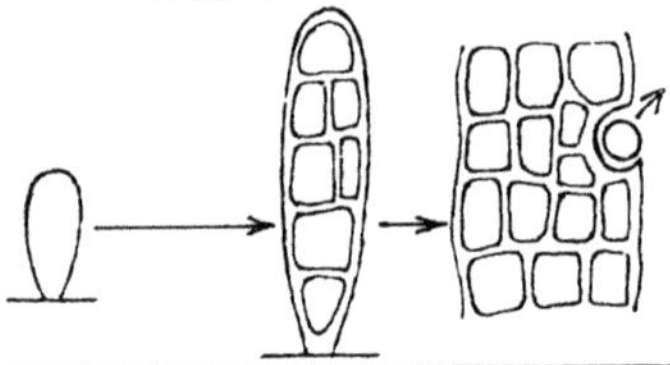 | Siphono-nema | daughter cells<br>+<br>planococci | daughter cells<br>+<br>plano-cytes | |
| 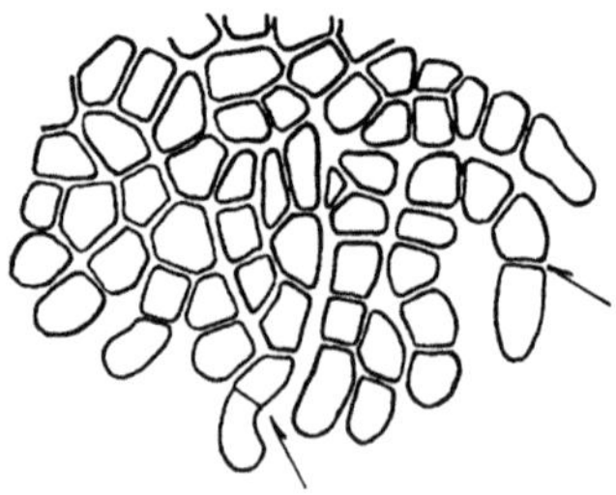 | Hydro-coccus | process:<br>cell division | daughter cells | Hydrococcaceae |

**Table 4.** Various types of cell division and reproduction in families of chroococcal cyanoprokaryotes. (After Komárek & Anagnostidis 1986.)

| Division pattern | Examples | Terms from literature | Accepted term | Fam. |
|---|---|---|---|---|
| 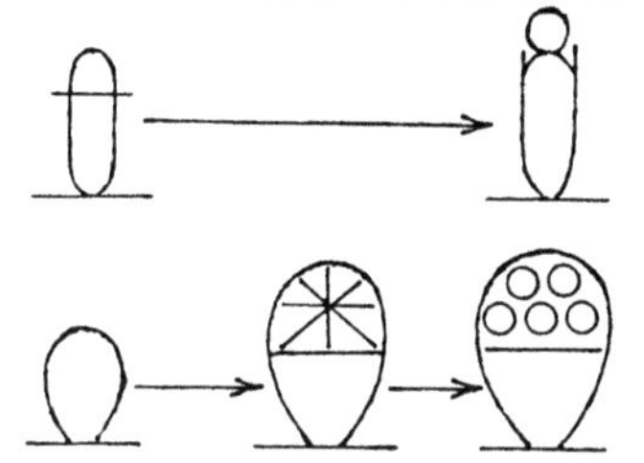 | Chamae-siphon<br>Chamae-calyx | process:<br>binary fission<br>budding<br>product:<br>exospores<br>endospores<br>buds | exocytes | **Chamaesiphon.** |
| 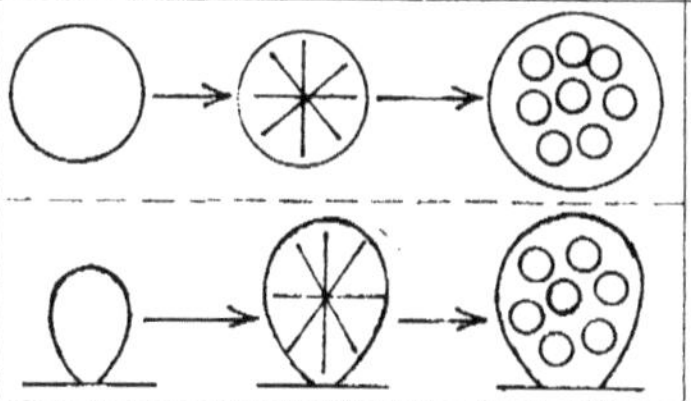 | Stanieria<br>Cyanocystis | process:<br>multiple fission<br>product:<br>endospores<br>baeocytes | baeocytes | **Dermocarpell.** |
| 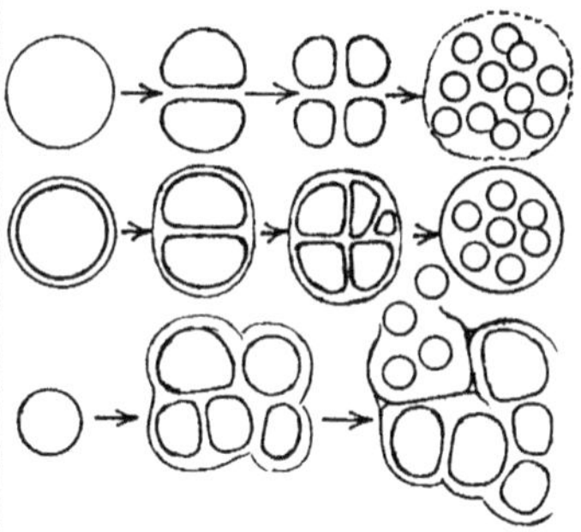 | Chroococ-cidium<br>Chroococci-diopsis<br>Xenococcus | process:<br>cell division<br>binary fission<br>multiple fission<br>product:<br>daughter cells<br>endospores<br>baeocytes | daughter cells + baeocytes (mono-cytes, plano-cytes) | **Xenococcaceae** |
| polarized growth<br>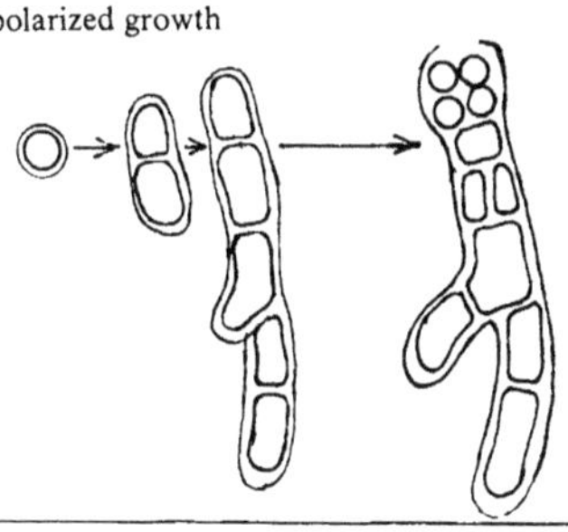 | Pleurocapsa<br>Hyella<br>Radaisia | process:<br>cell division<br>binary fission<br>multiple fission<br>product:<br>cells<br>endospores<br>baeocytes | daughter cells + baeocytes | **Hyellaceae** |

| Groups and clusters (WATERBURY & RIPPKA 1979) | no. of strains | width of cells | motility | trophic level | N-fixation (in anaerob.) | pigment syst. | halophily | max. temperature | mol. % GC-DNA | ecology |
|---|---|---|---|---|---|---|---|---|---|---|
| ***Cyanobacterium stanieri* (typ)** | 1 | | – | | – | PC | | | 39.0–40.5 | |
| *Synechococcus*-group (*Cyanobacterium*-cluster) | | | | | | | | | | |
| – «alcalic str.» | 1 | 1.7–2.3 | – | a | – | PC | – | | 39 • | alk. F |
| – «acidic str.» | 1 | 1.8–2.0 | – | a | – | PC | – | | 40.5 | acid. F |
| «*Cyanothece*»-group | | | | | | | | | | |
| – large | 1 | 5.0–6.0 | | a | + | PE | | | 41.2 | paddy f. F |
| – «halotolerant» | 3 | 4.0–5.0 | | (a) | (+) | PC | | 35 | 42.4 | B–S |
| – small | 2 | 3.0–3.5 | | a | + | PC | | | 48.6 | paddy f. F |
| – halobiotic | 1 | 3.5 | | a | – | PC | + | 43 | | B–M |
| *Synechococcus*-group (Marine-cluster «c») | | | | | | | | | | |
| – with PE | 1 | 1.9–2.1 | – | fac. h | + | PE | + | | 47.4 | pl. M |
| – autotr. | 1 | 1.2–1.5 | – | a | – | PC | – | | 48.5 | B–S |
| – halof., autotr. | 2 | 1.2–1.5 | – | fac. h | – | PC | – | 43 | 48.5–49.0 | pl. M |
| – halof., heterotr. | 1 | 1.8–2.0 | – | fac. h | – | PC | – | 39 | 49.3–49.5 | pl. M |
| *Synechococcus*-group (*Synechococcus*-cluster) | | | | | | | | | | |
| – motile, large | 1 | 3 | + | a | – | PC | – | 39 | 47.5–48.0 | F |
| – middle str. | 1 | 1.3–1.5 | – | a | – | PC | – | 37 | 50.2 | F |
| – small str. | 2 | 1.0–1.2 | – | a | PC | – | 43 | 54.8–56.0 | F | |
| – high DNA/DNA homol. | 3 | 1.0–1.2 | – | (a) | (–) | PC | – | 43 | 55.0–55.1 | F |
| – thermal | 3 | 1.2–1.4 | – | a | | PC | – | 53 | 52.0–53.6 | F term. |

| Groups and clusters (Waterbury & Rippka 1979) | no. of strains | width of cells | motility | trophic level | N-fixation (in anaerob.) | pigment syst. | halophily | max. temperature | mol. % GC-DNA | ecology |
|---|---|---|---|---|---|---|---|---|---|---|
| *Synechococcus*-group (Marine-cluster «a») | | | | | | | | | | |
| – motile | 3 | 0.9–1.7 | + | a | – | PE | + | 30 | 59.0–60.0 | pl. M |
| – immotile | 12 | 0.6–1.5 | – | a | – | PE | + | 30 | 55.0–61.0 | pl. M |
| *Synechococcus*-group (Marine-cluster «b») | | | | | | | | | | |
| – halobiotic | 1 | 0.9–1.1 | – | a | – | PC | + | | 63 | pl. M |
| – halotolerant | 3 | 0.8–1.4 | – | a | – | PC | – | (41) | 69.4–69.5 | pl. M |
| *Synechococcus*-group (*Cyanobium*-cluster) | | | | | | | | | | |
| – large F | 2 | 1.2–1.4 | – | a | – | PC | – | 35 | 65.7–68.4 | F |
| – brackish | 1 | 0.8–1.0 | – | a | – | PC | – | | 66.6 | B–S |
| – small F | 5 | 0.8–1.0 | – | a | – | PC | – | 35–37 | 66.3–71.4 | F |
| ***Cyanobium gracile* (typ)** | 1 | | – | | – | PC | – | 35–43 | 66.7–71.4 | |

**Table 5.** Characters of simple cyanoprokaryotic strains from the *Cyanobacterium/Synechococcus/Cyanobium* group, derived from Waterbury & Rippka (1989) and Rippka & Cohen-Bazire (1983). The main character for the clustering of strains is mol% GC-DNA.

biotopes numerous such new types are frequently found (particularly in a great variety of tropical and extreme biotopes). They do not correspond to any of known species described in the literature and authors commonly object to the description of many new taxa. The discovery, that the majority of tropical cyanoprokaryote species are not yet taxonomically defined, is unacceptable to many authors to such a degree, that they give them arbitrary species epithets available in the literature (even in cases where the substantial differences from the well-known species are evident). Such stable morphotypes are sometimes dominant in newly investigated biotopes; detailed studies also their ecological (and subsequently geographical) delimitation confirm.

At present, we know of numerous morphologically distinguishable taxonomic units (traditionally designated as species) occurring only under special ecological (microecological) conditions and distributed world-wide, or only restricted to certain regions. However, the same morphotype can be stable in one region, but variable in another. This fact (distribution in special biotopes over the world) is especially evident in species from extreme biotopes (*Cyanothece aeruginosa, Cyanobacterium minervae, Synechococcus lividus, Geitleria calcarea, Mastigocladus laminosus*), but the same phenomenon has been demonstrated for numerous species from «normal» (including planktic) biotopes, at least in certain regions (e.g., *Woronichinia naegeliana* – Čmiech & al. 1987, *Aphanizomenon flos-aquae* – Pechar & Kalina in prep., *Nostoc pruniforme* – Mollenhauer 1970, etc.). There exist freshwater planktic species, which are cosmopolitan, as well as substantial differences between planktic cyanoprokaryotic species from northern (e.g. Scandinavian), temperate and tropical water bodies. Numerous ecologically specialized, morphologically distinct and geographically delimited types (species) do exist.

Cyanoprokaryotes have a special speciation process, which is probably identical to bacteria (comp. Schopf in Pennisi 1994). Asexual cyanoprokaryotes evidently possess great adaptability, which has permitted continual adaptation to very variable environmental conditions during their long existence. It is almost impossible to speak about «species» in a similar sense to eukaryotic plants or animals. Such continually appearing modifications are characterized by ecophysiological deviations, later followed by morphological changes. They can be stable for a long time under constant conditions and form characteristic and so definable «species». However, populations also exist, which are composed of numerous, small deviations which in any given population possess more or less markedly deviant characteristics, and which can, moreover, stabilize after isolated in culture. This process explains the long existence of cyanoprokaryotes since the early Precambrian with continually changing, repeating and renewing types which can easily develop (also recently) in new occurring niches, and especially, the recent existence of recognizable taxonomic «units» differnig from one another (comp. Schopf in Pennisi 1994). Very wide mosaic of similar, but not quite identical «small» populations and strains is known as well as stabilized, morphologically clearlydelimited, specialized and recognizable types in ecologically distinct conditions (probably consisting of different genotypes). The present diversity of cyanoprokaryotes is therefore complicated and the taxonomic classification is extremely difficult. In nature all possible units exist, from genera composed only of small deviations at the population level (strains), to widely-distributed, morphologically clearly recognizable and ecologically distinct types (traditional «good» species).

This situation is complicated by the fact, that various characters in cyanoprokaryotes are phenotypically plastic (with various degree of stability), and express different «taxonomic values». E.g., there are cyanoprokaryotes with «typical» chromatic adaptation, in which the change in PC/PE ratio is dependent on environmental (light) conditions. Other types have a very stable phycobilin ratio under changing conditions (*Planktothrix agardhii*/*Planktothrix rubescens*). Between these two extreme examples, all possible modifications can appear, e.g. populations with a variable pigment ratio which stabilizes in culture, etc. The majority of ecophysiological and morphological features (e.g. presence of sheaths, size variation, etc.), used in cyanoprokaryotic taxonomy, shows the same diverse variability in different genera.

The category of «species» is conventional to a certain degree. In cyanoprokaryotes, all the three above mentioned concepts are not contradictory one to another, but show the complex continual diversification of cyanoprokaryotes inside only few (or numerous?) genotype entities. Hence, classification at all levels is necessary in experimental and ecological practice. Therefore, the definition, what is a species or strain (or any subspecific) category in cyanoprokaryotes, depends on our interpretation. The discussion in this field will surely continue in the future with increasing intensity. However, at present, we will keep the traditional species for our review stressing the morphological and ecological characters, which permit the recognition of the commonly and repeatedly occurring, ecologically delimited populations in natural biotopes. Of course, all species defined using this method contain numerous populations and strains, which can sometimes differ substantially in small but stable characteristics (morphological, biochemical, structural, ecological) and their classification is therefore difficult. The uniform designation of all such subspecific units as taxonomic forms (com. Elenkin 1936–1949, Hollerbach & al. 1953, and others) does not express infraspecific variation satisfactorily. The linkage of several traditional species to one genotype is probable, but not commonly accepted on the basis of present knowledge. However, the main goal of this book is the identification of cyanoprokaryote populations, strains and their clusters («species»). The manual should enable mainly the orientation in natural populations of cyanoprokaryotes; therefore, such stable units must be still respected, which really do exist in nature under specific ecological conditions, and which are important for ecological and technical-hydrobiological research.

We have tried to include all modern knowledge of cyanoprokaryotes into our systematic review so far as is possible. In our previous studies (Anagnostidis & Komárek 1985, 1988, 1990, Komárek & Anagnostidis 1986, 1989), we have defined cyanoprokaryotic units at the generic level, taking into consideration the data available from recent publications. The traditional cyanoprokaryotic species (*Chroococcales* in the 1st part) are summarized in this book.

# Ecology and distribution

The simple morphology of numerous cyanoprokaryotic populations results in the underestimation of the importance of ecology in relation to various cyanoprokaryote species. Morphologically similar populations often occur in different biotopes, however, their taxonomic identity is very problematic. The morphologically identical strains (e.g., mucilaginous colonies with randomly distributed spherical cells – *Aphanocapsa*-type) are referred from various localities (e.g., peaty waters, freshwater plankton, mountain streams, marine littoral, thermal springs, etc.), but their ecophysiological properties are very different. Unfortunately, their diversity is determinable only after long-term comparative studies and usually only after isolation in culture. Such studies are very rare and not very useful for ecologists, who need to be able to rapidly identify dominant populations. The result is, that several specific names exist, which are commonly used for all similar populations from very varied biotopes. *Microcystis pulverea, Aphanocapsa elachista, Aphanocapsa grevillei, Aphanocapsa pulchra, Aphanothece nidulans, Aphanothece halophytica* etc. are representatives of these «ubiquitous» species.

On the other hand, the numerous systematic – ecological studies of cyanoprokaryotic species show their delimited ecological properties and the restriction of their specific biotopes. The ecological diversity of cyanoprokaryotes indicates also the existence of morphologically and biologically clearly different, stable species with evidently identical or very similar ecology. This is obvious and easy recognizable of morphologically distinct types, but it was found also in simpler species, morphologically hardly distinguishable one from another. Dependence on salinity, heterotrophic/autotrophic metabolisms, thermophility, light dependence, etc. are important markers of taxonomic (and genetic?) identity. Autecological (i.e., ecophysiological) characteristics are, therefore, important characteristics of cyanoprokaryote taxa, and particularly in simple types the ecological properties are an important indication of their identity or diversity.

Of course, if we speak about species ecology, we must always keep in mind the microbiotopes. It means that the occurrence of species in apparently unusual ecological conditions can be explained by the close microconditions enveloping the cells. All cyanoprokaryotes are variable within certain limits that are dependent on ecological parameters. Numerous stages (status) of different species are morphologically identical; however, the unification of all such types into one large taxon only on the basis of their morphologically identical stages is simple and easy practice, but incorrect. More detailed experimental studies always demonstrate the diversification of the populations studied into numerous ecophysiologically different strains.

Geographic distribution is another factor dependent on ecological specificity. Because the cyanoprokaryotes are not ubiquitous, they can not be spread throughout the all world biotopes. Due to their ancient origin, cyanoprokaryotes have had time and possibilities to colonize all suitable biotopes. Numerous species have really cosmopolitan distributions, but only in areas that meet their ecological needs. It is particularly apparent in species, which only grow under very specialized conditions (e.g., in thermal waters with limited temperature range, peaty waters, hypersaline and heliothermal lakes, etc.). Species which are found in less extreme biotopes (plankton of eutrophic waters, etc.) are more widely distributed, but may also have restricted ecological niches.

Aside from cyanoprokaryote species with world-wide distribution, numerous species exist, which are restricted by their ecological properties to very restricted areas. Numerous tropical and subpolar species are restricted to clearly demarcated regions and the occurrence of endemic species is quite probable. There is especially great variability in some extreme biotopes, in which the development of specialized, endemic species (from various groups of plants and animals) is evident. The development of special stable cyanoprokaryotic taxa is to be expected. Mineral springs with various kinds of chemistry and temperature regimes, hypersaline localities with diverse salt concentrations, marine coasts from polar regions to warm tropical seas, volcanic lakes in different climatic regions, travertines and stromatolites of various kinds, diverse aerophytic biotopes, tropical swamps of various kinds, etc. belong to such biotopes.

It is interesting that there are regions with species not known elsewhere, i.e. with endemic species known only in certain countries. In Europe there are differences in both qualitative and quantitative parameters of the cyanoprokaryotic flora, not only from ecologically diverse biotopes (sea coast – soil algae – plankton of eutrophic reservoirs – mountain streams and wet rocks, etc.), but also between the wet Atlantic regions, Mediterranean countries and Nordic areas. It is worth noting, that from the Baltic area numerous species have been found unknown in other countries (freshwater as well as brackish water species), e.g., *Woronichinia karelica*, *W. compacta*, *Microcrocis sabulicola*, *Cyanodictyon iac*, *Coelosphaerium dubium* (in original sense), *Cyanothamnos plancticus*, *Nodularia litorea*, and others. This diversity is particularly distinct, if we compare fully (not only on the basis of morphological similarity) the cyanoprokaryotic populations from very different biotopes throughout the world. The cyanophyte flora from marine limestone coasts in Dalmatia and Greece, Persian Gulf, Californian coast, Pacific Islands, Caribbean Sea, etc. or benthic algae from rivers in Norway, the Alps, the Himalayas, or central Mexico can hardly be identical, in spite of several similar or identical species, which occur everywhere. If we consider all the possible biotopes, in which cyanoprokaryotes occur in the biosphere and the diversity of cyanoprokaryotes, which have been found in these various biotopes, the idea of ubiquitous species seems to be absurd. The ecophysiological properties and distribution (correctly determined!) belong, therefore, to the important characteristics of cyanoprokaryotic taxa.

# Glossary

## Cytomorphological terms

**aerotopes** – groups of gas vesicles, occurring mainly in many planktic species, seen in cells as brownish, refractive, irregular bodies; they are a facultative phenomenon, influencing buoyancy; they can appear throughout the whole cell volume or they can be localized in special regions of the cell protoplast (near poles or in central parts of cells); old incorrect term = «gas vacuoles» (they are not any type of vacuoles);
**akinetes** – resting cells, arising from one or more confluent cells after the storage of assimilates and diminishing of thylakoids; akinetes are larger than vegetative cells, enveloped by a thick, multilayered cell wall; they can germinate after a period of unfavourable conditions; the position of akinetes on a trichome, their morphology and type of germination can be characteristic for species (sometimes for genera);
**apical (terminal) cells** – terminal cells in polarized thallus, at the ends of pseudofilamentous rows of cells and in trichomes; their shape is often stable and important for species identification;
**baeocytes** – small reproductive cells, differentiating from mother cells after simultaneous or successive multiple fission, enveloped by firm sheaths, which splits or gelatinizes during the baeocyte liberation; old incorrect term = «endospores»; see also «nanocytes»;
**basal cells** – usually specifically modified cells by which the heteropolar thallus is attached to the substrate; in unicellular heteropolar cyanoprokaryotes they are sometimes modified morphologically (narrowed, elongated, widened, etc.); basal parts of cells or basal cells of a thallus sometimes have a specially adapted sheaths or pseudovaginae;
**binary fission** – simple division of a cyanoprokaryotic cell into two daughter cells which grow more or less to the original size (and sometimes original shape) of the mother cell before the next division;
**branching** – occurs in filamentous cyanoprokaryotes; the two main types are false and true branching (for explanation see the volume with filamentous types); see also «pseudobranching»;

**calyptra** – remnants of a gelatinous sheath or of a necridic cell, forming a thickened cap on the outer cell wall of an apical cell in filamentous cyanoprokaryotes;
**centroplasma** – inner, pale part of cyanoprokaryotic cell protoplasts (without thylakoids); homogeneous colouration of the whole cell indicates an other than parietal position of thylakoids;
**chromatoplasma** – usually outer, more intensely coloured part of the cyanoprokaryotic cell protoplasts, where thylakoids are situated; disinguishable chromatoplasma usually indicates a parietal arrangement of thylakoids;
**colony** – microscopic or macroscopic aggregates of cells or filaments, organized in a characteristic form; usually enveloped by clearly delimited or diffluent, distinct or fine mucilage of various consistency and colour; cells in colonies belong to several to many generations, sometimes with their own individual gelatinous envelopes; the morphological diversity of cells or polarity of colonies exists in several families (mechanisms influencing the diversification of cells in colonies are still unknown);
**colouration** – colouration of thallus depends on several types of pigments: photosynthetic pigments within cells, including chlorophyll *a* and phycobilins, the ratio of which results in cells being yellow-green, blue-green to olive-green, reddish or violet (the cell colour may be stable or be dependent on chromatic adaptation); sheath pigments and several carotenoids colour the envelopes and sheaths (red, violet, blue, or yellow-brown);
**dimensions** (cell and colony dimensions) – dimensions are given in µm, rarely in cm; in this book always in the form «length × width»;
**divarication** – see «pseudobranching»;
**envelopes** (gelatinous, mucilaginous) – a wide range of mucilaginous substances excreted from cells, which are not an integral part of a cell wall; envelopes around cells can be of diverse width, consistency and be clearly delimited or diffluent, stratified, and variously coloured by sheath pigments; they participate in colony formation and only differ from sheaths by having a more diffluent consistency; structure and colouration of the mucilaginous envelope sometimes is species and genus specific;
**exocytes** – reproductive cells which separate solitarily, in rows or in clusters from the apical part of heteropolar cells, which are usually attached to a substrate; characteristic for *Chamaesiphonaceae*; former incorrect term = «exospores».
**expulsing cells** – special type of monocytes (solitary reproductive cells), liberating expulsively and almost spontaneously from colonies under stress (or at the end of the vegetation period); characteristic particularly for the g. *Woronichinia*;
**filament** – trichome enveloped by a sheath, or only a trichome without sheath (broader term than the specialized cyanoprokaryotic «trichome»);
**gas vesicles** – small cylindrical, intracellular vesicles aggregated densely and parallel to each other in aerotopes; wall of gas vesicles is pure protein, content is gaseous; appearance and degradation of gas vesicles control the buoyancy of some (planktic) species;
**hairs** – apical ends of heteropolar trichomes, usually with narrowed and elongate cells (rarely with only elongate cells); hairs may not develop or may atrophy in waters rich in phosphate (Whitton 1987); they are lost during formation of hormogonia.
**heterocytes** – special cells arising from vegetative cells in distinct places in some filamentous cyanoprokaryotes (basal, intercalary); morphology: thick, several-layered cell wall, an imperfect photosynthetic apparatus (photosystem II is blocked); function: contain enzymatic system permitting the assimilation of gaseous nitrogen (active nitrogenase under anaerobic conditions); their frequency is influenced by N-metabolisms (in several species heterocytes can be missing

facultatively in environment rich in nitrogen); various nostocalean species behave differently from one another and, of course, reactions to nitrate and ammonia can differ; old incorrect term = «heterocysts» (they do not correspond to any definition of «cysts»);

**heteropolarity** – occurs in cells, trichomes, filaments and thallus, which are morphologically diversified into basal (usually attached) and apical (free) ends; heteropolarity is an important developmental phenomenon and is an accepted characteristics in cyanoprokaryotic taxonomy at the level of genera and families;

**hormogonia** – short sections of cyanoprokaryotic trichomes, separating from the original trichome after fragmentation or after formation of necridic cells or heterocytes, and serving for reproduction; within certain limits, character of hormogonia is constant in various species; hormogonia are liberated from sheaths in ensheathed types;

**intrathylakoidal spaces** – widened thylakoids, forming within cells vacuole-like spaces in several genera and species; they are well recognizable only in EM-fine sections of cells, but they sometimes look like «vacuolization» or «keritomy» under a light microscope; their origin and function is still not well explained and probably depends on the light regime;

**involution cells** – cells of an anomalous (monstrous) shape and size, different from cells, growing under suboptimal conditions («normal» in natural biotopes); they occur rarely in nature, commonly in culture, as a reaction on sublethal (stress) conditions; their form (filamentous, irregular) can be characteristic for various genera;

**keritomy** – net-like or striated plasma in cells (visible in light microscope); in principle keritomy results (i) from the thylakoid arrangement (usually radial, e.g. in *Cyanothece* or *Phormidium*), or (ii) from the appearance of widened intrathylakoidal spaces (e.g. in *Cyanothece, Chroococcidiopsis, Tychonema* and others);

**monocytes** – solitary cells, liberated from thallus (colony) and serving for reproduction without preceding cell division;

**multiple fission** – simultaneous or successive (in rapid sequence) division of cells into more than two or into numerous daughter cells, evidently as the result of repeated multiplication of genomes within a mother cell without cell division until after the end of genome multiplication;

**nanocytes** - small reproductive cells, differentiating from mother cells after simultaneous or successive multiple fission within fine, gelatinous slime; comp. also «baeocytes»;

**necridic (sacrified) cells** – dying cells; their death is an integral part of the life cycle, particularly in filamentous cyanoprokaryotes, they do not always die under the influence of limiting environmental conditions or at the end of the life cycle; their appearance precedes false branching, exocyte formation or hormogonia differentiation;

**necrids** (necroids) – see «necridic cells»;

**phycobionts** – cyanoprokaryotes (and algae) living symbiotically as the oxyphototrophic (plant) components of lichens; (the old term, not accepted in modern lichenology and phycology = «gonidia»);

**planocytes** – common term for solitary reproductive, motile cells (nanocytes, baeocytes, exocytes);

**pseudobranching** – apparently branched, divaricate thallus, without a physiological connection between cells; it appears in species in which the cells are arranged in rows, but not in trichomes with cells forming a physiological entity;

**pseudofilaments,** pseudofilamentous rows – rows of cells in thallus (colony), which are not distinctly connected in one physiological unit and do not represent one physiological entity;

**pseudovagina** – sheath in heteropolar cells or pseudofilaments, which opens only at one (apical) end;
**resting stages** (resting cells) - inactive cells, serving for overwintering or for survival unfavorable conditions, usually dormant for a particular period of a year; the designations are «arthrospores», «spores» or «akinetes», but the structure of envelopes and genesis of resting cells of coccal cyanoprokaryotes are probably different from those of akinetes in *Nostocales* (in which «akinetes» are a characteristic morphological feature); morphology of resting cells in coccal genera: slightly enlarged cells with numerous assimilate grains, enveloped by thick and firm (sometimes layered and intensely coloured) sheath;
**sheaths** – firm, thin or thick, morphologically distinct mucilaginous layers around cells and trichomes (rarely fasciculate trichomes), excreted from cells (not parts of a cell wall); sheaths often have a tube-shaped form (enveloping the trichome) open at both ends or closed (in several unicellular genera); sheaths can be lamellate and coloured by sheath pigments; presence, facultative presence, frequency, form and colouration of sheaths are among the taxonomic features used in species and to a lesser extent in generic taxonomy;
**stalks, strands** – mucilaginous excretions forming characteristic thin formations as a part of colonies; form, type and mode of divarication of stalks are stable and used for generic classification;
**thallus** – general term used in botany in several senses, referring to body type of algae; usually used in the sense of a colony (particularly macroscopic) of characteristic morphology, sometimes containing morphologically diverse types of cells, but without special organs of differentiated plant body (such as roots, stalks and leaves); in its original sense, thallus usually is flat, but in cyanoprokaryotes it is used to charcterize colonies of diverse shape;
**trichome** – row of cells (uni- or multiseriate), which are connected into one physiological unit (physiological transport exists between the cells, at least from a part); trichome is the traditional, special term for «filament» in cyanoprokaryotes without the enveloping sheath (not wholly suitable, but commonly used);
**vacuolization** – true vacuoles rarely appear in cyanoprokaryotes, probably only appear in apical cells of several filamentous species and in cells under stress; see also «keritomy» and «intrathylakoidal spaces».

## Ecological and biogeographic terms

**acidophilic** – living in (or preferring) biotopes with pH lower than 7;
**aerophytic** – living on air (in terrestrial habitats, on rocks and stones, bark of trees, etc.), needing only the atmospheric moisture (humidity) to sustain its life;
**alkaline** (water, substrate) - with pH higher than 7;
**atmophytic** - living in aerophytic localities, receiving moisture from cold or hot vapour (usually in vicinity of waterfalls or hot springs);
**backwaters** - stagnant or slowly flowing waters (oxbows, pools, small lakes) in the inundation zone of rivers;
**benthic** - living on the bottom of stagnant and flowing waters;
**biofilms** – see «mats»;
**biolithogenic** - precipitating carbonates in mucilage and in sheaths, usually participating in travertine and stromatolite formation;
**biotope** – locality (habitat) in which lives a defined community of organisms; biotope is characterized by the geographic width, height above sea level, climate, geological and pedological substrate, and by environmental abiotic and biotic factors; (syn.: ecotope);
**boreo-alpine distribution** - occurring in polar and subpolar regions and in high mountains;

**brackish** - living in marine coastal localities, usually with increased and variable salt content;
**calcification** - deposition of carbonates (sometimes combined with other mineral precipitates) in mucilage or in sheaths;
**calcifilic** - preferring limestone substrate (usually living in karstic regions, in water bodies in limestone areas, etc.);
**catharobic** - in clear water with very low salt and nutrient content and low trophic level;
**cavernicolous** - living in caves;
**corticolous** - living on the bark of trees;
**cosmopolitan** - distributed world-wide (but in cyanoprokaryotes usually in ecologically restricted localities);
**cryoseston, cryosestic** - living on the surface or below the surface of snow fields or of glaciers, usually in summer season in polar regions and in high mountains;
**cyanelles** - cyanoprokaryotic cells that live as endophytes or endozoids in cells of various eukaryotic plants and animals;
**dystrophic** - connected with *Sphagnum*-bogs, peaty lakes and moorland waters with low pH (less than 7) and with water usually brownish, coloured by humine acids; low trophic level;
**edaphic** – living in soil;
**endobiotic** – living within other organisms;
**endogloeic** - living in mucilage of other plant or animal organisms;
**endolithic** - living under the surface of stones and rocks; (the main types: **euendoliths** – actively dissolving and boring into the substrate, **cryptoendoliths** – living in small cavities and holes in stones, (**chasmoendoliths** – living in fissures);
**endophytic** - living within cells or in intercellular spaces of eukaryotic plant organisms;
**endozoic** - living within cells or in intercellular spaces of animals; usually in epidermal or skin layers of marine littoral invertebrates;
**epilithic** - attached to stones or rocks or growing on stony substrate and on stony walls (aerophytic or submerged);
**epilittoral** - highest zone of littoral above the water level, influenced continually or occasionally by waves (in splash zone);
**epipelic** - living on the muddy bottom of water bodies;
**epiphytic** - attached to or sessile on plant substrates (other cyanoprokaryotes, algae, mosses, vascular plants);
**epipsammic** - living on the sandy substrate (in **psammon**), submersed or on (or near) the surface of wet sand (subaerophytic);
**epizoic** - attached or sessile on animals or on parts of their body (also on the skin, hairs, etc.); commonly on the surface of animal shells (on shells boring species are common, usually classified also as endolithic types);
**eurytherm** - living in wide range of temperatures;
**euryhaline** - living in wide range of salt concentrations;
**eutrophic** - with high content of nutrients;
**freshwater** - living in various types of inland waters, without a high salt content;
**halobiotic** - living in localities with high or unusually high salt content;
**halophilic** - preferring localities with higher salt content;
**halotolerant** – living in habitats with normal and high salt concentration;
**heliothermal** - living in localities with constantly high temperature and heated by solar energy (mainly in subtropical and tropical regions);
**hypersaline** - with extremely high salt concentration (usually in coastal or desert swamps and drying lakes with increasing salt concentration and with various kind of salts, sometimes different from these ones which are usual in marine and oceanic biotopes);

**hypertrophic** - with extremely high content of nutrients, usually also connected with organic pollution;
**littoral** - part of coastal biotopes (of rivers, water reservoirs or seas), influenced by water level (contact zone) and together with associated shallow areas (bays, swamps) connected to the water body; upper littoral = above the average water level, lower littoral = below the average water level;
**marine** - occurring in coastal or pelagic sea or oceanic biotopes;
**mats (biofilms)** - more or less flat, usually macroscopic colonies on a substrate (usually submersed, or on wet soil); (Stal & Caumette 1994);
**mesohaline** - with higher salt content than in fresh waters, but lower than in sea water;
**mesotrophic** - naturally eutrophicated water with slightly increased salt content and trophic level;
**metaphytic** - free-living among other algae, submersed mosses or vascular plants, but not attached to them;
**mineral water bodies** - with high conductivity (with high content of carbonates, sulphates, chlorids, etc.): «soda-lakes», etc.;
**moorland waters** - usually acidic localities with pH lower than 7 and with much vegetation, often peaty or *Sphagnum*-bogs;
**nordic** - living in northern regions of temperate zone, and in polar and subpolar areas (in Europe including Fennoscandia, western parts of Siberia, Iceland, etc.);
**oligohaline** - living in biotopes with low salt concentration;
**oligotrophic** - with low trophic level;
**pantropical** - distributed throughout the whole tropical zone;
**peaty** - living in peaty, acidic localities, usually connected with *Sphagnum* (with pH usually lower than 7);
**periphytic** - attached, sessile on any submersed substrate;
**phycobiontic** - living as oxyphototrophic components in lichens («gonidial algae»);
**picoplanktic** – planktic species with solitary living cells, ± less than 2(3) µm in diameter;
**planktic** - free-floating in the plankton; (the commonly used term «planktonic» is not in agreement with the both ancient and modern Greek grammars);
**psammic** - living in psammon, i.e., in sandy wetted (sometimes periodically) or submersed biotopes;
**psychrophilic** - preferring cold biotopes;
**saline, salty, salinic** - living in marine, marine-coastal, less frequently in inland localities with high salt content (salts of various kinds, including sulphates, chlorides, etc.);
**saprobiotic, saprobionts** - living in localities with high content of organic matter;
**saprobity** - degree of pollution (catharobic – oligosaprobic – a- ß-mesosaprobic – polysaprobic waters);
**sapropelic** - living in sapropel, i.e., on muddy bottom rich in organic matter;
**splash zone** - part of upper littoral, influenced occasionally or continually by waves;
**stenotherm** - living within narrow temperature range (cold stenotherm = in low temperatures, thermophilic stenotherm = in high temperatures);
**stromatolites** - calcareous deposits of various and characteristic shapes, resulting from the biolithogenic activity of cyanoprokaryotes;
**subaerophytic** - living in aerophytic localities, but only after intermittent water supply, i.e., from rain, flowing water, dripping water, fog, etc. (usually in occasionally or seasonally dry localities);
**sulphuretum** - community of microorganisms (mainly bacteria and cyano-

bacteria-cyanoprokaryotes) that lives in localities with high sulphur content (usually in anaerobic conditions);
**supralittoral** - aerophytic localities above the littoral zone of lakes and seas, usually still influenced by the water body;
**terrestrial** - living on or below the surface of soil, or in biotopes connected with soil;
**thermal, thermobiotic, thermophilic** - living in biotopes with higher temperature, usually constantly over 20–30 °C; mainly in thermal springs and waters;
**thiobiocenoses** – communities of organisms living in localities with high sulphur content (bottom of deep lakes, thermal springs, combined with halophilic biotopes, etc.);
**travertines** - calcareous deposits and cascades resulting from the biolithogenic activity of cyanoprokaryotes in calcium-rich aquatic biotopes, usually in springs, streams, waterfalls;
**trophic level** - degree of nutrient level in biotopes (katharobic – oligotrophic – mesotrophic – eutrophic – hypertrophic waters);
**tychoplanktic** - planktic in deeper layers of water bodies; usually developing on the bottom and secondarily occurring in the plankton;
**waterbloom** - mass development of planktic species, forming macroscopically visible floating colonies, or water discolouration; a typical, macroscopically recognizable waterbloom is usually composed of species with gas vesicles (aerotopes);
**xerotherm** - preferring dry and warm localities.

## Acknowledgment

The authors wish to thank to everyone, who has helped us with the work of this book. We are particularly indebted to J. W. G. Lund (Ambleside) for the language revision and numerous important remarks regarding the text. Other colleagues we would like to thank are B. Büdel (Kaiserslautern), A. Couté (Paris), G. Cronberg (Lund), I. Dor (Jerusalem), A. Economou-Amilli (Athens), S. Golubić (Boston), J. Komárková (České Budějovice), L. Kováčik (Třeboň), B. Meyer (Plön), D. Mollenhauer (Lochmühle), T. Țafas (Athens), and K. Thomasson (Uppsala). We are also grateful to Mrs. D. Švehlová (Třeboň) for very valuable technical help.
We particularly appreciate the patience and support of our wives during the long preparation time for this book.

# Systematic part

Note: Several traditional chroococcal genera (comp. Geitler 1932) were recognized to be non cyanobacterial organisms, and they were excluded from the system of cyanoprokaryotes. They are, e.g., *Cyanotheca* Pascher 1914 (type: caulobacteria), *Dactylococcopsis* Hansgirg 1888 (type: green algae), *Marssoniella* Lemmermann 1900 (type: microsporidian genus *Gurleya*), *Tetrapedia* Reinsch 1867 (type: green algae), *Tetrarcus* Skuja 1934 (type: probably Rhodobacteria).

## 1. Family **Gloeobacteraceae** Komárek et Anagnostidis 1995

Cells solitary or in mucilaginous clusters with irregular outline and with irregularly agglomerated cells. Cells almost spherical or slightly elongate to rod shaped, nonmotile, either without mucilage or with a very narrow mucilaginous envelope or with distinct, wide and slightly stratified mucilaginous envelopes, which sometimes enclose small groups of cells; envelopes colourless or bluish. Cell content pale blue-green, reddish, violet, or almost colourless, without thylakoids and phycobilisomes, but with chlorophyll *a* and phycobilins (photosynthetic cyanoprokaryotic pigments). A single cytoplasmic membrane envelopes the protoplasma and pigments. Without involution cells.
Cell division in one direction in successive generations, in elongate cells (*Gloeobacter*) crosswise and perpendicular to their longitudinal axis. Cells grow into the original form before the next division.

Notes: Only one genus *Gloeobacter* is described. However, Andreoli & al. (1989) studied the ultrastructure of a picoplanktic species with more or less spherical or widely oval cells, 0.5–0.8 µm in diameter (long), and enclosed within a very

---

GLOEOBACTER
**Fig. 1.** *Gloeobacter violaceus*, form of mucilaginous colonies: **a** – from Rippka & al. (1974), **b** – from Waterbury & Rippka (1989); redrawn from microphotos.

CYANOBACTERIUM
**Fig. 2.** *Cyanobacterium stanieri*, outlines of cells within mucilaginous envelopes: **a** – from Rippka & Cohen-Bazire (1983), **b** – reference strain PCC 7202 of the «*Cyanobacterium*-cluster of the *Synechococcus*-group» from Waterbury & Rippka (1989), **c** – from Komárek & al. (1999).
**Fig. 3.** *Cyanobacterium cedrorum*, solitary cells: after Sauvageau (1892).

CYANOBIUM

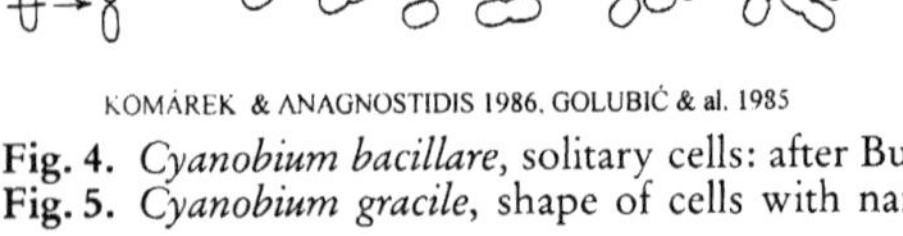


**Fig. 4.** *Cyanobium bacillare*, solitary cells: after Butcher (1952).
**Fig. 5.** *Cyanobium gracile*, shape of cells with narrow mucilaginous envelopes:

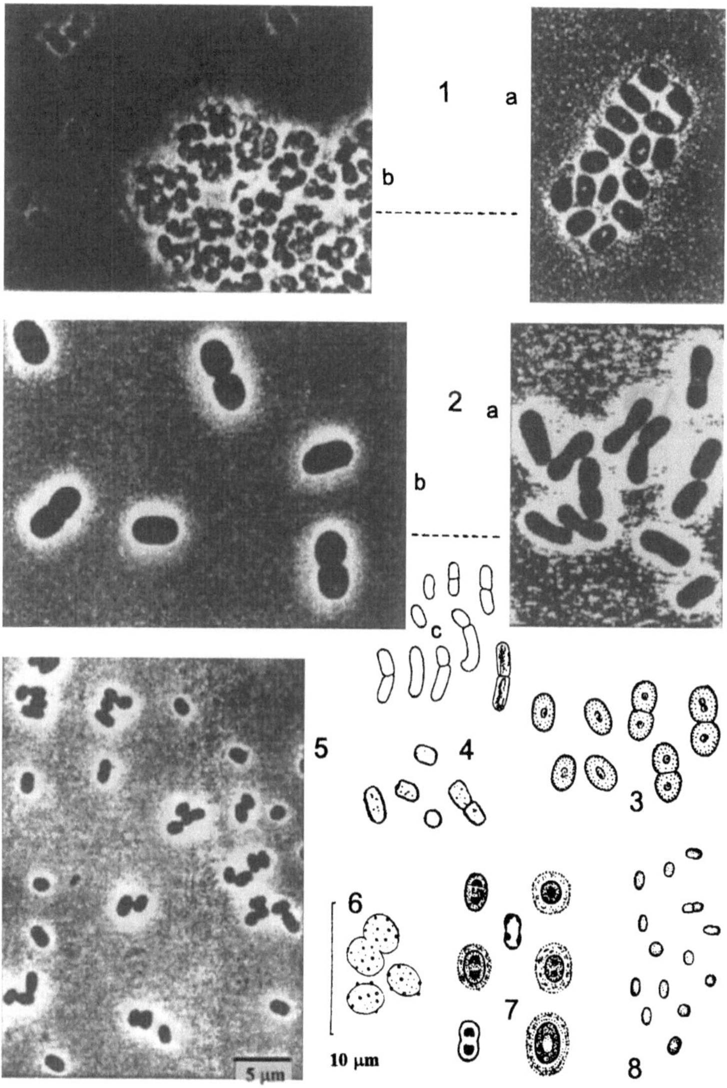


reference strain PCC 6307 of the «*Cyanobium*-cluster of the *Synechococcus*-group» from Waterbury & Rippka (1989), type species of the genus.

**Fig. 6.** «*Synechococcus*» *ferrunginosus*: after Wawrik (1979).

**Fig. 7.** *Cyanobium gaarderi*, solitary cells with fine mucilaginous envelopes: after Ålvik (1934).

**Fig. 8.** *Cyanobium parvum*, solitary cells: after Chu & al. (1991).

narrow mucilaginous envelope, from brackish waters of the lagoon S. Gilla near Cagliari (Italy – Sardinia). They consider several specimens to be probably cyanoprokaryotes without thylakoids. Their organisms should be classified into a special genus of *Gloeobacteraceae*, which differs from *Gloeobacter* in cell form and by the absence of distinct mucilaginous envelopes. Unfortunately, the description of Andreoli & al. is very short (cell structure is not clearly defined, pigment composition and type of cell division are unknown) and their population needs further study. Similar types found in small freshwater oligotrophic lakes in northern USA (NY) were investigated by Corpe & Jensen (1992).

## 1. **Gloeobacter** Rippka et al. 1974

Cells solitary or aggregated in irregular groups, widely oval or rod shaped, surrounded by narrow, delimited gelatinous envelopes, containing one or more ensheathed cells. Cells non motile.
Cell division by simple binary transverse fission, perpendicular to the long axis of the cell. Reproduction probably by liberation of solitary ensheathed cells or small groups of cells.
**Important references:** Rippka & al. (1974); Komárek & Anagnostidis (1986); Waterbury & Rippka (1989).

### 1. **Gloeobacter violaceus** Rippka et al. 1974 (fig. **1**)

*Gloeothece coerulea* sensu auct. (since *Gloeothece coerulea* Geitler really is synonymous with *Gloeobacter violaceus*, the correct name must be, according to both the Botanical and the Bacteriological Codes, «*Gloeobacter coeruleus*»).

Cells form irregular clusters, widely oval to rod shaped with rounded ends, with multilayered, bluish sheaths around the cells or groups of cells. Cells 2–3 × 1.5 µm, with almost colourless content and solitary granules (violet in the mass).
Occurrence: Described from a culture, isolated from a sample of epilithic cyanoprokaryotes from a limestone rock; Switzerland.

**Other possible species:**
– Another species of the genus *Gloeobacter* could be *Aphanothece caldariorum* (for description see p. 86–88); this presumption needs to be confirmed by electron microscopy.

# 2. Family **Synechococcaceae** Komárek et Anagnostidis 1995

Cells solitary or in mucilaginous colonies with cells arranged irregularly or more or less in one direction in a colony, sometimes forming pseudofilamentous rows. Mucilaginous colonies mainly structureless, amorphous, but genera also occur with spherical colonies and cells distributed at their periphery or at the ends or inside of mucilaginous stalks; several genera have special gelatinous envelopes around the cells. Cells rarely spherical, usually slightly or distinctly elongate, widely oval, oval, ellipsoidal, fusiform, cylindrical (rod like) to almost «filamentous».
Cell division always only in one plane in successive generations, usually perpendicular to the longitudinal axis of the cell (exception: cells of *Johannesbaptistia* divide perpendicular to the shorter axis).

Note: This family is defined by elongated cells and by cell division in one plane (perpendicular to longer cell axis) in successive generations. It is divided into two subfamilies according to ability to form irregular or filamentous involution cells. However, genera with very various ultrastructural features (e.g., thylakoid arrangements), occur in this family, which indicate wider diversity (e.g., *Cyanobium – Synechococcus – Cyanobacterium – Cyanothece*, etc.).

**Key to the subfamilies:**

**1a** Cell length/width ratio usually (1)1.2–3 : 1, rarely higher; cells solitary or irregularly arranged in formless or more or less spherical colonies, without typical pseudofilamentous formations; involution cells of irregular outline; facultative occurrence of nanocytes . . . . . . . . **1. Aphanothecoideae** (p. 37)

**1b** Cell length usually several times longer than cell width (except in pseudofilaments can be wider than long); cells cylindrical, rarely short oval, in colonies often arranged in one direction (sometimes in pseudofilamentous rows); involution cells long, cylindrical, «filamentous», with asymmetric cell division; without nanocytes . . . . . . . . . . . . . . **2. Synechococcoideae** (p. 101)

# 1. Subfamily **Aphanothecoideae** Komárek et Anagnostidis 1995

*Aphanothecoideae* Komárek et Anagnostidis 1986 sine lat. diagn.

Cells solitary or in mucilaginous colonies, usually irregularly arranged. Several genera have cells in the periphery of mucilaginous spheres or at the ends of or within gelatinous stalks, radiating from the colonial center. In several genera, cells are surrounded by their own, sometimes lamellate and coloured envelopes. Cells almost spherical, oval, widely oval or rod shaped; cell length/width ratio usually less than 3:1; involution cells (if present) irregular, not filamentous. Pseudofilaments are missing (indistinct and irregular rows of cells appear only in *Cyanodictyon* and *Lithococcus*).

Cell division by binary fission, only in one plane in successive generations, perpendicular to the longer cell axis (but sometimes the cells change very soon their position, during or just after division; in this case, the obligatory transversal cell division is not quite evident); cells divide into two isomorphic cells, growing to the original size before next division. Nanocytes facultatively present in some but not all species.

**Key to the genera:**

**1a** Cells solitary or in groups (but not forming mucilaginous colonies), more or less widely oval or (rarely) almost cylindrical, without mucilage or sometimes with narrow, fine and diffluent gelatinous envelopes around single cells . . . . . . . . . . . . . . . . . . . . . . . . . . . . . . . . . . . . . . . . . . . . . . . . . **2**

**1b** Cells in gelatinous colonies, usually with more or less distinct mucilaginous envelopes, almost spherical, oval or cylindrical . . . . . . . . . . . . . . . . . . **4**

**2a** Cells cylindrical to widely oval, with ± homogeneous contents, striated plasma, or with differentiated centro- and chromatoplasmic regions; cells mainly small, (0.4)1–12(30) µm long . . . . . . . . . . . . . . . . . . . . . . . . . **3**

**2b** Cells oval to widely oval, with reticulately keritomized plasmatic content (irregular arrangement of thylakoids); cells (6.2)7–45(100) µm long . . . . . . . . . . . . . . . . . . . . . . . . . . . . . . . . . . . . . . . . **3. Cyanothece**

**3a** Cells ± oval, (0.4)1–4.5(10) × 0.2–3(6) µm, with differentiated centro- and chromatoplasma (= peripheral arrangement of thylakoids) . **1. Cyanobium**

**3b** Cells cylindrical to widely oval, (2)3.7–12(30) × 1.7–7(20) µm, with homogeneous cell content, but often irregularly lengthwise «striated» (at higher magnifications) . . . . . . . . . . . . . . . . . . . . . . . . . . **2. Cyanobacterium**

Notes: The genera *Cyanobium* and *Cyanobacterium* are simple unicellular cyanoprokaryotes which were originally described from cultures without adequate morphological characters. About 50 strains were studied by biochemical and ecophysiological methods (Rippka & al. 1979; Rippka & Cohen-Bazire 1983; Waterbury & Rippka 1989) and different types were characterized under the designations of «clusters» (*Cyanobium, Cyanobacterium, Cyanothece* [not sensu Komárek 1976], *Synechococcus* and «marine clusters a – c»). This whole group contains several species, which can be divided into above defined genera also according to various types of thylakoid arrangement (Komárek & al. 1999): this diversity must be studied in complexity, particularly concerning cytomorphological variation, fine structure, molecular methods and comparison of more strains and natural populations. The review of both genera are modified according to Komárek & al. (1999).

**4a** Cells form spherical or irregular mucilaginous colonies . . . . . . . . . . . . . 5
**4b** Cells situated at or near the ends of mucilaginous, morphologically diverse stalks . . . . . . . . . . . . . . . . . . . . . . . . . . . . . . . . . . . . . . . . . . . 13
**5a** Cells distributed more or less irregularly or in irregular rows throughout the mucilage of the colonies (but not forming pseudofilaments) . . . . . . . . . . 6
**5b** Cells situated only on or near the surface of gelatinous spheres . . . . . . . 12
**6a** Cells arranged into more or less spherical or amorphous colonies, always irregularly, but with different densities of cells in different regions of a colony . . . . . . . . . . . . . . . . . . . . . . . . . . . . . . . . . . . . . . . . . . . 7
**6b** Cells arranged in mucilaginous, more or less spherical or irregular net-like colonies, at least partly in rows . . . . . . . . . . . . . . . . . . . . . . . . . 10
**7a** Colonies contain small ferric precipitates of diverse shape; cells almost spherical or widely oval, rarely short cylindrical; without their own distinct gelatinous envelopes; colonies planktic, more or less spherical (at least in young stages) or irregular . . . . . . . . . . . . . . . . . . . . . . . . . . . . . . . . 8
**7b** Colonies usually without ferric precipitates (but can be incrusted with lime); cells oval or rod shaped; sometimes (at least on the colonial periphery) the cells are surrounded by their own gelatinous envelopes (excluding planktic species); colonies mainly amorphous . . . . . . . . . . . . . . . . . . . . . . . . 9
**8a** Colonies more or less spherical with ferric precipitates in granular form; cells almost spherical or widely oval . . . . . . . . . . . . . . . . . . . **4. Cyanogranis**
**8b** Colonies irregular with ferric precipitates in the form of minute black rings or dot-like granules at cells; cells widely oval or shortly cylindrical, rounded at poles . . . . . . . . . . . . . . . . . . . . . . . . . . . . . . . . . . . **6. Cyanocatena**
**9a** Colonies with calcareous incrustations; participating in travertine formation; benthic in streams of limestone areas . . . . . . . . . . . . . . . . . **5. Lithomyxa**
**9b** Colonies without calcareous incrustations; not participating in travertine formation (with various ecology) . . . . . . . . . . . . . . . . . . . . . . . . . . . 14
**10a** Colonies irregularly reticulate, composed of mucilaginous strands in which the cells are embedded and are arranged more or less in rows . . . . . . . . 11
**10b** Colonies more or less spherical, planktic, with cells clustered in the center, from which short, sometimes indistinct rows radiate in all directions to the colony's periphery . . . . . . . . . . . . . . . . . . . . . . . . . . . . . **9. Radiocystis**
**11a** Free-living (planktic) or endogloeic in other planktic cyanoprokaryotes . . . . . . . . . . . . . . . . . . . . . . . . . . . . . . . . . . . . . . . . **7. Cyanodictyon**
**11b** Endolithic in calcareous rocks . . . . . . . . . . . . . . . . . . . . **8. Lithococcus**
**12a** Cells situated on the surface of gelatinous colonies, which are more or less

spherical, sometimes slightly irregular or composed of firm, mucilaginous, homogeneous spheres or stalks; epipelic and metaphytic
. . . . . . . . . . . . . . . . . . . . . . . . . . . . . . . . . . . . . . . . . **10. Epigloeosphaera**

**12b** Cells more or less localized in one layer under the surface of more or less spherical mucilaginous colonies (the mucilage extends beyond the cells); free floating . . . . . . . . . . . . . . . . . . . . . . . . . . . . . . **11. Lemmermanniella**

**13a** Colonies more or less spherical, free floating, with slightly curved cells attached transversely to the ends of more or less thin, gelatinous stalks (staining !), which are pseudodichotomously branched and radiate out from the center of the colonies . . . . . . . . . . . . . . . . . . . . . . **12. Cyanonephron**

**13b** Colonies in the form of thick, gelatinous, firm, rigid stalks, which are sometimes lamellate, attached to the substrate, with cells at or within their apices . . . . . . . . . . . . . . . . . . . . . . . . . . . . . . . . . . . . **13. Hormothece**

**14a** Cells completely (in planktic species) or, at least in the center of a colony, without their own gelatinous envelopes (non-planktic species, sometimes with coloured mucilage) . . . . . . . . . . . . . . . . . . . . . . **14. Aphanothece**

**14b** All cells enveloped by their own gelatinous, usually lamellate and often coloured envelopes (with the exception of special developmental stages)
. . . . . . . . . . . . . . . . . . . . . . . . . . . . . . . . . . . . . . . . . . . **15. Gloeothece**

## 1. **Cyanobium** Rippka et Cohen-Bazire 1983

*Synechococcus* sensu auct. post., sine typo; ?*Chroobacter* Pringsheim 1953

Cells solitary or in twos after division, not in colonies, small, usually about 1–2(4) µm long and about 1(–3) µm wide, without gelatinous envelopes, oval, ellipsoid to shortly rod-shaped, usually with well visible chromatoplasma. Thylakoids are arranged parietally along the cell walls.

Cell division by binary fission (obligatory by pinching ?), transverse to the long axis, resulting in two isomorphic daughter cells. Cells grow into the original shape and size before the next division.

Important references: Komárek (1976, sub *Synechococcus* p.p.); Chang (1980; sub *Synechococcus*); Rippka & Cohen-Bazire (1983); Komárek & Anagnostidis (1986); Waterbury & Rippka (1989); Albertano & al. (1996); Komárek (1996); Komárek & Cepák (1997); Komárek & al. (1999).

Notes: Other small picoplanktic species studied in recent years by numerous biologists (Y. Watanabe, Maeda, Kennaway, Albertano and others) belong to this genus, as well as various species and populations described under the generic name *Synechococcus* (from which they differ morphologically by more oval cell form, by lack of filament-like involution cells with asymmetric cell division and partly by the thylakoid arrangements. Rippka & Cohen-Bazire (1983) distinguished in *Cyanobium* three types at the species level from nine strains studied (but without valid taxonomic descriptions). Other important, but taxonomically not well defined populations are in the list of unrevised *Cyanobium* species.

**Key to the species:**

**1a** Marine, or living in salty habitats . . . . . . . . . . . . . . . . . . . . . . . . 2

**1b** Freshwater . . . . . . . . . . . . . . . . . . . . . . . . . . . . . . . . . . . . . . . . . 4

**2a** Cells 0.8–1.7 µm wide, red, usually intensely motile, in oceanic plankton
. . . . . . . . . . . . . . . . . . . . . . . . . . . . . . . . . . . . . . . . . . . **1. C. waterburyi**

**2b** Cells 1.2–1.5µm wide, reddish or blue-green, not intensely motile, in brackish or halophilous littoral habitats . . . . . . . . . . . . . . . . . . . . . . . . . . 3

**3a** Cells blue-green to slightly reddish, 1.5–2.8 × 1.2–1.5 µm . **2. C. gaarderi**

**3b** Cells blue-green, 1.7–4.5 × ±1.5 µm . . . . . . . . . . . . . . . . **3. C. bacillare**

4a Red cells, living in mucilage of planktic cyanoprokaryotes in lakes . . . . . 5
4b Blue-green cells . . . . . . . . . . . . . . . . . . . . . . . . . . . . . . . . . . . . . . 6
5a Cells 1–1.5 × 0.8 µm, (G/C cont. = 36) . . . . . . . . . . . . . . **4. C. rubescens**
5b Cells 1.2–2 × 0.8 µm, (G/C cont. = 70) . . . . . . . . . . . . . **5. C. virga-rosea**
6a Cells 0.4–4.5 µm long . . . . . . . . . . . . . . . . . . . . . . . . . . . . . . . . . 7
6b Cells 5–10 × 3–6 µm . . . . . . . . . . . . . . . . . . . . . . . . . . . **9. C. parvum**
7a Cells 0.2–0.4 µm wide, thermophilic . . . . . . . . . . . . . . . . . . . **6. C. gracile**
7b Cells 0.7–1.5 µm wide, not thermophilic . . . . . . . . . . . . . . . . . . . . . . . 8
8a Cells1.5–3 × 0.9–1.1 µm, planktic . . . . . . . . . . . . . . . . . **7. C. plancticum**
8b Cells 1.5–4.5 × 0.7–1.5 µm, epiphytic . . . . . . . . . . . . . . **8. C. diatomicola**

**1. Cyanobium waterburyi** Komárek 1999 (fig. **14**)

*Synechococcus* sp. sensu Waterbury et al. 1979

Solitary cells without mucilage, almost spherical, oval to rod shaped with rounded ends, red (not able to adapt chromatically), intensely actively motile, 1.8–2.2 × 0.8–1.7 µm, only dividing transversely (in one plane), with peripherally arranged thylakoids.Maximal temperature limit = 30 °C, G/C (% mol) = 59–60.
Occurrence: Marine, halophilic, autotroph, in oceanic plankton (down to 400 m depth!) of the Pacific Ocean (off the coast of Peru), Atlantic Ocean (northern Gulf Stream, Georges Bank, Sargasso Sea, Woods Hole Harbor) and Arabian/Persian Gulf.

**2. Cyanobium gaarderi** (Ålvik) Komárek et al. 1999 (fig. **7**)

*Synechococcus gaarderi* Ålvik 1934; *Cyanothece gaarderi* (Ålvik) Komárek 1976

Cells solitary or in pairs, not forming chain-like series, spherical or widely ellipsoidal, with indistinct, colourless, wide mucilage envelope, 1.5–2.8 × 1.2–1.5 µm.
Occurrence: Marine, planktic in brackish and salty water bodies near the sea coast; Norway.

**3. Cyanobium bacillare** (Butcher) Komárek et al. 1999 (fig. **4**)

*Synechococcus bacillaris* Butcher 1952

Cells solitary or in short chain-like formations, almost spherical, oval or cylindrical, with a blue-green, finely granulate content, 1.7–4.5 × more or less 1.5 µm.
Occurrence: Planktic in reservoirs containing seawater; England.

**4. Cyanobium rubescens** (Chang) Komárek et al. 1999

*Synechococcus rubescens* Chang 1980

Cells short, rod-shaped, with rounded ends, pink to reddish, 1–1.5 × 0.8 µm, without gelatinous envelopes, on agar forming short rows; thylakoids arranged concentrically; G/C (% mol) = 36.
Occurrence: Freshwater, epiphytic on the planktic, waterbloom- forming *Planktothrix rubescens*; Switzerland.

**5. Cyanobium virga-rosea** (Chang) Komárek et al. 1999

*Synechococcus virga-rosea* Chang 1980

Cells solitary or in pairs after division, forming short rows on agar, rod-shaped, rounded at the ends, without gelatinous envelopes, pink or pale red, 1.2–2 × 0.8 µm; thylakoids arranged concentrically; G/C (% mol) = 70.

Occurrence: Freshwater, epiphytic on the planktic, waterbloom- forming *Planktothrix rubescens*; Switzerland.

**6. Cyanobium gracile** Rippka et Cohen-Bazire 1983 (fig. **5**)

Cells oval, pale blue-green, without enveloping mucilage, solitary, with homogeneous content, 0.4–1.5(2.4) × ± 0.2–0.4 µm. Obligate photoautotroph. GC (% mol) = 66–71.
Occurrence: Described from cultures, thermophilic strains (Tmax = 35–41 °C), (six known strains).

**7. Cyanobium plancticum** (Drews) Komárek et al. 1999

*Synechococcus plancticus* Drews et al. 1961

Cells solitary, oval to rod-shaped, without mucilage (?), blue-green, 1.5–3 × 0.9–1.1 µm.
Occurrence: Freshwater, planktic in ponds containing waste waters, sometimes in masses; Czech Republic, Germany.

**8. Cyanobium diatomicola** (Geitler) Komárek et al. 1999 (fig. **9**)

*Synechococcus diatomicola* Geitler 1953

Cells solitary, in pairs or in short rows, one after another (especially after rapid, repeated division), later in agglomerations, without mucilage or enveloped by very fine, diffluent, colourless and indistinct slime (staining !). Cells short rod-shaped, straight or rarely slightly curved, rounded at the ends, with homogeneous, pale or bright blue-green content, (1.5) 2.3–3 (4.5) × 0.7–1.2 (1.5) µm. Thylakoids arranged parietally (concentrically in cross section).
Occurrence: Freshwater, submersed, epiphytic on diatom frustules and filamentous algae (*Oedogonium*) in clear waters (springs, swamps, littoral of peaty lakes); Austria, Switzerland, Slovakia (?), recorded also in China.

**9. Cyanobium parvum** (Migula) Komárek et. al 1999 (fig. **8**)

*Synechococcus parvus* Migula 1906; *Synechococcus minutus* W. West 1912; *Synechococcus cedrorum* Sauvageau sensu auct. (hydrophytic, metaphytic populations)

Cells solitary or in pairs, oval to cylindrical with rounded ends, blue-green, 5–10 × 3–6 µm.
Occurrence: Freshwater, metaphytic in small stagnant waters, pools, ditches, backwaters; sporadically distributed all over the temperate and subtropical world, possibly cosmopolitan.

**Species outside Europe:**
To the *Cyanobium*-group probably belong also species from other biotopes (outside Europe). E. g.,
- marine *Cyanobium oceanicum* (Hall et Claus 1964) Komárek et al. 1999;
- thermophilic *Cyanobium amethystinum* (Copeland) Komárek et al. 1999 (fig. **17**);
- thermophilic *Cyanobium eximium* (Copeland) Komárek et al. 1999 (fig.**15**);
- thermophilic *Cyanobium roseum* (Copeland) Komárek et al. 1999 (fig. **18**).

**Unrevised taxa:**
- *Synechococcus ferrunginosus* Wawrik1979 (fig. **6**): Cells solitary, without mucilage, ellipsoidal, with pale blue-green content, without recognizable chromato-

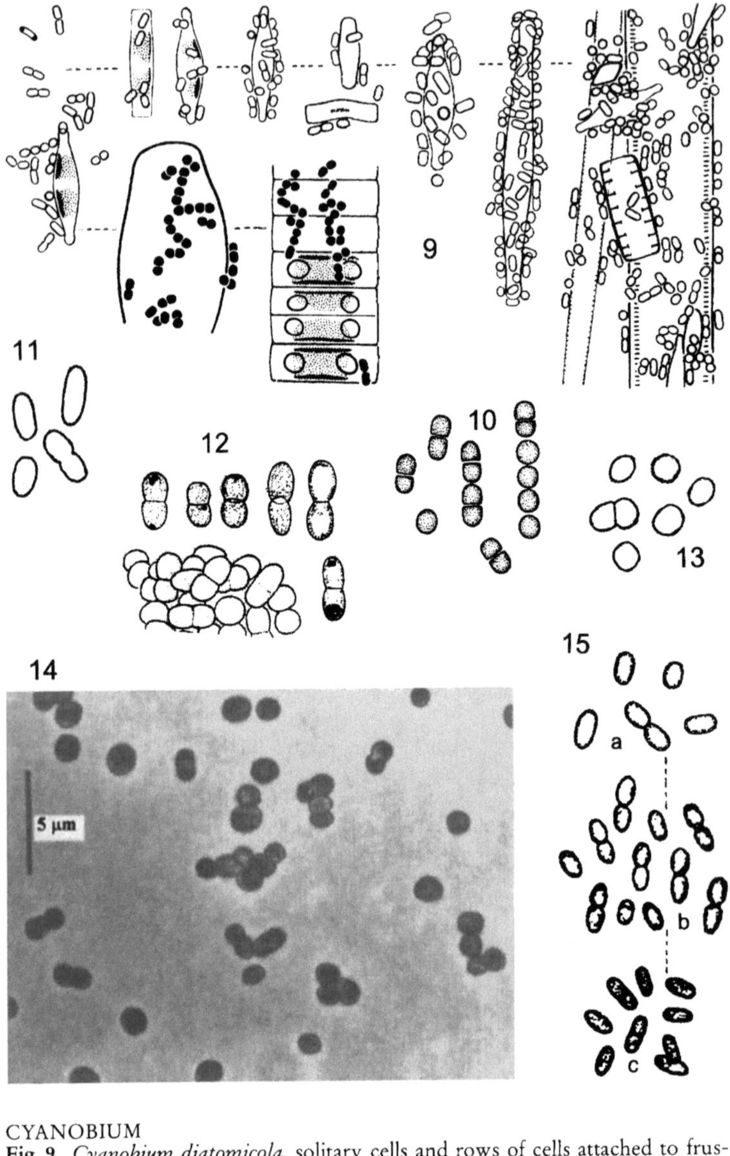


CYANOBIUM

**Fig. 9.** *Cyanobium diatomicola*, solitary cells and rows of cells attached to frustules of various diatoms: after Geitler (1953).

**Fig. 10.** *Cyanobium diatomicola* (sub *Synechococcus*), dividing cells and rows of cells: after Chu & al. ed. (1992).

and centroplasma, with fine ferric precipitates (in form of small solitary granules) on the cell surface; cells 3–4 × 2.5–3 µm. – Described from plankton of a small pond in time of melting of ice (March, temp. = 3 °C); Austria (Waldviertel).
– *Synechococcus vantieghemii* (Pringsheim) Bourrelly 1970 [Syn.: *Chroobacter vantieghemii* Pringsheim 1953] (fig. **11**): Cells solitary, rod-shaped, rounded at the ends, without mucilage, blue-green, 1.2–1.5 × 0.6–0.8 µm. – Isolated from a sample of mud; England.
– *Synechococcus* sp.(*cedrorum*) sensu Böcher 1949 (fig. **12**): Cells in irregular groups in the mucilage of other cyanoprokaryotes, in various *Oscillatoria* and *Pseudanabaena* mats, solitary or in twos after division, ellipsoid or oval, pale blue-green, usually with clearly differentiated centro- and chromatoplasma, sometimes with prominent polar granules (aerotopes?), 4–7 × 3–4 µm; motility not proved. – Freshwater, benthic (sapropelic) in a lake; Denmark.
– *Synechococcus* sp. sensu Kennaway in mscr. (fig. **13**): Cells solitary, almost spherical or subspherical to widely oval, without mucilage. Cell content more or less homogeneous, pale blue-green, without prominent granules; cells 0.8–1.2 µm in diameter. – Freshwater, planktic in clear lakes; UK (Wales).
– Andreoli & al. (1989a) found in the brackish lagoon of S. Gilla near Cagliari (Italy – Sardinia) a common picoplanktic cyanoprokaryote with elongated, narrow-oval cells and usually one peripheral thylakoid, with cells about 0.6–0.8 × 0.3–0.4 µm, without any visible mucilage around cells. According to the still insufficiently known features, this microorganism could belong in the vicinity of *Cyanobium*, but it needs to be reconsidered.
– Among cyanobacteria included in Bergey's Manual 1989 (elaborated by Castenholz & Waterbury) were presented 6 clusters corresponding more or less to *Cyanobium* species definable by traditional phycological methods, after the evaluation of only 51 strains. The mentioned clusters are characterized by the different G/C (% mol) – ratios, with additional ecophysiological criteria. The evaluation of these clusters according to the traditional taxonomic criteria see in Komárek (1996) (fig. **16**).
– Out of numerous picoplanktic morphotypes from small oligotrophic lakes, the ultrastructure of which is described in Corpe & Jensen (1992: lakes in NE. USA-NY, e.g. lake Arthur), at least those depicted in their fig. 4A, correspond morphologically to *Cyanobium*; the cells (0.8–0.9 × 0.3 µm) have few parietal thylakoids. Further, the authors present several other morphotypes, with more or less oval cells with varied dimensions and thylakoids oriented radially, irregularly, or only at one side of elongated cells (their figs. 1A – arrows, 2A-D, 3C, 4B-C; comp. *Cyanobacterium*).
– In literature populations morphologically similar to *Cyanobium* occur, the taxonomic position of which is still questionable, e.g. «*Synechococcus* cf. *elongatus*» from the domestic sewage in Jerusalem, Israel (Dor in Komárek 1976; fig. **19**),

---

**Fig. 11.** «*Synechococcus*» *vantieghemii*: after Pringsheim from Bourrelly (1970).
**Fig. 12.** «*Synechococcus*» sp.: after Böcher (1949).
**Fig. 13.** «*Synechococcus*» sp. sensu Kennaway: orig. from live material from nature.
**Fig. 14.** *Cyanobium waterburyi* («*Synechococcus*» sp. sensu Waterbury & al.), outline of cells from the photomicrograph (strain Syn-48): from Waterbury & al. (1979).
**Fig. 15.** *Cyanobium eximium*, solitary cells, dividing cells: **a** – after Copeland (1936), **b** – after Emoto & Yoneda (1941), **c** – after Dor (1967), (all sub *Synechococcus eximius*).

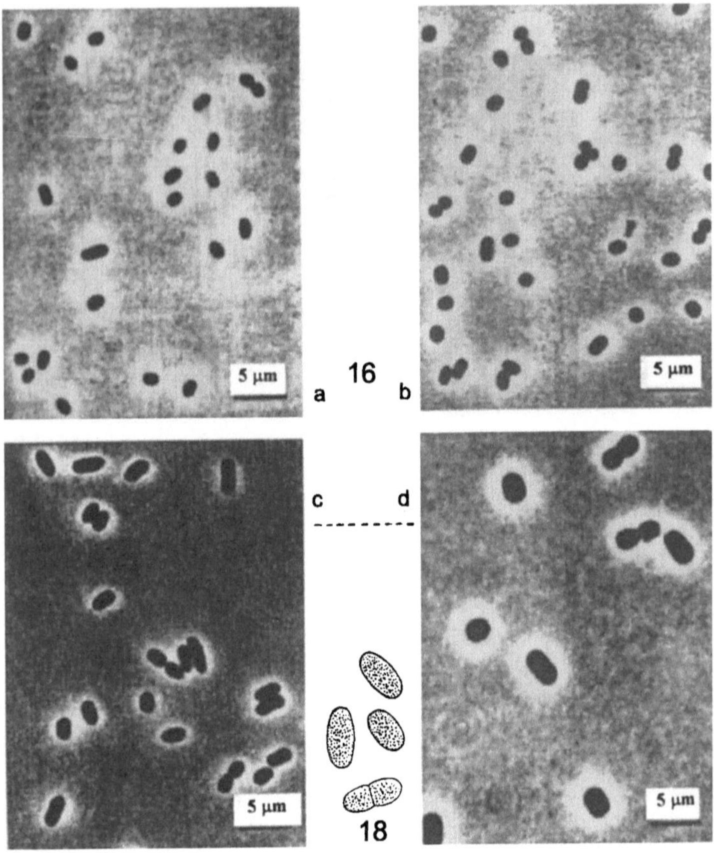


16

18

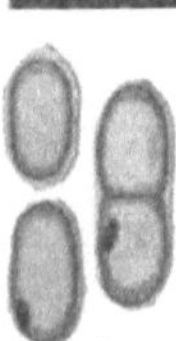

17

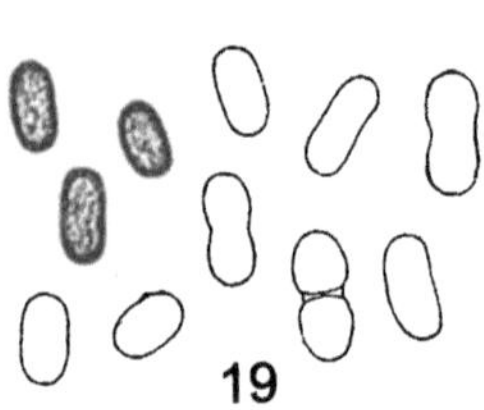

19

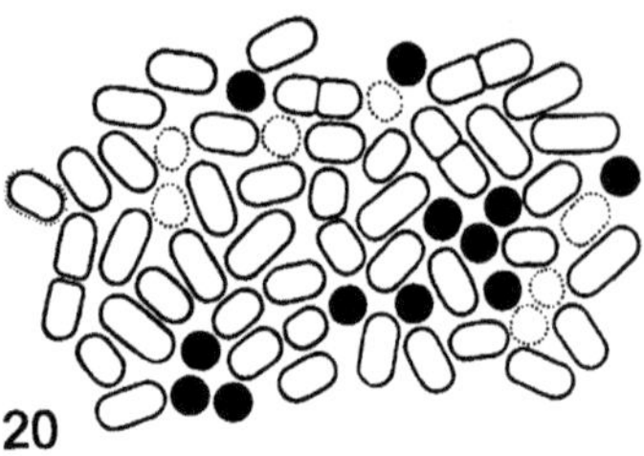

20

CYANOBIUM

**Fig. 16.** Examples of *Cyanobium*-like strains, presented in Waterbury & Rippka (1989, Bergey's Manual) as separate clusters: **a** – reference strain WH 8103 of «Marine-cluster A of the *Synechococcus*-group», **b** – reference strain WH 5701 of «Marine-cluster B of the *Synechococcus*-group», **c** – reference strain PCC 6301 of

or «*Synechococcus* sp.» growing epiphytic on *Potamogeton natans* in Austria (Geitler 1962; fig. 20).
- Freshwater picoplanktic populations of *Cyanobium*-type are known from oligotrophic to mesotrophic lakes of northern Germany, where their maxima occur usually in deeper layers (about 20 m) (Krienitz in mscr.). Cells are solitary, almost spherical or slightly oval, with 1-2(3) parietal thylakoids, division by pinching.
- Three picoplanktic populations of *Cyanobium*-like organisms were described by Albertano & al. (1996,1997a and 1997b) from the photic zone of the central Baltic Sea , designated like cyanobacterial cells < 3µm of «*Synechococcus*-type»: (a) rods 0.25-2.01 × 0.09-0.55 µm, (b) ellipsoids 0.75-5.67 × 0.75-1.20 µm, and (c) coccoids 1.22-5.13 × 1.22-3.03 µm. All three types correspond ± to marine-clusters sensu Waterbury & Rippka (1989). The parietal thylakoids, typical for *Cyanobium*, were demonstrated in two from these strains, one red and one green, in the poster of Albertano & al. (1997b). In both these strains was identified a narrow slimy layer around cells, with very interesting radial and tubular «spinae» under all culture conditions.
- Numerous data about oceanic and marine «*Synechococcus*»- populations (from subpolar to tropic areas) appeared in scientific papers in last years (comp., e.g., «Abstracts» 1997, e.g., Lantoine & Neveux, Vaulot, Wood & al., and others). The identity or diversity of all these populations is not studied yet. However, from information presented, numerous these populations correspond rather to the genus *Cyanobium* Rippka & Cohen-Bazire, than *Synechococcus*.

## 2. **Cyanobacterium** Rippka et Cohen-Bazire 1983

Cells solitary or in groups of twos after division, sometimes in irregular clusters, but never in colonies, without gelatinous envelopes, more or less cylindrical, rod-shaped with rounded ends up to widely oval, straight or slightly arcuate. Cell content homogeneous, without separation of centro- and chromatoplasma, but sometimes with lengthwise striation. Thylakoids situated in cells lengthwise; cell-wall S-layer of square lattice (p4) was identified in *C. minervae* and *C. cedrorum*. Cell division by binary fission (pinching), transversely to the longer cell axis, resulting in two isomorphic, rarely asymmetric daughter cells. Cells grow into the original shape and size before the next division.
**Important references:** Rippka & Cohen-Bazire (1983); Wilmotte & Stam (1984); Komárek & Anagnostidis (1986); Waterbury & Rippka (1989); Cepák (1993); Komárek (1996); Komárek & Cepák (1997); Komárek & al. (1999).

**Key to the species:**

| | | |
|---|---|---|
| 1a | Cells more or less cylindrical ............................ | 2 |
| 1b | Cells more or less oval ..................................... | 3 |
| 2a | Cells rounded-flattened at the ends, 2–3.4 µm wide ....... | **1. C. stanieri** |
| 2b | Cells rounded at the ends, 1.2–2(3) µm wide .... | **2. C. synechococcoides** |

«the *Synechococcus*-cluster of the *Synechococcus*-group», **d** – reference strain PCC 7002 of «Marine-cluster C of the *Synechococcus*-group».
**Fig. 17.** *Cyanobium amethystinum*: after Copeland (1936).
**Fig. 18.** *Cyanobium roseum*: after Copeland (1936).
**Fig. 19.** «*Synechococcus* cf. *elongatus*» : after Dor from Komárek (1976).
**Fig. 20.** «*Synechococcus* sp.»: after Geitler (1962).

**3a** Cells with ± homogeneous cell content, cells 5–12 × 2.8–4(5) µm; subaerophytic species . . . . . . . . . . . . . . . . . . . . . . . . . . . . **3. C. cedrorum**
**3b** Cells with lengthwise striate keritomization of chromatoplasma; from various water biotopes . . . . . . . . . . . . . . . . . . . . . . . . . . . . . . . . **4**
**4a** Thermal species, cells 4–10 × 3–4.5 µm . . . . . . . . . . . . . . . **4. C. minervae**
**4b** Freshwater or brackish species . . . . . . . . . . . . . . . . . . . . . . . . . . . **5**
**5a** Cells more or less cylindrical, 5–7 µm wide; metaphytic in moorland waters and pools . . . . . . . . . . . . . . . . . . . . . . . . . . . . . . . . **5. C. notatum**
**5b** Cells more or less oval, 10–20 µm wide . . . . . . . . . . . . . . . . . . . . . . . . **6**
**6a** Cells 9.5–17 × 10–12 µm, metaphytic in saline waters (littoral of Baltic Sea) . . . . . . . . . . . . . . . . . . . . . . . . . . . . . . . . . . . . . . . **6. C. diachloros**
**6b** Cells 17–30 × 12–20 µm, metaphytic in the littoral of northern freshwater lakes . . . . . . . . . . . . . . . . . . . . . . . . . . . . . **7. C. crassiusculum**

## 1. **Cyanobacterium stanieri** Rippka et Cohen-Bazire 1983 (fig. **2**)

Cells solitary or in twos after division, sometimes in small irregular clusters, without mucilage or with very narrow, diffluent layer around cells, more or less cylindrical, straight or slightly arcuate, 3.7–5.2(15) × 2–3.4 µm ; cell content pale blue-green, without granules, with ± homogeneous content, sometimes very indistinctly lengthwise, irregularly striated (immersion!). Division sometimes asymmetrical (under stress). GC (% mol) = 39–41. Non-motile.
Occurrence: Described from cultures (two known strains: PCC 7202, PCC 7502).

## 2. **Cyanobacterium synechococcoides** Komárek 1999 (fig. Table **2B/f**)

Cells solitary or aggregated in irregular clusters (mats), but without gelatinous envelopes. Cells cylindrical with more or less rounded ends, with blue-green, homogeneous or slightly lengthwise striated or finely granular content , 2–8(25) × 1.0–2(3) µm. Thylakoids arranged lengthwise.
Occurrence: Littoral of not polluted stagnant waters.

## 3. **Cyanobacterium cedrorum** (Sauvageau) Komárek et al. 1999 (fig. **3**)

*Synechococcus cedrorum* Sauvageau 1892; *Cyanothece cedrorum* (Sauvageau) Komárek 1976

Cells widely elliptical to cylindrical with rounded ends, solitary or in pairs, pale blue-green, 5–10 (12) × 2.8–4 (5) µm.
Occurrence: Subaerophytic, on the bark of trees and probably also on wet rocks; recorded mainly from warm areas of the northern temperate zone and sporadically from subtropical and tropical countries. Various concepts occur in literature, the hydrophytic populations particularly should be revised (some of them belong to *Cyanobium parvum*, see p. 41).

---

CYANOBACTERIUM
**Fig. 21.** *Cyanobacterium minervae*: **a-b** – drawings from Yellowstone Nat. Park after Copeland (1936), **c** – from Japan after Yoneda (1943), **d** – from Israel after Dor (1967), **e** – from China after Chu & al. (1992).
**Fig. 22.** Morphotypes of *Cyanobacterium minervae*, from thermal springs in Greece, after Anagnostidis (1961).

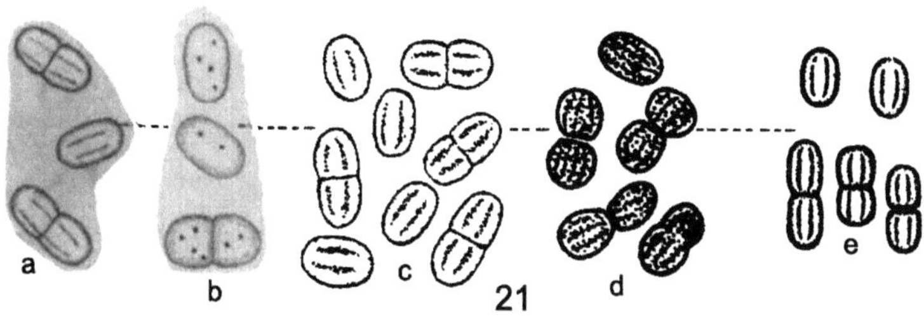


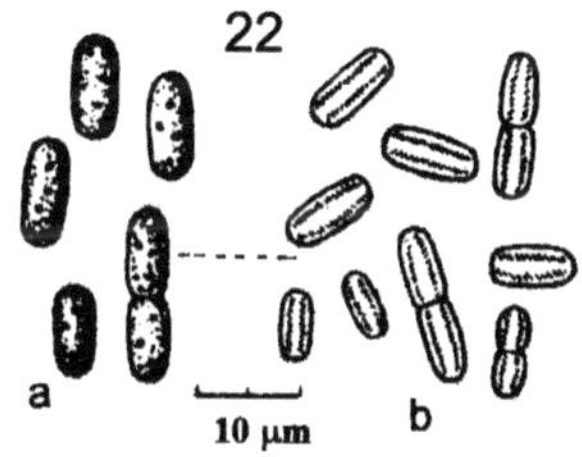


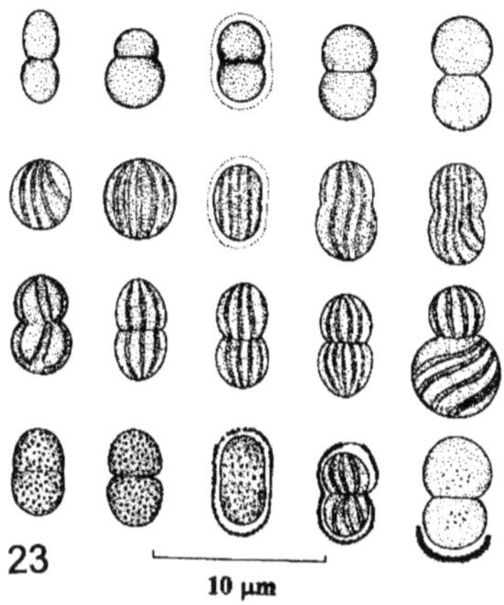


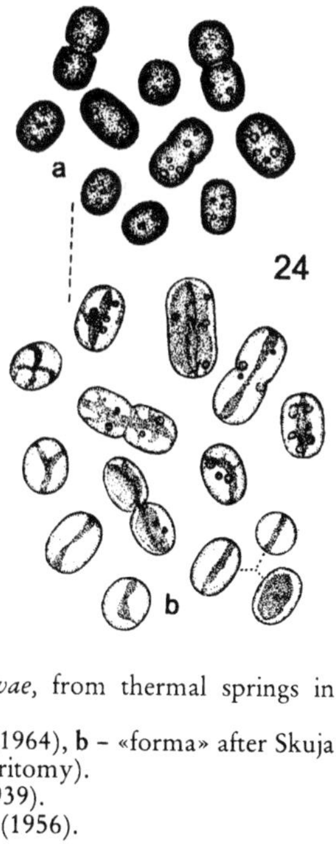


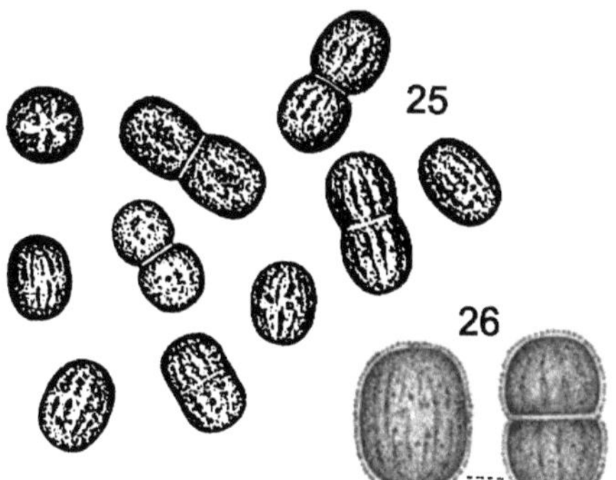


**Fig. 23.** Morphotypes of *Cyanobacterium minervae*, from thermal springs in Slovakia, after Hindák (1978).
**Fig. 24.** *Cyanobacterium notatum*: **a** – after Skuja (1964), **b** – «forma» after Skuja (1964, probably physiological state with visible keritomy).
**Fig. 25.** *Cyanobacterium diachloros*: after Skuja (1939).
**Fig. 26.** *Cyanobacterium crassiusculum*: after Skuja (1956).

**4. Cyanobacterium minervae** (Copeland) Komárek et al. 1999 (fig. **21**)

*Synechococcus minervae* Copeland 1936 ? incl. var. *maior* Copeland 1936; *Cyanothece minervae* (Copeland) Komárek 1976

Cells solitary or in pairs during division (pairs frequent but always separating before next fission), cell division sometimes asymmetric, sometimes in olive-green or blue-green masses; cells almost spherical or oval to oval-cylindrical, pale blue-green or olive green, usually with visible longitudinally keritomized (striated) chromatoplasma, (2) 4–10 × (2.2) 3–4.5 (5.2) µm. In populations from Slovakia very fine and thin mucilaginous sheaths were present (max. 0.8µm wide), rarely with ferric, brownish precipitates.
Occurrence: Obligate thermobiotic, mainly in alkaline thermal springs at temperatures of 23.5–64.1 °C; known from thermal springs all over the world; Europe: Greece, Slovakia, Iceland (?); the identity of populations from different regions is not certain (see, e.g., Anagnostidis 1961 from Greece – fig. **22**, Hindák 1978 from Slovakia – fig. **23**).

**5. Cyanobacterium notatum** (Skuja) Komárek et al. 1999 (fig.**24**)

*Synechococcus notatus* Skuja 1964; *Cyanothece notata* (Skuja) Komárek 1976

Cells solitary or two together during division, widely oval, broadly rounded at poles, deep blue-green or grey-green with a distinctly visible chromatoplasma, with homogeneous or slightly keritomized content, with few, large, sometimes indistinct granules, 6–12 × 5–7 µm, without mucilaginous envelopes.
Occurrence: Freshwater, metaphytic in pools, moorland waters and small lakes, among other algae and mosses; Sweden (Abisko).

**6. Cyanobacterium diachloros** (Skuja) Komárek et al. 1999 (fig. **25**)

*Synechococcus diachloros* Skuja 1939; *Cyanothece diachloros* (Skuja) Komárek 1976

Cells solitary or two together during division, widely oval or shortly cylindrical with rounded ends, blue-green or greyish olive-green, with finely granulate content and longitudinally striate chromatoplasma, with indistinct mucilaginous envelopes, 9.5–17 × 10–12 µm.
Occurrence: Halophilic, in slightly salty waters, known from the littoral region of the SE. area of Baltic Sea (Latvia).

**7. Cyanobacterium crassiusculum** (Skuja) Komárek et al. 1999 (fig. **26**)

*Synechococcus diachloros* var. *crassiusculus* Skuja 1956; *Cyanothece crassiuscula* (Skuja) Komárek et Anagnostidis 1995

Cells solitary or two together during division, widely oval, blue-green or olive-green, with longitudinally striate chromatoplasma, 17–30 × 12–20 µm, without mucilaginous envelopes.
Occurrence: Freshwater, metaphytic in the littoral of lakes (secondarily planktic); Sweden.

**Species outside Europe:**
– *Cyanobacterium formosanum* Komárek 1999 [syn.: *Synechococcus* sp. in Chou & Huang 1991]: Cells solitary or two together, oval, blue-green, without mucilage, (5.5)10.5–11.8 × 5.2–5.9 µm. – Recorded from rice fields in Taiwan (described from culture).

## 3. Cyanothece Komárek 1976

Cells solitary or in twos after division (but sometimes aggregated in groups, clusters and mats), without gelatinous envelopes or with a narrow (rarely wide), fine, hyaline, indistict gelatinous margin around single cells. Cells widely oval to almost cylindrical with widely rounded ends, with length/width ratio = (1)1.2–1.8(2.6) : 1. Cell content with net-like (reticulate) keritomization (= irregular to radial arrangement of thylakoids, intrathylakoidal spaces). S-layer with p2 symmetry was found in *C. aeruginosa*. Net-like type of nucleoids. Involution cells irregular (never filamentous as in *Synechococcus*); (fig. **27d, 30**). In several species motility was recorded. Cells always divide by binary fission (cleavage) into two morphologically equal, hemispherical daughter cells, in one plane, which is perpendicular to the longer cell axis in successive generations. Cells grow to the original size and shape before the next division takes place. Reproduction by cell division and succeeding separation of daugther cells.
**Important references:** Komárek (1976); Rippka & Cohen-Bazire (1983); Komárek & Anagnostidis (1986); Rippka & Waterbury (1989); Roussomoustakaki & Anagnostidis (1991); Cepák (1993, 1996, 1997); Komárek & Cepák (1998); Komárek & al. (1998).

**Key to the species:**

**1a** Metaphytic and planktic in heliothermal salty coastal (marine) waters, cells 7–15 × 6–13.5 µm . . . . . . . . . . . . . . . . . . . . . . . . . . . . . **1. C. halobia**
**1b** Metaphytic in moorland and peaty waters, raised bogs and on wet rocks, rarely in the littoral of lakes . . . . . . . . . . . . . . . . . . . . . . . . . . . . . . . **2**
**2a** Cells usually 10–50 × 10–38 µm . . . . . . . . . . . . . . . . . . . **2. C. aeruginosa**
**2b** Cells usually 30–70 (100) × (19) 28–52 µm . . . . . . . . . . . . . . **3. C. major**

### 1. Cyanothece halobia Roussomoustakaki et Anagnostidis 1991 (fig. 27)

Cells solitary, more or less motile, during division two together, roundly oval or subcylindrical with widely rounded ends, with a very thin, homogeneous, diffluent gelatinous enveloping layer, 7–12 (15) × (6) 7–12 (13.5) µm; cell content greyish olive-green, sometimes slightly (dirty) reddish or pale pinkish (carotenoids), rarely bright blue-green, with small scattered granules, without aerotopes, with distinctly reticulately keritomized chromatoplasma.
Occurrence: Halobiotic in heliothermal, coastal saltworks; Greece, Italy, Spain, probably with world-wide distribution. Similar populations were found in Yallahs ponds, Jamaica (Golubić 1980, sub *Synechocystis*/*Synechococcus* sp.).

### 2. Cyanothece aeruginosa (Nägeli) Komárek 1976 (fig. 28)

*Synechococcus aeruginosus* Nägeli 1849; *Synechococcus crassus* Archer 1867; *Synechococcus fuscus* Zeller 1873; *Synechococcus grandis* Playfair 1918; *Coccochloris aeruginosa* (Nägeli) Drouet et Daily 1952; *Anacystis aeruginosa* (Nägeli) Drouet et Daily 1956 pro parte typico

Cells solitary or in twos during division, oval to shortly cylindrical with widely rounded ends, blue-green, rarely yellowish, with finely granulate content and usually visible reticulately keritomized chromatoplasma, (9) 10–45 (50) × (7) 10–30 (36) µm. Sometimes a narrow, fine, colourless gelatinous layer around the cells.
Occurrence: Freshwater, in clear, cold moorland waters, raised bogs, on wet rocks and in swamps (pH below 7), rarely secondarily in cryoseston, from lowlands to mountains; probably cosmopolitan in ecologically corresponding biotopes, particularly in both temperate and circumpolar zones (whole of Eurasia, N. America,

Australia, Argentine, Kerguelen Islands, New Zealand, Antarctica, etc.); data from other biotopes (thermal waters, salines, crusts in dry areas, polluted waters) and from tropical regions are uncertain and usually related to other species.

### 3. **Cyanothece major** (Schröter) Komárek 1976 (fig. **31**)

*Synechococcus major* Schröter 1884; *Synechococcus major* f. *crassior* Lagerheim 1886; *Synechococcus major* var. *maximus* Lemmermann 1896; *Synechococcus aeruginosus* var. *maximus* (Lemmermann) Lemmermann 1907

Cells solitary or in twos during division (rarely aggregates of cells without mucilage), widely oval, blue-green, with granulated, reticulately keritomized content, (18 ?) 30–70 (100) × (19) 28–52 (76) µm. Narrow, but usually distinct, colourless, structureless layers of mucilage around the cells.

Occurrence: Freshwater, benthic and metaphytic in peaty waters, in the littoral of oligotrophic and dystrophic lakes, sometimes also subaerophytic on wet peaty soils; colder parts of the whole temperate world, including Arctic and Antarctic islands, subarctic Eurasia and northern Canada (Ellesmere island); in Europe in high mountain regions (Austria, Bulgaria, Czech Republic, Slovakia), probably with a boreo-alpine distribution.

**Species outside Europe:**

- *Cyanothece epiphytica* M. Watanabe et Komárek 1998: Cells 6.2–12.5 × 6.2–7.3 µm, epiphytic on mosses in cold, catharobic waters in Himalayas.
- *Cyanothece lineata* Komárek et Komárková-Legnerová 1999: Cells oval, 20–30.2 × 15.7–20.2 µm, described from the plankton of small mesotrophic lakes in Mexico.
- *Cyanothece shiloi* (Campbell et Golubić) Komárek et Anagnostidis 1995 [syn.: *Aphanothece shiloi* Campbell et Golubić 1985]: Cells 13–20 × 6–8 µm, with keritomized chromatoplasma, in a gelatinous homogeneous mass, described from a hypersaline lake (Sinai peninsula, Solar lake); (fig. **32**).

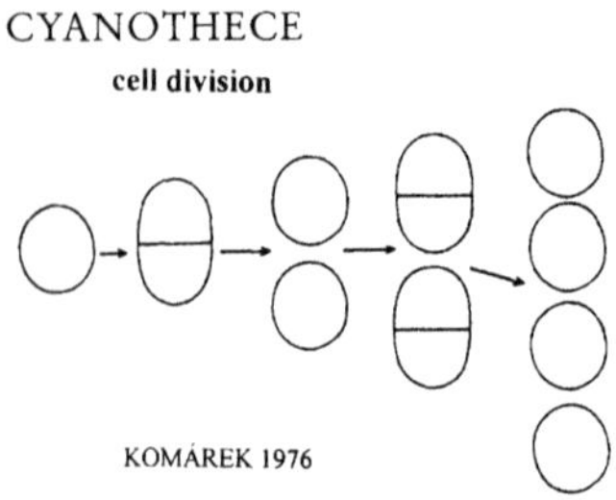


**Fig. 27.** *Cyanothece halobia:* **a-b** – variability in cell form and dividing cells, **c** – dividing cells; **d** – involution forms; all after Roussomoustakaki & Anagnostidis (1991).

**Fig. 28.** *Cyanothece aeruginosa*: **a** – after Geitler (1960); **b** – after Skuja (1956), **c** – after Geitler (1925), **d** – after Komárek in Fott (1956).

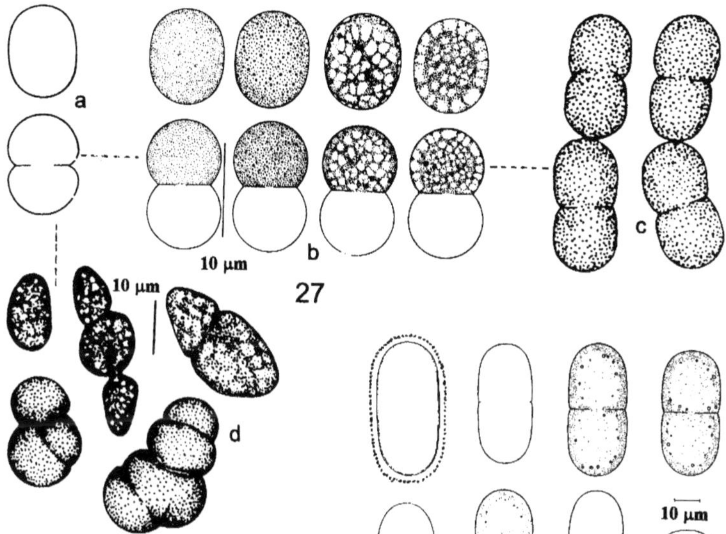


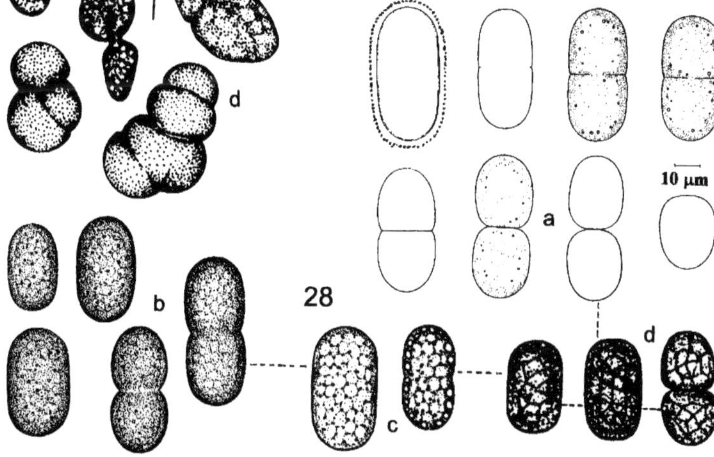


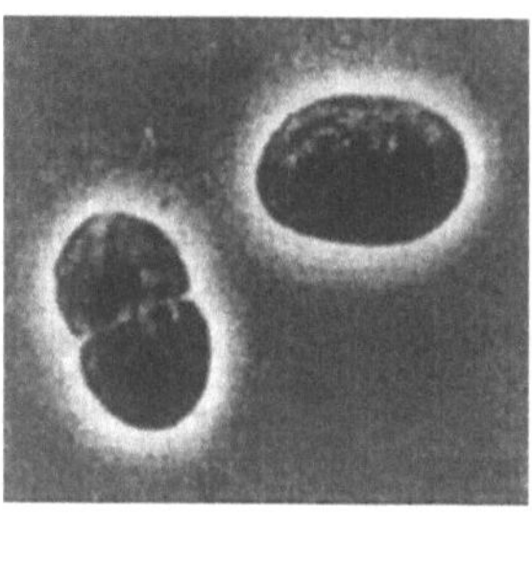


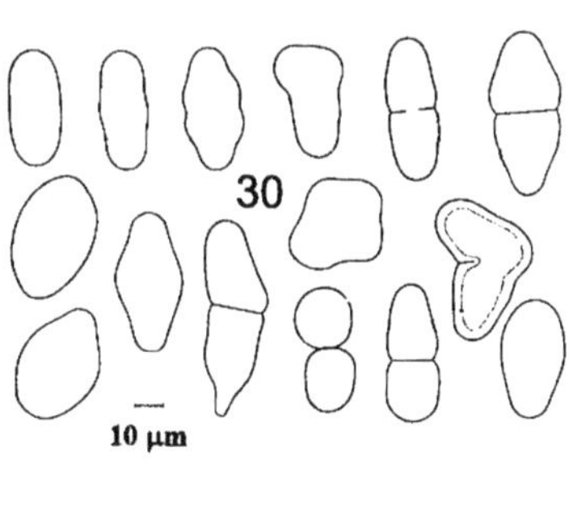


**Fig. 29.** Reference strain PCC 7424 «for the *Cyanothece*-group» from Waterbury & Rippka (1989, Bergey's Manual); probably identical with *Cyanothece aeruginosa*.
**Fig. 30.** Involution forms of *Cyanothece maior*, from Geitler (1960).

**Unrevised species:**
- *Synechococcus salinus* Frémy 1941: Cells 7.5–10 × 5 (more or less) µm, from salty biotopes; probably a *Cyanothece* species.
- *Cyanothece*-group sensu Rippka & Cohen-Bazire (1983): A broadly conceived taxonomic unit, based on the reference strain PCC 7424, isolated from a rice paddy in Senegal. Waterbury & Rippka (1989) characterized it as «unicellular coccoid to rod-shaped cyanobacteria that divide by transverse binary fission in a single plane. The cells are 3 µm or greater in diameter, contain photosynthetic thylakoids and lack structured sheaths». The morphology is insufficiently defined. Evidently this is a collective cluster, possibly with more taxonomic units from a different genus (see, e.g., *Cyanobacterium* or *Cyanobium*, Table **5**).
- *Cyanothece* sp. sensu Komárek 1989: Cells 4.6–7 µm long; from periodical road pools in Cuba (fig. **33**); comp. with *Cyanobacterium*.
- De Philippis & al. (1997) studied 15 strains designated «*Cyanothece*»-group. They found fatty acids composition to be a good marker for strain identification, and clear separation of strains from saline and alkaline environments. It is a pity, that the phenotype characterization is not included. The generic identification of these strains with the genus *Cyanobacterium* is also possible.

## 4. Cyanogranis Hindák 1982

Colonies free floating, microscopic, subspherical to irregular in outline, mucilaginous, composed of a cluster of irregularly aggregated cells embedded within fine, homogeneous, colourless and diffluent slime. Ferric precipitates occur the gelatinous colonies. Cells spherical or subspherical, sometimes widely oval.
Cell division only in the transverse plane, perpendicular to the longer axis of slightly elongate cells, in one plane in successive generations. Reproduction by fragmentation of colonies.
**Important references:** Wawrik (1963,1974, sub *Microcystis*); Heynig (1970, sub *Microcystis*); Hindák (1975, 1982, 1983); Hindák & Moustaka (1988).

**Key to the species:**
1a Cells up to 1.5 µm in diameter, greenish or bright blue-green; ferric precipitates mainly around or near the cells . . . . . . . . . . . . . . . **1. C. ferruginea**
1b Cells 4–6 µm in diameter, grey or pale blue-green; ferric precipitates aggregated mainly in the center of a colony . . . . . . . . . . . . . . . . . **2. C. basifixa**

### 1. Cyanogranis ferruginea (Wawrik) Hindák 1982 (fig.**35**)

*Microcystis ferruginea* Wawrik 1963; *Dactylococcopsis ferruginea* (Wawrik) Hindák 1975

Colonies free floating, more or less spherical to widely ellipsoidal, with a small number of cells (3–50, rarely more, up to 100), in sparse or dense clusters, usually up to 12 µm in diameter; cells sometimes enveloped by ferric, blackish precipitates, which are occasionally aggregated on the periphery of cell groups. Mucilage very fine, visible usually only after staining. Cells spherical to slightly oval, pale green to bright blue-green, without aerotopes, 0.6–1.5 × 0.4–1 (1.5) µm. Sometimes colonies lack precipitates.

---

CYANOTHECE
**Fig. 31.** *Cyanothece maior*: **a** – after Lemmermann from Starmach (1966), **b** – after Starmach (1973), **c** – after Skuja (1956, as «f. *maxima*»).

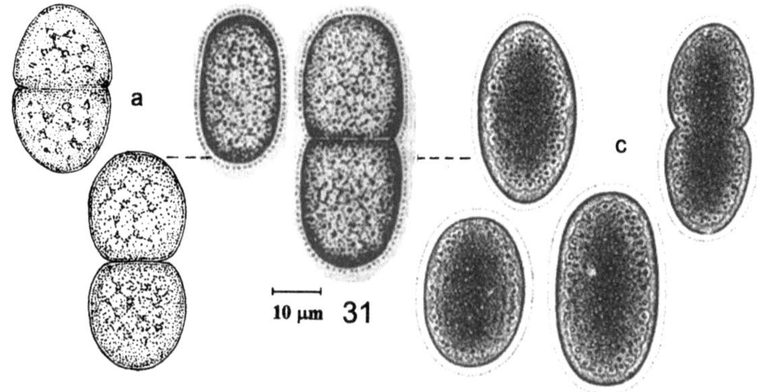


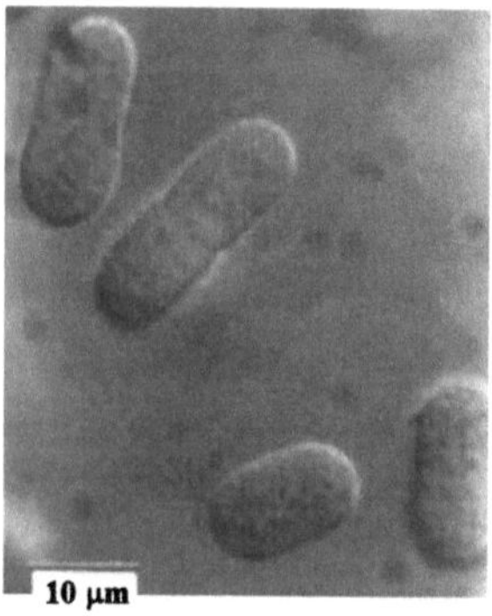


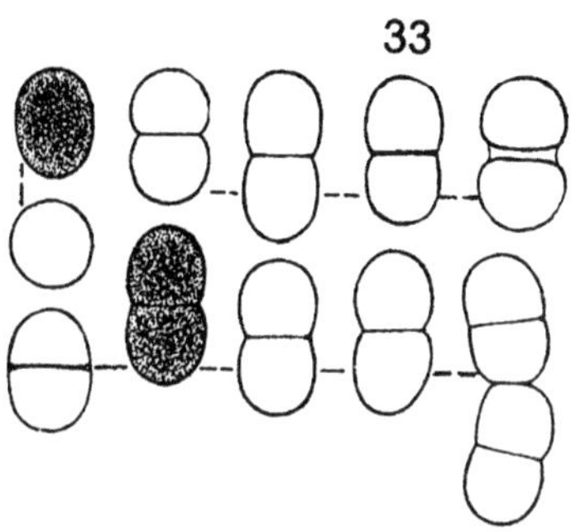


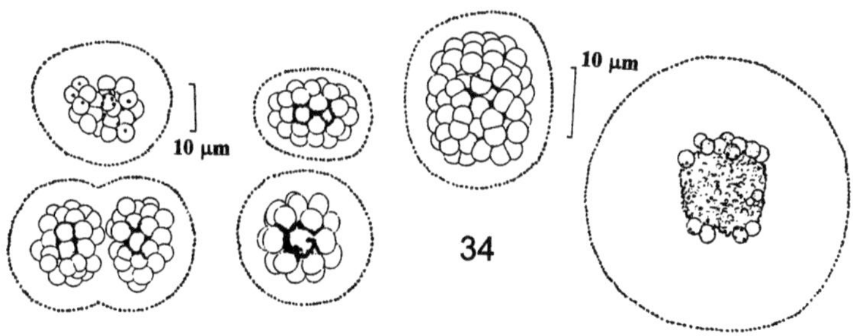


**Fig. 32.** *Cyanothece shiloi*, from hypersaline Solar lake (Egypt – Sinai peninsula): from Campbell & Golubić (1985).
**Fig. 33.** *Cyanothece* sp., from unpolluted temporary pools in roads in Cuba: after Komárek (1989).

CYANOGRANIS
**Fig. 34.** *Cyanogranis basifixa*: after Hindák (1982).

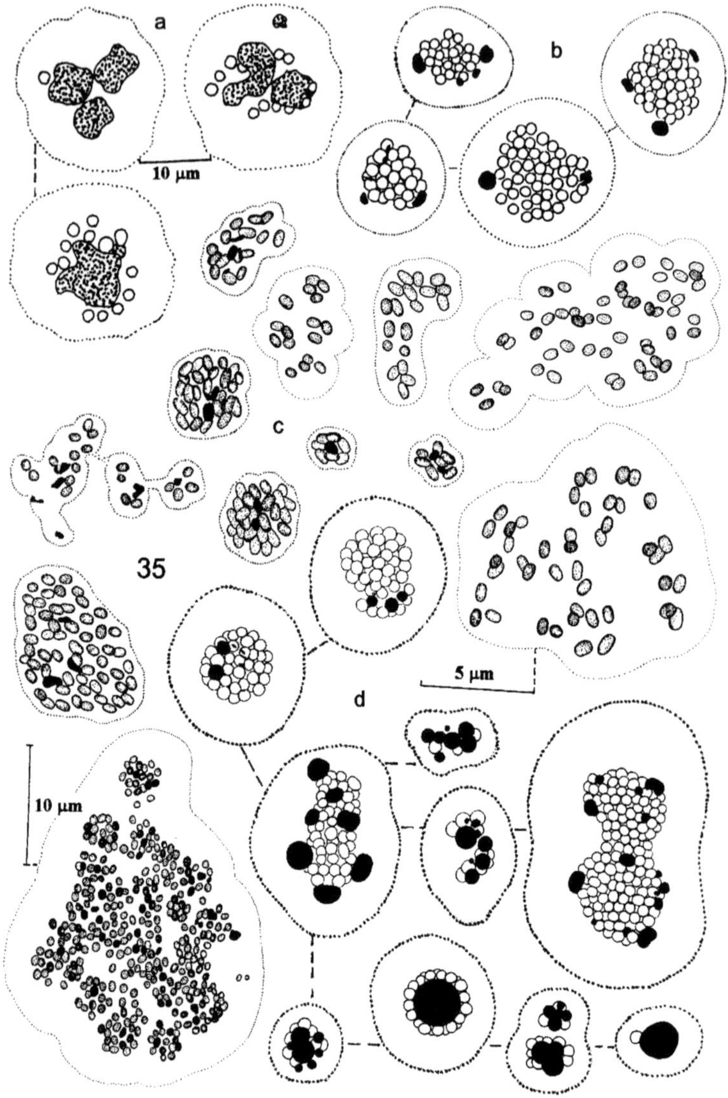


CYANOGRANIS

**Fig. 35.** *Cyanogranis ferruginea*: **a** - after Wawrik (1974, sub *Microcystis ferruginea*) from Austria, **b** - after Hindák & Moustaka (1988) from N. Greece, **c** - after Hindák (1975) from W. Slovakia with more elongated cells 0.6–1 x 0.4–0.6 µm, **d** - after Hindák (1982) from SW. Slovakia.

Occurrence: Freshwater, planktic in fishponds and pit-holes, usually in slightly alkaline water; known from central and S. Europe (Austria, Czech Republic, Germany, Greece, Slovakia).

### 2. **Cyanogranis basifixa** Hindák 1982 (fig. **34**)

Colonies free floating, spherical or oval, mainly with 20–40, rarely with more cells, with hyaline, homogeneous, colourless and usually diffluent mucilage, 30–60 µm in diameter; in the center of colonies usually there are spherical, brownish ferric precipitates in the form of large granules. Cells spherical to oval, with homogeneous, grey or pale blue-green content, without aerotopes, 4–6 µm in diameter.
Occurrence: Freshwater, planktic in small lakes and backwaters in the inundation zone of rivers; known only from Czech Republic (S. Bohemia) and S. Slovakia.

## 5. **Lithomyxa** Howe 1931

Colonies distributed over a substrate, crustose, usually heavily incrusted with lime, hard, primarily composed of a mucilaginous layer of homogeneous slime with irregularly arranged cells (never in rows), surrounded by their own gelatinous, diffluent envelopes. Cells more or less spherical, ovoid or short cylindrical, sometimes aggregated into small, 2-, 4- or few-celled, irregular groups.
Cell division probably in one plane in successive generations. Reproduction possibly by solitary, liberated cells. Insufficiently known, monospecific genus.
**Important references:** Geitler (1942).

**Species outside Europe:**
- *Lithomyxa calcigena* Howe 1931: Cells 0.3–1 µm wide; in streams, limestone regions, important biolithogenic and travertine forming species in N. America; possibly bacteria.

## 6. **Cyanocatena** Hindák 1975

Colonies free floating with cells irregularly arranged in very fine slime, microscopic, irregular, with varying number of cells; in young colonies cells sometimes are arranged in irregular rows. Mucilage structureless, colourless, diffluent at the margin; cells without their own gelatinous envelopes, but often with brownish or blackish rings (ferric precipitates). Cells cylindrical, elongate with rounded ends, sometimes slightly arcuate.
Cells divide by binary fission in one plane in successive generations, transversely to the longitudinal axis, often where the ring-like precipitations are located. Reproduction by disintegration of colonies. Monospecific genus.
**Important references:** Hindák (1975, 1982, 1983); Heynig (1979).

### 1. **Cyanocatena planctonica** Hindák 1975 (fig. **36**)

Colonies few-celled to many-celled, irregularly spherical, oval or irregular, with fine, hyaline, homogenous mucilage (staining!), up to 20 µm in diameter. Cells irregularly arranged, short, cylindrical-oval or cylindrical with rounded ends, straight or slightly arcuated, sometimes with 1–2 (3) ring-like ferric precipitates on the surface, grey or pale blue-green, with homogeneous content, without aerotopes, 1–2.5 (5) × 0.5–1.2 µm, after division almost spherical. Within the mucilage there sometimes are free ring-like precipitates.
Occurrence: Freshwater, planktic in eutrophic ponds and lakes, usually in sum-

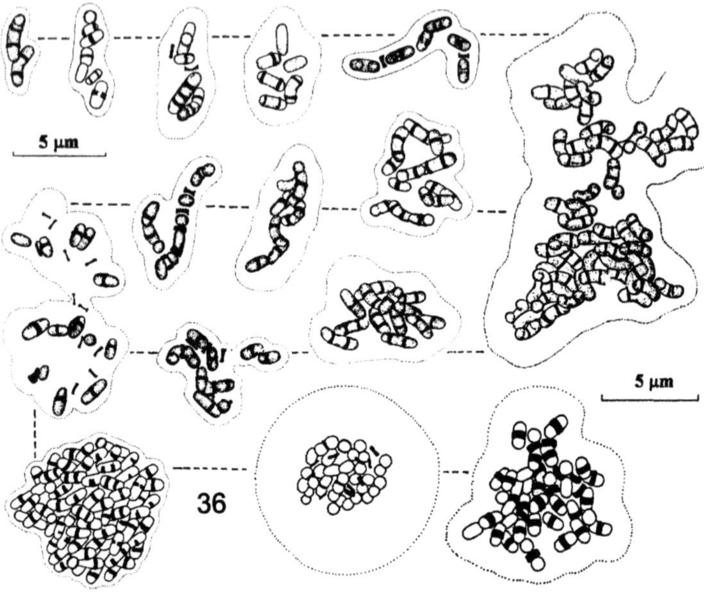


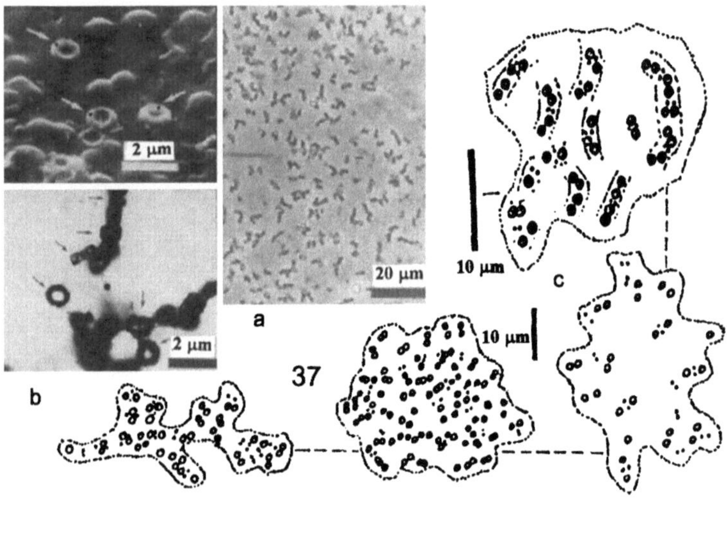


CYANOCATENA

**Fig. 36.** *Cyanocatena planctonica*, irregular colonies of varied size with cells and ferric precipitates in the form of brown rings on the cell surface and among cells: from Hindák (1975,1982).

mer or autumn; Austria, Czech Republik, Germany, Greece, Slovakia; possibly related to *Cyanodictyon imperfectum.*

**Species outside Europe:**
- *Cyanocatena bicudoi* Ramírez 1996: Small, only few-celled (2-celled) colonies, cells 5.7–6.2 × 2.5–3.7µm, rounded conical at the ends; planktic, described from Colombia (Munic. de Medellín).

## 7. **Cyanodictyon** Pascher 1914

*Sphaerodictyon* Geitler 1925

Colonies free floating, initially more or less spherical, soon becoming irregularly reticulate, flat or three-dimensional, often elongate, composed of gelatinous, irregularly divaricate and anastomozing strands, with uniseriate, in older stages multiseriate rows of cells. Mucilage structureless, colourless, usually indistinct and diffluent at the margin. Cells spherical or slightly elongate, without prominent inclusions.

Cell division always in one plane in successive generations, more or less perpendicularly to the mucilaginous strands; hemispherical divided cells grow to the original shape before the next division. Reproduction by disintegration of colonies.

**Important references:** Pascher (1914); Geitler (1925, 1932, 1942); Elenkin (1936, 1938, 1949); Cronberg & Weibull (1981); Hickel (1981); Cronberg (1988); Economou-Amilli & Spartinou (1991); Komárková-Legnerová & Cronberg (1994); Cronberg & Komárek (1994); Mayer (1994); (Lederer (1995).

**Key to the species:**

**1a** Cells more or less spherical . . . . . . . . . . . . . . . . . . . . . . . . . . . . . . . 2
**1b** Cells distinctly elongate to rod-shaped . . . . . . . . . . . . . . . . . . . . . . . . . 5
**2a** Cells less than 1 µm in diameter . . . . . . . . . . . . . . . . . . . . . . . . . . . . . 3
**2b** Cells more than 1 µm in diameter . . . . . . . . . . . . . . . . . . . . . . . . . . . . 4
**3a** Colonial nets irregular, indistinct, free-living, enveloped by common fine, colourless mucilage; cells very fine, more or less spherical or slightly elongate, 0.4–1 µm in diameter, with blackish ferric precipitations in the form of rings and granules . . . . . . . . . . . . . . . . . . . . . . . . . . **1. C. imperfectum**
**3b** Colonial nets always more or less spherical, small, growing endogloeic in mucilage of planktic *Anabaena* species; cells 0.5–1 µm in diameter . . . . . . . . . . . . . . . . . . . . . . . . . . . . . . . . . . . . . **2. C. endophyticum**
**4a** Colonies more or less globose or slightly elongate, strands usually with uniseriately arranged cells (at least two cells side by side in pairs); cells 1–1.5 µm in diameter . . . . . . . . . . . . . . . . . . . . . . . . **3. C. reticulatum**
**4b** Colonies usually elongate, strands in old colonies often multiseriate; cells 1.9–3.8 µm in diameter . . . . . . . . . . . . . . . . . . . . . . . . **4. C.tubiforme**
**5a** Colonies more or less three-dimensional . . . . . . . . . . . . . . . . . . . . . . . . 6
**5b** Colonies very small, composed of uniseriate, almost unbranched rows of elongate, rod-shaped cells, enclosed within fine or more or less firm sheaths; cells 1.5–2.4 × 0.2–0.5 µm . . . . . . . . . . . . . . . . . . . . . **8. C. filiforme**

---

CYANODICTYON

**Fig. 37.** *Cyanodictyon imperfectum*: **a** – photomicrograph of a colony from Cronberg & Weibull (1981), **b** – photomicrographs of ferric ring-shaped precipitates liberated from cells after Cronberg & Weibull (1981), **c** – irregular microcolonies from plankton after Komárková-Legnerová & Cronberg (1994).

**6a** Colonies irregular in outline, many-celled when old, with cells in one or more rows; cells approximately 1.5 × 0.8–1 µm . . . . . . . . . . . . . . . . . . . . 7
**6b** Colonies more or less spherical, with cells arranged in short, arcuate, indistinct, radial rows in the surface layer of colonial mucilage; cells 1.2–2 × 1–1.2 µm . . . . . . . . . . . . . . . . . . . . . . . . . . . . . . . . . . . . **7. C. iac**
**7a** Cells approximatelly 1.5 × 0.8–1 µm; planktic in lakes and ponds . . . . . . . . . . . . . . . . . . . . . . . . . . . . . . . . . . . . . **5. C. planctonicum**
**7b** Cells 2.5–4.5 × 1.5–2 µm; metaphytic in mountain raised bogs . . . . . . . . . . . . . . . . . . . . . . . . . . . . . . . . . . . . . **6. C. turfosum**

## 1. Cyanodictyon imperfectum Cronberg et Weibull 1981 (fig. 37)

Colonies free floating, microscopic, initially in the form of short, irregular rows with 2–8 cells, later (especially in mass development) in small, loose, net-like formations, in old colonies rather irregularly arranged, forming fragmented rows; length of solitary rows 5–10 µm; cells in touching pairs after division, later small spaces arise between them; sheaths fine, colourless, indistinct. Cells very small, irregularly spherical or slightly elongate (before division), pale blue-green, 0.4–0.8–1 µm in diameter; the cells have the ability to precipitate iron-oxide in the form of double dots or rings between cell pairs. Only one parietal thylakoid per cell.
Occurrence: Freshwater, planktic in mesotrophic or slightly eutrophic lakes, sometimes in masses, forming weak greenish discolourations; recorded from Germany, Greece, Sweden, and E. African great lakes, probably cosmopolitan; the generic classification should be revised.

---

CYANODICTYON

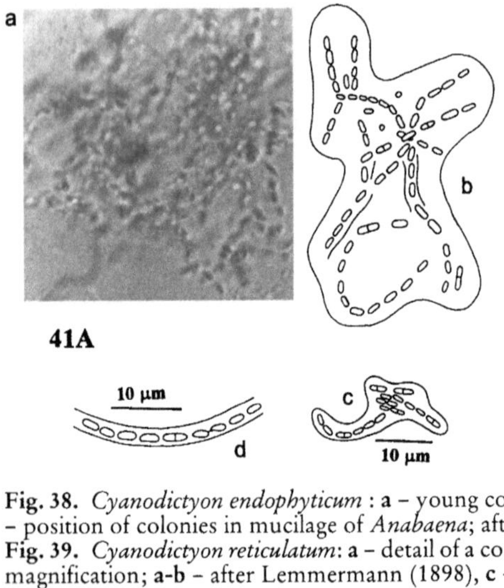


**Fig. 38.** *Cyanodictyon endophyticum* : **a** – young colonies, **b** – detail of a colony, **c** – position of colonies in mucilage of *Anabaena*; after Pascher (1914).
**Fig. 39.** *Cyanodictyon reticulatum*: **a** – detail of a colony, **b-c** – colonies at different magnification; **a-b** – after Lemmermann (1898), **c** – after Komárková-Legnerová & Cronberg (1994).
**Fig. 40.** *Cyanodictyon tubiforme*: **a** – characteristic shape of a colony, **b** – old

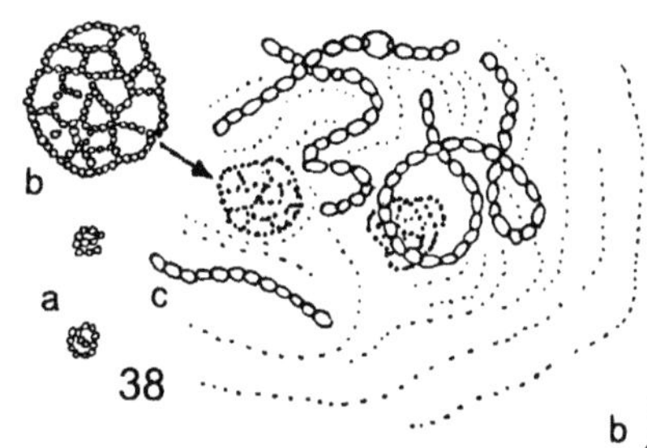


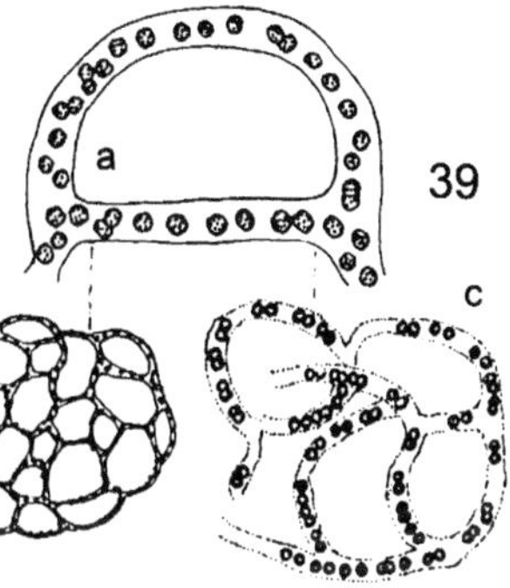


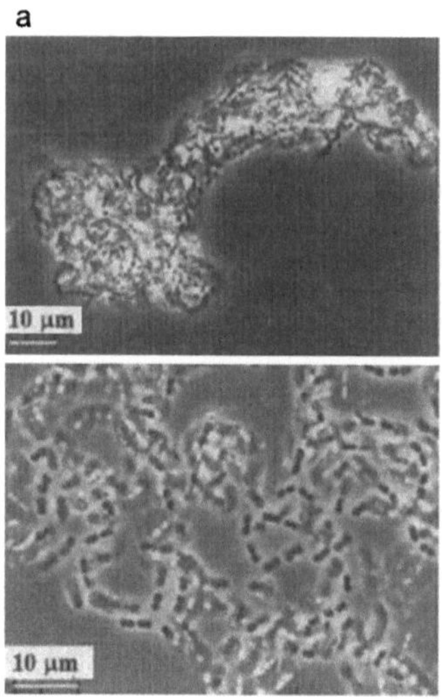


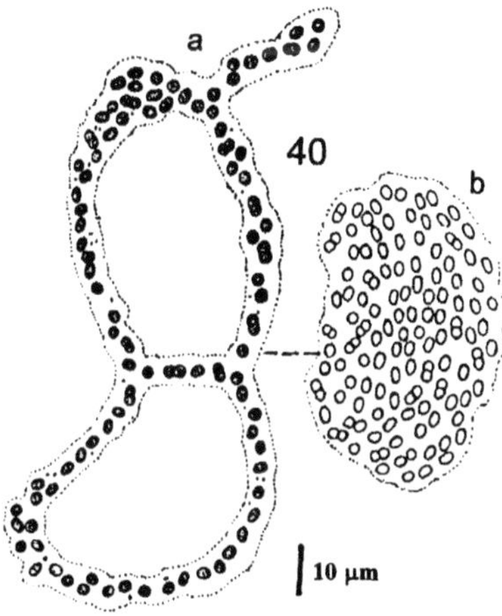


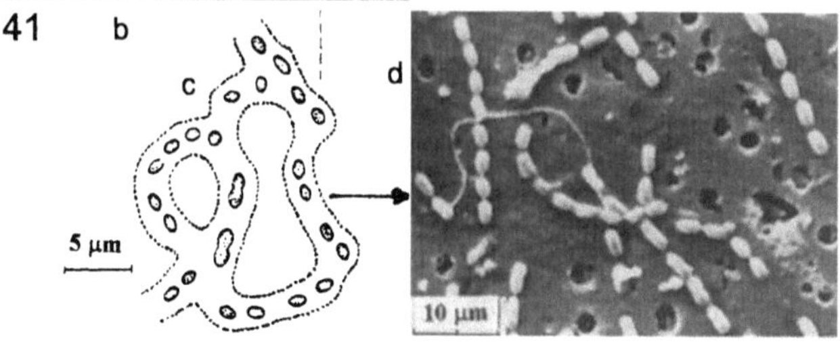


colony with numerous rows of cells; after Komárková-Legnerová & Cronberg (1994).

**Fig. 41.** *Cyanodictyon planctonicum*: **a-b** – shape of colonies, **c-d** – details of rows of cells; from Hickel (1981).

**Fig. 41A.** *Cyanodictyon turfosum*: **a-b** – old colonies, **c** – young colony, **d** – detail of pseudofilament; after Lederer (1995).

## 2. **Cyanodictyon endophyticum** Pascher 1914 (fig. **38**)

Colonies initially flat, later almost spherical, net-like (in form of closed spherical, irregular nets), with densely and regularly disposed cells in rows, usually single, rarely in pairs after division. Cells spherical, blue-green, 0.5–1(1.5) µm in diameter.
Occurrence: Freshwater, endogloeic, in mucilage of planktic *Anabaena* species; sporadic in central and E. Europe.

## 3. **Cyanodictyon reticulatum** (Lemmermann) Geitler 1925 (fig. **39**)

*Polycystis* (*Clathrocystis*) *reticulata* Lemmermann 1898; *Coelosphaerium reticulatum* (Lemmermann) Lemmermann 1910; *Sphaerodictyon reticulatum* (Lemmermann) Geitler 1925

Colonies free floating, more or less spherical or slightly elongate up to irregular, with scattered and irregularly arranged cells in simple gelatinous rows (strands); colonies irregularly reticulate with fine colourless, homogeneous mucilage (staining!); meshes of net 3–34 µm in diameter; cells in uniseriate strands, rarely in two parallel rows. Cells spherical, pale blue-green, without aerotopes, 1–1.5 µm in diameter.
Occurrence: Freshwater, planktic in mesotrophic or eutrophic lakes of central and northern parts of the whole temperate zone (Austria, Denmark, Germany, N. Greece, Russia, Ukraine, Canada), recorded also from eastern part of Baltic Sea.

## 4. **Cyanodictyon tubiforme** Cronberg 1988 (fig. **40**)

Colonies initially forming simple rows of spherical cells (slightly elongate before division); cells arranged irregularly in uniseriate, fine, diffluent, colourless mucilaginous strands, later variously divaricated and anastomozing, net-like, forming loops and circles respectively; in old colonies, cells shift after division and create irregular multiseriate, elongate, sometimes clathrate colonies, up to 400 µm long. Mucilage fine, colourless, homogeneous, diffluent. Cells spherical, hemispherical after divison, or widely oval to rod-shaped before division, with pale blue-green or greyish homogeneous content, without aerotopes, 2.2–3.8 × 1.9–2.2 (3) µm.
Occurrence: Freshwater, planktic in shallow eutrophic lakes; sporadically distributed in W. Baltic area ( N. Germany, S. Sweden).

## 5. **Cyanodictyon planctonicum** Meyer 1994 (fig. **41**)

*Cyanodictyon reticulatum* sensu Hickel 1981; ? *Aphanothece clathratiformis* var. *maior* Liebetanz 1925

Colonies free floating, initially more or less spherical, tridimensional, later elongate to irregular, reticulate, multicellular, up to 150 µm long; colonial mucilage colourless, fine (staining !), enveloping the whole net; cells within the colony form pseudofilamentous rows, initially with large distances between cells (7–35 µm), later cells closer to one another; mucilaginous sheaths visible after staining. Cells oval to almost rod-shaped, straight or sometimes slightly curved, with visible chromatoplasma, yellowish or pale blue-green, without aerotopes, 1.5 × 0.8–1 µm.
Occurrence: Freshwater, planktic in eutrophic lakes; Denmark, N. Germany, Canada.

## 6. **Cyanodictyon turfosum** Lederer 1995 (fig. **41A**)

Colonies microscopic, mucilaginous, irregular net-like, 50–200 µm in diameter, with irregular outline. Mucilage fine, colourless, diffluent. Cells rod-shaped, less

frequently ellipsoid, without aerotopes, pale blue-green, 2.5–4.5 × 1.5–2.0 μm, arranged into short, arcuated pseudofilaments.
Occurrence: Metaphytic in peaty raised bogs with acidic water (*Sphagnum* - pools); Czech Republic (Šumava Mts).

### 7. **Cyanodictyon iac** Cronberg et Komárek 1994 (fig. **42**)

Colonies microscopic, spherical or irregularly spherical, sometimes lobate, simple or composed of 2–3 or several subcolonies, up to 70 μm in diameter. Mucilage colourless, hyaline, diffluent. Cells almost spherical after division, later oval or widely oval, 1.2–2 × 1–1.2μm, with homogenous, pale blue-green content, arranged irregularly in colonies and indistinctly radially, or in short, arcuate, indistinct rows in the superficial layer of the colony. Short rows of cells sometimes have their own diffluent, narrow, hardly visible mucilaginous envelopes (staining!). Cells separate from one another soon after division.
Occurrence: Planktic in mesotrophic lakes; still known only from S. Sweden.

### 8. **Cyanodictyon filiforme** Komárková-Legnerová et Cronberg 1994 (fig. **43**)

Solitary pseudofilaments or groups of several pseudofilaments forming loose bundles resembling nets, free floating; pseudofilaments irregularly flexuous, curved in various directions, unbranched, composed of simple rows of elongate cells, enveloped by fine, mucilaginous, distinct, thin, colourless, tube-like sheaths. Cells arranged one after another, lying mostly in pairs or in short chains inside sheaths, fine, rod-shaped, pale blue-green, 1.5–2.4 × 0.2–0.5 μm; cells divide when they reach the length of about 2 μm. To some extent it resembles *Wolskyella* or *Planktolyngbya*.
Occurrence: Freshwater, planktic in mesotrophic and eutrophic lakes; known particularly from northern regions of the temperate zone (Sweden, Canada, etc.).

## 8. **Lithococcus** Ercegović 1925

Colonies endolithic, irregularly reticulate, composed of randomly divaricate and anastomozing, more or less limited and firm mucilaginous, filamentous strands, colourless, homogeneous, inside with uniseriate, irregular rows of cells, oriented parallel to the strands; in places with widened mucilage, the cells are irregularly distributed; without special envelopes around them. Cells spherical, ellipsoidal or oval, near to one another after division.
Cell division by binary fission, probably only in one plane in successive generations, perpendicular to the length of the strands. Reproduction possibly by liberation of solitary cells. Insufficiently known monospecific genus.
**Important references:** Ercegović (1925); Geitler (1932, 1942).

### 1. **Lithococcus ramosus** Ercegović 1925 (fig. **44**)

Colonies microscopic. Cells mainly ellipsoidal or oval, rarely spherical or hemispherical after division, bright blue-green, 0.8–2.2 μm in diameter.
Occurrence: Aerophytic, endolithic in calcareous rocks; known only from Croatia.

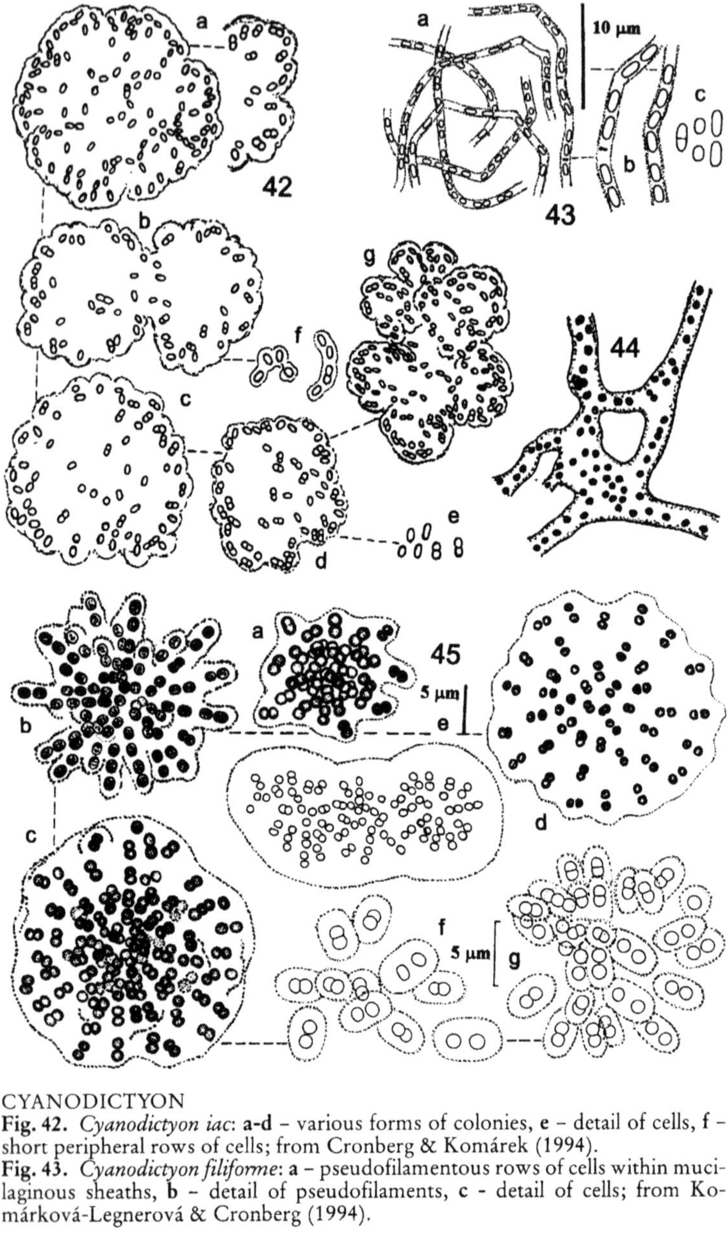


CYANODICTYON

**Fig. 42.** *Cyanodictyon iac*: **a-d** – various forms of colonies, **e** – detail of cells, **f** – short peripheral rows of cells; from Cronberg & Komárek (1994).

**Fig. 43.** *Cyanodictyon filiforme*: **a** – pseudofilamentous rows of cells within mucilaginous sheaths, **b** – detail of pseudofilaments, **c** – detail of cells; from Komárková-Legnerová & Cronberg (1994).

## 9. **Radiocystis** Skuja 1948

Colonies free floating, more or less spherical to irregular, microscopic, with fine, colourless, diffluent gelatinous envelopes, with cells more or less irregularly grouped in the colonial center and radiating in all directions in more or less uniseriate rows; cells in radial rows slightly distant from one another or (after division) in pairs, widely oval or almost spherical, facultatively with aerotopes. Cells divide transversely in one plane in successive generations, in later stages perpendicularly to the radiating rows. Reproduction by disintegration of mucilaginous colonies.

**Important references:** Skuja (1948); Starmach (1966); Hindák & Moustaka (1988); Komárek & Komárková-Legnerová (1993); Hindák (1996).

**Key to the species:**

1a Cells with gas vesicles, spherical to spherical-oval, 2.5–3.8 × 1.7–2.6 µm ........................................ **1. R. geminata**

1b Cells without gas vesicles, short cylindrical, only after division almost spherical, 4–5 × 2.5–3 µm .................... **2. R. aphanothecoidea**

### 1. **Radiocystis geminata** Skuja 1948 (fig. **45**)

Colonies planktic, microscopic, more or less spherical, slightly elongate or composed of subcolonies, up to 70 (350) µm in diameter; mucilage fine, colourless, with irregular diffluent margin. Cells spherical to oval, pale or bright blue-green, facultatively with aerotopes, 2.5 (3) – 4.8 (5) × 2.7–3.6 (4.8) µm.

Occurrence: Freshwater, planktic in mesotrophic and slightly eutrophic lakes over both temperate zones; very similar populations (taxonomically identical ?) are commonly recorded from tropical stagnant water bodies. A smaller morphotype (cells 2–3 × 2 µm) was found in the lake Volvi, Greece (Hindák & Moustaka 1988).

### 2. **Radiocystis aphanothecoidea** Hindák 1996

Colonies free floating, planktic, microscopic, spherical, oval or irregular, 15–50 µm in diameter, with cells more or less radially arranged from the center to the periphery; colonial mucilage homogeneous, hyaline, structureless, (2)5–20 µm overpassing the cells. Cells shortly cylindrical with rounded ends, in twos after division or solitary, 4–5 × 2.5–3 µm, usually with own mucilaginous envelopes, which are up to 1 µm wide, tough, hyaline, but conspicuous; cell content homogeneous, pale blue-green, with some refractive granules, without gas vesicles.

Occurrence: Freshwater, planktic in shallow lakes and oxbow lakes; known only from the inundation zone of Danube river and its tributaries, in the section from Austria to Hungary and South and S-E. Slovakia.

---

LITHOCOCCUS

**Fig. 44.** *Lithococcus ramosus*: after Ercegović (1925).

RADIOCYSTIS

**Fig. 45.** *Radiocystis geminata*: **a-c** – colonies after Skuja (1948), **d-e** – colonies after Komárková-Legnerová & Cronberg (1994), **f-g** – colonies from Lake Volvi (Greece) after Hindák & Moustaka (1988).

**Species outside Europe:**
- *Radiocystis elongata* Hindák 1996: Colonies to 60 µm in diameter, cells spherical-oval to short cylindrical, 2–5 × 1.8–2(2.5) µm, without gas vesicles. Described from lake plankton in central Canada (Manitoba, South Indian Lakes).
- *Radiocystis fernandoi* Komárek et Komárková-Legnerová 1993: Cells (6)7–8 × 5–6 (7) µm. Tropical species (Brazil, Indonesia, Srí Lanka, S. Africa).
- *Radiocystis magadzae* Komárek et Cronberg 1999: Cells oval, (1)2.4–3 × 1–1.2(2.5) µm. Described from S. Africa.

## 10. **Epigloeosphaera** Komárková-Legnerová 1991

Colonies mucilaginous, more or less spherical, oval or elongate, microscopic, solitary, in pairs or compound, forming irregular groups where colonies of different size are tightly grouped together, free floating in the plankton, metaphyton, or living in sapropel; cells distributed irregularly and distant from each other on the surface of the primary colonial mucilage (hyaline gelatinous spheres or cylinders), without their own gelatinous envelopes. Mucilage of primary colonies homogeneous, delimited, smooth on the surface. Cells oval or rod-shaped, small, short, without aerotopes or prominent granules.
Cell division by transverse binary fission into two equal daughter cells, perpendicularly to the long axis. Reproduction by the disintegration of compound and single colonies, even into unicellular fragments.
**Important references:** Zalessky (1926, sub *Aphanothece*); Komárková-Legnerová (1991); Komárek et Cronberg (1998).

### 1. **Epigloeosphaera glebulenta** (Zalessky) Komárková-Legnerová 1991 (fig. **46**)

*Aphanothece glebulenta* Zalessky 1926

Solitary colonies up to 35 µm in diameter, with colourless, homogeneous and distinctly delimited smooth surface (staining!); cells are situated on the colonial

EPIGLOEOSPHAERA

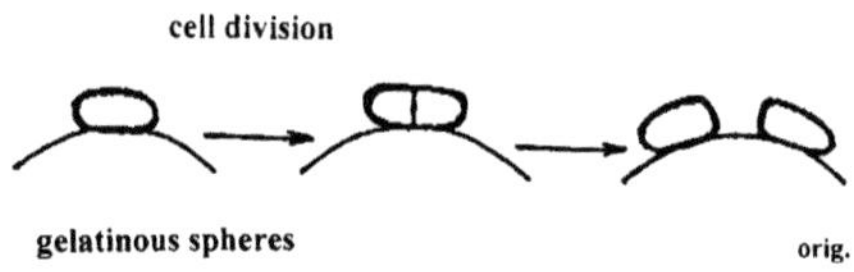


**Fig. 46.** *Epigloeosphaera glebulenta*: **a-e** – after Komárková-Legnerová (1991), **f** – iconotype of Zalessky (1926, sub *Aphanothece glebulenta*).

LEMMERMANNIELLA

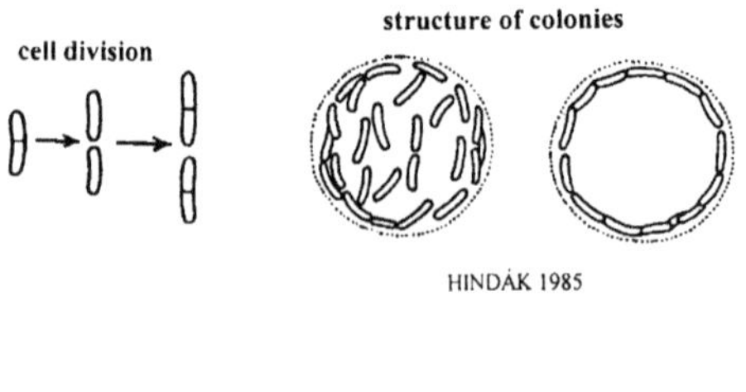

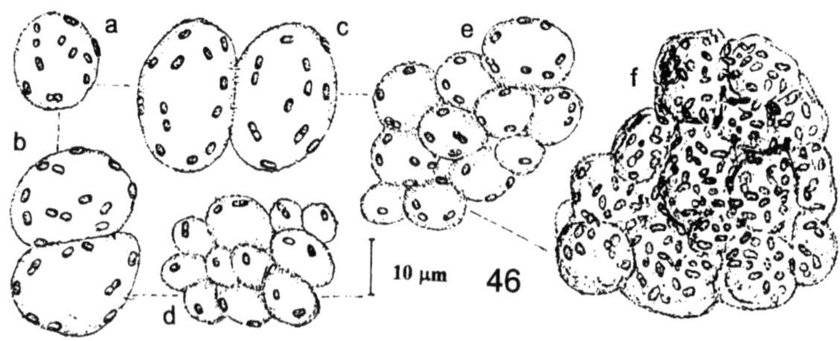


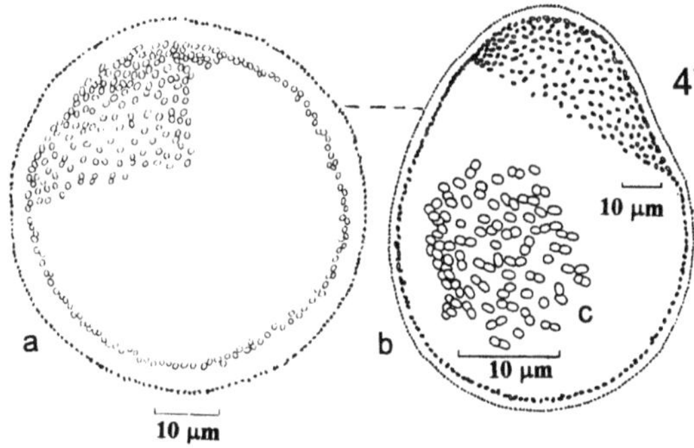


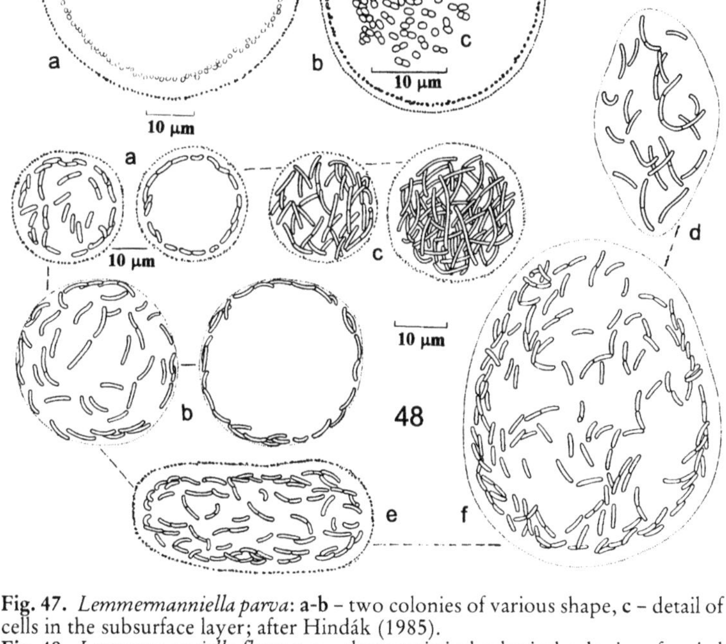


**Fig. 47.** *Lemmermanniella parva*: **a-b** – two colonies of various shape, **c** – detail of cells in the subsurface layer; after Hindák (1985).
**Fig. 48.** *Lemmermanniella flexa*: **a-c** – characteristical spherical colonies of varied size and cell density, **d-f** – irregular colonies; after Hindák (1985).

surface (from outside), distant one from another, oval to rod-shaped with rounded ends, pale blue-green, 1.5–4 × 0.8–2 µm.
Occurrence: Freshwater, benthic, epipelic in clear northern lakes, facultatively in plankton (solitary or forming small compound colonies); N. Russia (near Sankt Petersburg), Sweden, Canada.

**Species outside Europe:**
– *Epigloeosphaera filamentosa* Komárek et Cronberg 1998: Elongate and divaricate mucilaginous colonies, cells 2.8–4.2 × 2.8–3.2 µm; swamps in South Africa.

## 11. **Lemmermanniella** Geitler 1942

*Lemmermannia* Elenkin 1933

Colonies microscopic, in form of mucilaginous spheres, simple or compound, with several spherical subcolonies, sometimes slightly irregular oval, free floating in plankton or metaphyton, consisting of fine, homogeneous, colourless mucilage; beneath the surface is a layer of irregularly, more or less tangentially arranged cells, sometimes overlapped by a thin, but distinctly delimited layer of colonial slime. Cells oval or rod-shaped, up to several times longer than wide, usually slightly distant one from another or in small clusters, pale blue-green, without aerotopes. Without stalks in the colonial center.
Cell division by transverse binary fission into two equal parts, perpendicular to the longer cell-axis, in one plane in successive generations. Cells become more or less the original size before the next division. They soon separate from one another after division and usually shift to the parallel position; rarely remaining in short rows. Reproduction by «budding» of daughter spherical colonies from the surface of the old mother colony, or by disintegration of colonies into small clusters of cells or even solitary cells.
**Important references:** Elenkin (1933); Geitler (1942); Komárková-Legnerová & Cronberg (1985); Hindák (1985).

**Key to the species:**
**1a** Cells more or less oval, 1–1.8 × 0.8–1 µm . . . . . . . . . . . . . . . . **1. L. parva**
**1b** Cells rod-shaped, usually over 1.5 × longer than wide . . . . . . . . . . . . . . **2**
**2a** Cells (0.7) 1.1–3.7 (4.3) × 0.5–1.6 µm, straight . . . . . . . . . . . . **2. L. pallida**
**2b** Cells 3–15 × 0.8–1 µm, usually slightly curved . . . . . . . . . . . . . **3. L. flexa**

### 1. **Lemmermanniella parva** Hindák 1985 (fig. **47**)

Colonies free floating, spherical, oval or broadly irregular oval, multicellular. Mucilaginous envelopes hyaline, structureless, 55–120 (180) µm in diameter. Cells solitary or in twos after division, never forming pseudofilamentous rows nor groups, arranged irregularly in one layer just beneath the periphery of colonies, or the mucilage forms a narrow gelatinous layer around cells; cells very short cylindrical or oval, straight, pale blue-green or grey, 1–1.5 (1.8) × 0.8–1 µm.
Occurrence: Originally freshwater, described from eutrophic rivers (Danube) or fishponds (SW. Slovakia), but commonly occurs in brackish waters in northern Europe (e.g., known also from bays of the Baltic Sea and in the Netherlands).

### 2. **Lemmermanniella pallida** (Lemmermann) Geitler 1942 (fig. **49**)

*Coelosphaerium pallidum* Lemmermann 1898; *Lemmermannia pallida* (Lemmermann) Elenkin 1933

Colonies free floating, spherical or with small spherical daughter colonies lying on

the surface of an old mother colony, multicellular, to 85 (138) µm in diameter. Mucilage colourless, delimited, overlapping the cell layer by a thin gelatinous layer (up to 7 µm thick), sometimes indistinctly striated. Cells irregularly but more or less densely distributed in a surface layer, isolated or in small parallel groups (secondarily shifted after division); cells rod-shaped, cylindrical, straight, with rounded ends, pale blue-green, (0.7) 1.1–3.7 (4.3) × 0.5–1.6 µm.
Occurrence: Freshwater, planktic in large, slightly eutrophic lakes; known only from the Baltic area (N. Germany, NW. Russia, S. Sweden) and maybe from SE. Russia (Volga river) and central Asia (lake Zaisan). Records from Argentine (Tierra del Fuego) obviously concern another species (a real *Coelosphaerium*).

### 3. **Lemmermanniella flexa** Hindák 1985 (fig. **48**)

Colonies free floating, spherical or irregularly oval or elliptical, multicellular. Mucilaginous envelopes hyaline, structureless, 20–70 µm in diameter. Cells solitary, in twos after division, or in short (up to 4-celled) pseudofilamentous rows, irregularly arranged in one layer just beneath the periphery of the colonial mucilage; cells narrow, rod-shaped, long cylindrical to sausage-shaped, slightly bent or arcuate with rounded ends, pale blue-green or grey, (3) 7–9 (15) × 0.8–1 µm.
Occurrence: Freshwater, planktic in the summer and autumn plankton; known only from a gravel-pit lake with irregular inflow of thermal water, SW. Slovakia.

## 12. **Cyanonephron** Hickel 1985

Colonies free-living, more or less irregularly spherical, globose or ellipsoidal, sometimes composed of subcolonies, with a system of mucilaginous, pseudodichotomously branched and structureless stalks, radiating from the center of colonies (staining!); common gelatinous material surrounding the whole colony not present. Cells more or less elongate, solitary and crosswise at the ends of divaricate stalks, more or less on the colonial periphery, without aerotopes, with parietally (concentrically) localized thylakoids.
Cells divide by binary fission, transversely to their longitudinal axis and perpendicularly to the colonial surface, always in one plane in successive generations. Cells reach the original size before the next division. Reproduction by disintegration of colonies. Monotypic genus.
**Important references:** Hickel (1985); Komárek & Anagnostidis (1986).

### 1. **Cyanonephron styloides** Hickel 1985 (fig. **50**)

Colonies free floating, 12–29 µm (specimens from Finland to more than 100 µm) in diameter, composed of very fine, indistinct, colourless stalked gelatinous material; semicircular subgroups of 4–6–8 cells are formed within the colonies. Cells cylindrical, rod-shaped to kidney-shaped, always slightly arcuate (with the convex side facing the outside of the colony), with homogeneous, grey or pale yellowish-green content, without distinct inclusions, 2.3–4.5 × 0.8–1.2 µm (specimens from Finland 2–5.5 × 0.8–1.2 µm), usually covered with a thin layer of indistinct, colourless slime (staining !).
Occurrence: Freshwater (also brackish ?), planktic in shallow, hypertrophic lakes near the sea-shore (sometimes connected to the sea); sporadic in Baltic region (Finland, N. Germany, S. Sweden), but recently recorded also from Africa (Zimbabwe ?).

## 13. Hormothece Jao 1944

Colonies microscopic to macroscopic, mucilaginous, irregular, attached to a substrate, composed of thick, firm (tough), irregularly or more or less parallely and radially arranged gelatinous, colourless stalks, usually indistinctly or dis-

LEMMERMANNIELLA

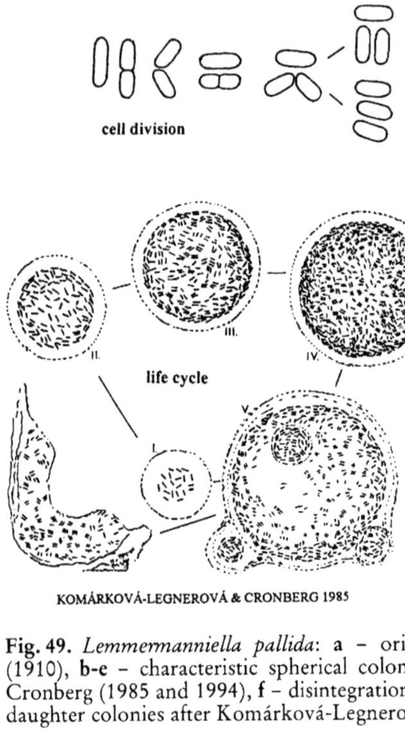


**Fig. 49.** *Lemmermanniella pallida*: **a** – original drawing after Lemmermann (1910), **b-e** – characteristic spherical colonies after Komárková-Legnerová & Cronberg (1985 and 1994), **f** – disintegration of old colony and reproduction by daughter colonies after Komárková-Legnerová & Cronberg (1985).

CYANONEPHRON

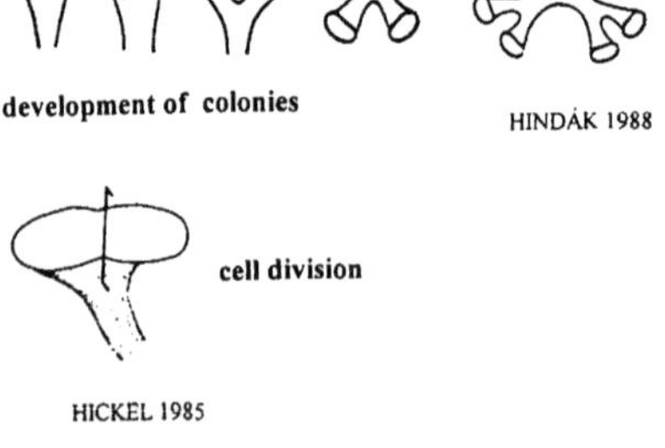

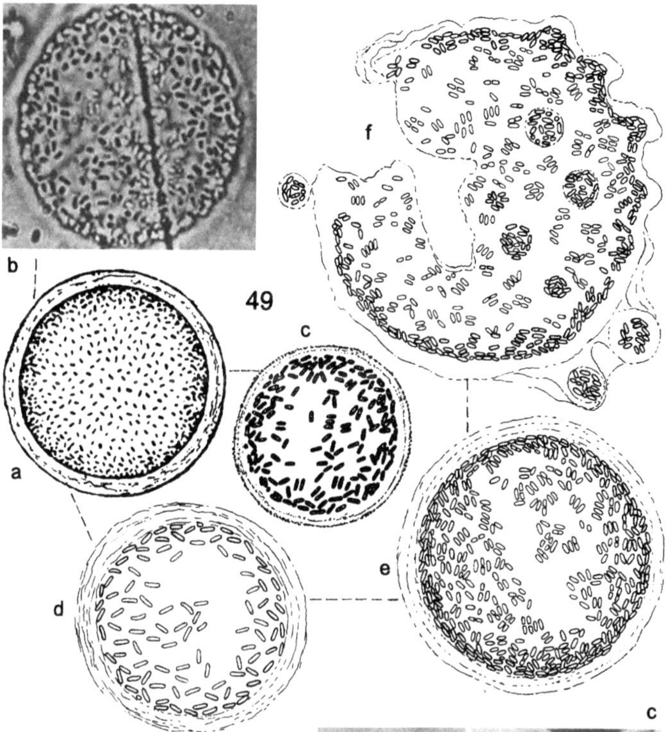


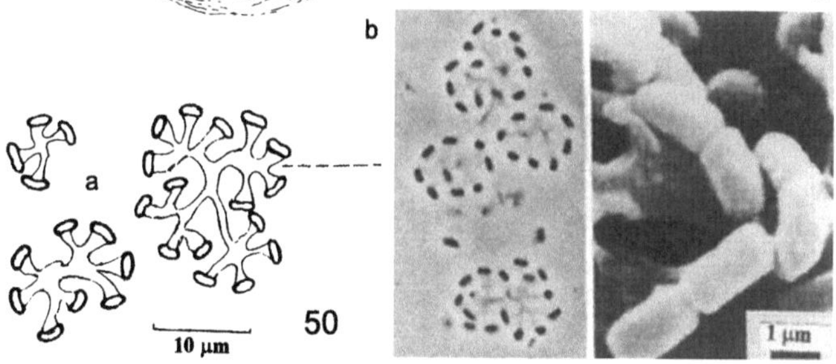


**Fig. 50.** *Cyanonephron styloides*: **a-b** – colonies of varied size, **c** – morphology of dividing cells; after Hickel (1985).

tinctly transversally lamellose, sometimes changing into an amorphous mass in the basal regions; cells solitary, in irregular, short rows with the cells distant one from another or in few-celled groups within the ends of the stalks. Cells oval or shortly rod-shaped with rounded ends.
Cell division by binary fission, always only in one plane in successive generations, perpendicular to the longitudinal cell axis; cells grow into the original size and shape before the next division. Reproduction probably by liberation of cells from colonies.
**Important references:** Geitler (1933, sub *Cyanostylon*); Margalef (1946, sub *Cyanostylon*); Jao (1944); Ley (1947); Komárek & Anagnostidis (1995).

### 1. Hormothece banyolensis (Margalef) Komárek et Anagnostidis 1995 (fig. 52)

*Cyanostylon banyolensis* Margalef 1946

Colonies microscopic, attached to the stony substrate, compound from short, ± radially arranged, thick, finely transversely stratified, colourless, slightly refractive and hyaline strands with 1–2(4) cells situated more or less at the top of strands, to 30 µm long and 7–11 µm wide. Cells cylindrical, after transversal division almost spherical, rounded at both ends, blue-green, 4–5 × 2–3 µm. Cells are enveloped by *Gloeocapsa*-like, stratified envelopes, situated often obliquely to the axis of the strand.
Occurrence: Freshwater, on wet rocks, among *Schizothrix* above the water level of lakes; known only from Spain (Banolas).

**Species outside Europe:**
- *Hormothece cylindrocellulare* (Geitler) Komárek et Anagnostidis 1995: [syn.: *Cyanostylon cylindrocellulare* Geitler 1933]: Colonies macroscopic, forming layers, composed of thick, mucilaginous, colourless strands, with pale blue-green, oval cells within their ends, up to 6 × 2–3 µm; waterfalls in Indonesia (Geitler & Ruttner 1935), recorded also from wet rocks in Brasil-SP (Sant'Anna & al. 1991); (fig. **53**).
- *Hormothece geitleriana* Azevedo et Sant'Anna 1994: Colonies gelatinous, green, stalks radial, firm, hyaline, lamellated; cells oblong, 5.9–7.6 × 5.1–5.9 µm, blue-green; on moistened soil in culture, Brazil.
- *Hormothece jaoi* Ley 1947: Colonies microscopic with transversally striated, thick, colourless mucilaginous stalks, within the ends of which are situated 1–2 cylindrical, blue-green cells, 5 × 2.2–2.5 µm; wet rocks in China (N. Kwangtung).
- *Hormothece rupestris* Jao 1944 [syn.: *Cyanostylon rupestris* (Jao) Bourrelly 1970]: Type species, colonies microscopic, with densely lamellated and pseudodichotomously branched, colourless stalks; cells 8–10 × 5–5.5 µm; wet calcareous rocks in China (Kwangsi); (fig. **51**).
- *Hormothece starmuehlneri* (Bourrelly) Komárek et Anagnostidis 1995 [syn.: *Cyanostylon starmuehlneri* Bourrelly 1984]: Colonies attached, mucilaginous, macroscopic, hemispherical, yellow-pinkish, up to 3 mm in diameter, composed from radially arranged, pseudodichotomously branched, mucilaginous, striated stalks, at ends with 1–2(8) ellipsoid cells, approximately 3.5 × 2.5 µm; freshwater, in lake littoral, New Caledonia.

## 14. Aphanothece Nägeli 1849 nom. cons.

*Coccochloris* Sprengel 1807 nom. rejic.; *Anacystis* Meneghini 1837 nom. rejic., non «*Anacystis nidulans*» sensu auct.; *Microcystis* Unterabt. *Anacystis* Kützing 1846; *Polycystis* sect. *Anacystis* Hansgirg 1892; *Cyanogastrum*, Schiller 1956 incl.

Colonies multicellular, micro- to macroscopic (up to several centimeters in diameter), with cells arranged irregularly, loosely or densely in common mucilage, gelatinous, spherical or (usually) amorphous, greenish, blue-green or brown. Common mucilage fine, colourless, homogeneous and diffluent (subg. **Anathece**), or distinct, delimited, more or less smooth on the surface, structureless (subg. **Cyanogastrum**), or in marginal parts lamellate and yellowish or yellow-brown (rarely reddish), with cells in their own indistinct or distinct and/or lamellose, delimited envelopes (subg. **Aphanothece**). Cells widely oval to cylindrical (rod-shaped) with rounded ends, rarely oval-fusiform, straight or slightly arcuate, sometimes with clearly visible chromatoplasma; in a few planktic species solitary aerotopes occur facultatively.

Cell division by binary fission (mainly cleavage) in one plane in successive generations, perpendicular to the longitudinal cell axis; in several species the daughter cells shift immediately after the division from the longitudinal position, giving the impression of «oblique» or «longitudinal» division. In several species facultative nanocyte production has been recorded. Reproduction by fragmentation of colonies, sometimes even into solitary cells.

**Important references:** Geitler (1932, 1942, 1960); Komárek & Anagnostidis (1986); Campbell & Golubić (1985); Kováčik (1988).

**Key to the species:**

1a Colonies microscopic (only exceptionally macroscopic if they are in aggregates), irregular, with colourless, homogenous, diffluent slime; cells without their own mucilaginous envelopes, usually very small (about 1 µm wide, rarely wider); planktic species (with the exception of the epizoic *A. protohydrae*); subg. **Anathece** . . . . . . . . . . . . . . . . . . . . . . . . . . . . . 2

1b Colonies microscopic, later usually macroscopic, often with coloured slime in marginal parts; cells rod-shaped, cylindrical with rounded ends or oval to widely oval, occasionally with their own mucilaginous envelopes near the colonial surface, usually more than 1 µm wide; not euplanktic species, they live mainly benthic, epipelic, metaphytic or aerophytic, only secondarily sometimes floating on the water surface (large colonies); subg. **Aphanothece** . . . . . . . . . . . . . . . . . . . . . . . . . . . . . . . . . . . . . . . . . . . . . . 11

Note: Members of the subg. **Cyanogastrum** are known only from tropical countries; (examples see p. 89–91, fig. **81–83**).

2a Width of cells smaller than 1 µm (exceptionally up to 1.6 µm) . . . . . . . . 3

2b Width of cells larger than (0.8) 1 µm . . . . . . . . . . . . . . . . . . . . . . . . . . 7

3a Cells widely oval (sometimes subspherical) to cylindrical, cell length/width ratio less to 2 : 1 . . . . . . . . . . . . . . . . . . . . . . . . . . . . . . . . . . . . 4

3b Cells rod-shaped, cell length/width ratio always higher than 2 : 1 (comp. also **5b**, *A. nebulosa*) . . . . . . . . . . . . . . . . . . . . . . . . . . . . . . . . . . . 6

4a Cells in colonies distributed irregularly, forming scattered cloudy agglomerations within a colonial mucilage, surrounded by other regularly and sparsely distributed cells; cells 1–2.2(2.8) × 0.8–1(1.6) µm; facultatively with solitary aerotopes . . . . . . . . . . . . . . . . . . . . . . . . . . . . . . **1. A. minutissima**

4b Cells more or less regularly distributed throughout the colonies, or forming clusters with aggregated cells within common slime . . . . . . . . . . . . . . . 5

5a Cells densely arranged, but regularly distributed, with distinct spaces between them, 1–2 × 0.5–1 µm . . . . . . . . . . . . . . . . . . . . **2. A. bachmannii**

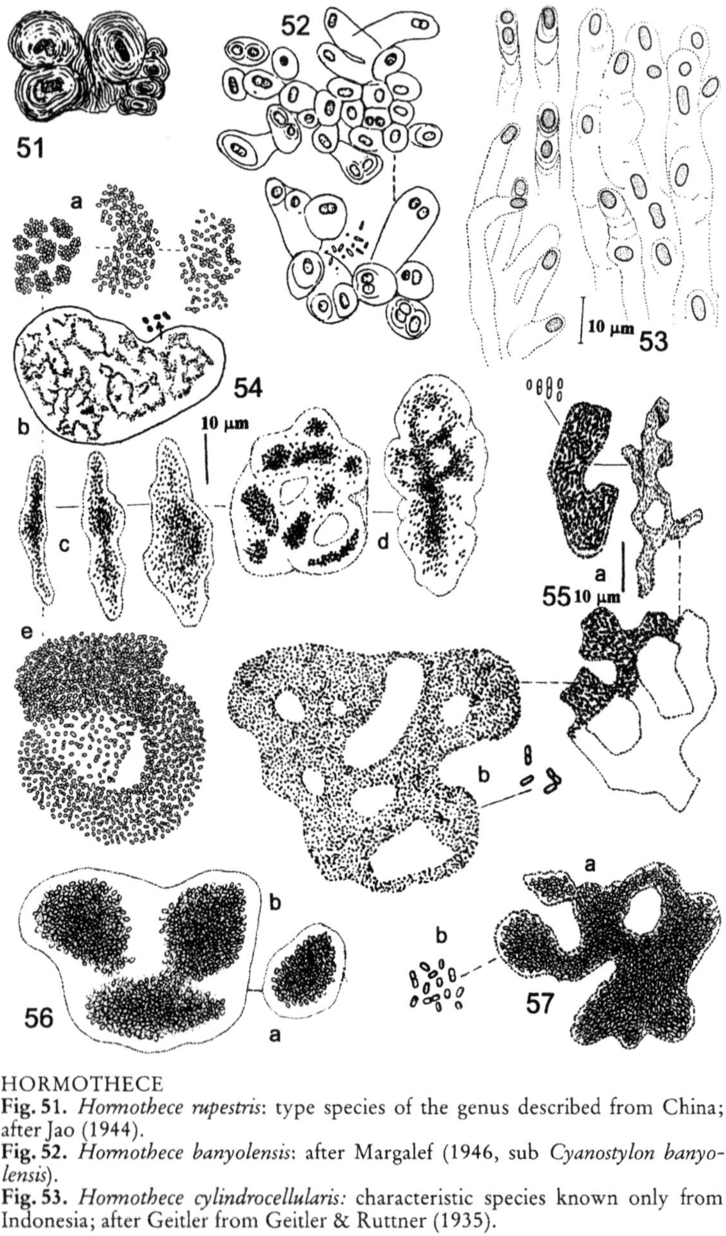


HORMOTHECE

**Fig. 51.** *Hormothece rupestris*: type species of the genus described from China; after Jao (1944).

**Fig. 52.** *Hormothece banyolensis*: after Margalef (1946, sub *Cyanostylon banyolensis*).

**Fig. 53.** *Hormothece cylindrocellularis:* characteristic species known only from Indonesia; after Geitler from Geitler & Ruttner (1935).

**5b** Cells very densely aggregated in separate clusters within common mucilage, 0.5–1(2.5) × 0.3–0.4 µm . . . . . . . . . . **3. A. nebulosa**

**6a** Colonies irregular, often flattened, sometimes clathrate; cells 0.8–3.5(4.5) × 0.4–1(2) µm without polar granules . . . . . . . . . . **4. A. clathrata**

**6b** Colonies more or less spherical, never clathrate; cells 2–6 × 0.3–0.4 µm, with small solitary polar granules . . . . . . . . . . **5. A.biceps**

**7a** Colonies epizoic or endogloeic . . . . . . . . . . **8**

**7b** Planktic or benthic (metaphytic), free-living freshwater species . . . . . . . . **9**

**8a** Colonies epizoic on *Protohydra* in the marine littoral; cells 2–3 × 0.8–1.7 µm . . . . . . . . . . **6. A. protohydrae**

**8b** Solitary cells or groups of cells endogloeic in other planktic cyanoprokaryotes (*Woronichinia, Chroococcus,* etc.) . . . . . . . . . . **7. A. endophytica** (See also the species **7a. A. desikacharyi**)

**9a** Colonies irregular, with more or less loosely arranged cells (with distinct spaces between cells), always microscopic, cells 2–3.5 × 1–1.5(2) µm, without aerotopes . . . . . . . . . . **8. A. smithii**

**9b** Colonies more or less spherical, with very densely arranged cells and clearly delimited margin (diffluent only in old colonies), sometimes several colonies united to form microscopic or to macroscopically visible agglomerations **10**

**10a** Cells intensely green, without gas vesicles, 1.6–4 × 0.8–1.2 µm **9. A. floccosa**

**10b** Cells pale blue-green, facultatively with aerotopes, 2.8–6.5(8.5) × 1–3(4.5) µm . . . . . . . . . . **10. A. elabens**

**11a** From marine or salty biotopes; micro- or macroscopic colonies . . . . . . **12**

**11b** From various freshwater and aerophytic biotopes; mainly in macroscopic colonies . . . . . . . . . . **15**

**12a** Cells oval or cylindrical, but always more or less straight . . . . . . . . . . . **13**

**12b** Cells cylindrical, always distinctly arcuate, more or less 12 × 4–5 µm . . . . . . . . . . **17. A. curvata**

**13a** Colonies small, on calcareous rocks in the marine epilittoral; Mediterranean species . . . . . . . . . . **15. A. marina**

**13b** Colonies macroscopic, in inland and coastal salt marshes, swamps and pools . . . . . . . . . . **14**

---

APHANOTHECE subg. ANATHECE

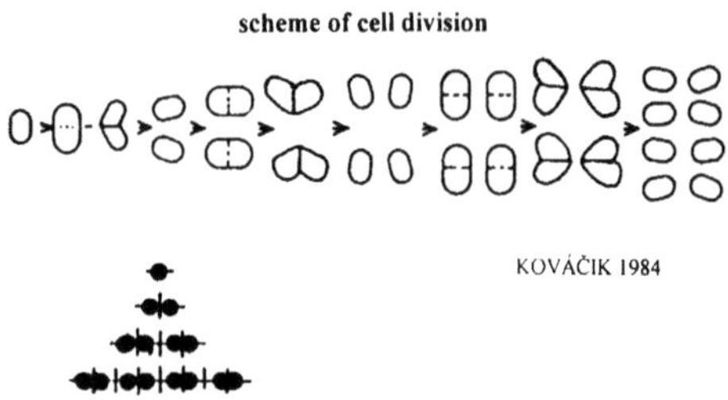


**Fig. 54.** *Aphanothece minutissima* (planktic): **a,e** - after Nygaard (1949), **b** - after Bachmann (1921), **c-d** - after Komárková-Legnerová & Cronberg (1994).
**Fig. 55.** *Aphanothece bachmannii* (planktic): **a** – after Komárková-Legnerová & Cronberg (1994), **b** – after Bachmann (1923).
**Fig. 56.** *Aphanothece nebulosa*: after Skuja (1964).
**Fig. 57.** «*Aphanothece clathrata* var. *rosea*»: after Skuja (1948).

14a Cells more or less oval, 4–7 × 3.4–4.6 µm . . . . . . . . . . . . **16. A. utahensis**
14b Cells subspherical, oval to rod-shaped, 6–15.4 × 3.2–5 µm . . **18. A. salina**
15a Thermal species . . . . . . . . . . . . . . . . . . . . . . . . . . . . . . . . . . . . . **16**
15b From non-thermal biotopes . . . . . . . . . . . . . . . . . . . . . . . . . . . . . **17**
16a Cells 4–7 × 2.5–4 µm, protoplasts usually with several granules . . . . . . . . . . . . . . . . . . . . . . . . . . . . . . . . . . . . **12. A. thermicola**
16b Cells 5.5–12 × 3.5–5 µm, usually with more or less homogeneous content . . . . . . . . . . . . . . . . . . . . . . . . . . . . . . . . . . . . **20. A. bullosa**
17a Freshwater, benthic, epipelic and metaphytic, only secondarily floating, sometimes on the water surface . . . . . . . . . . . . . . . . . . . . . . . . . . **18**
17b Aerophytic, subaerophytic and terrestrial species . . . . . . . . . . . . . . . . **21**
18a Colonies amorphous, usually bright blue-green; in moorland and peaty waters, acidic swamps and on wet peaty soil . . . . . . . **19. A. microscopica**
18b Colonies more or less irregularly spherical, diffluent, or composed of subcolonies, usually with distinctly delimited margin, in oligotrophic to mesotrophic or slightly eutrophic small water bodies, ponds and lake littoral **19**
19a Cells in very densely packed groups, usually surrounded by brownish-violet envelopes; epilithic in katharobic or oligotrophic subarctic lakes . . . . . . . . . . . . . . . . . . . . . . . . . . . . . . . . . . . . **11. A. atro-crustacea**
19b Cells in colonies more or less regularly distributed, not in groups, but single cells in marginal parts sometimes with their own, diffluent, colourless gelatinous envelopes; in mesotrophic up to eutrophic waters . . . . . . . . **20**
20a Colonies microscopic to macroscopic (up to 2 cm in diameter), grey or pale blue-green; cells pale grey-green, 1.6–4.5 × 1–1.9 µm; epipelic and metaphytic in clear (not polluted), small water bodies, pools, backwaters, cold water aquaria, usually with plenty of water plants . . . . . . **13. A. nidulans**
20b Colonies up to several cm in diameter, green, blue-green or olive-green; cells blue-green, 3.8–9(11) × 2–5(6.8) µm; benthic, epipelic, later free floating in mesotrophic or slightly eutrophic small ponds, lakes and basins . . . . . . . . . . . . . . . . . . . . . . . . . . . . . . . . . . . . **21. A. stagnina**
21a Cells oval, maximally 2.5 × longer than wide . . . . . . . . . . . . . . . . . . . . **22**
21b Cells rod-shaped, straight or slightly curved, always more than 3 × longer than wide (with the exception of nanocyte production) . **22. A. caldariorum**
22a Cells 1.5–4.5 × 1–2.3 µm; colonies amorphous, green or olive-green, large, macroscopic, on wet stony substrates; envelopes around cells facultative, usually colourless and diffluent . . . . . . . . . . . . . . . . . . . . **14. A. saxicola**
22b Cells over 3.8 µm long and 2–8 µm wide; colonies usually brownish; cells near the colonial margin always with their own, coloured gelatinous envelopes . . . . . . . . . . . . . . . . . . . . . . . . . . . . . . . . . . . . . . . . . . . . . . . . **23**
23a Cells 3.4–8 × 2–4.8 µm, on wet limestone rocks . . . . . . . **23. A. castagnei**
23b Cells 5.5–10(16) × 3.2–8 µm, usually in soil (mainly alkaline), rarely on stony substrates . . . . . . . . . . . . . . . . . . . . . . . . . . . . . . . . . **24. A. pallida**

### 1. **Aphanothece minutissima** (W. West) Komárková-Legnerová et Cronberg 1994 (fig. **54**)

*Microcystis* sp. sensu Bachmann 1911; *Microcystis minutissima* W. West 1912; *Aphanothece pulverulenta* Bachmann 1921

Colonies free floating, microscopic (to 140 µm in diameter), gelatinous, subspherical, elongate or irregular, with irregularly and more or less densely aggregated cells but of variable density throughout colony («cloudy clusters»), sometimes clathrate; mucilage fine, usually wide, colourless, homogeneous, with a more or less visible, diffluent margin. Cells widely oval or shortly rod-shaped with rounded ends, pale blue-green, yellowish green or greyish, with homoge-

neous content, without gas vesicles (facultatively ?), 1–1.5–2(?2.8) × 0.8–1(1.6) µm.
Occurrence: Freshwater, planktic from oligotrophic, dystrophic or mesotrophic to eutrophic lakes, ponds and pools; central to subpolar regions of temperate zones. Populations from tropical lakes should be revised (genotype identity?).

### 2. **Aphanothece bachmannii** Komárková-Legnerová et Cronberg 1994 (fig. **55**)

*Aphanothece clathrata* var. *brevis* Bachmann 1923

Colonies microscopic, free floating, sometimes more than 100 µm in diameter, irregular, lobate up to clathrate, with loosely or densely, more or less equally and regularly distributed cells (sometimes in indistinct, parallel rows) throughout colony; young colonies usually flat, sometimes only with approximately one layer of cells; mucilage very fine, colourless, indistinct, structureless, diffluent; colonies easily disintegrate. Cells elongate, short oval to cylindrical with rounded ends, pale blue-green or greyish, 0.8–2 × 0.5–1 µm.
Occurrence: Freshwater or slightly halophilic (different ecotypes?), planktic in large, mesotrophic lakes, ponds and reservoirs; temperate zones, in Europe commonly distributed, known also from Baltic Sea. Records from tropical lakes need further study and their identity with northern populations should be confirmed.

### 3. **Aphanothece nebulosa** Skuja 1964 (fig. **56, 59**)

*Aphanothece nebulosa* f. *elongata* Skuja 1964 incl.

Colonies microscopic, free-living, more or less spherical, globose or irregular, usually 20–300 µm in diameter, with homogeneous, colourless, hyaline mucilage, indistinctly delimited and sometimes diffluent, surrounding very dense aggregates of cells with a gelatinous margin. Cells shortly cylindrical or oval with rounded ends, sometimes clearly rod-shaped (f. *elongata*), without their own mucilaginous envelopes, pale or bright blue-green, with homogeneous content, 0.5–1(2.5) × 0.3–0.4 µm.
Occurrence: Freshwater, epipelic or secondary free floating (planktic) in clear lakes, among other algae; N. Sweden (Abisko).

### 4. **Aphanothece clathrata** W. et G.S. West 1906 (fig. **58**)

*Aphanothece gracilis* Schiller 1954

Colonies free floating, microscopic (up to 400 µm in diameter), gelatinous, initially irregularly spherical or lenticular, later irregular, elongate, more or less flat and sometimes clathrate, with a large number of irregularly and more or less densely aggregated cells. Mucilage fine, homogeneous, colourless or indistinctly yellowish, diffluent at the margin. Cells fine, always rod-shaped or very slightly spindle-shaped, rarely slightly arcuate, 2–4-times longer than wide, pale blue-green or greyish green, without gas vesicles, (0.8)1.5–3.5(4.5) × 0.4–1(2) µm.
Occurrence: Freshwater, planktic in mesotrophic or eutrophic lakes, ponds, pools and rivers, sometimes causing a slight greenish colouration of water, usually in summer; common, over the whole temperate zone, records from tropical regions need confirmation. Identity with halophilic populations (e.g., from Baltic Sea) is debatable.

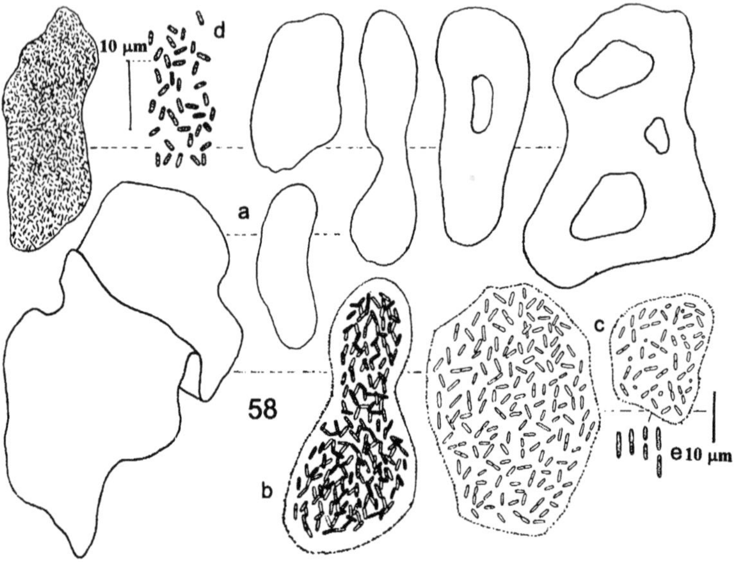


58

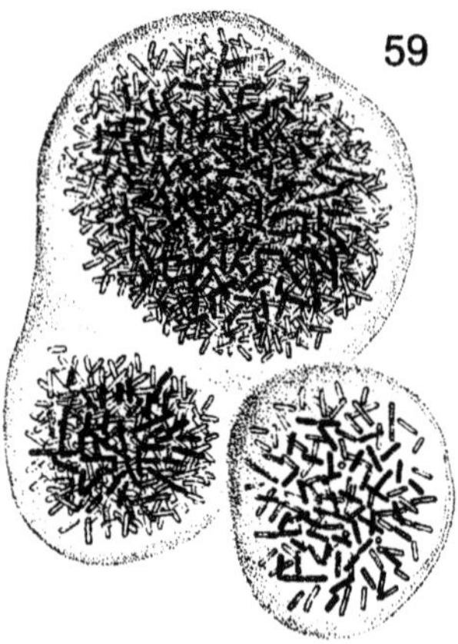

59

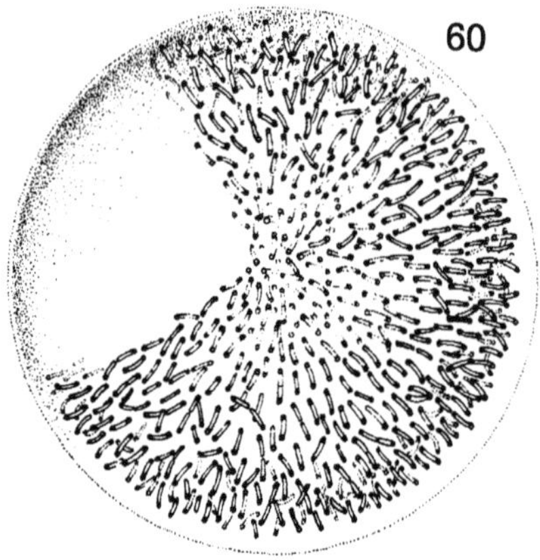

60

APHANOTHECE subg. ANATHECE

**Fig. 58.** *Aphanothece clathrata* (planktic): **a** – form of colonies, **b-c** – colonies with characteristical arrangement of cells, **d-e** – detail of rod-shaped cells; **a, d** – after Komárek (1958), **b** – after G.M. Smith (1920), **c, e** – after Komárková-Legnerová & Cronberg (1994).

**Fig. 59.** *Aphanothece nebulosa*: after Skuja (1964, sub *A. nebulosa* f. *elongata*).

**Fig. 60.** *Aphanothece biceps*: after Skuja (1964).

### 5. **Aphanothece biceps** Skuja 1964 (fig. **60**)

Colonies microscopic, free floating, more or less spherical, up to 150 µm in diameter (rarely more ?), distinctly delimited, with densely aggregated and more or less radially and subserially arranged cells; mucilage colourless, hyaline, forming an approximately 2–5 µm wide margin around the cell clusters. Cells elongate cylindrical, straight or slightly curved, with rounded ends, without their own mucilaginous envelopes, with pale blue-green or bluish content, with solitary, polar refractive granules, 2–6 × 0.3–0.4 µm.
Occurrence: Freshwater, epipelic or secondary planktic, usually sporadic among other algae; N. Sweden (Abisko).

### 6. **Aphanothece protohydrae** Häyrén in Luther 1923

Colonies mucilaginous, microscopic, amorphous, irregular, colourless, not clathrate, with densely arranged cells. Mucilage colourless, diffluent. Cells pale blue-green, 2–3 × 0.8–1.7 µm.
Occurrence: In salty waters, epizoic on the surface of *Protohydra leuckartii*; Finland, littoral in the Baltic Sea (near Tvärminne).

### 7. **Aphanothece endophytica** (W. et G. S. West) Komárková-Legnerová et Cronberg 1994 (fig. **61**)

*Aphanothece nidulans* var. *endophytica* W. et G. S. West 1912; *A. saxicola* f. *endophytica* (W. et G. S. West) Elenkin 1938, *Borzia endophytica* (W. et G. S. West) Anagnostidis et Komárek 1988

Cells solitary or in groups, sometimes arranged in irregular, short rows, situated within the mucilage of various planktic cyanoprokaryotes, without their own mucilaginous envelopes; cells distant from one another. Cells short oval or short cylindrical with rounded ends, with grey-bluish or pale blue-green content, (1)2.0–2.7(4.5) × (0.8)1.3–1.8 µm.
Occurrence: Endogloeic in mucilage of other planktic cyanoprokaryotes (*Woronichinia, Microcystis, Coelosphaerium, Chroococcus*, etc.), in clear lakes; a similar species (clusters with more cells distant one from another) occurs epiphytically and endogloeically in periphytic algae; Austria, Greece, England, Poland, Scotland, Slovakia, Sweden, USA, probably more widely distributed, particularly in northern Europe.

During the press of this book was described the following endogloeic species:

### 7a. **Aphanothece desikacharyi** Hindák 1995

Cells solitary or in pairs after division, or in groups, short cylindrical or subspherical (young), rounded at the ends, 1–1.8 × 1–1.2 µm, with a gelatinous envelope. Cell content pale blue-green, homogeneous, without aerotopes and conspicuous granules.
Occurrence: Endogloeic, often on the surface of host cells in colonies of *Microcystis aeruginosa*, or sometimes forming a distinct superficial layer on the host colony; described from the plankton of the river Danube near Bratislava (SW. Slovakia).

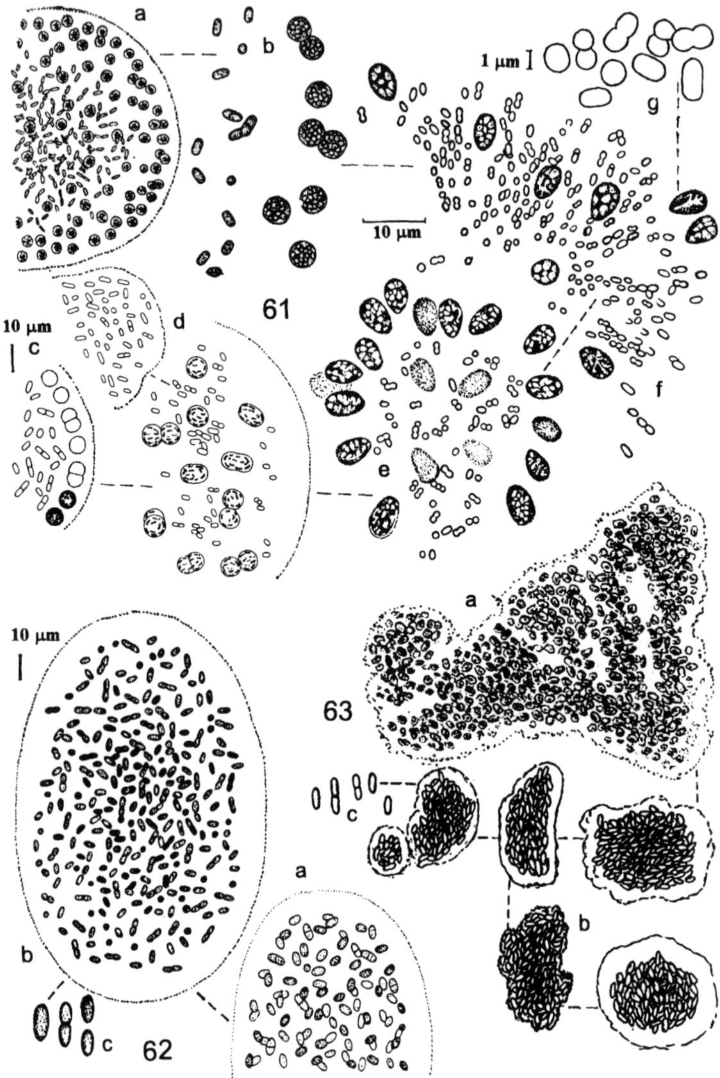


APHANOTHECE subg. ANATHECE

**Fig. 61.** *Aphanothece endophytica* (endogloeic): **a-d** - in mucilage of *Microcystis* colonies, **e-f** - within colonies of *Woronichinia naegeliana*, **g** - detail of cells from colonies of *Woronichinia naegeliana;* a - after West from Kondrateva & al. (1984), **b** - after G.M. Smith (1920), **c** - after Komárková-Legnerová & Cronberg (1994), **d** - after Hindák & Moustaka (1988), **e-g** - after Wojciechowski (1971).

**8. Aphanothece smithii** Komárková-Legnerová et Cronberg 1994 (fig. **62**)

*Aphanothece nidulans* sensu G.M. Smith 1920 et auct. post. (planktic populations)

Colonies microscopic, more or less spherical or irregular in outline, sometimes elongate, free floating, with more or less regularly (usually not very densely) distributed cells; colonial mucilage colourless, usually overlapping all the cells by a narrow, hyaline, sometimes diffluent margin. Cells more or less oval to cylindrical, straight, pale blue-green or grey-blue, sometimes with minute aerotopes, (1.5)2–3.5 × 1–1.5(2) µm.

Occurrence: Freshwater, planktic in mesotrophic and slightly eutrophic, usually large water bodies (lakes, reservoirs, ponds); both temperate zones, common in colder (northern) regions (Canada, Scandinavia); known also from bays of Baltic Sea.

**9. Aphanothece floccosa** (Zalessky) Cronberg et Komárek 1994 (fig. **63**)

*Microcystis floccosa* Zalessky 1925

Colonies small, more or less spherical, microscopic, later aggregated into macroscopically visible, irregular clusters. Individual colonies usually with distinct, colourless, not structured, but on the surface firmly delimited slime, forming a wide margin around the cells; cells very densely aggregated. Cells thin, rod-shaped, intensely green coloured, without aerotopes, (1.6)3–4 × 0.8–1.2(2.4) µm.

Occurrence: Epipelic and metaphytic in benthos and littoral of clear, oligotrophic or mesotrophic lakes; known from N. Europe (Finland, N. Russia, Sweden), rarely from central Europe.

**10. Aphanothece elabens** (Brébisson in Meneghini) Elenkin 1938 (fig. **65**)

*Micraloa elabens* Brébisson in Meneghini 1842; *Microcystis elabens* (Brébisson) Kützing 1846; *Polycystis elabens* (Brébisson) Kützing 1849; *Microcystis elabens* var. *maior* Bachmann 1921, incl.; *Microcystis aphanothecoides* Zalessky 1926; *Aphanothece elabens* var. *minor* Nygaard 1929, incl.; *Coccochloris elabens* (Brébisson) Drouet et Daily 1956

Colonies microscopic, spherical or flat, later macroscopic (aggregate colonies), blue-green or olive-green. Mucilaginous envelopes homogeneous or slightly concentrically lamellate, usually distinctly delimited, rarely diffluent. Cells oval to cylindrical, olive-green, facultatively (in plankton) with aerotopes, 2.8–6.5 × 1–2(3) µm. Species with many misinterpretations in the literature. All populations from the temperate zone with small cells (3–6 × 1–2 µm) belong probably to the typical *A. elabens* (incl. var. *minor*) (fig. **65**), whereas the tropical populations with cells 4–8.5 × 2–4 µm belong to the species *A. conglomerata* Rich 1933 (fig. **64**). Taxonomic and ecological revisions are needed.

---

**Fig. 62.** *Aphanothece smithii*: **a** – after G.M. Smith (1920, sub *A. nidulans*), **b-c** – after Komárková-Legnerová & Cronberg (1994).

**Fig. 63.** *Aphanothece floccosa*: **a** – large colony after Zalessky (1926, sub *Microcystis floccosa*, **b,c** – small colonies and detail of cells after Cronberg & Komárek (1994).

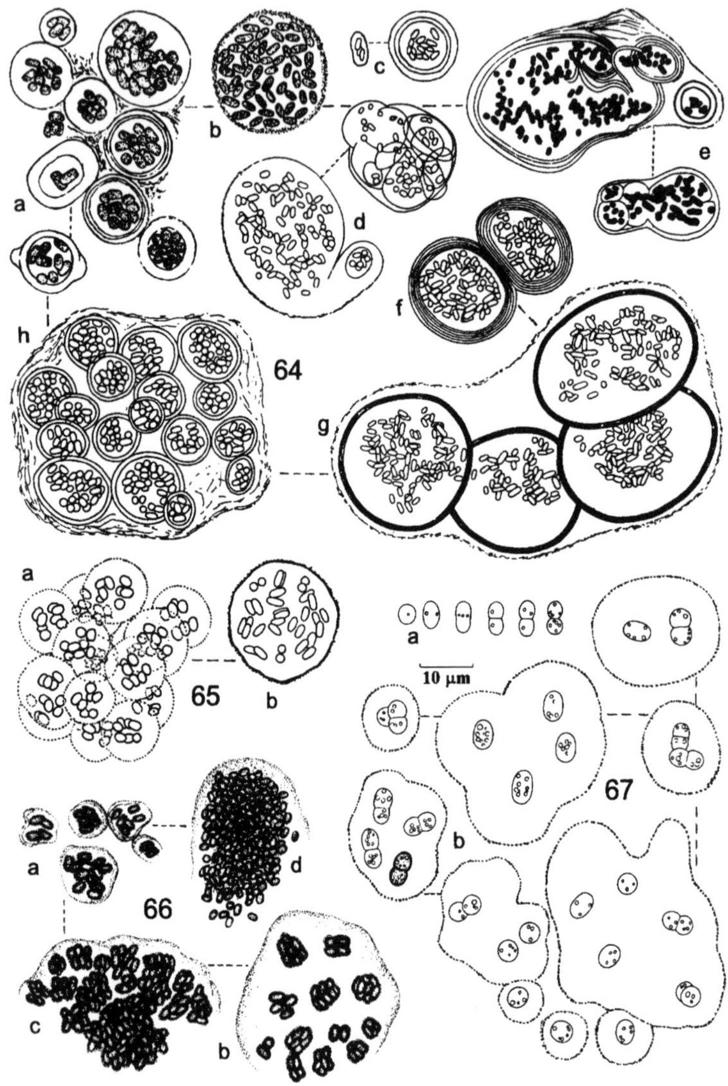


APHANOTHECE subg. ANATHECE

**Fig. 64.** *Aphanothece conglomerata*: **a** – after Rich (1933) from S. Africa, **b** – after Frémy (1930, sub *Microcystis elabens*) from Africa, **c-h** – after Nygaard (1920, sub *Microcystis elabens* var. *minor*) from Indonesia.

**Fig. 65.** *Aphanothece elabens*, European morphotypes with smaller cells: **a** – after

Occurrence: Freshwater, epipelic and metaphytic, later sometimes free-floating in clear lakes; occurs in temperate zones, but recorded commonly from tropical regions (as pantropical species ?), occasionally occurs up to subpolar regions (comp. Kondrateva & al. 1984); Europe: Baltic states, Hungary, Russia, Ukraine.

### 11. **Aphanothece atro-crustacea** Skuja 1964 (fig. **66**)

Colonies mainly microscopic, irregular spherical, later irregular, subspherical or flattened, macroscopic, mucilaginous to crust-like, blackish violet, sessile on submersed stones and rocks. There are mucilaginous envelopes around the separate cells or cell groups (with 2-4-8 or more cells), initially hyaline, later brown-violet of varying intensity, and confluent; common mucilage colourless or slightly violet, irregular and sometimes diffluent at the margin. Cells in old colonies densely and irregularly aggregated, more or less oval to cylindrical with rounded ends, with grey-blue, homogeneous content, 2–4 × 1.5–2 µm.
Occurrence: Freshwater, benthic, epilithic among other algae in littoral of clear northern (subarctic) lakes; N. Sweden.

### 12. **Aphanothece thermicola** Hindák 1978 (fig. **67**)

Colonies microscopic, mucilaginous, only few-celled (usually with 4–8 cells), up to 50 µm in diameter, rarely multicellular, with irregularly scattered cells and a conspicuous, irregular, colourless and structureless mucilage envelope 10–15 µm wide; cells without own visible gelatinous envelopes. Cells widely oval to oval-cylindrical, distant one from another, only after division in two, with blue-green content, usually with several granules, 4–7 × 2.5–4 µm.
Occurrence: Periphytic in a small open basin with thermal water (49.5 °C); known only from thermal springs, W. Slovakia (comp. *A. conferta* Richter, p. 91).

### 13. **Aphanothece nidulans** Richter in Wittrock & Nordstedt 1884 (fig. **68**)

*Aphanothece saxicola* f. *nidulans* (Richter) Elenkin 1938; non *Anacystis nidulans* sensu auct.

Colonies initially microscopic, forming spherical mucilaginous aggregates among other algae, later macroscopic, firm, more or less spherical or irregular (to 2 cm in diameter), grey-green, yellow-green or pale brownish, with more or less densely arranged cells. Mucilage colourless, fine and confluent inside the colonies, at the margin delimited. Cells oval up to cylindrical, straight or very slightly arcuate, pale grey-blue, yellow-green to blue-green, in marginal parts sometimes with their own indistinct, diffluent, colourless, not lamellate envelopes, with homogeneous content, (1.6)2.5–4(4.5) × (0.8) 1–1.9 µm.

---

Teiling (1942, corresponding probably to the typical *A. elabens*) from Sweden, **b** – after Hegewald & al. (1975) from Hungary.
APHANOTHECE subg. APHANOTHECE
**Fig. 66.** *Aphanothece atro-crustacea*: **a-d** – original drawings of Skuja (1964) from N. Sweden.
**Fig. 67.** *Aphanothece thermicola* (thermal springs): **a** – detail of cells, **b** – microscopic colonies; after Hindák (1978) from culture.

Occurrence: Freshwater, in benthos, periphyton and metaphyton of small stagnant waters, not polluted, clear water bodies, swamps, pools, backwaters, coldwater aquaria, rarely secondarily free floating; in temperate zones distributed, records from tropical countries (E. Africa, Brazil, India, Srí-Lanka) should be checked. Planktic populations (G.M.Smith 1920, Guarrera & al. 1987, Hindák & Moustaka 1988) with microscopic colonies from lakes belong mainly to *A. smithii*. Microscopic colonies from metaphyton and periphyton of swamps and littoral of freshwaters (similar to *A. lemnae* Chu, described from China) possibly represent special morphotypes or a separate species. *A. nidulans* var. *thermalis* Hansgirg 1892 (with cells 2–3.2 × 0.8–1.2 µm) from thermal springs in Europe, Asia and Africa, probably represents a separate species.

### 14. Aphanothece saxicola Nägeli 1849 (fig. 69)

*Aphanothece subachroa* Hansgirg 1892

Colonies microscopic, later macroscopic, gelatinous, amorphous, almost colourless (grey-green) up to intensely green, spreading over wet substrates (stones, rarely wood), with more or less densely packed cells. Mucilage colourless or slightly yellow-brownish at the margin; cells sometimes with indistinct, colourless individual mucilaginous envelopes near the colonial surface. Cells oval to cylindrical, with more or less homogeneous content with visible chromatoplasma, blue-green or olive-green, (1.5)2–4.5(6 ?) × 1–2.3 µm.
Occurrence: Aerophytic and subaerophytic on wet stony, mainly non calcareous substrates, rocks and walls (greenhouses), rarely on wet wood; northern temperate zone, sporadic, particularly in regions with wet climate; data from hydrobiotopes (and/or tropical regions) should be checked (comp. Uherkovich & Rai 1979 from Rio Negro, Brazil; Senna 1994 from Brazil; Romo & al. 1995 from W. Africa; Branco & al.1996 from Brazilian mangroves; etc.).

### 15. Aphanothece marina (Ercegović) Komárek et Anagnostidis 1995 (fig.70)

*Synechococcus marinus* Ercegović 1932

Colonies microscopic, amorphous, more or less colourless. Common mucilage fine, colourless. Cells oval to subcylindrical, rounded at both poles, usually surrounded by their own, homogeneous, confluent envelopes (to 10 µm wide), with blue-green or blackish-green, homogeneous content, 7–10 × 2–4 µm.
Occurrence: Marine, in epilittoral among other epilithic algae on limestone seashores; Croatia (Dalmatia), Greece (Aegean Sea, Ionian Sea).

---

APHANOTHECE subg. APHANOTHECE

**Fig. 68.** *Aphanothece nidulans* (benthic, metaphytic): **a** – detail of cells, **b** – old macroscopic colonies from freshwater biotopes, after Komárek (1975); **c** – small colony – morphotype after Hegewald & al. (1975).

**Fig. 69.** *Aphanothece saxicola* (aerophytic): **a** – small colony, **b** – cells with individual envelopes, after Nägeli (1849); **c** – old macroscopic, mucilaginous, aerophytic colonies on a stony substrate, orig.

**Fig. 70.** *Aphanothece marina* (marine coasts): cells enveloped by individual gelatinous envelopes; after Ercegović (1932).

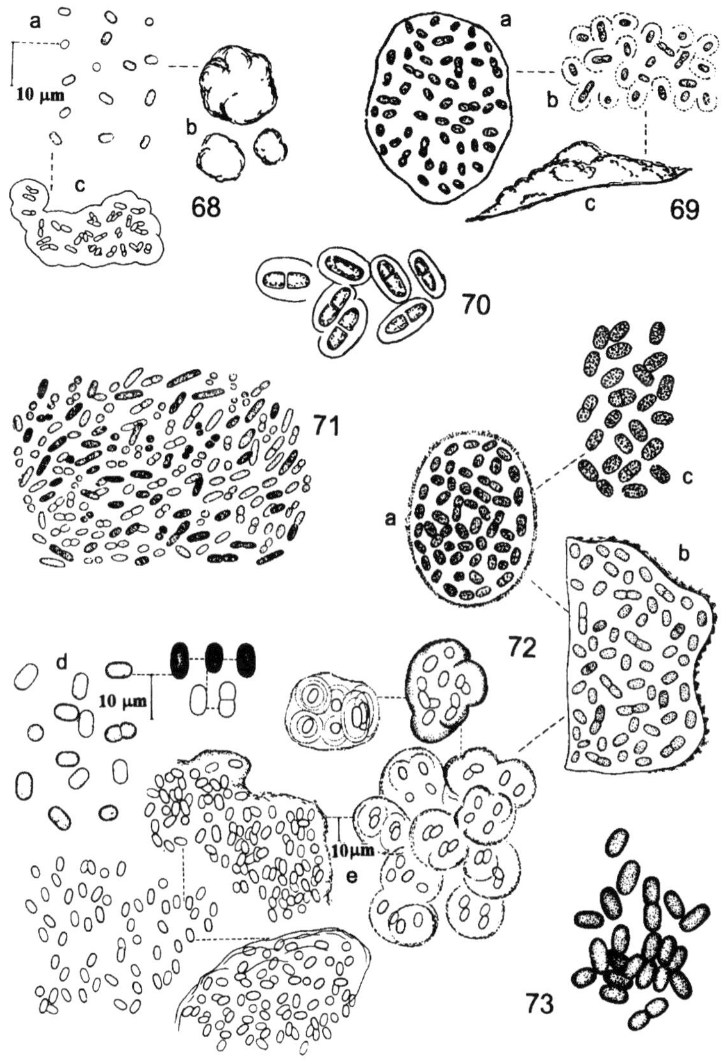


**Fig. 71.** *Aphanothece salina* (halophytic): inner part of a colony; after Elenkin & Danilov (1915).

**Fig. 72.** *Aphanothece microscopica* (benthic in moorland waters and swamps): **a** – small colony, **b** – marginal part of a colony, **c-d** – detail of cells, **e** – various types of benthic colonies; **a-b** – after Nägeli (1849), **c** – after Frémy (1930), **d-e** – after Komárek (1975).

**Fig. 73.** *Aphanothece microscopica* «fa. *endophytica*»: from Skuja (1948).

### 16. **Aphanothece utahensis** Tilden 1898

*Aphanothece packardii* (Farlow) Setchell 1917

Colonies amorphous, spreading, macroscopic, soft, sometimes thin-membranous, blue-green or brown, up to 6 cm in diameter. Cells almost spherical to shortly wide oval, singly or in pairs in common mucilage, 4–7 × 3.4–4.6 µm.
Occurrence: Brackish, benthic and metaphytic or even floating in coastal swamps and in the littoral of inland salty lakes; recorded occasionally from warmer regions of temperate zones (Ukraine – Crimea, USA).

### 17. **Aphanothece curvata** Lagerheim 1883

Colonies spreading over substrates, flat, macroscopic, incrusted with lime, with more or less densely aggregated cells. Cells cylindrical, arcuate, with blue-green content, circa 12 × 4–5 µm. Insufficiently known species.
Occurrence: Marine, on old leaves of *Zostera* and remains of plants; Baltic Sea (Sweden).

### 18. **Aphanothece salina** Elenkin et Danilov 1915 (fig. **71**)

Colonies large, usually elongate, forming a mucilaginous, irregular blue-green or olive-green mass on a substrate, to 9 cm in diameter, with densely, homogeneously and randomly arranged cells, separated from one another or in pairs. Mucilage more or less colourless, structureless without special envelopes around cells. Cells widely oval, ellipsoidal to elongate rod-shaped, pale blue-green, often intensely granular, 6–15.4 × 3.2–5 µm. Facultative nanocyte production probable.
Occurrence: In salty inland and coastal water bodies, lakes and swamps; coasts of Black Sea (Ukraine – Crimea), central Asia.

### 19. **Aphanothece microscopica** Nägeli 1849 (fig. **72, 73**)

Colonies mucilaginous, bright blue-green, initially microscopic, spherical or oval with more or less delimited mucilage, later amorphous, macroscopic, with irregular and diffluent margin, up to 2 mm in diameter, with more or less densely located cells. Mucilage colourless, homogeneous, later diffluent. Cells widely oval to cylindrical, sometimes with their own colourless, simple envelopes at the colonial margin, intensely blue-green, rarely pale green or olive-green, usually with visible chromatoplasma, (3.2) 4–9(10) × 3–4.8(6) µm.
Occurrence: Freshwater, benthic and epiphytic in moorland waters, peaty swamps, pools or in littoral of oligotrophic lakes (sometimes in masses), rarely subaerophytic on moist peaty soils, secondary in plankton of clear water bodies in both temperate zones. Data from tropical regions (Africa, India) probably concern another species; f. *endophytica* Skuja (1948) is probably only an ecomorphosis (fig. **73**).

### 20. **Aphanothece bullosa** (Meneghini) Rabenhorst 1865 (fig. **74**)

*Microcystis bullosa* Meneghini 1837

Colonies macroscopic, mucilaginous, irregular, amorphous with firm surface, greyish, yellowish-green or olive-green, to 15 cm in diameter. Mucilage colourless, structureless, with irregularly arranged cells, sometimes surrounded by their

---

APHANOTHECE subg. APHANOTHECE
**Fig. 74.** *Aphanothece bullosa* (thermophilic): **a-b** – parts of old colonies, **c** – young, microscopic colony, **d-g** – marginal parts of large colonies, **h** – details of

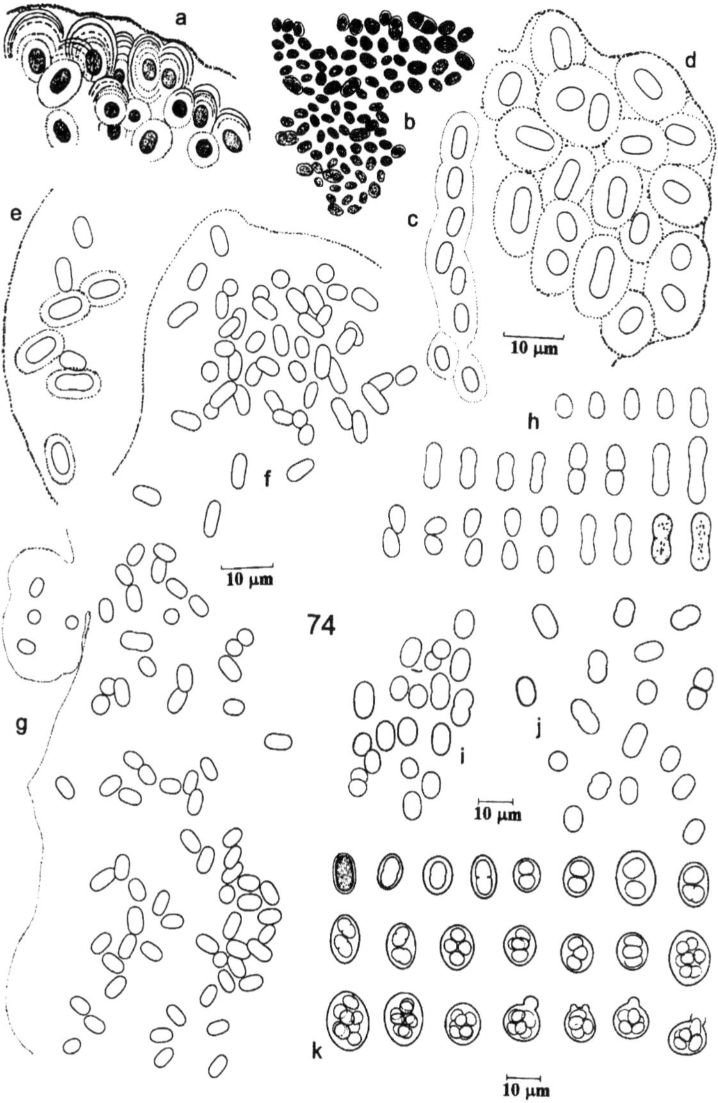


dividing cells, **i-j** – parts of colonies described as «var. *maior*» from Indonesia, **k** – formation of nanocytes in «var. *maior*»; **a** – from Bourrelly & Manguin (1952), **b** – after Frémy (1930), **c-h** – after Hindák (1978), **i-k** – from Geitler & Ruttner (1935). [Only the figs **c-h** illustrate the European material from Slovakia.]

own gelatinous envelopes near the colonial margin. Cells oval to elongate cylindrical with rounded ends, with more or less homogeneous content, sometimes finely granular, pale blue-green or olive-green, (3)5.5–12 × (2.5)3.5–5 µm. Cell division probably by pinching.
Occurrence: Subaerophytic and atmophytic at edges of thermal springs, on rocks and stones wetted by thermal waters (from 20 °C up to 50 °C), rarely submerged, attached to the substrate or (secondarily) free floating in warm water; thermal springs, probably all over the world, recorded from Europe (Greece, Italy, Slovakia), Africa, Asia (Indonesia, Japan, Russia-Kamchatka) and central America.

## 21. **Aphanothece stagnina** (Sprengel) A. Braun in Rabenhorst 1863 (fig. **75**)

*Coccochloris stagnina* Sprengel 1807; *Aphanothece prasina* A. Braun in Rabenhorst 1863; *Aphanothece piscinalis* Rabenhorst 1865; *Aphanothece mooreana* (Harvey) Lagerheim 1883; *Aphanothece tuberculata* (Areschoug) Forti 1907; *Microcystis roeseana* (De Bary) Elenkin 1938 (?)

Colonies initially small, later macroscopic, gelatinous, with distinct, firmly delimited margin, spherical or irregularly elongate and lobate, usually consisting of subcolonies, to 4(6) cm in diameter, bright blue-green or dark olive-green, in the center sometimes with small calcite crystals, with more or less densely, but in the colonial center loosely dispersed cells. Mucilage around subcolonies and of the individual cell envelopes colourless, delimited at the beginning of the life cycle, later more diffuse and confluent. Cells oval to widely cylindrical with rounded ends, on the colonial margin sometimes with their own indistinct colourless gelatinous envelopes, with homogeneous, pale to bright blue-green cell content, 3.8–9(11) × (2)3–5(6.8) µm. Nanocyte production recorded.
Occurrence: Freshwater, benthic, epipelic, sometimes attached to submerged water plants, later free floating on the water surface, occasionally in masses, usually in the smaller eutrophic bodies; common all over the temperate zones, but with very distinct and delimited ecology and characteristic life cycle (data from Argentine see in Halperin 1964, Guarrera & al. 1987, etc.); however, numerous records from tropical countries (and from the southern hemisphere) should be compared with temperate populations for identity (e.g. Frémy 1930, Halperin & al. 1974, Guarrera 1986, Werner 1988); data from salty waters are questionable and need confirmation (comp. Zhuang & al. 1993).

## 22. **Aphanothece caldariorum** Richter 1880 (fig. **76**)

*Aphanothece microspora* (Meneghini) Rabenhorst 1863; *Aphanothece trentepohlii* (Mohr) Grunow in Rabenhorst 1865; *Aphanothece muralis* (Tomaschek) Lemmermann 1907; *Gloeothece caldariorum* (Richter) Hollerbach in Elenkin 1938

Colonies macroscopic, mucilaginous, amorphous, irregular, sometimes forming a gelatinous, formless mass on a substrate, greyish, greenish or pale violet. Mucilage

---

APHANOTHECE subg. APHANOTHECE
**Fig. 75.** *Aphanothece stagnina* (epipelic, later free-floating): **a-b** – various forms of macroscopic colonies (almost 1: 1), **c-e** – microscopic spheroidal colonies, **f-g** – marginal parts of old colonies with details of cells, **h** – sparse distribution of cells within a colony; **a** – after Woloszynska from Huber-Pestalozzi (1938), **b** – after Frémy (1930), **c-d** – after G.M. Smith (1920, **d** – sub *A.prasina*), **e** – after Werner & Rosa (1992) from Brasil, **f** – after Komárek in Fott (1959), **g** – after Komárková-Legnerová & Cronberg (1994), **h** – after Forbes.

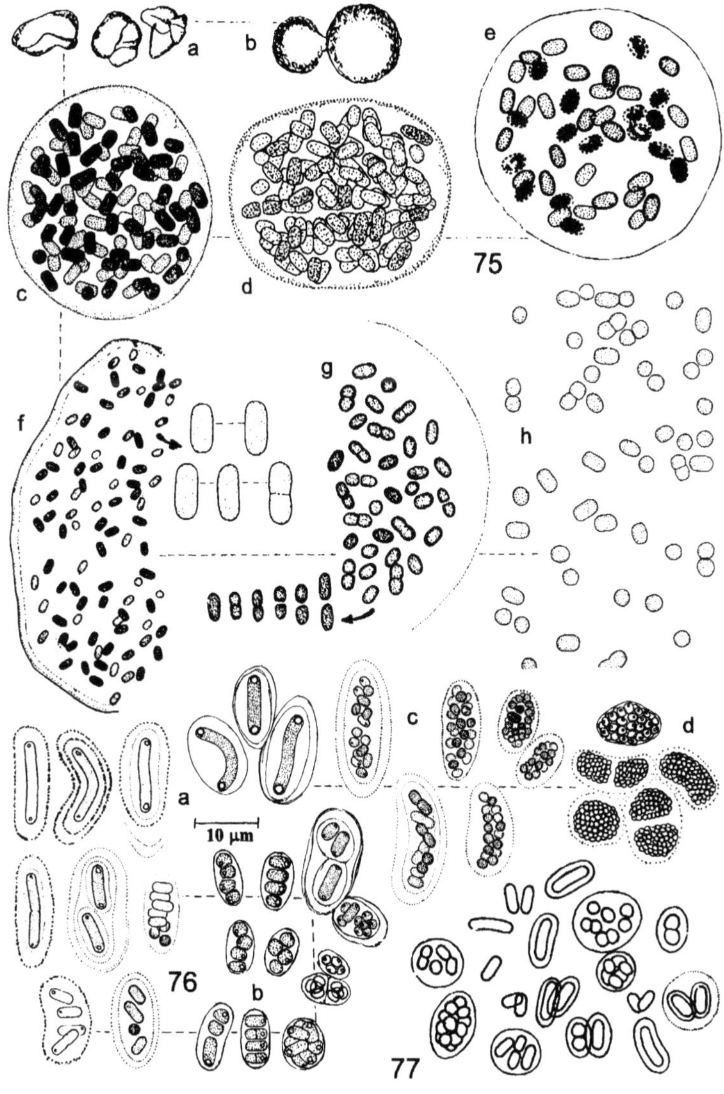


**Fig. 76.** *Aphanothece caldariorum* (subaerophytic): **a** – individual cells with their mucilaginous envelopes and dividing cells, **b-d** – cells dividing in nanocytes; after Geitler (1925 and 1960).
**Fig. 77.** *Aphanothece* cf. *caldariorum*: vegetative cells and nanocyte formation; after Geitler (1932, sub *Aphanothece microspora*).

colourless, homogeneous; cells on the colonial periphery with their own gelatinous, hyaline, structureless or slightly concentrically lamellate envelopes. Cells cylindrical, rod-shaped, rarely straight, usually slightly arcuate or sigmoid, with pale blue-green, yellowish, greenish-violet or greyish content, sometimes with prominent, large, refractive granules at both poles, (1.2)4–7.5(12) × (0.6)1.2–2(3) µm. Facultative nanocyte production (0.6–1.2 µm in diameter). Possibly different from other *Aphanothece* species (polar granules, nanocyte formation, thylakoids ?).
Occurrence: Aerophytic, on wet rocks, walls, in greenhouses, atmophytic near thermal springs; widely distributed, probably cosmopolitan (?).

### 23. Aphanothece castagnei (Brébisson) Rabenhorst 1865 (fig.78)

*Gloeothece heufleri* Grunow in Rabenhorst 1865?

Colonies macroscopic, mucilaginous, irregularly spherical to amorphous, firm on the surface, dirty blue-green, olive-green or (mainly) brown, with more or less densely and irregularly arranged cells, near the colonial margin regularly surrounded by their own, usually slightly lamellate and yellowish brown envelopes. Cells oval or cylindrical with widely rounded ends, blue-green with visible narrow chromatoplasma, (3.8)4–8 × (2)3–4.8(5.5) µm.
Occurrence: Aerophytic, on wet rocks and mosses, common particularly in mountains; probably cosmopolitan, common in European temperate zone; records from aquatic biotopes (even thermal springs, thiobiocoenoses, mangroves, salty waters – Baltic Sea and Atlantic sea coast, planktic in tropical lakes, etc.) are evidently incorrect.

### 24. Aphanothece pallida (Kützing) Rabenhorst 1863 (fig. 79)

Colonies micro- or macroscopic, mucilaginous, fine, diffluent or with delimited margin, amorphous, almost colourless (greyish), later blue-green or brown, with loosely (in the colonial center) to densely arranged cells. Mucilage colourless, structureless, but cells near the colonial surface usually with their own concentrically lamellate and intensely yellow-brown envelopes. Cells long oval to cylindrical, with pale grey-blue, blue-green or olive-green, fine granular content, (5.6)7.5–10.5(16) × (3.2)5–8 µm.
Occurrence: Terrestrial, in soils and on periodically wetted stony walls; sporadic, probably cosmopolitan.

**Species outside Europe:**
- *Aphanothece baccilloidea* Gardner 1927: Cells 1.6–2 × 0.8–1 µm, subaerophytic, in limestone depressions, Cuba, Porto Rico, USA – Florida.
- *Aphanothece botryosa* Obuchova 1965.

---

APHANOTHECE subg. APHANOTHECE
**Fig. 78.** *Aphanothece castagnei* (aerophytic, yellow-brown envelopes): **a** – marginal part of a colony, **b** – cells with well developed mucilaginous envelopes, **c** – part of a young colony, **d** – inner part of a colony (sub. «*Anacystis rupestris*»), **e** – variation of cells; **a-b,e** = orig., **c** – after Nägeli (1849), **d** – from G.M. Smith (1950, sub *Anacystis rupestris*).
**Fig. 79.** *Aphanothece pallida* (yellow-brown envelopes): **a** – after Geitler (1932), **b** – after M.Watanabe & Komárek (1988).
**Fig. 80.** *Aphanothece karukerae* (marine, red cells): **a** – macroscopic attached colony, **b** – detail of cells; after Lami from Kosinskaja (1948).

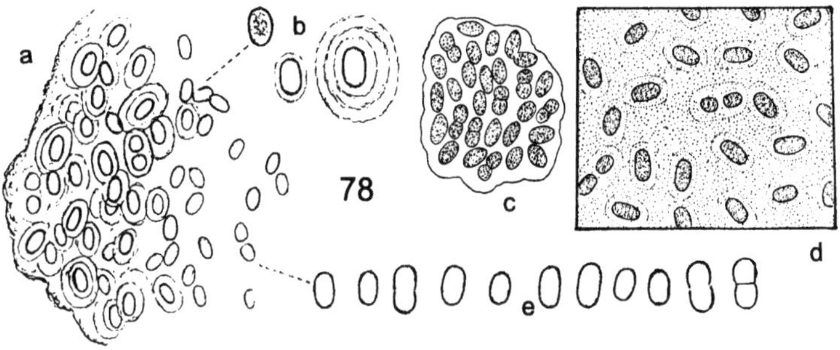


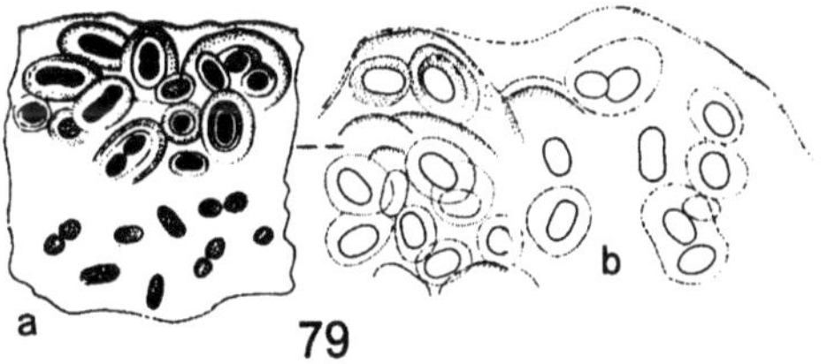


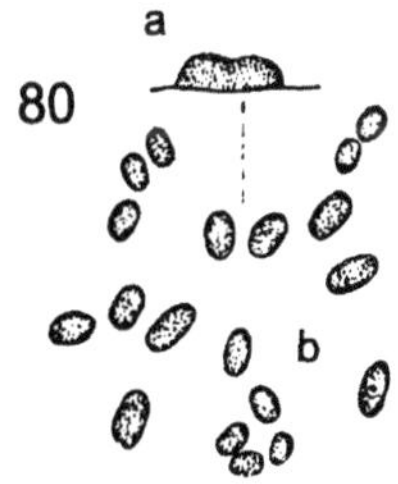


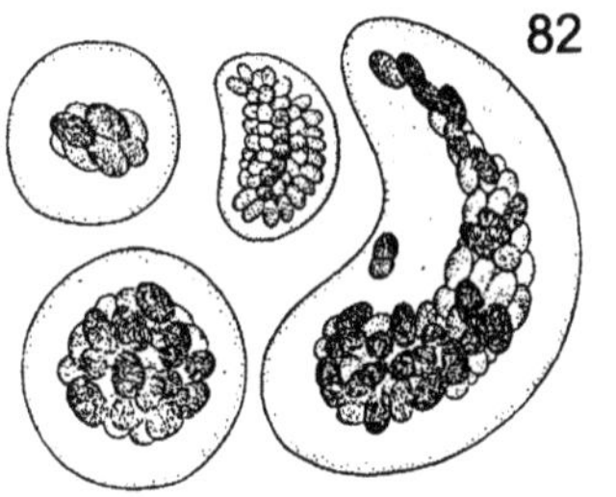


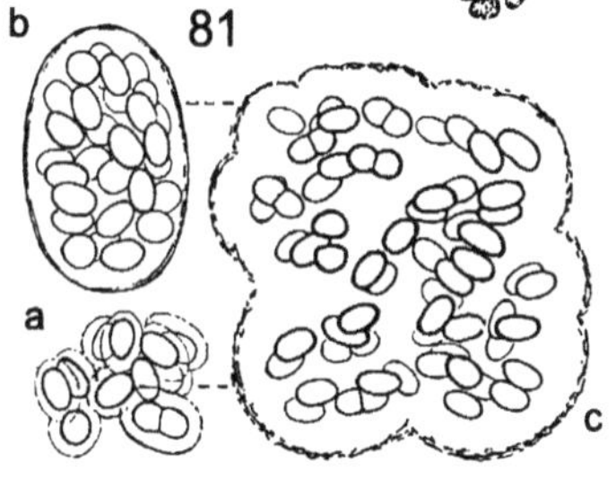


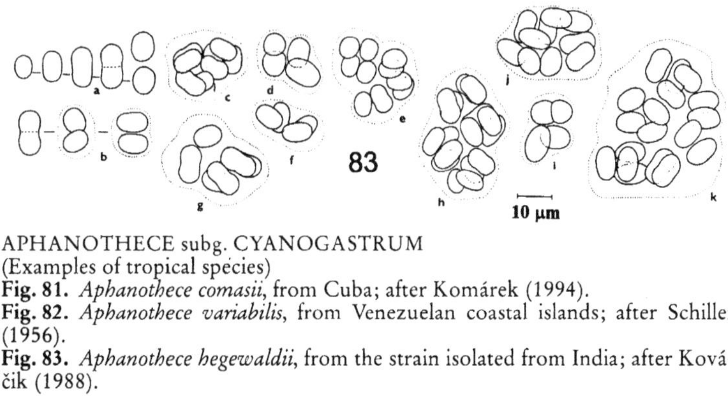


APHANOTHECE subg. CYANOGASTRUM
(Examples of tropical species)
**Fig. 81.** *Aphanothece comasii*, from Cuba; after Komárek (1994).
**Fig. 82.** *Aphanothece variabilis*, from Venezuelan coastal islands; after Schiller (1956).
**Fig. 83.** *Aphanothece hegewaldii*, from the strain isolated from India; after Kováčik (1988).

– *Aphanothece cohenii* Campbell et Golubić 1985: Cells 3–9 × 2–4 μm (in culture 3.5–5.5 × 2–2.5 μm), amorphous gelatinous mats in shallow hypersaline biotopes, Egypt (Sinai).
– *Aphanothece comasii* Komárková-Legnerová et Tavera 1996: From the subg. *Cyanogastrum*, with cells aggregated within spheroidal colonies; cells oval, (2.3)3–5 × (1.8)2–2.5(4) μm; metaphytic and planktic in freshwater tropical swamps and lakes (Cuba, eastern coastal areas of Mexico); (fig. **81**).
– *Aphanothece conglomerata* Rich 1932: Well defined tropical and subtropical metaphytic to planktic species with cells 4–8.5 × 2–4 μm, with aerotopes and distinct mucilaginous colonies; sometimes forming rich, composed colonial groups and waterblooms; often identified as a «large type of *A. elabens*», present in tropical countries (subg. *Cyanogastrum*?); described from S. Africa, known also from Venezuela and southern USA (Florida); (fig. **64**).
– *Aphanothece cylindracea* (Gardner) Komárek et Anagnostidis 1995 [syn.: *Anacystis cylindracea* Gardner 1927]: Cells 10–14 × 6–7.5 μm, microscopic colonies on rocks, Porto Rico.
– *Aphanothece gardneri* de Toni 1936 [syn.: *A. gelatinosa* Gardner 1927]: Cells oval, 8–10 × 4.9–5.2 μm, described from hot springs in China (Fukien).
– *Aphanothece goetzei* Schmidle 1901: Cells 7 × 1.7 μm, described from thermal springs in Africa.
– *Aphanothece hegewaldii* Kováčik 1988: From the subg. *Cyanogastrum*; cells 7–16 (18) × 5–8(10) μm, metaphytic and planktic in water bodies, probably pantropical; (fig. **83**).
– *Aphanothece karukerae* Lami 1938: Cells red, 9–13 × 6–6.5 μm, in hemispherical red mucilaginous colonies on coral reefs; marine, Algeria, Guadeloupe, probably common also in Mediterranean Sea; (fig. **80**).
– *Aphanothece krumbeinii* Campbell et Golubić 1985: Cells 7.5–15.5 × 5–6 μm, yellow-orange pinnacle mats in shallow hypersaline biotopes, Egypt (Sinai); probably widely distributed in similar biotopes.
– *Aphanothece lemnae* Chu 1952: Cells 4.2–6.8 × 3.2–4.4 μm, growing in small, more or less spherical colonies (60–80 μm in diameter), epiphytic on *Lemna trisulca* (?) in China; possibly initial stages of *A. microscopica*.
– *Aphanothece microscopica* var. *granulosa* Gardner 1927: Cells 7–8 × 4.5–5 μm, aerophytic on rocks, Porto Rico; probably a special species.
– *Aphanothece minor* Frémy 1941: Small gelatinous, subspherical, blue-green colonies with loosely arranged cells, 3 × 1.5 μm; known from saline biotopes in Antilles.
– *Aphanothece opalescens* Gardner 1927: Cells 2–3.2 × 1–1.4 μm, compact mucilaginous colonies in metaphyton of water bodies, Porto Rico.
– *Aphanothece sacrum* (Suringar) Okada 1953: Cells 6–7 × 3.5–4 μm, large macroscopic colonies on stones in running waters, E. Asia (Hirose & Hirano 1981); edible. Sometimes synonymized with *A. stagnina* (Compère 1970), but ecologically and also morphologically (colonies, arrangement of cells) very different.
– *Aphanothece stagnina* var. *nemathece* Frémy 1930: Colonies elongated, up to 1 mm long (or more), up to 35 μm wide, cells 6–7 × 3–4 μm, pale blue-green; metaphytic in littoral of freshwaters, described from Africa (Gabon).
– *Aphanothece stratus* Komárek et Cronberg 1999: Planktic species from the subg. *Anathece*, cells circa (0.7)1 × 0.6–0.8 μm, pale grey-blue, in irregular colonies; described from Lake Kariba (Zimbabwe).
– *Anacystis thermalis* Copeland 1936: Cells 3–6 × 3–3.7 μm, spherical colonies in thermal water (circa 42 °C), USA (Yellowstone); probably a separate thermophilic *Aphanothece* species from the subg. *Cyanogastrum*.
– *Aphanothece uliginosa* Taylor 1928: Irregular colonies up to1.5 μm in diameter,

with irregularly distributed cells, in homogeneous mucilage; cells spherical to oval, 3.8 × circa 2.8 µm; described from pools in the mountains of British Columbia, Canada.
- *Aphanothece variabilis* (Schiller) Komárek 1995 [syn.: *Cyanogastrum variabile* Schiller 1956]: Cells oval, (3.7?)4.2–10 × 2.5–7 µm; known from coastal swamps in Caribbean Islands and coasts, probably pantropical species (known also from Brasil); (fig. **82**).

**European species needing revision:**
- *Aphanothece rufescens* Hansgirg 1892: Macroscopic, formless, brownish colonies, cells ellipsoid, 2.5–3.5 µm wide, subaerophytic on wet walls or wood.
- *Aphanothece clathrata* var. *rosea* Skuja 1948: Cells 1.3–1.5 × 1 µm, densely agglomerated in irregular colonies, planktic in lakes, Sweden; rhodobacteria ?; (fig. **57**).
- *Aphanothece conferta* Richter in Hauck & Richter 1892: Cells 4.5–5 × 3.5 µm; originally described as an atmophytic, thermophilic species (comp. *A. thermicola*), in later concepts cited only as aerophytic from warm countries (e.g., on bark of trees); (Tilden 1910; Desikachary 1959).
- *Aphanothece cretaria* Frémy 1942: Cells pale blue-green or yellowish, 3.5–9 × 2.5–3 µm, in homogeneous, structureless mucilage forming hemispherical nests about 1 mm in size; orange, epilithic (aerophytic) colonies in holes of limestone rocks, NE. France. Not confirmed.
- *Aphanothece gelatinosa* (Hennings) Lemmermann 1907: Insufficiently described; sensu Prescott (1951) cells 4–4.5 × 3.8–4 µm, in spherical colonies, benthic in water bodies (calcareous areas).
- *Aphanothece halophytica* Frémy in Hof et Frémy 1933: Halobiotic, with diverse views of its species identity (Yopp & al. 1978; Golubić 1980; Dor & Hornoff 1985; Campbell & Golubić 1985; Rippka & Waterbury 1989); a collective species, which actually represents a questionable cluster of euryhaline and eurythermal taxa; for discussion, see Golubić (1980), Montoya & Golubić (1991) and Roussomoustakaki & Anagnostidis (1991). Comp. also, e.g., Garcia-Pichel & al. (1997), Ventura & al. (1997), and others.
- *Aphanothece heterospora* Rabenhorst 1864: Cells –20 × 9–10 µm, described from lake plankton in Germany (Schleswig), not recorded again in its original sense; (Geitler 1932).
- *Aphanothece longior* Naumann 1921: Cells –10.5 × 1 µm, described from a stream in Sweden, not unambiguously recorded again; sensu Hirose & Hirano (1977) = probably *Rhabdoderma.*
- *Aphanothece margaritacea* (Kützing) Forti 1907: Cells globose, olive-green, 3.7–4.5 µm in diameter (?); marine, epiphytic on *Polysiphonia*, Adriatic Sea; insufficiently known.
- *Aphanothece naegelii* Wartmann in Rabenhorst 1861: Insufficiently known, various concepts (Geitler 1932; Skuja 1949; Watanabe & Komárek 1988; and others).
- *Aphanothece nostocopsis* Skuja 1934: Cells 4.5–10(15) × 3–5 µm, often in pseudofilamentous rows, in colonies occur extremely large cells, in dystrophic waters (Latvia); certainly belongs to another genus.
- *Aphanothece pallida* sensu Guarrera et al. 1986: From Argentine (Tierra del Fuego), probably a species of the subg. *Cyanogastrum.*
- *Aphanothece rubra* Liebetanz 1925: Cells rod-shaped, straight or arcuate, 1.5–3.6 × 1–1.2 µm, red in amorphous, structurless, reddish colonies, subaerophytic, halobiotic in wet margins of salty swamps and lakes; described from brackish waters in Poland.

## 15. **Gloeothece** Nägeli 1849

Colonies usually small, composed of sheathed cells or groups of cells (2-, 4- or more-celled); cells obligatorily enclosed within their own gelatinous envelopes and sometimes surrounded by common mucilage, usually in groups and irregularly distributed in large agglomerations (micro- to macroscopic mucilaginous or «granular» thallus). Cells and groups of cells enclosed in hyaline or concentrically lamellate, colourless or intensely bluish, violet or yellow-brown envelopes, delimited at the margin. Cells widely oval to rod-shaped, rounded at the ends, pale blue-green, blue-green, olive-green or violet.
Cell division transverse to the longitudinal axis, in one plane in successive generations; cells grow in the original size and shape before next division; they produce their own envelopes just after division. Reproduction by disintegration of colonial aggregates and (in several species) by the liberation of cells after gelatinization of envelopes. Nanocyte production recorded in a few species.
**Important references:** Geitler (1932, 1942); Hollerbach in Elenkin (1938); Skuja (1937, 1949, 1964); Desikachary (1959); Starmach (1966); Rippka & al. (1979); Rippka & Waterbury (1989).

**Key to the species:**

**1a** Cell envelopes colourless, but sometimes distinctly stratified and slightly yellowish . . . . . 2
**1b** Cell envelopes distinctly and usually intensely (at least at the colonial margin) coloured . . . . . 9
**2a** Hydrophytic, living in various aquatic biotopes . . . . . 3
**2b** Aerophytic, subaerophytic and atmophytic . . . . . 5
**3a** Cells to 1 μm wide; envelopes around cells more or less homogeneous, not or indistinctly stratified; benthic or secondarily planktic . . . . . 4
**3b** Cells 7.5–8.8 × 4.5–6 μm; envelopes usually stratified; metaphytic . . . . . **9. G. membranacea**
**4a** Cells 0.8–2 × 0.4–0.6 μm, more or less straight; epipelic and planktic (tychoplanktic) in clear northern lakes . . . . . **1. G. subtilis**
**4b** Cells 3–6 × 0.6–0.8 μm, clearly arcuate; epipelic in clear lakes . . . . . **2. G. abiscoensis**
**5a** Cells rod-shaped, with length/width ratio always greater than 3 : 1 . . . . . 6
**5b** Cells more or less oval, with length/width ratio less than 3 : 1 . . . . . 7
**6a** Gelatinous envelopes around cells simple, not stratified or with only 2-(3) μm wide colourless layers; cells 10–18 × 1.5–2.5 μm . . **5. G. linearis**
**6b** Gelatinous envelopes around cells distinctly concentrically lamellate; cells 5–6 (10) × circa 1μm . . . . . **6. G. violacea**
**7a** Envelopes not or indistinctly stratified . . . . . 8
**7b** Envelopes clearly concentrically lamellate . . . . . **10. G. tepidariorum**
**8a** Cells 1.6–2.3 μm wide, envelopes usually confluent, indistinctly delimited . . . . . **7. G. confluens**
**8b** Cells 2.5–4.5 μm wide, envelopes distinctly delimited . . . . . **8. G. palea**
**9a** Envelopes distinctly blue or dark blue-violet (at least partly) . . . . . 10
**9b** Envelopes mainly yellow, orange or yellow-brown . . . . . 11
**10a** Cells 1.2–4 × 0.5–2 μm . . . . . **4. G. cyanochroa**
**10b** Cells circa 15 × 7.5–8.6 μm . . . . . **14. G. ustulata**
**11a** Cells 1.5–2.5 × 1–1.7 μm, benthic on stony substrates in nordic lakes . . . . . **3. G. incerta**
**11b** Cells more than 6 × 3 μm large . . . . . 12
**12a** Strictly aerophytic, envelopes yellow, orange or yellow-brown . . . . . 13

**12b** Subaerophytic, atmophytic, rarely submerged, envelopes colourless or yellow, cells 6.5–12.5 × 4.5–5.5 µm . . . . . . . . . . . . . . . . . . **11. G. fusco-lutea**
**13a** Envelopes pale orange, colonies orange, on mosses and limestone wet rocks; cells 8–11 × 6–8 µm . . . . . . . . . . . . . . . . . . . . . . . . . . **12. G. dubia**
**13b** Envelopes yellow to yellow-brown, colonies pale or dirty olive-green or brownish, on wet rocks; cells 6.5–13(16) × 3.6–6 (10) µm **13. G. rupestris**

## 1. **Gloeothece subtilis** Skuja 1964 (fig. **84**)

Colonies microscopic, mucilaginous, free floating, spherical or slightly irregular to amorphous, with irregularly disposed cells. Common mucilage delimited, colourless, homogeneous or sometimes slightly concentrically lamellate, cells and small groups of cells (2–4-celled) with their own, fine, colourless, slightly lamellate envelopes (clearly visible after staining). Cells shortly cylindrical with rounded ends up to oblong, sometimes slightly curved, with homogeneous, pale blue-green content, without aerotopes or prominent inclusions, 0.8–2(4.5) × 0.4–0.6(1) µm.
Occurrence: Freshwater, in mesotrophic lakes with alkaline sediments, epipelic and free in benthos among other algae (metaphytic), sometimes secondarily in plankton; Sweden.

## 2. **Gloeothece abiscoensis** (Skuja) Komárek et Anagnostidis 1995 (fig. **86**)

*Gloeothece vibrio* var. *abiscoensis* Skuja 1964

Colonies microscopic, more or less ellipsoidal or irregularly rounded, among other algae, up to 60 × 40 µm large. Mucilage colourless, envelopes around cells wide, simple, not lamellate, to 12 µm wide. Cells cylindrical, elongate, arcuate to coiled or sigmoid, solitary or in pairs, blue-green, 3–6 × 0.6–0.8 µm.
Occurrence: Freshwater, benthic in water bodies with calcareous sediments; N. Sweden (Abisko).

## 3. **Gloeothece incerta** Skuja 1964 (fig. **87**)

Colonies mainly micro-, rarely macroscopic, mucilaginous, amorphous, yellow-brown, forming crust-like mats on rocky and stony substrates, composed of subcolonies with several cells in smaller groups (up to 16 µm in diameter). Envelopes around cells hyaline or yellow-brown, distinctly lamellate, sometimes diffluent at the margin. Cells oval or oval-cylindrical, pale blue or olive-green, with distinct, visible chromatoplasma, homogeneous or slightly granular, 1.5–2.5 × 1–1.5–1.7 µm.
Occurrence: Freshwater, submerged, on stones and rocks, among other algae, in littoral of clear nordic lakes; N. Sweden (Abisko). Records from marine littoral (Brazil, *Gloeothece* cf. *incerta* sensu Sant'Anna & al. 1995) concern surely another species.

## 4. **Gloeothece cyanochroa** Komárek 1995 (fig. **88,89,90**)

*Gloeothece coerulea* sensu Golubić & Campbell 1979 (non *Gloeothece coerulea* Geitler = ? syn. to *Gloeobacter violaceus*)

Groups of cells, in small, usually microscopic, irregular, fine colonies. Envelopes around cells hyaline, but in marginal parts intensely bluish or dark blue-violet, slightly lamellate (1–2-4 layers). Cells oval up to short cylindrical with rounded

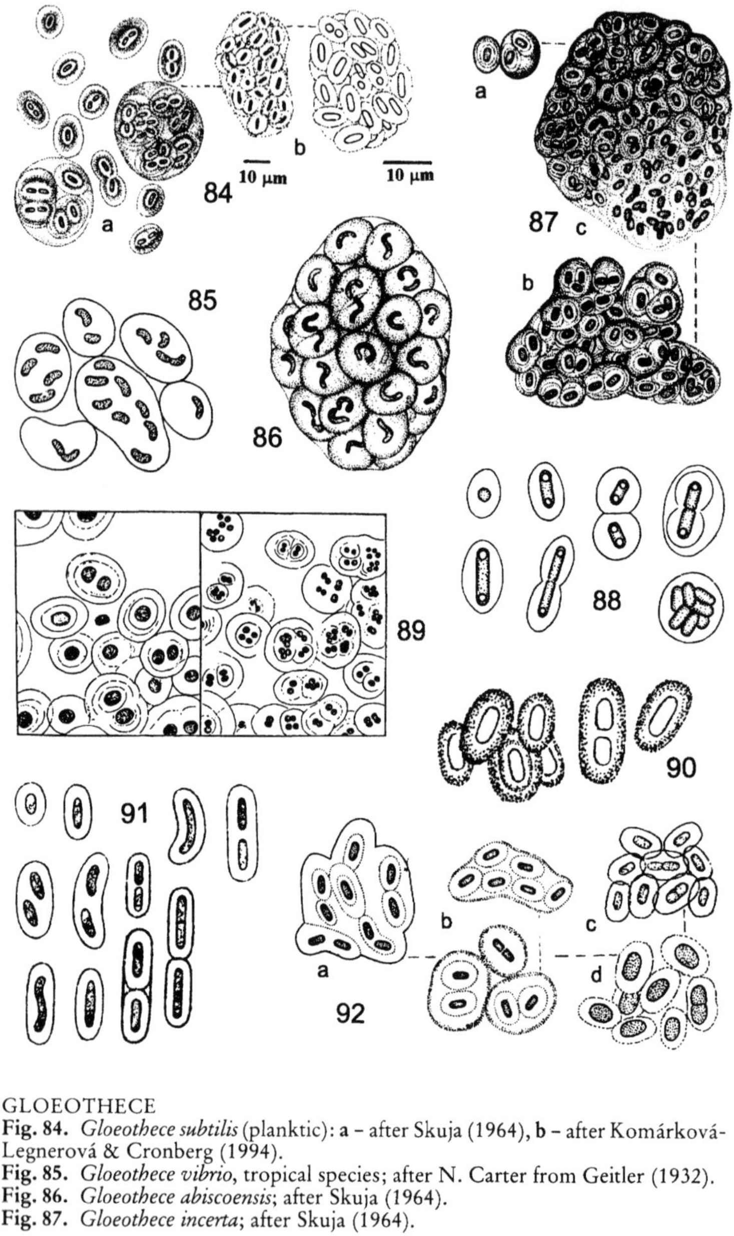


GLOEOTHECE

**Fig. 84.** *Gloeothece subtilis* (planktic): **a** – after Skuja (1964), **b** – after Komárková-Legnerová & Cronberg (1994).

**Fig. 85.** *Gloeothece vibrio*, tropical species; after N. Carter from Geitler (1932).

**Fig. 86.** *Gloeothece abiscoensis*; after Skuja (1964).

**Fig. 87.** *Gloeothece incerta*; after Skuja (1964).

ends, with pale blue-green, more or less homogeneous content, without prominent granules, 1.2–3(4) × 0.5–1.2(2) µm.
Occurrence: Aerophytic and subaerophytic on wet granitic and non-calcareous rocks, in shallow ditches, among algae and detritus; Norway, Svalbard, N. Sweden, probably also in the Alps.

### 5. **Gloeothece linearis** Nägeli 1849 (fig. **91**)

Solitary sheathed cells or small irregular groups of cells in microscopic colonies, united sometimes into gelatinous, olive-green to reddish macroscopic aggregates. Envelopes around cells wide, colourless, simple, not lamellate, inconspicuous or distinctly delimited, only rarely confluent; common mucilage sometimes diffluent. Cells cylindrical, rod-shaped, straight, slightly curved or sigmoid with rounded ends, with pale blue-green, homogeneous content, without granules, 10–18 × 1.5–2.5 µm.
Occurrence: Aerophytic, on wet rocks and walls, at low pH, usually in mountains (data from plankton concern *Rhabdoderma lineare* and *R. compositum* !); known mainly from the Alps and central Europe.

### 6. **Gloeothece violacea** Rabenhorst 1865

Sheathed groups of cells joined to spreading, fine, thin, gelatinous, purple-violet, later up macroscopic colonies. Envelopes around cells distinctly lamellate, colourless or slightly violet in marginal colonial parts, thick, up to 12.5 µm wide. Cells long cylindrical, straight or slightly curved, 5–6(10) × circa 1 µm. Insufficiently known species.
Occurrence: Aerophytic, in caves in dim light intensity; known only from S. France.

### 7. **Gloeothece confluens** Nägeli 1849 (fig. **92**)

Cells solitary or in pairs (rarely more, united together irregularly by common envelopes), in irregular, amorphous, pale yellowish, greenish or slightly reddish, gelatinous, microscopic to macroscopic aggregates. Envelopes around individual cells more or less wide, colourless, not lamellate, diffluent at the margin and confluent. Cells short cylindrical with rounded ends, with pale blue-green, homogeneous content, (2.2) 5.5–7.5 × 1.6–3 µm.
Occurrence: Aerophytic, on wet rocks, stony walls, among mosses, rarely on soil surface; over the whole temperate world, but recorded also from Brazil (Sant' Anna & al. 1991); identity of various populations need confirmation.

---

**Fig. 88.** *Gloeothece* cf. *cyanochroa*; after Geitler (1927; sub *Gloeothece coerulea*, which was synonymized with *Gloeobacter violaceus*).
**Fig. 89.** Nanocyte formation in «*Gloeothece coerulea*»; from Geitler (1932); morphology of cells is different from the type (comp. fig. 88); the relation (identity?) to *Gloeobacter violaceus* is unclear.
**Fig. 90.** *Gloeothece cyanochroa* (aerophytic, epilithic): typical morphology from the Svalbard populations (blue envelopes); orig.
**Fig. 91.** *Gloeothece linearis* (aerophytic): after Nägeli (1849).
**Fig. 92.** *Gloeothece confluens*: **a** – after West from Kondrateva & al. (1984), **b** – after Hollerbach from Kondrateva & al. (1984), **c** – after Nägeli from Starmach (1966, iconotype), **d** – after Bourrelly & Denizot (1967).

## 8. **Gloeothece palea** (Kützing) Rabenhorst 1865 (fig. **93**)

Rarely sheathed solitary cells, usually small, regular colonies of 2–4 (many) cells, united into gelatinous, dirty blue-green macroscopic agglomerations. Envelopes around cells distinctly delimited, more or less wide, not lamellate or only with 1–2 indistinct layers, colourless or slightly yellowish. Cells oval to cylindrical, blue-green, but sometimes very pale, with homogeneous content, 3.8–13.5 × 2.5–4.5 μm.
Occurrence: Aerophytic and atmophytic, on wet rocks and walls, among mosses, in the vicinity of thermal springs, rarely on wet soils; known from the whole European temperate zone; aquatic or various tropical habitats recorded from Burma, Brazil, etc., probably concern other species.

## 9. **Gloeothece membranacea** (Rabenhorst) Bornet 1892 (fig. **94**)

*Aphanothece membranacea* Rabenhorst 1865

Small cell clusters joined by common mucilage into gelatinous, sometimes spherical or membranous, olive-green or brownish colonies reaching macroscopic dimensions (several mm in diameter, *Nostoc*-like). Envelopes around cells colourless, lamellate, only sometimes diffluent at the margin. Cells more or less oval or short cylindrical with widely rounded ends, blue-green, 7–8.8 × 4–6 μm.
Occurrence: Freshwater, submerged, metaphytic in swamps, rarely atmophytic or on intensely wetted rocks, in rocky ditches and pools, sporadic, from lowlands to mountains; known mainly from Europe, but probably distributed over the whole warmer areas of the temperate and subtropical zones.

## 10. **Gloeothece tepidariorum** (A. Braun) Lagerheim 1883 (fig. **95**)

*Gloeothece rupestris* var. *tepidariorum* (A.Braun) Hansgirg 1891

Small colonies of sheathed cells grouped together into irregular agglomerations of up to macroscopic size; colonies dirty greyish-green or olive-green. Envelopes around cells colourless, intensely concentrically lamellate. Cells pale blue-green, with more or less homogeneous content, 8–15 × 5–6.2(9 ?) μm.
Occurrence: Aerophytic and atmophytic, on rocks and wet walls, common in glasshouses; distributed in central Europe.

## 11. **Gloeothece fusco-lutea** Nägeli 1849 (fig. **96**)

Small, few-celled (usually 4- or 8-, rarely up to 32-celled) colonies, grouped together in mucilaginous, fine, blue-green to brown aggregate colonies of microscopic or macroscopic size, spreading over a substrate. Envelopes around cells colourless or yellow to brownish, usually slightly lamellate (primary and secondary envelopes). Cells oval to cylindrical, pale blue-green with more or less homogeneous content, 6.5–12.5 × 4.4–5.5 μm.

---

GLOEOTHECE

**Fig. 93.** *Gloeothece palea*: **a** – after Skuja (1964); **b** – after Bourrelly & Denizot (1967, false sub «*G. pallida*»).

**Fig. 94.** *Gloeothece membranacea*: one example of very rare drawings of this species, after Watanabe & al. (1979) from Papua – New Guinea.

**Fig. 95.** *Gloeothece tepidariorum*: **a** – after Geitler (1932), **b** – after Lemmermann from Geitler (1925).

**Fig. 96.** *Gloeothece fusco-lutea* (benthic, metaphytic, yellow-brown envelopes):

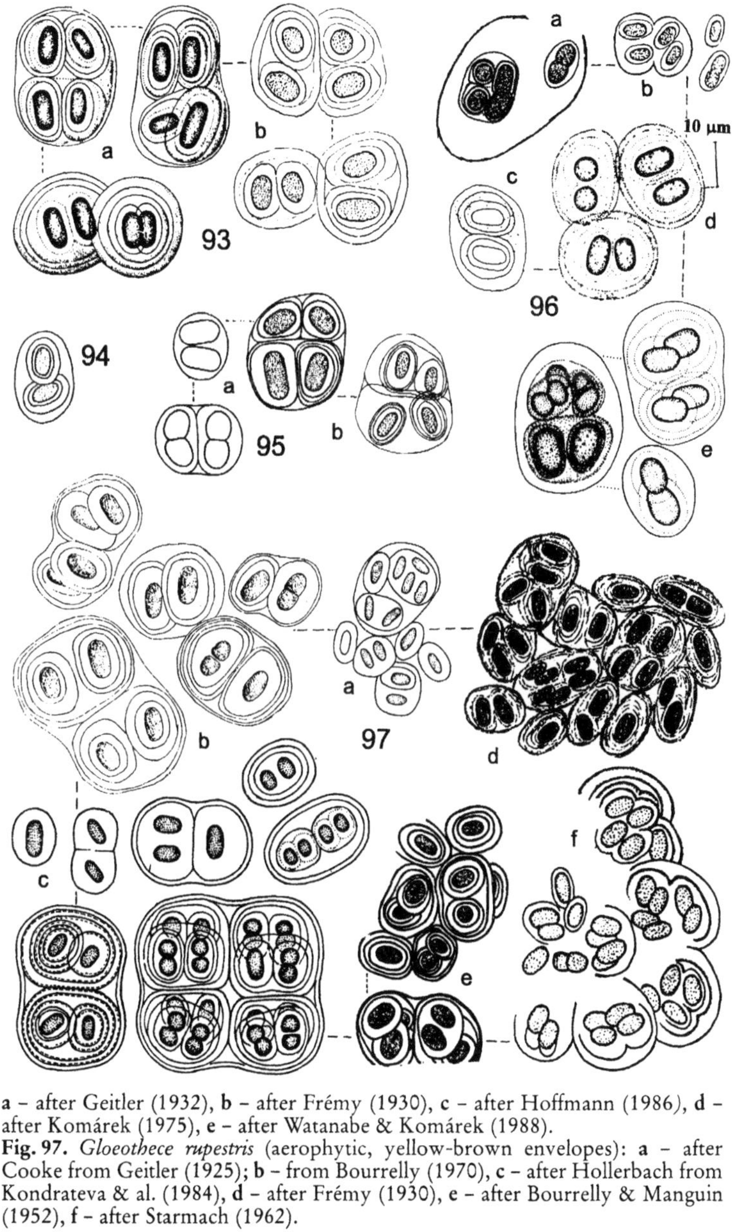


**a** – after Geitler (1932), **b** – after Frémy (1930), **c** – after Hoffmann (1986), **d** – after Komárek (1975), **e** – after Watanabe & Komárek (1988).

**Fig. 97.** *Gloeothece rupestris* (aerophytic, yellow-brown envelopes): **a** – after Cooke from Geitler (1925); **b** – from Bourrelly (1970), **c** – after Hollerbach from Kondrateva & al. (1984), **d** – after Frémy (1930), **e** – after Bourrelly & Manguin (1952), **f** – after Starmach (1962).

Occurrence: Subaerophytic or freshwater, submerged, in metaphyton of clear and acidic pools and swamps, among water plants and algae, rarely on and under intensely wetted rock walls, in small ditches and pools, from lowland to mountains; known from the whole European temperate zone; identifications from tropical regions are uncertain.

## 12. **Gloeothece dubia** (Wartmann) Geitler 1932 (fig. **99**)

Small, 2–4-celled colonies, joined by common mucilage into gelatinous masses; colonies sometimes macroscopic, mucilaginous, intensely orange, irregularly subspherical or hemispherical to more or less spread out over a substrate. Envelopes around cells slightly lamellate or not lamellate, occasionally slightly granulate, pale orange to reddish, more intensely coloured at the colonial margin, sometimes diffluent. Cells widely oval, blue-green or pale greenish violet, 8–11 × 6–8 µm.
Occurrence: Aerophytic on mosses and wet rocks in limestone areas (on alkaline substrates); sporadic throughout the temperate zone in Europe (also in Greenland).

## 13. **Gloeothece rupestris** (Lyngbye) Bornet in Wittrock et Nordstedt 1880 (fig. **97**)

*Gloeothece rupestris* f. *maxima* (W. West) Hollerbach in Elenkin 1938 incl.

Colonies small with 2–8 to 32 cells, enveloped by common sheaths (rarely also solitary sheathed cells), irregular, more or less oval, mucilaginous, usually 25–40 µm in diameter, forming agglomerations reaching macroscopic size, with regularly distributed cells. Gelatinous envelopes around cells and cell groups more or less wide, colourless to yellow or yellow-brown, distinctly concentrically lamellate and delimited, rarely almost unstratified. Cells ellipsoidal or oval to short cylindrical with rounded ends, pale blue-green or olive-green, 5–9(13–16) × 3.6–4-6(10) µm.
Occurrence: Aerophytic and atmophytic, on wet rocks and walls, sometimes on mosses, mainly in mountains, but in numerous morphotypes, from which several were described as subspecific taxa (fig. **98**); known from the whole European and Asian temperate zones; possibly of cosmopolitan distribution, but with distinct ecology (comp., e.g., records from Brazil – SP; Sant'Anna & al. 1991); populations from various tropical countries (e.g., from Philippines, Umezaki & Modelo 1987) have very different ecology and type of colonies.

## 14. **Gloeothece ustulata** Beck-Mannagetta 1929 (fig. **100**)

Colonies usually 2–4-celled, irregularly spherically elongate, more or less ellipsoidal, to 50 µm long and sometimes slightly flattened, grouped into irregular aggregations. Envelopes around cells hyaline, only with indistinct and rare

---

GLOEOTHECE
**Fig. 98.** Morphotypes of *Gloeothece rupestris*: **a** – «var. *maior*» after Geitler (1960), **b** – «var. *maior*» after Gonzalves & Kamat (1958), **c** – «var. *maxima*» after Gonzalves & Kamat (1958), **d** – «*Gloeothece* cf. *rupestris*» after Böcher (1949) from mud of lakes.

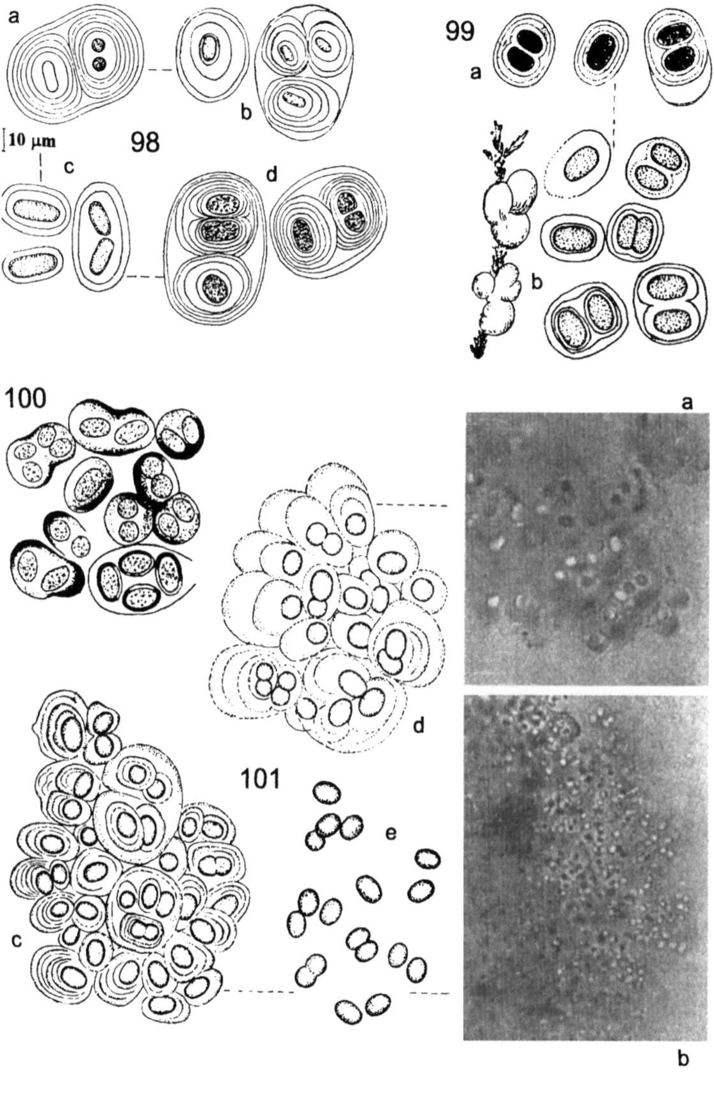


**Fig. 99.** *Gloeothece dubia*: **a** - after Wille from Geitler (1932), **b** - from Starmach (1966).
**Fig. 100.** *Gloeothece ustulata*: after Beck from Geitler (1932).
**Fig. 101.** *Gloeothece heufleri*: **a-b** - micrographs from Vodeničarov (1961) from Bulgaria, **c-e** - after Prichod'kova from Kondrateva & al. (1984).

lamellation, colourless or slightly bluish to violet and blackish violet (dark violet colouration is usually more concentrated at one side of envelopes). Cells ellipsoidal to oval, blue-green, with homogeneous, slightly granulate content, circa 15.5 × 7.5–8.6 µm. Cells are liberated from colonies by gelatinization of the mucilage.
Occurrence: Aerophytic, on wet rocks in the Alps (over 1100 m); Austria (Carinthia).

**Species outside Europe:**
- *Gloeothece baileyana* Schmidle 1896: Cells 12–20 × 8–12 µm, described from Australia; revision needed.
- *Gloeothece endochromatica* Gardner 1927: Cells 1.5–2.4 × 0.9–1.2 µm, with dark violet envelopes, aerophytic on limestone, described from Porto Rico.
- *Gloeothece filiformis* Sant'Anna et al. 1991: Long, cylindrical cells in stratified envelopes, 5–20 × 1.8–2.5 µm, arranged in rows; aerophytic on damp rocks, described from Brasil; comp. with *Gloeobacter* (thylakoids ?).
- *Gloeothece interspersa* Gardner 1927: Cells 7–7.5 × 3.6–4 µm, with wide colourless envelopes, aerophytic, described from Porto Rico.
- *Gloeothece kriegeri* Frémy 1932: Cells more or less 2–2.5 × 1.5 µm, with hyaline colourless envelopes, described from New Guinea.
- *Gloeothece lunata* W. et G.S.West 1894.
- *Gloeothece opalothecata* Gardner 1927: Cells 2.2–5 × 2–2.5 µm, with wide colourless envelopes, aerophytic on limestone, described from Porto Rico; maybe *Aphanothece*.
- *Gloeothece parvula* Gardner 1927: Cells 2.4–4.5 × 1.2–1.5 µm, with colourless envelopes, aerophytic, described from Porto Rico.
- *Gloeothece prototypa* Gardner 1927: Cells 2–3 × 0.8–1.2 µm, with narrow colourless envelopes, aerophytic on limestone, described from Porto Rico; according to Hoffmann (1986) also among mosses, found in Louxembourg (identity with Gardner's type should be revised).
- *Gloeothece rhodochlamys* Skuja 1949: Cells 3–5.5 × 2–3 µm, with red envelopes, aerophytic, known from Burma and New Caledonia. Similar species was found in the marine supralittoral zone in Brazil (SP).
- *Gloeothece rupestris* var. *maior* Geitler 1933: Cells –16 × –6 µm, with lamellate colourless or yellow-brown sheaths, aerophytic, described from Indonesia; possibly a separate species.
- *Gloeothece samoensis* Wille 1913: Cells more or less (5)-8 × (3)4–5 µm, with colourless, lamellate envelopes, aerophytic, described from Samoa Islands and known from other Pacific islands; records from Europe probably belong to other species.
- *Gloeothece tophacea* Skuja 1937: Cells –4.5 × 1.2–1.5 µm, with not lamellate colourless envelopes, subaerophytic on lime tufa, described from China.
- *Gloeothece vibrio* N. Carter 1922: Cells 2–5(7) × 1–1.5(2) µm, distinctly curved, with colourless envelopes, described from rocks in New Caledonia; (Bourrelly 1984); (fig. **85**).

**Uncertain European species:**
- *Gloeothece cystifera* (Hassall) Rabenhorst 1865: Incomplete description.
- *Gloeothece distans* Stizenberger in Rabenhorst 1860: Described from moorland waters near Konstanz; (Vodeničarov 1962: Rila Mts., Bulgaria).
- *Gloeothece goeppertiana* (Hilse) Forti in de Toni 1907: Recorded originally from soils, later in various concepts (Geitler 1932; Copeland 1936; Claus 1962).
- *Gloeothece heufleri* Grunow in Rabenhorst 1865: Described from Austria, later found in Caucasus (Voronichin 1924) and in Rhodopi Mountains, Bulgaria

(Vodeničarov 1961), aerophytic on calcareous rocks; further study is necessary (structure of colonies corresponds rather with *Chlorogloea*; comp. also *Aphanothece castagnei*); (fig. **101**).
– *Gloeothece monococca* (Kützing) Rabenhorst 1865: According to Geitler (1932) probably *Chroothece*.

## 2. Subfamily Synechococcoideae

Cells solitary without mucilage or forming mucilaginous colonies; in colonies rarely irregularly arranged, usually oriented in one direction or forming isopolar pseudofilamentous rows of cells (rarely in heteropolar thallus). Mucilaginous envelopes lacking or homogeneous, mainly structureless, colourless, diffluent at the margin, or forming tube-like sheaths. Cells widely oval, oval, fusiform, short cylindrical or (mainly) rod-shaped to elongate cylindrical; facultatively (but commonly in cultures) filamentous involution cells are present.
Cell division only in one direction in successive generations, perpendicular to the long axis of the cell, symmetrically or asymmetrically (usually in cultures, especially in involution cells); without nanocytes.

**Key to the genera:**

**1a** Solitary cells without mucilage or with very fine, thin gelatinous envelopes around single cells; sometimes in groups, but not in compact colonies . . **2**
**1b** Cells in mucilaginous colonies, sometimes incrusted . . . . . . . . . . . . . . . **3**
**2a** Cells more or less fusiform, straight, long, always several times longer than wide . . . . . . . . . . . . . . . . . . . . . . . . . . . . . . . . . . . . . **2. Myxobaktron**
**2b** Cells cylindrical, sometimes slightly curved or sigmoid, two to several times longer than wide . . . . . . . . . . . . . . . . . . . . . . . . . **5. Synechococcus**
**3a** Cells randomly distributed or scattered throughout mucilaginous colonies, but sometimes oriented by long axis more or less in one direction . . . . . **4**
**3b** Cells more or less in uniseriate or multiseriate rows, usually in elongate, sometimes tube-like formations with fine or firm sheaths; rows of cells (pseudofilamentous colonies) live solitary, or arranged irregularly, parallely or radially into irregular, flat (stratified) or spherical (radially arranged) colonies . . . . . . . . . . . . . . . . . . . . . . . . . . . . . . . . . . . . . . . . . **6**
**4a** Cells widely oval to cylindrical, in macroscopic, compact, delimited, firm gelatinous colonies; periphytic or metaphytic . . . . . . . . . . . . **1. Dzensia**
**4b** Cells elongate, cylindrical or fusiform, sometimes curved, in microscopic colonies with fine, colourless, homogeneous, diffluent slime; free-floating . **5**
**5a** Cells fusiform or at least with slightly tapering and pointed-rounded poles . . . . . . . . . . . . . . . . . . . . . . . . . . . . . . . . . . . . . . . . . **3. Rhabdogloea**
**5b** Cells cylindrical with rounded ends . . . . . . . . . . . . . . . **4. Rhabdoderma**
**6a** Cells mainly oriented with their longer axes congruent with the long axis of the row of cells; they are arranged in one to more rows one after another or slightly irregularly (in few genera some cells are sometimes slightly displaced from the straight to a crosswise position, creating «zig-zag» formations) . **7**
**6b** Cells obligatorily in uniseriate pseudofilamentous colonies (rarely pseudo-dichotomously divaricate), slightly distant from one another, always with longer axis perpendicular to that of the pseudofilaments (cells «shorter than wide»; cell division also proceeds transversely to the pseudofilamentous colonies, i.e., parallel to the longer cell dimension) . . . . . . . . . . . . . . . . . . . . . . . . . . . . . . . . . . . . **12. Johannesbaptistia**
**7a** Colonies more or less spherical, very fine, free-floating, with irregular, bush-

like, divaricate rows of rod-shaped cells, radiating from the colonial center . . . . . . . . . . . . . . . . . . . . . . . . . . . . . . . . . . . . . . . . **6. Cyanothamnos**

7b Colonies not spherical (but sometimes hemispherical); solitary pseudofilaments, irregular strata or parallely oriented rows of cells in one layer . . . 8

8a Cells always oriented in the direction of the elongate, tube-like colony, more or less one after another . . . . . . . . . . . . . . . . . . . . . . . . . . . . . . . . 9

8b Cells in elongated colonies in distinct rows, which are often arranged parallely, but sometimes displaced slightly obliquely or transversely to the long axis of the colony (in «zig-zag» position); colonies form mats («strata») . . . . . . . . . . . . . . . . . . . . . . . . . . . . . . . . . . . . . . . . . . . . . . . 10

9a Cells in distinct uniseriate pseudofilaments, cylindrical, usually elongate, within a narrow or slightly widened, sheath-like envelope . . **7. Wolskyella**

9b Cells irregularly multiserially arranged (but oriented solely in one direction) in wide, delimited tube-like sheaths, not in distinct pseudofilamentous rows . . . . . . . . . . . . . . . . . . . . . . . . . . . . . . . . . . . . . . . . . . . **8. Bacularia**

10a Colonies mucilaginous and elongate, spreading over a substrate, irregularly arranged; in soil . . . . . . . . . . . . . . . . . . . . . . . . . . . . . . . **9. Tubiella**

10b Colonies forming hemispherical, later flattened or amorphous strata on a substrate . . . . . . . . . . . . . . . . . . . . . . . . . . . . . . . . . . . . . . . . 11

11a Colonies crust-like, with vertically and parallely, densely arranged pseudofilaments; cells without their own gelatinous envelopes; benthic in creeks . . . . . . . . . . . . . . . . . . . . . . . . . . . . . . . . . . . . . . **10. Rhodostichus**

11b Colonies soft, hemispherical or flattened, with radial rows of cells; cells with their own gelatinous envelopes, usually within sheaths; benthic . . . . . . . . . . . . . . . . . . . . . . . . . . . . . . . . . . . . . **11. Pseudoncobyrsa**

## 1. **Dzensia** Voronichin 1929

Colonies mucilaginous, microscopic to macroscopic, more or less spherical, oval or ovoid, sometimes formed from hemispherical subcolonies, up to 1 cm in diameter, with distinct and delimited, firm pellicular surface. Cells almost spherical or oval up to long cylindrical, without their own gelatinous envelopes, in young colonies irregularly distributed, later in more or less regular, parallel and radially oriented rows, surrounded by scarcely visible (staining !) tube-like gelatinous sheaths.

Cell division (according to Voronichin's orig. fig.) by binary fission (pinching), in one direction in successive generations, perpendicular to the longer axis (origin of radial rows); reproduction by disintegration of colonies.

**Important references:** Voronichin (1929); Hof & Frémy (1933); Elenkin (1938); Geitler (1942); Hollerbach & al. (1953); Bourrelly (1970, 1985); Komárek & Anagnostidis (1986).

### 1. **Dzensia salina** Voronichin 1929 (fig. **102**)

Colonies free-floating, macroscopic, more or less spherical, 1–10 mm in diameter, blue-green or brownish, soft but with smooth and tough or slightly warty surface, with densely arranged cells. Mucilage colourless; radial tubes hardly visible. Cells spherical (particularly at high rates of division), widely oval to long cylindrical, sometimes with their own gelatinous envelopes, 2.5–4 µm in diameter or up to 6.8–10(17) × 3.5–4 µm.

Occurrence: Halobiotic, in inland and coastal saline lakes, sometimes in masses; Russia (W. Siberia), probably widely distributed in salty central Asian, SE. European and Mediterranean lakes and saltworks.

**Unrevised species:**
- In the littoral of clear S. Norwegian lakes occurs a species which corresponds to the generic diagnosis of *Dzensia*: Mucilaginous colonies with oval to rod-shaped cells, arranged in more or less radial rows, with their own mucilaginous envelopes (Lindström in litt.).
- *Entophysalis weningeri* Bourrelly 1984: Colonies spherical, composed of subcolonies, several mm in diameter; cells oval to rod-shaped, to 5(–10) × 1.8–2 µm, with their own envelopes, radially arranged; epilithic in a lake littoral (in masses), described from New Caledonia.

## 2. **Myxobaktron** Schmidle 1904

Solitary cells without mucilage or with very fine, narrow individual gelatinous layer, straight or slightly arcuate, several times longer than wide, fusiform, gradually tapering to the ends and pointed or pointed-rounded. Cell content homogeneous, blue-green, olive-green or yellowish, in one planktic species with aerotopes.
Cell division transverse, perpendicular to the longitudinal axis, in the same plane in successive generations, into two, more or less isomorphic daughter cells soon separating from one another (rarely asymmetrically).
**Important references:** Geitler (1932, 1942); Potts (1980, sub *Dactylococcopsis*); Walsby & al. (1983, sub *Dactylococcopsis*); Komárek & Anagnostidis (1986); Hindák (1992).

**Only species outside Europe:**
- *Myxobaktron hirudiforme* G.S. West 1912: Cells 30–62 × 1.8–2 µm, metaphytic, known from a pool in Angola (G.S. West 1912; Geitler 1932, 1942; Huber-Pestalozzi 1938); (fig. **103**).
- *Myxobaktron salinum* Walsby in Komárek et Anagnostidis 1986 sec. Walsby et Hindák 1992: Cells (12)25–80 (dividing stages up to 150) µm long, (2)4–6(8) µm wide, with aerotopes, from the heliothermal saline Solar Lake (Egypt, Sinai), (Potts 1980 sub *Dactylococcopsis* sp.; Walsby & al. 1983 sub *Dactylococcopsis salina*; Komárek & Anagnostidis 1986, Hindák 1992a: strain CCAP 1417/1); records from heliothermal saltworks of Greece (Anagnostidis & Roussomoustakaki 1991) referred to *Myxobaktron* sp. probably belong to this species; (fig. **104**). Similar populations were found in plankton of salty Yallahs ponds in Jamaica (Golubić 1980).
- *Myxobaktron usterianum* Schmidle 1904: Cells 20–200 × 2 µm, blue-green, brackish from the estuary of a river in the Philippines; type species.

**Unrevised species:**
- *Myxobaktron palatinum* Schmidle 1905: Insufficently described.
- *Dactylococcopsis acicularis* var. *grandis* Frémy 1930: Long, fusiform cells, up to 140 × 5 µm; metaphytic in marshes, Gabon.
- *Dactylococcopsis libyca* Marchesoni 1946: Cells 110–140(200 before division) × 2.5–3 µm, from the salty Murzuc Lake in central Libya; probably *Myxobaktron* species.
- *Dactylococcopsis magna* Bharadwaja 1963: Cells 250–350 × 8–10 µm, planktic in India (Imphal); probably *Myxobaktron* species.
- Kufferath (1929) described from Costa Rica an aerophytic organism with solitary, narrow, spindle-shaped cells (16–20 × 1 µm) as a monospecific cyanoprokaryote -*Cyanocloster muscicola*; description and figure are more like a unicellular chlorophyte.

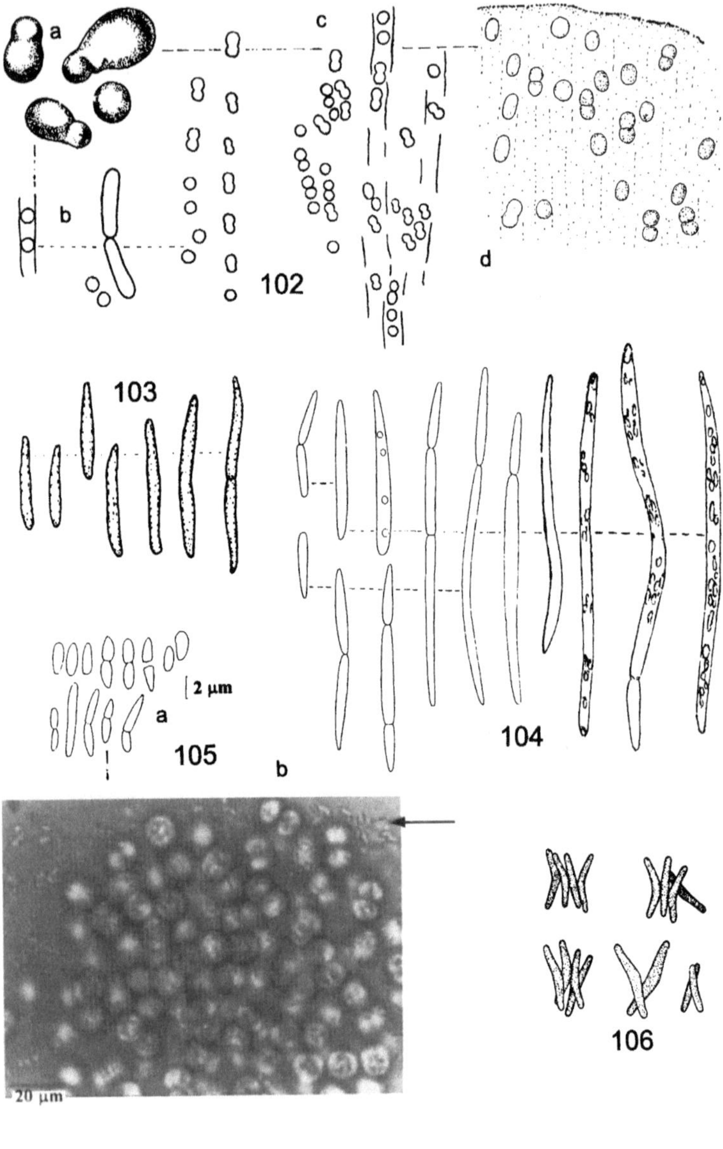
a
b
c
d
102
103
104
2 µm
a
105
b
20 µm
106

DZENSIA

**Fig. 102.** *Dzensia salina*: **a** – macroscopic colonies, **b** – various cell types, partly dividing, **c-d** – parts of colonies; **a-c** – after Voronichin (1929), **d** – from Bourrelly (1970).

MYXOBAKTRON

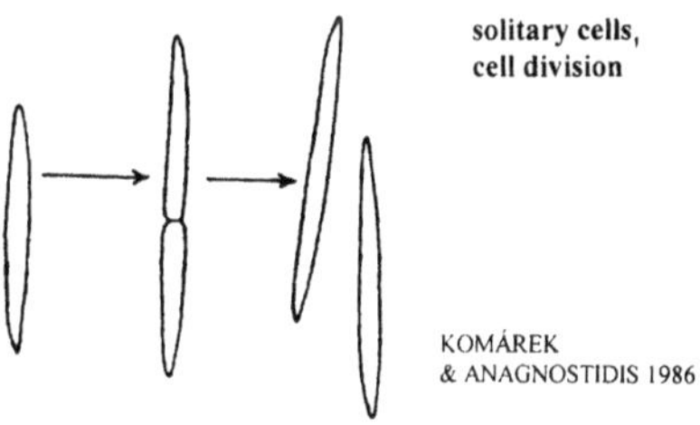


**Fig. 103.** *Myxobaktron hirudiforme*: after G.S. West from Huber-Pestalozzi (1938).
**Fig. 104.** *Myxobaktron salinum*: after Hindák (1992).

RHABDOGLOEA

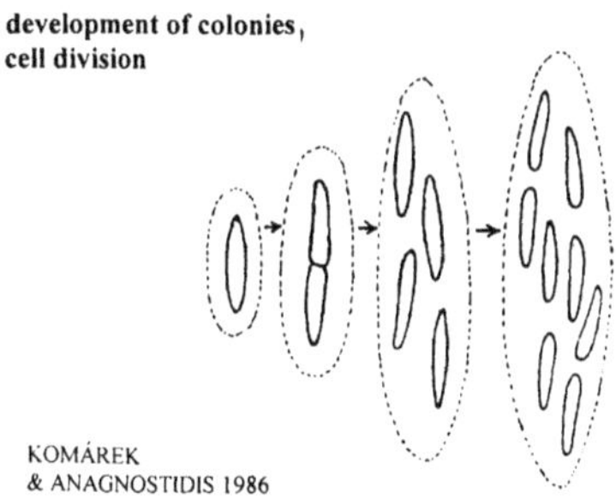


**Fig. 105.** *Rhabdogloea minuta* (endogloeic): **a** – variation in cell form and dividing cells, **b** – arrangement of cells at the surface of a *Microcystis* colony; from Hickel (1991).
**Fig. 106.** *Rhabdogloea scenedesmoides* (planktic): groups of cells in plankton, after Nygaard (1949, sub *Dactylococcopsis scenedesmoides*).

## 3. **Rhabdogloea** Schröder 1917

*Dactylococcopsis* Hansgirg sensu R. et F. Chodat 1925 et auct. post., not Hansgirg 1888

Colonies mucilaginous, microscopic (rarely macroscopic), irregular, usually free-floating, with cells distant from one another, situated irregularly (more or less parallel in young colonies) in fine, diffluent (rarely limited), colourless, structureless mucilage; number of cells usually small (sometimes only 1–3 cells in a colony). Cells elongate, narrow, spindle-shaped, ellipsoidal or more or less cylindrical with short attenuate and acute or acute-rounded poles, without aerotopes, straight or slightly arcuate or sigmoidally coiled; around the cells or their small groups thereof sometimes a fine individual, homogeneous envelope. In cultures (*R. smithii/ellipsoidea*) pseudofilamentous formations and filamentous involution cells (like in *Synechococcus*) often with asymmetrical cell division.
Cell division by binary fission, transverse to the long axis into two more or less isomorphic daughter cells; the cells separate from one another before the next division. Reproduction by disintegration of colonies.
**Important references:** Geitler (1932, 1942, sub *Dactylococcopsis*); Komárek (1969a, 1969b, sub *Dactylococcopsis*); Bourrelly (1970, 1985 sub *Aphanothece*); Komárek (1983); Hindák (1984, 1992); Komárek & Anagnostidis (1986); Lederer (1995b).

**Key to the species:**

**1a** Cells with rounded (or acute-rounded) ends . . . . . . . . . . . . . . . . . . . . . . 2
**1b** Cells with clearly acute ends . . . . . . . . . . . . . . . . . . . . . . . . . . . . . . . 6
**2a** Cells endogloeic (in the slime in planktic cyanoprokaryotes), 2.3-(7) × 0.8–1.1 µm . . . . . . . . . . . . . . . . . . . . . . . . . . . . . . . **1. R. minuta**
**2b** Cells planktic, over 1.2 µm wide . . . . . . . . . . . . . . . . . . . . . . . . . . . . 3
**3a** Cells symmetrical lengthwise, elongate spindle-shaped with rounded or acute-rounded ends, sometimes curved or sigmoid . . . . . . . . . . . . . . . . 4
**3b** Cells clearly asymmetrical lengthwise with one side more convex than the other, with clearly rounded ends; cells usually 10–20 µm long . . . . . . . . 5
**4a** Cells straight, almost cylindrical, at poles shortly attenuate and widely rounded; metaphytic in moorland waters . . . . . . . . . . . . . . **2. R. linearis**
**4b** Cells straight, slightly arcuate or sigmoid, narrow spindle-shaped, at poles acute-rounded; planktic in large, clear lakes (rarely in small water bodies) over the whole temperate zone . . . . . . . . . . . . . . . . . . . . **3. R. smithii**
**5a** Cells irregularly distributed in wide, more or less spherical, homogeneous, colourless mucilage, 18–20 × 3.5–5 µm; planktic in lakes **4. R. planctonica**
**5b** Cells more or less parallely arranged in twos or fours, forming «scenedesmoid» groups (rarely solitary cells or over 4 together), without mucilage (invisible without staining ?); planktic in lakes . . . . . **5. R. scenedesmoides**
**6a** Cells more or less straight, in amorphous mucilaginous colonies, spindle-shaped or ellipsoidal, 6–10 × 1.5–2.5 µm; in cryoseston . . **6. R. hungarica**
**6b** Cells clearly lunate, arranged in spherical mucilage, circa 5.4 × 1.3 µm; planktic in lakes . . . . . . . . . . . . . . . . . . . . . . . . . . . . **7. R. elenkinii**

### 1. **Rhabdogloea minuta** Hickel 1991 (fig. **105**)

Colonies indistinct, confluent with the slime of the host cyanoprokaryote (*Microcystis*); cells solitary or in small groups on the surface of the colonial mucilage of the host, more or less distant from one another or parallely or irregularly distributed in clusters. Cells cylindrical, oval, elliptical or fusiform, with slightly attenuate and acute-rounded (bluntly pointed) ends (rounded only after division),

straight or slightly bent, sometimes little irregular, pale blue-green or greyish, without gas vesicles; cell content homogeneous, without gas vesicles, thylakoids localized parietally; cell dimensions: (1.8)2.3–5(7) × (0.5)0.8–1.2 µm. Short several-celled pseudofilaments occur rarely.
Occurrence: In brackish water, endogloeic in mucilage of planktic *Microcystis (M. flos-aquae)* colonies; decribed from N. Germany – Schleswig/Holstein (Schlei-Fjord). The same species was identified by Hindák (1995) from a fishpond near Bratislava, Slovakia, also in a mucilage of *M. flos-aquae.*

### 2. **Rhabdogloea linearis** (Geitler) Komárek 1983 (fig. **107**)

*Dactylococcopsis linearis* Geitler 1935

Colonies elliptical, oval or irregular, free-living, sometimes several joined together, composed of groups of (2)4–8–16 (to more than 20) cells oriented more or less in one direction within the slime; mucilage fine or tough, colourless, hyaline, homogeneous, delimited at the margin, but sometimes hardly visible, more or less widened around cells; cell pairs sometimes lying at right angles to one another in one subcolony. Cells straight or indistinctly bent, rod-shaped, shortly tapered to the ends and widely rounded, ± linear, blue-green or olive-green, with small solitary granules, (6)10–12 × 2.5–6 µm. Specimens with shorter cells (var. *serians* and var. *curta* sensu Teiling 1946) belong evidently in the variation range of typical *R. linearis* (comp. Hindák 1992).
Occurrence: Freshwater, in moorland waters and peaty lakes, particularly in mountains; Austria (Alps), Bulgaria, Czech Republic (Šumava Mts.), Slovakia (High Tatra Mts.).

### 3. **Rhabdogloea smithii** (R. et F. Chodat) Komárek 1983 (fig. **108, 109, 110**)

*Dactylococcopsis rhaphidioides* Hansgirg sensu G.M. Smith 1920; *Dactylococcopsis smithii* R. et F. Chodat 1925

Colonies microscopic, more or less ellipsoidal or irregular, with a small number or up to more than thirty widely separated cells, more or less oriented in one direction within the colony. Mucilage colourless, homogeneous, diffluent. Cells irregularly spindle-shaped, usually slightly curved or sigmoid, tapering to acute-rounded ends, pale blue-green or greyish green, (3.5)9–13 (involution cells to over 25) × 1.4–3(3.5) µm.
Occurrence: Freshwater, mainly planktic in large, cold, clear (oligotrophic to mesotrophic) lakes, rarely in clear small water bodies, found also in eastern Baltic Sea; over both temperate zones, data from tropical countries (Africa, Brazil, India) concern possibly another species (*R. ellipsoidea* ?; ecological difference recorded from swamps in tropics).

### 4. **Rhabdogloea planctonica** (Teiling) Komárek 1983 (fig. **111**)

*Dactylococcopsis planctonica* Teiling 1942

Colonies microscopic, mucilaginous, more or less spherical, usually with a large number of irregularly disposed cells. Mucilage homogeneous, colourless, delimited. Cells more or less straight or slightly arcuate, subcylindrical or asymmetrical with one side more convex than the other, narrowed and rounded at the ends, with greenish content and with solitary small refractive granules, 18–20 × 3.5–5 µm; a form with smaller cells (8.2–11.6 × 1.2–3 µm) was described by Fedorov (1992) from Russia (W. Siberia).
Occurrence: Freshwater, planktic in lakes; Sweden.

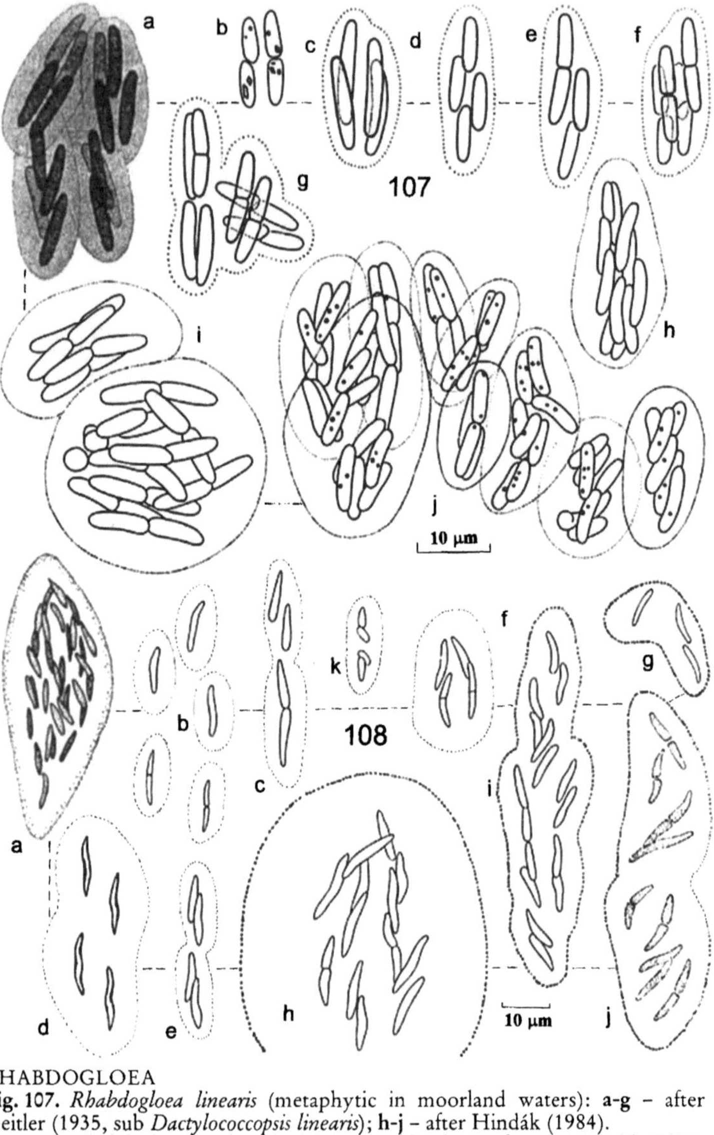


RHABDOGLOEA

**Fig. 107.** *Rhabdogloea linearis* (metaphytic in moorland waters): **a-g** – after Geitler (1935, sub *Dactylococcopsis linearis*); **h-j** – after Hindák (1984).

**Fig. 108.** *Rhabdogloea smithii* (planktic): **a** – old colony, after G.M.Smith (1920, sub *Dactylococcopsis raphidioides*), **b** – solitary cells with gelatinous envelopes, **c-g** – few-celled colonies, **h-j** – old colonies; **b-i** – after Hindák (1984, sub *Rhabdogloea ellipsoidea*), **j** – after Teiling (1946, sub *Dactylococcopsis ellipsoidea*).

**5. Rhabdogloea scenedesmoides** (Nygaard) Komárek et Anagnostidis 1995 (fig. **106**)

*Dactylococcopsis scenedesmoides* Nygaard 1949

Colonies free-floating, microscopic, with more or less parallely («scenedesmoidally») arranged cells, in twos or four in one row, sometimes slightly alternating or crosswise disposed. Mucilage not recorded, but probably present (staining !). Cells pale blue-green, slightly granulate, elongate, rod-shaped or spindle-shaped, slightly arcuate and asymmetric (one side convex, the other slightly concave), gradually narrowed to the ends and rounded, sometimes inconspicuously capitate, 12.5–21.5 × 2–3.5 µm.

Occurrence: Freshwater, planktic in large lakes, usually recorded at rather low temperatures (below 12 °C), recorded also from winter plankton (data from strongly eutrophic reservoirs are probably incorrect; Starmach 1966); northern temperate zone, of very sporadic occurrence.

**6. Rhabdogloea hungarica** (Kol) Hindák 1984 (fig. **112**)

*Dactylococcopsis hungarica* Kol 1955; *Dactylococcopsis caucasica* Kiss 1959; *Gloeothece transsylvanica* Kol sensu Garrick 1965

Colonies gelatinous, irregular, colourless, diffluent, microscopic, spherical when young, later elongate or amorphous, 20–300 µm in diameter. Mucilage colourless, diffluent. Cells fusiform with more or less bluntly pointed ends, 6–10 × 1.5–3.5 µm.

Occurrence: In cryoseston, causing blue colouration of snow in mountains; sporadically in the northern hemisphere, Hungary (Bükk Mts.), Caucasus, USA.

**7. Rhabdogloea elenkinii** (Roll) Komárek et Anagnostidis 1995 (fig. **113**)

*Dactylococcopsis elenkinii* Roll 1928

Colonies microscopic, more or less spherical or subspherical, with a large number of scarcely or densely and irregularly disposed cells within slime. Mucilage colourless, structureless, more or less delimited. Cells spindle-shaped, crescent-shaped, arcuate, gradually tapered to the ends and pointed, circa 5.4 × 1.3 µm.

Occurrence: Freshwater, planktic in rivers; E. Europe (Belorussia, Russia, Ukraine). A morphotype with cells 2–3 × 1–1.2 µm is recorded from lake Volvi, Greece (fig. **113c**).

**Species outside Europe:**

- *Rhabdogloea brasilica* Azevedo et Kováčik 1996: Microscopic gelatinous colonies with numerous cells, cells 2.3–12.1 × 1.4–3 µm, ellipsoid with conical-rounded ends; subaerophytic, on moisted and shadowed walls near the sea shore, Brazil (SP).
- *Rhabdogloea ellipsoidea* Schröder 1917 [syn.: *Dactylococcopsis ellipsoidea* (Schröder) Geitler 1942, *Aphanothece ellipsoidea* (Schröder) Bourrelly 1970]: Morphologically similar to *R. smithii*, differs ecologically, probably distributed only in swamps of the tropical zone; the relations of both mentioned species need to be resolved; (Bourrelly 1961, Hindák 1984).
- *Rhabdogloea subtropica* Hindák 1984: Fusiform, straight cells, 5–10 × 1.2–1.8 µm, tropical species known only from Cuba; (fig. **114**).

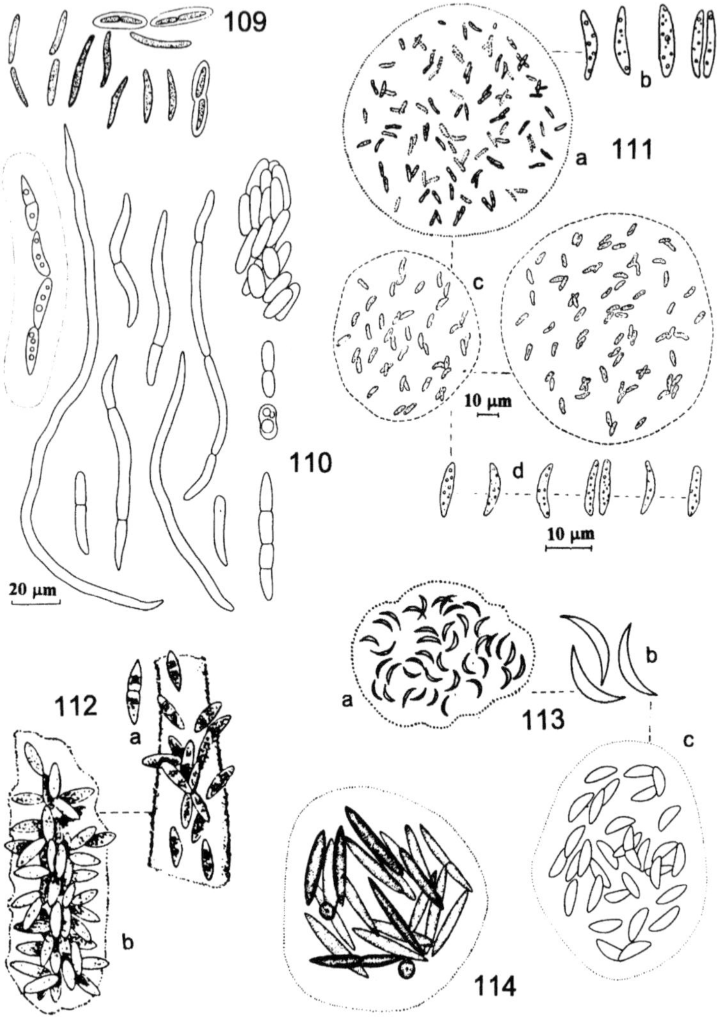


RHABDOGLOEA

**Fig. 109.** *Rhabdogloea smithii*: solitary cells after Chodat from Geitler (1932, sub *Dactylococcopsis smithii*).

**Fig. 110.** *Rhabdogloea smithii*, anomalous filamentous (involution) cells from culture: after Hindák (1984).

**Fig. 111.** *Rhabdogloea planctonica* (planktic): **a,c** – spherical planktic colonies, **b,d** – solitary cells; **a-b** – after Teiling (1942, sub *Dactylococcopsis planctonica*), **c-d** – after Fedorov (1992, sub *Dactylococcopsis planctonica* f. *minor*).

**Unrevised species:**
- Many other species have been described in the genus *Dactylococcopsis*, most of which belong to green algae (*Monoraphidium, Ankistrodesmus, Selenastrum*) and diatoms (*Nitzschia*).
- *Dactylococcopsis antarctica* Fritsch 1912: On surface of mats of other cyanoprokaryotes in Antarctica; probably a chroococcalean species (comp. Broady 1982), which needs further study.
- *Dactylococcopsis pectinatellophila* W. West 1911: Widely and irregularly ellipsoid cells with pale blue-green content, 8–13 × 3–5.5 µm; possibly a tropical (Burma) *Rhabdogloea* species (W. West & Annandale 1911).
- *Dactylococcopsis planctonica* f. *minor* Fedorov 1969: Differs from the type by substantially smaller cells (8.2–11.6 × 1.2–3 µm), known from lakes of W. Siberia; probably a separate *Rhabdogloea* species.
- *Rhabdogloea* sp. sensu Anagnostidis & Roussomoustakaki 1991: Found in heliothermal coastal saltworks in S. Greece.

## 4. **Rhabdoderma** Schmidle et Lauterborn 1900

*Gloeocapsa* subg. *Rhabdoderma* (Schmidle et Lauterborn) Elenkin 1923; *Catella* Ålvik 1934 incl.

Cells in small microscopic, irregularly oval or elongate-irregular mucilaginous colonies, usually distant from one another or in short pseudofilamentous rows, irregularly and widely spaced, but often oriented in one direction in a colony. Mucilage fine, colourless, homogeneous, sometimes diffluent (staining !); cells rarely with individual fine, indistinct gelatinous envelopes. Cells cylindrical, rod-shaped, straight, arcuate or sigmoid, with rounded ends, sometimes several times longer than wide; occasionally are present long, filamentous involution cells or unevenly long cells in one colony.
Cells only divide transversely to the longer axis, symmetrically or asymmetrically; daughter cells sometimes remain joined together in short pseudofilamentous rows. Reproduction by disintegration of colonies.
**Important references:** Geitler (1932, 1942); Komárek (1958, 1970a); Healey (1985); Komárek & Anagnostidis (1986).

**Key to the species:**

| | | |
|---|---|---|
| **1a** | Planktic in freshwater lakes | 2 |
| **1b** | Non-planktic species | 3 |
| **2a** | Cell length/width ratio greater than 3:1 | **1. R. lineare** |
| **2b** | Cells short rod-shaped, 4–8 × 3–3.5 µm | **5. R. compositum** |
| **3a** | In cryoseston, cells 6–12 × 1–2.5 µm | **4. R. transsylvanicum** |
| **3b** | Not in cryoseston, metaphytic and epiphytic in various hydrobiotopes | 4 |
| **4a** | Metaphytic in freshwaters biotopes | 5 |
| **4b** | Metaphytic and subaerophytic in salty biotopes | 6 |

**Fig. 112.** *Rhabdogloea hungarica* (cryosestic): **a** – after Kol (1955, sub *Dactylococcopsis hungarica*), **b** – after Kiss (1959, sub *Dactylococcopsis caucasica*).
**Fig. 113.** *Rhabdogloea elenkinii* (planktic): **a-b** – after Roll (1928, sub *Dactylococcopsis elenkinii*), **c** – morphotype from the lake Volvi, Greece, after Hindák & Moustaka (1988).
**Fig. 114.** *Rhabdogloea subtropica*, described from Cuba: after Hindák (1984).

**5a** Cells long cylindrical, usually clearly arcuate, circa 10 × 1 µm, in moorland waters . . . . . . . . . . . . . . . . . . . . . . . . . . . . . . . . **2. R. vermiculare**

**5b** Cells cylindrical, distinctly sigmoid, circa 5 × 1 µm; in small streams . . . . . . . . . . . . . . . . . . . . . . . . . . . . . . . . . . . . . . . **3. R. irregulare**

**6a** Cells reddish, metaphytic in coastal salty water bodies, 2.5–7.5 × 1.5–2.5 µm (these dimensions do not cover involution cells) . . . . . . . . . **6. R. rubrum**

**6b** Cells pale blue-green, metaphytic and subaerophytic in coastal salty waters, up to 6 × 3 µm . . . . . . . . . . . . . . . . . . . . . . . . . . . . . . . **7. R. curtum**

## 1. Rhabdoderma lineare Schmidle et Lauterborn 1900 (fig. 115, 116)

*Rhabdoderma lineare* var. *spirale* Woloszyńska 1913; *Rhabdoderma gorskii* Woloszyńska 1917 incl.; *Aphanothece viennensis* Schiller 1954; *Synechococcus linearis* (Schmidle et Lauterborn) Komárek 1970

Colonies small, few-celled, later sometimes multicellullar, with cells oriented more or less in one direction or irregularly arranged. Mucilage scarcely visible, fine, homogeneous, colourless, sometimes diffluent at the margin; single cells without or sometimes with very fine individual hyaline envelopes. Cells long cylindrical, rod-shaped, straight, slightly arcuate or clearly sigmoid, occasionally in pseudofilamentous rows, pale blue-green or grey-green, with fine granular content, (3) 4–10 (22) × (0.8) 1.5–3 (3.5) µm. In colonies sometimes the cells are of different lengths, occasionally dividing asymmetrically.

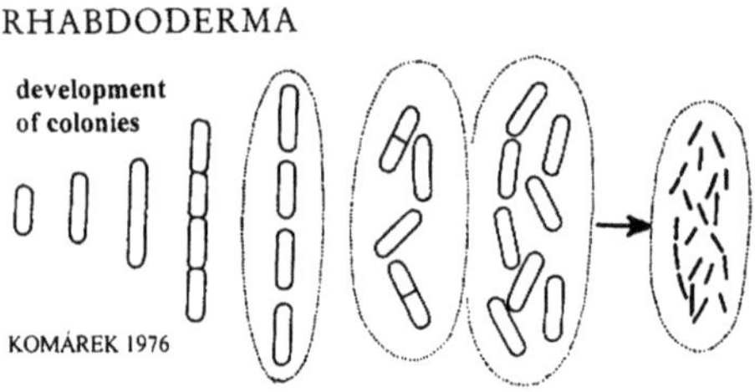


119
b

**Fig. 115.** *Rhabdoderma lineare* (only planktic!): **a** – iconotype after Schmidle from Starmach (1966), **b** – after G.M. Smith (1920), **c** – after Wojciechowski (1971), **d** – variability of cells and colonies after Komárek (1958).

**Fig. 116.** Morphotypes of *Rhabdoderma lineare*, described as separate taxa: **a** – after Woloszyńska from Geitler (1932, sub *Rhabdoderma gorskii*), **b-c** – after Lundberg (1931, sub *Rhabdoderma gorskii* var. *spiralis*).

**Fig. 117.** *Rhabdoderma vermiculare* (from moorland waters): after Fott (1952).

**Fig. 118.** *Rhabdoderma irregulare* (from metaphyton in streams): iconotype after Naumann (1921, sub *Spirillopsis irregularis*).

**Fig. 119.** *Rhabdoderma transsylvanicum* (cryosestic): **a** – after Kol (1955, sub *Gloeothece transsylvanica*), **b** – after Kol (1968, sub *Gloeothece transsylvanica*).

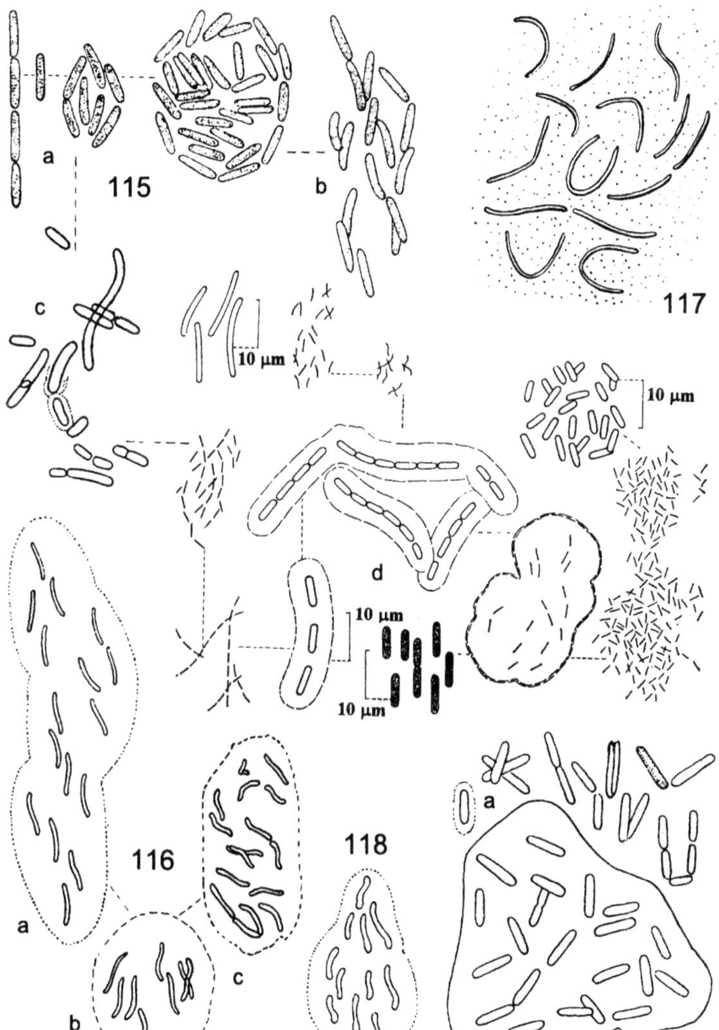
a
115
b
117
c
10 µm
10 µm
d
10 µm
10 µm
116
a
b
c
118
a
119

Occurrence: Freshwater, planktic in oligo- to mesotrophic water bodies, common in temperate zones; sporadical records from tropical countries (Indonesia) probably concern another species.

## 2. Rhabdoderma vermiculare Fott 1952 (fig. 117)

Cells long cylindrical, narrow, solitary or irregularly disposed and widely spaced in common mucilage, slightly or intensely arcuate or sigmoid, indistinctly attenuated at the ends, with homogeneous, pale blue-green content, circa 7–10 × –0.7–1 µm (however, the cells are up to 20-times longer than wide judging from drawings); mucilage very fine, irregular, not structured, colourless, but distinct, up to 600 µm in diameter, without special envelopes around cells.
Occurrence: Freshwater, acidophilic, metaphytic in pęaty bogs, moorland waters and raised bogs; known only from Czech Republic (Šumava Mts.) and Slovakia (W. Carpathians).

## 3. Rhabdoderma irregulare (Naumann) Geitler 1925 (fig. 118)

*Spirillopsis irregularis* Naumann 1921

Colonies more or less spherical or irregular, microscopic, mucilaginous. Mucilage hyaline, colourless, homogeneous, delimited. Cells distant from one another, cylindrical, sigmoid, circa 2–5 × 1–1.5 µm. Insufficiently known species.
Occurrence: Freshwater, metaphytic and periphytic, described from a stream in Sweden, found very rarely in small mountain streams in Czech Republik (Moravia -Beskydy Mts.), and recorded also from backwaters of the river Guadalete in Spain (Gonzalez Guerrero 1946).

## 4. Rhabdoderma transsylvanicum (Kol) Komárek et Anagnostidis 1995 (fig. 119)

*Gloeothece transsylvanica* Kol 1955

Colonies microscopic, irregular, mucilaginous, dirty olive-green or reddish-violet, with sparsely and irregularly arranged cells. Mucilage fine, colourless, homogeneous, diffluent, sometimes indistinct. Cells cylindrical, rod-shaped with rounded ends, straight, bluish, with homogeneous content, 6–12 × 1–2.5 µm.
Occurrence: In cryoseston on summer snowfields in mountains (bluish colouration); E. Carpathians (Roumania). Wawrik (1987) described *Rhabdoderma sertosum* from autumn melt water in Austria (Waldviertel), with very similar morphology (cells 10–12 (13) × 2–2.5 µm) to *R. transsylvanicum* (comp. fig. 120).

## 5. Rhabdoderma compositum (G.M.Smith) Fedorov 1967 (fig. 121)

*Gloeothece linearis* var. *composita* G.M.Smith 1920; *Rhabdoderma lineare* f. *compositum* (G.M.Smith) Hollerbach in Elenkin 1938; *Gloeothece linearis* f. *unicellularis* Hollerbach in Elenkin 1938; *Gloeothece linearis* Nägeli sensu auct. sine typo (planktic populations)

Colonies small, microscopic (mainly 20–65 × 10–18 µm), few-, (1)2–4 (12)-celled, with cells usually more or less in one lengthwise row (one after another), but often some of them slightly displaced from the colonial axis; after division the two cells remain near to one another; several small colonies often irregularly aggregated together. Mucilage hyaline, colourless, delimited, rarely indistinct, oval to cylindrical, enveloping and copying the outline of the cell row. Cells oval to shortly cylindrical with widely rounded ends, straight, blue-green or pale blue-green,

with more or less homogeneous content, 4–8 × 3–3.5 μm, with individual, distinct, hyaline gelatinous envelopes.
Occurrence: Freshwater, planktic in large clear lakes; temperate zone: central and northern Europe, northern Russia (W. Siberia), Canada, USA.

**6. Rhabdoderma rubrum** (Ålvik) Komárek et Anagnostidis 1995 (fig. **122**)

*Catella rubra* Ålvik 1934

Colonies small, microscopic, elongate, mucilaginous, with cells in separate rows, in which they are slightly distant from one another, or (after division) in pseudofilamentous formations, enveloped by hyaline, finely delimited, hardly visible mucilage, which exceeds the cell width 2–4x; number of cells from one to several in a colony; rows of cells are straight or slightly curved, in later stages multicellular; long cells sometimes divide simultaneously into several daughter cells. Cells oval to cylindrical with rounded ends, with red content, 2.5–7.5 × 1.5–2.5 μm; elongate, filament-like and irregular involution cells very often occur (cultures).
Occurrence: In salty water bodies in the coastal marine regions; Norway (described from cultures).

**7. Rhabdoderma curtum** (Setchell) Komárek et Anagnostidis 1995 (fig. **123**)

*Synechococcus curtus* Setchell in Collins et al. 1907

Cells solitary with wide gelatinous envelopes or in mucilaginous colonies, sometimes joined into large aggregates, macroscopically visible. Mucilage fine, translucent, homogeneous, colourless. Cells cylindrical (according to drawings rather fusiform), usually irregularly arranged, pale blue-green, circa 6 × 3 μm; (according to the cell shape belongs to *Rhabdogloea*).
Occurrence: Marine, in heliothermal salty coastal pools and swamps, and in warm marine littoral zones, benthic or free- floating; described from USA (California), in Europe known from Greece and Portugal; probably more widely distributed throughout the world (e.g., SW. Pacific coasts – Nauru Islands).

**Species outside Europe:**
- *Rhabdoderma tenuissimum* Komárek et Kling 1991: Several-celled planktic colonies with very narrow rod-shaped cells (2–7.8 × 0.5–0.6 μm), described from large E. African lakes; (fig. **124**).
- *Rhabdoderma zygnemicolum* Copeland 1936: Cells oval to cylindrical, straight, 3–3.5 × 1.9–2.4 μm, epiphytic on *Zygnema* filaments, described from acidic thermal waters in Yellowstone National Park, USA, at temperatures of 32.9–43.80 °C; (fig. **125**).

**Unrevised taxa:**
- *Rhabdoderma bellyeense* Hortobágyi 1944: Small, free-floating colonies with short rows of 2–4 cylindrical cells with rounded ends, 14–14.3 × 4.7–4.9 μm, slightly distant from one another, enveloped by wide (12.5–13 μm), fine, colourless, more or less delimited mucilage; planktic in one pond in Hungary (Baranya), not confirmed.
- *Rhabdoderma cavanillesiana* Gonzalez 1946: Solitary cells or pseudofilamentous formations with oval or irregular-oval cells, 6–33 × 5–7 μm, blue-green content and aerotopes; described from saline habitats near the estuary of the river San Pedro, Spain; comp. with *Romeria*.

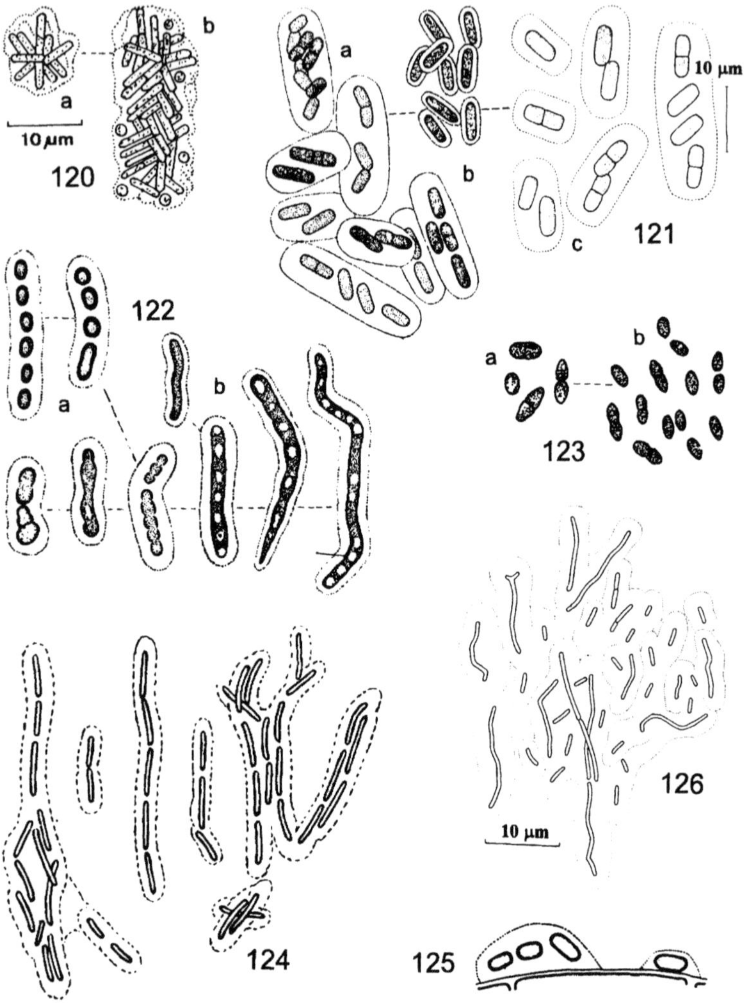


RHABDODERMA

**Fig. 120.** *Rhabdoderma sertosum* (cryosestic): after Wawrik (1987).
**Fig. 121.** *Rhabdoderma compositum* (planktic): **a-b** – after G.M. Smith (1920, sub *Gloeothece linearis* var. *composita*), **c** – after Fedorov (1967).
**Fig. 122.** *Rhabdoderma rubrum* (halophilic, red cells): **a** – colonies, **b** – anomalies (involution cells) from culture; after Ålvik (1934, sub *Catella rubra*).
**Fig. 123.** *Rhabdoderma curtum* (halophilic): **a** – after Setchell in Collins & al. (1907, sub *Synechococcus curtus*), **b** – after Kosinskaja (1948, sub *Synechococcus curtus*).

- *Rhabdoderma lineare* var. *minor* Emoto et Hirose 1939 : Cells 4–7 × 0.7–1 µm; described from thermal springs in Japan (Hirose & Hirano 1977); probably of species rank.
- *Rhabdoderma salinum* Tutin 1940: Planktic species, described from Bolivia (Titicaca region); not found again, according to Drouet & Daily (1956) possibly bacteria (?).
- *Rhabdoderma surchanense* Ergašev 1980: Very poorly described, not identifiable.
- Guarrera (1986) distinguished in specimens from Argentine (Tierra del Fuego) *R. gorskii* (with cells 14–30 × 2–5.8 µm) and *R. sigmoideum* (cells 9.6 × 3.6 µm).
- *Rhabdoderma* sp. in Hindák & Moustaka (1988): Planktic, cells 1.5–15 × 0.6–0.8 µm, rarely with Y-shaped cell ends; in summer plankton of lake Volvi, Greece; (fig. **126**).

## 5. Synechococcus Nägeli 1849

*Lauterbornia* Pringsheim 1968 incl.

Cells solitary or grouped in microscopic or macroscopic, irregular clusters, but not forming mucilaginous colonies; cells sometimes in short series of pseudofilamentous formations with 2–4 (20) cells. Mucilage absent or very fine, colourless, homogeneous, diffluent, around single cells. Cells long oval or cylindrical, sometimes several times longer than wide, straight, arcuate or sigmoid, 1.5 up to more than 20 µm long and 0.4–6 (11) µm wide, with parietal thylakoids; fine subunit arrays with p4 symmetry were found in few *Synechococcus* strains. Particularly in cultures there can arise long filamentous involution forms.
Cells divide by binary fission, transversely (always in one plane in successive generations, perpendicular to the longitudinal axis) into two isomorphic or different (after asymmetric binary fission) daughter cells, which occasionally remain joined for longer period in pairs (obligatory by cleavage ?). Reproduction by cell division; (fig. **127**).
**Important references:** Geitler (1932, 1942, 1960); Copeland (1936); Anagnostidis (1961); Bailey-Watts & al. (1968); Pringsheim (1968b); Padmaja & Desikachary (1968, 1969); Komárek (1970, 1976); Padmaja (1972); Rippka & al. (1979); Wilmotte & Stam (1984); Komárek & Anagnostidis (1986); Waterbury & Rippka (1989); Cepák (1993).

**Key to the species:**

| | | |
|---|---|---|
| **1a** | Species from thermal biotopes | **4. S. bigranulatus** |
| **1b** | Species from other biotopes | 2 |
| **2a** | Species from various hydrobiotopes | 3 |
| **2b** | Aerophytic and subaerophytic species | **11** |
| **3a** | Planktic species | 4 |
| **3b** | Metaphytic, epiphytic or endophytic | 6 |

Note: See also two endogloeic species, **1a S. endogloeicus** and **1b S. epigloeicus.**

| | | |
|---|---|---|
| **4a** | Cells blue-green | **5** |

**Fig. 124.** *Rhabdoderma tenuissimum* (planktic in tropical lakes): after Komárek & Kling (1991).
**Fig. 125.** *Rhabdoderma zygnemicolum* (epiphytic, thermophilic, described from USA): after Copeland (1936).
**Fig. 126.** *Rhabdoderma* sp. from plankton of the lake Volvi, Greece: after Hindák & Moustaka (1988).

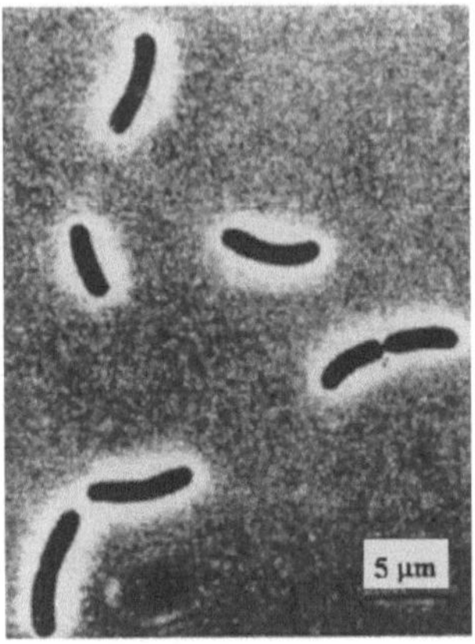


127

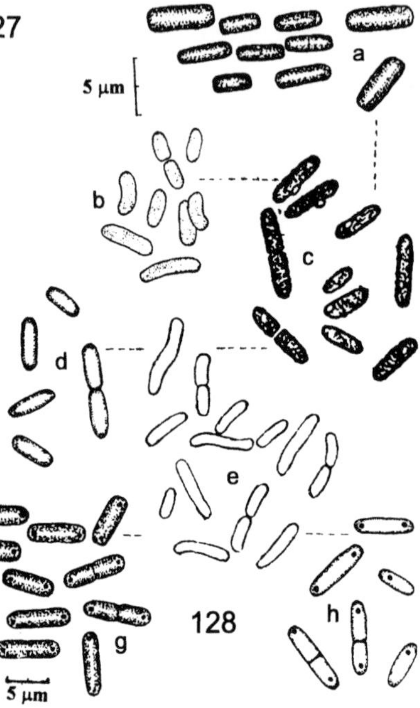


128

129

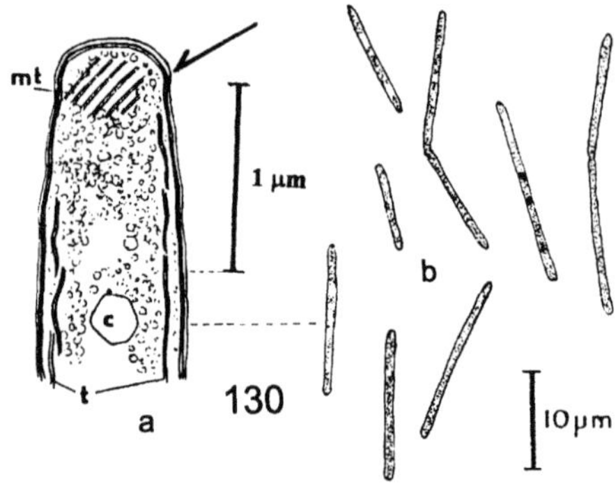


130

4b Cells red, cylindrical, circa 3–11 × 0.8–2.3 μm ...... **2. S. rhodobaktron**
5a Cells short rod-shaped, more or less rounded at both ends, 1.5–8.5 (- 35, invol. forms) × 0.4–1.3 μm ...... **1. S. nidulans**
5b Cells elongate cylindrical, slightly capitate (widened) at both ends, 4–40 × 0.7–0.9 μm ...... **3. S. capitatus**
6a Freshwater ...... 7
6b Saltwater ...... 10
7a Cells blue-green, benthic and metaphytic in swamps and waters with hydrophytes (see also endogloeic species: **1a S. endogloeicus** and **1b S. epigloeicus.** ...... 8
7b Cells reddish or violet, in springs and fens ...... 9
8a Cells 2.7–8 × 2.3 μm, mainly in swamps ...... **7. S. ambiguus**
8b Cells 6–13 × 4–6 μm, benthic in lakes ...... **8. S. mundulus**
9a Cells to 11 × 2.5–5.5 μm, in springs, cell content more or less violet ...... **9. S. violaceus**
9b Cells to 36 × 3.6–6 μm, in fens, cell content reddish . **10. S. roseo-purpureus**
10a Cells 6–12 × 2–3.5 μm, in heliothermal saltworks and on marine beaches; Black Sea and Mediterranean Sea ...... **6. S. salinarum**
10b Cells 6–24 × 4.5–6 μm; Baltic Sea littoral ...... **11. S. subsalsus**
11a Cells 2–9 (in cultures up to 68) × 1.2–3 μm, on wet soil and stones, wet lake epi- and supralittoral, rarely under water in springs ...... **5. S. elongatus**
11b Cells wider than 3 μm, on wet rocks and rock walls ...... 12
12a On calcareous rocks, cells pale blue-green, 6–14 × 3.5–5.5 μm, in Mediterranean countries ...... **12. S. sciophilus**
12b Not on calcareous rocks, cells blue-green or brownish, up to 33 × 5–11 μm, central Europe ...... **13. S. brunneolus**

---

SYNECHOCOCCUS

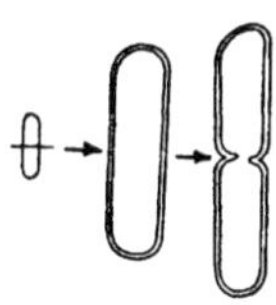

KOMÁREK & ANAGNOSTIDIS 1986

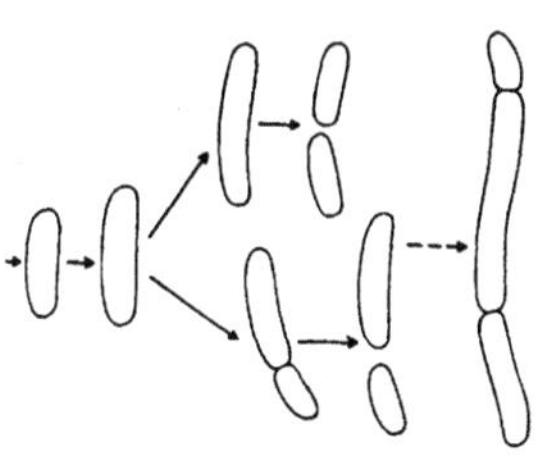


KOMÁREK 1976

**Fig. 127.** Outlines of cells of the typical thermophilic *Synechococcus*-strain PCC 6716, designated as member «of the *Synechococcus*-cluster of the *Synechococcus*-group»; from Waterbury & Rippka (1989, Bergey's Manual).
**Fig. 128.** *Synechococcus bigranulatus* (populations from thermal springs, corresponding to «thermophilic *Synechococcus elongatus*»): **a** – after Anagnostidis (1961, sub *S. elongatus* fa. *thermalis*), **b** – after Dor (1967, sub *Synechococcus elongatus* var. *vestitus*), **c** – after Emoto & Hirose (1942, sub *Synechococcus elongatus* fa. *indefinitus*), **d** – after Copeland (1936, sub *Synechococcus elongatus* var. *vestitus*), **e** – after Skuja (1933, *Synechococcus bosshardii*), **f** – after Skuja (1933, iconotype of *Synechococcus bigranulatus*), **g** – after Anagnostidis (1961, sub *Synechococcus elongatus* var. *amphigranulatus*), **h** – after Copeland (1936, iconotype of *Synechococcus elongatus* var. *amphigranulatus*).
**Fig. 129.** *Synechococcus rhodobaktron* (planktic, red): orig.
**Fig. 130.** *Synechococcus capitatus* (planktic): **a** – EM fine section of the cell end (redrawn from Bailey-Watts & al. 1968), **b** – solitary cells from Bailey-Watts & Komárek (1991).

**1. Synechococcus nidulans** (Pringsheim) Komárek in Bourrelly 1970 (fig. **131, 132**)

«*Anacystis nidulans* Drouet» sensu Kratz & Myers 1955 et auct. post. [non *Anacystis nidulans* (Richter) Drouet et Daily 1952 = *Aphanothece nidulans* Richter in Wittrock et Nordstedt 1884]; Lauterbornia nidulans Pringsheim 1968; *Synechococcus leopoliensis* (Raciborski) Komárek 1970 sine typo

Cells solitary, free-floating, short oval to rod-shaped, without mucilage, with more or less homogeneous, pale blue-green content, straight or sometimes slightly sigmoid or arcuate, 1.5–8.5 × 0.4–1.3(2.2) µm. Filamentous involution cells up to 35 µm long often arise (particularly in cultures).

Occurrence: Freshwater, planktic in small water bodies, ditches, pools, rarely in ponds and large lakes, sometimes in masses; known from temperate zones. Important model experimental species cultivated in various strains in numerous laboratories throughout the world, usually under the name «*Anacystis nidulans*», less «*Agmenellum quadruplicatum*» (the both generic names are nomenclatorically and taxonomically incorrect).

---

SYNECHOCOCCUS

**Fig. 131.** *Synechococcus nidulans*: **a** – variation in cell morphology at different temperatures (over 30°C numerous filamentous involution cells appear), **b-c** – involution cells under suboptimal conditions in culture; **a-b** – after Komárek (1970), **c** – from Pringsheim (1968).

**Fig. 132.** *Synechococcus nidulans*, populations from nature: **a** – after Komárek (1989, sub *Synechococcus* cf. *nidulans* from Cuba), **b** – after Komárková-Legnerová & Cronberg (1994).

**Fig. 132A.** *Synechococcus endogloeicus*: after Hindák (1995).

**Fig. 132B.** *Synechococcus epigloeicus*: after Hindák (1995).

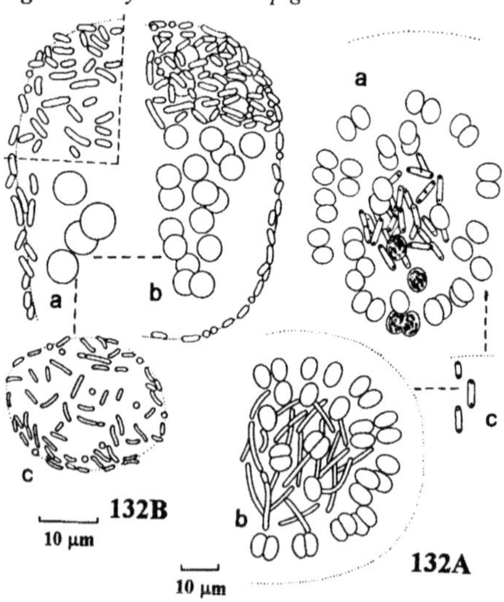


**132B**

**132A**

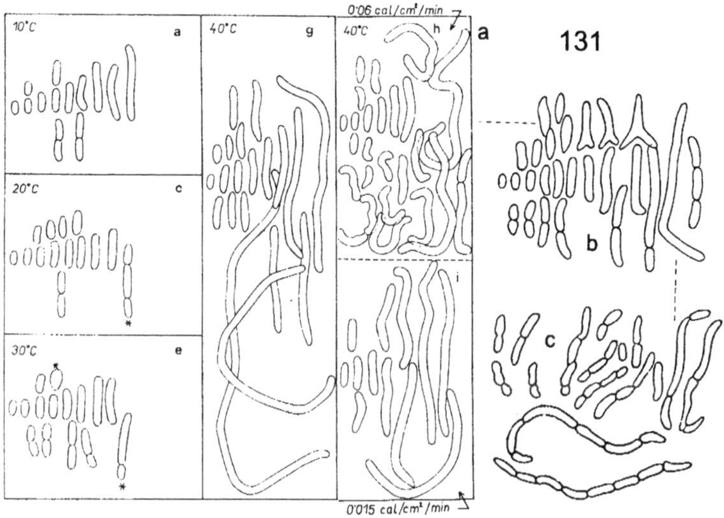


131

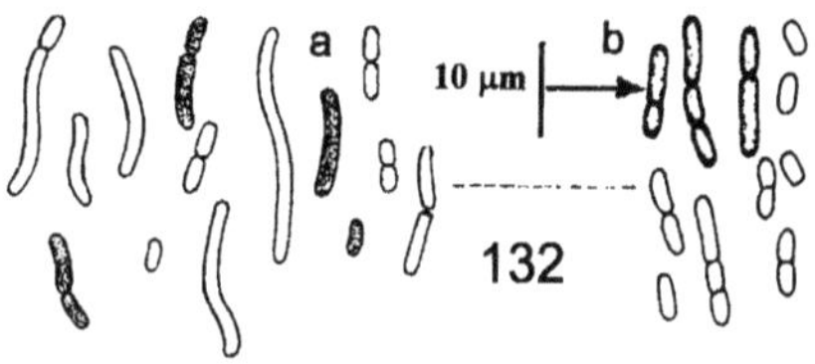


132

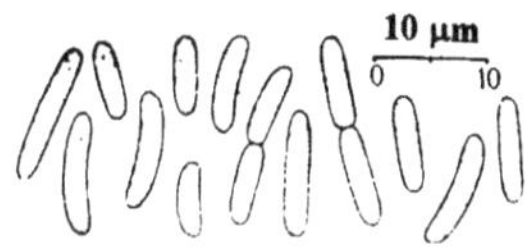


133

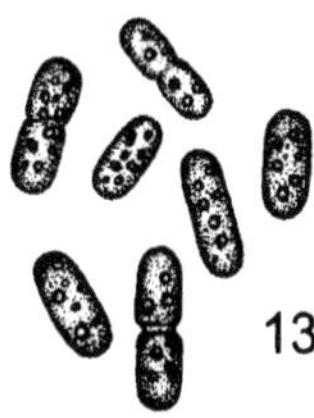

134

135

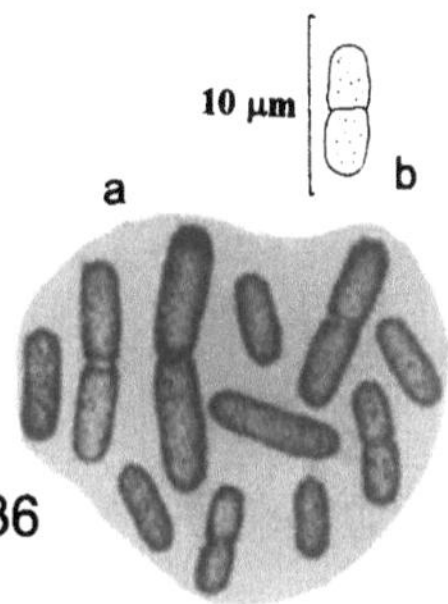


136

**Fig. 133.** *Synechococcus salinarum*: after Komárek (1956).
**Fig. 134.** *Synechococcus mundulus*: after Skuja (1964).
**Fig. 135.** *Synechococcus subsalsus*: after Skuja (1939).
**Fig. 136.** *Synechococcus sciophilus*: **a** – after Skuja (1970), **b** – after Hoffmann (1986).

During the press of this book were described the following endogloeic species:

### 1a. **Synechococcus endogloeicus** Hindák 1995 (fig. **132A**)

Cells solitary, or in pairs after division, rarely 3–4 together in rows, cylindrical with rounded ends, straight or slightly arcuated, 4–15 × 1.8–2 μm. Sometimes appear unequally long daughter cells after asymmetrical cross-like cell division. Cell content pale blue-green, homogeneous without aerotopes, sometimes with solitary granules at poles.
Occurrence: Endogloeic in the mucilage of the planktic species *Woronichinia naegeliana*, usually in the inner parts of a colony; described from the Lake Loch Lochmond (Scotia).

### 1b. **Synechococcus epigloeicus** Hindák 1995 (fig. **132B**)

Cells solitary or in pairs after division, cylindrical with rounded ends, straight or very slightly arcuated, 2.5–4.5 × 1.5–2 μm, in pairs sometimes of different length (after occasional asymmetrical cell division). Cell content pale blue-green, homogeneous, without aerotopes and granulation.
Occurrence: Epigloeic on the colonial surface of *Microcystis aeruginosa* in plankton of eutrophic lakes; described from the eutrophic sand pit lake near Košice (SE. Slovakia).

### 2. **Synechococcus rhodobaktron** Komárek et Anagnostidis 1995 (fig. **129**)

Cells solitary, free-floating, rod-shaped, cylindrical, rounded at the ends, without mucilage, in twos after division, more or less straight or slightly arcuated, constantly with reddish content, 3–9 (11) × 0.8–2.3 μm.
Occurrence: Freshwater, planktic in large lakes; known from central Japan and central Europe; probably in corresponding biotopes over the whole temperate world.

### 3. **Synechococcus capitatus** Bailey-Watts et Komárek 1991 (fig.**130**)

*Synechococcus* sp. in Bailey-Watts et al. 1968

Cells solitary or two together; sometimes (in cultures) short chain-like formations with up to 4 cells arise. Cells more or less straight or slightly curved, rod-shaped, at the ends slightly capitate (widened), pale green or almost bluish, with finely granulate content, without mucilage, 4–40 × 0.7–0.9 μm. Involution filament-like cells recorded, particularly in cultures.
Occurrence: Freshwater, planktic in large, cold lakes; Scottland (Loch Leven), Scandinavia, Japan, occasionally in cold, not polluted reservoirs in central Europe.

### 4. **Synechococcus bigranulatus** Skuja 1933 (fig. **128**)

*Synechococcus elongatus* f. *thermalis* Geitler 1935; *Synechococcus elongatus* var. *amphigranulatus* Copeland 1936; *Synechococcus elongatus* var. *vestitus* Copeland 1936; *Synechococcus elongatus* f. *indefinitus* Emoto et Hirose 1942; *Synechococcus amphigranulatus* (Copeland) Yoneda 1943; *Synechococcus elongatus* var. in Castenholz 1969

Cells solitary or forming irregular mucilaginous clusters on a substrate; cell aggregates arsenic green, dull blue-green, black-green or olive-green (at higher

temperatures); cells rod-shaped, straight or (rarely) slightly curved, with rounded poles, sometimes (after division) two together, pale or bright blue-green, in nature often with solitary polar granules at both ends (not commonly so in cultures), usually with distinct chromatoplasma, (2) 3–9 × (1.2) 1.5–3 μm. Involution forms. Slow motility recorded.
Occurrence: Thermal springs, sometimes forming fine mats on the substrate (stones), also flocculant, in temperatures from (20)30 to 75.8°C (probably there are different temperature-dependent types); known from thermal springs all over the world. Used also as a thermophilic model strain in laboratories (usually under designation «thermophilic *Synechococcus elongatus*»).

### 5. **Synechococcus elongatus** (Nägeli) Nägeli 1849 (fig. **137**)

*Synechococcus parvulus* Nägeli 1849; *Synechococcus racemosus* Wolle 1881; *Synechococcus geitleri* De Toni 1936

Cells solitary or in clusters on substrate, sometimes in chain-like formations with up to 4 cells. Cells rarely oval, mostly cylindrical, straight or slightly irregularly curved, with blue-green, homogeneous or finely granulate content (rarely with small solitary granules), occasionally with very thin mucilage layer around cells, 2–9 (–68, in cultures) × 1.2–3 μm. Filamentous involution forms known.
Occurrence: Subaerophytic on wet soil, walls of aquaria, wet stony epi- and supralittoral of lakes and pools, edges of wells, etc., rarely under water in springs or in littoral of lakes; all over the temperate zones. Records from tropical countries and/or other biotopes should be revised.

### 6. **Synechococcus salinarum** Komárek 1956 (fig. **133**)

Cells solitary or two together after division, cylindrical with rounded ends, straight or slightly arcuate, without mucilage, with pale blue-green, more or less homogeneous content, 6–12 × 2–3.5 μm.
Occurrence: Marine, in heliothermal saltworks, salty swamps, on wet saline soil and in psammon of shallow sandy beaches; Bulgaria, Greece, probably more distributed.

### 7. **Synechococcus ambiguus** Skuja 1937

Cells solitary or in pairs after division, oval to cylindrical, straight or slightly curved, with rounded ends, with very fine and indistinct mucilaginous layer (staining !). Cells with homogeneous or finely granulate content, pale blue-green, 2.7–8 × 2–3 μm.
Occurrence: Freshwater, metaphytic in swamps and ponds with water plants and in lake littorals, from clear to highly eutrophic waters (different ecotypes ?); described from China (Yünan), but distributed probably over the whole temperate and subtropical zones.

### 8. **Synechococcus mundulus** Skuja 1964 (fig. **134**)

Cells solitary or two together after division, oval to cylindrical with rounded poles, in the middle part sometimes slightly narrowed, grey-blue to blue-green with distinct chromatoplasma, without mucilaginous envelopes; cell content homogeneous or with several large, conspicuous, refractive granules, 6–13 × 4–6 μm.
Occurrence: Freshwater, benthic in lakes among other algae; N. Sweden (Abisko).

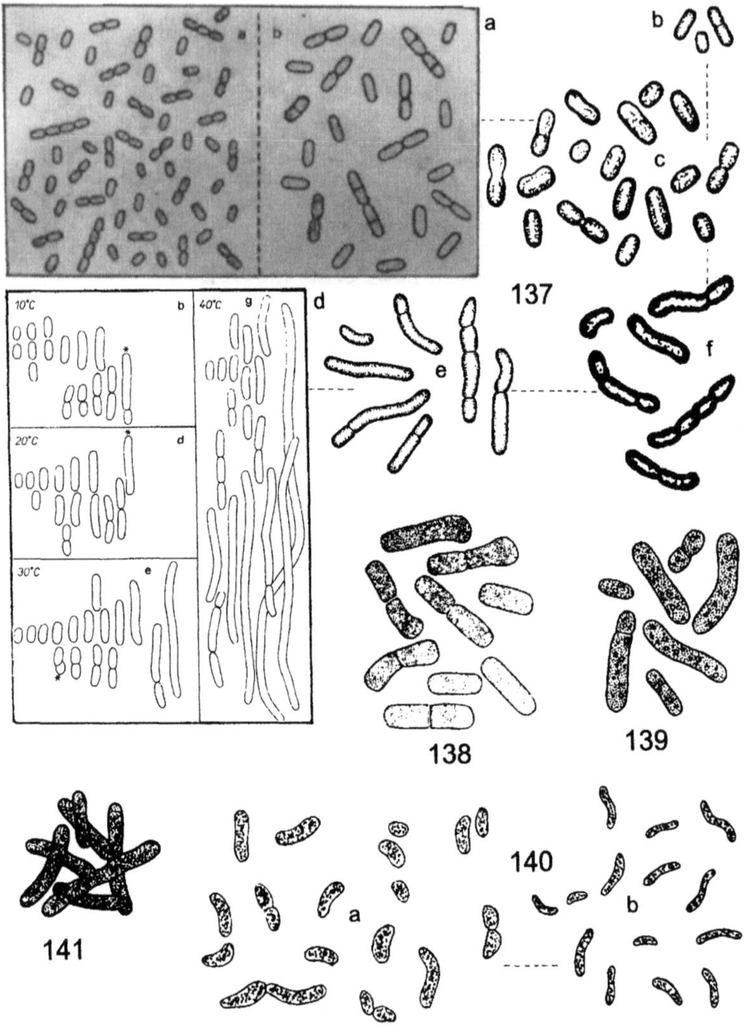


SYNECHOCOCCUS

**Fig. 137.** *Synechococcus elongatus*: **a** – after Nägeli (1849), **b-c** – characteristical cell form, **d** – cell morphology at various temperatures (over 30°C there are numerous filamentous involution cells), **e-f** – involution cells from natural populations; **b,e** – from Geitler (1925), **d** – from Komárek (1970), **c,f** – after Skuja (1937).

**Fig. 138.** *Synechococcus koidzumii* (thermal springs in Japan): after Yoneda (1943).

**Fig. 139.** *Synechococcus rayssae* (thermal springs in Israel): after Dor (1967).

### 9. **Synechococcus violaceus** Grunow in Rabenhorst 1868

Cells usually in fine, pale violet mats on a substrate, oval, at the poles roundly truncate, with pale bluish or slightly violet content, up to 11 × 2.5–5.5 µm. Insufficiently known species.
Occurrence: Freshwater, metaphytic in springs; Austria (near Wagrain).

### 10. **Synechococcus roseo-purpureus** G.S. West 1899

Cells solitary or two together, oval to subcylindrical with rounded poles, with pinkish purple or reddish content, sometimes with several granules, up to 36 (?) × 3.6–6 µm. Insufficiently known species.
Occurrence: Freshwater, metaphytic, together with *Mougeotia*, described from a fen; recorded only from England (Cambridgeshire).

### 11. **Synechococcus subsalsus** Skuja 1939 (fig. **135**)

Cells solitary or two together after division, long elliptical to cylindrical with rounded poles and pale blue-green, homogeneous or finely granulate content with distinctly visible chromatoplasma, sometimes enveloped by thin, fine, individual, colourless mucilaginous envelopes. Cells pale blue-green, 6–24 × 4.5–6 µm.
Occurrence: Brackish, littoral of Baltic Sea; Lithuania (near Riga).

### 12. **Synechococcus sciophilus** Skuja 1964 (fig. **136**)

Cells solitary or two together after division, oval to cylindrical with rounded poles, straight or slightly curved, pale grey-blue, grey-brown to almost colourless, with homogeneous and very fine granulate content, 6–14 × 3.5–5.5 µm.
Occurrence: Aerophytic, among algae on moist calcareous rocks, soil, or in caves; Italy, Luxembourg. *S. sciophilus* was published also from Switzerland (Hindák 1983), but from dystrophic water (probably another species).

### 13. **Synechococcus brunneolus** Rabenhorst 1863

Cells long oval up to cylindrical with wide rounded ends and blue-green or brownish content with visible chromatoplasma, up to 33 × 5–11 µm. Insufficiently described species.
Occurrence: Aerophytic, on wet, not calcareous rocks; recorded from central Europe and Japan.

**Species outside Europe:**
- *Synechococcus carcerarius* Norris 1967: Cells 8–10 × 3–5 µm; marine, endobiotic in protists and diatoms in Indian Ocean.
- *Synechococcus intermedius* Gardner 1927: Cells 12–14 × 6–7 µm, intensely blue-green; aerophytic on bark of trees, also in water on leaves and wood (?), described from Porto Rico; comp. *Cyanobium*.
- *Synechococcus koidzumii* Yoneda 1943: Cells 5–16 × 2.5–4.8 µm; thermal springs in Japan (Yoneda 1943; Hirose & Hirano 1981) (fig. **138**); may be also in USA,

---

**Fig. 140.** *Synechococcus sigmoideus* (planktic in lakes): after Moore & Carter from Geitler (1932; **b** – sub «fa. *minor*»).
**Fig. 141.** *Synechococcus sulphuricus*: from Israel; after Dor (1967).

India and Israel: probably identical with *S. rayssae* Dor 1967 (fig. **139**) and *S. vulcanus* var. *bacillarioides* Copeland 1936.
- *Synechococcus lividus* Copeland 1936: Cells long rod-shaped, straight or curved, 5–15 (32) × 1.2–2.2 µm; thermobiotic species at high temperatures (usually more than 50°C); several taxa were described, which are probably morphotypes and/or ecotypes of *S. lividus* (var. *curvatus* Copeland 1936, var. *siderophilus* Copeland 1936, var. *non-granulatus* Emoto et Hirose 1939, *S. arcuatus* Copeland 1936, *S. vescus* Copeland 1936, *S. vescus* var. *non-granulatus* Emoto et Hirose 1940, *S. sublividus* Emoto et Yoneda 1940, *S. praelongus* Emoto et Yoneda 1940). *S. bosshardii* Skuja 1932 (described from hot springs in high central Asia) belongs possibly also to this group; in that case it seems justified to conserve the epithet «*lividus*». *S. lividus* is known from numerous hot springs all over the world (Japan, New Zealand, USA, Zimbabwe, etc.), but not yet recorded from Europe; (fig. **142**).
- *Synechococcus marinus* Jao 1948: Cells ± oval, 9–11 × 6.5–8.5 µm, from San Juan island, scattered among algae on floating remnants of seaweeds and animals; comp. *Cyanobium* or *Cyanothece*.
- *Synechococcus minutissimus* Negoro 1944: Cells narrow, rod-shaped, slightly arcuate, 4.5–10 × 0.9–1 µm, from thermal springs in Japan; (Hirose & Hirano 1981).
- *Synechococcus sigmoideus* (Moore et Carter) Komárek 1970 [syn.: *Rhabdoderma sigmoideum* Moore et Carter 1923]: Cells short cylindrical, usually sigmoid, 4–13 × 1.5–3 µm; planktic in lakes, USA (N. Dakota), (Moore & Carter 1923; Geitler 1932; Komárek 1976) (fig. **140**); data from rice-fields from Greece (Tsangridis 1982) probably concern another species.
- *Synechococcus sulphuricus* Dor 1967: Cells 6–24 × 1.8–2.4 µm, with yellowish content; described from thermal springs in Israel; (fig. **141**).
- *Synechococcus vulcanus* Copeland 1936: Cells 4–17 × 2–3 µm, with green to yellowish content, thermobiotic in hot springs (always over 40°C); morphologically and ecologically very similar also are *S. viridissimus* Copeland 1936 and *S. caldarius* Okada 1939; known from USA, Japan and India (?).
- Takamura & al. (1993) presented 4 *Synechococcus* species isolated from plankton of Japanese lakes, without taxonomic definition; two types are probably identical with *S. capitatus* and *S. rhodobaktron*, but one blue-green species with cells 2.3–8.7 × 0.5–0.7 µm represents probably a still unknown species.

**Unrevised European taxa:**
- *Synechococcus arcuatus* var. *calcicolus* Fjerdingstad 1965; (false Messikommer): Arcuate, kidney-shaped cells with pale olive-green, homogeneous content, 8 × 3.2 µm; described from limestone rocks (aerophytic, «Tintenstriche») in the «Alpes maritimes», S. France; surely different from the typical *S. arcuatus*, which is described from thermal springs in USA (in 55–64.1°C).

---

SYNECHOCOCCUS
**Fig. 142.** *Synechococcus lividus*: **a-d** – selected morphotypes described by Copeland (1936) as different taxa from Yellowstone Nat. Park (**a** – typical *S. lividus*, **b** – sub var. *siderophilus*, **c** – sub var. *curvatus*, **d** – sub *Synechococcus vulcanus*), **e** – part of a mat after Emoto & Yoneda (1941) from hot springs of Japan.

CYANOTHAMNOS
**Fig. 143.** *Cyanothamnos planctícus*: **a** – detail of cells, **b** – planktic colony, **c** – part of a colony; after Cronberg (1991).

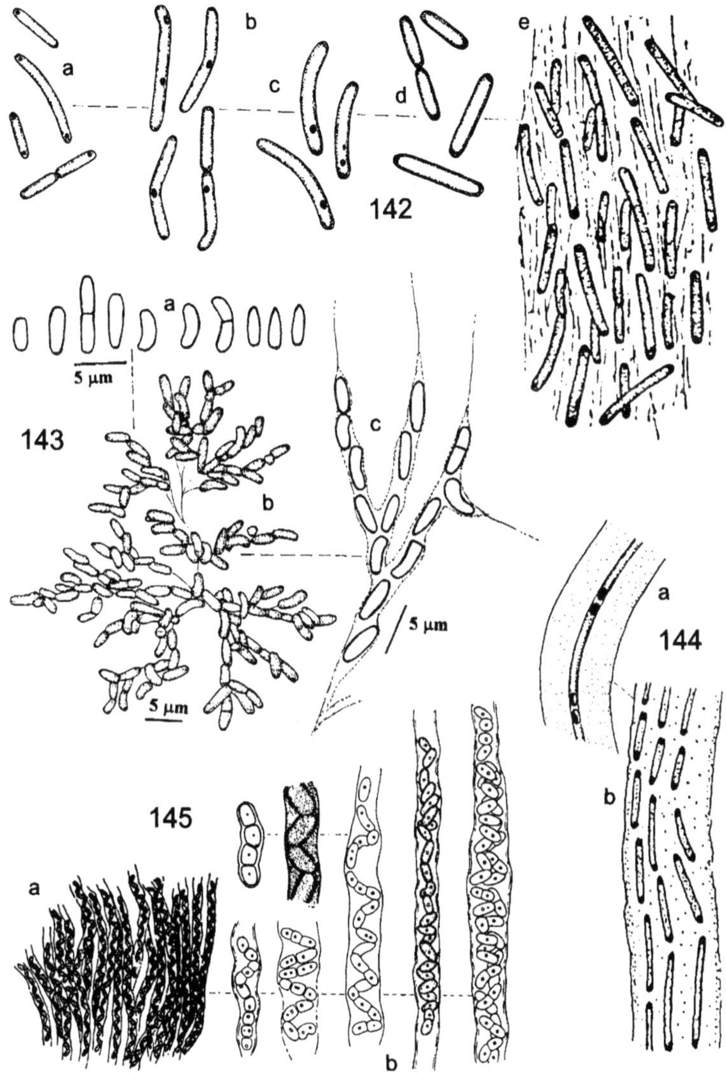


WOLSKYELLA

**Fig. 144.** *Wolskyella filamentosa*, described from S. Africa: after Claus (1963).

TUBIELLA

**Fig. 145.** *Tubiella elenkinii:* a – part of a colony (mat), **b** – details of pseudofilaments; after Hollerbach (1935).

– *Synechococcus roseo-persicinus* Grunow in Rabenhorst 1863: Cells red, probably a rhodobacterium.
– *Synechococcus* cf. *rayssae* (false *raysii*) Dor sensu Hindák 1983: From wet granitic rocks near the lake Morskie Oko, High Tatra Mts., Poland; evidently different species from thermal *S. rayssae.*
– Cyanelles, living endobiotically within cells of the rhizopod *Paulinella chromatophora* Lauterborn, correspond cytomorphologically and by the type of division to the genus *Synechococcus.* However, they never were found free living, and they occur probably exclusively as endosymbionts in *Paulinella*; (see, e.g., Lukavský & Cepák 1992).

## 6. **Cyanothamnos** Cronberg 1991

Colonies free-living, more or less spherical in outline, composed of a system of irregular and irregularly divaricate mucilaginous, indistinctly tube-like stalks radiating to all sides from the colonial center, containing short rows of rod-like cells, arranged more or less one after another; the appearance of a colony is like a densely branched bush with radially oriented branches. False branching of stalks arises after a displacement of divided cells from the original direction, by the following cell division they lie in a new direction and by formation of a new sheath. Mucilaginous stalks colourless, thin, fine, sometimes diffluent, enveloping cell rows; the slime sometimes overlaps the end cells and extends in thin threads. Cells oval to rod-shaped with rounded ends, without aerotopes. Cell division by binary fission, in the same plane in successive generations, perpendicular to the longitudinal axis; cells sometimes diverge from one another slightly after division and reach the original size and shape before next fission. Reproduction by disintegration of colonies. Monotypic genus.
**Important reference:** Cronberg (1991).

### 1. **Cyanothamnos plancticus** Cronberg 1991 (fig. **143**)

Colonies free-floating, more or less spherical and radially bush-like, slightly elongate, 25–55 × 15–35 µm, rarely composed of two subcolonies. Radial pseudofilamentous cell rows (usually with 4–6 cells), falsely branched and divaricate, are kept together by a thin, colourless, diffluent mucilage (staining !). The number of cells per colony varies greatly, they can be packed in the cell rows in old stages; the cells sometimes disappear in the colonial center. Cells elongate oval to cylindrical, rod-shaped, straight or sometimes slightly curved, with rounded ends, in the periphery of colonies more or less bluntly pointed, with pale green content without gas vesicles or prominent granules, 4–5 × 2–2.4 µm.
Occurrence: Freshwater, planktic in eutrophic lowland lakes; S. Sweden.

## 7. **Wolskyella** Claus 1963

Solitary pseudofilaments, simple, unbranched, to 12 µm wide, composed from thin or wide sheaths and (inside) from 1–4 straight, uniseriate rows of cylindrical cells; cells situated one after another, close to or distant from one another (if more rows are within sheaths, they are arranged parallel to one another). Sheaths smooth, structureless and homogeneous, colourless, or thin, firm on the marginal surface. Cells rod-shaped, long cylindrical, more or less rounded at the ends, usually several times longer than wide.
Cell division only transversely to the longitudinal axis, in the same plane in successive generations; reproduction by fragmentation of pseudofilaments and liberation of cells from sheaths (planocytes).

**Important references:** Claus (1963); Ehresmann & Claus (1968); Bourrelly (1970, 1985).

**Species outside Europe:**
- *Wolskyella filamentosa* Claus 1963: In sheaths 1–4 rows of cells, freshwater, S. Africa; (fig. **144**).
- *Wolskyella solitaria* Ehresmann et Claus 1968: Solitary pseudofilaments with sheaths, marine, USA (Long Island Sound).

## 8. Bacularia Borzì 1905

*Bacillosiphon* Copeland 1936

Colonies mucilaginous, elongate, in the form of tubular sheaths, more or less tapering and (sometimes) pointed (rarely rounded) or open at both ends; slime homogeneous, fine, colourless, enveloping numerous, more or less lengthwise and parallely to the longitudinal axis (axial core) oriented cells, but without regular arrangement (not in rows or in pseudofilaments); cells slightly distant from one another, only after division two together for a short time. Colonies sometimes joined into an amorphous, clathrate thallus. Cells cylindrical, rod-shaped, with rounded ends.
Cell division crosswise (transversely) to the long axis into two isomorphic daughter cells, which reach the original length before next division. Reproduction by liberation of solitary cells (planocytes ?) from the ends of tube-like sheaths and by disintegration of a colony.
**Important references:** Copeland (1936, sub *Bacillosiphon*); Geitler (1942); Frémy (1949); Bourrelly (1970, 1985).

**Key to the species:**
**1a** Cells 2.2–3.4 µm long, periphytic and planktic in stagnant waters, W. Siberia . . . . . . . . . . . . . . . . . . . . . . . . . . . . . . . . . . . . . . **1. B. vermicularis**
**1b** Cells 10–14 µm long, epilithic on marine coasts, Mediterranean region . . . . . . . . . . . . . . . . . . . . . . . . . . . . . . . . . . . . . . **2. B. caerulescens**

### 1. Bacularia vermicularis (Fedorov) Komárek et Anagnostidis 1995 (fig. 146)

*Rhabdoderma vermiculare* Fedorov 1969

Colonies microscopic, elongate, tube-like or slightly flattened; colonial mucilage homogeneous, colourless, delimited, but sometimes scarcely visible; mucilaginous sheaths sometimes polarized (one end of an elongate colony is more rounded after liberation from the substrate). Cells cylindrical, straight, pale blue-green, 2.2–3.4 × 0.8–1.5 µm.
Occurrence: Freshwater, peri- and metaphytic, attached to various substrates in lake littoral and in swamps, later secondarily planktic; Russia (W. Siberia, Omsk region).

### 2. Bacularia caerulescens Borzì 1905

*Aphanothece caerulescens* (Borzì) Geitler 1932 non A. Braun 1863

Colonies amorphous, sometimes irregularly clathrate, consisting of cylindrical tube-like, distinct, delimited, colourless sheaths, enveloping the cells. Cells long cylindrical, straight, pale blue, 10–14 × 2–3 µm.
Occurrence: Marine, on rocky coasts; Italy (Sicily).

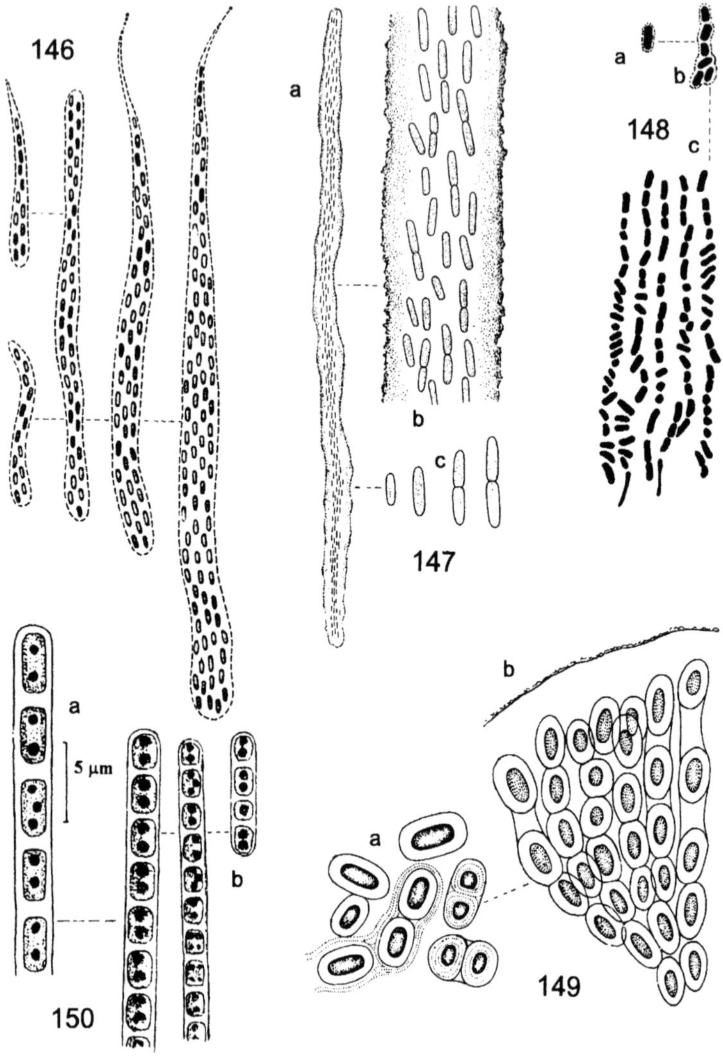


BACULARIA

**Fig. 146.** *Bacularia vermicularis*: after Fedorov (1969, sub *Rhabdoderma vermiculare*).

**Fig. 147.** *Bacularia thermalis*: after Frémy (1949).

RHODOSTICHUS

**Fig. 148.** *Rhodostichus expansus*: **a** – initial one-celled stage, **b** – young pseudofila-

**Species outside Europe:**

- *Bacularia gracilis* Komárek 1995: Cells rod-shaped, pale blue-green, 3.6–7 × 0.5–0.8 µm; described from metaphyton of alkaline lakes in Cuba.
- *Bacularia indurata* (Copeland) Frémy 1949 [syn.: *Bacillosiphon induratus* Copeland 1936]: Cells 7–15 × 1.8–2.2 µm, sheaths intensely incrusted by lime, strictly thermobiotic species, known only from hot springs (62–70° C) of USA (Yellowstone).
- *Bacularia thermalis* Frémy 1949: Filamentous, gelatinous colonies, up to 100 µm wide, attenuating to the apex, colourless or yellow-green; cells rod-shaped, straight or slightly arcuate, pale blue-green, with homogeneous content, 5–20 × circa 2.5 µm; described from thermal springs in Zaire (Ilehe); (fig. **147**).

## 9. **Tubiella** Hollerbach 1935

Colonies in the form of an amorphous, gelatinous mass, spreading over the substrate, composed of more or less cylindrical tubes, densely and more or less parallely arranged, sometimes wider in some sections than others, unbranched; sheaths fine and scarcely visible or with distinct margin, rounded or truncate at the ends, later open. Within the sheaths a uniseriate row (rarely 2–3 rows) of usually «zig-zag» disposed cells, rarely in straight, wavy or screw-shaped formations. Cells elongate, oval to cylindrical, enveloped by very fine, colourless, scarcely visible gelatinous envelopes (staining !), slightly distant from or touching one another.

Cell division transverse, in the same plane in successive generations, perpendicular to the longitudinal axis; reproduction by solitary cells liberated from gelatinous tubes. Monospecific genus.

**Important references:** Hollerbach (1935); Geitler (1942); Hollerbach & al. (1953); Bourrelly (1970, 1985); Kondrateva & al. (1984).

### 1. **Tubiella elenkinii** Hollerbach 1935 (fig. **145**)

Tube-like colonies blue-green, mucilaginous, thin, spreading over a substrate, 3.3–8.3 µm wide; cells initially in uniseriate rows, later forming irregular pseudofilaments, usually with «zig - zag» or screw-shaped, rarely irregular cell arrangement; in each tube usually only one row of cells, rarely two to three. Cells long oval or cylindrical (mainly 1.5–2.5 × longer than wide), with rounded ends and scarcely visible individual, diffluent, colourless gelatinous envelopes; cell content blue-green with usually clearly visible chromatoplasma, with 1 (2–3 - more) refractive granules of varying size, usually localized in cell center; cells (2.5) 4.2–5.2 (6.2) × (1.7) 2–2.5 (3) µm, sometimes involution cells up to 12 µm long are present.

---

ment, **c** - part of thallus with pseudofilamentous arrangement of cells; after Geitler & Pascher in Geitler (1931).

PSEUDONCOBYRSA

**Fig. 149.** *Pseudoncobyrsa lacustris*: **a** – unicellular or few-cellular stages after Skuja (1930), **b** - part of a thallus, after Kirchner in Schröter & Kirchner (1896, sub *Oncobyrsa lacustris*).

PALIKIELLA

**Fig. 150.** *Palikiella elegans*: after Claus (1962).

Occurrence: Terrestrial, isolated from sandy soils; known only from Russia (vicinity of Sankt Petersburg).

## 10. **Rhodostichus** Geitler et Pascher in Geitler 1931

Colonies widely spreading, consisting of one layer of erect, parallely and more or less densely arranged rows of cells, vertically oriented to the substrate and enveloped by common, fine mucilage and indistinct mucilaginous tube-like envelopes; cells in rows, slightly distant from one another, uniseriately organized one after another or (in short sections) oriented at right angles to the row. Mucilage in the basal parts of colonies more dense than elsewhere (there it forms an irregular net-like structure, seen from above), in the upper parts fine, homogeneous. Cells cylindrical, with rounded poles (never spindle-shaped).
Cell division transversely to the longitudinal axis, in the same plane in successive generations. Reproduction probably by the liberation of solitary cells. The classification into the *Entophysalidaceae* or *Hydrococcaceae* (near *Cyanodermatium*) is possible, considering the structure of colonies and basal cells (see also Bourrelly 1970).
**Important references:** Geitler (1931, 1942); Bourrelly (1970, 1985).

### 1. **Rhodostichus expansus** Geitler et Pascher in Geitler 1931 (fig. **148**)

Colonies flat, spreading over a stony substrate, crust-like, dull red, up to 50 µm thick; erect cell rows enveloped by common colourless, fine, structureless mucilage, dense near the pseudofilament bases, scarce near the colonial surface. Cells oval, short or long cylindrical, rod-shaped, straight or slightly curved, rounded at both ends, of varying length (also in one row), with pale reddish or pink (sometimes almost colourless), more or less homogeneous content, 2–3 (4) × 0.6 µm.
Occurrence: Freshwater, benthic on stony substrates in streams, in low light intensity (forest zone); Austria (Carinthia, Tirol).

## 11. **Pseudoncobyrsa** Geitler 1925

Colonies mucilaginous, more or less polarized, attached to stony substrates, gelatinous, more or less hemispherical or flattened, (to 2 or more mm high), with cells situated in distinctly «branched», divaricate, radially or vertically oriented rows, enveloped by gelatinous, tubular and confluent sheaths, and, moreover, cells with individual spherical, distinct, delimited or diffluent, non-stratified envelopes (staining !). The whole colony is embedded in homogeneous mucilage with distinct margin. Cells more or less spherical, ellipsoidal or cylindrical with rounded ends, sometimes up to 25 µm long.
Cell division probably only perpendicularly to the long axis of the radially oriented cells. Reproduction imperfectly described, probably by solitary cells. (Since the cell division proceeds irregularly in more than two planes, the genus belongs to the *Entophysalidaceae*).
**Important references:** Geitler (1925c, 1932, 1942); Skuja (1930).

### 1. **Pseudoncobyrsa lacustris** (Kirchner) Geitler 1925 (fig. **149**)

*Oncobyrsa lacustris* Kirchner in Schröter et Kirchner 1896

Colonies hemispherical, gelatinous, soft, later amorphous, to 2 mm in diameter, green, blue-green or dirty orange, with radially oriented, separate cells (only after division near to one another); the rows of cells are distinct in young colonies or at

the colonial margin, later arranged irregularly. Cells ellipsoidal or almost cylindrical, 15–25 (36) × 10–14 µm, with distinct individual, 3–6 µm thick colourless envelopes.
Occurrence: Freshwater, benthic or in littoral of clear lakes; Germany (Lake Constance), Estonia; little known species.

**Unrevised species (outside Europe):**
- *Chlorogloea microcystoides* sensu Bourrelly 1984: Colonies consisting of mucilaginous, colourless stalks, parallely arranged; within the stalks are rows of spherical, subspherical or oval cells, 2–3 µm in diameter, enveloped by their individual sheaths; in running water in New Caledonia.

## 12. **Johannesbaptistia** De Toni 1934

*Cyanothrix* Gardner 1927 (non *Cyanothrix* Schmidle 1897); *Heterohormogonium* Copeland 1936
Solitary pseudofilaments, unbranched or rarely laterally or pseudodichotomously divaricate (and occasionally anastomozing), straight or slightly wavy, consisting of one row of transversally located narrow-oval cells, situated perpendicularly to the longitudinal axis of the pseudofilaments. Mucilaginous envelopes tube-like, colourless, structureless, slightly diffluent or more or less distinct at the margin, rounded at the ends. Clear spaces are visible between cells, only after division the daughter cells remain together. Cells discoid, in lateral view transversally narrow-oval, pale grey-blue or blue-green with fine granular content without gas vesicles, apical cell rounded. Pseudofilaments sometimes contain numerous necridic cells.
Cell division in one plane, perpendicular to the longitudinal axis of pseudofilaments (i.e., transversely to the longer axis of cells); reproduction by disintegration of pseudofilaments and by facultative liberation of solitary cells.
**Important references:** Gardner (1927a, sub *Cyanothrix*); Taylor (1928); Copeland (1935, sub *Heterohormogonium*); Drouet (1936b, 1938); Daily (1946); Kiselev (1940, sub *Cyanothrix*); Iyengar & Desikachary (1946); Hollerbach & al. (1953); Drouet & Daily (1956); Desikachary (1959); Umezaki (1961); Srinivasan (1962); Bourrelly (1970, 1985); Islam (1985); Wenzel & de Halperin (1987).

### 1. **Johannesbaptistia pellucida** (Dickie) Taylor et Drouet 1938 (fig. **151**)

*Hormospora pellucida* Dickie 1874
Cells arranged in one row in solitary, unbranched, short or long cylindrical uniseriate pseudofilaments, to 300–600 (2000) µm long, (9) 13–20 (22) µm wide, more or less straight, wavy or irregularly slightly coiled; cells enveloped by wide mucilaginous, colourless and hyaline, homogeneous or rarely indistinctly lamellate envelopes, usually indistinct at the margin or (rarely) firm and distinct; solitary cells do not possess their own gelatinous envelopes (or rarely they have indistinct sheaths); cells distant, slightly apart from one another, sometimes there are necridic cells. Cells discoid or rounded lenticular, in colonial rows shorter than long, 1.5–5.5 (6.5) × (2) 6.6–8.3 (10) µm; apical cells rounded conical. Cell content pale blue-green or olive-green, homogeneous or finely granular.
Occurrence: Benthic and metaphytic among other algae in heliothermal freshwaters, mesohaline to saline (to 60‰) water biotopes, sometimes component of mats in salty seapages, perhaps also tychoplanktic in lakes; originally described from Brazil, known from the whole tropical zone, and also from Argentine, Caspian Sea, Balkan Peninsula, France, Greece, S. USA, etc.

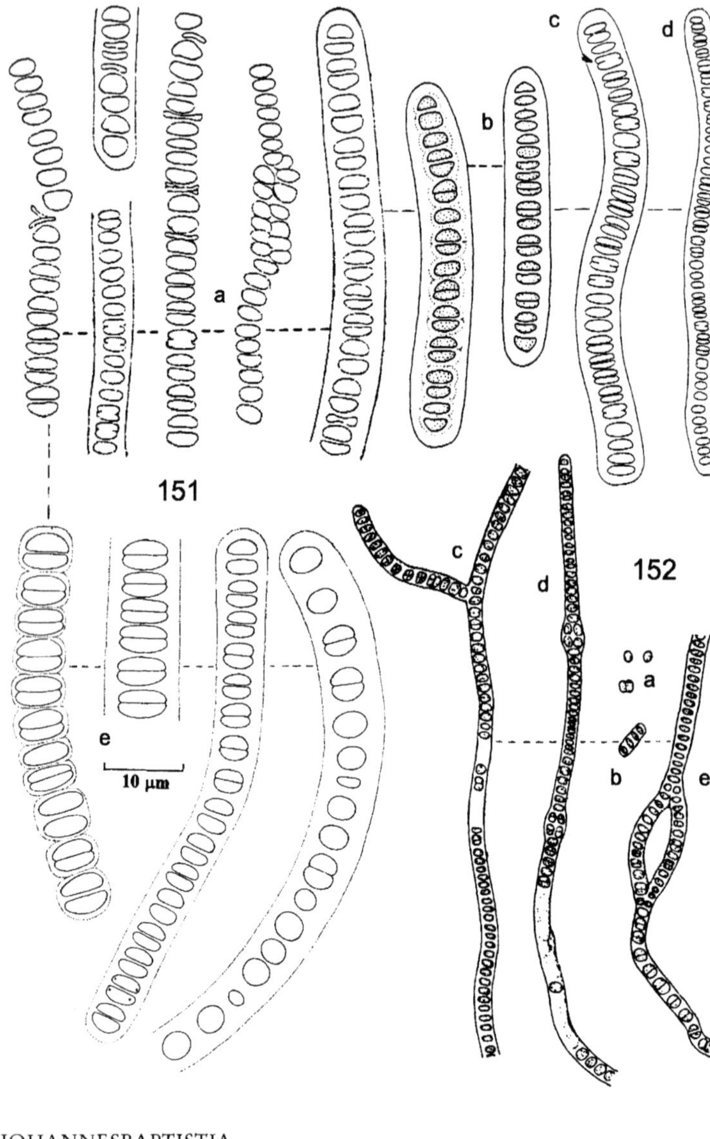


JOHANNESBAPTISTIA

**Fig. 151.** *Johannesbaptistia pellucida*, morphotypes according to various authors: **a** – after Kiselev from Hollerbach & al. (1953, sub *Cyanothrix gardneri* f. *caspica*), **b** – after Umezaki (1961, sub *J. pellucida*), **c** – after Gardner (1927, sub *Cyanothrix*

**Species outside Europe:**

- *Johannesbaptistia schizodichotoma* (Copeland) Komárek et Anagnostidis 1995 [syn.: *Heterohormogonium schizodichotomum* Copeland 1936]: Pseudofilamentous colonies ± 2 µm wide, to 1 mm long; cells spherical or compressed spherical, 1–2.5 × 1.2–1.6 µm; described from thermal mineral springs (28–42° C, pH = 6.5) in USA (Yellowstone); (fig. **152**).

**Unrevised species:**

Numerous taxa were described from the genus *Johannesbaptistia*, but all have to be revised. Both morphological and ecological diversities are remarkable. Drouet & Daily (1956) accepted only one variable species, *J. pellucida* with wide distribution. The described but still unrevised taxa are as follows:

- *Cyanothrix anabaeniformis* Kiselev 1940 [syn.: *C. gardneri* f. *anabaeniformis* (Kiselev) Kiselev 1947]: Pseudofilaments short to several mm long, cells 1.5–2.5 × 2.5–3.5 µm; benthic in shallow heliothermal lagoons with high salinity (35–60‰), described from NE. part of Caspian Sea.
- *Cyanothrix caspica* Kiselev 1940 [syn.: *C. gardneri* f. *caspica* (Kiselev) Kiselev 1947]: Pseudofilaments short to several mm long, straight or flexuous, cells 2.5–4 × 5.5–10 µm; benthic in shallow heliothermal lagoons with low salinity (up to 20‰), described from NE. part of Caspian Sea.
- *Cyanothrix cavanillesii* Gonzalez Guerrero 1946: Pseudofilaments curved, 8–50-celled; cells pale blue-green, spherical or subspherical, often two together, 2–3 × 4–5.5 µm; described from a muddy habitat in the estuary of the river Guadalete, Spain.
- *Cyanothrix gardneri* (Frémy) Kiselev 1947 [syn.: *Johannesbaptistia gardneri* Frémy 1935]: The common name (later synonym) used originally by Frémy (1935) for both the Gardner's species *C. primaria* and *C. willei*. The Kiselev's recombination was used later by himself (1947) and Hollerbach et al. (1953) with further amplification of this taxon (incl. *C. caspica, C. anabaeniformis* and *Nodularia ?fusca* Taylor 1945); this wide concept corresponds in principle with that of Drouet & Daily (1956; *Johannesbaptistia pellucida*) and Srinivasan (1962).
- *Cyanothrix primaria* Gardner 1927 [syn.: *Johannesbaptistia primaria* (Gardner) De Toni 1934]: Pseudofilaments about 300 µm long, straight or more or less flexuous, 18–22 µm wide; cells 3.6–5 × 10–15 µm; envelopes thick, firm, hyaline; described from Porto Rico (metaphyton of freshwater lakes).
- *Cyanothrix willei* Gardner 1927 [syn.: *Johannesbaptistia willei* (Gardner) De Toni 1934]: Pseudofilaments about 200 µm long, 9–11 µm wide, cells 3.5–4.5 × 6–6.5 µm; envelopes almost transparent, about 2 µm thick; described from Porto Rico (metaphytic among algae in freshwater lakes).
- *Geminella scalariformis* f. *marina* G. S. West in Collins & Hervey 1917: The type form belongs to green algae, but the f. *marina* is probably identical with *Johannesbabtistia*.
- *Heterohormogonium* sp. in Karim 1968.
- *Johannesbaptistia pellucida* sensu Wenzel et Halperin 1987: Pseudofilaments (6) 7–8 µm wide, cells 2.5–3 × 3.5–4 (5) µm; Argentine (Tierra del Fuego).
- *Johannesbaptistia whitfordii* Thomas et Gonzalves 1965: Pseudofilaments more than 5 mm long and 20 µm wide, cells greenish-blue, disc-shaped, 2.5–3 (4) ×

---

*willei*), **d** – after Gardner (1927, sub *Cyanothrix primaria*), **e** – after Hindák (1988, sub *J. pellucida*).

**Fig. 152.** *Johannesbaptistia schizodichotoma*, from Yellowstone thermal springs: after Copeland (1936, sub *Heterohormogonium schizodichotomum*).

10.8–12 µm, mostly in pairs; mucilaginous envelope hyaline, more or less indistinct. Among filaments of *Spirogyra* at 32–40°C, described from thermal springs, W. India (Tuwa).
– *Johannesbaptistia* spp. in Potts 1980: Cells 7.4 µm (wide ? ); found in shallow water mats in Solar Lake and other saline coastal pools and ponds in Egypt (Sinai Peninsula).

Notes to the family **Synechococcaceae:**
Into the family *Synechococcaceae* (subg. *Synechococcoideae*), belongs probably also the genus
**Palikiella** Claus 1962, with two insufficiently defined species, *P. hungarica* Claus (type species) and *P. elegans* Claus (fig. **150**), described from wetted wood and calcareous stones in a cave Abaliget, Hungary (not found again): Solitary pseudofilaments composed from uniseriately arranged cells slightly distant from one another, and firm, colourless sheaths, closed at both ends. Cells bluish, with distinguishable centro- and chromatoplasma, and few prominent granules.

# 3. Family **Merismopediaceae** Elenkin 1933

Coccal, living as solitary cells or in irregular, flat (tabular, consisting of one layer of cells), or spherical mucilaginous colonies. Mucilage mainly colourless, structureless, but in spherical colonies sometimes with the development of a special central stalk system; around cells sometimes and facultatively simple, fine, individual gelatinous envelopes. Cells spherical, obovate, oval or rod-shaped.
Cell division by binary fission, always in two planes perpendicular to one another (in elongated cells crosswise and lengthwise); cells reach more or less the original form and size before next division; nanocyte formation occurs very rarely. Reproduction by disintegration of colonies, rarely by solitary cells.

**Key to the subfamilies:**
**1a** Cells solitary or in irregular or tabular (in one layer) colonies . . . . . . . . . . . . . . . . . . . . . . . . . . . . . . . . . . . . . . **1. Merismopedioideae**
**1b** Spherical colonies with cells arranged radially and more or less peripherally, sometimes with central mucilaginous stalk system . . . . . . . . . . . . . . . . . . . . . . . . . . . . **2. Gomphosphaerioideae** (p. 192)

## 1. Subfamily **Merismopedioideae**

*Holopediaceae* Elenkin 1933 incl.; *Beckiaceae* Elenkin 1933 incl.; *Planimetrinales* Starmach 1966

Cells solitary or in colonies. Colonies microscopic, rarely visible macroscopically, usually free-living, rarely attached to the substrate, irregular or flat (with cells in one, more or less tabular layer, irregularly arranged or in perpendicular rows). Cells spherical, rarely obovate, oval or elongate to rod-shaped, in several species surrounded by individual gelatinous envelopes. In the cell wall of numerous species of the *Merismopedioideae* with simple morphology (from the genera *Aphanocapsa* and *Synechocystis*) a special hexagonal «S-layer» (p6) was found (fig. **154**; Šmarda & al. 1979, Šmarda 1991).
Cell division by binary fission in two planes perpendicular to each other in successive generations. Reproduction by disintegration of colonies.

**Key to the genera:**

1a Solitary, free-living cells . . . . . . . . . . . . . . . . . . . . . . . . **1. Synechocystis**
1b Cells in colonies or in groups, sometimes attached to the substrate . . . . . 2
2a Colonies mucilaginous, irregular, formless, with irregularly arranged cells . . . . . . . . . . . . . . . . . . . . . . . . . . . . . . . . . . . . . . . . . . **2. Aphanocapsa**
2b Colonies always flattened, but sometimes wavy or arcuate, usually consisting of a single cell layer, free-living, or rarely attached to the substrate . . . 3
3a Colonies sessile on filamentous algae, consist of a single layer of cells; cells divide perpendicularly to the substrate . . . . . . . . . . . . . . . **8. Mantellum**
3b Colonies free-living, not attached to algae (but sometimes growing on stones or rocks) . . . . . . . . . . . . . . . . . . . . . . . . . . . . . . . . . . . . . . . 4
4a Cells elongate, oriented by their longer axis perpendicularly to the plane of the colony . . . . . . . . . . . . . . . . . . . . . . . . . . . . . . **7. Microcrocis**
4b Cells spherical; if slightly elongate, they are arranged with their longer axis in the plane of the colony . . . . . . . . . . . . . . . . . . . . . . . . . . . . . . 5
5a Cells irregularly arranged in the plane of the colony . . . . . . . . . . . . . . . 6
5b Cells arranged in the plane of the colony in perpendicular rows . . . . . . . 7
6a Colonies tabular, flat . . . . . . . . . . . . . . . . . . . . . . . . . . . **3. Coccopedia**
6b Colonies irregular in outline, lobate, wavy or sometimes globular arcuate . . . . . . . . . . . . . . . . . . . . . . . . . . . . . . . . . . . . . . . . . . **4. Pannus**
7a Cells spherical or hemispherical after division, rarely more or less rounded rectangular . . . . . . . . . . . . . . . . . . . . . . . . . . . . . . **5. Merismopedia**
7b Cells elongate and slightly ovoid, forming characteristic rectangular groups within the colonies . . . . . . . . . . . . . . . . . . . . . . . . . . **6. Cyanotetras**

## 1. Synechocystis Sauvageau 1892

Cells solitary, spherical, after division hemispherical and for a short time two together, never forming colonies, without mucilage or (mainly) with narrow, fine, colourless and usually diffluent and indistinct mucilaginous envelopes. Cells spherical or widely oval. In several strains the cell walls contain a special «S-layer» with characteristical hexagonal (p6) substructure (Šmarda et al. 1979; Vaara 1982; etc.) (fig. **154**). The stability, function and taxonomic significance of this layer is not yet clear. Different but stable pigment ratios in related species (and populations) have been found (Kováčik in litt.).

Cell division by binary fission (mainly by pinching) into two morphologically equal, hemispherical daughter cells, which reach the original globular shape before next division; cells always divide in two perpendicular planes in successive generations (fig. **153**). If the mucilaginous envelopes are formed around cells, they split together with dividing cells.

**Important references:** Geitler (1960); Anagnostidis (1961); Starmach (1966); Komárek (1976); Šmarda & al. (1979); Vaara (1982); Kováčik (1983, 1988); Komárek & Anagnostidis (1986); Hernandéz-Mariné & al. (1990).

**Key to the species:**

1a Cells usually smaller than 1 µm in diameter, rarely up to 1.4 µm . . . . . . . 2
1b Cells greater than 1 µm in diameter, usually more than 1.5 µm . . . . . . . . . 5
2a Aerophytic species from wet calcareous rocks . . . . . . . . . **1. S. primigenia**
2b Species from submerged biotopes, mainly from the sapropel of lakes and tychoplankton . . . . . . . . . . . . . . . . . . . . . . . . . . . . . . . . . . . . . . 3
3a Cells without mucilaginous envelopes . . . . . . . . . . . . . . . . . . . . . . . . 4
3b Cells with gelatinous envelopes, 0.7–1 µm in diameter . . . . **3. S. bourrellyi**
4a Cells 0.7–0.8 µm in diameter . . . . . . . . . . . . . . . . . . . . . . **2. S. parvula**
4b Cells 1–1.4 µm in diameter . . . . . . . . . . . . . . . . . . . . . . **4. S. diplococca**

5a Epiphytic species, attached to the surface of or living within the mucilage of other algae . . . . . . . . . . . . . . . . . . . . . . . . . . . . . . . . . . . . 6
5b Free-living or epizoic (endozoic) species . . . . . . . . . . . . . . . . . . . . . . . 8
6a Endogloeic in mucilage of planktic cyanoprokaryotes, cells 1–2.9 µm in diameter . . . . . . . . . . . . . . . . . . . . . . . . . . . . . . . **5. S. endobiotica**
6b In sheaths and mucilage of non-planktic algae . . . . . . . . . . . . . . . . . . . . 7
7a In sheaths of *Phormidium*-like cyanoprokaryotes; cells more or less 1.5 µm in diameter . . . . . . . . . . . . . . . . . . . . . . . . . . . . . . . . . . **6. S. minima**
7b In mucilage of aerophytic algae on limestone rocks; cells 2.5–3.5 µm in diameter . . . . . . . . . . . . . . . . . . . . . . . . . . . . . . . . . **11. S. pevalekii**
8a Species from mineral and thermal springs (or similar biotopes) . . . . . . . . 9
8b Species free-living or epizoic in diverse freshwater, brackish and saline water biotopes . . . . . . . . . . . . . . . . . . . . . . . . . . . . . . . . . . . . . . . . 10
9a Cells 1.2–2 µm in diameter, with clearly recognizable chromatoplasma, without individual gelatinous envelopes . . . . . . . . . . . . . . **7. S. thermalis**
9b Cells 1.6–2.9 µm in diameter, with homogeneous cell content, with individual gelatinous envelopes . . . . . . . . . . . . . . . . . . . . . . **8. S. minuscula**
10a Free-living species; mainly with blue-green cells . . . . . . . . . . . . . . . . . 11

SYNECHOCYSTIS

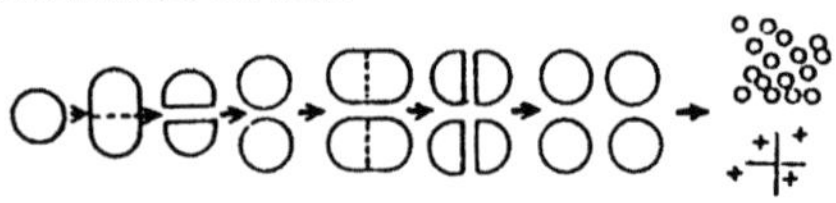

**cell division**
**(in two perpendicular planes)**

KOVÁČIK 1988

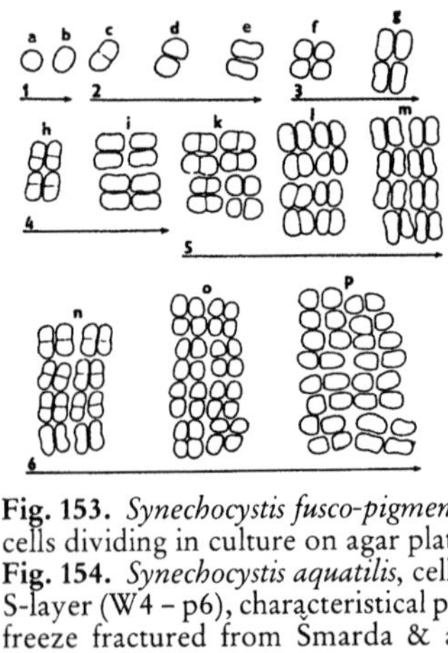


**cell division**
**on agar plate**

KOMAREK 1970

**Fig. 153.** *Synechocystis fusco-pigmentosa*, strain Hindák 1965/23: microphotos of cells dividing in culture on agar plate; (orig. Kováčik).
**Fig. 154.** *Synechocystis aquatilis*, cell wall layers with visible crystalline hexagonal S-layer (W4 – p6), charącteristical particularly for various *Synechocystis* strains: **a** – freeze fractured from Šmarda & al. (1979), **b** – scheme from Karlsson & al. (1983).

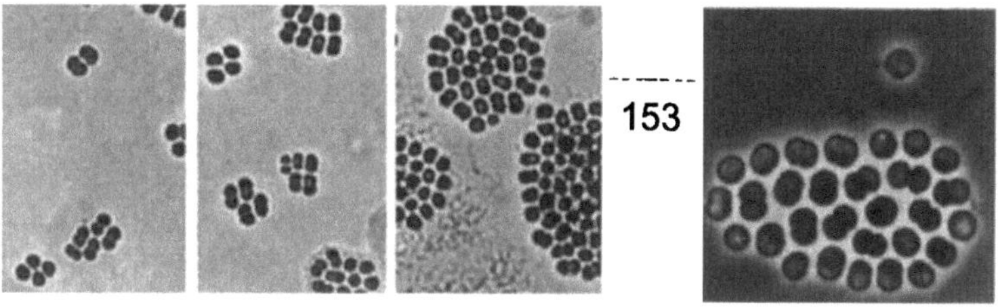

153

a

154

a

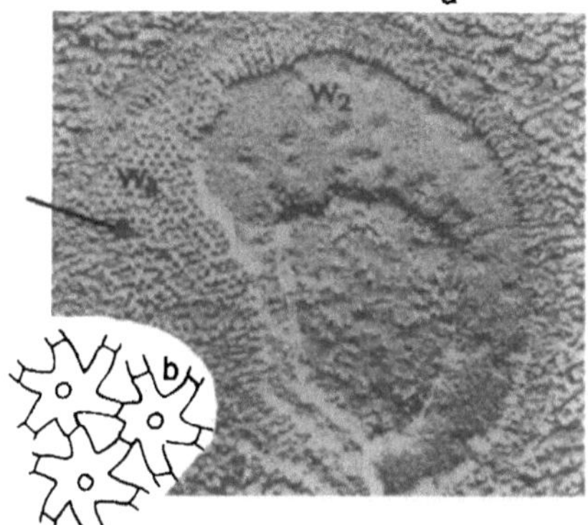


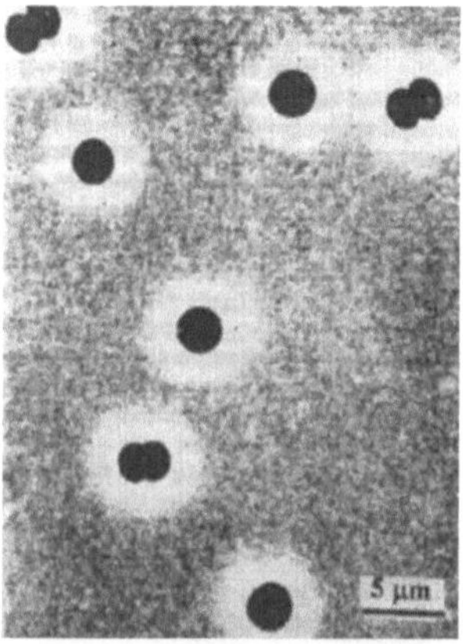


a

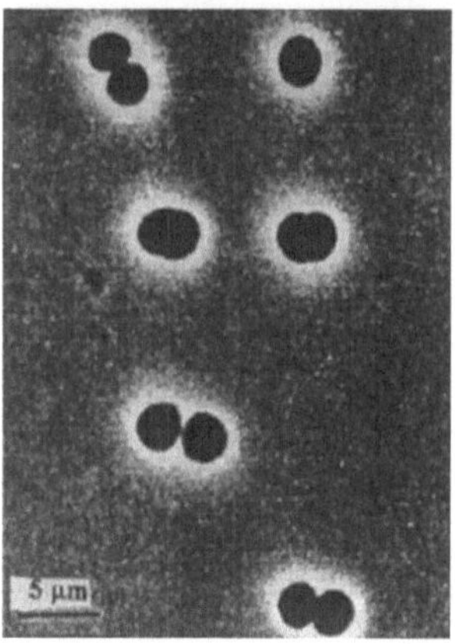


b

c

155

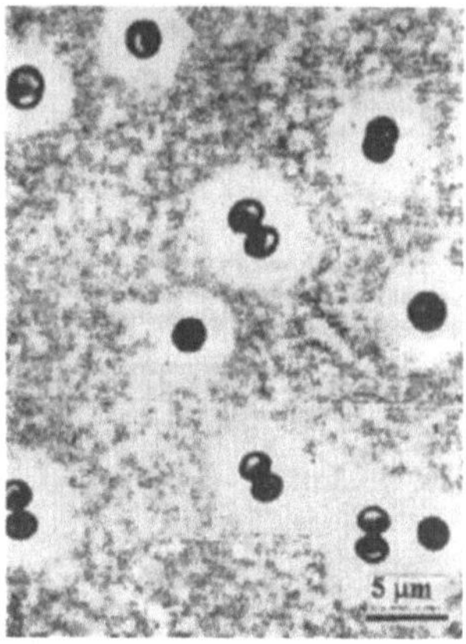


**Fig. 155.** Three examples of various *Synechocystis* clusters, differing in GC-ratios and ecology (according to strain origin) from Waterbury & Rippka (1989, Bergey's Manual): **a** – reference strain (WH 8501) of the «Marine-cluster of the *Synechocystis*-group», **b** – reference strain (PCC 6308) of the «low GC-cluster of the *Synechocystis*-group», **c** – reference strain (PCC 6714) of the «high GC-cluster of the *Synechocystis*-group».

10b Epi- and endozoic species (on marine ascidians); red cells . **14. S. trididemni**
11a Cell diameter always less than 10 μm . . . . . . . . . . . . . . . . . . . . . . . . . . **12**
11b Cell diameter (7) 10–29 (41) μm . . . . . . . . . . . . . . . . . . . . . . . . . . . . . **15**
12a Cells (1.8) 3–6 (7) μm in diameter . . . . . . . . . . . . . . . . . . . . . . . . . . . . **13**
12b Cells 6–9.6 μm in diameter . . . . . . . . . . . . . . . . . . . . . . . . . . . . . . . . . **14**
13a Cells (1.8) 2.2–4.5 μm in diameter; in metaphyton and plankton of biotopes with higher salinity, brackish waters and inland saline water bodies . . . . . . . . . . . . . . . . . . . . . . . . . . . . . . . . . . . . . . . **9. S. salina**
13b Cells (3) 4–7 μm in diameter; in plankton and metaphyton of freshwater biotopes . . . . . . . . . . . . . . . . . . . . . . . . . . . . . . . . **10. S. aquatilis**
14a Species from small pools under wet stone walls and rocks, subaerophytic; cells 7.5–8 μm in diameter . . . . . . . . . . . . . . . . . . . . . . . . **12. S. buzasii**
14b Species from metaphyton and plankton of saline lakes; cells 6.4–9.6 μm in diameter . . . . . . . . . . . . . . . . . . . . . . . . . . . . . . . . . . **13. S. crassa**
15a Cells (7) 10–15 μm in diameter; planktic in northern clear lakes and in moorland waters . . . . . . . . . . . . . . . . . . . . . . . . **15. S. septentrionalis**
15b Cells 14–29 (41) μm in diameter; planktic and in metaphyton of moorland or peaty lakes . . . . . . . . . . . . . . . . . . . . . . . . . . . . . . . **16. S. sallensis**

## 1. **Synechocystis primigenia** Gardner 1927 (fig. **156**)

Cells perfectly globular, without visible gelatinous envelopes, pale blue-green, 0.7–1 μm in diameter.
Occurrence: Aerophytic on calcareous rocks and walls; described from Porto Rico, but probably with cosmopolitan distribution.

## 2. **Synechocystis parvula** Perfiliev 1923

Cells globular, solitary or two together, with pale blue-green content and very thin, colourless mucilaginous envelope (?), 0.7–0.8 μm in diameter.
Occurrence: Sapropelic and tychoplanktic in lakes; recorded from Russia (near Sankt Petersburg, central Asia, W. Siberia) and Bulgaria (lake Urdino).

## 3. **Synechocystis bourrellyi** Komárek 1976

*Synechocystis minima* Bourrelly 1953

Cells globular to widely elliptic, solitary, in great quantity (in culture) deep blue-green, with distinct chromatoplasma, embedded in mucilaginous envelopes, 0.7–1 μm in diameter.
Occurrence: Described from cultures (Paris).

## 4. **Synechocystis diplococca** (Pringsheim) Bourrelly 1970 (fig. **157**)

*Tetrachloris diplococcus* Pringsheim 1953

Cells globular, solitary or (after division) in twos, pale blue-green, without mucilage, 1–1.4 μm in diameter (during division to 2.4 μm long).
Occurrence: Known from cultures, from benthic sample of mud (England?), later recorded from fissures of rocks and stones influenced by steam exhalations of thermal volcanic fumaroles (Greece).

**5. Synechocystis endobiotica** (Elenkin et Hollerbach) Elenkin et Hollerbach in Elenkin 1938 (fig. **158**)

*Aphanocapsa endophytica* G.M. Smith 1920 (?); *Synechococcus endobioticus* Elenkin et Hollerbach 1923; *Microcystis endophytica* (G.M. Smith) Elenkin 1938

Cells solitary or in small groups, globular, with greyish or pale blue-green, homogeneous content; cells are usually situated near the colonial surface, but their occurrence inside the colony is also known; around the cells fine, gelatinous envelopes, confluent with the mucilage of the host cyanoprokaryotes; cells (1)1.8–3 µm in diameter, before division elongated, up to 4 µm in diameter.
Occurrence: Endogloeic in mucilage of freshwater planktic cyanoprokaryotes (*Chroococcus, Microcystis, Woronichinia*); sporadically in diverse localities over the whole temperate zone.

**6. Synechocystis minima** Voronichin 1927

Cells solitary, globular, pale blue-green, 1.5 µm in diameter.
Occurrence: In sheaths of *Phormidium corium*; known only from N. Caucasus and W. Siberia.

**7. Synechocystis thermalis** Copeland 1936 (fig. **159**)

Cells globular with blue-green homogeneous content, with distinct chromatoplasma, without individual mucilaginous envelopes around cells; cells (1.2) 1.5–1.8 (2) µm in diameter.
Occurrence: In acidic (?) thermal springs, probably only at high temperatures (28.2 ?) 62–69 °C; recorded from hot springs in USA, Japan, India and Greece; the relation to *S. minuscula* should be elucidated.

**8. Synechocystis minuscula** Voronichin 1926 (fig. **160**)

Cells solitary, globular, with blue-green homogeneous content, without visible chromatoplasma, with very thin gelatinous individual hyaline envelopes, 1.6–2.5 (2.9) µm in diameter.
Occurrence: In alkaline mineral and thermal springs and pools, sometimes also in saline swamps, probably in temperatures of up to about 50 °C (records for up to 72 °C are suspect); recorded sporadically from diverse localities in Europe, Asia and N. America. Identity with tropical populations (Brazil; Senna 1988) should be revised.

**9. Synechocystis salina** Wislouch 1924 (fig. **161**)

*Synechocystis aquatilis* var. *minor* Geitler 1935; *S. aquatilis* f. *salina* (Wislouch) Komárek 1976

Cells spherical, solitary or two together, pale blue-green, with homogeneous content, (1.8)2.2–4(4.5) µm in diameter; in cultures up to 6.5 µm diameter is recorded.
Occurrence: Planktic and metaphytic in salty and alkaline waters, sometimes in mass development (green colour of water); known from the whole Europe and central Asia.

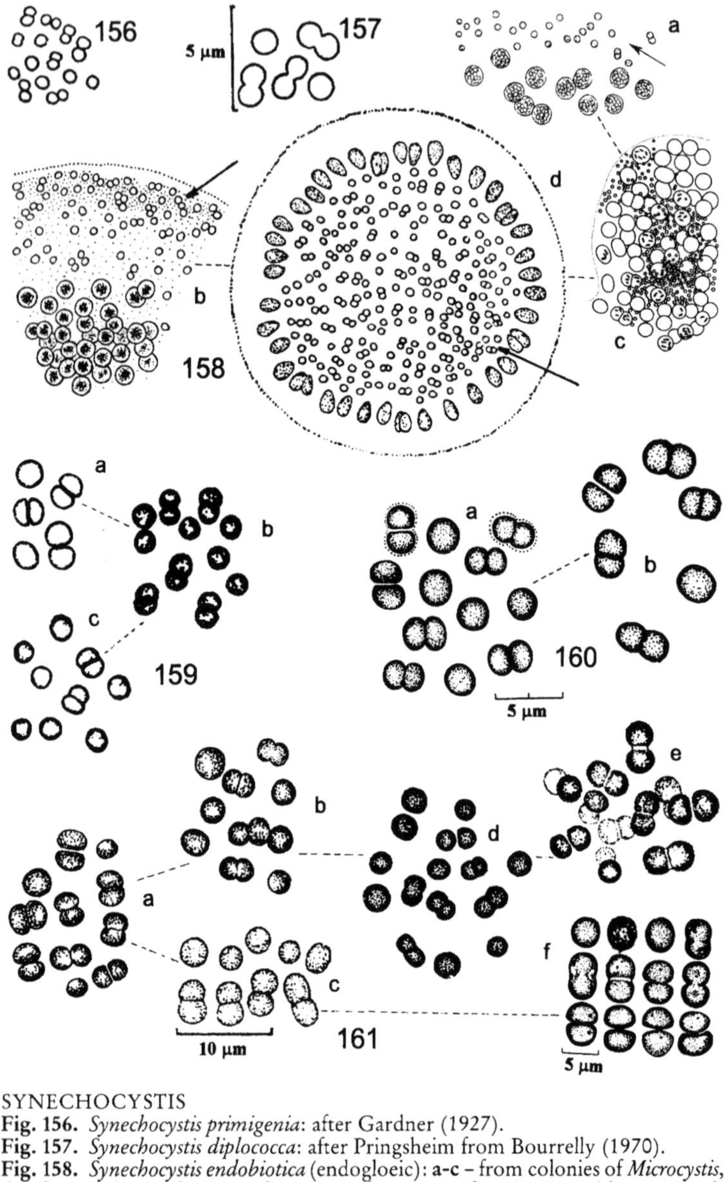


SYNECHOCYSTIS
**Fig. 156.** *Synechocystis primigenia*: after Gardner (1927).
**Fig. 157.** *Synechocystis diplococca*: after Pringsheim from Bourrelly (1970).
**Fig. 158.** *Synechocystis endobiotica* (endogloeic): **a-c** – from colonies of *Microcystis*, **d** – from colony of *Woronichinia naegeliana*; **a** – after G.M. Smith (1920, sub *Aphanocapsa endophytica*), **b** – after Skuja (1926, sub *Aphanocapsa endophytica*),

## 10. **Synechocystis aquatilis** Sauvageau 1892 (fig. **154, 162**)

Cells globular or widely oval, solitary or two together, with pale blue-green content, sometimes with thin individual mucilaginous envelopes (?), (3)4.5–7(7.8 ?) µm in diameter; according to several authors the cells are motile.
Occurrence: Planktic and metaphytic in small freshwater bodies and ponds, usually slightly polluted or with a relatively high content of nutrients; probably cosmopolitan, all over Europe, but not frequently. Records from aerophytic habitats are wrong.

## 11. **Synechocystis pevalekii** Ercegović 1925 (fig. **163**)

Cells solitary, spherical, after division hemispherical, with blue-green homogeneous content, 2.4–3.5 µm in diameter.
Occurrence: Subaerophytic, in mucilage of epilithic algae, on calcareous rocks; described from Croatia, but more widely distributed, particularly in central and S. Europe; records from aquatic biotopes are questionable.

## 12. **Synechocystis buzasii** Palik 1948 (fig. **164**)

Cells solitary, spherical, pale blue-green, without mucilaginous envelopes, 7.5–8 µm in diameter.
Occurrence: In mineral and warm (49 °C) springs; Hungary (Balint spring).

## 13. **Synechocystis crassa** Voronichin 1929

Cells spherical, solitary or in mucilaginous groups with up to 20 cells (?) with individual fine, diffluent and indistinct gelatinous envelopes, 6.4–10 µm in diameter.
Occurrence: In salty and mineral lakes; distributed sporadically all over the Russia and central Asia.

## 14. **Synechocystis trididemni** Lafargue et Duclaux 1979

*Synechocystis didemni* var. *rosea* Lewin 1975

Cells spherical, red, 7.5–14 µm in diameter, without any mucilaginous envelopes. Thylakoids (3–5) arranged in peripheral bands, parallely to the cell wall.
Occurrence: Marine, epizoic in Ascidians; tropical Atlantic, probably also widely distributed in the Mediterranean Sea (W. part, Adriatic Sea); about variability of Mediterranean morphotypes see Hernandéz-Mariné & al. (1990).

---

c – after Hindák & Moustaka (1988, sub *Aphanocapsa endophytica*), **d** – after Elenkin & Hollerbach from Starmach (1966).
**Fig. 159.** *Synechocystis thermalis*: **a** – after Copeland (1936), **b** – after Emoto & Yoneda (1940), **c** – from Chu & al. edit. (1991).
**Fig. 160.** *Synechocystis minuscula*: **a** – after Anagnostidis (1961), **b** – from Chu & al. edit. (1991).
**Fig. 161.** *Synechocystis salina*: **a-b** – after Anagnostidis (1961, **a** – sub *S. aquatilis*), **c** – after Komárek (1976), **d** – after Skuja from Starmach (1966), **e** – after Hortobágyi (1963), **f** – after Hindák (1968); (usually sub *S. aquatilis* var./f. *salina*).

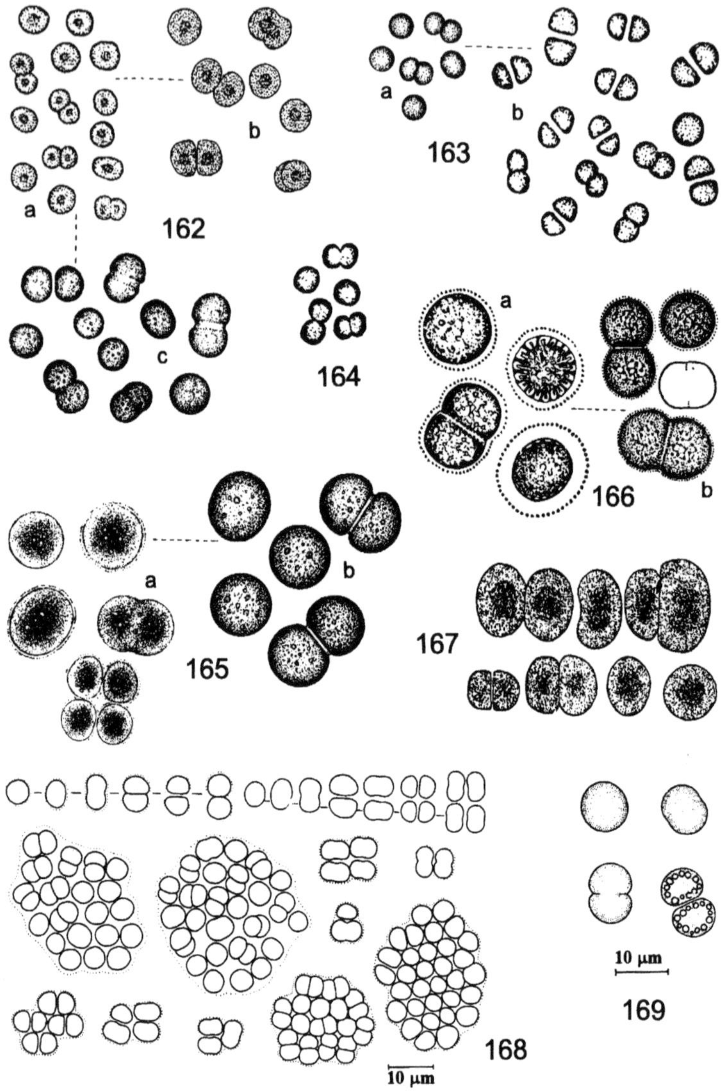


SYNECHOCYSTIS

**Fig. 162.** *Synechocystis aquatilis*: **a** – after Sauvageau (from Starmach 1966), **b** – after G.M.Smith (1950), **c** – after Skuja (1956).

**Fig. 163.** *Synechocystis pevalekii*: **a** – after Skuja (1964), **b** – redrawn iconotype of Ercegović from Kondrateva & al. (1984).

**15. Synechocystis septentrionalis** Skuja 1956 (fig. **165**)

Cells spherical or widely elliptical, after division more or less hemispherical and two or four together, pale blue-green or greyish with solitary granules with very thin, fine and diffluent individual envelopes, (7) 10–15 µm in diameter.
Occurrence: Planktic in clear lakes and peaty waters; N. Sweden, Finland, Switzerland.

**16. Synechocystis sallensis** Skuja 1929 (fig. **166, 167**)

*Synechococcus euryphyes* Beck-Mannagetta 1926 ?

Cells spherical, solitary or two together after division, with pale blue-green content, with broad or narrow mucilaginous, fine and colourless envelopes, (12) 14–30 µm in diameter.
Occurrence: Metaphytic and secondarily tychoplanktic in swamps, lowland and upland moorland waters and in the littoral of clear, cold lakes, among water plants and mosses, usually at low pH; Sweden, Baltic states, northern Russia, Greenland, central Asia and Carpathian mountains (Slovakia, Roumania, Hungary ?), probably also S. Argentine (Tierra del Fuego). The specimens from Brazil (Senna 1992) belong probably to another species (morphology, ecology).

**Species outside Europe:**
- *Synechocystis consortia* Norris 1967: Marine, endobiotic in dinoflagellates.
- *Synechocystis fuscopigmentosa* Kováčik 1988: Tropical swamps (Cuba, SE. Asia, etc.); (fig. **153, 168**).
- *Synechocystis limnetica* Popovskaja 1968: Planktic in lake Baikal.
- *Synechocystis major* (Geitler) Komárek et Anagnostidis 1995 [syn.: *Synechocystis crassa* var. *major* Geitler 1933]: Cells 7.5–10 µm in diameter, known only from metaphyton of tropical thermal waters; (Geitler & Ruttner 1935); (fig. **169**).
- *Synechocystis planctonica* Proškina-Lavrenko 1951: Planktic in Azovian Sea, Sivas Bay.
- *Synechocystis* – group, «a. Marine-cluster» sensu Waterbury & Rippka 1989 (Bergey's Manual): Obligatory photoautotroph, immotile, with phycoerythrin predominating (reddish cells), thermophilic (26–32 °C); isolated from plankton of tropical Atlantic Ocean; probably a separate *Synechocystis* species not yet validly defined; (fig. **155a**).

Note: *Synechocystis didemni* Lewin 1975 was recognized as a chloroprokaryotic species (Prochlorophytes).

---

**Fig. 164.** *Synechocystis buzasii*: after Palik (1949).
**Fig. 165.** *Synechocystis septentrionalis*: **a** – after Skuja (1956); **b** – after Skuja (1964).
**Fig. 166.** *Synechocystis sallensis*: **a** – after Skuja (1948), **b** – after Skuja from Hollerbach & al. (1953).
**Fig. 167.** *Synechocystis sallensis?* (= *Synechococcus euryphyes*): after Beck-Mannagetta (1926).
**Fig. 168.** *Synechocystis fusco-pigmentosa*, from agar-culture: after Kováčik (1988).
**Fig. 169.** *Synechocystis major*: after Geitler (1933).

**Unrevised taxa:**

- *Aphanocapsa montana* Cramer sensu Lewin et Robinson 1979: Spherical cells (2–4 µm in diameter) living in hollow hairs of polar bears in zoological gardens (USA, Germany, Uruguay); very probably a *Synechocystis* species.
- *Synechocystis spirogyrae* Skvorcov 1946.
- *Synechocystis willei* Gardner 1927: Cells 3.5–3.8 µm in diameter; metaphytic in pools in Porto Rico; probably identical with *S. aquatilis*.
- Waterbury & Rippka (1989, Bergey's Manual) recorded several different *Synechocystis* clusters differing by GC-ratios and by their origin (ecology); (fig. **155**).
- Andreoli & al. (1989) described insufficiently three or four types of picoplanktic cyanoprokaryotes from brackish waters in Valles Pozzatini near the Po river delta (Italy – Venetia) with rounded cells, peripherally localized thylakoids and more or less narrow mucilaginous envelopes, which could be classified into the genus *Synechocystis*. However, the type of division is not recorded and the correct taxonomic position will have to resolved be in the future: (i) Spherical cells 1–1.8 µm in diameter, with distinct, delimited, 0.1–0.2 µm wide mucilaginous envelope; (ii) subspherical to oval cells, 0.6–1.2 µm long, enveloped by irregular, electron-dense gelatinous sheath in form of small, separated gelatinous clusters (and probably sometimes grouped together in small colonies); (iii) spherical cells about 0.4–0.5 µm in diameter, with fragmented peripheral thylakoid and very narrow, indistinct, diffluent mucilaginous envelope.
- Another picoplanktic, cyanoprokaryotic population of Andreoli & al. (1989) was recorded from the brackish lagoon of S. Gilla (Italy – Sardinia); the cells were spherical, 0.4–0.7 µm in diameter, with one peripheral thylakoid and an about 0.01–0.05 µm wide mucilaginous envelope with interesting hexagonal structure (reminiscent of the «S-layer» ?); insufficiently described, nothing is known about cell division, revision is needed.
- Corpe & Jensen (1992) described the fine structure of several morphotypes from the picoplankton of the small oligotrophic Lake Arthur in NE. USA: (i) Fig. 3A – cells more or less 0.7 µm in diameter, thylakoids in groups, oriented parallel to cell wall, division probably by pinching; (ii) Figs. 1B, 4A upper, 5A – group of spherical or widely oval cells, circa 0.5–0.6 or 0.7–0.8 × 0.6 µm large, dividing by pinching, with parietally or radially disposed thylakoids (variable character ? ); (iii) Fig. 3B upper – cells only about 0.3 µm in diameter, with few parietal thylakoids, division by pinching. All these types correspond probably to the traditional genus *Synechocystis* (life cycles ?).
- Takamura & al. (1993) described two *Synechocystis*-like types from plankton of Japanese lakes: (i) cells spherical, reddish, 0.6–1.2 µm in diameter; (ii) cells spherical, blue-green, 0.6–0.9 µm in diameter. Without taxonomic evaluation.

## 2. **Aphanocapsa** Nägeli 1849

Colonies microscopic, more or less spherical, irregular or flat, in several species macroscopic, irregular, usually with numerous, sparsely or densely, irregularly arranged cells; colonial mucilage mainly colourless, more or less homogeneous with indistinct margin, or (particularly in macroscopic colonies) firm, formless, but clearly delimited; cells without individual gelatinous envelopes or occasionally (in firm colonies) with narrow, diffluent individual envelopes. Cells more or less spherical, after division hemispherical, without aerotopes. Cell wall S-layer of hexagonal lattice (identified in few small-celled species).
Cell division by binary fission in two perpendicular planes in successive generations (mainly cleavage); cells remain sometimes in groups together after division (in twos or tetrads). Reproduction by colony disintegration. In a few species facultative nanocyte production is recorded.

**Important references:** Geitler (1932, 1942); Frémy (1934); Feldmann (1937); Kosinskaja (1948); Padmaja (1972); Komárek & Anagnostidis (1986); Kováčik (1988).

**Key to the species:**

**1a** Aerophytic or subaerophytic species . . . . . . . . . . . . . . . . . . . . . . . . . . . . . 2
**1b** Species living in various hydrobiotopes . . . . . . . . . . . . . . . . . . . . . . . . . . . 5
**2a** Cells 1–1.5 µm in diameter, colonies formless, yellowish-brown . . . . . . . . . . . . . . . . . . . . . . . . . . . . . . . . . . . . . . . **6. A. fusco-lutea**
**2b** Cells more than 2 µm in diameter . . . . . . . . . . . . . . . . . . . . . . . . . . . . . . . 3
**3a** Cells 2–3 (4.5) µm in diameter . . . . . . . . . . . . . . . . . . **12. A. muscicola**
**3b** Cells always more than 4 µm in diameter . . . . . . . . . . . . . . . . . . . . . . . 4
**4a** Cells 4–7 µm in diameter; on limestone wet rocks and walls **19. A. parietina**
**4b** Cells 7.5–9.5 µm in diameter; on granitic wet rocks or wet peaty soils . . . . . . . . . . . . . . . . . . . . . . . . . . . . . . . . . . . . . . . **20. A. testacea**
**5a** Planktic species . . . . . . . . . . . . . . . . . . . . . . . . . . . . . . . . . . . . . . . . . . 6
**5b** Species from other water biotopes . . . . . . . . . . . . . . . . . . . . . . . . . . . 12
**6a** Cells densely arranged in colonies . . . . . . . . . . . . . . . . . . . . . . . . . . . . . 7
**6b** Cells loosely distributed; the spaces between cells greater than the diameter of one or two cells . . . . . . . . . . . . . . . . . . . . . . . . . . . . . . . . . . . . . . . 8
**7a** Cells 1–2 µm in diameter, joined into compact, more or less spherical, rarely irregular colonies; cell masses blue-green or yellowish-green **3. A. incerta**
**7b** Cells circa 1 µm in diameter, in irregular colonies; cell masses pale grey or blue-green . . . . . . . . . . . . . . . . . . . . . . . . . . . . . . . . . **4. A. holsatica**
**8a** Cells 0.5–0.7 (1.2) µm in diameter, pale blue-green, in irregular colonies . . . . . . . . . . . . . . . . . . . . . . . . . . . . . . . . . . . . . . **2. A. delicatissima**
**8b** Cells over 1.2 µm in diameter . . . . . . . . . . . . . . . . . . . . . . . . . . . . . . . . 9
**9a** Cells very sparsely and irregularly distributed within more or less spherical colonies, 1.3–2 µm in diameter . . . . . . . . . . . . . . . . . . . . **9. A. elachista**
**9b** Cells regularly distributed in colonies, over 1.5 µm in diameter . . . . . . 10
**10a** Cells 1.5–2 (2.4) µm in diameter . . . . . . . . . . . . . . . . . . . . . . . . . . . . 11
**10b** Cells 2–3 µm in diameter, pale blue-green, regularly distributed in irregular colonies . . . . . . . . . . . . . . . . . . . . . . . . . . . . . . . . . **15. A. planctonica**
**11a** Cells 1.5–2 µm in diameter, in small, irregular, cloud-like colonies; mucilaginous margin wavy . . . . . . . . . . . . . . . . . . . . . . . . . . . . . **10. A. nubilum**
**11b** Cells 1.6–2 (2.4) µm in diameter, in irregular colonies with diffluent and scarcely visible margin . . . . . . . . . . . . . . . . . . . . . . . . . . **11. A. conferta**
**12a** Freshwater species . . . . . . . . . . . . . . . . . . . . . . . . . . . . . . . . . . . . . . 13
**12b** Species from brackish, marine or inland salty biotopes . . . . . . . . . . . . . 19
**13a** Species from thermal springs, cells 2.2–4.5 µm in diameter . **16. A. thermalis**
**13b** Not in thermal waters . . . . . . . . . . . . . . . . . . . . . . . . . . . . . . . . . . . . . 14
**14a** Epilithic in running water or in splash zone of lake littoral . . . . . . . . . . 15
**14b** In stagnant waters, periphytic or benthic . . . . . . . . . . . . . . . . . . . . . . . 17
**15a** Colonies with ferric precipitates; cells 2–2.5 µm in diameter; benthic in streams . . . . . . . . . . . . . . . . . . . . . . . . . . . . . . . . . . **13. A. sideroderma**
**15b** Colonies not with ferric precipitates . . . . . . . . . . . . . . . . . . . . . . . . . . . 16
**16a** Cells 0.8–4 µm in diameter, benthic (epilithic) in calcareous streams, usually in mountain regions . . . . . . . . . . . . . . . . . . . . . . . . . . **17. A. rivularis**
**16b** Cells 2.5–4 µm in diameter, in clear, mountain streams (epilithic or in detritus), on non-calcareous, usually slightly acidic substrates, on stony margins of wells, etc. . . . . . . . . . . . . . . . . . . . . . . . . . **18. A. fonticola**
**17a** Epiphytic on submerged plants, in mesotrophic waters; colonies irregular, cells pale, greyish blue-green, 1–2.2 µm in diameter . . . . . **5. A. parasitica**

17b Benthic species . . . . . . . . . . 18
18a Gelatinous, microscopic to macroscopic, more or less spherical colonies, epipelic and metaphytic in clear, small water bodies; cells 1.5–2.5 µm in diameter, pale blue-green . . . . . . . . . . **21. A. hyalina**
18b Amorphous colonies, benthic in moorland waters, swamps or on acidic (peaty) substrates; cells 3–6 µm in diameter, bright blue-green . . . . . . . . . . **22. A. grevillei**
19a Cells less than 2 µm in diameter . . . . . . . . . . 20
19b Cells more than 2 µm in diameter . . . . . . . . . . 22
20a Cells 0.4–0.5 µm in diameter, in an irregular, gelatinous mass; in brackish pools on the N. Atlantic European coasts . . . . . . . . . . **1. A. marina**
20b Cells 1–2 µm in diameter . . . . . . . . . . 21
21a Cells 0.8–1.2 µm in diameter, pale blue-green, in a gelatinous mass; benthic (sapropelic) in salty inland lakes and pools in central Asia and SE. Russia . . . . . . . . . . **7. A. salina**
21b Cells 1–1.6 µm in diameter, pale blue-green or olive-green, a slightly elongate; on shells and limestone rocks in Mediterranean region . . . . . . . . . . **8. A. concharum**
22a Cells 2–6 µm in diameter . . . . . . . . . . 23
22b Cells more than 6 µm in diameter . . . . . . . . . . 26
23a Cells reddish, 2–3 (5) µm in diameter, slightly elongate; epizoic on fish near the French coast (Banyuls) . . . . . . . . . . **14. A. feldmannii**
23b Cells more or less blue-green; free-living species . . . . . . . . . . 24
24a Colonies attached to small stones or growing among sand particles; cells bright blue-green, 3–4.5 µm in diameter . . . . . . . . . . **23. A. reinboldii**
24b Colonies not living in psammon . . . . . . . . . . 25
25a Cells 3–6 µm in diameter, in more or less spherical, ellipsoidal or irregular colonies with more or less firm, distinct margin; benthic in shallow salty pools and swamps, N. parts of Atlantic European coasts . . **24. A. litoralis**
25b Cells 4–6 µm in diameter, in amorphous, gelatinous colonies; on calcareous rocks in the splash zone of the Mediterranean coasts . . . . . . . . **25. A. orae**
26a Cells 6–8 µm in diameter, blue-green or olive-green; amorphous colonies in benthos of inland and coastal salty waters . . . . . . . . . . **26. A. salinarum**
26b Cells 6–12 µm in diameter, reddish; epizoic or in organic deposits on Mediterranean marine coasts . . . . . . . . . . **27. A. raspaigellae**

## 1. **Aphanocapsa marina** Hansgirg 1890 (fig. 170)

*Microcystis marina* (Hansgirg) Kosinskaja 1948; *Anacystis marina* (Hansgirg) Drouet et Daily 1948 p.p.

Colonies small, more or less formless, gelatinous, greyish-blue, sometimes united into a dark blue-green mass. Cells spherical, commonly in twos (after division), pale blue-green, sometimes with individual thin, colourless gelatinous envelopes, 0.4–0.5 µm in diameter.

---

APHANOCAPSA

**Fig. 170.** *Aphanocapsa marina*: **a-b** – after Kosinskaja (1948), **c** – after Frémy (1929).

**Fig. 171.** *Aphanocapsa delicatissima* (planktic): **a** – after W. et G.S.West (1912), **b** – after West from Kondrateva & al. (1984), **c** – after G.M.Smith (1920), **d** – after Komárková-Legnerová & Cronberg (1994).

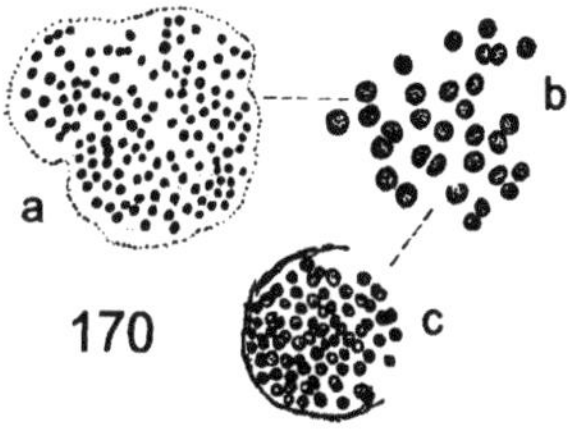


170

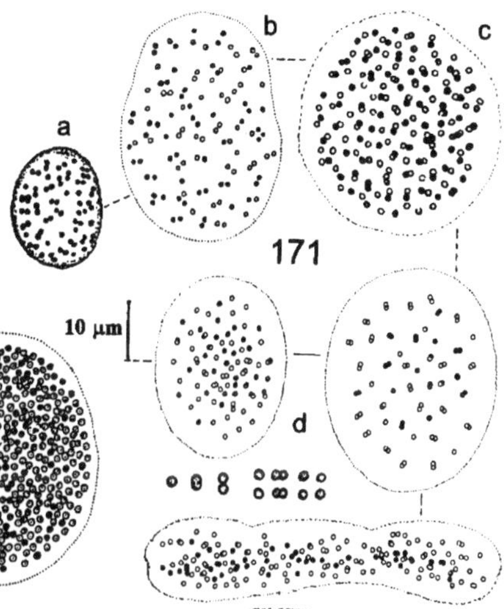


171

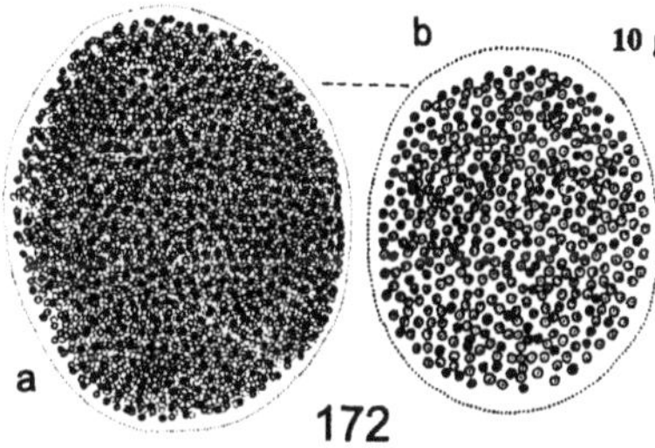


172

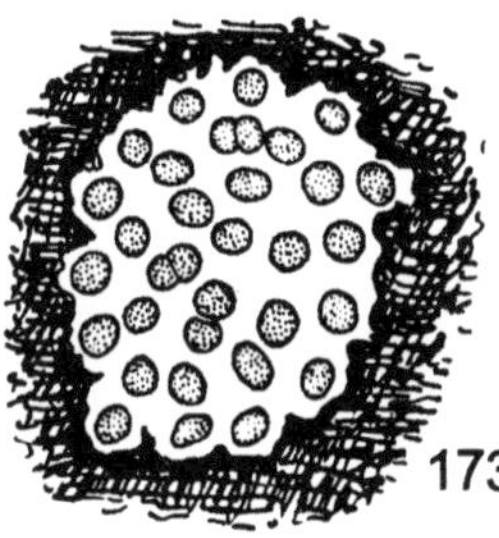

173

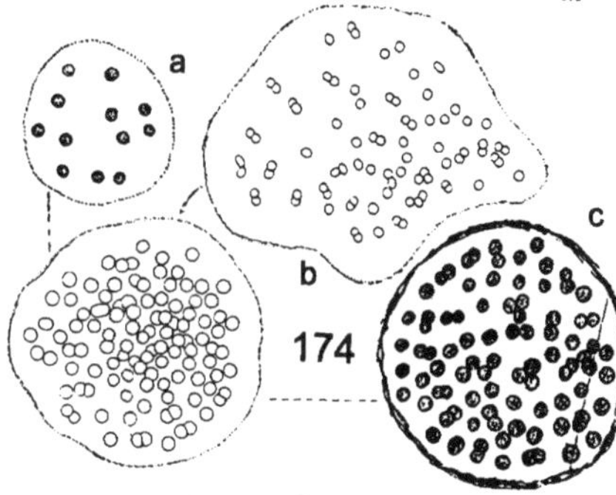


174

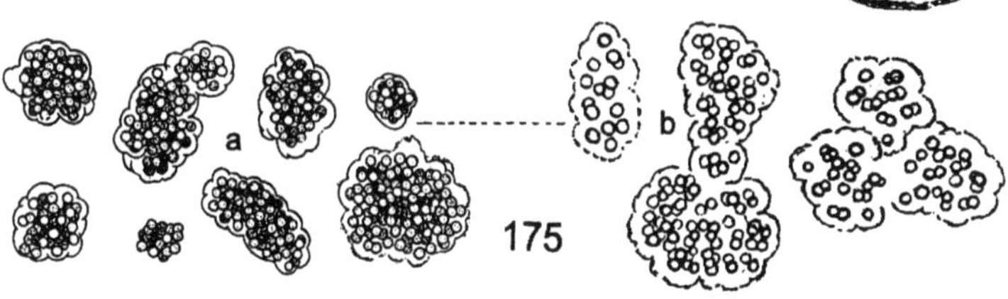


175

**Fig. 172.** *Aphanocapsa incerta* (planktic): **a** – after G.M.Smith (1920), **b** – after Kosinskaja from Kondrateva & al. (1984).
**Fig. 173.** *Aphanocapsa concharum*: after Frémy (1934, sub *Aphanocapsa endolithica* var. *marina*).
**Fig. 174.** *Aphanocapsa elachista* (planktic): **a** – after W. et G. S.West from Starmach (1966), **b** – after Hindák & Moustaka (1988), **c** – after Frémy (1930).
**Fig. 175.** *Aphanocapsa nubilum* (planktic): **a** – after Nygaard (1949, sub *Microcystis pulverea* var. *racemiformis*), **b** – after Komárek & Kling (1991).

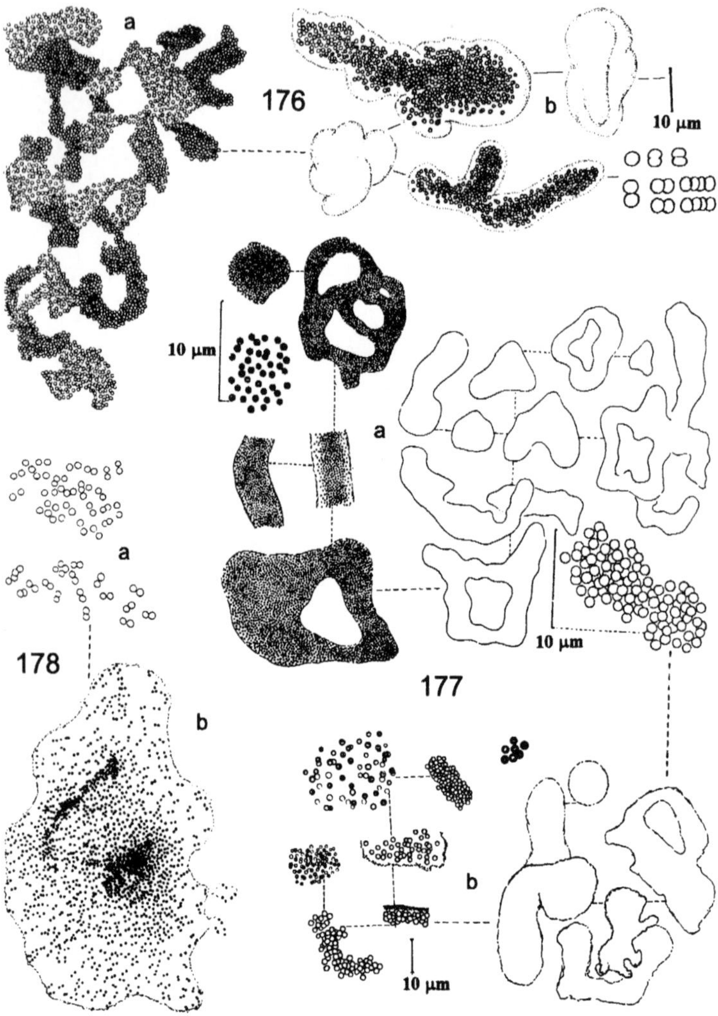


APHANOCAPSA

**Fig. 176.** *Aphanocapsa holsatica* (planktic): **a** – after Nygaard (1949, sub *Microcystis holsatica*), **b** – after Komárková-Legnerová & Cronberg (1994).
**Fig. 177.** *Aphanocapsa holsatica*: **a** – after Komárek (1958, sub *Diplocystis incerta*), **b** – after Komárek (1975, sub *Microcystis incerta*).
**Fig. 178.** *Aphanocapsa* cf. *holsatica*: after Hindák & Moustaka (1988); cells seem to be arranged in rows (cell division in one plane?).

Occurrence: Marine and brackish, in pools with salty water and on wet rocks in the intertidal zone; Atlantic coasts of Europe (Norway, France – Chausey Islands), less common in Mediterranean region (Greece – Aegean Sea).

## 2. Aphanocapsa delicatissima W. et G.S. West 1912 (fig. 171)

*Microcystis pulverea* f. *delicatissima* (W. et G.S. West) Elenkin 1938; *Microcystis delicatissima* (W. et G.S. West) Starmach 1966

Colonies planktic, spherical, elliptical or irregular, up to 50 µm in diameter, with more or less loosely arranged, evenly spread out cells; mucilage colourless or slightly yellowish, homogeneous, diffluent. Cells minute, spherical, grey or pale blue-green, without aerotopes, 0.5–0.8 (1.2) µm in diameter.
Occurrence: Planktic in lakes and ponds; probably cosmopolitan, but recorded mainly from temperate regions; Europe: widely distributed.

## 3. Aphanocapsa incerta (Lemmermann) Cronberg et Komárek 1994 (fig. 172)

*Polycystis incerta* Lemmermann 1899; *Microcystis incerta* (Lemmermann) Lemmermann 1907; *Microcystis pulverea* var. *incerta* (Lemmermann) Crow 1923; *Microcystis pulverea* f. *incerta* (Lemmermann) Elenkin 1938; *Anacystis incerta* (Lemmermann) Drouet et Daily 1952

Colonies microscopic, spherical, sometimes slightly flattened, rarely somewhat irregular, with a great number of irregularly and densely aggregated cells; the agglomerations of cells are yellowish-green. Mucilage colourless or slightly yel-

**Fig. 179A.** *Aphanocapsa parasitica*: **a** – after Tarnavschi (1956, sub *Microcystis pulverea* f. *parasitica*), **b** – «*Microcystis parasitica* f. *dinobryonis*» after Wawrik (1976).

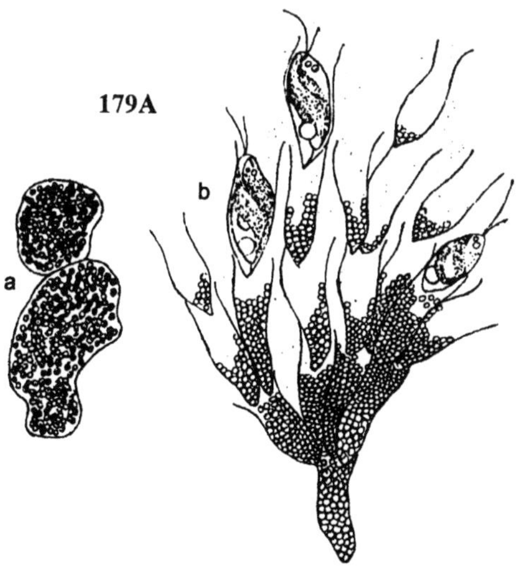

low, structureless, diffluent and indistinct at the margin (in culture completely diffluent and cells are solitary). Cells spherical with finely granulate yellowish-green or pale blue-green content, without aerotopes, 0.5–2 (2.7) µm in diameter.
Occurrence: Commonly in plankton of eutrophic water bodies; cosmopolitan; (often misinterpreted as *A. elachista*). Data from other biotopes (subaerophytic) are evidently wrong.

**4. Aphanocapsa holsatica** (Lemmermann) Cronberg et Komárek 1994 (fig. **176, 177, 178**)

*Clathrocystis holsatica* Lemmermann 1903; *Microcystis holsatica* (Lemmermann) Lemmermann 1907

Young colonies more or less spherical, later irregular, lobate, elongate or clathrate, up to more than 300 µm in diameter, with usually clearly visible colourless mucilage of irregular outline; more or less densely aggregated cells, having in the mass a pale greyish blue-green colour. Cells spherical, pale grey-blue or greyish blue-green, about 1 µm in diameter.
Occurrence: In plankton of eutrophic freshwater bodies (also in Baltic Sea), very common; cosmopolitan.

**5. Aphanocapsa parasitica** (Kützing) Komárek et Anagnostidis 1995 (fig. **179A, 179**)

*Microcystis parasitica* Kützing 1843; *Anacystis parasitica* (Kützing) Trevisan 1848; *Microcystis pulverea* f. *parasitica* (Kützing) Elenkin 1938

Colonies irregular, microscopic, diffluent, attached to a substrate, with more or less densely aggregated small cells. Mucilage colourless, diffluent. Cells spherical, circa 1–2 (2.2 ?) µm in diameter, pale greyish-blue-green.
Occurrence: Epiphytic on submerged plants in swamps and littoral of not very polluted ponds and lakes; in Europe widely distributed, data for other regions uncertain. Wawrik (1976) described *Microcystis parasitica* f. *dinobryonis* from empty scales of *Dinobryon sertularia*, Austria (fig. **179 A-b**); the relation of this cyanoprokaryote to *A. parasitica* should be confirmed.

---

APHANOCAPSA
**Fig. 179.** *Aphanocapsa parasitica* (epiphytic): after Fritsch from Hollerbach & al. (1953).
**Fig. 180.** *Aphanocapsa conferta* (planktic): **a** – after W. et G. S. West (1912), **b** – after G.M.Smith (1920), **c-d** – after Komárková-Legnerová & Cronberg (1994, dimensions of **d** correspond rather to *A. planctonica*).
**Fig. 181.** *Aphanocapsa muscicola* (aerophytic): **a** – after Wille from Hollerbach & al. (1953, sub *Microcystis muscicola*), **b** – after Hoffmann (1986, sub *Microcystis muscicola*), **c** – after Hoffmann (1986, sub *Microcystis montigena*), **d** – after Banerji from Desikachary (1959, sub *Aphanocapsa montana*).
**Fig. 182.** *Aphanocapsa feldmannii* (marine): after Frémy from Kosinskaja (1948).
**Fig. 183.** Planktic *Aphanocapsa* sp. sensu Teiling (1944), corresponding to *A. conferta*.

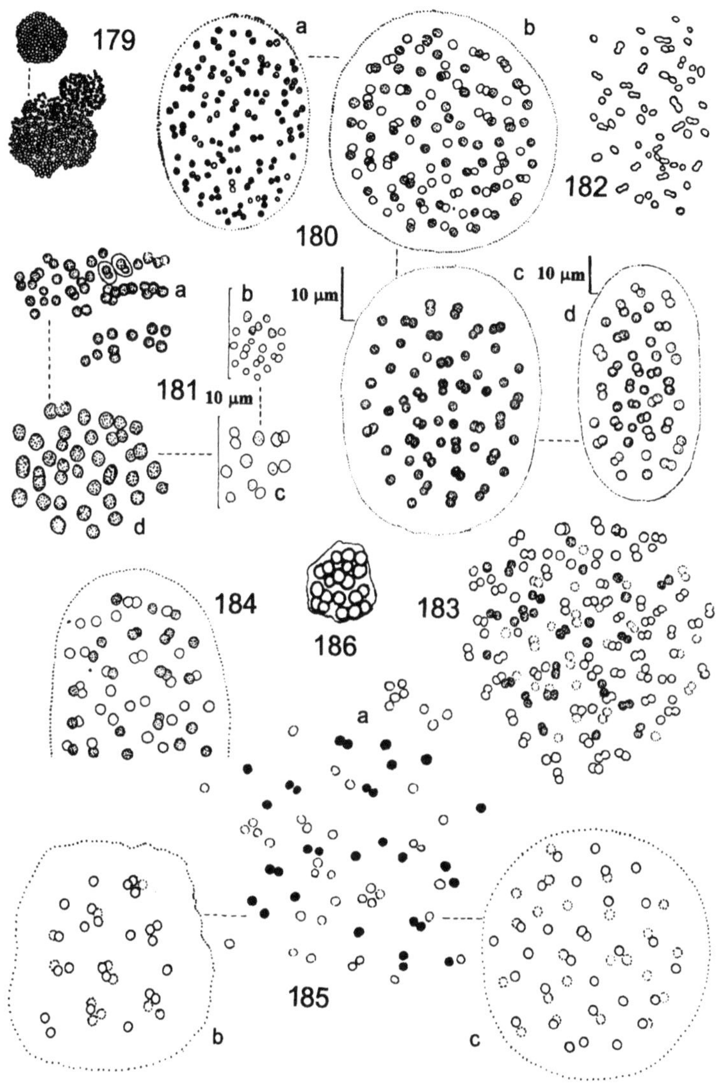


**Fig. 184.** *Aphanocapsa planctonica* (planktic): after G.M. Smith (1920).
**Fig. 185.** *Aphanocapsa koordersii* (planktic), known from tropical countries: a – after Ström (1923), **b** – after Yacubson (1974), **c** – after Yacubson (1974, sub *A. planctonica*).
**Fig. 186.** *Aphanocapsa fonticola*: after Kann (1978).

### 6. **Aphanocapsa fusco-lutea** Hansgirg 1892 (fig. **189**)

*Microcystis hansgirgiana* Elenkin 1938

Colonies mucilaginous, formless, diffluent, dirty yellow-green, greyish up to yellowish-brown, attached to a substrate, sometimes macroscopic. Cells spherical, blue-green or yellowish, sparsely or densely distributed within the slime, 1–1.5 µm in diameter, usually with individual, diffluent gelatinous envelopes.
Occurrence: On wet soil, wet stone walls and rocks, walls of glasshouses, often near thermal springs; sporadical in the Eurasian temperate zone.

### 7. **Aphanocapsa salina** Voronichin 1929 (fig. **199**)

*Microcystis salina* (Voronichin) Elenkin 1938

Colonies macroscopic, formless, gelatinous, later forming an amorphous mass with loosely disposed cells, which are solitary or in twos (after division); in young colonies cells more clustered. Cells spherical, pale blue-green, 0.8–1.2 µm in diameter.
Occurrence: Benthic (sapropelic) in salty lakes and swamps, sometimes in masses, commonly in salty thiobiocoenoses; SE. Europe, Ukraine, Russia (also inland salines, W. Siberia, Central Asia).

### 8. **Aphanocapsa concharum** Hansgirg 1890 (fig. **173**)

*Aphanocapsa endolithica* var. *marina* Frémy 1934; *Microcystis concharum* (Hansgirg) Kosinskaja 1948

Colonies mucilaginous, thin, fine, usually diffluent and amorphous, dark blue-green, forming strata on calcareous rocks. Common mucilage colourless, structureless. Cells spherical or spherical-ellipsoidal, pale blue-green or olive-green, 1–1.6 µm in diameter (before division elongate and to 3 µm long).
Occurrence: Marine, on mollusc shells and algae; Mediterranean coasts (Adria), French coasts.

### 9. **Aphanocapsa elachista** W. et G.S. West 1894 (fig. **174**)

*Microcystis pulverea* f. *elachista* (W. et G.S. West) Elenkin 1938; *Microcystis elachista* (W. et G.S. West) Starmach 1966 sec. Compère 1967

Colonies microscopic, spherical or oval, to 100 µm in diameter, with very sparsely disposed solitary cells, after division in twos. Mucilage homogeneous, scarce, colourless, usually diffluent. Cells spherical, grey or pale blue-green, without aerotopes, 1.3–2 µm in diameter.
Occurrence: Planktic in eutrophic waters, very common in tropical countries, rare in warm regions of the temperate zone; Europe: occasionally in warmer areas or during the summer season in large water bodies. Records from other biotopes (e.g., bark of trees in W. Africa, Romo & al. 1995) belong evidently to another species.

### 10. **Aphanocapsa nubilum** Komárek et Kling 1991 (fig. **175**)

*Microcystis pulverea* var. *racemiformis* Nygaard 1949; *Microcystis pulverea* f. *racemiformis* (Nygaard) Hollerbach 1955

Colonies small, microscopic, free floating, irregularly spherical, botryoidal or cloud-shaped, multicellular, up to 56 µm in diameter, rarely composed from subcolonies. Mucilage colourless, diffluent, more or less following the outer margin of the cluster of cells, up to 3 µm wide around marginal cells, sometimes

distinctly visible. Cells spherical, more or less densely aggregated, irregularly distributed, pale blue-green, 1.2–2 µm in diameter, without gas vesicles.
Occurrence: Planktic in clear lakes, ponds and swamps, sporadic, but probably cosmopolitan; Europe: rarely found in lakes.

**11. Aphanocapsa conferta** (W. et G.S. West) Komárková-Legnerová et Cronberg 1993 (fig. **180, 183**)

*Aphanocapsa elachista* var. *conferta* W. et G.S. West 1912; *Microcystis pulverea* f. *conferta* (W. et G.S. West) Elenkin 1938

Colonies planktic, free floating, microscopic to 80 µm in diameter, more or less spherical or irregular, with scattered, more or less evenly densely arranged cells. Mucilage delicate, homogeneous, colourless, structureless, indistinct at the margin. Cells spherical, pale blue-green or greyish, sometimes in twos, without aerotopes, 1.5–2 (2.4) µm in diameter.
Occurrence: Planktic in eutrophic water bodies; commonly in temperate zones; sporadic records from tropical countries need confirmation (*A. koordersii* ?).

**12. Aphanocapsa muscicola** (Meneghini) Wille 1919 (fig. **181**)

*Coccochloris muscicola* Meneghini 1843; *Aphanocapsa montana* Cramer 1862; *Aphanocapsa naegelii* Richter 1884; *Aphanocapsa minima* Migula 1933; *Microcystis muscicola* (Meneghini) Elenkin 1938; *Microcystis montigena* Hoffmann 1986

Colonies colourless, yellowish to dark blue-green or olive-green, formless, sometimes united into an amorphous gelatinous mass with densely packed cells; cells sometimes surrounded by individual gelatinous envelopes. Cells spherical, blue-green or yellowish-green, 2–4 (4.5) µm in diameter.
Occurrence: Aerophytic, on wet rocks, wood and soil; over the whole temperate zone.

**13. Aphanocapsa sideroderma** Naumann 1922

*Aphanocapsa siderosphaera* Naumann 1922 incl.

Colonies microscopic, mucilaginous, with ferric precipitates, with more or less densely arranged cells. Cells widely oval (?), yellowish-green, sometimes with vacuole-like inclusions, 2–2.5 µm in diameter.
Occurrence: On stones in streams; known only from S. Sweden.

**14. Aphanocapsa feldmannii** Frémy 1933 (fig. **182**)

*Microcystis feldmannii* (Frémy) Kosinskaja 1948

Colonies small, amorphous, without distinct mucilage, reddish, growing in the epidermal layer of marine littoral sponges. Cells more or less densely and irregularly aggregated, spherical or elongate before division (pinching !); spherical, 2–3 µm in diameter, before division up to 5 µm long, with reddish homogeneous content; individual thin gelatinous layers around cells. Probably belongs to *Synechocystis*.
Occurrence: Marine, epizoic in the epidermis of sponges *Hircinia variabilis* and *Petrosia ficiformis* var. *dura* and others (depth down to 20 m); along the coasts of W. Mediterranean Sea (France – Banyuls, Algeria), Portugal, also found in Antilles and Guadeloupe.

**15. Aphanocapsa planctonica** (G.M. Smith) Komárek et Anagnostidis 1995 (fig. **184**)

*Aphanocapsa elachista* var. *planctonica* G.M. Smith 1920; *Microcystis pulverea* f. *planctonica* (G.M. Smith) Elenkin 1938

Colonies microscopic, more or less irregular, with homogeneous, colourless mucilage, rarely diffluent, usually distinct and clearly delimited. Cells regularly and more or less loosely distributed, spherical, greyish blue-green, 2–3 µm in diameter.
Occurrence: Common in clear mesotrophic lakes of N. and S. America (USA, Canada, Argentine); Europe: probably in northern regions. Data from aerophytic habitats are evidently incorrect, records from tropical regions concern rather *A. koordersii.*

**16. Aphanocapsa thermalis** Brügger 1863 (fig. **187**)

*Aphanocapsa thermalis* (Kützing) Brügger 1863 sensu Brügger non sensu Kützing; *Microcystis thermalis* (Brügger) Claus 1959; ? *Microcystis thermalis* Yoneda 1938

Colonies mucilaginous, irregular, formless, with homogeneous colourless mucilage, usually with densely packed cells (they touch one to another or have spaces up to 1/2 cell diameter between them), blue-green. Cells spherical or slightly oval, 2.2–4.5 µm in diameter, bright blue-green, sometimes with individual fine, diffluent mucilaginous envelopes (esp. in peripheral parts of colonies). Probably facultatively nanocyte production.
Occurrence: Subaerophytic (and in running hot water), on wet rocks in thermal springs (in temperatures of up to 68.7 °C); common in thermal springs of northern hemisphere, known also from Argentine and New Zealand.

**17. Aphanocapsa rivularis** (Carmichael) Rabenhorst 1865 (fig. **188**)

*Aphanocapsa anodontae* Hansgirg 1892; *Aphanocapsa endolithica* Ercegović 1925; *Aphanocapsa endolithica* var. *rivulorum* Geitler 1927; *Microcystis anodontae* (Hansgirg) Elenkin 1938; *Microcystis calciphila* Schwabe 1947; *Microcystis anodontae* var. *rivulorum* (Geitler) Kann 1978

Colonies microscopic, mucilaginous, amorphous, blue-green, with densely aggregated cells, up to 40 µm in diameter, later sometimes forming a large gelatinous mass or an irregular green coating on stony substrates. Cells spherical or widely and irregularly oval, blue-green, 0.8–4 (6 ?) µm in diameter (different morphotypes with diverse cell diameter).

---

APHANOCAPSA
**Fig. 187.** *Aphanocapsa thermalis* (thermal springs): **a** – after Wenzel & Halperin (1991) from Argentine, **b** – after Palik (1949) from Hungary, **c** – after Dor (1967) from Israel, **d** – after Anagnostidis (1961) from Greece.
**Fig. 188.** *Aphanocapsa rivularis* (benthic in streams): **a** – after Padmaja (1972) from India, **b** – after Geitler (1927) from Austria, **c** – after Frémy (1930) from Africa.
**Fig. 189.** *Aphanocapsa fusco-lutea*: after Naumann from Starmach (1966, sub *Microcystis hansgirgiana*).
**Fig. 190.** *Aphanocapsa parietina*: after Nägeli (1849).
**Fig. 191.** *Aphanocapsa testacea*: after Nägeli (1849).

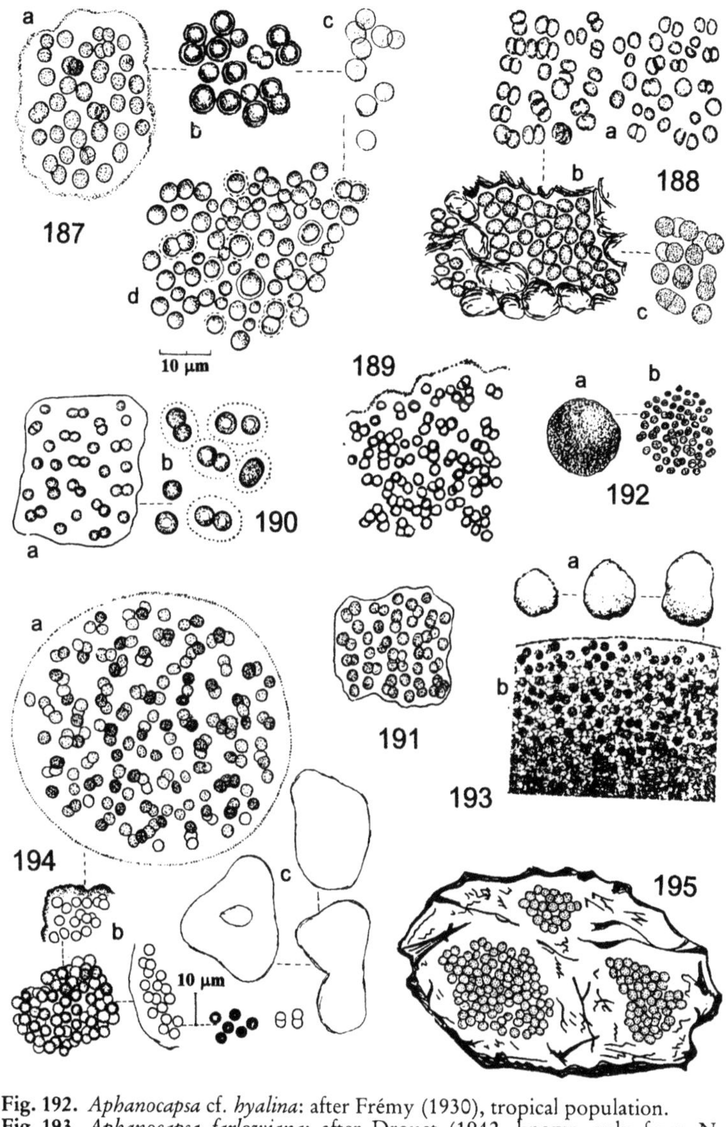


**Fig. 192.** *Aphanocapsa* cf. *hyalina*: after Frémy (1930), tropical population.
**Fig. 193.** *Aphanocapsa farlowiana*: after Drouet (1942, known only from N. America).
**Fig. 194.** *Aphanocapsa grevillei*: **a** – after G.M.Smith (1925), **b-c** – after Komárek (1975).
**Fig. 195.** *Aphanocapsa reinboldii*: after Frémy from Kondrateva & al. (1984).

Occurrence: Epi- or endolithic on or in calcareous stones or shells in streams, mainly in mountain areas; widely distributed, particularly in limestone mountains of the temperate zones; records from saline waters (comp. *A. endolithica* var. *marina* Frémy 1914) and from plankton are surely wrong, those from tropical countries need confirmation.

**18. Aphanocapsa fonticola** Hansgirg 1890 (fig. **186**)

*Microcystis fonticola* (Hansgirg) Kann 1978

Colonies fine, mucilaginous, formless, with more or less densely aggregated cells, dark blue-green. Cells spherical, elongate before division, blue-green, (1.5) 2.5–4 µm in diameter.
Occurrence: In clear springs and streams, mainly in mountain areas (mainly in non-calcareous biotopes; similar types from limestone areas probably belong to *A. rivularis*); Europe: common.

**19. Aphanocapsa parietina** Nägeli 1849 (fig. **190**)

*Aphanocapsa virescens* (Hassall) Rabenhorst 1865; *Aphanocapsa flava* (Kützing) Rabenhorst 1865; *Microcystis parietina* (Nägeli) Elenkin 1938

Colonies large, sometimes forming an amorphous mass with more or less loosely distributed cells, yellowish-green, dirty blue-green or dark green and brownish olive-green; cells usually with individual gelatinous, diffluent envelopes. Cells spherical, usually pale blue-green, rarely yellowish, olive-green or brownish, usually with visible chromatoplasma and/or solitary granules, (3) 4–7 µm in diameter.
Occurrence: On wet limestone rock walls and wet alkaline soils; Europe: sporadic; data from plankton concern other species.

**20. Aphanocapsa testacea** Nägeli 1849 (fig. **191**)

*Microcystis testacea* (Nägeli) Elenkin 1938

Colonies small, simple or forming a macroscopic, gelatinous, yellowish-brown or dirty green or reddish mass, with more or less densely distributed cells, which are solitary or in twos, sometimes with individual mucilaginous envelopes. Cells spherical or oval, yellowish or brownish-green, 7.5–9.5 µm in diameter.
Occurrence: On wet granitic rocks or on peaty soil, near swamps etc.; sporadically over the whole temperate zone of Eurasia.

**21. Aphanocapsa hyalina** (Lyngbye) Hansgirg 1892 (fig. **192**)

Colonies more or less spherical or irregularly spherical, gelatinous, blue-green or dirty olive-green, micro- to macroscopic (to 5 cm in diameter), with loosely or more or less densely arranged cells, free living; sometimes forms macroscopic diffluent gelatinous mass. Mucilage thick, gelatinous, not structured, colourless. Cells spherical or oval before division, polygonal-rounded in the colonial center, pale blue-green, 1.5–2.5 µm in diameter.
Occurrence: Freshwater, benthic, epipelic, in clear swamps, peaty waters and small stagnant water bodies; known from central Europe, mainly from mountains (but also from N. Germany); similar rich populations (probably a separate species), sometimes with somewhat larger cells, are known from littoral of lakes in Africa (Lake Nyassa; lakes in Nigeria).

**22. Aphanocapsa grevillei** (Berkeley) Rabenhorst 1865 (fig. **194**)

*Microcystis grevillei* (Berkeley) Elenkin 1938

Colonies more or less amorphous-spherical, later irregular, to 1 cm in diameter, sometimes composed of subcolonies, mucilaginous, green, with irregularly and densely aggregated cells. Cells spherical or slightly oval, with bright blue-green content without aerotopes, sometimes with distinct chromatoplasma, 3.5–6 µm in diameter.

Occurrence: Commonly benthic in moorland waters and acidic swamps, occasionally on peaty soil and on granitic substrates (wet rocks); sporadic all over Europe, probably over the whole temperate zone; planktic records concern other species (mainly *Microcystis smithii*), records from thermal biotopes and tropical regions need revision; (comp. fig. **196**).

**23. Aphanocapsa reinboldii** (Richter) Komárek et Anagnostidis 1995 (fig. **195**)

*Anacystis reinboldii* Richter in Reinbold 1889; *Microcystis reinboldii* (Richter) Forti 1907

Colonies small, irregular, flattened, attached to particles of sand, to 60 µm in diameter, later forming small subspherical or amorphous clusters separating from the substrate, with densely aggregated cells (sometimes very tightly). Cells spherical, rarely slightly elongate or flattened laterally, pale blue-green, rarely (in old colonies) olive-yellowish, 3–4 µm in diameter.

Occurrence: Benthic in inland and coastal salty shallow water bodies or sea beaches with sandy bottoms, always halophilic; common in Baltic Sea, but known also from other European sandy coasts (including Black Sea).

**24. Aphanocapsa litoralis** (Hansgirg) Komárek et Anagnostidis 1995

*Polycystis litoralis* Hansgirg in Foslie 1890; *Microcystis litoralis* (Hansgirg) Forti in de Toni 1907

Colonies spherical, ellipsoidal or rarely irregular, usually 30–60 µm in diameter, often consisting of subcolonies, with clearly visible and distinctly delimited colonial mucilage, with densely aggregated cells. Cells spherical or oval, pale blue-green, without aerotopes, 3–5 µm in diameter, but sometimes distinctly longer before division.

Occurrence: In salty swamps and pools, usually together with marine algae, but also in inland saline waters; probably cosmopolitan, in Europe known from Norway, Baltic Sea and northern Atlantic coasts and Portugal; records from tropical paddy fields (Tiwari 1975) or from plankton in saline water (? var. *natans* Wille 1914) concern evidently other species.

**25. Aphanocapsa orae** (Kosinskaja) Komárek et Anagnostidis 1995 (fig. **198**)

*Aphanocapsa litoralis* Hansgirg 1892; *Microcystis halophila* (nomen nudum ?); *Microcystis orae* Kosinskaja 1948

Colonies macroscopic, mucilaginous, formless, firm, dark yellowish-brown to blackish-green. Cells spherical or subspherical, sparsely or densely grouped within colonies, blue-green or yellowish, 4–6 µm in diameter.

Occurrence: Halophilic on wet rocks and walls, sometimes polluted (marine ports); Atlantic (France, Portugal) and Mediterranean coasts of Europe (Croatia – Istria, France, Spain), Canary Islands, coasts of S. Pacific (Nauru, Samoa, Ellice,

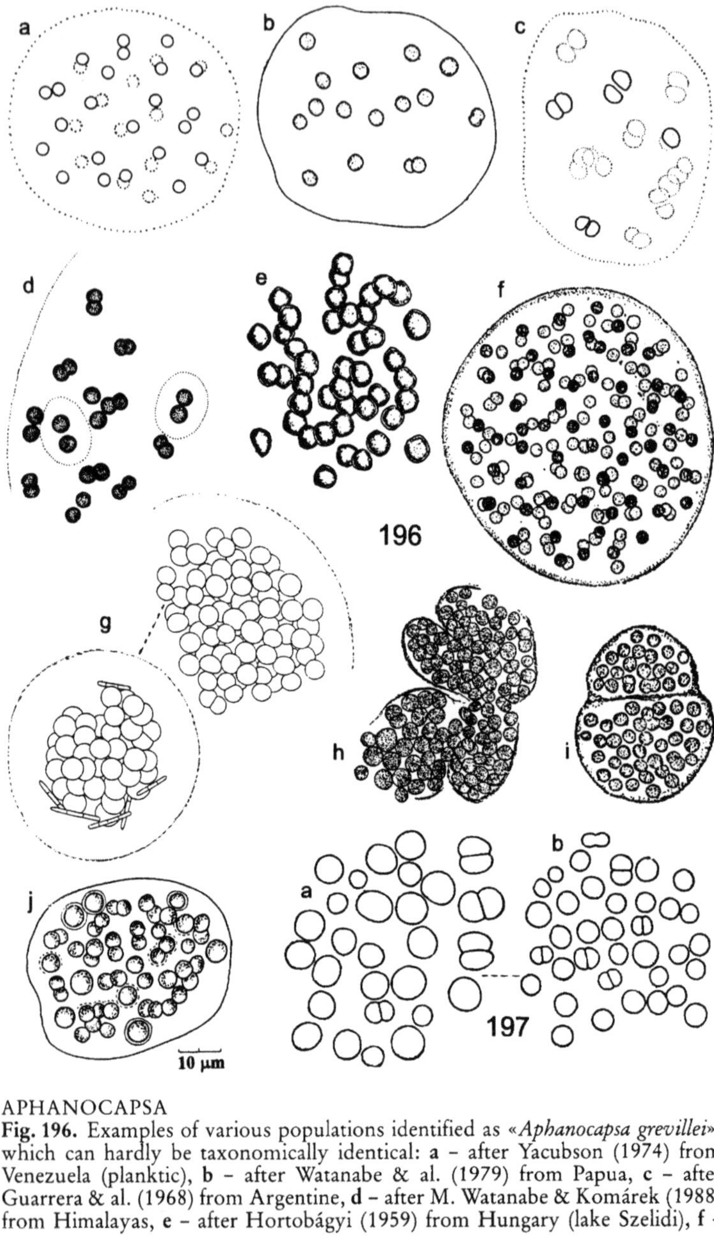


APHANOCAPSA

**Fig. 196.** Examples of various populations identified as «*Aphanocapsa grevillei*», which can hardly be taxonomically identical: **a** – after Yacubson (1974) from Venezuela (planktic), **b** – after Watanabe & al. (1979) from Papua, **c** – after Guarrera & al. (1968) from Argentine, **d** – after M. Watanabe & Komárek (1988) from Himalayas, **e** – after Hortobágyi (1959) from Hungary (lake Szelidi), **f** –

Gilbert Islands, Chile - Bay of Concepción); «*A. litoralis*» sensu Guarrera & al. 1980 from Argentine (Tierra del Fuego) belongs probably to *A. rivularis*.

## 26. Aphanocapsa salinarum Hansgirg 1887

Colonies mucilaginous, formless, small, with slimy diffluent margin. Cells spherical, sometimes in groups in 2–4, pale blue-green or olive-green, with wide individual, diffluent, colourless mucilaginous envelopes, 6–8 (10) µm in diameter.
Occurrence: Salty inland and coastal swamps and soils; known from central and S. Europe. Recorded also from Galapagos Islands.

## 27. Aphanocapsa raspaigellae (Hauck) Frémy 1933 (fig. 197)

*Chroococcus (?) raspaigellae* Hauck 1888; *Microcystis raspaigellae* (Hauck) Kosinskaja 1948

Colonies small, formless, reddish, without distinct mucilage, irregularly distributed in the epidermal layer of various marine sponges. Cells more or less densely and irregularly aggregated, spherical, with more or less homogeneous red content (in formol purple), (4) 6–12 µm in diameter; thin individual gelatinous envelopes around single cells.
Occurrence: In the tissues of epidermis of marine littoral sponges (*Raspaigella, Hircinia, Sycon, Halichondria*) and in organic detritus, usually associated with *Aphanocapsa feldmannii;* probably along the whole Mediterranean Sea coast; also recorded from Portugal, Caribbean Sea and tropical Pacific Ocean.

**Main species outside Europe:**
- *Aphanocapsa albida* Zeller 1873: Cells agglomerated in mucilaginous colonies; marine, described from Burma, not revised.
- *Aphanocapsa arctica* Whelden 1947: Microscopic, more or less spherical, hyaline colonies (to 90 µm in diameter), with tightly packed spherical cells, 8–10 µm in diameter; described from Canadian Arctic.
- *Aphanocapsa benaresensis* Bharadwaja 1935: Freshwater, colonies more or less spherical, hollow, to 1.5 cm in diameter, hyaline or olive-green, cells 4–6.2 µm in diameter, with narrow individual sheaths; benthic in stagnant water, ponds, Bangladesh, India; (Desikachary 1959, Islam & Nahar 1967).
- *Aphanocapsa botryoides* Copeland 1936: Small spherical to formless colonies united into botryoidal clusters, with pink to pinkish green cells, 1.8–2.1 µm in diameter; described from thermal springs, USA, smaller than *A. thermalis*; (fig. 200).
- *Aphanocapsa crassa* Ghose 1927: Brownish gelatinous colonies with sparsely distributed cells, 7–12 µm in diameter; epiphytic on submerged plants, described from a pond near Rangoon, Burma.

---

after G.M.Smith (1920) from USA (planktic, see *Microcystis smithii*), **g** - after Hindák & Moustaka (1988) from Greece (planktic), **h-i** - after Frémy (1930a, 1930b) from Africa, **j** - after Anagnostidis (1961) from Greece (thermal waters).
**Fig. 197.** *Aphanocapsa raspaigellae* (marine): after Frémy from Kosinskaja (1948, two forms from different hosts).

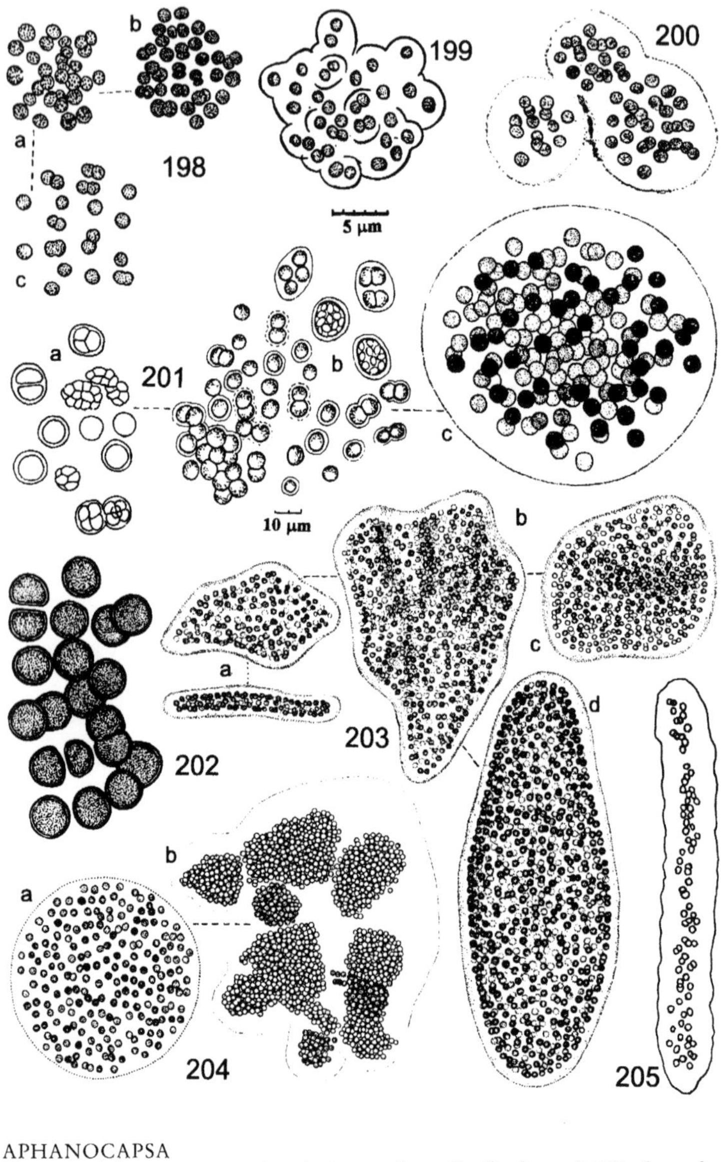


APHANOCAPSA

**Fig. 198.** *Aphanocapsa orae* (marine): **a** – from Desikachary (1959), **b** – after Frémy from Kosinskaja (1948), **c** – after Bourrelly & Denizot (1967).
**Fig. 199.** *Aphanocapsa salina* (halophilic): from Golubić (1960).

– *Aphanocapsa cumulus* Komárek et Cronberg 1999: Large cloudy colonies with densely arranged cells, 0.6–0.9 µm in diameter; described from ponds near Lake Kanta, Zimbabwe.
– *Aphanocapsa farlowiana* Drouet et Daily 1942: Macroscopic colonies, benthic and in littoral of small lakes of N. America (Canada, USA), sometimes in masses; acc. to Bourrelly (1949) also in France (lake du Bourget); (fig. **193**).
– *Aphanocapsa howei* Collins 1920: Dark blue-green mucilaginous colonies with spherical or rarely slightly fusiform cells, 9–15 µm in diameter, with pale blue-green content; benthic in salt lake, Bahamas.
– *Aphanocapsa intertexta* Gardner 1927: Aerophytic on rocks, cells 2.4–2.6 µm in diameter; described from Porto Rico.
– *Aphanocapsa koordersii* Strøm 1923: Colonies more or less spherical or irregular, cells 2–2.8 (3) µm in diameter, commonly planktic in tropical lakes; (Geitler 1932, Huber-Pestalozzi 1938, Islam & Nahar 1967, Yacubson 1974); (fig. **185**).
– *Aphanocapsa protea* (Copeland) Komárek et Anagnostidis 1995 [syn.: *Microcystis protea* Copeland 1936]: Small colonies mixed with other cyanoprokaryotes, with cells 1–1.3 µm in diameter; thermal species from USA.
– *Aphanocapsa tolliana* Copeland 1936: Microscopic colonies with loosely arranged cells, 0.6–0.8 µm in diameter; thermal springs of USA (Copeland 1936), Argentine (Wenzel & Halperin 1991) and Israel (Dor 1967).
– *Aphanocapsa venezuelae* Schiller 1952: Free-floating microscopic colonies, cells sparsely disposed within mucilage, 2.4–2.6 µm in diameter; swamps, Los Aves islands, Venezuela.
– *Aphanocapsa zanardinii* (Hauck) Hansgirg 1889: Marine, described from Sarawak.
– *Microcystis ovalis* Hollenberg 1939: Spherical or oval colonies with distinct hyaline envelope; cells tightly packed, spherical, pale blue-green or olive-green, without aerotopes, 1.5–2 µm in diameter; described from cultures of marine algae, USA (S. California).
– *Microcystis splendens* Hollenberg 1939: Marine, described from California, USA.
– *Microcystis thermalis* Yoneda 1938: Large colonies in thermal springs, probably a separate *Aphanocapsa* species, known from Japan; (Emoto & Hirose 1952).
– The genus *Aphanocapsa* is very diversified and many morphotypes occur (probably at the species level), particularly in tropical biotopes. Many of these

**Fig. 200.** *Aphanocapsa botryoides* (from thermal springs): after Copeland (1936) from Yellowstone Nat. Park, USA.
**Fig. 201.** Example of various concepts of *Aphanocapsa biformis*: **a** – from Geitler (1932), **b** – after Anagnostidis (1961), **c** – from Huber-Pestalozzi (1938, after G.M.Smith 1920, sub *A. rivularis*, probably *Microcystis smithii*); specimens **a** and **b** correspond rather to the genus *Chroococcidiopsis*.
**Fig. 202.** *Aphanocapsa sesciacensis*: after Frémy (1928).
**Fig. 203.** *Aphanocapsa* sp. sensu Skuja (1964; sub *Microcystis complanata*).

**205A**

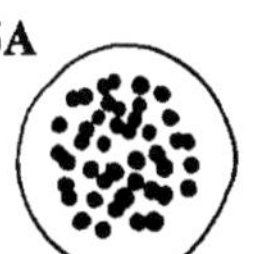

**Fig. 204.** Example of various concepts of *Microcystis pulverea* (probably various *Aphanocapsa*-species): **a** – after Kosinskaja from Kondrateva & al. (1984), **b** – after Nygaard (1940).
**Fig. 205.** *Aphanocapsa* sp.: after Lemmermann (1910, sub *Microcystis stagnalis* var. *pulchra*).
**Fig. 205A.** *Aphanocapsa nivalis*: after Lagerheim from Kol (1968).

types were incorrectly identified and need taxonomic validation or new description.

**Unrevised European species:**

- *Aphanocapsa biformis* A. Braun in Rabenhorst 1876: Numerous concepts, type unclear; (comp. fig **201**).
- *Aphanocapsa brunnea* (A. Braun) Nägeli 1849: Brownish macroscopic colonies, cells 4.5–5.5 µm in diameter, aerophytic (recorded also from tree trunks in India); in various concepts (Hansgirg 1886, Tilden 1910, Biswas 1925, Geitler 1932, Skuja 1949, Desikachary 1959, etc.).
- *Aphanocapsa depressa* (Berkeley) Rabenhorst 1865: Macroscopic colonies.
- *Aphanocapsa distans* Zalessky 1926: Sapropelic, NW. Russia.
- *Aphanocapsa endolithica* var. *violascens* Ercegović 1925: Cells 0.8–3 µm in diameter, with pale violet content; freshwater, endolithic in calcareous rocks, Croatia.
- *Aphanocapsa ferruginea* Beck-Mannagetta 1931: Cells 6.9–10.3 µm in diameter, in yellow-brown mucilage; in moorland waters, Austria.
- *Aphanocapsa fusco-lutea* forma sensu Anagnostidis 1961: Thermal, differing from type in cell colour, dimensions and ecology, Greece.
- *Aphanocapsa le-jolisii* (Thuret) Frémy 1924 [syn.: *Cagniardia adriatica* (Kützing) Trevisan 1848, *Aphanothece le-jolisii* Thuret 1905]: Marine, epilithic, similar to *A. orae*, cells 5–6 µm in diameter, violet; Croatia – Dalmatia, France.
- *Aphanocapsa lewisii* Keefe 1927: Cells 0.7–1 µm in diameter, described from a freshwater pool, in various concepts (Geitler 1932, Golubić 1967a).
- *Aphanocapsa litoralis* var. *macrococca* Hansgirg 1892: Cells 6–10 µm in diameter, various concepts (Geitler 1932, Frémy 1934, Kosinskaja 1948, Beljakova 1987, Umezaki & Modelo 1987); sensu Kaas (1985) probably a special species; described from Croatia (Dalmatia).
- *Aphanocapsa naegelii* Richter 1884: Mucilaginous greyish green or blue brownish amorphous colonies with cells 2.5–5 µm in diameter; aerophytic on rocks and stony walls, recorded sporadically from the whole Europe and (usually) from tropical countries (Seckt 1922: Argentine, Frémy 1930b: Madagascar, etc.).
- *Aphanocapsa nivalis* Lagerheim 1894: Spherical colonies up to 150 µm in diameter, with homogeneous, colourless or pale grey-violet mucilage, cells 6–7 µm in diameter, blue-green, loosely arranged; described from cryoseston in Svalbard; Kol (1928) reports from Tatra Mts. (Slovakia) similar form with cells 3–3.5 µm in diameter; (fig. **205A**).
- *Aphanocapsa pulchra* sensu auct.: Original *A. pulchra* is planktic (see under *Microcystis smithii*); the name was used in many various concepts (G.M.Smith 1920, Seckt 1922, Rayss 1944, Gonzalves & Kamat 1958, Golubić 1960, 1967a, Rino 1969, Tiwari 1975, and others).
- *Aphanocapsa richteriana* Hieronymus in Hauck et Richter 1892: Macroscopic mats (indefinite strata), cells 2–3 µm in diameter, closely aggregated one to another; recorded from pools and swamps, also in greenhouses; possibly identical with *A. muscicola* (p. 155); data from USA see in Daily (1942).
- *Aphanocapsa roberti-lamii* Frémy 1934: Macroscopic, reddish violet strata, with red-violet spherical cells, 1–3 µm in diameter, with thin, firm individual envelopes; described from walls of aquaria containing sea water, France (Atlantic coast), after Golubić (1960) near Split, Croatia – Dalmatia; according to figures rather *Chroococcidiopsis*.
- *Aphanocapsa roeseana* De Bary 1870 [syn.: *Microcystis roeseana* (De Bary) Elenkin 1938]: Colonies macroscopic, gelatinous, cells 5–8 µm in diameter; described from Germany, later records are taxonomically diverse in different authors (Geitler 1932, Desikachary 1959). Recently recorded in original sense from littoral of water biotopes in Hungary (A. Schmidt, pers. comm.).

- *Aphanocapsa rufescens* Hansgirg 1892: see *Aphanothece*.
- *Aphanocapsa sesciacensis* Frémy 1928: Cells 10–16 µm in diameter; marine (Adriatic Sea, France – Atlantic coast, Turkey – near Izmir), probably synonymous to *A. zanardinii* (Hauck) Hansgirg 1892; revision is needed (Geitler 1932, Frémy 1934, Kosinskaja 1948, Dural & al. 1997); (fig. **202**).
- *Aphanocapsa smaragdina* (Hauck) Hansgirg 1889: Marine from Adria, unclear species.
- *Aphanocapsa violacea* Grunow 1865: Not revised since description, probably *Gloeocapsa*.
- *Aphanocapsa* sp. sensu Ardré 1969: Cells 3.5–8 µm in diameter, pale green (preserved in formol), epiphytic on *Codium*; marine , coasts of Portugal.
- *Microcystis argentea* Schiller 1954: Microscopic, flat irregular colonies with spherical cells in one plane in colourless, diffluent mucilage, cells 2.4–3 µm in diameter, without gas vesicles; described from a pond in Austria (near Vienna).
- *Microcystis clarionensis* (Setchell et Gardner) de Toni 1938: Marine, described from Clarion islands, Mexico.
- *Microcystis complanata* Skuja 1964: Microscopic, always flattened colonies to 400 µm long, cells spherical, 1–1.5 µm in diameter, without gas vesicles, similar to *A. incerta;* described from N. Sweden (Abisko), not recorded since description; (fig. **203**).
- *Microcystis exigua* Zalessky 1926: Belongs to *A. incerta*-group.
- *Microcystis fusco-lutea* (Hansgirg) Forti 1907: Aerophytic on wet walls and rocks, cells 3–4.4 µm in diameter, yellow-brown mucilage; probably an *Aphanocapsa*-species.
- *Microcystis minuta* Kylin 1937: Marine (Baltic Sea), with 1.5–2 × 1.5 µm, reddish cells, attached to the substrate, probably a valid *Aphanocapsa* species (Lindstedt 1943, Kosinskaja 1948). Lindstedt (1943) described two taxonomic varieties, var. *aeruginosa* and var. *violacea*.
- *Microcystis protogenita* (Biasoletto) Cooke 1882: Similar to *A. incerta* with very densely arranged cells; not recorded later.
- *Microcystis pulverea* (Wood) Forti in de Toni 1907: Used in many different taxonomic concepts (G.M. Smith 1920, Nygaard 1926, Geitler 1932, Buell 1938, Palik 1949, Rayss 1944, Nygaard 1949, Hollerbach & al. 1953, Desikachary 1959, Tarnogradskij 1959, Kondrateva 1959, Vialov & Zdobnikova 1961, Vodeničarov 1962, Starmach 1966,1973, Tiwari 1975, Kondrateva & al. 1984, Hindák & Moustaka 1988, Senna 1992, etc.); probably a collective species; (fig. **204**).
- *Microcystis stagnalis* Lemmermann 1903 [syn.: *Polycystis pallida* Lemmermann 1898]: Commonly used name for populations in temperate zones but in different interpretations; the original concept not clear, revision is needed (the same is valid for *M. stagnalis* var. *pulchra* Lemmermann 1910); (fig. **205**).
- *Palmella mediterranea* Kützing 1843: Marine, described from Italy, possibly an *Aphanocapsa* species.
- *Palmella oceanica* Crouan 1867: Marine, described from Atlantic coast of France, possibly an *Aphanocapsa* species.
- *Pleurocapsa minor* «status adultus» sensu Uherkovich (1961) from the Tisza river, Hungary, represents probably an *Aphanocapsa* species.
- To the genus *Aphanocapsa* probably belong two provisionarily described *Microcystis* species from thermal springs in Puyehue, Chile (*M. enteromorpha*) and Masutomi, Japan (*M. eucapsiformis*) by Schwabe (1944).

## 3. **Coccopedia** Troickaja 1922

Colonies microscopic, mucilaginous, flat, tabular, with cells irregularly distributed in one layer (without forming rows perpendicular to one another). Mucilage

colourless, homogeneous, with irregular outline. Cells spherical, without individual gelatinous envelopes. Cell division in two planes, more or less perpendicular to one another; cells separate from one another soon after division. Reproduction by colony disintegration.
**Important references:** Hollerbach & al. (1953); Kiseleva (1931); Kondrateva & al. (1984).

**Key to the species:**
1a Cells 1.5–2 µm in diameter . . . . . . . . . . . . . . . . . . . . . . . . **1. C. limnetica**
1b Cells 2.3–4 µm in diameter . . . . . . . . . . . . . . . . . . . . . **2. C. turkestanica**

### 1. **Coccopedia limnetica** Troickaja 1922

Colonies flat, microscopic to 250 µm in diameter, sometimes elongate in one direction and in outline more or less oval but never lobate, with irregularly loosely or densely distributed cells. Cells spherical, pale blue-green, 1.5–2 µm in diameter, before cell division slightly elongate, after division sometimes in groups; spaces between cells sometimes up to 10 µm.
Occurrence: Small muddy water bodies, metaphytic or planktic; eastern Europe (Estonia, Ukraine).

### 2. **Coccopedia turkestanica** E. Kiseleva 1931

Colonies microscopic, flat, to circa 100 µm diameter. Cells spherical, pale blue-green, irregularly distributed, 2.3–4 µm in diameter.
Occurrence: Benthic in paddy fields, lakes and periodic pools; known from E. Europe (Ukraine), Russia (W. Siberia), and central Asia.

## 4. **Pannus** Hickel 1991

Colonies microscopic, free floating, hollow, in young stages irregularly spherical with holes, later more irregular in outline, flattened, wavy or hemispherical, arcuate, clathrate or composed of subcolonies, in old stages usually elongate and lobate, with one to three layers of more or less radially or irregularly, densely distributed cells. Around colonies fine, colourless, homogeneous, watery (diffluent at the margin) mucilage. Cells spherical, with or without gas vesicles (two different genera?).
Cell division in two planes in successive generations, perpendicularly to the surface of the colony into two equal parts. Reproduction by colony disintegration.
**Important references:** Hickel (1991); Hindák (1993).

**Key to the species:**
1a Cells without gas vesicles . . . . . . . . . . . . . . . . . . . . . . . . . . . . . . . . 2
1b Cells with gas vesicles (with visible aerotopes) . . . . . . . . . . . . . . . . . . . 3
2a Colonies in young stages spherical, later lobate to clathrate; cells 1–1.5 µm in diameter . . . . . . . . . . . . . . . . . . . . . . . . . . . . . . **1. P. spumosus**
2b Colonies flat, irregular, never clathrate; cells 1.5–2 µm in diameter . . . . . . . . . . . . . . . . . . . . . . . . . . . . . . . . . . . . **2. P. planus**
3a Cells (and colonies) reddish, 1.5–3 µm in diameter; usually living in neuston (reddish water bloom near water surface ?) . . . . . . . . . . . . . **3. P. leloupii**
3b Cells blue-green, 2–3 µm in diameter; microscopic colonies in plankton . . . . . . . . . . . . . . . . . . . . . . . . . . . . . . . . **4. P. microcystiformis**

**1. Pannus spumosus** Hickel 1991 (fig. **206**)

Colonies to 180 × 20 µm diameter, more or less spherical, hollow, later irregularly lobate with holes, with densely arranged cells in one layer in the colonial surface. Cells spherical, pale blue-green, without gas vesicles, surrounded by individual fine, colourless, indistinct mucilaginous envelopes, 1–1.5 µm in diameter.
Occurrence: Planktic, described from brackish bays and reservoirs; known from N. Germany (Schleswig-Holstein, Schlei Fjord). Data from the river Danube near Bratislava (Slovakia) and from fishponds in SW Slovakia (Hindák 1993) probably refer to another species (fig. **207**).

**2. Pannus planus** Hindák 1993 (fig. **208**)

Colonies microscopic, free floating, small, flat, never clathrate, oval to irregularly oval or more or less cylindrical in outline, 30–70 µm in diameter, with cells arranged randomly in one layer; colonial mucilage indistinct, hyaline, homogeneous, structureless. Cells spherical, 1.5–2 µm in diameter, solitary or after division in twos; cell content homogeneous, pale blue-green, without gas vesicles. Possibly belongs to *Coccopedia.*
Occurrence: Freshwater, planktic in small reservoirs (but not forming water blooms), described from a gravel-pit lake; W. Slovakia.

**3. Pannus leloupii** (Kufferath) Hindák 1993 (fig. **210**)

*Coelosphaerium leloupii* Kufferath 1950

Colonies flat or globose, microscopic, clathrate and lobate, to 200 µm long, sometimes composed of subcolonies, with cells arranged in one layer on the colonial surface. Cells more or less spherical, 1.5–3 µm in diameter, with reddish cell content, with gas vesicles (visible aerotopes).
Occurrence: Described from neuston (water bloom ?) in one pond in Belgium; not confirmed.

**4. Pannus microcystiformis** Hindák 1993 (fig. **211**)

*Microcystis* sp. sensu Hindák et Moustaka 1988

Colonies microscopic, free floating, in young stages few-celled, later multicellular, flat, oval or irregular in outline, sometimes clathrate and composed of small subcolonies, to 200 µm diameter (or larger), with irregularly and densely arranged cells; mucilaginous envelopes fine, homogeneous, hyaline, structureless. Cells spherical, 2–3 µm in diameter, in twos after division, arranged irregularly in a layer 1 – (2–3) cell thick in the plane of colony, sometimes in irregular rows or chains; cell content pale blue-green or grey, with conspicuous aerotopes.
Occurrence: Described from the autumn and winter plankton of Lake Volvi, N. Greece; not forming a water bloom. Not confirmed.

**Unrevised species:**

– *Coelosphaerium compositum* Liebetanz 1925: Similar colonial morphology to *P. spumosus*: colonies 10–20 µm, compound colonies to 80 µm in diameter; cells 1–1.5 µm in diameter; from metaphyton in brackish waters in Poland (Baltic coast). Probably identical with *Pannus spumosus*; (fig. **209**).

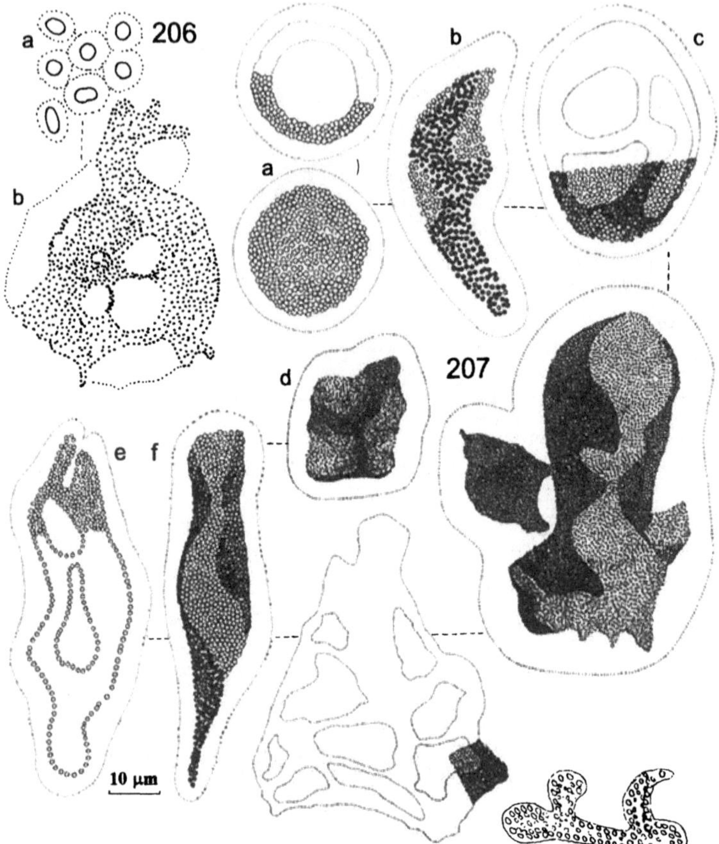


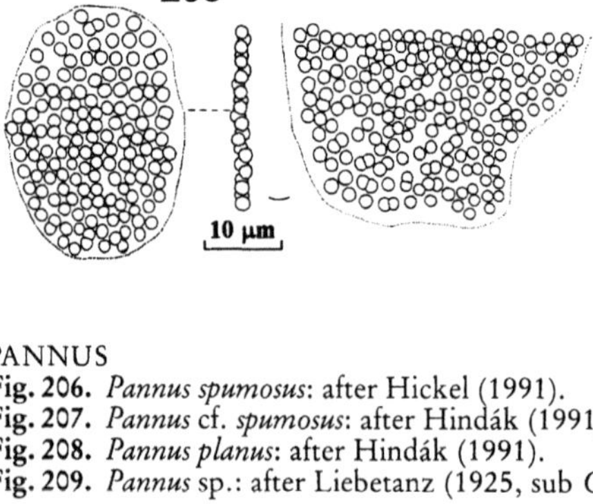


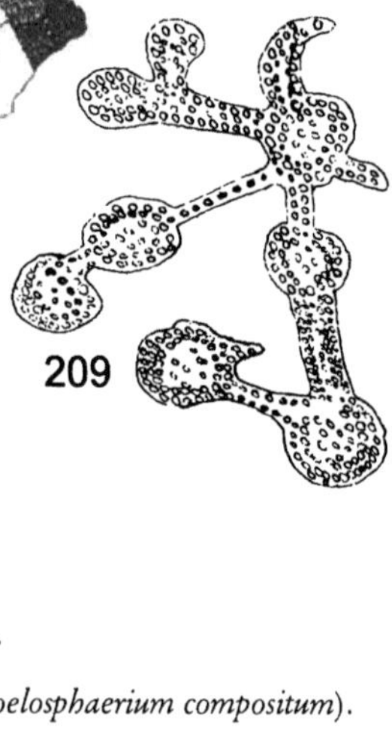


PANNUS

**Fig. 206.** *Pannus spumosus*: after Hickel (1991).
**Fig. 207.** *Pannus* cf. *spumosus*: after Hindák (1991).
**Fig. 208.** *Pannus planus*: after Hindák (1991).
**Fig. 209.** *Pannus* sp.: after Liebetanz (1925, sub *Coelosphaerium compositum*).

## 5. Merismopedia Meyen 1839

*Agmenellum* Brébisson 1839; *Gonidium* Ehrenberg in Meneghini 1849; *Merismopedium* sensu auct.; *Merismopedia* subg. *Pseudoholopedia* Ryppowa 1925; *Pseudoholopedia* (Ryppowa) Elenkin 1938 incl.

Colonies microscopic, rarely (in large species) macroscopic, free -living, tabular, with one layer of densely or loosely arranged cells in the plane of the colony, square or rectangular, later sometimes with irregular outline, flat or wavy; large colonies sometimes contorted or composed of several subcolonies. Colonies consist of few to very many (in several species over 4000) cells; cells arranged in perpendicular rows, distant or close together. Colonial mucilaginous envelopes colourless, structureless, hyaline, fine, usually with indistinct narrow or wide margin; cells sometimes with individual gelatinous envelopes (in several species facultatively so). Cells spherical or widely elliptical (when they are elongate or oval, the longer axis lies in the plane of the colony), after division hemispherical, with homogeneous, pale or bright blue-green or reddish (violet, pink) content; in a few species in central parts of cells appear refractive bodies or aerotopes. The parietal arrangement of thylakoids was found in all up to now studied members of *Merismopedia.*

Cell division (binary fission) into two equal daughter cells, in two planes perpendicularly to one another and to the plane of the colony in successive generations; cells do not change their alignment after division (reason for regular perpendicular rows of cells) and reach the original size before next division. Reproduction by disintegration of colonies.

Notes : Small-celled *Merismopedia* species with reddish cell content can be easily misidentified as *Thiopedia* (*Rhodobacteria*). - Palinska & al. (1996) isolated from a salinic microbial mats in Norderney Island (Germany) five strains, which corresponded morphologically to three species, *M. elegans, M. glauca* and *M. punctata.* They proved the evolutionary relationship to *Synechocystis* (one strain) and the genotype similarity of all other studied strains by the 16S rDNA sequencing. The stability of different phenotypes, ecotypes, and variability of various species need the further study and confirmation. The identity of, e.g., very variable *M. glauca* from saline biotopes (sensu Palinska & al. 1996) and *M. glauca* from acidic peaty waters should be also revised.

**Important references:** Geitler (1932, 1942, 1960); Bourrelly (1970, 1985); Hindák (1982, 1992b); Kondrateva & al. (1984); Komárek & Anagnostidis (1986); Kováčik (1988); Palinska & al. (1996).

**Key to the species:**

**1a** Cells with aerotopes . . . . . . . . . . 2

**1b** Cells without aerotopes . . . . . . . . . . 4

**2a** Cells with 1 – (2) aerotopes, slightly elongate, 3–5 µm in diameter . . . . . . . . . . **1. M. insignis**

**2b** Cells with more than 2 aerotopes, cell diameter usually smaller than 3 µm . 3

**3a** Cells 2–3 µm in diameter; in large water bodies (lakes, resevoirs) . . . . . . . . . . **2. M. trolleri**

**3b** Cells 1.3–2 µm in diameter; in mesotrophic ponds, lakes and reservoirs . . . . . . . . . . **3. M. marssonii**

**4a** Cells smaller than 2 (2.5) µm in diameter . . . . . . . . . . 5

**4b** Cells greater than (2) 2.5 µm in diameter . . . . . . . . . . 10

**5a** Colonies attached to small stones and sandy grains in psammon; cells (0.7) 1.5–2 µm in diameter . . . . . . . . . . **4. M. affixa**

**5b** Colonies not attached to the substrate, usually metaphytic . . . . . . . . . . 6

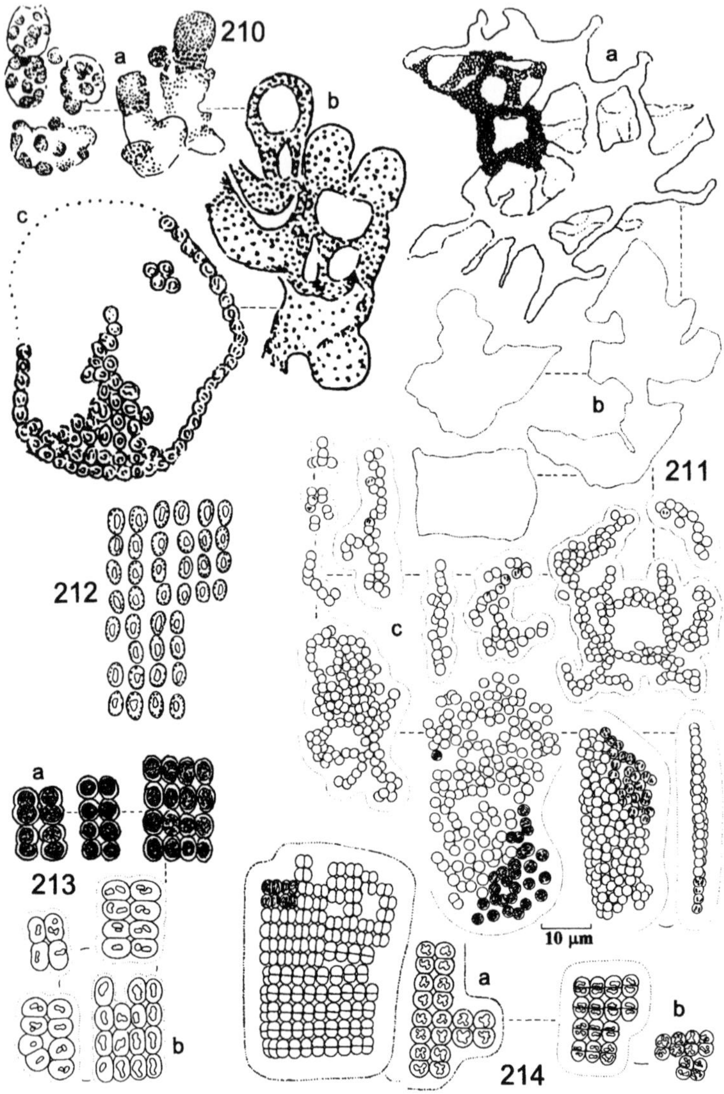


PANNUS

**Fig. 210.** *Pannus leloupii*: **a** – colonies, **b** – part of a colony, **c** – detail of a colony after Kufferath (1950).

**Fig. 211.** *Pannus microcystiformis*: **a** – colony, **b** – outline of colonies, **c** – detail of colonies of varied age; after Hindák & Moustaka (1988, sub *Microcystis* sp.).

**6a** Colonies with sparsely distributed cells, spaces between cells are several times greater than the cell diameter . . . . . . . . . . . . . . . . **5. M. danubiana**
**6b** Colonies with densely arranged cells, spaces mostly less than two cell diameters . . . . . . . . . . . . . . . . . . . . . . . . . . . . . . . . . . . . . . . . . . 7
**7a** Subaerophytic on wet rocks and stony walls, small quadrate colonies, cells up to 1 µm in diameter . . . . . . . . . . . . . . . . . . . . . . . . . . **6. M. minima**
**7b** Planktic and metaphytic in hydrobiotopes . . . . . . . . . . . . . . . . . . . . . . . 8
**8a** Cells 0.4–1.6 (2) µm in diameter, in small, tabular colonies, planktic species 9
**8b** Cells circa 2 µm in diameter, with individual mucilaginous envelopes, in many-celled colonies, metaphytic in Arctic swamps . . . . . . . **9. M. arctica**
**9a** Cells 0.4–1 µm in diameter, in quadrate colonies . . . . **7. M. warmingiana**
**9b** Cells 0.5–1.6 (2.4 ?) µm in diameter, in rectangular colonies **8. M. tenuissima**
**10a** Cells mainly (2) 2.5–4.5 (6.5) µm in diameter . . . . . . . . . . . . . . . . . . 11
**10b** Cells always over 5 µm in diameter . . . . . . . . . . . . . . . . . . . . . . . . . . . 18
**11a** Colonies with sparsely distributed cells, with large spaces between them, cells 2.4–4.2 µm in diameter . . . . . . . . . . . . . . . . . . . . . . . **12. M. duplex**
**11b** Colonies without large spaces between cells (less than two cell diameters) 12
**12a** Cells 2–3.5 µm in diameter, planktic species . . . . . . . . . . . . . . . . . . . . 13
**12b** Cells more than 3 µm in diameter, planktic and metaphytic species . . . . 14
**13a** In few-celled colonies (usually with 3–16, rarely up to 32 cells), mainly somewhat irregularly arranged («incomplete colonies»); cells 2–3 µm in diameter . . . . . . . . . . . . . . . . . . . . . . . . . . . . . . . . . **10. M. hyalina**
**13b** Colonies regular, plate-shaped, usually with 16, 32 or more cells; cells (2 ?) 2.5–3.5 µm in diameter . . . . . . . . . . . . . . . . . . . . . . . . **11. M. punctata**
**14a** Colonies with ferric precipitates close to the cells (in granular black form), cells 3–6 × 2.8–5 µm; planktic in small water bodies (backwaters, pools in inundation zone near rivers) . . . . . . . . . . . . . . . . . . . . **13. M. ferrophila**
**14b** Colonies without ferric precipitates, usually metaphytic species in various hydrobiotopes . . . . . . . . . . . . . . . . . . . . . . . . . . . . . . . . . . . . . . . . . 15
**15a** Colonies simple, usually with (8) – 16–64 – (128) cells . . . . . . . . . . . . 16
**15b** Colonies usually compound together and multicellular, wavy, with more than (32) 64 cells; cells 3–4 µm in diameter . . . . . . . . . . . **15. M. revoluta**
**16a** Thermal species, cells 3–4 µm in diameter . . . . . . . . . . . **14. M. thermalis**
**16b** Not in thermal biotopes, cells 3–6 (6.5) µm in diameter . . . . . . . . . . . . 17
**17a** Freshwater, common in swamps and small water bodies . . . **16. M. glauca**

MERISMOPEDIA

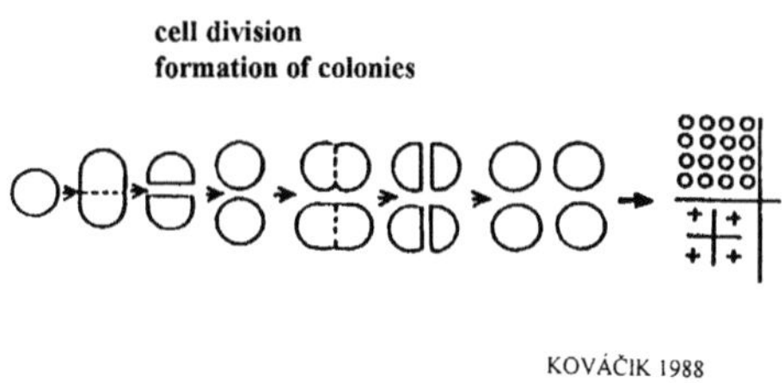


**Fig. 212.** *Merismopedia insignis*: after Škorbatov (1923).
**Fig. 213.** *Merismopedia trolleri*: **a** – after Bachmann (1921), **b** – after Hindák (1992).
**Fig. 214.** *Merismopedia marssonii*: after Hindák (1992).

**17b** Marine, coastal swamps and pools . . . . . . . . . . . . . **17. M. mediterranea**
**18a** Cells pink or red, rarely greenish, marine coasts, cells 6–13 µm in diameter . . . . . . . . . . . . . . . . . . . . . . . . . . . . . . . . . . . . . **18. M. litoralis**
**18b** Cells more or less blue-green . . . . . . . . . . . . . . . . . . . . . . . . . . . **19**
**19a** Cells with a more or less quadrangular-rounded outline, circa 6.5 × 4.4 µm; in moorland waters . . . . . . . . . . . . . . . . . . . . . . . . . . **21. M. angularis**
**19b** Cells spherical, oval or (after division) hemispherical . . . . . . . . . . . . . . 20
**20a** Colonies large, many-celled; cells 4.5–11 (–14) µm in diameter . . . . . . . **21**
**20b** Colonies few-celled (usually not more than 16–32 – celled); cells 10–17 µm in diameter . . . . . . . . . . . . . . . . . . . . . . . . . . . . . . . . **22. M. smithii**
**21a** Colonies wavy, composed of separated flat plates; cells spherical or hemispherical, blue-green, 4.5–7.8 µm in diameter . . . . . . . . **19. M. convoluta**
**21b** Colonies more or less flat; cells slightly elongate, bright blue-green, 5–14 × 5–11 µm . . . . . . . . . . . . . . . . . . . . . . . . . . . . . . . **20. M. elegans**

## 1. **Merismopedia insignis** Škorbatov 1923 (fig. **212**)

*Merismopedia jeanelii* Bachmann 1938; *Merismopedia glauca* f. *insignis* (Škorbatov) Geitler 1925 sec. Gruia 1963

Colonies small, plate-like, with widely separated cells, circa 2–4 µm from one another. Cells slightly elongate, oval, with aerotopes (usually with only one central aerotope), with pale blue-green content, 3–5 (5.5) × 3–4.5 (4.8) µm.
Occurrence: Planktic in freshwater lakes and rivers, known from widely separated localities in Eurasia and S. Africa (not rare).

## 2. **Merismopedia trolleri** Bachmann 1920 (fig. **213**)

Colonies small, usually (4) 8–16-celled, with rather densely arranged cells, sometimes in groups. Mucilage hyaline, forming a circa 1 µm wide margin around the cells. Cells spherical or widely oval, after division hemispherical, sometimes surrounded by individual gelatinous envelopes, with pale blue-green or reddish cell content and several (1–3) aerotopes (?), 2–3 µm in diameter.
Occurrence: Planktic in freshwater lakes all over the temperate zone, not very common; sporadic records from tropical countries should be checked.

## 3. **Merismopedia marssonii** Lemmermann 1900 (fig. **214**)

Colonies flat to slightly curved, more or less rectangular with up to 128 (> 200) densely arranged cells. Mucilage distinct, hyaline, to 3 µm beyond cells. Cells spherical or subspherical after division, with blue-green content and up to three aerotopes in the central part of cells, 1.3–2 (3.3 ?) µm in diameter.
Occurrence: Planktic in freshwaters (eutrophic lakes and ponds over the Eurasian temperate zone) but sporadic; records from tropical countries (from India with 4-celled colonies and slightly larger cells; E. Africa – lake Victoria) probably represent other species.

---

MERISMOPEDIA
**Fig. 215.** *Merismopedia affixa*: after Richter (1895).

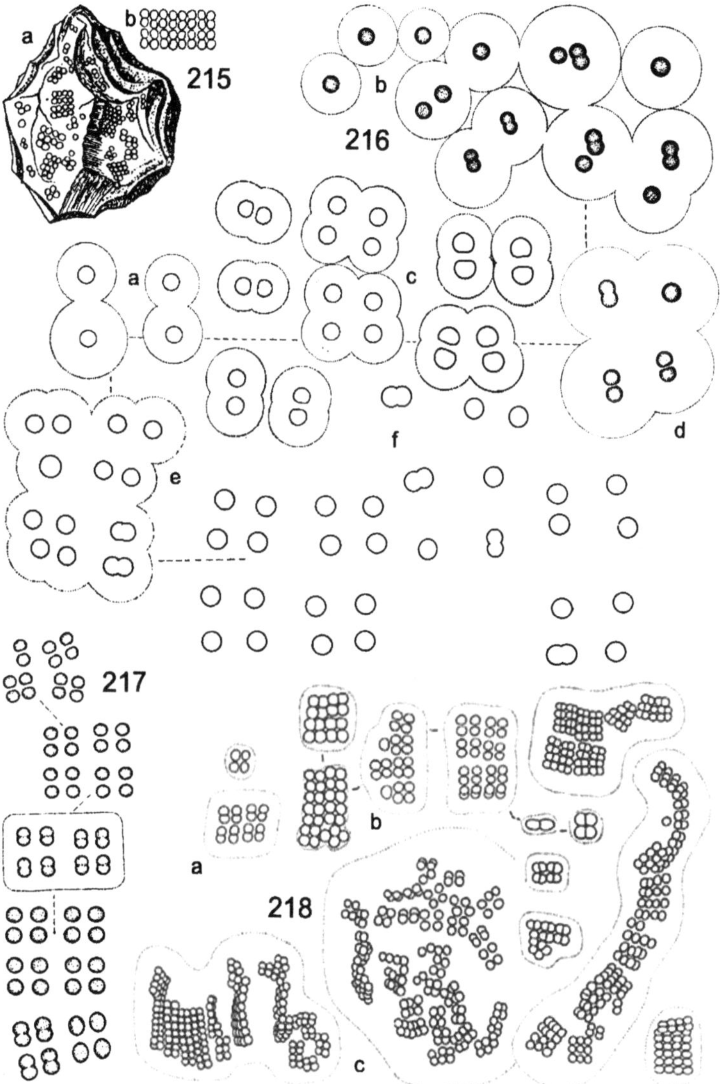


**Fig. 216.** *Merismopedia danubiana*: after Hortobágyi (1974).
**Fig. 217.** *Merismopedia* cf.*minima*: after Nygaard (1979, sub *M. tenuissima*), morphologically and ecologically (clear water biotopes in Greenland) corresponding rather with *M. minima*.
**Fig. 218.** *Merismopedia warmingiana*: after Hindák (1992, sub *M. minima*).

**4. Merismopedia affixa** Richter 1895 (fig. **215**)

Colonies attached to small stones or sandy particles, with 4–32 cells densely arranged in one plane in perpendicular rows, but sometimes not with a regular quadrate outline, rarely irregular. Cells spherical or slightly «polygonal» in outline, pale blue-green, (0.7) 1.5–2 µm in diameter.
Occurrence: Marine littoral, beaches of Baltic Sea; records from India concern another species.

**5. Merismopedia danubiana** Hortobágyi 1974 (fig. **216**)

Colonies with very sparsely distributed cells; in young colonies the gelatinous envelopes confluent, in old ones the cells with individual mucilaginous envelopes; colonies with 2–32 cells; around cells wide colourless slime; cells not always in one plane (?). Cells spherical, slightly elongate before division, without aerotopes, with blue-green content, 1.5–2.7 µm in diameter.
Occurrence: Known only from reservoirs near Budapest, Hungary.

**6. Merismopedia minima** Beck 1897 (fig. **217**)

Colonies small, quadrate, with 4–32 cells, with more or less densely arranged cells. Cells spherical, greyish or pale blue-green, sometimes with individual gelatinous envelopes, 0.4–0.8 (1.2) µm in diameter.
Occurrence: Subaerophytic on wet rocks, mainly in mountains, rarely metaphytic in clear mountain springs, pools and ponds; in corresponding biotopes over the whole temperate zones up to subpolar regions and in high mountains but sporadic; planktic records probably concern *M. warmingiana.*

**7. Merismopedia warmingiana** Lagerheim 1883 (fig. **218**)

*Merismopedia minima* sensu auct. (aquatic populations)

Colonies regular, flat, quadrate, microscopic with more or less densely arranged 4–16 (64) cells (sometimes grouped in tetrads within colony) and composed of subcolonies, easily disintegrating. Mucilage fine, diffluent, sometimes almost invisible (staining !), usually wide (circa 2 µm or wider) around cells. Cells 0.5–1 (1.2) µm in diameter, spherical, slightly elongate before division, pale blue-green.
Occurrence: Planktic in eutrophic water bodies, sometimes fertilized or polluted, common particularly in coastal, slightly brackish and saline ponds and bays; probably cosmopolitan; Europe: frequently occurring (occasionally in masses), but not commonly distributed.

**8. Merismopedia tenuissima** Lemmermann 1898 (fig. **219**)

Colonies flat, tabular, usually rectangular (not exactly quadrate), sometimes slightly wavy or forming an aggregation of subcolonies with (8) 16 (up to > 100) cells, with more or less densely aggregated cells. Mucilage colourless, hyaline, distinct, diffluent, sometimes individual gelatinous, hardly visible envelopes are formed around cells. Cells spherical or oval, after division hemispherical, with homogeneous, pale grey-blue or blue-green content, 0.4–1.6 (2–2.5 ?) µm in diameter.
Occurrence: Common in stagnant eutrophic freshwaters, mainly in fertilized fishponds, less common in lakes, also in brackish waters, particularly in the warm season of the year; probably cosmopolitan, in Europe common.

**9. Merismopedia arctica** (Kosinskaja) Komárek et Anagnostidis 1995

*Merismopedia punctata* f. *arctica* Kosinskaja 1933

Colonies large, multicellular, with more or less densely arranged cells in perpendicular rows, with a distinct margin of colonial slime mirroring that of the marginal cells. Cells usually surrounded by individual gelatinous, colourless envelopes, more or less spherical or hemispherical, 2–3.1 µm in diameter.
Occurrence: Freshwater, metaphytic in swamps and pools, small lakes, rarely in running waters; northern subpolar regions (tundra), Arctic islands of northern Europe and Asia, but recorded also from central Asia.

**10. Merismopedia hyalina** (Ehrenberg) Kützing 1845 (fig. **220**)

Colonies usually small, 4–8 (16)-celled, rarely with larger number of cells (in the case of composed colonies), flat, usually with cells in short rows, sometimes with irregularities («lacking» cells). Mucilage colourless, but often distinct, overlapping the cells up to 4 µm; individual envelopes around cells rarely visible. Cells spherical, hemispherical after division, with pale greyish or blue-green content, (1) 2–3 (3.5 ?)µm in diameter.
Occurrence: Planktic in mesotrophic and eutrophic freshwaters, probably also in brackish water, common; cosmopolitan, in Europe widely distributed.

**11. Merismopedia punctata** Meyen 1839 (fig. **222**)

*Merismopedia kuetzingii* Nägeli 1849; *Merismopedia convoluta* f. *minor* Wille 1922; *Merismopedia baumanii* Kufferath 1942; *Agmenellum quadruplicatum* sensu auct. (after Drouet & Daily 1956 p.p.)

Colonies tabular, flat, up to 64-celled (rarely more), usually in regular rows with more or less loosely arranged cells, rarely composed of subcolonies. Mucilage distinct or diffluent, colourless. Cells spherical, widely oval or hemispherical, pale blue-green, without individual gelatinous envelopes, (2) 2.5–3.6 (4 ?) µm in diameter.
Occurrence: Commonly planktic and metaphytic in mesotrophic freshwater habitats, but similar morphotypes often recorded also in brackish water over the world (f. *minor* Lagerheim 1883, other taxon ?); cosmopolitan but more widely distributed in tropical and warmer areas; Europe: sporadically.

**12. Merismopedia duplex** Playfair 1918 (fig. **221**)

*Merismopedia cyanea* Playfair 1918; *Merismopedia distans* Fedorov 1969; *Merismopedia distans* Messikommer 1971

Colonies small, microscopic, free floating, with 8–32 cells, with large spaces between cells (more than 2 cell diameters), but in regular perpendicular rows. Mucilage very fine, diffluent, colourless, indistinct. Cells spherical or oval with pale blue-green content, (2.4) 2.8–4.2 µm in diameter.
Occurrence: Freshwater, metaphytic in swamps, moorland and peaty waters, rarely secondarily planktic in lakes; cosmopolitan (?), but sparsely distributed in corresponding biotopes; Europe: Switzerland, Russia (W. Siberia), probably more widely distributed.

**13. Merismopedia ferrophila** Hindák 1982 (fig. **223**)

Colonies small, free, plate-like or oval in outline, 2–64-celled (rarely more) with fine, diffluent, colourless and structureless, hyaline mucilage, with ferric precipitates on the surface or in the neighbourhood of cells (precipitates in shape of

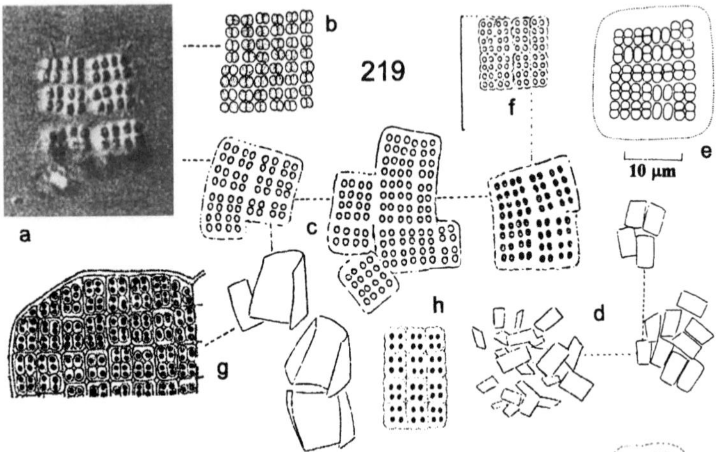


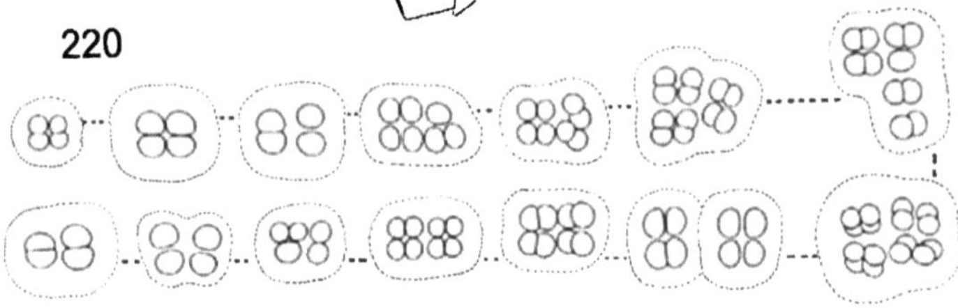


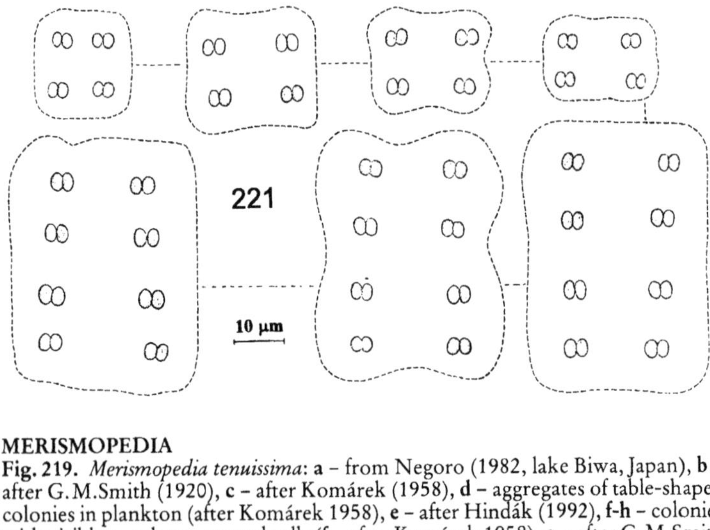


MERISMOPEDIA

**Fig. 219.** *Merismopedia tenuissima*: **a** – from Negoro (1982, lake Biwa, Japan), **b** – after G.M.Smith (1920), **c** – after Komárek (1958), **d** – aggregates of table-shaped colonies in plankton (after Komárek 1958), **e** – after Hindák (1992), **f-h** – colonies with visible envelopes around cells (**f** – after Komárek 1958), **g** – after G.M.Smith

brown, spherical or irregular bodies, 0.5–3 µm in diameter); cells sometimes in tetrads or in groups of 8–16 cells close together; mucilaginous envelopes hyaline, surrounding cells by 3–8 µm wide margin; cells spherical or widely oval, hemispherical after division, with grey or pale blue-green content, 3–5 (6) × 2.8–3.5 (5) µm.
Occurrence: In summer and autumn plankton of stagnant and flowing waters (ponds, gravel-pit lakes, rivers) in inundation zone of the river Danube; known only from S. and W. Slovakia.

### 14. Merismopedia thermalis Kützing 1843

*Agmenellum thermale* (Kützing) Drouet et Daily 1956 pro typo (non sensu auct. post.)

Colonies small, tabular, flat, with 16–32–64 cells (rarely more), with densely arranged cells, sometimes a little irregular in outline. Mucilage colourless, scarcely visible. Cells spherical or oval before division, pale blue-green, (2 ?) 3–4 µm in diameter.
Occurrence: Thermal waters, in metaphyton or free floating; known from Eurasia.

### 15. Merismopedia revoluta Askenasy 1894

Colonies large, with several hundred cells, sometimes composed of more flat or wavy subcolonies, with more or less densely packed cells in perpendicular rows. Cells spherical, pale blue-green, 3–4 (before division –7) µm in diameter.
Occurrence: Freshwater, metaphytic in small water bodies with hydrophytes; Europe: sporadic, known from a few localities.

### 16. Merismopedia glauca (Ehrenberg) Kützing 1845 (fig. **225**)

*Merismopedia aeruginea* Brébisson in Kützing 1849 ?; *Merismopedia nova* Wood 1872; *Agmenellum thermale* sensu auct.

Colonies flat, quadrate, usually 64-celled (rarely more), with more or less densely and regularly arranged cells, sometimes consisting of subcolonies. Mucilage colourless, sometimes diffluent but distinct, slightly extending beyond marginal cells (2–4 µm). Cells spherical or widely oval before divison, hemispherical after

---

from Geitler (1932), **h** – after Komárková-Legnerová & Cronberg (1992), **i** – after Fedorov (1967, sub *M. minima*).

**219i**

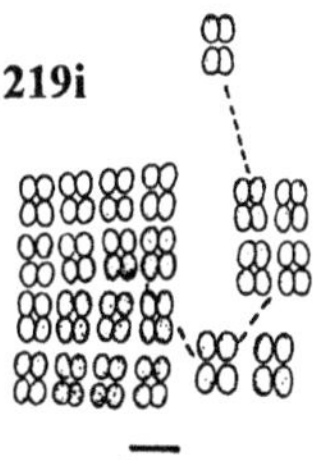


**Fig. 220.** *Merismopedia* cf. *hyalina*: after Hindák (1992, sub *M. punctata*); population corresponding rather to *M. hyalina*.
**Fig. 221.** *Merismopedia duplex*: after Fedorov (1969, sub *M. distans*).

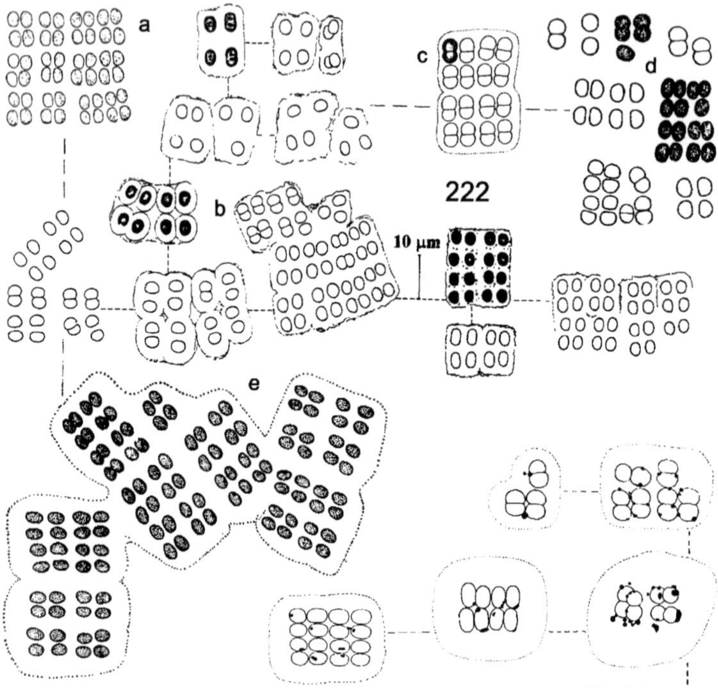


222

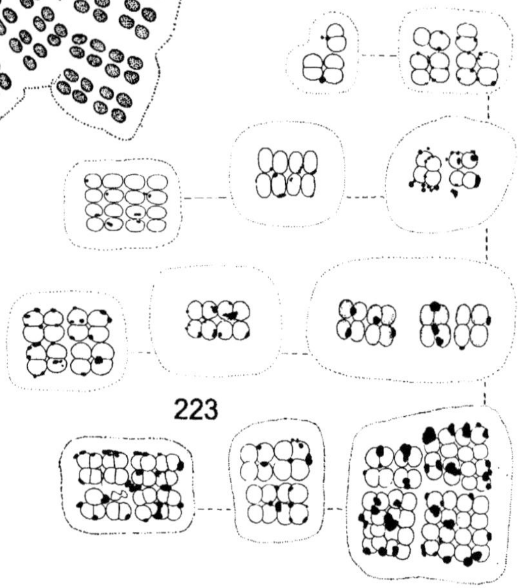

223

224

MERISMOPEDIA

**Fig. 222.** *Merismopedia punctata*: **a** – after G.M.Smith (1920), **b** – after Komárek (1975), **c** – after Hindák (1992), **d** – after Komárek (1989), **e** – after G.M.Smith from Kondrateva & al. (1984).

**Fig. 223.** *Merismopedia ferrophila*: after Hindák (1989).

**Fig. 224.** *Merismopedia mediterranea*: after Kosinskaja (1948, sub *M. glauca* f. *mediterranea*).

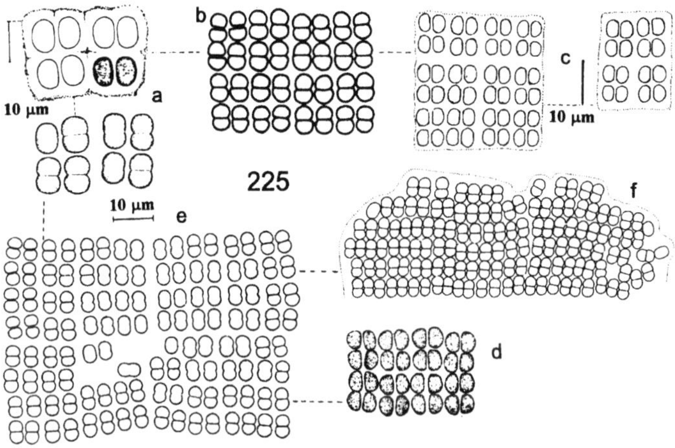


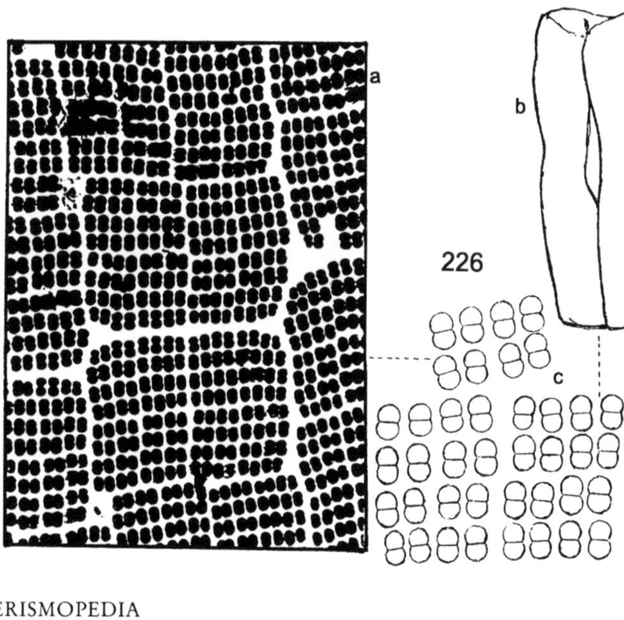


MERISMOPEDIA

**Fig. 225.** *Merismopedia glauca*: **a-d** – small colonies, **e-f** – parts of old colonies (more like *M. convoluta*); **a** – after Komárek (1975), **b** – after Werner & Rosa (1992), **c** – after Komárková-Legnerová & Cronberg (1994), **d** – after G.M.Smith (1920), **e** – after Geitler (1960), **f** – after Hindák (1989).

**Fig. 226.** *Merismopedia convoluta*: **a** – part of a compound colony, **b** – shape of a large colony, **c** – detail of cell arrangement; **a** – after Pascher from Geitler (1960), **b-c** – orig. Komárek.

division, with homogeneous, pale blue-green content (specimens with pink cells were also recorded – deep water forms ?), (2.8) 3–6 µm in diameter.
Occurrence: Freshwater, metaphytic and benthic in swamps, stagnant waters, littoral of mainly unpolluted ponds and lakes, facultatively planktic; cosmopolitan distribution, common in Europe. Morphologically similar populations from saline waters belong probably into the variation range of another species (? var. *amethystina* (Lagerheim) Lagerheim ex Forti 1907; *Merismopedia aeruginea* f. *ovata* Wille 1924).

### 17. **Merismopedia mediterranea** Nägeli 1849 (fig. **224**)

*Merismopedia glauca* f. *mediterranea* (Nägeli) Collins 1910

Colonies flat with many cells (usually 64 and more), regularly and densely arranged in perpendicular rows. Mucilage colourless, usually distinct at the margin. Cells spherical, slightly oval before division, with grey blue-green content (specimens with pink cells were recorded – special taxon ?), 3–6.5 µm in diameter.
Occurrence: Marine littoral, salty swamps and pools; known from Mediterranean region, Portugal and coasts all over the temperate and subtropical zones.

### 18. **Merismopedia litoralis** (Oersted) Rabenhorst 1865

*Erythroconis litoralis* Oersted 1842; *Merismopedia glauca* subsp. *amethystina* Lagerheim 1883; *Merismopedia elegans* var. *mandalensis* Wille 1900; *Merismopedia massartii* Kufferath 1942

Colonies small, flat, usually with about 32 cells (rarely more), with irregular outline. Mucilage colourless, distinct, cells rarely spherical, usually distinctly elongate, greenish or with reddish content, 6–13 × 6–10 µm or more or less 7 µm in diameter. (Unrevised species !).
Occurrence: Recorded from the marine littoral, on *Zostera* leaves, etc.; known from all European marine coasts, mainly in North Sea, Mediterranean and Black Sea regions.

### 19. **Merismopedia convoluta** Brébisson in Kützing 1849 (fig. **226**)

*Merismopedia willei* Gardner 1927; *Pseudoholopedia convoluta* (Brébisson) Elenkin 1938 pro typo (non sensu Elenkin)

Colonies very large, flat, wavy (rolled up) or arcuate plates with many cells (to over 560), to several mm in diameter (then macroscopically visible), sometimes composed of large quadrate subcolonies, with regularly arranged cells in perpendicular rows. Mucilage colourless, with distinct margin. Cells perfectly spherical (after division hemispherical), blue-green, 4–5.2 µm in diameter.
Occurrence: Freshwater, metaphytic in stagnant and flowing waters, usually with submerged hydrophytes; probably cosmopolitan, but sporadically distributed, mainly in tropical countries; Europe: known from several localities.

### 20. **Merismopedia elegans** A. Braun in Kützing 1849 (fig. **227**)

Colonies few- to many-celled (16–4000), in outline usually rectangular, wavy, with cells more or less densely arranged in perpendicular rows (in old colonies with a large number of cells sometimes slightly irregularly arranged). Mucilage more or less firm, with distinct margin up to 10 µm beyond the cells. Cells widely oval, rarely almost spherical, after division hemispherical, with homogeneous or finely granular content, bright blue-green or greenish, 5–9 × (4) 5–7 µm.

Occurrence: Freshwater, metaphytic and epipelic in benthos of springs, swamps, pools, pond littorals, rivers, occasionally tychoplanktic, usually in slightly acidic waters; probably cosmopolitan, sporadically in Europe; records from saltwaters need confirmation (they are similar morphologically, but belong probably to another species).

## 21. **Merismopedia angularis** Thompson 1938 (fig. **228**)

Colonies microscopic, tabular, flat, small to large, with up to 32 (or more, - 256) cells, in a thin, firm gelatinous envelope. Mucilage usually distinct, hyaline, colourless, extending slightly beyond the cell clusters. Cells very densely arranged in perpendicular rows, in outline rounded rectangular, sometimes slightly alternating in rows, bright blue-green, 4–6.5 (7) × 4–4.4 µm.
Occurrence: Freshwater, metaphytic in acidic swamps and moors; sparsely distributed over the whole northern temperate zone up to subpolar regions (Svalbard), in Europe less frequently, usually in mountains (Alps, Šumava Mts., etc.). Species known from nature, but its stability was proved in culture.

## 22. **Merismopedia smithii** de Toni 1939 (fig. **229**)

*Merismopedia elegans* var. *major* G.M. Smith 1920; *M. major* (G.M. Smith) Geitler 1925

Colonies small, few-celled (usually with not more than 16–32 cells), flat. Cells more or less densely packed, spherical or oval before division, 10–14 (17) µm in diameter, with bright blue-green or greenish, homogeneous content, without prominent granules, without individual envelopes. Mucilage colourless, diffluent, usually indistinct.
Occurrence: Freshwater, metaphytic species, growing in cold swamps and moorland waters; distributed over northern USA and Canada, in Europe known from N. Sweden (Abisko), other European records probably concern other species and should be checked (e.g., Austria - Lunzer See, etc.).

**Species outside Europe:**
- *Merismopedia aeruginea* Brébisson in Kützing sensu Frémy 1930: Colonies elongate, with 4–64 cells, cells about 5 µm in diameter; usually synonymized with *M. glauca*, but accepted by Frémy (1930) from tropical Africa and by Diaz & Wenzel (1993) from Andes, Argentine, as a separate species.
- *Merismopedia cantonensis* Chu 1987: With small cells and wavy, rolled up colonies, metaphyton and plankton of unpolluted swamps, littoral of lakes, etc., in tropical countries (S. China, S. and central Africa).
- *Merismopedia gardneri* (Collins) Setchell in Gardner 1906: Pools on Pacific coast of N. America (Gardner 1906, Kosinskaja 1948).
- *Merismopedia remota* Hirose et Chung in Chung 1970: Freshwater, described from Youngnam, Korea.
- *Merismopedia sinica* Ley 1947: Rectangular cells with aerotopes; described from China.
- *Merismopedia tenuissima* var. *polyedrica* Guarrera in Guarrera et al. 1986: Colonies tabular, arranged perpendicularly to one another, cells circa 2 µm in diameter; described from Argentine (Tierra del Fuego).

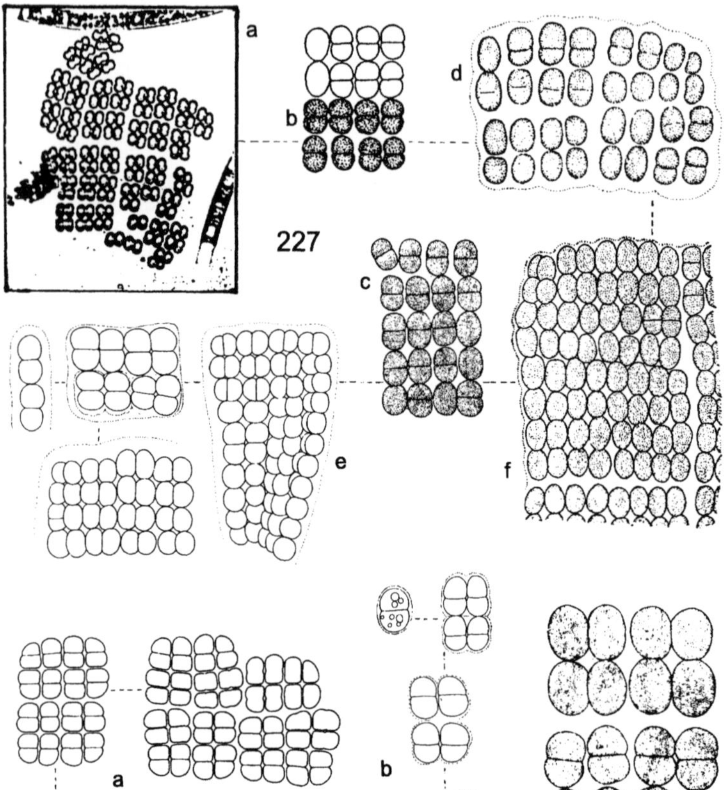


MERISMOPEDIA

**Fig. 227.** *Merismopedia elegans*: **a** – from Negoro (1982), **b** – after Teiling (1944), **c** – after G.M.Smith (1920), **d** – after Kosinskaja from Kondrateva & al. (1984), **e** – after Hindák (1992), **f** – after G.M.Smith (1950).

**Fig. 228.** *Merismopedia angularis*: **a** – after Geitler (1960, sub «*M. glauca* aus einer alter Rohkultur»), **b** – after Hindák (1992, sub «*Merismopedia*? sp.»), **c** – after Thompson (1938).

**Unrevised European species:**
- *Merismopedia chondroidea* Wittrock in Wittrock & Nordstedt 1878: Cells reddish, springs near Uppsala, Sweden.
- *Merismopedia paludosa* Bennet 1886: Cells rectangular, described from England.
- *Merismopedia tranquilla* (Ehrenberg) Trevisan 1845: Probably identical with *M. punctata.*
- *Merismopedia quadruplicata* (Meneghini) Trevisan 1845: Various concepts; the name «*Agmenellum quadruplicatum* (Meneghini) Brébisson» (Drouet & Daily 1956) was usually used for *M. punctata* and *M. glauca*; strains in world collections designated by that name contain mainly *Synechococcus nidulans.*
- *Merismopedia thermalis* Kützing 1843 sensu auct. post.: Originally thermal species (comp. p. 177); the later nomenclatural combination «*Agmenellum thermale* (Kützing) Drouet et Daily 1956» was used in a very wide concept; several authors identified *M. glauca, M. mediterranea* and *Microcrocis marina* by that name.
- *Merismopedia* sp. sensu Hindák 1992: From a peat bog lake near Luzern, Switzerland, belongs probably to *M. angularis*; (fig. **227b**).
- *Pseudoholopedia convoluta* var. *subsalina* Proškina-Lavrenko 1950:
- *Synechocystis* sp. sensu Castenholz & Waterbury 1989 (Bergey's Manual) belongs probably to the genus *Merismopedia*: low GC-cluster, freshwater, obligatory photoautotroph, facultatively gliding motility of cells, facultatively high percentage of phycoerythrin, growing in flat sheets.

## 6. Cyanotetras Hindák 1988

Colonies microscopic, free floating, flat, tabular, one-layered, few-celled, surrounded by wide, homogeneous mucilaginous envelope. Cells oval, arranged in twos or flat tetrads. Cell content homogeneous, with solitary granules.
Cell division into two equal parts, in two perpendicular planes in successive generations; reproduction by fragmentation of colonies.
**Important references:** Hindák (1988).

### 1. Cyanotetras fusca Hindák 1988 (fig. 230)

Colonies 2–4 – 16-celled, with rectangular arrangement of cells, which are sometimes slightly irregularly shifted from one another. Mucilaginous envelopes wide, surrounding the flat colony by a rounded or oval, wide (6–13 µm) margin, brownish with very fine Fe-precipitations. Cells broadly oval, pale or bright blue-green, without aerotopes, 1.5–2.5 × 1.2–1.8 µm.
Occurrence: Described from a small forest pool near Bratislava, SW. Slovakia.

**228**

c

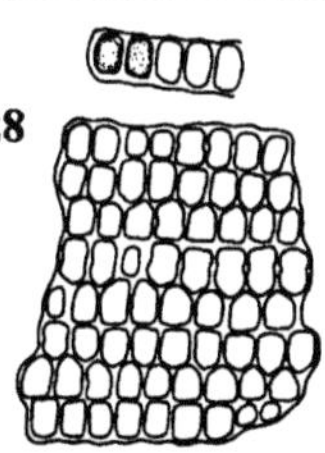

**Fig. 229.** *Merismopedia smithii*: after G.M.Smith (1920, sub *M. elegans* var. *maior*).

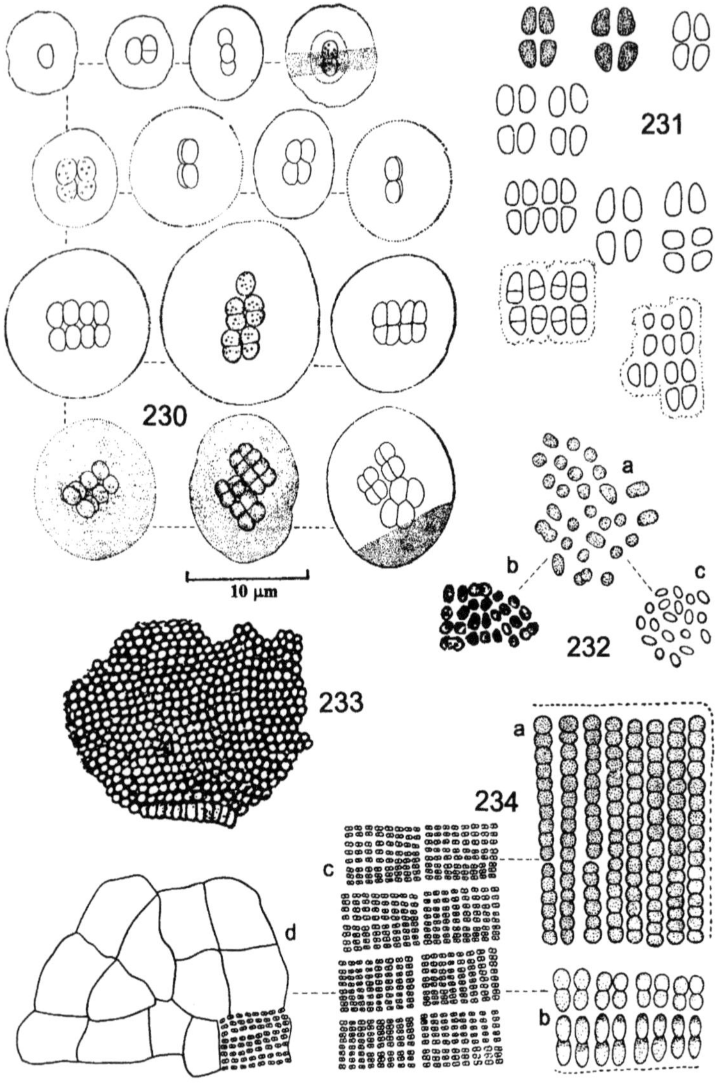


CYANOTETRAS

**Fig. 230.** *Cyanotetras fusca*: after Hindák (1988).

**Fig. 231.** *Cyanotetras crucigenielloides*, known from central America: after Komárek (1994).

**Species outside Europe:**
- *Cyanotetras crucigenielloides* Komárek 1995: Colonies 2–4–(32)-celled, cells 1.5–2.8 × 1–2.4 µm, pale blue-green, with very fine slime, always without ferric precipitates; freshwater, metaphytic, known from central America (Cuba, Mexico); (fig. **231**).
- *Cyanotetras aerotopa* Komárek et Komárková-Legnerová 1999: Freshwater, planktic, colonies 2–8-celled, cells circa 5 µm long, with aerotopes; described from Mexico.

## 7. Microcrocis Richter 1892

*Merismopedia* subg. *Holopedium* Lagerheim 1883; *Holopedium* Lagerheim 1893; *Beckia* Elenkin 1933 incl.

Colonies microscopic to macroscopically visible, flat, tabular, free-living, mucilaginous, composed of one layer of densely packed cells, sometimes composed of subcolonies; cells sometimes arranged in perpendicular rows (usually in young colonies), later quite irregularly (in several species rows of cells rarely are formed); mucilage colourless, cells without individual mucilaginous envelopes. Cells elongate, ellipsoidal, oval or rod-shaped, with rounded ends, oriented with their longer axis perpendicular to the plane of colony (during the life cycle stages with rounded cells occur). Thylakoids (in *M. geminata*) are wavy and disposed through the whole cell content (Chang in litt.).

Cells divide lengthwise in two perpendicular planes in successive generations (i.e., lengthwise with the longer cell axis) and perpendicularly to the plane of colony; cells separate and slightly shift apart from one another after division. Reproduction by disintegration of colonies. During the vegetation cycle of several species there are various morphological (*«Synechocystis-»*, *«Merismopedia-»*, typical *«Microcrocis-»*) stages.

**Important references:** Geitler (1932, sub *Holopedium*; 1942); Skuja (1948); Fott (1972); Niell & Anadon (1978, sub *Agmenellum*); Komárek & Anagnostidis (1986); Frank & Landman (1988).

**Key to the species:**

1a Cells in colonies (seen from above) more or less rounded in outline, or (old colonies) slightly polygonal; in young colonies (rarely during the main period of the life cycle) cells sometimes in perpendicular rows (subg. **Microcrocis**) . . . . . . . . 2

1b Cells in outline clearly polygonal, particularly in old colonies (subg. **Beckia**) . . . . . . . . 7

2a Cells to 6 (7) µm long . . . . . . . . 3

2b Cells (6) 7–19 µm long . . . . . . . . 5

3a Cells 2–4.5 (?) × 1.5–3 µm, tychoplanktic . . . . . . . . **1. M. irregularis**

3b Cells more than 4.6 µm long . . . . . . . . 4

4a Freshwater, benthic, cells 5.4–7 × 3.5–5 µm . . . . . . . . **2. M. obvoluta**

---

MICROCROCIS

**Fig. 232.** *Microcrocis irregularis*: **a-b** – after Lagerheim (1883), **c** – from Chu & al. (1992).

**Fig. 233.** *Microcrocis obvoluta*: after Tiffany (1934).

**Fig. 234.** *Microcrocis gigas*: **a-b** – parts of a colony, **c** – part of an old compound colony, **d** – compound colony with subcolonies; after Ryppowa from Kondrateva & al. (1984).

4b Marine psammon, cells 5–6 × 3–4 (5) µm . . . . . . . . . . . . . **3. M. sabulicola**
5a Marine, cells 6–19 × 4–7 µm (in typical stages) . . . . . . . . . . . **6. M. marina**
5b Freshwater metaphyton and tychoplankton . . . . . . . . . . . . . . . . . . . . . **6**
6a Cells in colonies mainly in rows, 7.6–12 × 4–6 µm . . . . . . . . . . **4. M. gigas**
6b Cells mainly irregularly arranged, 12–16 × 3.4–6 (7) µm . . **5. M. geminata**
7a Mucilage around cells with numerous granules (precipitates ?), colonies up to more than 60 µm in diameter . . . . . . . . . . . . . . . . . . . . . . . . **7. M. bella**
7b Mucilage without granules, colonies up to more than 200 µm in diameter . . . . . . . . . . . . . . . . . . . . . . . . . . . . . . . . . . . . . . . **8. M. granulata**

## 1. **Microcrocis irregularis** (Lagerheim) Geitler 1942 (fig. **232**)

*Merismopedium* (subg. *Holopedium*) *irregulare* Lagerheim 1883; *Holopedium irregulare* sensu auct. (false *«Holopedia irregularis»* Lagerheim)

Colonies microscopic, plate-shaped, sometimes wavy or rolled up, with irregular outlines. Cells loosely or densely irregularly arranged, only in young colonies in irregular rows, slightly elongate, ellipsoidal or cylindrical, pale blue-green, 2–4.5 (?) × (1 ?) 1.5–3 µm.
Occurrence: Freshwater, planktic (mainly tychoplanktic) in lakes, known also from aquaria; very sporadic over the whole northern temperate zone; records from Africa need confirmation.

## 2. **Microcrocis obvoluta** (Tiffany) Frank et Landman 1988 (fig.**233**)

*Holopedium obvolutum* Tiffany 1934; *Microcrocis dietelii* f. *obvoluta* (Tiffany) Fott 1972; *Microcrocis* sp. sensu Frank et Landman 1988 incl.

Colonies plate-shaped, with more than 100 (to 200–400) cells, usually more than 100 µm in diameter, cells densely packed in one plane, not in rectangular rows. Cells elongate oval, rounded at both poles, from above (top view) rounded or slightly rounded-polygonal, sometimes slightly widened in the equatorial part, pale blue-green, (4.6) 5.4–7 × 3–5 (6.8) µm.
Occurrence: Freshwater, benthic in pools and ponds, usually with sandy bottom; known from a few localities in N. America and Europe (Netherlands).

## 3. **Microcrocis sabulicola** (Lagerheim) Geitler 1942 (fig. **235**)

*Merismopedium* (subg. *Holopedium*) *sabulicolum* Lagerheim 1883; *Holopedia sabulicola* (Lagerheim) Kirchner 1900

Colonies initially few-celled, attached to sand particles, later free-living in epipsammon; young colonies with cells arranged more or less in perpendicular rows (*«Merismopedia»* stage), later irregularly and densely packed together, more than several hundreds cells in one flat, sometimes wavy colony, irregular in outline. Cells oval, bright blue-green, (3) 5–6 × 3–4 (5) µm.
Occurrence: Marine and brackish species, commonly on sandy beaches of western Baltic Sea (submersed), records also from salty marshes in England; data from Spain should be revised.

## 4. **Microcrocis gigas** (Ryppowa) Komárek et Anagnostidis 1995 (fig. **234**)

*Merismopedia gigas* Ryppowa 1925; *Pseudoholopedia convoluta* (Brébisson) Elenkin 1938 sine typo

Colonies flat, regular, with densely arranged cells in perpendicular rows, sometimes composed of several more or less quadrate subcolonies, later to 1 cm in

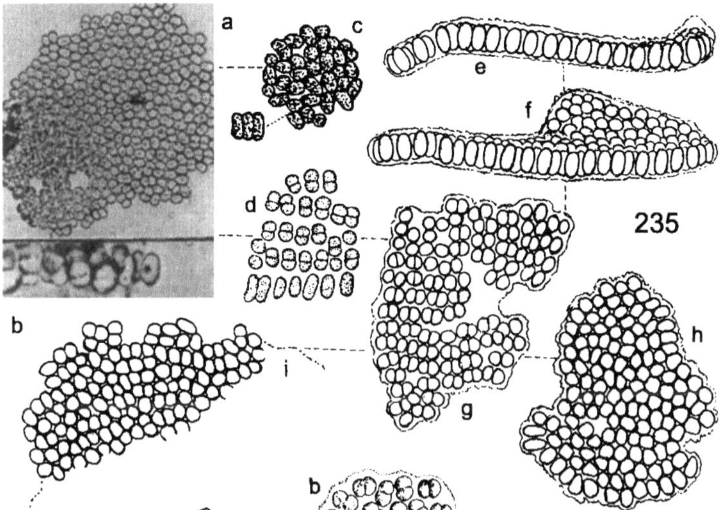


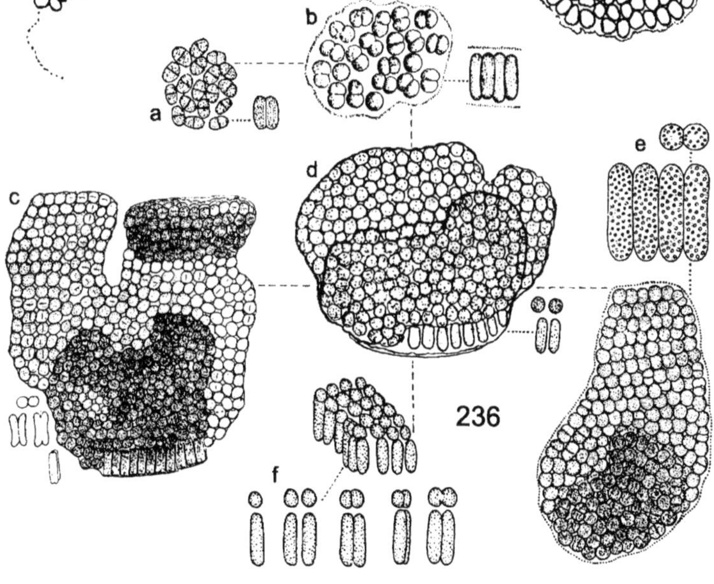


MICROCROCIS

**Fig. 235.** *Microcrocis sabulicola*: **a-b** – from Lindstedt (1943), **c** – iconotype after Lagerheim (1883), **d** – after Kosinskaja (1948), **e-i** – orig. from Baltic Sea.

**Fig. 236.** *Microcrocis geminata*: **a** – after Lagerheim (1883), **b** – after Bourrelly (1970), **c** – after Richter (1892, sub *Microcrocis dietelii*), **d** – after Kosinskaja from Kondrateva & al. (1984, sub *Holopedia geminata*), **e** – after Fott (1972, sub *Microcrocis dietelii*), **f** – after G.M.Smith (1950).

diameter and irregular in outline; rows becoming slightly irregular in old colonies. Cells elongate, ellipsoidal to rod-shaped, without individual gelatinous envelopes, bright blue-green or olive-green, 7.6–12 × 4–6 µm.
Occurrence: Brackish water, metaphytic; sporadic in the temperate zone of E. Europe (Roumania); records from other European and N. American countries need confirmation. «*Pseudoholopedia convoluta*» sensu Uherkovich (1979) from Rio Negro, Brasil, belongs surely in the g. *Merismopedia.*

### 5. **Microcrocis geminata** (Lagerheim) Geitler 1942 (fig. **236**)

*Merismopedia (subg. Holopedium) geminata* Lagerheim 1883; *Microcrocis dietelii* Richter 1892; *Holopedia geminata* (Lagerheim) Lagerheim 1893; *Holopedium dietelii* (Richter) Migula 1905

Colonies flat, tabular, sometimes wavy and rolled up, with irregular and slightly lobate margin, to 3 mm in diameter; cells densely and irregularly arranged (in young colonies sometimes in indistinct rows); old colonies sometimes with numerous cells (1800–3000). Cells rod-shaped with rounded ends, blue-green, 12–16 × 3.4–6 (7) µm.
Occurrence: Freshwater, benthic, metaphytic or occasionally free-floating in pools, swamps and littoral of lakes; scattered localities known from all over the northern temperate zone and from Argentine (Tierra del Fuego).

### 6. **Microcrocis marina** (Lagerheim) Komárek et Anagnostidis 1995 (fig. **237**)

*Merismopedia elegans* var. *marina* Lagerheim 1883; *Agmenellum thermale* (Kützing) Drouet et Daily sensu Niell et Anadon 1978

Colonies flat, regularly tabular or irregular, with cells arranged sparsely or densely in perpendicular rows or later irregularly; old colonies contain over 2500 cells. Cells very variable, in some developmental stages spherical or widely oval, later rod-shaped, oriented with their longer axis to the plane of colony, with pale green, homogeneous or slightly granular content, 6–19 × 4–7 µm; cells often divide asynchronously in a colony.
Occurrence: Brackish, sporadic along the northern and western European coasts (France, Spain, Sweden).

### 7. **Microcrocis bella** (Beck-Mannagetta) Komárek et Anagnostidis 1995 (fig. **238**)

*Holopedium bellum* Beck-Mannagetta 1929; *Beckia bella* (Beck-Mannagetta) Elenkin 1938; *Microcrocis bella* (Beck-Mannagetta) Geitler in Skuja 1964 (sine basionymo)

Colonies microscopic, flat, with up to 90 cells (more than 60 µm in diameter), of irregular outline. Colonial mucilage between cells scarcely visible, but provided with «prominent granules». Cells very densely packed, in outline clearly polygonal, cylindrical, at the poles conical-rounded, on the periphery rounded, bright blue-green, circa 7 × 3.4–4 µm.
Occurrence: Stagnant waters in Austria (Carinthia) and Sweden (Abisko).

---

MICROCROCIS
**Fig. 237.** *Microcrocis marina*, various stages: from Niell & Anadon (1978, sub *Agmenellum thermale*).

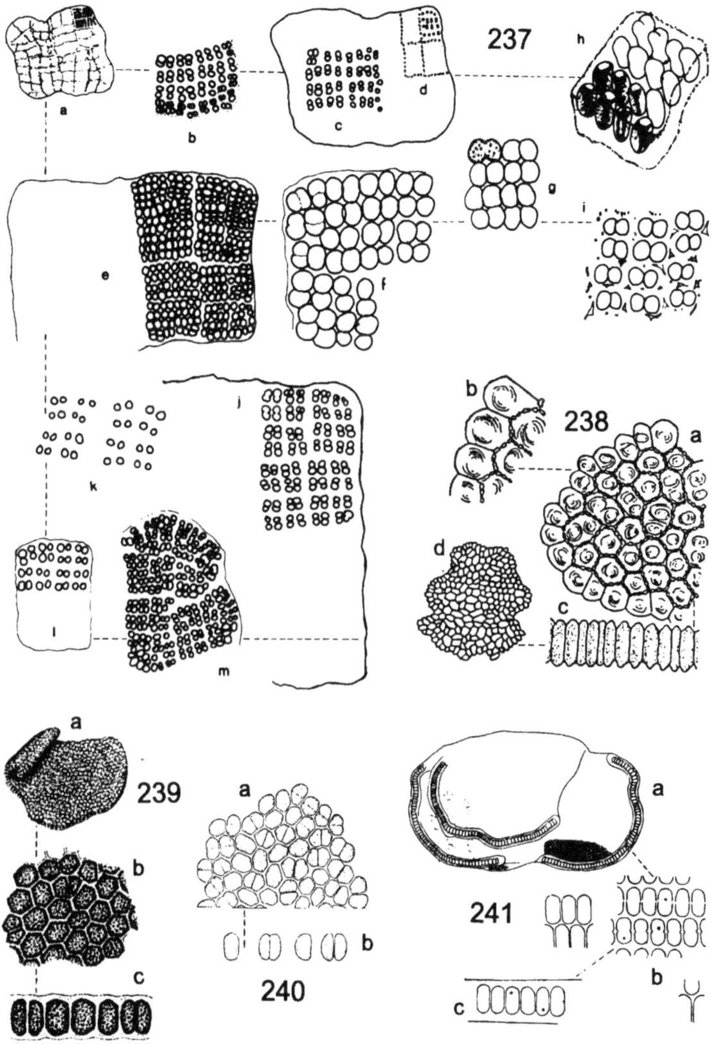


**Fig. 238.** *Microcrocis bella*: **a** – part of a colony, **b** – detail of marginal cells, **c** – elongated cells from the side view, **d** – colony; after Beck-Mannagetta (1929).
**Fig. 239.** *Microcrocis granulata*: **a** – colony, **b** – detail of cells from the culture of a colony, **c** – elongated cells from the side view; after Skuja (1948).
**Fig. 240.** *Microcrocis pulchella*: after Buell (1938).
**Fig. 241.** *Microcrocis* sp.: **a** – outline of a colony with rows of cells from side view, **b-c** – detail of cells; after Pevalek (1925, sub *Holopedia dednensis*).

**8. Microcrocis granulata** (Skuja) Skuja in Starmach 1966 (fig. **239**)

*Holopedium granulatum* Skuja 1948

Colonies flat, tabular, wavy or rolled up, to more than 200 µm in diameter. Cells very densely packed, polygonal in outline, cylindrical, more or less rounded or conical-rounded at the poles, with blue-green, finely granular content, 6.5–8 × 3.5–4.7 µm.

Occurrence: Freshwater, planktic (mainly tychoplanktic) in lakes, known only from Sweden.

**Species outside Europe:**

– *Microcrocis pulchella* (Buell) Geitler 1942: Freshwater, known from N. America (Buell 1938, Geitler 1942); (fig. **240**).

– *Microcrocis solea* (Komarenko) Komárek et Anagnostidis 1995 [syn.: *Beckia solea* Komarenko 1975]: From the subg. *Beckia*, described from plankton of the river Chroma, Yakutia, Russia (Komarenko & Vasilieva 1975).

**Unrevised species:**

– *Microcystis* sp. sensu Frank & Landman 1988: Cells 5.4–6.7 µm long, 4.6–6.8 / 3.0–4.2 µm wide, with two lateral swellings; from the Lake Vechten, the Netherlands; (fig. **242**).

– *Holopedia dednensis* Pevalek 1925: (fig. **241**).

– *Holopedia mutisiana* Guerrero 1946: Colonies microscopic, tabular, quadrate, cells 9–17 × 3–5 µm, arranged in one plane, mucilage colourless; planktic in swamps of the estuary of the river San Pedro, near Cádiz, Spain; not recorded since description, very probably a stage of *Microcrocis marina*; (fig. **243**).

– *Holopedia clementei* Guerrero 1946: Colonies microscopic, 4–50-celled, mucilage colourless, cells regularly spaced from one another, 10–13 × 3–5 µm; planktic in backwater swamps of the river Guadalete, near Cádiz, Spain; not recorded since description, very probably a stage of *Microcrocis marina*; (fig. **244**).

## 8. **Mantellum** Dangeard 1941

Small, more or less spherical cells, solitary or in groups (in one layer), attached on the substrate (usually epiphytic on other algae and filamentous cyanoprokaryotes), forming one-layered colony; cells are usually arranged in perpendicular rows on a substrate (sometimes indistinct; result of cell division in two perpendicular planes). Cells more or less spherical, usually without distinct mucilaginous envelopes (but with basal gelatinous discs ?).

Cells divide perpendicularly to the substrate and probably in two planes perpendicular to one another (deduced from the arrangement of cells, but according to

---

MICROCROCIS

**Fig. 242.** *Microcrocis* sp. sensu Frank & Landman: after Frank & Landman (1988).

**Fig. 243.** *Microcrocis* sp. : after Guerrero (1946, sub *Holopedia mutisiana*).

**Fig. 244.** *Microcrocis* sp.: after Guerrero (1946, sub *Holopedia clementei*).

MANTELLUM

**Fig. 245.** *Mantellum adherens*: **a-c** – cells attached to filamentous algae, **d** – solitary cells with visible centro- and chromatoplasmic regions and thin mucilaginous

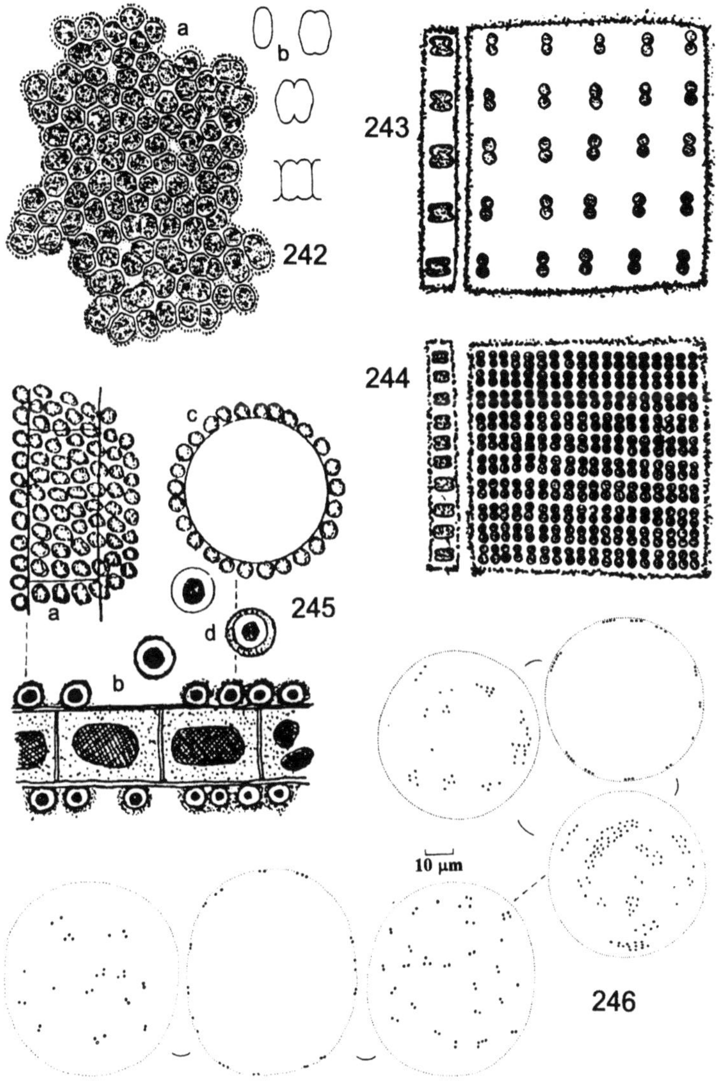


envelopes; after Dangeard (1941), in the iconotype numerous rod-like bacteria are present, attached radially in the slime around cells.

COELOSPHAERIUM

**Fig. 246.** *Coelosphaerium punctiferum*: after Hindák & Moustaka (1988, sub *C. minutissimum*).

the original diagnosis in three planes ? ). Insufficiently known genus, probably with more species.
**Important references:** Dangeard (1941).

### 1. **Mantellum adherens** Dangeard 1941 (fig. **245**)

Cells spherical or slightly oval before division, 2.5–3 µm in diameter, with blue-green or pale pinkish, homogeneous content; central part of cells is pale (almost colourless). Around cells very fine, diffluent mucilage.
Occurrence: Marine, epiphytic on algae and filamentous cyanoprokaryotes in intertidal zone of Atlantic coast; France, England (Channel Islands).

**Species outside Europe:**
- *Mantellum himalayense* M. Watanabe et Komárek 1995: Freshwater, epiphytic on algae and mosses in cold, katharobic waters in Himalayas, E. Nepal.
- *Mantellum rubrum* Tavera et Komárek 1996: From slightly salinic, inland volcanic lake, deep-water form, epiphytic on filamentous algae, with reddish cells; described from Mexico.

**Unrevised species:**
- To the genus *Mantellum* possibly belong several «small» *Xenococcus* species reproducing by cell division perpendicularly to the substrate, without baeocytes, especially from marine littoral biotopes.

Notes to the family *Merismopediaceae:* The uncertain genus *Chroostipes* Pascher 1914: Solitary rod shaped cells dividing lengthwise; this type of cell division occurs in cyanoprokaryotes only in merismopedioid types (*Microcrocis*); however, this genus with one species *C. linearis* has never been satisfactorily described nor revised (Geitler 1925, 1932, 1942); probably a eubacterium. – Another uncertain genus is *Pilgeria* Schmidle 1901 with a type species *P. brasiliensis*: Free-floating spheroidal colonies with densely aggregated cells in the peripheral layer, slightly radially elongated. Described from Brasil. Since 1901 not found again; (Geitler 1932, Bourrelly 1970).

## 2. Subfamily **Gomphosphaerioideae** (Elenkin) Komárek et Hindák 1988

*Coelosphaeriaceae* Elenkin 1933 incl.; *Gomphosphaeriaceae* Elenkin 1933 incl.; *Woronichiniaceae* Elenkin 1933 incl.

Cells located on the periphery of more or less spherical, oval or irregularly spherical, free floating or, rarely, sessile colonies, usually radially oriented, in one or more layers; colonies microscopic, rarely macroscopic (visible as a «pulverous» waterbloom), sometimes composed of several subcolonies, mucilaginous, enveloped usually by distinct or indistinct (staining !) gelatinous, colourless mass. Cells arranged radially in one or more peripheral layers, more or less distant one from another or densely agglomerated, spherical, ellipsoidal, oval or obovate, after division sometimes cordiform, rarely almost cylindrical with rounded ends; in several genera attached to the ends of thin or thick mucilaginous, simply or pseudodichotomously divaricate stalks, or covered by special thin, gelatinous envelopes within thick stalks. The stalk system has radial structure in the center of the colony and sometimes becomes diffluent in the course of the life cycle.
Cell division by binary fission in two planes in successive generations, perpendicular to one another and to the surface of a colony. Reproduction by the disintegra-

tion of colonies and, in some genera, by the liberation of solitary cells from the colony.

**Key to the genera:**

1a Cells arranged more or less at the periphery of mucilaginous spherical colonies, without any central stalk system, but sometimes with irregular gelatinous clusters in the center of colonies . . . . . . . . . . . . . . . . . . . . . . . **2**

1b Cells arranged in colonies radially, usually in the peripheral layer or displaced slightly from one another, during the main period of the life cycle attached to the ends of thin or within the ends of thick radial stalks, commencing from the colonial centre . . . . . . . . . . . . . . . . . . . . . . . . . . . . . . . . . . . . **4**

2a Cells spherical or hemispherical (after division), always more or less distant from one another, usually in one (rarely in two) layers on the periphery . **3**

2b Cells usually slightly (widely) oval, distant from one another in young colonies, more densely packed in old colonies, radially arranged in one or several layers on the colonial periphery; in the centre of young colonies sometimes gelatinous clusters with mucilaginous protuberances . . . . . . . . . . . . . . . . . . . . . . . . . . . . . . . . . . . . . . . . **3. Coelomoron**

3a All adult spherical cells of the same size, without individual mucilaginous envelopes . . . . . . . . . . . . . . . . . . . . . . . . . . . . . . . **1. Coelosphaerium**

3b Adult cells sometimes of variable size, always surrounded by their own mucilaginous envelopes . . . . . . . . . . . . . . . . . . . . . **2. Coelosphaeriopsis**

4a Cells attached to the ends of thin, pseudodichotomously divaricate stalks, usually clearly visible without staining, without enveloping mucilage around cells (but sometimes in very fine common colonial slime); cells radially oriented and usually radially displaced; colonies spherical, free floating . . . . . . . . . . . . . . . . . . . . . . . . . . . . . . . . . . . . . . . . . **4. Snowella**

4b Cells located at the ends or within more or less wide mucilaginous stalks, which are sometimes invisible without staining; colonies free floating . . . **5**

5a Colonies irregularly spherical, often composed of subcolonies, free-living; cells usually slightly elongate, situated clearly radially at the ends of stalks (sometimes visible only after staining), which are of the same width as or thinner than the cell . . . . . . . . . . . . . . . . . . . . . . . . . . . . . . . . . . . . **6**

5b Colonies oval, rarely more or less spherical, free floating; cells spherical or hemispherical, situated within the ends of wide, fine stalks, which are usually thicker than the cell diameter . . . . . . . . . . . . . . . . . . **7. Siphonosphaera**

6a Cells to 6 µm long (rarely up to 7 µm), spherical, oval or obovoid, joined to the ends of indistinctly visible, gelatinous, simple stalks, which are densely and parallely arranged from the colonial center, about as wide as cells, divaricate only after cell division (sometimes visible as radially layered central gelatinous mass); cells in old colonies densely packed in the peripheral layer; reproduction commonly by the liberation of solitary cells **5. Woronichinia**

6b Cells usually over 6 µm long, obovoid, slightly club-shaped or cordiform (characteristic persistent division stages), joined to the ends of colourless gelatinous, wide (but never as wide as the cell), pseudodichotomously branched stalks; the gelatinous matrix of stalks often envelopes also cells, forming thin layer around them; between cells always remarkable «spaces»; reproduction only by the disintegration of colonies . . **6. Gomphosphaeria**

## 1. **Coelosphaerium** Nägeli 1849

*Coelocystis* Nägeli in Kützing 1849

Colonies microscopic, more or less spherical or oval, sometimes composed of subcolonies, rarely irregular when old, free-living, usually floating in plankton,

enveloped by fine, colourless, usually not sharply defined, sometimes indistinct mucilage; mucilage limited to just around the peripheral layer of cells or forming a gelatinous layer (margin) around cells. Cells more or less in one layer on the colonial periphery, spherical or hemispherical after division, usually slightly or distinctly distant from one another, with or without aerotopes. Mucilage homogeneous, fine, never structured, without any stalks in the colonial center.
Cell division in two planes in successive generations, perpendicular one to another and more or less so to the colonial surface. Reproduction by disintegration of colonies.
**Important references:** Geitler (1932); Huber-Pestalozzi (1938); Komárek (1958); Skuja (1956, 1964); Hindák (1988); Komárek & Hindák (1988); Komárek & Komárková-Legnerová (1992).

**Key to the species:**

**1a** Cells without aerotopes (groups of gas vesicles) . . . . . . . . . . . . . . . . . . 2
**1b** Cells with aerotopes . . . . . . . . . . . . . . . . . . . . . . . . . . . . . . . 7
**2a** Cells circa 1 µm (up to 1.4 µm) in diameter . . . . . . . . . . . . . . . . . . . . 3
**2b** Cells larger than 1.2 µm in diameter . . . . . . . . . . . . . . . . . . . . . . . . 4
**3a** Cells irregularly and sparsely distributed on the periphery of colonies, sometimes very distant from one another, just at the slime margin; tropical and subtropical species . . . . . . . . . . . . . . . . . . . . . . . . **1. C. punctiferum**
**3b** Cells more or less regularly distributed in the peripheral layer of colonies, mucilage slightly extending beyond the cell layer; known from northern Europe . . . . . . . . . . . . . . . . . . . . . . . . . . . . . . **2. C. minutissimum**
**4a** Cells 1.2–1.6 µm in diameter; in northern clear or mesotrophic lakes (rarely also in mountains of the middle temperate zone), mainly with circumpolar distribution . . . . . . . . . . . . . . . . . . . . . . . . . . . . **4. C. subarcticum**
**4b** Cells always over 1.8 µm in diameter . . . . . . . . . . . . . . . . . . . . . . . . 5
**5a** Colonies always spherical (or widely oval), or composed of few subcolonies, to 100 µm in diameter; cells 1.8–5.0 µm in diameter, with homogeneous blue-green content, usually with visible chromatoplasma . . . . . . . . . . . 6
**5b** Old colonies elongate and irregular, usually more than 140 (even to more than 1000) µm in length; cells 2.4–5.7 µm in diameter, with 1–2 refractive granules in plasma . . . . . . . . . . . . . . . . . . . . . . . . . . **7. C. limnicolum**
**6a** Cells 1.8–4 µm in diameter, arranged slightly irregularly in the peripheral layer . . . . . . . . . . . . . . . . . . . . . . . . . . . . . . . **5. C. kuetzingianum**
**6b** Cells 2–5.5 µm in diameter, arranged distinctly in one peripheral layer . . . . . . . . . . . . . . . . . . . . . . . . . . . . . . . . . . . **6. C. aerugineum**
**7a** Cells 1.3–1.5 µm in diameter . . . . . . . . . . . . . . . . . . . . . . . . **3. C. natans**
**7b** Cells 5–7 µm in diameter . . . . . . . . . . . . . . . . . . . . . . . . . **8. C. dubium**

## 1. **Coelosphaerium punctiferum** Komárek et Komárková-Legnerová 1992 (fig. **246** )

Colonies free floating, spherical to broadly oval, 2.5–70 (- 80) µm in diameter, with cells irregularly and very sparsely distributed just beneath the periphery of the fine, hyaline colonial mucilage, distant from one another, sometimes in small loose and flat groups. Cells tiny, spherical, ca. 1 µm in diameter, with pale greyish blue-green homogeneous content, without gas vesicles.
Occurrence: Planktic in ponds and lakes in tropical and subtropical (in summer) regions; probably pantropical; Europe: Mediterranean countries.

## 2. **Coelosphaerium minutissimum** Lemmermann 1900 (fig. **247**)

Colonies free-living, spherical or oval, usually 20–30 µm, rarely up to 30–170 µm (f. *definitum* Skuja 1964) in diameter, with cells more or less regularly arranged in one layer near the peripheral surface, with 4–10 (13) µm broad external gelatinous margin; mucilage fine, colourless, indistinct, but clearly delimited. Cells spherical, after division hemispherical, pale blue-green, without gas vesicles, to 1.4 µm in diameter.
Occurrence: Planktic in clear or slightly polluted, sometimes also brackish water biotopes in northern Europe. Tropical populations concern another species.

## 3. **Coelosphaerium natans** Lemmermann 1900

Colonies free floating, more or less spherical, with cells more or less densely arranged near the colonial surface, sometimes slightly displaced to one another, in up to three «layers», with very sparse and thin outer gelatinous envelope; mucilage fine, colourless, diffluent. Cells spherical, 1.3–1.5 µm in diameter, with aerotopes.
Occurrence: Planktic in stagnant waters, mainly lakes, known from northern Germany, Baltic states and NW. Russia.

## 4. **Coelosphaerium subarcticum** Komárek et Komárková-Legnerová 1992 (fig. **248**)

Colonies free floating, spherical or oval, sometimes composed of 2 (3) hemispherical subcolonies, mucilaginous, to 110 µm in diameter, with cells distributed irregularly in one layer near the colonial surface; mucilage fine, colourless, indistinct. Cells spherical or hemispherical after division, 1.2–1.6 µm in diameter, with pale blue-green or greyish, more or less homogeneous content, without gas vesicles.
Occurrence: Planktic in northern lakes of temperate up to circumpolar zone. Europe: Rarely in central regions, commonly in northern states; occurs also in Canada.

## 5. **Coelosphaerium kuetzingianum** Nägeli 1849 (fig. **251, 252**)

*Coelocystis kuetzingiana* Nägeli in Kützing 1849

Colonies free floating, spherical or widely oval, up to 100 µm in diameter, sometimes consisting from 2 subcolonies, with relatively densely arranged cells in one or two layers (or slightly irregularly) very near the colonial surface (the cells are sometimes slightly shifted one to another). Slime colourless, indistinct. Cells spherical or hemispherical after division, 1.8–3(4?) µm in diameter, pale blue-green or olive-green, slightly granular, without aerotopes.
Occurrence: Planktic in mesotrophic water bodies; not very common, but possibly with cosmopolitan distribution, frequently in tropical regions.

## 6. **Coelosphaerium aerugineum** Lemmermann 1898 (fig. **249,258**)

Colonies free floating, spherical, up to about 100 µm in diameter, sometimes consisting of 2–3 hemispherical subcolonies with irregularly and sparsely or densely aggregated cells (however, they never touch one another), arranged in one layer near the surface of the homogeneous, colourless, diffluent or delimited gelatinous envelope (which is up to 5 µm wide). Cells spherical or hemispherical after division, (2)2.5–4(5.5) µm in diameter, usually bright blue-green, slightly granular, without gas vesicles. This species was often misinterpreted as *C. kuetzin-*

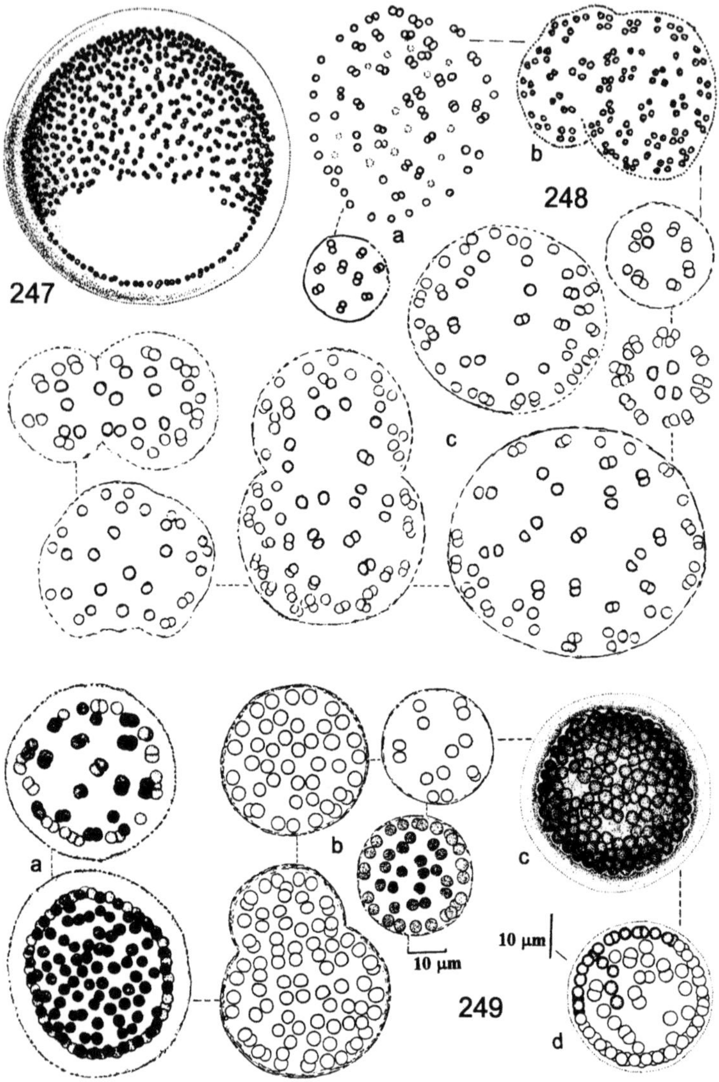

247
248
a
b
c
249
a
b
c
d
10 µm
10 µm

*gianum* by later authors (G.M.Smith 1920, Komárek 1958, etc.) in contrast to the original description of Nägeli.
Occurrence: In slightly eutrophic to mesotrophic water bodies, mainly planktic, never in masses; known (in typical form) mainly from temperate zone of the northern hemisphere (but also from Argentine); numerous floristic data from tropical countries exist, which must be revised.

## 7. **Coelosphaerium limnicolum** Lund 1942 (fig. **253**)

Colonies large, hollow, with more or less densely and irregularly disposed cells in the peripheral layer (sometimes with two or three layers of cells), either perfectly spherical or oval oblong, later more or less irregular, usually 141–880 × 78–343 µm as much as 1010 µm in diameter (original data); mucilaginous envelopes thick (8.3–21.6 µm wide around marginal cells), sometimes containing a few subcolonies. Cells spherical to broadly oblong containing one or two prominent granules, 2.4–5.7 µm in diameter.
Occurrence: Submerged, metaphytic, not recorded in plankton; till now found only in the littoral of a pond in England (Surrey).

## 8. **Coelosphaerium dubium** Grunow in Rabenhorst 1865 (fig. **254**)

Colonies exactly spherical or widely oval, sometimes composed of 2–3 subcolonies, to 150 µm in diameter with more or less densely arranged cells in one layer; the colourless colonial mucilage distinct but diffluent, not refractive at the

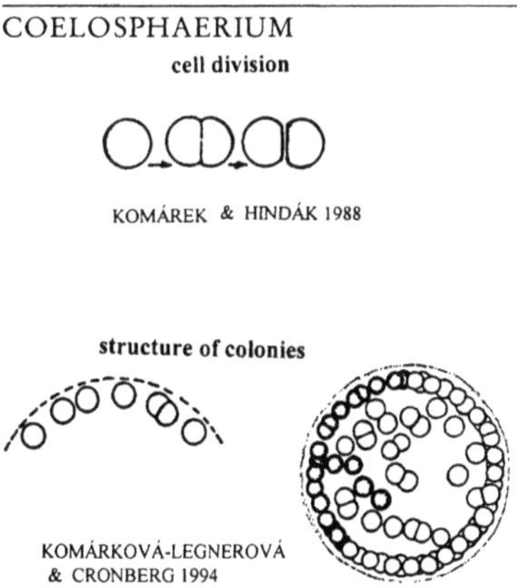


**Fig. 247.** *Coelosphaerium minutissimum*: after Skuja (1964, sub *C. minutissimum* f. *definitum*).
**Fig. 248.** *Coelosphaerium subarcticum*: **a** – after Bachmann (1907, sub *C. kuetzingianum*), **b** – after Tikkanen (1986, sub *C. minutissimum*), **c** – after Komárek & Komárková-Legnerová (1992).

**Fig. 249.** *Coelosphaerium aerugineum*: **a** – after G.M.Smith (1920), **b** – after Komárek (1958), **c** – after Skuja (1956), **d** – after Komárková-Legnerová & Cronberg (1994); (all sub *C. kuetzingianum*).

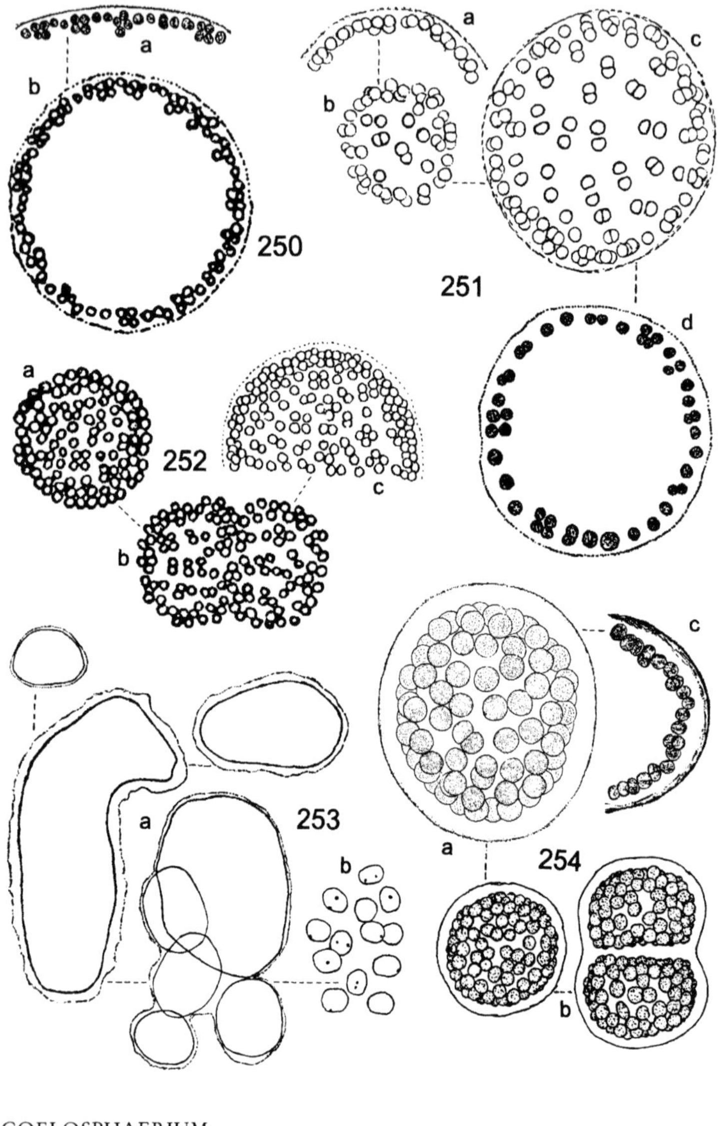


COELOSPHAERIUM

**Fig. 250.** *Coelosphaerium confertum*: after Frémy (1930), tropical population from Africa.

**Fig. 251.** *Coelosphaerium kuetzingianum*: **a-c** – after Komárek & Komárková-Legnerová (1992) from Canadian lakes, **d** – after Frémy (1930) from Africa.

margin, extending to 8 µm beyond the cellular layer, firm, not stratified. Cells spherical, pale blue-green with aerotopes, 5–7 µm in diameter.
Occurrence: Sparsely distributed over the temperate zone in lakes and ponds, also in slightly brackish biotopes; the tropical records are questionable. Europe: Recently found only in Baltic region (N. Germany). Data (Guarrera & al. 1986) from Argentine (Tierra del Fuego) probably concern another species.

**Species outside Europe:**
- *Coelosphaerium confertum* W. et G.S. West 1896: Cells 1.8–2.4 µm in diameter; mainly tropical, European records not confirmed (see Kosinskaja in Kondrateva et al. 1984 sub *C. kuetzingianum*); description: see Frémy (1930), Huber-Pestalozzi (1938), etc. C. cf. *confertum* with cells 1.8–3 µm is mentioned by Komárek & Komárková-Legnerová (1992) from Canadian lakes; (fig. **250, 251**).

**Unrevised taxa:**
- *Coelosphaerium anomalum* (Bennet) de Toni et Levi 1888 sec. Hansgirg 1890: Moorland waters, swamps, small ponds and lakes in Eurasia; very problematic and unrevised species; reported from Ukraine without further documentation; (Kondrateva & al.1984); (fig. **255**).
- *Coelosphaerium geitleri* Schiller 1954: In a pond of Austria (?). Incomplete description.
- *Coelosphaerium goetzei* Schmidle 1901: Cells 4–10 µm in diameter; stagnant waters in Africa, recorded also from Brazil, probably pantropic; (Frémy 1930); (fig. **256**).
- *Coelosphaerium goetzei* var. in Skuja 1948: Epipelic and planktic in lakes of Sweden; revision necessary; (fig. **257**).
- *Coelosphaerium kuetzingianum* sensu Schwabe 1944: From thermal springs in Chile, colonies up to 250 µm in diameter, cells 3.7–4.3 µm in diameter, more or less densely arranged.
- *Coelosphaerium kuetzingianum* f. in Skuja 1956: Colonies 30–100 µm, cells 3.5–4 µm in diameter, planktic in lake Erken, Sweden, (fig. **258**).
- *Coelosphaerium?* sp. in Golubić 1980: Halophilic species, cells spherical, 1.5–2 µm in diameter, hollow spherical colonies, known from Yallahs ponds, Jamaica.

## 2. **Coelosphaeriopsis** Lemmermann 1900

Colonies microscopic, more or less spherical, up to 500 µm in diameter, free-living, enveloped by colourless, fine, homogeneous mucilage. Cells spherical or oval before division, situated more or less in the colonial periphery in one layer, each surrounded by its own, narrow, delimited gelatinous envelope, with more or less homogeneous or slightly granular content.

---

**Fig. 252.** *Coelosphaerium kuetzingianum*: after Nägeli (1849) from Kondrateva & al. (1984); this figure is the iconotype of *C. kuetzingianum*.
**Fig. 253.** *Coelosphaerium limnicolum*: **a** – irregular colonies and aggregates of colonies, **b** – detail of cells; after Lund (1942).
**Fig. 254.** *Coelosphaerium dubium*: **a** – typical colony after G.M.Smith (1920), **b** – after Schmula from Geitler (1932), **c** – after Frémy (1930) from Africa (probably a separate species).

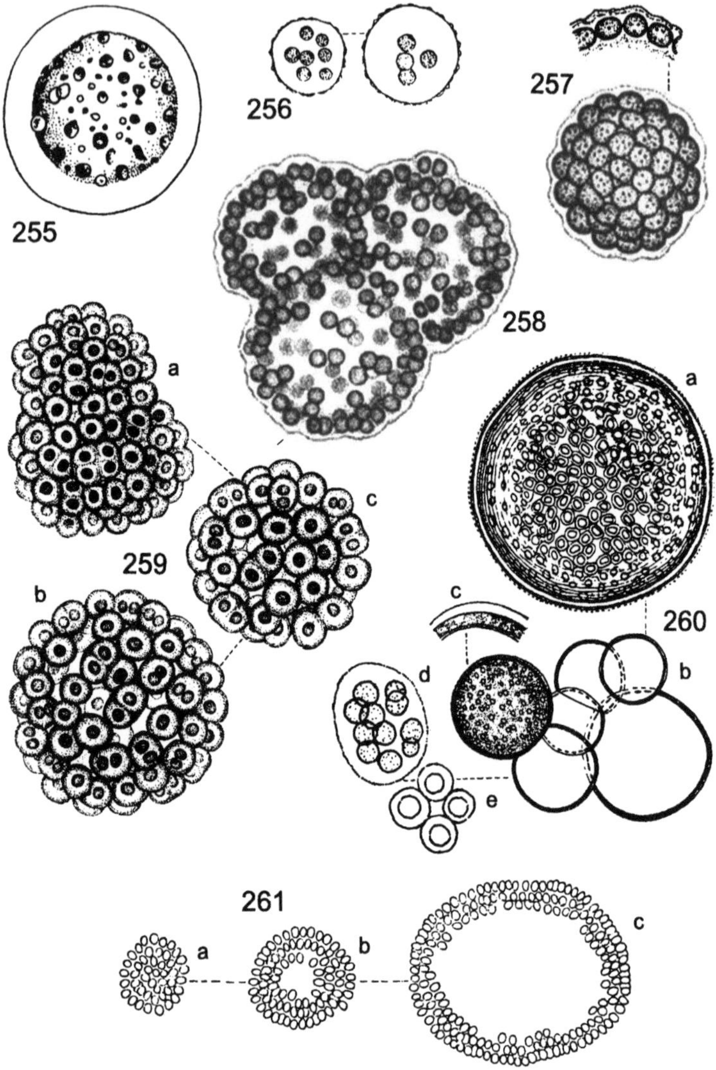


COELOSPHAERIUM

**Fig. 255.** *Coelosphaerium anomalum*: after Bennet from Kondrateva & al. (1984); the shape of cells and structure of the colony are unclear.

**Fig. 256.** *Coelosphaerium goetzei*: after Schmidle from Frémy (1930), African specimens.

Cell division by fission, probably in two perpendicular planes in successive generations; reproduction probably by disintegration of colonies.
**Important references:** Geitler (1932, 1942); Skuja (1964, sub *Coelosphaerium*); Bourrelly (1970, 1985).

## 1. **Coelosphaeriopsis chlamydocystis** (Skuja) Komárek et Anagnostidis 1995 (fig. **259**)

*Coelosphaerium chlamydocystis* Skuja 1964

Colonies free floating or living among other algae, spherical or oval, 20–60 µm in diameter, with cells aggregated in the periphery of more or less hollow colonies; each cell surrounded by its own gelatinous envelope, common mucilage absent. Cells spherical or slightly elongate (broadly oval), without envelopes 1.7–2.5 µm to 4 × 3 µm in diameter; gelatinous envelope colourless, not or slightly stratified, sometimes slightly yellowish, distinct, delimited, about 3 µm wide; cell content pale blue-green or olive-green, homogeneous or slightly granular without gas vesicles.
Occurrence: Known only from lakes of N. Sweden (Abisko).

**Species outside Europe:**
- *Coelosphaeriopsis halophila* Lemmermann 1899 [syn.: *Coelosphaerium halophilum* (Lemmermann) Geitler 1925]: Coastal lakes with saline water in Pacific islands (Laysan); (Huber-Pestalozzi 1938); (fig. **260**).

# 3. **Coelomoron** Buell 1938

Colonies microscopic, more or less spherical or irregular spherical, sometimes compound, free floating, enveloped by fine, colourless, diffluent, very often indistinct, rarely distinct delimited mucilage. Cells in young colonies randomly and more or less sparcely disposed, diverging slightly radially from one another, in old colonies concentrated in the peripheral layer or up to three layers thick. In the center of colonies is a condensed cluster of slime, from which radiate irregular, fine and diffluent stalks or protuberances, which very soon disappear (staining !); in the center of old colonies fine slime. Cells slightly elongate, situated more or less radially in the outer layers.
Cell division in two planes in successive generations, perpendicular one to another

---

**Fig. 257.** *Coelosphaerium goetzei* var. in Skuja: after Skuja (1948) from Sweden.
**Fig. 258.** *Coelosphaerium kuetzingianum* fa. in Skuja: after Skuja (1956) from Sweden.

COELOSPHAERIOPSIS
**Fig. 259.** *Coelosphaeriopsis chlamydocystis*: after Skuja (1964, sub *Coelosphaerium chlamydocystis*).
**Fig. 260.** *Coelosphaeriopsis halophila*: **a** – spherical colony, **b** – group of colonies, **c** – detail of a colonial margin, **d** – young colony, **e** – detail of cells; after Lemmermann from Huber-Pestalozzi (1938), Geitler (1942) and Kosinskaja (1948).

COELOMORON
**Fig. 261.** *Coelomoron regulare*: after Buell (1938).

and more or less to the colonial surface; reproduction by disintegration of colonies.
**Important references:** Geitler (1942); Komárek (1989); Komárková-Legnerová & Tavera (1996); Senna & al.(1998).

### 1. Coelomoron pusillum (Van Goor) Komárek 1988 (fig. 262)

*Coelosphaerium pusillum* Van Goor 1924; *Coelosphaerium collinsii* Drouet et Daily 1942; *Gomphosphaeria pusilla* (Van Goor) Komárek 1958

Colonies microscopic, more or less spherical or irregularly oval, sometimes composed of two subcolonies, 15–20 (- 30) µm in diameter, with cells more or less radially arranged in the periphery, slightly distant and displaced from one another, sometimes 2–4 near together, later aggregated; often only with 10–30 cells. Mucilaginous envelopes colourless, diffluent, sometimes recognizable only after staining, extending circa 2.5 µm beyond the cell layers; dense mucilage in the center of young colonies radiates out in poorly visible string-like protuberances. Cells widely oval or obovoid (tropical populations), pale blue-green, very finely granulate, without gas vesicles, 2.2–4.5 × 1.8–4 µm.
Occurrence: In plankton of stagnant or slowly flowing eutrophic waters; known from the temperate zone of Europe and N. America (frequently) and from the whole tropical region (special morphotype ?).

---

COELOMORON
**Fig. 262.** *Coelomoron pusillum*: **a** – after van Goor (1924, sub *Coelosphaerium pusillum*), **b-d** – after Komárek (1958, sub *Gomphosphaeria pusilla*).
**Fig. 263.** *Coelomoron microcystoides*: after Komárek (1988).
**Fig. 264.** *Coelomoron* sp. div.: **a** – after West from Geitler (1925, sub *Microcystis chroococcoidea*) from Antarctica (!), **b** – after Virieux (1912, sub *Chroococcus rochei*), **c** – after Nygaard (1949, sub *Microcystis chroococcoidea* var. *minor*).

SNOWELLA

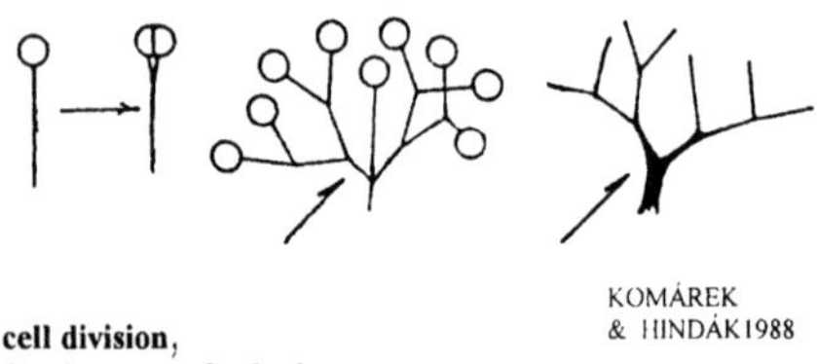


**cell division, development of colonies**

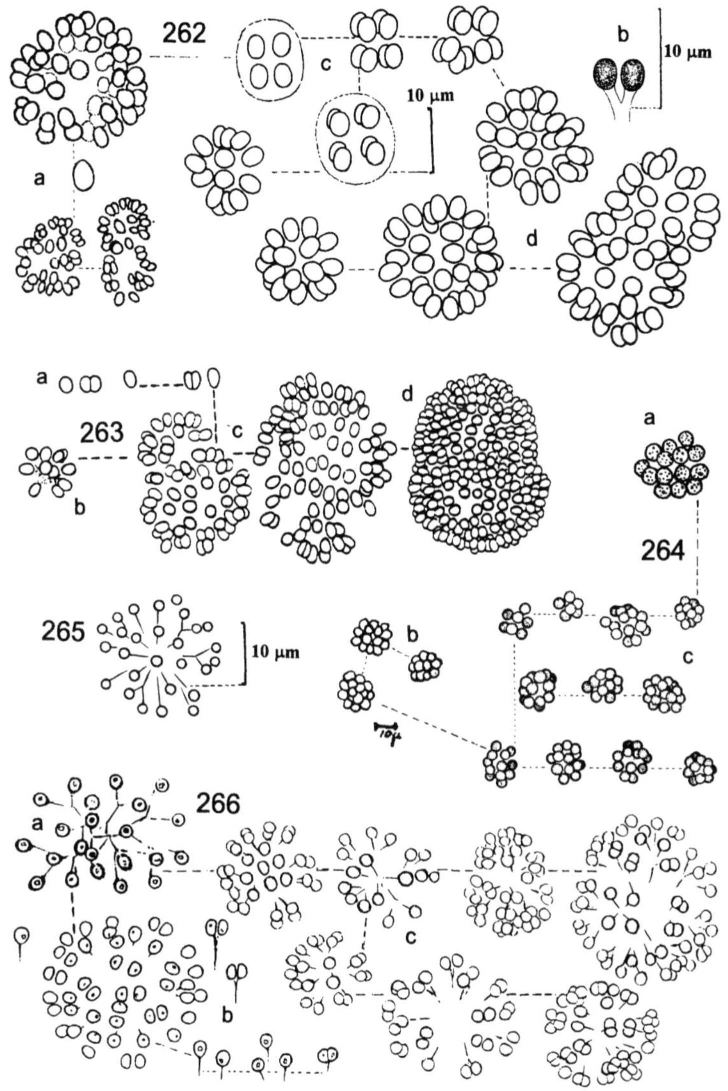


**Fig. 265.** *Snowella atomus*: after Komárek from Komárek & Hindák (1988).
**Fig. 266.** *Snowella septentrionalis*: **a** – from Komárek & Hindák (1988), **b** – after Komárek & Hindák (1988), **c** – form of colonies after Komárek & Komárková-Legnerová (1992).

**Species outside Europe:**
– *Coelomoron microcystoides* Komárek 1988: Colonies to 80 µm in diameter, cells 1.6–3.6 × 0.9–3 µm; in metaphyton and plankton of water bodies with rich higher vegetation, tropical species (Cuba, Mexico, S. Africa); (fig. **263**).
– *Coelomoron minimum* (Bernard) Komárková-Legnerová et Tavera 1996 [syn.: *Microcystis minima* Bernard 1909]: Colonies to 60 µm in diameter, cells 2.2–3.2 × 2–2.8 µm, with aerotopes; rare planktic species from tropical lakes (Malaysia, Mexico).
– *Coelomoron regulare* Buell 1938: With up to three peripheral layers of cells; among other algae in stagnant waters, known only from N. America; (Geitler 1942); (fig. **261**).
– *Coelomoron vestitum* Komárek 1988: Colonies to 50 µm in diameter, surrounded usually by firm envelope, cells 1–6.5 × 0.8–3.5 µm; in natural and alkaline swamps, known only from Cuba.
– *Coelomoron tropicale* Senna et al. 1998: Colonies subspherical, to 16.8 µm in diameter, cells almost spherical, pale greyish blue-green, 1.8–3 × 1.4–3 µm; metaphytic and planktic in slightly mesotrophic waters in SE. Brazil.
– *Coelosphaerium pallidum* sensu Prescott 1951: In *Sphagnum*-bogs, N. America; probably a *Coelomoron* species.

**Unrevised species:**
- *Chroococcus rochei* Virieux 1911: Colonies more or less spherical, 16–22 µm in diameter, usually with 8–24 radially arranged, widely oval to spherical cells, which are pale green, 4–4.5 µm in diameter, described originally from freshwater pools in France; (fig. **264b**). *Microcystis chroococcoidea* var. *minor* Nygaard (1949) from Denmark is probably identical with this species (fig. **264c**); comp. *Chroococcus minimus*, p. 286.
– *Microcystis chroococcoidea* W. et G. S.West 1911: Colonies ±spherical, with 6–24 cells, to 33 µm in diameter; cells pale yellowish blue-green, 4–7 µm in diameter; described from a salty lake in Antarctica, very probably a *Coelomoron* species; (fig. **264a**).

## 4. **Snowella** Elenkin 1938

Colonies more or less spherical or irregularly oval, rarely compound, free floating, with distinct or indistinct (staining !), homogeneous, colourless and wide mucilaginous envelopes. Within colonies there is a more or less stable system of fine, thin, thread-like mucilaginous stalks radiating from the colonial center, more or less pseudodichotomously divaricated and sometimes fasciculate, widened in the center. Cells spherical or slightly elongate, joined to the ends of stalks during the whole life cycle, distant from one another. If the cells are densely disposed on the periphery of old multicellular colonies, there always remain at least narrow spaces between them; cells usually not strictly in one peripheral layer, but slightly radially shifted from one another.
Cell division only by binary fission in two planes in successive generations, perpendicular to one another and to the colony surface; disintegration of colonies even into solitary cells.
**Important references:** G.M. Smith (1920, sub *Gomphosphaeria*); Geitler (1932, sub *Gomphosphaeria*); Elenkin (1938); Hollerbach & al. (1953); Komárek & Anagnostidis (1986, sub *Gomphosphaeria*); Komárek & Hindák (1988); Komárek & Komárková-Legnerová (1992).

**Key to the species:**

1a Cells spherical . . . . . . . . . . . . . . . . . . . . . . . . . . . . . . . . . . 2
1b Cells slightly elongate, drop-shaped, or obovoid . . . . . . . . . . . . . . . . . . 5
2a Cells without gas vesicles or prominent inclusions within protoplasts . . . 3
2b Cells with 1–2 aerotopes (refractive granules ?), 1.2–3.4 µm in diameter, colonies more or less spherical, up to 50 µm in diameter; nordic species . . . . . . . . . . . . . . . . . . . . . . . . . . . . . . . . . . **3. S. septentrionalis**
3a Cells less than 2.4 µm in diameter . . . . . . . . . . . . . . . . . . . . . . . . . . 4
3b Cells 2.4–4 µm in diameter . . . . . . . . . . . . . . . . . . . . . . . . **5. S. litoralis**
4a Cells 0.6–1.4 µm, colonies up to 25 µm in diameter . . . . . . . **1. S. atomus**
4b Cells 1.2–2.4 µm, colonies with many cells, up to 60 µm in diameter . . . . . . . . . . . . . . . . . . . . . . . . . . . . . . . . . . **2. S. arachnoidea**
5a Cells with one to several aerotopes (refractive granules ?) in the center, always sparsely disposed in the colony (also if multicellular); colonies slightly irregular, to 50 µm in diameter . . . . . . . . . . . . . . . . . . . . . . . **4. S. fennica**
5b Cells always without gas vesicles, sparsely or (in old colonies) more densely disposed on the periphery of colonies; colonies more or less spherical or oval, up to 80 µm in diameter . . . . . . . . . . . . . . . . . . . . . . . **6. S. lacustris**

## 1. **Snowella atomus** Komárek et Hindák 1988 (fig. **265**)

Colonies free floating, solitary, more or less spherical, to 25 µm in diameter, with randomly disposed cells, joined to thin mucilaginous stalks, radiating from the center of the colonies. Cells spherical, after division hemispherical, with homogeneous, pale grey-blue content, 0.6–1.4 µm in diameter.
Occurrence: Planktic and tychoplanktic in mesotrophic lakes and large reservoirs; central and northern Europe.

## 2. **Snowella arachnoidea** Komárek et Hindák 1988 (fig. **267**)

Colonies free floating, solitary, spherical, oval or composed of subcolonies, up to 60 µm in diameter, with many, sparsely disposed cells, joined to thin, pseudodichotomously and frequently branched stalks, usually widened in the colonial center. Cells spherical, after division hemispherical, with homogeneous, pale grey-blue content, 1.2–2.4 µm in diameter.
Occurrence: Planktic in large oligotrophic to mesotrophic lakes; known only from S. Austria (Tirol).

## 3. **Snowella septentrionalis** Komárek et Hindák 1988 (fig. **266**)

Colonies free floating, solitary, mostly spherical, to 50 µm in diameter, multicellular, with sparsely disposed cells joined to the ends of thin mucilaginous stalks, radiating from the center of the colony. Cells spherical, after division hemispherical, with pale olive-green or yellow-green content with 1–2 distinct aerotopes (refractive granules ?), 1.2–3.4 µm in diameter.
Occurrence: Planktic and metaphytic in lakes and ponds in northern regions of temperate zone; Finland, Sweden, Canada, found also in eastern Baltic Sea.

## 4. **Snowella fennica** Komárek et Komárková-Legnerová 1992 (fig.**268**)

Colonies free floating, solitary, irregularly spherical or elongate, up to 50 µm in diameter, multicellular, with cells distant from one another and always radially displaced, joined to the ends of thin, mucilaginous, pseudodichotomously branched stalks, radiating from the center of the colony. Old cells slightly

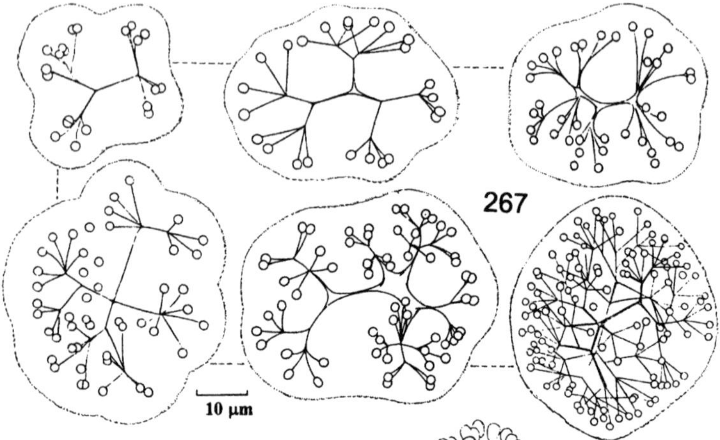


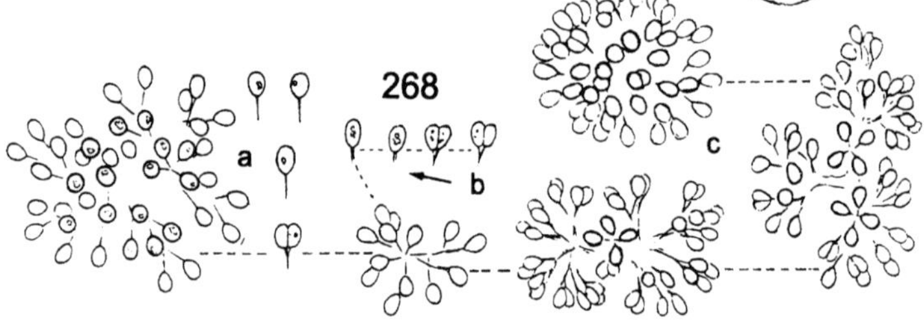


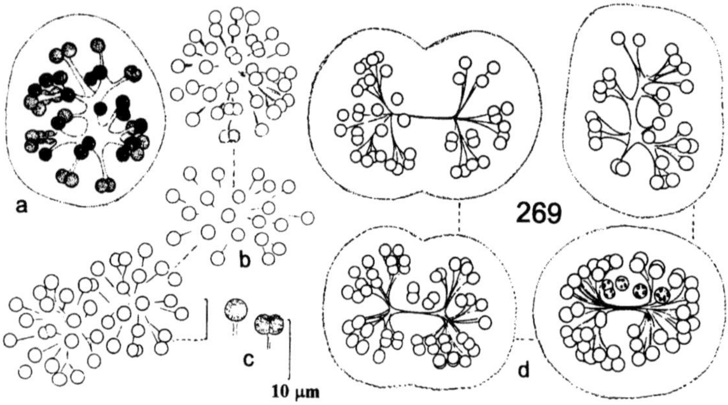


SNOWELLA

**Fig. 267.** *Snowella arachnoidea*: from Komárek & Hindák (1988).

**Fig. 268.** *Snowella fennica*: **a** – colony and detail of cells after Komárek & Hindák (1988, sub *S. septentrionalis* p.p.), **b-c** – after Komárek & Komárková-Legnerová (1992).

elongate, drop-shaped or more or less elliptical, after division slightly asymmetrical, with one or two aerotopes, 2.8–4.2 × (1.2) 2.6–2.8 µm.
Occurrence: Planktic in nordic lakes (N. Europe, Canada), eastern Baltic Sea and in lakes in Austria (Salzkammergut, special morphotype with up to 250 cells in a colony); possibly also in S. Argentine.

### 5. **Snowella litoralis** (Häyrén) Komárek et Hindák 1988 (fig. **269**)

*Gomphosphaeria litoralis* Häyrén 1921

Colonies spherical, mainly to 30 µm, rarely to 95 µm in diameter, always with cells distant from one another. Cells precisely spherical, after division hemispherical with pale greyish blue-green content, 2.4–4 µm in diameter.
Occurrence: Planktic in lakes and large reservoirs, common in the whole temperate zone; also in slightly salty water (coastal parts of Baltic Sea).

### 6. **Snowella lacustris** (Chodat) Komárek et Hindák 1988 (fig. **270**)

*Gomphosphaeria lacustris* Chodat 1898; *Coelosphaerium lacustre* (Chodat) Ostenfeld 1907

Colonies microscopic, spherical or ellipsoidal, sometimes compound, to 80 µm in diameter, with cells disposed sparsely in young and densely in old colonies. Mucilaginous envelopes fine, diffluent, stalks thin, usually clearly visible without staining. Cells elliptical or obovoid, pale grey-blue or blue-green, without gas vesicles, (1.5 ?) 2–4 × 1.5–3.5 µm.
Occurrence: Planktic in mesotrophic or eutrophic stagnant waters, lakes, ponds and reservoirs, known also from eastern Baltic Sea, common but never in masses; probably cosmopolitan. In Austrian lakes was found *S. lacustris*-like species with many-celled colonies and larger cells with aerotopes; similar type was recorded from Argentine (Tierra del Fuego) sub «*Gomphosphaeria aponina* var. *delicatula* fa.».

**Species outside Europe:**
- *Snowella rosea* (Snow) Elenkin 1938: Planktic (mostly tychoplanktic) in N. American lakes; (G.M. Smith 1920).

## 5. **Woronichinia** Elenkin 1933

Colonies more or less spherical or irregular oval, commonly composed of subcolonies, free-living, usually with a narrow, colourless enveloping mucilaginous layer. In the center of colonies is a system of radially and more or less parallely oriented, simple, unbranched stalks, which split along their whole length after cell division; stalks thick, usually as wide as the width of cells, commonly densely packed together and so causing the radial lamellation of the central colonial mucilage (staining !); the stalk system sometimes diffluent within old colonies. Cells rarely almost spherical, usually slightly elongate, widely oval, oval or obovoid, separating from one another immediately after division, but in old colonies very densely, radially aggregated in the peripheral layer.

---

**Fig. 269.** *Snowella litoralis*: **a** – after G.M.Smith (1920, sub *Gomphosphaeria lacustris*), **b-c** – after Komárek (1958, sub *Gomphosphaeria rosea* p.p.), **d** – from Komárek & Hindák (1988).

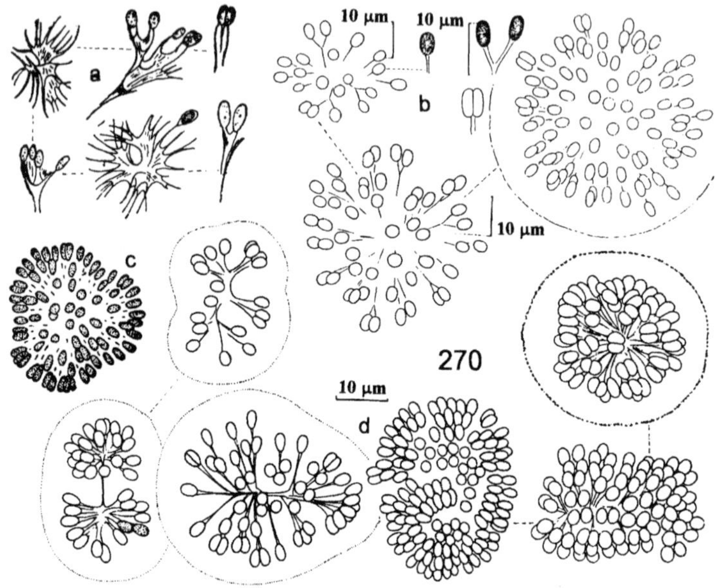


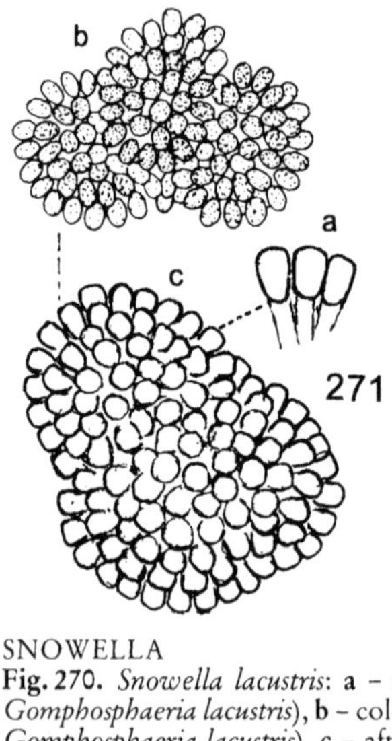


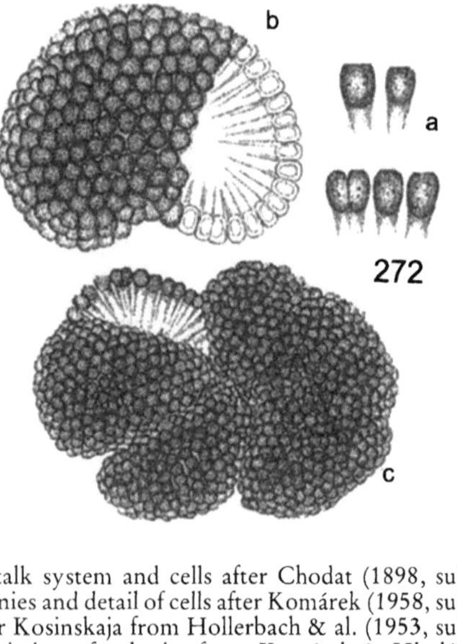


SNOWELLA

**Fig. 270.** *Snowella lacustris*: **a** – stalk system and cells after Chodat (1898, sub *Gomphosphaeria lacustris*), **b** – colonies and detail of cells after Komárek (1958, sub *Gomphosphaeria lacustris*), **c** – after Kosinskaja from Hollerbach & al. (1953, sub *Gomphosphaeria lacustris*), **d** – variation of colonies from Komárek & Hindák (1988).

Cell division in two planes perpendicular to one another in successive generations, and perpendicular to the colony surface; reproduction by disintegration of colonies and especially by liberation of solitary cells (sometimes spontaneously) from the colony (expulsing cells).

**Important references:** Geitler (1932, sub *Coelosphaerium*); Elenkin (1933); Huber-Pestalozzi (1938, sub *Coelosphaerium*); Hollerbach & al. (1953); Komárek (1958, 1984, sub *Gomphosphaeria*); Skuja (1964, sub *Gomphosphaeria*); Komárek & Anagnostidis (1986, sub *Gomphosphaeria*); Cmiech & al. (1987, sub *Gomphosphaeria*); Komárek & Hindák (1988); Komárek (1989); Komárek & Komárková-Legnerová (1992); Cronberg & Komárek (1994).

**Key to the species:**

1a Cells commonly with numerous aerotopes; cells 4.5–7 × 1.5–5 µm, colonies to more than 100 µm in diameter; cosmopolitan species . . **1. W. naegeliana**

1b Cells without aerotopes, only occasionally with a few aerotopes . . . . . . . 2

2a Cells more than 3 µm long (or in diameter) . . . . . . . . . . . . . . . . . . . . . . 3

2b Cells less than 3 µm long (or in diameter) . . . . . . . . . . . . . . . . . . . . . . 8

3a Cells pink or reddish, widely oval to almost spherical; planktic in lakes . . . . . . . . . . . . . . . . . . . . . . . . . . . . . . . . . . . . . . . . . . . . **7. W. fusca**

3b Cells always pale greyish-blue or blue-green . . . . . . . . . . . . . . . . . . . 4

4a Cells narrow, obovoid, circa 2x longer than wide, 3–6 × 1.2–2 µm; colonies up to 30 (- 50) µm in diameter . . . . . . . . . . . . . . . . . . . . . . **6. W. karelica**

4b Cells widely obovoid or oval, less than 1.8x longer than wide . . . . . . . . 5

5a Cells clearly obovoid, at the apical end obtuse rounded, 3–5.6 × (1.5) – 3.4 µm; colonies to 80 µm in diameter . . . . . . . . . . . . . . **2. W. compacta**

5b Cells oval or widely oval to almost spherical . . . . . . . . . . . . . . . . . . . . 6

6a Cells 4–6 µm long, colonies to 80 µm in diameter; benthic in more or less clear water bodies . . . . . . . . . . . . . . . . . . . . . . . . . . . . . . . . . . . . . 7

6b Cells 5.5–7 × 3.5–4.5 µm, olive-green, colonies to 130 µm in diameter; metaphytic in small water bodies . . . . . . . . . . . . . . . . . . . . **3. W. robusta**

---

WORONICHINIA

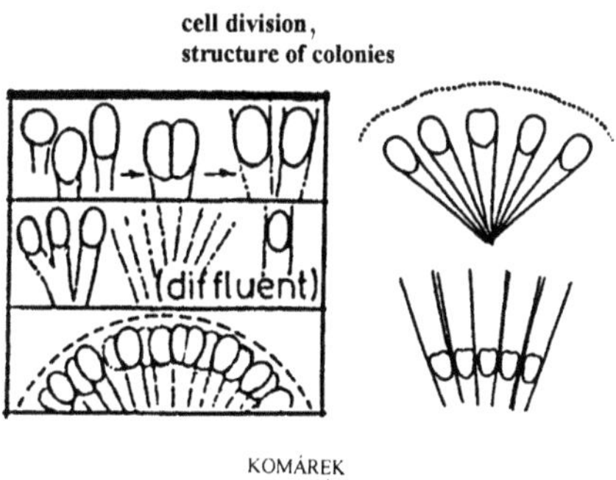


Fig. 271. *Woronichinia compacta*: **a** – detail of cells, **b–c** colonies; **b** – after Tikkanen (1986, sub *Gomphosphaeria compacta*), **a,c** – orig.
Fig. 272. *Woronichinia robusta*: **a** – detail of cells, **b-c** – colonies; after Skuja (1964, **a-b** – sub *Gomphosphaeria robusta*, **c** – sub *G. robusta* var. *meiocystis*).

7a Cells oval to widely oval, 4–6 × 1.5–3.5 µm, blue-green; benthic in clear reservoirs with sandy bottom . . . . . . . . . . . . . . . . . . . . . . **4. W. ruzickae**
7b Cells widely oval to almost spherical, 5–6 × 3.5–5 µm, with pale blue-green to olive-green, finely granulated content; epipelic on mud of ponds . . . . . . . . . . . . . . . . . . . . . . . . . . . . . . . . . . . . . . . . . . **5. W. botrys**
8a Cells slightly, but distinctly elongate . . . . . . . . . . . . . . . . . . . . . . . . . 9
8b Cells spherical . . . . . . . . . . . . . . . . . . . . . . . . . . . . . . . . . . . . . . . . 10
9a Colonies with cells densely packed in the peripheral layer; cells mainly oval, 2.5–3 × 1.6–2 µm; planktic in clear nordic lakes . . . . . . . . . **8. W. elorantae**
9b Colonies with cells distant from one another; cells oval, 1.4–3 × 1–1.8 µm; alkaline lakes in Hungary . . . . . . . . . . . . . . . . . . . . . . . . . **9. W. radians**
10a Cells 2–2.5 µm, colonies up to 45 µm in diameter; peaty pools **10. W. tenera**
10b Cells 1–1.5 µm, colonies up to 75 µm in diameter; planktic and benthic in lakes . . . . . . . . . . . . . . . . . . . . . . . . . . . . . . . . . . . **11. W. delicatula**

## 1. Woronichinia naegeliana (Unger) Elenkin 1933 (fig. 273)

*Coelosphaerium naegelianum* Unger 1854; *Gomphosphaeria naegeliana* (Unger) Lemmermann 1907

Colonies microscopic, spherical, ellipsoid, kidney-shaped or spherical-irregular (lobate) up to 180 (- 275 ?) µm in diameter, sometimes composed of subcolonies, with more or less densely and radially arranged cells in the peripheral monolayer, attached at the ends of gelatinous tubular stalks, which are of the same width as the cells; stalk system forms typical radial lamellation in the colonial center, diffluent in old colonies; mucilaginous envelope around the colony colourless or slightly yellowish, sometimes distinctly delimited, radially lamellate and extending beyond the cell layer; cells surrounded sometimes with individual, narrow gelatinous envelopes. Cells obovoid or ellipsoid, blue-green with numerous aerotopes, (3 ?) 5–7 × (1.5) 2.5–3.5 (5) µm. Reproduction often by disintegration of colony into solitary cells («expulsing cells», planocytes). Thylakoids peripheral. Variable species; different morphotypes, the taxonomic values of which need resolution (see e.g., Cronberg & Komárek 1994). Ultrastructure described by Cmiech & al. (1987).

Occurrence: Commonly planktic in eutrophic lakes and ponds, sometimes forming water blooms; temperate zones, in Europe and N. America up to northern regions; tropical records need revision.

## 2. Woronichinia compacta (Lemmermann) Komárek et Hindák 1988 (fig. 271)

*Gomphosphaeria lacustris* var. *compacta* Lemmermann 1899; *Gomphosphaeria compacta* (Lemmermann) Ström 1923; *Gomphosphaeria lacustris* f. *compacta* (Lemmermann) Elenkin 1938; ? *Coelosphaerium holopediforme* Schmidle 1901; ? *Gomphosphaeria robusta* var. *meiocystis* Skuja 1964 incl.

Colonies free-living, irregularly spherical or oval, up to 80 µm in diameter, later often consisting of subcolonies; cells densely, radially aggregated in the peripheral

---

WORONICHINIA

**Fig. 273.** *Woronichinia naegeliana*: **a** – outlines of colonies (after Kovalenko from Kondrateva & al. 1984), **b** – young colonies (after G.M.Smith 1920, Komárek & Hindák 1988, Komárek & Komárková-Legnerová 1992 and Komárková-Legner-

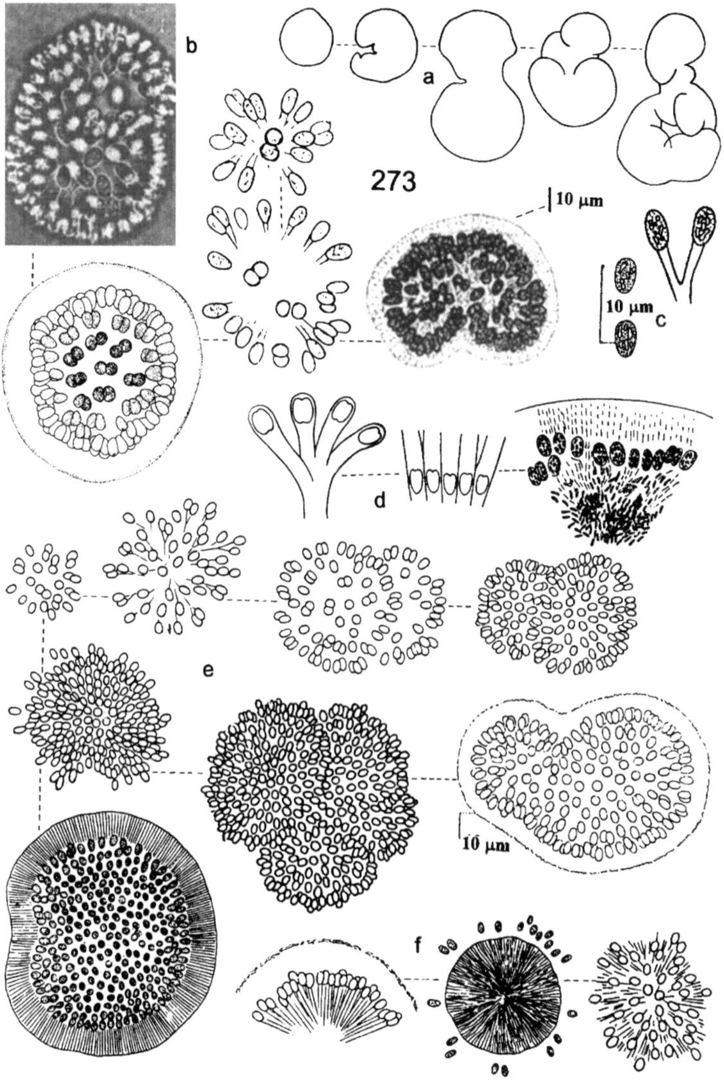


ová & Cronberg 1994), **c** - detail of cells (after Komárek 1958), **d** - position of cells within mucilaginous stalks (after Voronichin from Kondrateva & al. 1984 and Bachmann 1907), **e** - variation in colonies of various age (after Komárek 1958 and Elenkin from Kondrateva & al. 1984), **f** - arrangement of cells with visible stalk system, expulsion of cells and a colony after liberation of solitary cells (after Komárek 1958 and Elenkin from Kondrateva & al. 1984).

layer; mucilaginous envelope wide, colourless, diffluent at the margin; gelatinous stalks developed during the whole life cycle (staining !). Cells obovoid, sometimes rounded triangular in outline with obtuse-truncate poles, rarely oval, greyish blue-green, without aerotopes, 3–6 × (1.5) – 3.5 µm.
Occurrence: Planktic in lakes of the temperate zone, especially in northern regions; Europe: common particularly in Baltic region, including Baltic Sea.

**3. Woronichinia robusta** (Skuja) Komárek et Hindák 1988 (fig. **272**)

*Gomphosphaeria robusta* Skuja 1964

Colonies free-living, more or less spherical, subspherical, kidney-shaped or compound (lobed), 30–70–130 µm in diameter, with many grouped, radially oriented cells, with enveloping hyaline mucilage, slightly or clearly visible, extending up to 12 µm beyond the cells. Cells at the ends of hyaline, thick, after cell division pseudodichotomously divaricate stalks, radiating from the colonial center, oval, slightly obovate or obovate-cylindrical, with flattened and rounded apical ends, 5.5–7 × 3.5–4.5 µm; cell content homogeneous, olive-green or olive-bluish, with clearly distinct chromatoplasma, sometimes with small solitary granules.
Occurrence: Epipelic among benthic algae, mosses and water plants on muddy bottom of small water bodies or lake littorals; N. Sweden.

**4. Woronichinia ruzickae** Komárek et Hindák 1988 (fig. **274**)

*Gomphosphaeria compacta* sensu Komárek 1958

Colonies solitary, mainly irregularly spherical or oval, later compound, to 80 µm in diameter, multicellular, with cells radially and densely disposed in the peripheral layer, joined to thick stalks, radiating parallely from the center of colonies. Cells oval with homogeneous, pale blue-green content, 4–6 × 1.5–3.4 µm.
Occurrence: Benthic and metaphytic in oligotrophic ponds, swamps and pools with sandy bottom and sometimes submerged water plants; Czech Republic (S. Bohemia), Austria (Waldviertel).

**5. Woronichinia botrys** (Skuja) Komárek et Hindák 1988 (fig. **275**)

*Gomphosphaeria botrys* Skuja 1964

Colonies solitary, free floating, irregularly spherical, sometimes composed from several subcolonies, 20–60 (- 80) µm in diameter, with densely packed cells in the peripheral layer, enveloped by an indistinct, colourless, diffluent mucilaginous

---

WORONICHINIA

**Fig. 274.** *Woronichinia ruzickae*: **a** – colony from the original locality (from Komárek & Hindák 1988), **b** – young colonies (from Komárek 1975), **c** – old colony, **d-e** – detail of cells (**c-e** – after Komárek 1958, sub *Gomphosphaeria compacta*).

**Fig. 275.** *Woronichinia botrys*: **a** – cells, **b** – colonies; after Skuja (1964, sub *Gomphosphaeria botrys*).

**Fig. 276.** *Woronichinia karelica*: **a** – detail of cells, **b** – young colonies, **c** – old colony; **a-b** – after Komárek & Komárková-Legnerová (1992), **c** – after Kosinskaja from Hollerbach & al. (1953, sub *Gomphosphaeria lacustris* f. *compacta*).

**Fig. 277.** *Woronichinia fusca* (red cells): **a** – detail of cells, **b** – colonies; after Skuja (1948, sub *Gomphosphaeria fusca*).

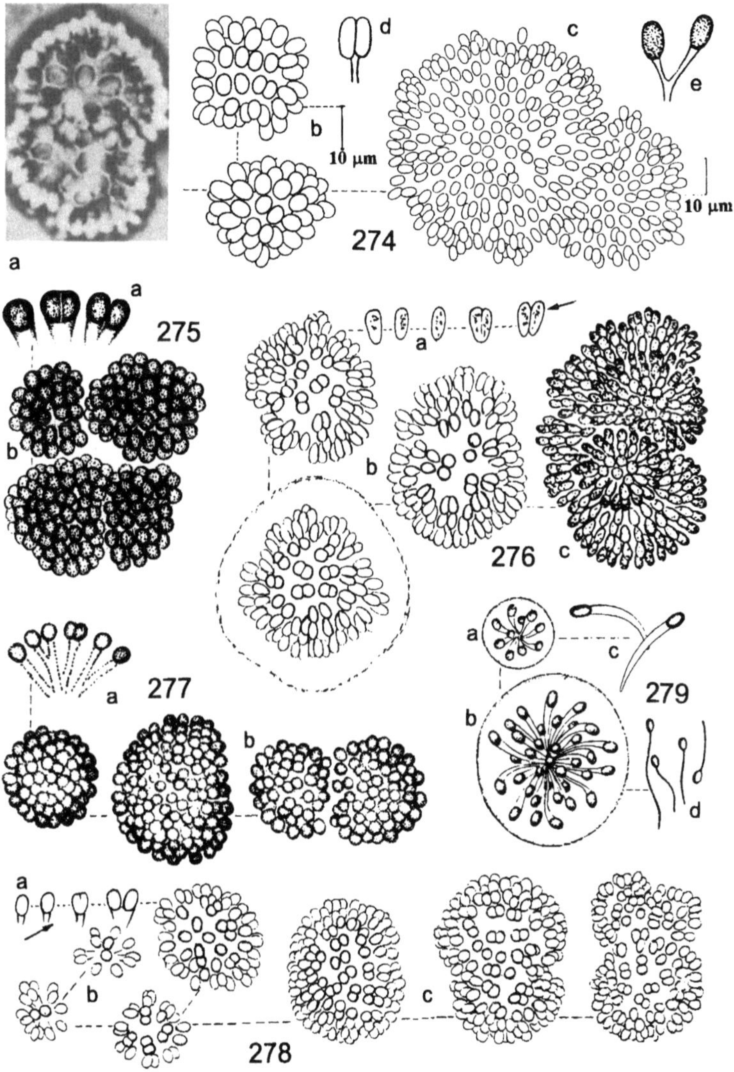


**Fig. 278.** *Woronichinia elorantae* (pale blue-green, olive-green, or yellowish cells): **a** – detail of cells, **b** – young colonies, **c** – old colonies; after Komárek & Komárková-Legnerová (1992).

**Fig. 279.** *Woronichinia radians*: **a** – young colony, **b** – colony, **c-d** – detail of cells attached to mucilaginous strands; after Hortobágyi (1957, sub *Gomphosphaeria radians*).

envelope. Cells widely obovate or oval, radially oriented, after division widely obovate-cordiform, 5–6 × 3.5–5 µm; cell content blue-green or olive-green, finely granulate, without gas vesicles.
Occurrence: In metaphyton and benthos among algae of small, clear (oligotrophic) water bodies; N. Sweden.

**6. Woronichinia karelica** Komárek et Komárková-Legnerová 1992 (fig. **276**)

Colonies solitary, irregularly spherical or oval, later composed of subcolonies, multicellular, up to 50 µm in diameter, with radially and densely packed cells in the peripheral layer; cells joined to the faintly visible, colourless, thick stalks radiating from the colonial center. Cells more or less elongate, obovate, 3–6 × 1.5–2 µm, with a homogeneous, pale blue-green content.
Occurrence: Planktic in oligo- or mesotrophic lakes and ponds of northern part of the whole temperate zone (N. Europe, N. Russia, Japan-Hokkaido, Canada), known also from eastern Baltic Sea.

**7. Woronichinia fusca** (Skuja) Komárek et Hindák 1988 (fig. **277**)

*Gomphosphaeria fusca* Skuja 1948

Colonies spherical, oval or slightly irregular, up to 50 µm in diameter, with very fine, colourless, scarcely visible mucilaginous envelope; gelatinous stalks visible after staining. Cells widely oval to almost spherical, densely packed in the peripheral layer, 3–5 × 2.5–4 µm; cell content pink or slightly violet, homogeneous or finely granulate.
Occurrence: In summer and autumn plankton of mesotrophic large lakes; Sweden, Switzerland.

**8. Woronichinia elorantae** Komárek et Komárková-Legnerová 1992 (fig. **278**)

Colonies solitary, irregularly spherical or oval, usually composed of subcolonies, compact, multicellular, with radially, densely packed cells in the peripheral layer, to 20 (38) µm in diameter; cells joined to the ends of thick, colourless stalks, almost invisible without staining, radiating from the center of the colonies. Cells oval or widely obovate, with homogeneous blue-green content, without gas vesicles, 2.5–3 × 1.6–2 µm.
Occurrence: In plankton and metaphyton of small water bodies and lake littoral with water plants; northern areas of temperate zone (Finland, Canada, Japan-Hokkaido); possibly also in littoral parts and bays of Baltic Sea.

**9. Woronichinia radians** (Hortobágyi) Komárek et Hindák 1988 (fig. **279**)

*Gomphosphaeria radians* Hortobágyi 1957

Colonies more or less spherical with 10–40 more or less radially oriented, slightly elongate cells; mucilage colourless, homogeneous, indistinctly delimited; unbranched gelatinous, colourless stalks radiating from the colonial center, separated from one another and widening continually to the ends (width of stalks at the ends reaches the cell width). Cells oval, with pale blue-green, more or less homogeneous content, without aerotopes, 1.4–3 × 1–1.8 µm.
Occurrence: Known only from Hungary, planktic in the alkaline lake Szelidi. Needs confirmation (bacteria ?).

**10. Woronichinia tenera** (Skuja) Komárek et Hindák 1988 (fig. **280**)
*Gomphosphaeria tenera* Skuja 1964

Colonies small, more or less irregularly spherical, oval or lobate, 8–100-celled, sometimes compound, in the center more or less hollow, 10–45 µm in diameter, surrounded by a very delicate, hyaline, scarcely visible, to 4 µm wide mucilaginous envelope; cells joined to the ends of thick, but very fine stalks, more or less densely packed in the peripheral layers. Cells more or less spherical or (after division) widely spherical-obovate, 2–2.5 µm in diameter, with a blue-green or (rarely) olivaceous, homogeneous content, without granules or aerotopes.
Occurrence: Among coccal and filamentous cyanoprokaryotes and algae in benthos of small water biotopes, mainly moorland waters and in *Sphagnum*-pools; known only from N. Sweden (Abisko).

**11. Woronichinia delicatula** (Skuja) Komárek et Hindák 1988 (fig. **281**)
*Gomphosphaeria delicatula* Skuja 1964

Colonies free floating, small, spherical, subspherical, oval or lobate, with cells forming more or less hollow sphere, 20–75 µm in diameter, with very fine, diffluent and usually invisible (or absent ?) common mucilaginous envelope, but with a distinct system of radially oriented, straight, gelatinous stalks, radiating from the colonial center; cells situated more or less in the colonial periphery in one layer or slightly displaced from one another in the radial direction, not very densely packed. Cells spherical or widely oval, 1–1.5 µm in diameter, with pale blue-green, homogeneous or finely granulate content, without gas vesicles.
Occurrence: Freshwater, free floating or among other algae in benthos of oligotrophic lakes; N. Sweden (Abisko).

**Species outside Europe:**
- *Woronichinia fremyi* (Komárek) Komárek et Hindák 1988: Tropical, planktic species with numerous small aerotopes; (Komárek 1984, 1988); (fig. **282**).
- *Woronichinia klingae* Komárek et Komárková-Legnerová 1992: Planktic in central Canadian lakes.
- *Woronichinia kuselae* M. Watanabe et Komárek 1994: Periphyton of streams and pools in central Himalayas.

**Unrevised species:**
- *Woronichinia hungarica* Hortobágyi 1939: Probably a morphotype of *W. naegeliana*; (fig. **283**).
- *Gomphosphaeria wichurae* (Hilse in Rabenhorst) Drouet et Daily 1952: Probably a morphotype of *W. naegeliana.*

## 6. Gomphosphaeria Kützing 1836

Colonies spherical or irregular, commonly composed of subcolonies, free-living, sometimes enveloped by indistinct, fine and diffluent mucilage; from the center of colony radiates a system of more or less thick, gelatinous stalks, which are, however, always thinner than the cell width; stalks are widened at the ends and envelope individual cells with a narrow layer; the stalk system is sometimes diffluent in the middle of the colony. Cells are always elongate, radially situated at the ends of stalks, obovate or club-shaped, after division remain and grow parallely joined together for a long part of the cell cycle, forming the characteristic cordiform shape. Solitary cells or cordiform couples are always slightly distant from one another with remarkable spaces between them, sometimes slightly radially displaced from one another.

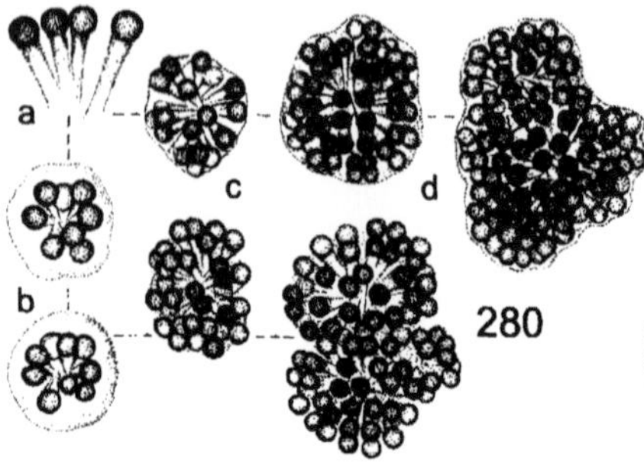


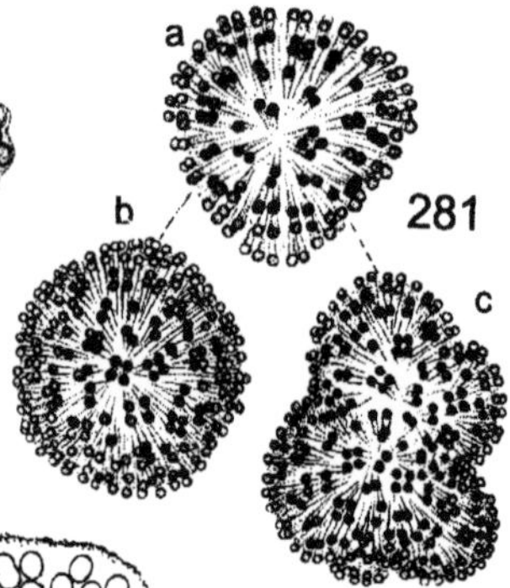


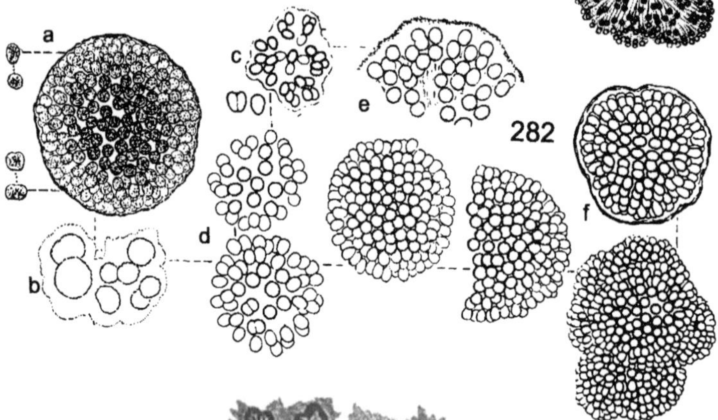


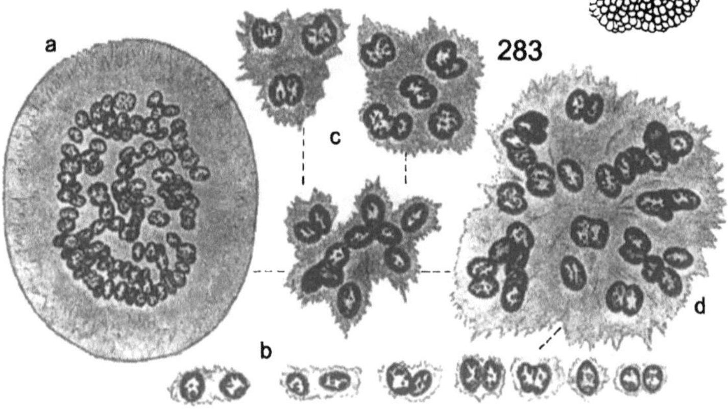


WORONICHINIA

**Fig. 280.** *Woronichinia tenera*: **a** – detail of cells, **b-c** – young colonies, **d** – old colonies; after Skuja (1964, sub *Gomphosphaeria tenera*).

**Fig. 281.** *Woronichinia delicatula*: after Skuja (1964, sub *Gomphosphaeria delicatula*).

Cell division in two planes in successive generations, perpendicular to one another and to the colony surface; the daughter cells divide completely in the later period of a cell cycle. Reproduction by the colony disintegration.
**Important references:** Geitler (1932, 1942); Huber-Pestalozzi (1938); Komárek & Anagnostidis (1986); Komárek & Hindák (1988); Komárek (1989); Komárek & Komárková-Legnerová (1992).

**Key to the species:**

**1a** Colonies to 40 μm in diameter, cells to 10 μm long; planktic species .... **2**
**1b** Old colonies more than 40 μm in diameter, cells 8–15 μm long; mainly metaphytic species .... **4**
**2a** Cells pinkish or violet, 6–9 μm long .... **1. G. lilacea**
**2b** Cells blue-green or olive-green .... **3**
**3a** Cells 7–10 × 2–4 (cordiform cells –5.6) μm .... **2. G. virieuxii**
**3b** Cells 4.8–7.2 (- 10) × 2.8–7.5 (cordiform cells –11) μm .... **3. G. natans**
**4a** Cells 8–12 × 4–6.5 (cordiform cells –9) μm, pale blue-green, with distinct individual gelatinous envelopes; freshwater swamps .... **4. G. aponina**
**4b** Cells 10.4–15 × 4–6 (cordiform cells –10) μm, bright blue-green, with indistinct individual envelopes; coastal and inland salty waters **5. G. salina**

## 1. **Gomphosphaeria lilacea** Virieux 1916 (fig. **284**)

Colonies microscopic, many-celled, solitary, sometimes composed of subcolonies; stalks colourless, relatively thin, colonies enveloped by diffluent mucilage. Cells obovoid, without gas vesicles, pinkish violet, 6–9 × 3–4 μm. According to the original description rather a *Woronichinia* species (Virieux 1916: «l'aspect de *G. Nägeliana*»).
Occurrence: Freshwater, planktic in lakes; described and known only from France (Lac de Ratay).

## 2. **Gomphosphaeria virieuxii** Komárek et Hindák 1988 (fig. **285**)

*Gomphosphaeria aponina* var. *delicatula* Virieux 1916

Colonies free floating, irregularly spherical, sometimes compound, to 25–35 μm in diameter, with more or less sparsely arranged cells, surrounded by diffluent, colourless gelatinous envelope. Cells obovoid or cordiform, narrowed at the proximal end, sometimes with visible individual narrow envelopes around cells, with pale or bright blue-green, more or less homogeneous or slightly granulate content, 7–10 × 2–4 (5.6) μm.
Occurrence: Planktic in clear or mesotrophic lakes, usually large ones, in temperate zones; Europe: France (Jura lakes), northern regions, also known from Caspian Sea.

---

**Fig. 282.** *Woronichinia fremyi*, planktic, with aerotopes (= groups of gas vesicles), from tropical plankton: **a** – colony with detail of cells, **b** – group of colonies in common mucilage, **c-d** – young colonies with detail of cells, **e** – part of a colony with visible stalk system, **f** – old colonies; **a-b** – after Komárek (1989), **c-f** – after Komárek (1984, sub *Gomphosphaeria fremyi*).
**Fig. 283.** *Woronichinia hungarica*: **a** – colony, **b** – detail of cells and initial stages of colony formation, **c-d** – young colonies; after Hortobágyi (1939).

Note: This description and cited figures of G. *virieuxii* correspond to concepts of later authors (e.g. Huber-Pestalozzi 1938); according to the original description, *G. aponina* var. *delicatula* Virieux 1916 is characterized by pale, very narrow cells (colonies 60–80 µm in diameter); the difference of this taxon from our *G. virieuxii* is probable; (fig. **286**).

### 3. **Gomphosphaeria natans** Komárek et Hindák 1988 (fig. **287**)

Colonies solitary, free floating, irregularly spherical or slightly elongate, to 30 µm in diameter (excluding mucilage), with wide mucilaginous, hyaline envelopes, with radially disposed, more or less densely grouped cells; mucilaginous stalks wide, hyaline, irregularly pseudodichotomously branched, at the ends widened and enveloping cells with a narrow layer. Cells widely obovoid, after division widely cordiform, with grey-blue or olive-green content, sometimes with sparsely distributed granules (one to several), (4.2) 4.8–7.2 (10) × (2.8) 3.8–7.5 (8) µm.
Occurrence: Planktic in oligotrophic or mesotrophic lakes, known from northern USA, Canada, and perhaps from S. Argentine; occurrence in Europe (Austria, France, S. Slovakia) is questionable; (fig. **288**).

---

GOMPHOSPHAERIA

HINDÁK 1988

**cell division, development of colonies**

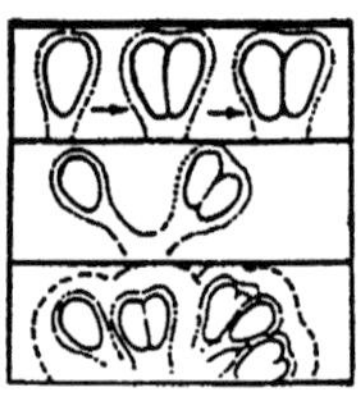

KOMÁREK & HINDÁK 1988

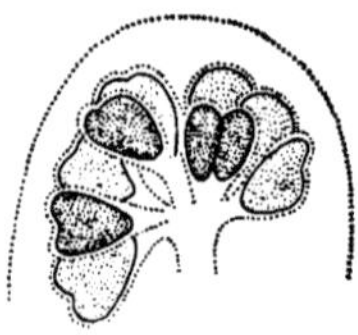

G.M. SMITH 1920

**Fig. 284.** *Gomphosphaeria lilacea*: after Virieux (1916), the original drawing corresponds rather to the genus *Woronichinia* or *Snowella*.
**Fig. 285.** *Gomphosphaeria virieuxii* in the present sense (sensu auct. post): **a** – after Proškina-Lavrenko & Makarova (1968, sub *G. aponina* var. *delicatula*), **b** – after Kosinskaja from Hollerbach & al. (1953, sub *G. aponina* var. *delicatula*), **c** – after Nygaard (1932, sub *G. aponina* var. *delicatula*), **d-e** - after Komárek & Komárková-Legnerová (1992).

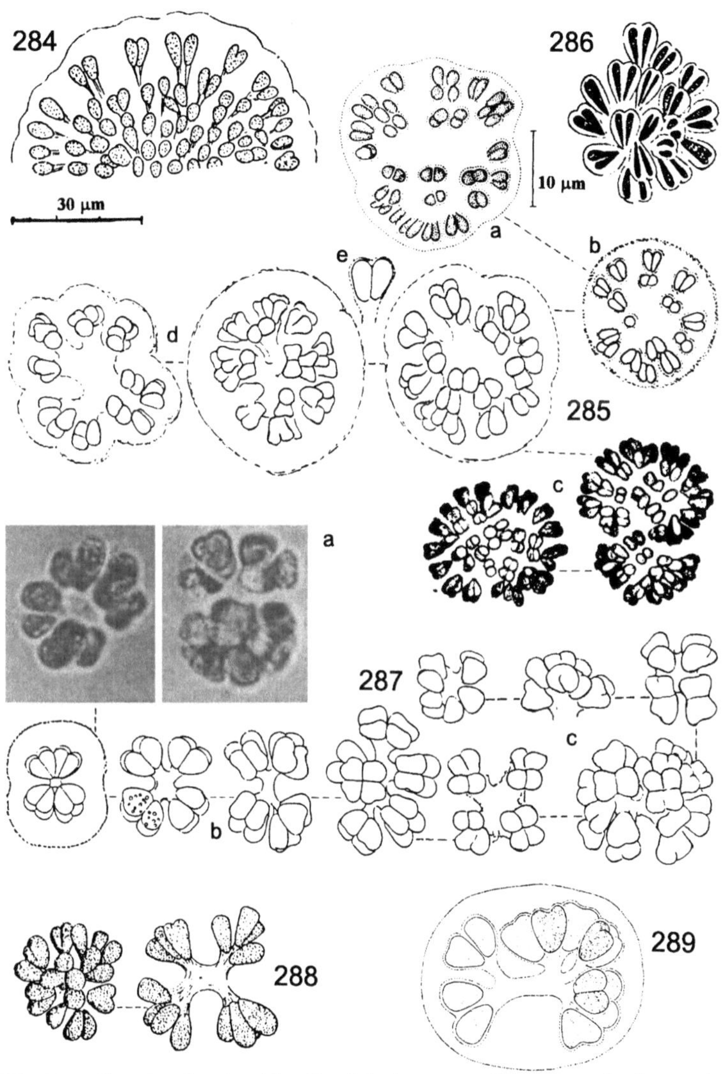


**Fig. 286.** *Gomphosphaeria virieuxii*, original concept: iconotype of «*G. aponina* var. *delicatula*» after Virieux (1916).
**Fig. 287.** *Gomphosphaeria natans*, from Canadian localities: **a-b** – after Komárek & Hindák (1988), **c** – after Komárek & Komárková-Legnerová (1992).
**Fig. 288.** *Gomphosphaeria* cf. *natans*, from France: after Virieux (1916, sub *G. aponina* f. *limnetica*).
**Fig. 289.** *Gomphosphaeria cordiformis*: after G.M. Smith (1920).

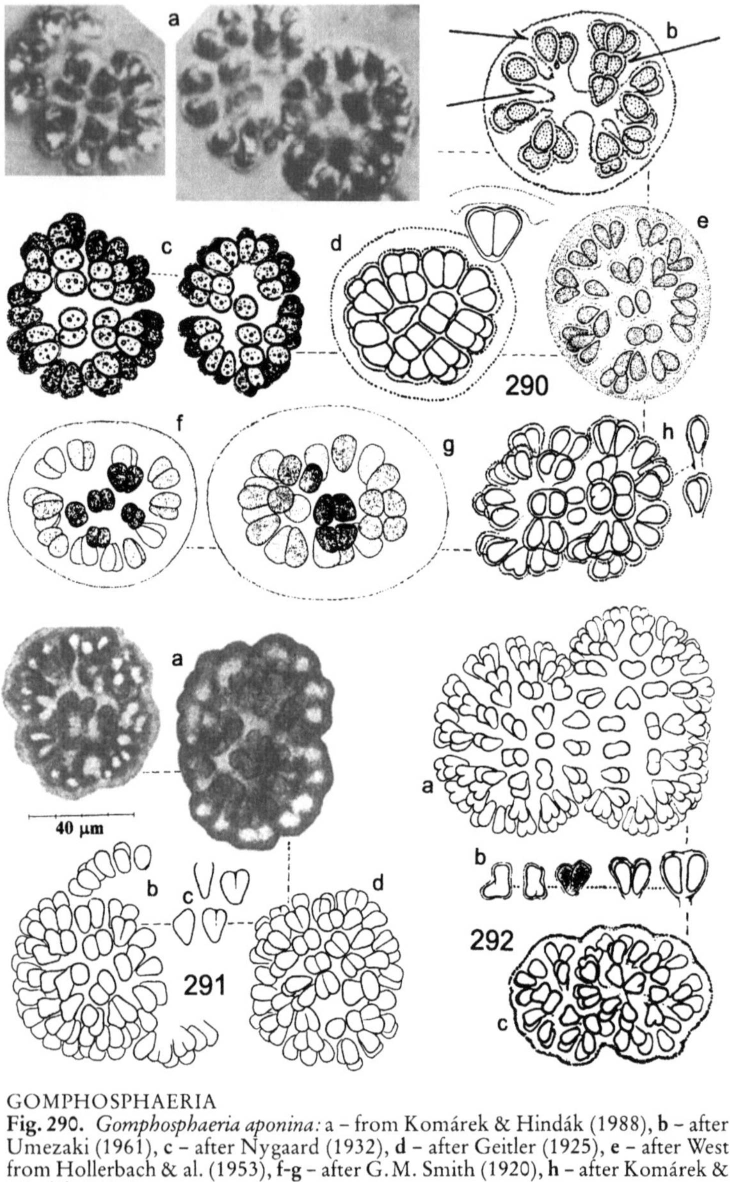


GOMPHOSPHAERIA

**Fig. 290.** *Gomphosphaeria aponina:* a – from Komárek & Hindák (1988), **b** – after Umezaki (1961), **c** – after Nygaard (1932), **d** – after Geitler (1925), **e** – after West from Hollerbach & al. (1953), **f-g** – after G.M. Smith (1920), **h** – after Komárek & Hindák (1988).

**Fig. 291.** *Gomphosphaeria salina*: **a** – from Uherkovich (1965, sub *G. aponina)*

**4. Gomphosphaeria aponina** Kützing 1836 (fig. **290**)

Colonies microscopic, irregularly spherical or elliptical, sometimes compound, to 95 µm in diameter, with sparsely or densely arranged cells, sometimes 2–4 near one to another; mucilaginous envelopes colourless, fairly thick, homogeneous, mucilaginous stalks more or less thick; special colourless gelatinous layers usually visible around cells. Cells obovoid or slightly club-shaped, during division cordiform, usually pale blue-green, rarely olive-green or yellowish, with homogeneous or finely granulate content, without aerotopes, (7.4)8–12 × (3.7)4–6.5 (cordiform cells -9) µm.

Occurrence: Among algae and metaphytic in oligotrophic and mesotrophic swamps and in littoral of freshwater lakes, springs, rarely tychoplanktic and planktic; commonly but sporadically distributed in temperate and subtropic zones (comp., e.g., Werner 1988) of both hemispheres, records from typical tropical biotopes probably concern other species (usually *G. multiplex*).

**5. Gomphosphaeria salina** Komárek et Hindák 1988 (fig. **291**)

*Gomphosphaeria aponina* sensu auct. (Uherkovich 1965, etc.)

Colonies microscopic, solitary, irregularly spherical or weakly elongate, to 110 µm long, with radially disposed cells; stalks mucilaginous, colourless, wide, hyaline, sometimes diffluent in the center of colonies and not well recognizable, irregularly pseudodichotomously divaricate, widened at the ends, enveloping the cells by a thin gelatinous layer; special layers around cells usually invisible without staining. Cells obovoid or slightly club-shaped, after division more or less cordiform, with homogeneous, brightly blue-green, yellowish green or olive-green content, 10.4–15 × 4–6 (10) µm.

Occurrence: Metaphytic and among other algae in marine coastal and inland salty swamps and pools, sporadic, all over the Europe; possibly cosmopolitan.

**Species outside Europe:**

- *Gomphosphaeria multiplex* (Nygaard) Komárek 1989: Cells 8.3–13.2 × 5.6–13.2 µm; coastal and inland swamps in tropical areas; (Frémy 1930, sub *G. aponina*; Huber-Pestalozzi 1938); (fig. **293**).
- *Gomphosphaeria okawangae* Komárek et Cronberg 1999: Metaphytic in swamps in Botswana (Buffalo fence).

---

from a «Soda - lake» in Hungary, **b-d** - after Komárek & Hindák (1988), **e** - after Setchell & Gardner (1919, sub *G. aponina*).

291

e

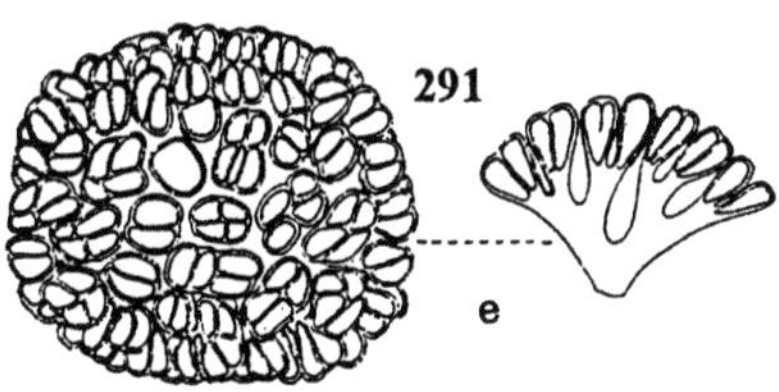

**Fig. 292.** *Gomphosphaeria semen-vitis,* from central America: after Komárek (1988).

– *Gomphosphaeria semen-vitis* Komárek 1989: Metaphytic, secondary planktic in littoral of lakes and in swamps, tropical species (Cuba, USA-Florida); (fig. **292**).
– *Gomphosphaeria* sp. sensu Ling et al. 1989: Documented and described without taxonomic definition (evidently a separate species), from Dalhousie springs, central Australia.

**Unrevised species:**
– *Gomphosphaeria cordiformis* (Wille) Hansgirg 1886: Recorded from swamps of the temperate zone including Europe, not revised in recent years; various concepts, see in G.M. Smith (1920), Geitler (1932), Skuja (1964), Guarrera & al. (1986); (fig. **289**).
– *Gomphosphaeria gessneri* Schiller 1956: Described from hypersalinic pools in Venezuela, not recorded again.

---

GOMPHOSPHAERIA
**Fig. 293.** *Gomphosphaeria multiplex*, tropical species: **a** – after Nygaard (1926, sub *G. aponina* var. *multiplex*), **b-c** – after Komárek (1989).

SIPHONOSPHAERA

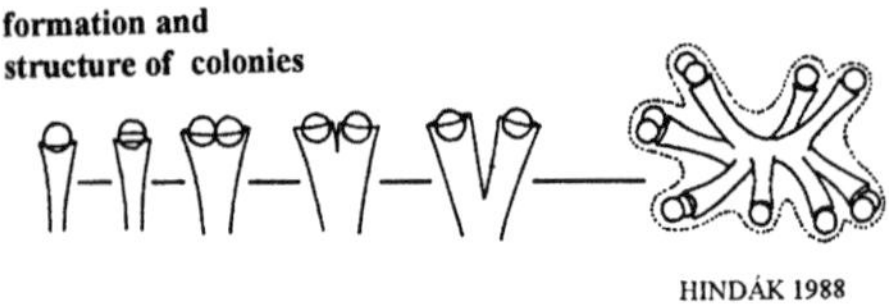


**Fig. 294.** *Siphonosphaera pannonica:* after Hindák (1988).

MICROCYSTIS

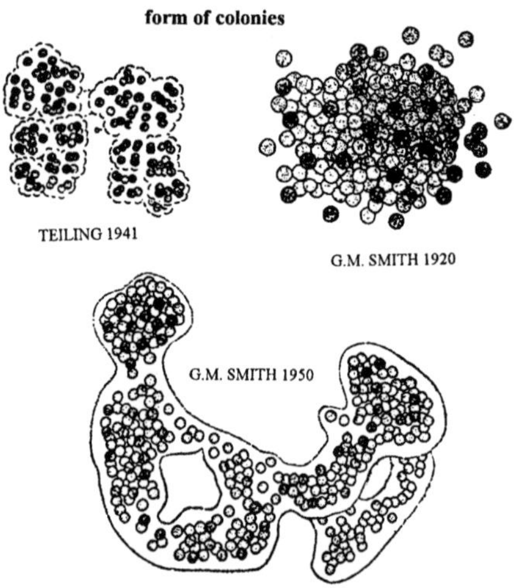

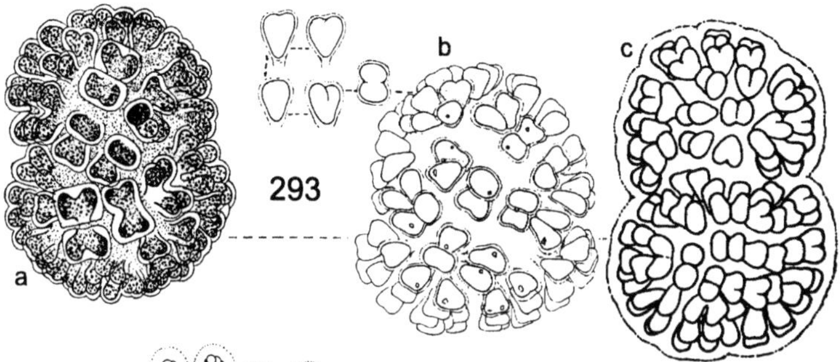


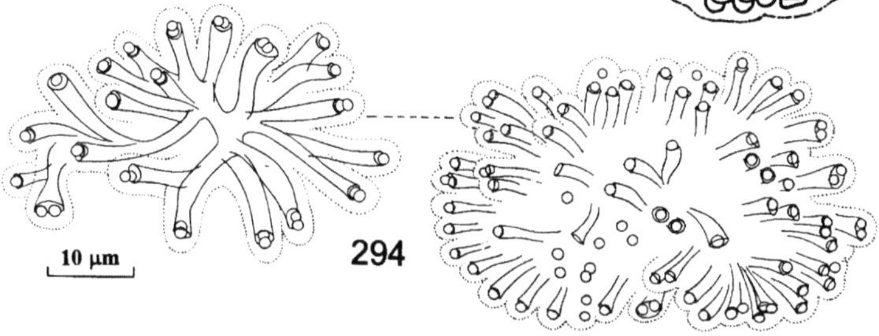


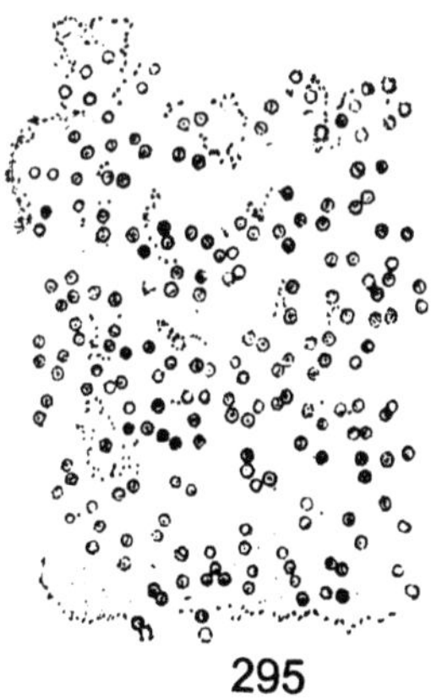


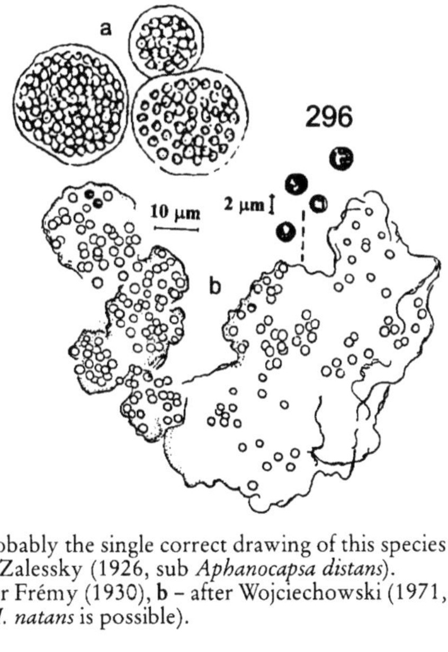


**Fig. 295.** *Microcystis* cf. *natans*: probably the single correct drawing of this species (each cell with 1–2 aerotopes) by Zalessky (1926, sub *Aphanocapsa distans*).
**Fig. 296.** *Microcystis firma*: **a** – after Frémy (1930), **b** – after Wojciechowski (1971, identity of this population with *M. natans* is possible).

## 7. **Siphonosphaera** Hindák 1988

Colonies microscopic, free floating, more or less spherical, composed of radially and slightly irregularly divaricate thick, tube-like, colourless stalks, originating from the colonial center. Cells narrower than stalks, situated within the end of stalks; the whole stalk system with cells is enveloped by the common, fine, diffluent, colourless mucilage; stalks slightly distant from one another. Cells in strings solitary or grouped in twos after division, geminate.
Cell division in two perpendicular planes in successive generations, reproduction by colony disintegration.
**Important references:** Hindák (1988).

### 1. **Siphonosphaera pannonica** Hindák 1988 (fig. **294**)

Colonies oval to irregular, multicellular, 30–50 µm in diameter; mucilaginous envelopes hyaline, colourless, surrounding the cells with a 1.5–2 µm wide gelatinous margin. Cells solitary or in twos after division, within the end part of radial tube-like stalks (sheaths); colonial mucilaginous envelopes follow the stalk system, fine, colourless (staining !). Cells spherical, 1–1.8 µm in diameter, with pale yellow-green content, without conspicuous inclusions or aerotopes.
Occurrence: In plankton of the lake Balaton, Hungary.

# 4. Family **Microcystaceae** Elenkin 1933

*Gloeocapsaceae* Elenkin 1933 incl.; *Stereometrinales* Starmach 1966

Coccal, colonial; cells arranged irregularly or in three-dimensional perpendicular rows in mucilaginous colonies, with or without individual gelatinous envelopes. Cells spherical, after division hemispherical, with or without aerotopes. Mucilage homogeneous or in form of special gelatinous, often stratified envelopes around cells.
Cell division regularly in three planes perpendicular one to another in successive generations, daughter cells grow into the original form and size before next division. Reproduction by disintegration of colonies. Facultative nanocyte production.

**Key to the genera:**

**1a** Cells without individual gelatinous envelopes, distributed within common fine, structureless and colourless mucilage . . . . . . . . . . . . . . . . . . . . . . **2**
**1b** Cells or their groups with individual gelatinous, distinct envelopes; small colonies, usually aggregated in granular strata . . . . . . . . . . . . . . . . . . . . **3**
**2a** Cells irregularly arranged in colonies, clustered irregularly in more or less spherical or formless planktic colonies; with aerotopes . . . **1. Microcystis**
**2b** Cells in three-dimensional, cubic colonies, arranged more or less in perpendicular rows (disturbed sometimes in old colonies); without aerotopes . . . . . . . . . . . . . . . . . . . . . . . . . . . . . . . . . . . . . . . . **4. Eucapsis**
**3a** With gelatinous, more or less spherical envelopes, often stratified and coloured (red, blue, yellow, brown), in irregular aggregates; mainly epilithic and epiphytic species . . . . . . . . . . . . . . . . . . . . . . . . . . . . . **2. Gloeocapsa**
**3b** In small, densely clustered, cubic or polyhedral subcolonies with cells surrounded by thin individual envelopes and widened, clearly delimited, firm slime, usually colourless; mainly living submersely . **3. Chondrocystis**

## 1. Microcystis Kützing ex Lemmermann 1907 nom. cons.

*Diplocystis* Trevisan 1848 nom. rejic.; *Anacystis* Meneghini sensu Kützing 1849; *Polycystis* Kützing 1849 nom. rejic.; *Clathrocystis* Henfrey 1856 nom. rejic.; *Microhaloa* Kützing sensu Rabenhorst 1865; non *Microcystis* Kützing 1833 (nom. rejic.)

Colonies micro- or macroscopic, free floating, spherical, oval, lobate to irregular, or elongate, in several species irregularly clathrate, sometimes composed of subcolonies or clustered together, with irregularly, sparsely or (usually) densely arranged cells in common mucilage. Mucilage fine, colourless, usually homogeneous or only indistinctly lamellate, diffluent or distinct and delimited, in several species with refractive outline (surface); around individual cells gelatinous envelopes never present. Cells spherical or hemispherical after division, with gas vesicles (joined into distinct aerotopes). Hexagonal substructure (p6) of S-layers was found in cell walls of several *Microcystis* strains.

Cell division by binary fission in three perpendicular planes in regular «cubic» arrangement; cells grow into the original size and shape before next division. Reproduction by disintegration of colonies into small clusters of cells or even into solitary cells.

**Important references:** Crow (1923a); Geitler (1932); Huber-Pestalozzi (1938); Teiling (1941, 1942); Komárek (1957, 1958, 1991); Desikachary (1959); Almestrand (1961); Starmach (1966); Kondrateva (1968b); Reynolds & al. (1981); Parker (1982); Kondrateva & al. (1984); Komárek & Anagnostidis (1986); Watanabe & al. (1986); Doers & Parker (1988); Hofbauer & Jüttner (1988); Grilli Caiola & al. (1991); Kato & al. (1991).

**Key to the species:**

**1a** Cells less than 3.2 μm in diameter . . . . . . . . . . . . . . . . . . . . . . . . . . . . . **2**

**1b** Cells mainly over 3 μm in diameter . . . . . . . . . . . . . . . . . . . . . . . . . . . . . **4**

**2a** Colonies more or less spherical to irregular, with more or less irregularly and evenly but not very densely distributed small cells (circa 1.5 μm in diameter) . . . . . . . . . . . . . . . . . . . . . . . . . . . . . . . . . . . . . . . . **1. M. natans**

**2b** Cells at least for part of the life cycle gathered into dense subspherical clusters . . . . . . . . . . . . . . . . . . . . . . . . . . . . . . . . . . . . . . . . . . . . . . **3**

**3a** Cells 0.8–2.5 μm in diameter; colonies more or less spherical, sometimes a few united together . . . . . . . . . . . . . . . . . . . . . . . . . . . **2. M. firma**

**3b** Cells 2–3.2 μm in diameter; colonies composed of more or less spherical subcolonies each with wide mucilaginous, indistinct margin, later disintegrating into amorphous mass of cells . . . . . . . . . . . . **3. M. ichthyoblabe**

**4a** Colonies more or less spherical, without holes, with sparsely arranged cells (distance between cells usually > than the cell diameter); aerotopes solitary, occurring facultatively . . . . . . . . . . . . . . . . . . . . . . . . . . . **4. M. smithii**

**4b** Colonies more or less irregular, at least in later stages with more or less densely packed cells; usually with numerous aerotopes . . . . . . . . . . . . **5**

**5a** Colonies composed of packet-like subcolonies (rarely solitary), sometimes aggregated together, not or only indistinctly clathrate («perforate») . . . . **6**

**5b** Colonies irregular, usually elongate, lobate and clathrate when old, reaching macroscopic size . . . . . . . . . . . . . . . . . . . . . . . . . . . . . . . . . . . . . **9**

**6a** Mucilage with indistinct margin, diffluent, not exceeding cells (or only up to 1 μm); cells very densely packed, (2.5) – 3.5–4.8 μm in diameter; colonies usually irregular-spherical in outline . . . . . . . . . . . . . . . . **5. M. flos-aquae**

**6b** Mucilage distinct or indistinct, but always distinctly exceeding the outline of cell cluster by more than 1 μm wide margin . . . . . . . . . . . . . . . . . . . **7**

**7a** Mucilaginous margin diffluent, but wide, distinct, with radial, hyaline gelatinous protuberances («connections»; staining !); cells 5–6 µm in diameter, colonies more or less spherical in outline, usually aggregated together . . . . . . . . . . . . . . . . . . . . . . . . . . . . . . . . . . . . . . . . **6. M. botrys**

7b Mucilaginous margin without radial, hyaline protuberances . . . . . . . . . . 8

8a Mucilage forms around more or less spherical or subspherical colonies (cell clusters) wide margin with diffluent or indistinctly delimited outline; colonies sometimes composed of subcolonies, separated by slime . . . . . . . . . . . . . . . . . . . . . . . . . . . . . . . . . . . . . . **7. M. novacekii**

8b Mucilage more or less copies outline of the packet-like clusters of cells (often more or less irregularly cubic), circa 1–3 µm wide, with delimited, slightly refractive surface; old colonies usually composed of irregularly aggregated packet-like subcolonies . . . . . . . . . . . . . . . . . . . . . . . . . . **8. M. viridis**

9a Colonies enveloped by diffluent, sometimes indistinct slime . . . . . . . . . . . . . . . . . . . . . . . . . . . . . . . . . . . . . . **9. M. aeruginosa**

9b Colonies enveloped by distinctly delimited mucilage, usually slightly refractive on the margin and smooth on the surface; colonies intensely lobate and perforate . . . . . . . . . . . . . . . . . . . . . . . . . . . . . **10. M. wesenbergii**

## 1. **Microcystis natans** Lemmermann ex Skuja 1934 (fig. **295**)

Colonies microscopic, spherical, elongate, later irregular, usually not clathrate or at most indistictly clathrate, 40–200 µm in diameter, with irregularly, evenly, not very densely packed cells (or more aggregated near the colonial surface). Mucilage colourless, homogeneous, fine, visible usually after staining, exceeding cells by an to 4 µm wide margin. Cells spherical or slightly elongate (widely oval) before division, with aerotopes (usually with one central aerotope), circa 1.5 µm (up to ± 3 µm?) in diameter.
Occurrence: Freshwater, planktic in lakes and large water bodies; known from the north of Europe, but probably sporadically distributed over the cold regions of the whole temperate zone. Identity with populations from Argentine (Tierra del Fuego) with clathrate colonies and cells up to 2.4 µm in diameter is debatable.

## 2. **Microcystis firma** (Kützing) Schmidle 1902 (fig.**296**)

Colonies microscopic, spherical or more or less irregular spherical, never clathrate, with densely packed cells, with indistinct mucilaginous envelopes, which do not exceed the cell clusters to form a distinct margin; sometimes few colonies are united together. Cells spherical, with gas vesicles (visible aerotopes), (0.8)2–3.7(4.8) µm in diameter.
Occurrence: Freshwater, planktic in stagnant waters, sporadically; known from N. Europe (also in Baltic Sea); identity of sporadic records of this species from the Mediterranean region and Africa (e.g., Compère 1967, 1970, etc.) with the original concept needs confirmation.

## 3. **Microcystis ichthyoblabe** Kützing 1843 (fig. **297**)

Colonies more or less irregular-spherical, usually composed of small groups of cells grouped together (subcolonis), embedded in wide, fine, homogeneous, colourless, diffluent gelatinous envelopes, sometimes joined together; in old stages the subcolonies disintegrate into homogeneous and amorphous mass of cells, in which sometimes remain small groups of aggregated cells. Cells spherical, 2–3.2 µm in diameter. Known toxic strains.
Occurrence: Freshwater, planktic in mesotrophic or slightly eutrophic, but not

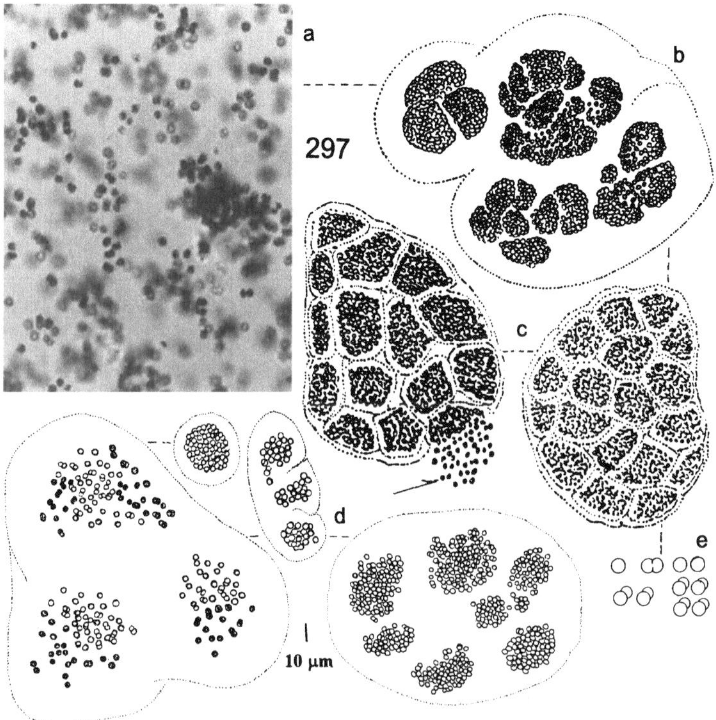


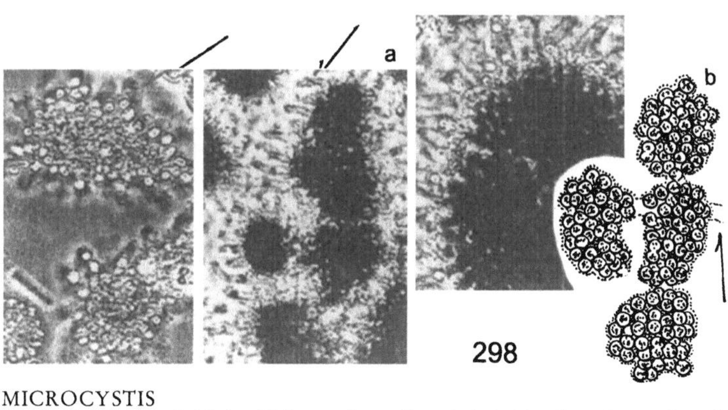


MICROCYSTIS
**Fig. 297.** *Microcystis ichthyoblabe*: **a** – from Komárek (1991) from Japan, **b** – after Teiling (1942) from Sweden, **c** – after Kützing (1845), **d-e** – after Komárková-Legnerová & Cronberg (1992) from Sweden, (**e** – cell division in three planes).
**Fig. 298.** *Microcystis botrys*: **a** – orig. photos of Cronberg, **b** – iconotype of Teiling (1942).

polluted lakes, sometimes forming water blooms; more in northern regions of the temperate zone, probably not occurring in tropical countries.

**4. Microcystis smithii** Komárek et Anagnostidis 1995 (fig. **299**)

*Aphanocapsa pulchra* (Kützing) Rabenhorst 1865; *Microcystis grevillei* sensu G.M. Smith 1920 (planktic form); *Microcystis grevillei* f. *pulchra* (Kützing) Elenkin 1936 sensu auct. post.; non *Microcystis pulchra* Flotow 1842 (= *«Pleurococcus», Chlorophyceae*)

Colonies free floating, more or less spherical or slightly irregular, later gelatinous, never clathrate, with scattered, rarely densely arranged cells; mucilage fine, colourless, more or less distinctly delimited or diffuse, homogeneous, without refractive margin; cells arranged more or less sparsely and regularly throughout the colony, singly or in pairs after division, spherical, blue-green, bright blue-green or olive-green, facultatively with one or several brownish aerotopes in each cell, 3.2–5.6 µm in diameter.
Occurrence: Freshwater, planktic in clear lakes; N. America, Argentine (Tierra del Fuego); N. Europe, Poland (near Lublin), N. Greece, data about planktic or not planktic *«M. pulchra»* or *«M. grevillei»* from other biotopes and areas (Africa, marine coasts, etc.) evidently concern other species.

**5. Microcystis flos-aquae** (Wittrock) Kirchner 1898 (fig. **300, 301**)

*Microcystis prasina* (Wittrock) Lemmermann 1904 ?; *Microcystis aeruginosa* f. *flos-aquae* (Wittrock) Elenkin 1938

Colonies microscopic, rarely macroscopic, pulverous (then irregular, compound), compact, more or less spherical or irregular in outline, not lobate, with very densely aggregated cells, only rarely with indistinct holes in old colonies; mucilaginous envelopes do not exceed the margin of the cell clusters (or at most to 1 µm), fine, hyaline, colourless, diffluent. Cells spherical, with aerotopes, (2.5) 3.5–4.8 (5.6 ?) µm in diameter. Without neurotoxins, with isopropylthio-compounds. Further characteristics (particularly in comparison with *M. aeruginosa* see in Doers & Parker 1988 and Hofbauer & Jüttner 1988).
Occurrence: Freshwater, planktic, usually together with other planktic algae and cyanoprokaryotes, sometimes part of water blooms, in mesotrophic and slightly eutrophic water bodies; in the whole temperate zone, particularly in northern regions; cosmopolitan ? .

**6. Microcystis botrys** Teiling 1942 (fig. **298**)

Colonies free floating, more or less spherical, with cells irregularly and very densely aggregated in mucilaginous, spherical groups. Mucilage around colonies wide, gelatinous, fine, colourless, sometimes indistinct but structured, enveloping clusters of cells and extending distinctly beyond their margin, usually by more than 1 µm, later diffluent; in the colonial mucilage occur indistinct (staining, phase contrast !), more or less wide, radially oriented semiglobose or tubular gelatinous protuberances, sometimes connecting the subcolonies. Cells spherical, with aerotopes, 5–6 µm in diameter.

---

MICROCYSTIS
**Fig. 299.** *Microcystis smithii*: **a** – after G.M. Smith (1920, sub *Aphanocapsa pulchra*), **b** – after G.M.Smith (1920, sub *Aphanocapsa grevillei*), **c** – after G.M. Smith

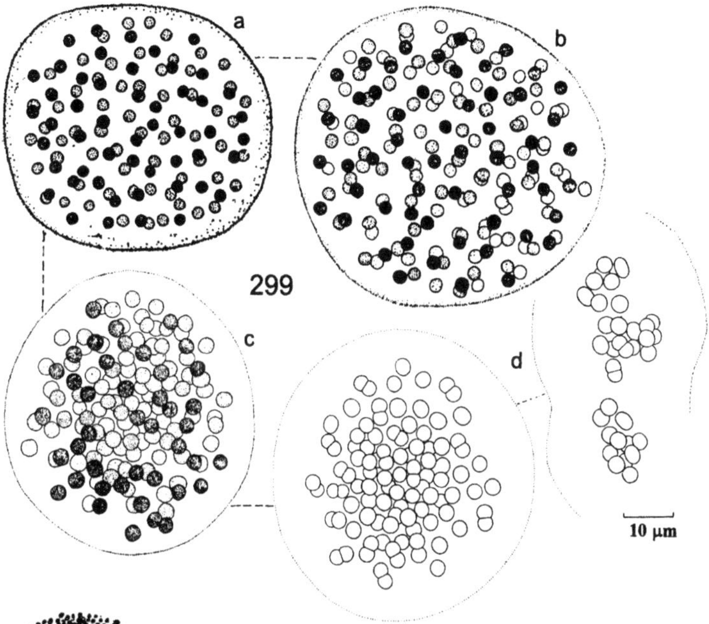


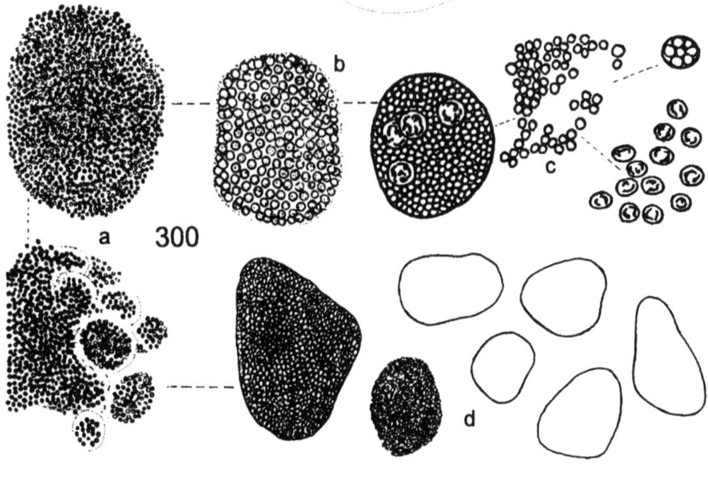


(1920, sub *Aphanocapsa rivularis*), **d** – after Hindák & Moustaka (1988, sub *Aphanocapsa* cf. *pulchra*).

**Fig. 300.** *Microcystis flos-aquae*: **a** – after Crow from Huber-Pestalozzi (1938), **b** – after West from Huber-Pestalozzi (1938, sub *M. marginata*), **c** – after Canabaeus from Huber-Pestalozzi (1938, the interpretation of nanocyte formation is questionable), **d** – after Komárek (1958, form of colonies).

Occurrence: Planktic in lakes and in slightly brackish waters; known mainly from the Baltic region (Sweden, E. Baltic Sea), but records from other countries also occur (central Europe, E. Africa!).

**7. Microcystis novacekii** (Komárek) Compère 1974 (fig. **302**)

*Anacystis marginata* (Meneghini 1837 sine typo, sensu Rabenhorst 1865 et auct. post.; *Microcystis marginata* (Meneghini) Kützing 1846 excl. typo; *Microcystis aeruginosa* f. *marginata* (Meneghini) Elenkin 1938 sine typo; *Diplocystis novacekii* Komárek 1958; *Anacystis montana* f. *montana* Drouet et Daily 1956 p.p.

Colonies microscopic, later rarely reaching macroscopic size (composed of subcolonies), lenticular or almost spherical, irregularly spherical or slightly elongate or with wavy outline, compact, with packed cells, without holes and not lobate, in old stages composed of several clustered subcolonies, with a large number of densely aggregated cells; several solitary cells around the cell clusters. Mucilage colourless or pale yellowish, sometimes irregularly and slightly concentrically stratified, delimited at the margin, not diffluent, but without a refractive outline; the wide margins around central cell clusters reach to 15 µm width. Cells spherical, with aerotopes, (2.4) 3–5.5 (6) µm in diameter.
Occurrence: Freshwater, planktic in mesotrophic and eutrophic water bodies (lakes, ponds, reservoirs), sometimes forming or participating in water blooms; common in the whole tropical zone, occasionally (especially in summer seasons) in temperate zone.

**8. Microcystis viridis** (A. Braun in Rabenhorst) Lemmermann 1903 (fig. **303**)

*Polycystis viridis* A. Braun in Rabenhorst 1862; *Microcystis aeruginosa* f. *viridis* (A. Braun) Elenkin 1938; *Anacystis cyanea* (Kützing) Drouet et Daily 1952 p.p. excl. typo; *Diplocystis viridis* (A. Braun) Komárek 1958

Colonies free floating, microscopic, packet-like, more or less rounded or cubic-rounded, clearly three-dimensional, later macroscopic (to 3100 µm long), composed of packet-like subcolonies, irregularly spherical or elongate, with densely aggregated cells. Mucilage colourless, more or less narrow or slightly distant from cell clusters (more or less copying the margin of cell groups) and delimited by slightly refractive outline, usually 5 – (10) µm wide; mucilage never forms a wide spherical margin around cell clusters. Cells spherical, with aerotopes, (3) 3.5–7 (7.9–8.4 ?) µm in diameter. Well characterized species morphologically as well as according to isozyme composition. Occurs in two (or more) size modifications. Toxic.
Occurrence: Freshwater, planktic, sporadically in slightly eutrophic lakes and ponds, sometimes forming water blooms; cosmopolitan distribution, the genetic identity of tropical and temperate populations should be checked (populations from Tierra del Fuego are obviously different).

---

MICROCYSTIS
**Fig. 301.** *Microcystis flos-aquae*: from Komárek (1991).
**Fig. 302.** *Microcystis novacekii*: **a** – shape of colonies, **b** – detail of a colonial

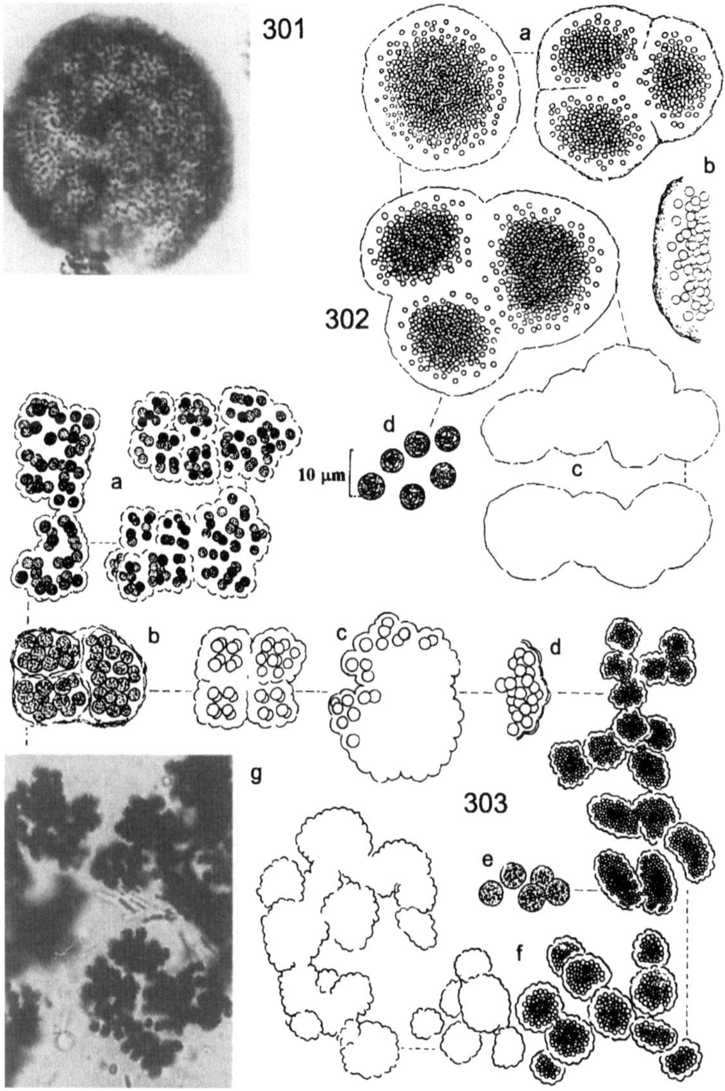


margin, **c** – outline of compound colonies, **d** – detail of cells; after Komárek (1958).

**Fig. 303.** *Microcystis viridis*: **a-c** – detail of colonies, **d** – detail of colonial margin , **e** – detail of cells, **f-g** – compound colonies; **a** – after Teiling (1941), **b** – after Frémy (1930), **c-f** – after Komárek (1958), **g** – from Komárek (1991).

**9. Microcystis aeruginosa** (Kützing) Kützing 1846 (fig. **304**)

*Micraloa aeruginosa* Kützing 1833 sine typo; *Diplocystis aeruginosa* (Kützing) Trevisan 1848; *Clathrocystis aeruginosa* (Kützing) Henfrey 1856; *Microcystis ochracea* (Brand) Forti 1907 ? ; *Anacystis cyanea* (Kützing) Drouet et Daily 1952 p.p. excl. typo; and numerous others.

Colonies mucilaginous, microscopic, more or less spherical, lenticular or feebly elongate, later macroscopic, irregular, lobate, distinctly elongate, usually (40)600–900(1000) µm, rarely up to 8 mm long, usually net-like clathrate with distinct holes, or composed of subcolonies, with irregular outline and with a large number of irregularly arranged and ± densely packed cells. Mucilage colourless, structureless, diffluent, sometimes forming a distinct, but not very wide margin around cells (usually to 5–8 µm). Cells spherical, sometimes slightly elongate before division, pale blue-green, with numerous aerotopes, (3?) 4–6 (9.4) µm in diameter. Rare nanocyte formation (?). Very polymorphic species, occurring in numerous modifications and seasonal status (Kondrateva 1968b; Reynolds & al. 1981); the genetic diversity follows also from the isozyme analysis (Kato et al. 1991). Toxic populations and strains.

Occurrence: Fresh- and brackish waters, planktic in eutrophic water bodies (lakes, fishponds, reservoirs), sometimes forming heavy water blooms, common; cosmopolitan distribution with exception of polar and subpolar regions. The genotype identity of different modifications from different areas has not been established.

**10. Microcystis wesenbergii** (Komárek) Komárek in Kondrateva 1968 (fig. **305**)

*Coelosphaerium dubium* Grunow in Rabenhorst sensu auct. post. (Hollerbach et al. 1953), excl. typo; *Microcystis aeruginosa* Kützing sensu Teiling 1941; *Diplocystis wesenbergii* Komárek 1958

Colonies spherical when young, later elongate and richly lobed and clathrate, with distinct holes, sometimes composed of subcolonies, reaching macroscopic size (to 6 mm long); cells arranged randomly, rarely densely, in young colonies rather near the colonial surface, but also over the whole colonial content. Mucilage colourless, structureless, with distinctly delimited, smooth, firm, continuous, not diffluent but with refractive outline; gelatinous margin extends 3–6 µm beyond the cell aggregations. Cells spherical, sometimes slightly elongate before division, with distinct aerotopes, 4–7 (8.5–10 ?) µm in diameter. Occurs in several size and morphological modifications, the taxonomic value of which needs elucidation (comp. Cronberg & Komárek 1994). Possibly toxic populations occur, but only non-toxic strains were found in Japan (Kato & Watanabe 1993).

Occurrence: Freshwater, common, planktic in eutrophic lakes and ponds, mixed with other planktic cyanoprokaryotes, rarely dominant in water blooms; with cosmopolitan distribution with the exception of subpolar regions, in various morphotypes (possibly other similar species exist, which belong to various genera).

**Species outside Europe:**

– *Microcystis bengalensis* Banerji 1936: Large, filamentous colonies with limited, indistinctly stratified mucilage and cells in groups, 3.5–6 µm in diameter. Tropical Asia.

– *Microcystis comperei* Komárek 1984: Spherical microscopic colonies with smooth, delimited mucilage (later diffluent), cells 4.5–5.2 µm in diameter. Planktic (sometimes forming water blooms) in eutrophic and hypertrophic water bodies;

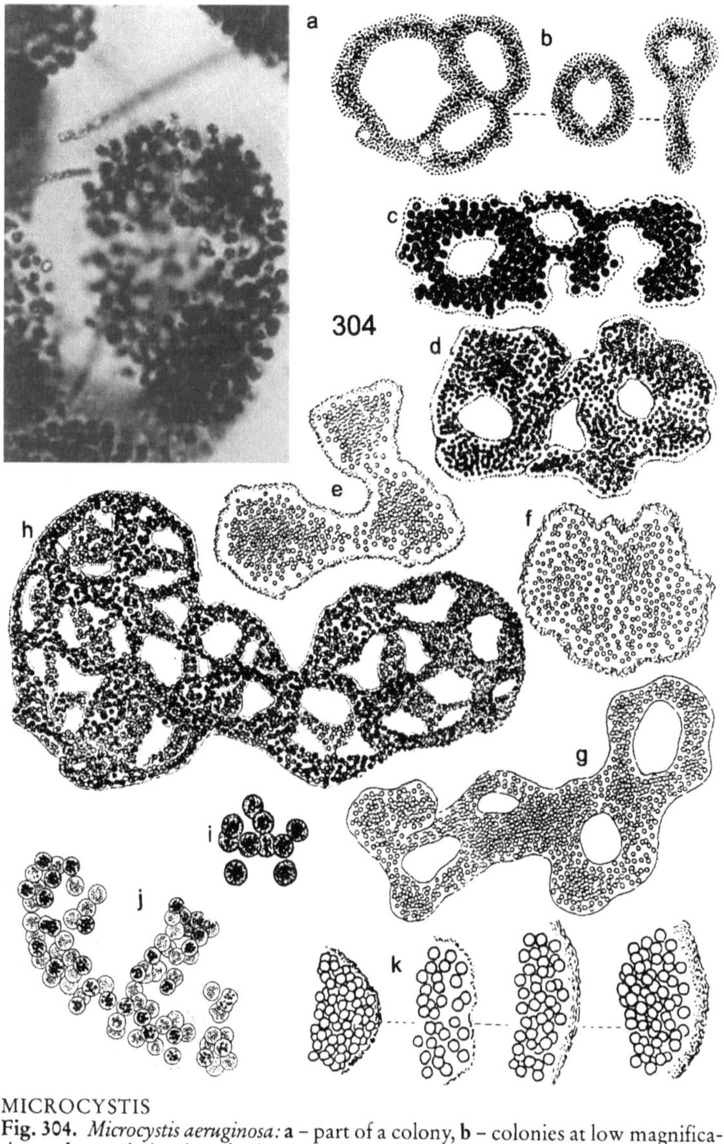


MICROCYSTIS

**Fig. 304.** *Microcystis aeruginosa:* **a** – part of a colony, **b** – colonies at low magnification, **c-h** – variation in colonies, **i-j** – detail of cells, **k** – variation in mucilaginous colonial margin; **a** – from Komárek (1991), **b** – after Kirchner from Kondrateva & al. (1984), **c** – after Teiling (1941), **d** – after Crow from Hollerbach & al. (1953), **e-g,i,k** – after Komárek (1958), **h** - after Kosinskaja from Hollerbach & al. (1953, subf. *sphaerodictyoides*), **j** – after G.M.Smith (1920).

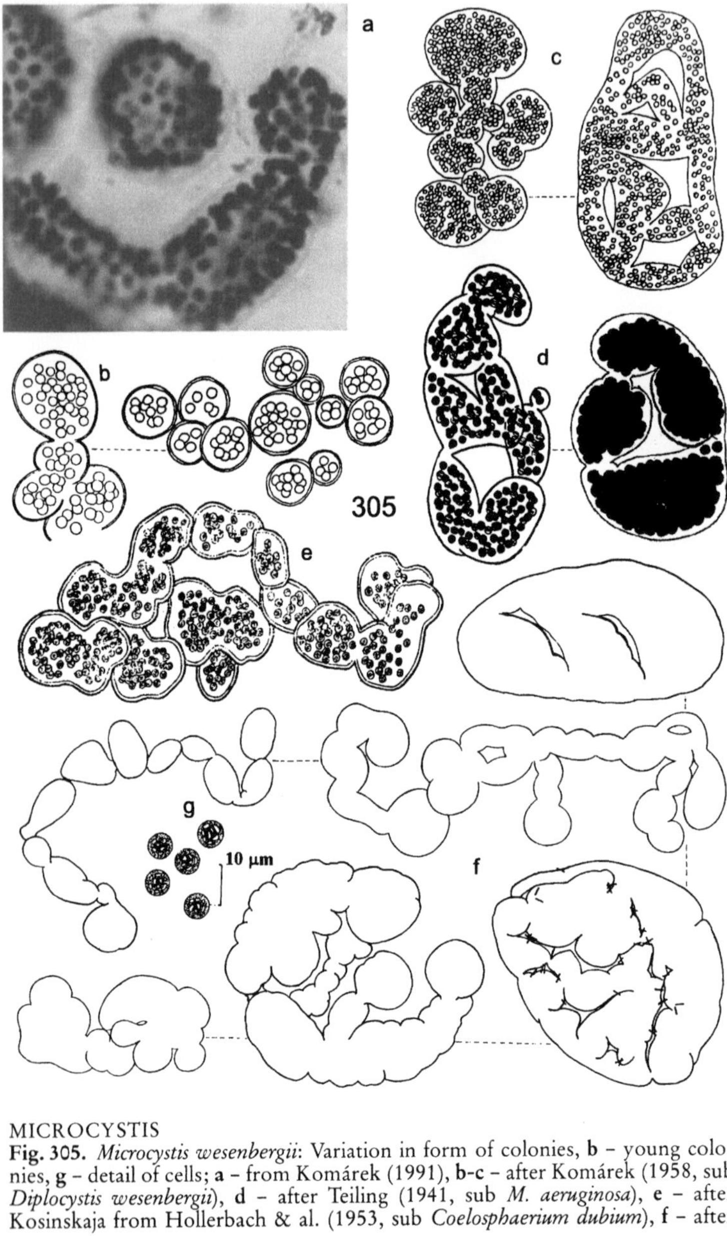


MICROCYSTIS

**Fig. 305.** *Microcystis wesenbergii*: Variation in form of colonies, **b** – young colonies, **g** – detail of cells; **a** – from Komárek (1991), **b-c** – after Komárek (1958, sub *Diplocystis wesenbergii*), **d** – after Teiling (1941, sub *M. aeruginosa*), **e** – after Kosinskaja from Hollerbach & al. (1953, sub *Coelosphaerium dubium*), **f** – after

Caribbean region. Possibly identical with *M. supersalsa* Schiller 1956 from Venezuela.

– *Microcystis densa* G.S. West 1909: Cylindrical colonies up to 1400 μm long, cells about 4 μm in diameter, densely arranged; described from E. Africa (Albert Lake), revision is needed.

– *Microcystis lamelliformis* Holsinger 1954: Colonies microscopic, usually flattened, spherical or irregular with fine, diffluent, scarcely visible mucilage, cells randomly but homogeneously arranged, 3–4 (5) μm in diameter. Planktic in freshwaters of tropical countries; originally described from Asia (Srí Lanka), but widely distributed also in Africa, tropical America, etc., forming heavy water blooms (toxic). Related to *M. ichthyoblabe*.

– *Microcystis maxima* Bernard 1908: Colonies large, spherical, to 1 mm in diameter, mucilage hyaline, diffluent, but clearly visible, cells (3) 4–5 (8) μm in diameter, elongate before division. Planktic in lakes and ponds, tropical Asia (data from Caspian Sea surely concern another species).

– *Microcystis pallida* (Farlow) Lemmermann 1910: Several taxonomic concepts, usually *Aphanocapsa*-like populations; according to Farlow's original description probably *Aphanothece*; revision needed.

– *Microcystis protocystis* Crow 1923: Colonies irregular with sparsely disposed cells, mucilage very fine, hyaline, diffluent, widened around cells, cells (3) 3.5–6.5 μm in diameter. Described from plankton in Srí Lanka, but probably one of the commonest pantropical species.

– *Microcystis pseudofilamentosa* Crow 1923: Colonies elongate, pseudofilamentous, to 1 mm long, composed of subcolonies, with diffluent, indistinct mucilage and cells unevenly distributed, sometimes in dense groups; cells 3–7 μm in diameter. Tropical Asia, but reported also from the Argentine (Tierra del Fuego).

– *Microcystis robusta* (Clark) Nygaard 1925: Colonies spherical, later irregular, disintegrating, elongate and lobate, sometimes clathrate, mucilage delimited, refractive, cells without gas vesicles (?), 6–9.5 μm in diameter. Sparsely over the whole tropical zone, probably one from modifications of «*Microcystis wesenbergii*» (without visible aerotopes ?). The populations from Tierra del Fuego (Guarrera & al. 1986) belong possibly to another species.

– *Microcystis scripta* (Richter) Geitler 1925: Probably synonymous to *M. aeruginosa* or *M. pseudofilamentosa*; comp. also *M. punctata* Schiller 1954. Recorded mainly from tropical regions.

– *Microcystis supersalsa* Schiller 1956: Young colonies spherical with delimited, later diffluent mucilage at the margin, cells without gas vesicles (?); described from saline pools of coastal areas in Venezuela, together with *M. densa* Schiller 1956 (it is later homonym to *M. densa* G.S.West 1909, and represents probably the old colonies of *M. supersalsa*).

**Unrevised taxa:**

– *Microcystis* sp. div.: Variable European populations of *Microcystis* with widened mucilaginous envelopes, which are diffluent at the margin, are usually identified as *M. aeruginosa*, *M. flos-aquae*, *M. marginata*, or they belong to *M. novacekii*. Similar species is also *M. lutescens* Schiller 1954. Their revision is needed; (fig. **306**).

– *Microcystis aphanothecioides* f. *minor* Hortobágyi 1951: Described from mineral lakes in Hungary; probably *Aphanothece*-species from the subg. *Cyanogastrum*.

---

Teiling (1941, sub *M. aeruginosa*) and Komárek (1958, sub *Diplocystis wesenbergii*), **g** – after Komárek (1958, sub *Diplocystis wesenbergii*).

- *Microcystis aeruginosa* var. *elongata* Rao 1927: Probably a separate species from tropical Asia, with elongate colonies and cells 2–2.8 µm in diameter.
- *Microcystis argentea* Schiller 1954: Cells 2.4–3 µm in diameter, without aerotopes (*Aphanocapsa* ?).
- *Microcystis elongata* Desikachary 1959: Probably synonymous with *M. bengalensis* Banerji.
- *Microcystis prasina* (Wittrock) Lemmermann sensu Frémy 1930: Cells 3.5–4.5 µm in diameter, aggregated densely in colonies with delimited mucilage, not overlapping the cell cluster; possibly the tropical morphotype (or convergent species) of *M. flos-aquae* (?).
- *Microcystis ramosa* Bharadwaya 1935: Probably synonymous with *M. pseudofilamentosa* Crow.
- *Microcystis rosea* Kufferath 1942: Reddish cells, 2–3 µm in diameter; diagnosis insufficient.
- «*Microcystis*-cluster» sensu Castenholz & Waterbury is the only *Microcystis* taxon included in Bergey's Manual (1989): With gas vesicles (joined in aerotopes), producing toxins and ß-cyclocitral.

## 2. **Gloeocapsa** Kützing 1843

Colonies micro-, later macroscopic (rarely only microscopic), usually multicellular, mucilaginous, amorphous, epilithic or epiphytic (rarely free floating or metaphytic), composed of small groups of irregularly arranged cells, enveloped by wide, usually stratified individual gelatinous envelopes, joined together into a formless mass. Cells and their groups always surrounded by wide gelatinous sheaths, concentrically lamellate (lamellation distinct or scarcely visible), in several species intensely or partly coloured by sheath pigments (gloeocapsin, scytonemin), yellow, yellow-brown, orange, red, blue or violet; gelatinous envelopes do not copy exactly the cell shape. Gelatinous envelopes usually delimited, only in some stages of the life cycle or under special conditions diffluent (also sometimes in cultures). Cells always spherical or rarely indistinctly elongated and oval, after division hemispherical, usually with pale blue-green, homogeneous content, sometimes with solitary granules; gas vesicles only in one planktic species.
Cell division by binary fission within gelatinous envelopes, in three perpendicular planes in successive generations; daughter cells shift in mucilage irregularly and produce their own gelatinous envelopes soon after division (= formation of lamellate sheaths). Cells grow to the original size and shape before next division (difference from *Chroococcaceae*). Reproduction by liberation of groups of cells or solitary cells (with or without envelopes) from the colonial mass and by disintegration of colonies into small gelatinous clusters containing several cells. In some species is known nanocyte production. *Aphanocapsa*-stages (with diffluent sheaths, resulting in arrangement of cells in more or less homogeneous slime) and resting stages (cells enveloped by thick, firm sheaths, incorrectly designated as «akinetes» = «Dauerzellen») develop under unfavourable conditions.

Note: Various *Gloeocapsa* species grow in morphologically very diverse stages, in dependence on seasonal and local environmental changes. The various status (st. «familiaris», «simplex», «lamellosus», «coloratus», «perdurans», etc.) were described in numerous species. The knowledge of seasonal variability of the whole population is therefore necessary to the correct identification.
**Important references:** Brand (1900); Geitler (1932); Nováček (1929, 1934); Jaag (1945); Skuja (1964); Golubić (1967a,c); Abdelahad & Bazzichelli (1991); Komárek (1993).

**Key to the species:**

1a Envelopes colourless (without sheath pigments, but exceptionally coloured brownish by Fe-precipitates) . . . . . 2
1b Envelopes coloured by sheath pigments (yellow, orange, red, blue or violet) . . . . . 14
2a Cells without envelopes always < 3 μm in diameter . . . . . 3
2b Cells without envelopes more than (2) 3 μm in diameter . . . . . 6
3a Envelopes with Fe-precipitates (yellow-brown), but without sheath pigments; cells with aerotopes, 1.5–3 μm in diameter; planktic . . . . . **1. G. siderochlamys**
3b Envelopes without Fe-precipitates; cells without aerotopes; not a planktic species . . . . . 4
4a In thermal springs, cells 1.8–3 μm in diameter . . . . . **2. G. gelatinosa**
4b Not in thermal springs, subaerophytic on wet rocks, cells 0.7–3 μm in diameter . . . . . 5
5a Envelopes around cells wide, hyaline, spherical, not lamellate; on more or less acidic (not calcareous) rocks . . . . . **3. G. punctata**
5b Envelopes indistinctly or distinctly lamellate, sometimes slightly granular; on limestone rocks . . . . . **4. G. aeruginosa**
6a Cells 1.2–6 μm in diameter; thermal species . . . . . 7
6b Cells 2–12 μm in diameter; not thermal, aerophytic species (on wet walls, rocks, in soil, etc.) . . . . . 8
7a Cells 3.7–6 μm in diameter . . . . . **5. G. arenaria**
7b Cells 1.2–1.6 μm in diameter . . . . . **6. G. thermophila**
8a Cells strictly spherical, 2–8 μm in diameter . . . . . 9
8b Cells slightly elongate, 6–12 μm in diameter . . . . . **13. G. decorticans**
9a Submersely in swamps and pools, cells 4–5 μm in diameter . **7. G. granosa**
9b Aerophytic or subaerophytic . . . . . 10
10a On wet salty soils, margins of shallow salinic pools, etc.; cells 3–6 μm in diameter . . . . . **11. G. salina**
10b On wet rocks and walls . . . . . 11
11a On sandstone rocks; cells 3–6 μm in diameter, envelopes around cells not lamellate . . . . . **10. G. coracina**
11b On other than sandstone substrates . . . . . 12
12a Envelopes around cells more or less spherical, always distinctly and densely concentrically lamellate . . . . . **12. G. caldariorum**
12b Envelopes around cells not or indistinctly lamellate . . . . . 13
13a Cells blue-green, 2–5 μm in diameter; commonly distributed on wet rocks . . . . . **8. G. atrata**
13b Cells olive-green and pinkish or violet, 2–6 μm in diameter; on calcareous rocks . . . . . **9. G. bituminosa**
14a Envelopes intensely blue or violet (at least in part of a colony) . . . . . 15
14b Envelopes yellow, yellow-brown, orange, or red in different intensities; never blue or blue-violet . . . . . 19
15a Cells less than 3(3.5) μm in diameter, subaerophytic . . . **14. G. compacta**
15b Cells more than 3 μm in diameter . . . . . 16
16a Envelopes blue to bright blue-violet, transparent, outer layers wide, usually almost colourless; aerophytic, mainly on calcareous rocks . . . . . 17
16b Envelopes dirty blue, blue-violet, blackish violet, dark to almost not transparent; on various substrates (rocks, wood) . . . . . 18
17a Cells 3–5 μm in diameter; on wet rocks from lowland to mountains, often in tropical countries . . . . . **17. G. nigrescens**
17b Cells 4–6 (8) μm in diameter; on calcareous rocks in high mountains . . . . . **18. G. alpina**

18a On wet wood, cells 3–6 µm in diameter . . . . . . . . . . . . . **16. G. lignicola**
18b On rocks and walls, usually aerophytic, from lowland to mountains, probably on indifferent (rather alkaline substrates; cells 3–4 (6.8) µm in diameter . . . . . . . . . . . . . . . . . . . . . . . . . . . . . . . . . . . . . . . . . . . . **15. G. violascea**
19a Envelopes red . . . . . . . . . . . . . . . . . . . . . . . . . . . . . . . . . . . . . . . . . 20
19b Envelopes orange, yellow-orange, yellow or yellow-brown . . . . . . . . 26
20a In thermal springs, subaerophytic or atmophytic; cells 1–2.6 µm in diameter . . . . . . . . . . . . . . . . . . . . . . . . . . . . . . . . . . . . . . . . . . . . **19. G. thermalis**
20b Not in thermal springs, cells always over (1.6) 2 µm in diameter . . . . . 21
21a Submersed, epilithic in littoral of cold nordic lakes . . . . . **20. G. tornensis**
21b Subaerophytic . . . . . . . . . . . . . . . . . . . . . . . . . . . . . . . . . . . . . . . . . 22
22a Cells (1.6) 2–4 (4.5) µm in diameter . . . . . . . . . . . . . . . . . . . . . . . . 23
22b Cells 4–7 µm in diameter . . . . . . . . . . . . . . . . . . . . . . . . . . . . . . . . 24
23a In soils and moors, on wet wood, etc.; acidophilic, cells 1.6–4 µm in diameter . . . . . . . . . . . . . . . . . . . . . . . . . . . . . . . . . . . . . . . . . . . **21. G. haematodes**
23b On wet, not calcareous rocks from lowlands to mountains, cells 2–3 (4.5) µm in diameter . . . . . . . . . . . . . . . . . . . . . . . . . . . . . . . . . . . **22. G. rupicola**
24a Outer envelopes not extremely widened . . . . . . . . . . . . . . . . . . . . . . 25
24b Outer envelopes very widened, more or less spherical, only slightly reddish or colourless, inner envelopes intensely red; cells 4–8 µm in diameter; high but not calcareous mountains . . . . . . . . . . . . . . . . . . . . . . **25. G. ralfsii**
25a Cells 4–6.5 µm in diameter, pale blue-green; on wet, usually non-calcareous rocks in high mountains . . . . . . . . . . . . . . . . . . . . . . . **24. G. sanguinea**
25b Cells 3.5–9 µm in diameter, olive-green; aerophytic on serpentine and not calcareous rocks, from lowland to mountains . . . . . . . . **23. G. novacekii**
26a Envelopes orange . . . . . . . . . . . . . . . . . . . . . . . . . . . . . . . . . . . . . . . 27
26b Envelopes yellow or yellow-brown . . . . . . . . . . . . . . . . . . . . . . . . . 28
27a Cells 2.5–3.5 µm in diameter . . . . . . . . . . . . . . . . . . . . . **26. G. reicheltii**
27b Cells 6–8 µm in diameter . . . . . . . . . . . . . . . . . **27. G. shuttleworthiana**
28a Cells less than 3 µm in diameter . . . . . . . . . . . . . . . . . . . . . . . . . . . . 29
28b Cells more than 3 µm in diameter . . . . . . . . . . . . . . . . . . . . . . . . . . . 30
29a Epilithic on non-calcareous rocks; cells 1.5–2 µm in diameter . . . . . . . . . . . . . . . . . . . . . . . . . . . . . . . . . . . . . . . . . . . **28. G. fusco-lutea**
29b Epilithic and chasmolithic on calcareous rocks; cells 0.8–3 µm in diameter . . . . . . . . . . . . . . . . . . . . . . . . . . . . . . . . . . . . . . . . . . . . **29. G. biformis**
30a On marine coasts, in splash zone on rocks, cells 3–6 µm in diameter . . . . . . . . . . . . . . . . . . . . . . . . . . . . . . . . . . . . . . . . . . . . . . **30. G. deusta**
30b Subaerophytic on inland rocks and walls; common species . . . . . . . . . 31
31a Cells 3–5 µm in diameter; on wet rocks in mountains up to the alpine zone . . . . . . . . . . . . . . . . . . . . . . . . . . . . . . . . . . . . . . . . . **31. G. kuetzingiana**
31b Cells (5) 6–8 (11) µm in diameter; on wet rocks and walls from lowland to mountains (mainly in forest zone) . . . . . . . . . . . . . . . . **32. G. rupestris**

**1. Gloeocapsa siderochlamys** (Skuja) Starmach 1966 (fig. 307)

*Chroococcus siderochlamys* Skuja 1948

Colonies microscopic, free floating, with 1–4 irregularly aggregated cells. Cells spherical or oval (before division), pale blue-green to almost colourless, with single aerotope, 1.5–2.9 µm in diameter, surrounded by wide, rounded envelopes. Mucilaginous envelopes concentrically lamellate, fine, yellow-brownish (coloured by Fe-precipitates, but without sheath pigments), more intensely coloured inside the colony, in outside layers almost colourless (according to original drawings).
Occurrence: Described from plankton of the lake Siggeforasjøn, Sweden. Not confirmed.

## 2. **Gloeocapsa gelatinosa** (Meneghini) Kützing 1843 (fig. **308**)

Colonies micro-, later macroscopic, forming greyish, olive-green to yellowish mucilaginous, thin or warty mats, composed from aggregated subcolonies. Mucilaginous envelopes not very wide, colourless, delimited, enveloping solitary cells or groups of cells, almost hyaline when young, later indistinctly to intensely stratified, 5.6–10 µm (with more cells up to 25 µm) in diameter. Cells spherical, blue-green, 1.8–3 µm in diameter.
Occurrence: Freshwater, submersely and atmophytic in warm mineral springs (usually up to 50 °C); known from central and SE. Europe and from Israel, Africa (Angola, Zaire), E. Asia (Indonesia – Java, Japan) and USA; findings from other biotopes (Angola, Antilles) need confirmation.

## 3. **Gloeocapsa punctata** Nägeli 1849 (fig. **309**)

Colonies micro- or macroscopic, forming dark grey mucilaginous, warty mass, composed of numerous subcolonies. Mucilaginous envelopes wide, spherical, (3.5) 7–17 (50) µm in diameter, only in old colonies slightly irregular, delimited, colourless, hyaline, not lamellate or only with 1–2 hardly visible concentric layers, surrounding 1–2 – 4, rarely more irregularly (densely or sparsely) disposed cells («*Aphanocapsa*»-stage), usually slightly distant from one another. Cells spherical, pale blue-green, 0.7–3 µm in diameter.
Occurrence: Subaerophytic, on wet rocks and walls from lowlands to mountains; geographical distribution unknown, common in central Europe, probably widely distributed.

## 4. **Gloeocapsa aeruginosa** Kützing 1843 (fig. **310**)

Colonies crusty, granular or gelatinous, sometimes forming amorphous macroscopic mass, blue-green or greyish olive-green, composed of more or less spherical, many-celled subcolonies, 16–50 µm in diameter. Mucilaginous envelopes colourless, delimited, more or less inflated, indistinctly striate. Cells small, spherical, blue-green, 2–3 µm in diameter.
Occurrence: Aerophytic, on wet, mainly limestone rocks; described from central Europe, nothing is known about its distribution, possibly cosmopolitan, commonly recorded in tropical countries. Probably in different concepts of authors (see figs. **309a-e**); comp. also Guarrera & al. (1987).

## 5. **Gloeocapsa arenaria** (Hassall) Rabenhorst 1865

Colonies small, gelatinous, olive-green to yellow-green, composed of oval or nearly spherical subcolonies to 43 µm in diameter. Mucilaginous envelopes around cells wide, colourless, indistinctly lamellate (with «diffluent» layers), 6–17 µm wide. Cells spherical, blue-green or brownish, (3) 3.3–6 µm in diameter.
Occurrence: In thermal and mineral springs; known from Europe, Japan, Tibet and USA (data from China – Tibet, Japan and USA with different ecology – aerophytic – very probably concern another species).

## 6. **Gloeocapsa thermophila** (Wood) Claus 1959

*Chroococcus thermophilus* Wood 1869
Colonies microscopic, gelatinous, greenish, composed mostly of few-celled nearly cubic, rarely ellipsoid subcolonies. Subcolonies usually contain 2–4 (- 8) cells with individual envelopes, to 15 (25) µm in diameter; sometimes, especially

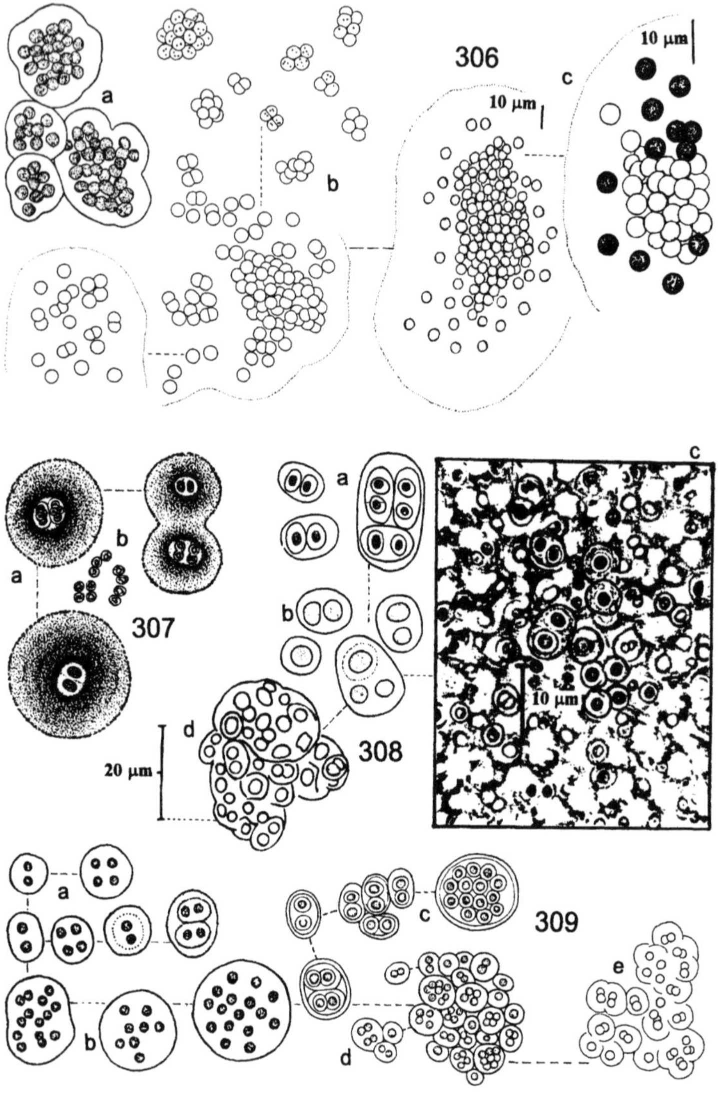


MICROCYSTIS

**Fig. 306.** Examples of *Microcystis* populations with various wide, colourless mucilaginous envelopes around clusters of cells: **a** – after Kirchner from Geitler (1925, sub *M. flos-aquae*), **b** – after Hindák & Moustaka (1988, sub *M.* cf. *pulverea*), **c** – after Komárková-Legnerová & Cronberg (1992, sub *M. aeruginosa*).

at the margin, there are wider subcolonies, which consist of solitary cells with individual cell envelopes, embedded within common, diffluent, colourless mucilage («*Aphanocapsa*»-stage ?). Mucilaginous envelopes around subcolonies wide, colourless, almost translucent, not lamellate. Mucilaginous individual envelopes around cells spherical, more or less narrow (usually narrower than cell diameter), fine, colourless, not lamellate, 0.8–2.5 µm wide. Cells spherical, pale greenish, 1.2–1.6 µm in diameter.
Occurrence: In thermal springs, over 30 °C; originally described from USA (California), later recorded also from Europe: Greece (Langadas) and Hungary (Bükkszek, Hajdúszoboszló).

**7. Gloeocapsa granosa** (Berkeley) Kützing 1845 (fig. **311**)

*Gloeocapsa fenestralis* Kützing 1843; non *G. granosa* sensu Geitler 1925

Colonies micro- or macroscopic, forming greenish or grey-green mucilaginous mass, composed of numerous subcolonies. Mucilaginous envelopes wide, spherical, colourless or pale grey, later (surrounding more cells) more or less irregular, delimited, not or indistinctly stratified (with few layers), up to 11 µm in diameter, sometimes enveloping more cells together and wider. Cells exactly spherical, pale blue-green, 4–5 µm in diameter; nanocytes known, 2.5 µm in diameter.

GLOEOCAPSA

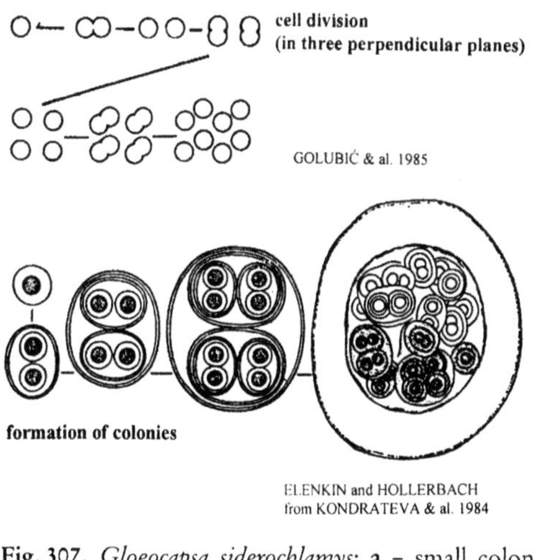


**Fig. 307.** *Gloeocapsa siderochlamys*: **a** – small colonies, **b** – solitary cells; after Skuja (1948).
**Fig. 308.** *Gloeocapsa gelatinosa* (envelopes colourless): **a** – after Frémy (1930), **b-c** – after Dor (1967), **d** – after Duvigneaud & Symoens (1949).
**Fig. 309.** *Gloeocapsa punctata* (envelopes colourless): **a, d** – typical small colonies, **b** – old colonies, **c** – colonies with visible envelopes around cells; **a-d** – after Kützing, Nägeli, Kukk and Hollerbach from Hollerbach & al. (1953), Starmach (1966) and Kondrateva & al. (1984); **e** – planktic population from Hungary after Hegewald & al. (1975), which belongs probably to another species.

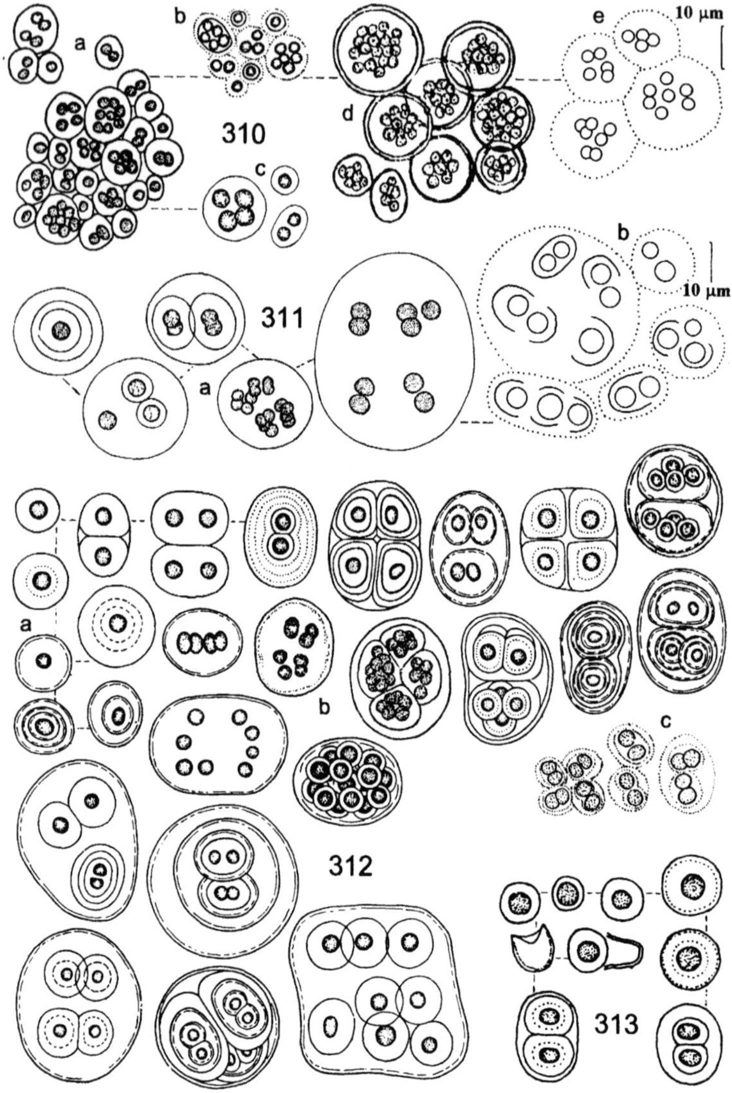


GLOEOCAPSA

**Fig. 310.** *Gloeocapsa aeruginosa* (envelopes colourless): **a** – after Cooke from Geitler (1925), **b** – after Thompson (1938), **c** – after Palik (1960), **d** – after Biswas (1926), **e** – after Yacubson (1974). Various and seemingly different concepts.

Occurrence: Freshwater, benthic, epiphytic and metaphytic among water plants in clear pools and swamps, submersed or in drying pools, sporadic; known mainly from central and N. Europe, the distribution is unknown, probably more distributed over temperate zones.

**8. Gloeocapsa atrata** Kützing 1845 (fig. **312, 314**)

*Gloeocapsa montana* Kützing 1843 (later homonym to *G. montana* (Lightfoot) Wille 1918 = *Ulva montana* Lightfoot 1777 = basionym of *Anacystis montana* (Lightfoot) Drouet et Daily 1952 = collective species of the more or less whole genus *Gloeocapsa* sensu Drouet et Daily 1956; on the type *Ulva montana* is based also *Palmella alpicola* Lyngbye 1819 = *Gloeocapsa alpicola* (Lyngbye) Bornet in W. et G.S. West 1903); *Gloeocapsa polydermatica* Kützing sensu auct. post. p.p. (typus: sec. Van Den Hoek 1963 = *Coccomyxa*); *Chroococcus varius* sensu auct. post.

Colonies usually macroscopic, in the form of dark grey or dirty grey-green mass, composed of many subcolonies, sometimes confluent. Subcolonies contain usually 2–4 – more cells with individual envelopes and unified together with envelopes of the 2nd and higher generations. Mucilaginous envelopes irregularly spherical, oval, ellipsoid, to 14.5 (40) µm in diameter, colourless or slightly bluish, usually finely but distinctly lamellate, in diffluent (marginal) parts unstratified. Cells spherical or indistinctly elongate (before division), pale blue-green or blue-green, 2–6 (8 ?) µm in diameter.

Occurrence: Subaerophytic, on wet rocks, walls, among mosses, on edges of wells, rarely on wet soil (data from thermal springs are incorrect); common in central and northern Europe (revised localities), about the real geographic distribution nothing is known (many incorrect identifications); probably over the whole temperate zone, including S. America. Numerous data from tropical regions must be revised.

**9. Gloeocapsa bituminosa** (Bory) Kützing 1849 (fig. **313**)

*Chroococcus bituminosus* (Bory) Hansgirg 1892; *Chroococcus caldariorum* Hansgirg 1892

Colonies flat, irregular, mucilaginous, slightly granular, dark brown, blackish-green to dirty black, rarely black-violet, usually in form of macroscopic, large, irregular gelatinous spots, composed of small, 2–8-celled subcolonies. Mucilaginous envelopes more or less spherical, fine, wide, colourless, homogeneous or finely (sometimes indistinctly), densely layered, delimited, 3–15 µm in diameter, enveloping usually 1–2 (4) cells. (Several authors write about blackish-brown sheaths; relation of such populations to *G. bituminosa* is questionable). Cells more

---

**Fig. 311.** *Gloeocapsa granosa* (envelopes colourless or pale greyish): **a** – after Geitler (1925), **b** – after Yacubson (1974, different from the type?).
**Fig. 312.** *Gloeocapsa atrata* (envelopes colourless or slightly bluish): **a** – young one- or two-celled colonies, **b** – variation of old compound colonies and various stages in the life cycle; **a-b** – after Wille and Hollerbach from Hollerbach & al. (1953) and Kondrateva & al. (1984), sub *G. montana* (incl. f. *fenestralis*) and *G. varia* (excl. typical *Chroococcus varius*); **c** – «*Gloeocapsa varia*» after Prescott from Starmach (1966, diverse concept from the type).
**Fig. 313.** *Gloeocapsa bituminosa*: after Hollerbach from Starmach (1966).

or less spherical, blue-green, pinkish, reddish or almost violet, 2–6 μm in diameter. Occasional nanocyte formation.
Occurrence: Subaerophytic, on wet, calcareous rocks, walls in glasshouses; known in the present sense only from Europe, other data must be checked. (*G. bituminosa* is not well defined and needs more study).

### 10. **Gloeocapsa coracina** Kützing 1843 (fig. **315**)

Colonies crustaceous, gelatinous, slippery. Mucilaginous envelopes around cells more or less spherical, colourless (sometimes slightly violet or bluish, but never intensely blue or blue-violet), up to 6 μm wide, not lamellate or distinctly stratified, sometimes laterally uniting. Cells spherical with homogeneous, pale blue-green content, 3–6 μm in diameter.
Occurrence: Subaerophytic, typical of wet sandstone rocks; rare in Czech Republic (N. Bohemia), Germany (S. Saxony), several times recorded also from tropical countries (India).

### 11. **Gloeocapsa salina** Hansgirg 1893 (fig. **316**)

Colonies micro- or (usually) macroscopic, gelatinous, amorphous, widely spread over a substrate, dirty-yellow, rarely slightly orange or yellow-brown, mainly composed of 2–4 – 8-celled subcolonies, 12–24 μm in diameter. Mucilaginous envelopes around cells colourless, distinctly lamellated. Cells yellow-green or yellowish, 3–6 μm in diameter.
Occurrence: Halophilic, on wet salty soils, on edges of wet drying salinic swamps and marshes, in shallow salty pools, etc.; inland and coastal saline localities in central Europe and in the Mediterranean region.

### 12. **Gloeocapsa caldariorum** Rabenhorst 1865 (fig. **317**)

*Gloeocapsa polydermatica* Kützing sensu auct. post. p.p. (typus: sec. Van Den Hoek 1963 = *Coccomyxa*).

Colonies micro- or macroscopic, composed of aggregated, but usually not diffluent, to 39 μm wide subcolonies, which are spherical or oval and contain usually 1–4 cells. Mucilaginous envelopes thick, delimited (only very rarely diffluent under special conditions), colourless, finely or (usually) distinctly densely stratified. Cells spherical, blue-green, 3–4.5 (8) μm in diameter.
Occurrence: Subaerophytic, on wet rocks, walls, in glasshouses, on diverse stony and glass substrates, rarely on soil; commonly distributed in central Europe, possibly cosmopolitan, geographic distribution should be reconsidered on the basis of correct determinations.

---

GLOEOCAPSA
**Fig. 314.** *Gloeocapsa atrata:* after Wille from Geitler (1932, sub *G. montana* and *G. atrata*).
**Fig. 315.** *Gloeocapsa coracina*: after Bharati & Bongale (1975), possibly different from the original Kützing's type from European sandstone rocks.
**Fig. 316.** *Gloeocapsa salina*: after Hansgirg from Geitler (1925, drawing in which lamellation of envelopes is not demonstrated).
**Fig. 317.** *Gloeocapsa caldariorum*: **a** – after Wille from Geitler (1932, sub *G.*

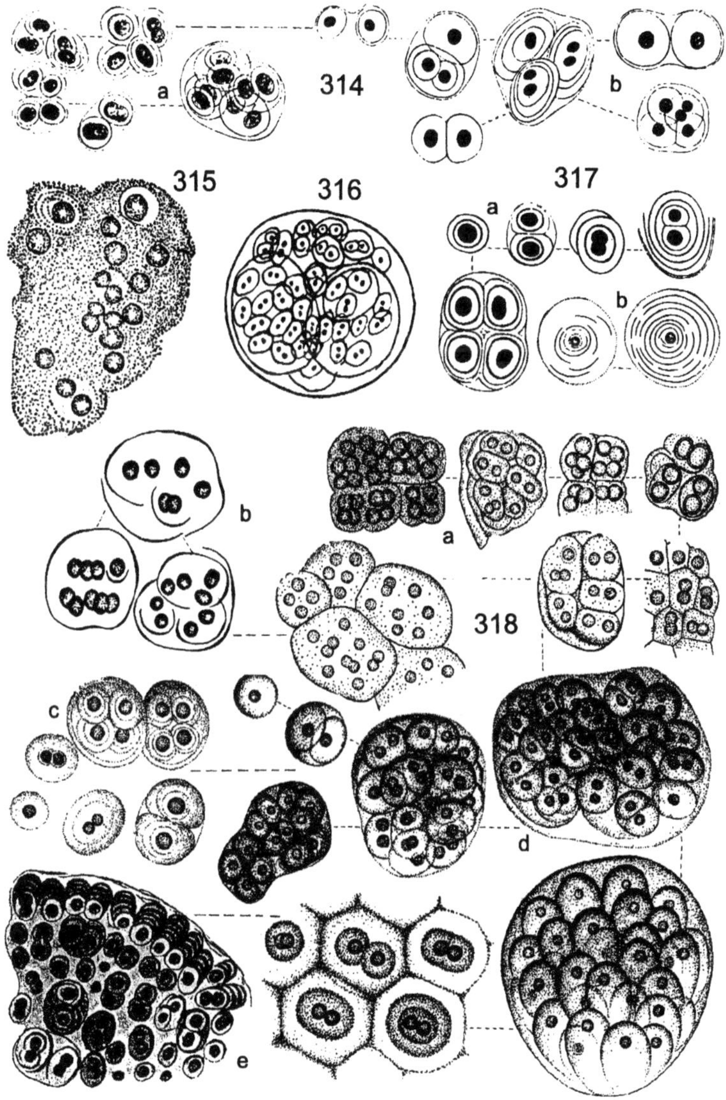


*polydermatica*, redrawn from the type exsiccate), **b** – after West from Geitler (1932).

**Fig. 318.** *Gloeocapsa compacta*: **a** – after Geitler from Kondrateva & al. (1984), **b** – after Golubić (1967), **c** – after Skuja (1964), **d** – after Skuja (1964), **e** – after Nováček (1934).

**13. Gloeocapsa decorticans** (A. Braun) Richter in Wille 1925 (fig. **319**)

Colonies small, composed of more or less regular subcolonies with 1–8 cells; subcolonies usually a slightly elongate, more or less oval, to 30 µm long. Mucilaginous envelopes colourless, delimited, wide, usually distinctly lamellate. Cells spherical or slightly ellipsoidal-oval, blue-green, 6–9.5 µm in diameter, during the cell division to 12 µm long.
Occurrence: Subaerophytic, on wet rocks, walls, known also from caves, data «from stagnant waters» should be checked; through Europe (and USA ?) sporadic, the distribution not known.

**14. Gloeocapsa compacta** Kützing 1845 (fig. **318**)

Colonies micro- or macroscopic, irregular, granular, firm, composed of tightly grouped subcolonies of irregular oval, ovoid, sack-like or irregular outline (flattened at points of contact); subcolonies contain 2 to more loosely or densely arranged cells with or without individual envelopes. Mucilaginous envelopes wide, colourless when young, later intensely blue to blue-violet, reddish violet or brown-violet, sometimes more coloured on one side than another, distinct, delimited, unstratified and homogeneous inside, or with indistinct, to 8.5 µm wide, very fine single, intensely coloured envelopes around cells (up to 3.5 µm in diameter). Sometimes groups of colonies enveloped by a firm sheath. Cells spherical, blue-green, (1.5)2–2.5 (3.5) µm in diameter. Aggregations of subcolonies are reminiscent of *Chondrocystis*.
Occurrence: Subaerophytic, on wet (usually calcareous) rocks and (rarely) bark of trees, from lowlands to mountains; common in high mountains of central Europe, Scandinavia, Italy and Spain. Probably wider distributed, but the current localities are not well known. Data from sea shores belong evidently to other species.

**15. Gloeocapsa violascea** (Corda) Rabenhorst 1865

*Protococcus violasceus* Corda in Sturm 1833 (typus = ?); (false «*violacea*»)

Colonies dirty and dark blue-green to blackish, granular, gelatinous when wet, micro- to macroscopic, irregular, composed of numerous, more or less spherical subcolonies; subcolonies not flattened laterally, contain usually several to many irregularly arranged cells, surrounded by a system of not very wide envelopes. Mucilaginous envelopes delimited or sometimes diffluent at the margin, indistinctly or distinctly stratified, intensely dirty blue, blue-violet or blackish violet (in old colonies very dark, sometimes not transparent), occasionally slightly reddish violet. Cells pale blue-green, 2.8–4 (6.8) µm in diameter.
Occurrence: Aerophytic, on walls (usually not calcareous, but known also from serpentine and limestone rocks), from lowlands to mountains, usually in rather shady localities and in cavities in rocks; known from central Europe (Austria, Czech Republic, Germany), the wider distribution needs confirmation.

---

GLOEOCAPSA
**Fig. 319.** *Gloeocapsa decorticans* (colourless envelopes): **a** – after Wille from Geitler (1932, redrawn from the type exsiccate of A. Braun), **b** – after Skuja (1939).

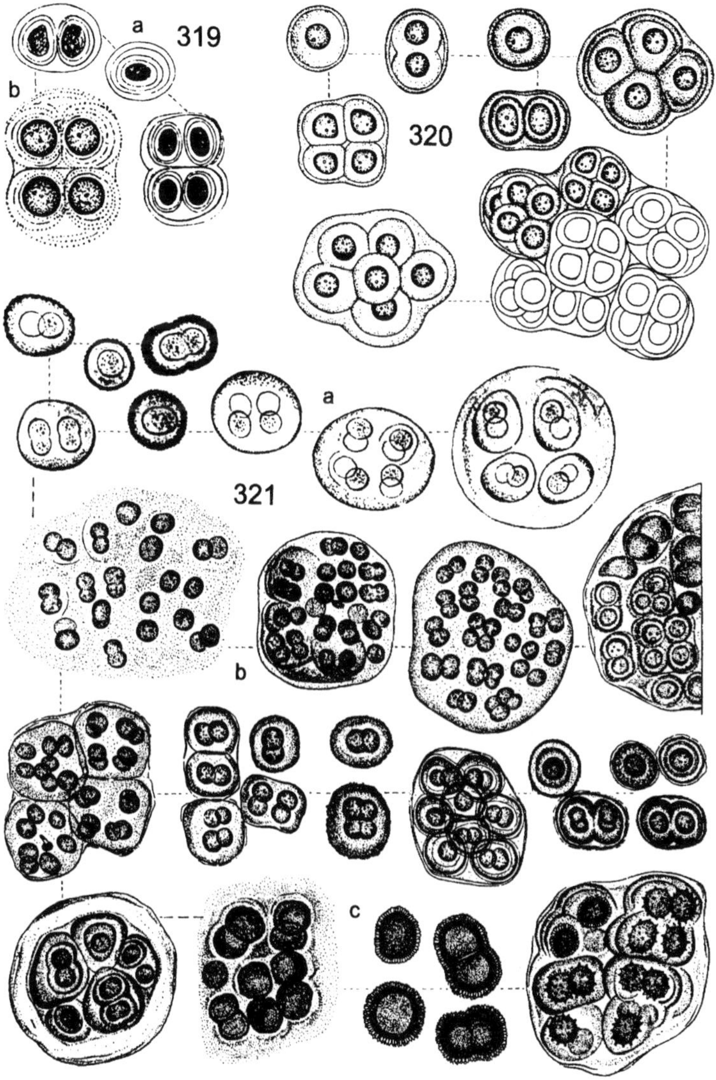


**Fig. 320.** *Gloeocapsa lignicola* (bluish or dark blue-violet sheaths): after Hollerbach from Kondrateva & al.(1984, sub *G. alpina* f. *lignicola*).
**Fig. 321.** *Gloeocapsa* cf. *lignicola* from stony (serpentine) substrate: **a** – simple colonies, **b** – stages in the life cycle, **c** – resting stages; after Nováček (1934 and from Geitler 1960, sub *G. alpina* sp. aff.).

**16. Gloeocapsa lignicola** Rabenhorst 1865 (fig. **320**)

*Gloeocapsa violacea* Kützing 1846; *Gloeocapsa alpina* f. *lignicola* (Rabenhorst) Hollerbach in Elenkin 1938

Colonies crusty, granular, blackish. Subcolonies usually 2–4-celled, rarely with more cells joined together with wide outer colourless or bluish to very intensely coloured, blue-violet, usually not or slightly striated sheath layers. Cells blue-green, (2.2) 3–6 μm in diameter.
Occurrence: Aerophytic, on wet woods and bark of trees, mainly in mountains (identity of similar populations from wet rocks and stony substrates should be confirmed), (fig. **321**); central Europe; relations to *G. alpina* are unclear.

**17. Gloeocapsa nigrescens** Nägeli in Rabenhorst 1857 (fig. **322**)

Colonies microscopic, spherical, 2 to many-celled, sometimes aggregated in irregular macroscopic, granular, blackish mass; subcolonies composed of groups of spherical cells, surrounded by individual envelopes, which are distant from cells and spherical in outline, not lamellate, hyaline, only rarely consisting of envelopes of different generations, inside violet (not reddish) of varying intensity, outside colourless; colonies to about 140 μm in diameter. Cells spherical, blue-green, 3–5 μm in diameter, sometimes there are *Aphanocapsa* stages.
Occurrence: Aerophytic, on wet calcareous and serpentine rocks in lowlands and mountains, usually in localities with low illumination and high air humidity; distribution in Europe unknown (often misinterpreted as other species), common especially in warm and tropical countries. Numerous localities are cited by Daily (1942) from USA.

**18. Gloeocapsa alpina** (Nägeli) Brand 1900 (fig. **323**)

*Gloeocapsa ambigua* Nägeli in Kützing 1849 (false «Kirchner»); *G. fuscolutea* Kirchner 1878

Colonies microscopic, 2–8-celled, but frequently aggregated in macroscopic, irregular, colonial granular, blackish mass. Solitary subcolonies arising often from groups of irregularly disposed cells, with individual, wide mucilaginous envelopes (regularly spherical in outline), which are finely granulate, lamellate (rarely not lamellate), blue to dark violet (not reddish); irregular groups of cells with coloured sheaths (such subcolonies to 40 μm in diameter) are sometimes surrounded by wide, more or less spherical, colourless or only slightly bluish, homogeneous, delimited envelopes, to 300 μm in diameter. Cells spherical, blue-green, 4–6 (8) μm in diameter. Sometimes *Aphanocapsa*-stages occur (without striated individual envelopes) and nanocytes (2.5–3.5 μm in diameter). Resting cells to 20 μm in diameter, with thin, firm, blackish violet sheaths.
Occurrence: Subaerophytic, mainly on calcareous rocks in mountains, rarely in lowlands, forming black spots and stripes on limestone walls, probably also on wood and secondarily on snow fields, rarely in caves; in central Europe (Alps, Carpathians) common, probably widely distributed in corresponding biotopes throughout the world.

---

GLOEOCAPSA
**Fig. 322.** *Gloeocapsa nigrescens* (blue envelopes): **a-c** – after Daily (1942), **d** – after Frémy (1930).
**Fig. 323.** *Gloeocapsa alpina* (blue or blue-violet envelopes): **a** – after Geitler

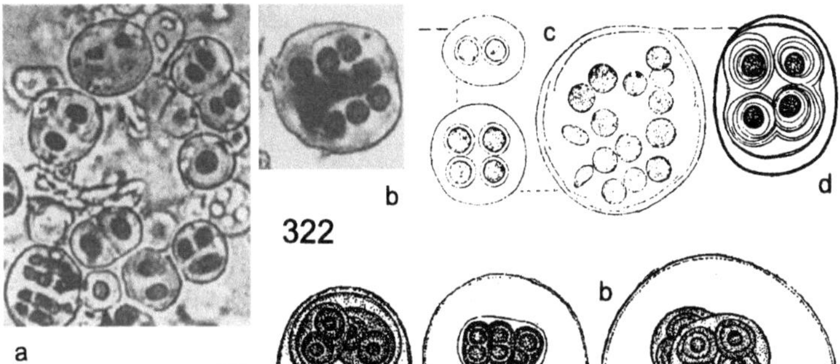


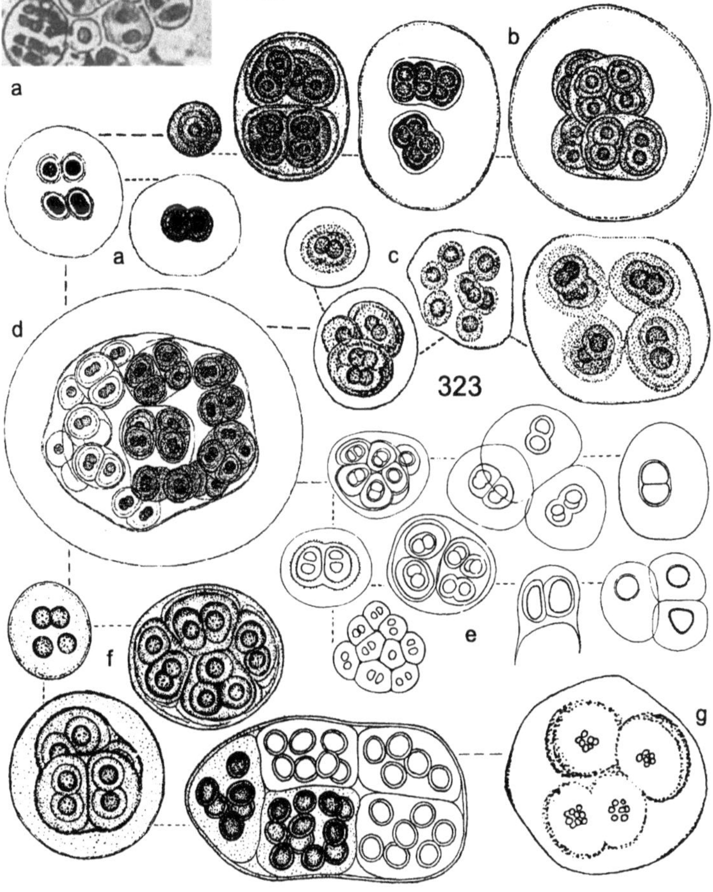


(1932), **b** – after Hollerbach from Kondrateva & al. (1984), **c** – after Starmach (1966, «typical stages»), **d** – after Geitler (1932), **e** – after Geitler (1925, at the right: three resting stages), **f** – after Hollerbach from Kondrateva (1984, sub *G. alpina* f. *ambigua*, at the right: resting stages), **g** – after Starmach (1966, degenerative cells).

## 19. **Gloeocapsa thermalis** Lemmermann 1905 (fig. **324**)

? *Gloeocapsa tepidariorum* A. Braun in Rabenhorst 1852

Colonies microscopic, composed of 2–8-celled subcolonies, less frequently grouped in macroscopically visible, more or less compact gelatinous clusters. Mucilaginous envelopes spherical or irregular, colourless or later reddish or brownish red, often granular, distinctly lamellate (usually with 3–4 layers), in single cells wider, up to 8 µm. Cells spherical, pale to bright blue-green, rarely bluish coloured, 1–2.6 µm in diameter.

Occurrence: In thermal springs; originally described from Hawaii (hot spring in the vicinity of volcano Mauna Kea), later recorded also from Europe: Greece (Nigrita), France (Chaudesaigues) and Hungary (Margaretheninsel, Bálintquelle) mostly atmophytic, together with *Mastigocladus laminosus* and *Leptolyngbya laminosa*. Taxonomic revision is needed. (*G. thermalis* Kützing 1843, described from thermal springs in Italy, has priority against *G. thermalis* Lemmermann, but the Lemmermann's species is often recorded and well known; the identity and nomenclature of both taxa has to be solved).

## 20. **Gloeocapsa tornensis** Skuja 1964 (fig. **325**)

Colonies small, micro- or macroscopic, gelatinous, subcrustaceous, thin, red-brownish to blackish, with solitary cells or groups of cells, all surrounded by their own, wide, more or less spherical, gelatinous, intensely red-brown, stratified or not stratified, not diffluent envelopes (usually wider than cell diameter). Cells spherical or slightly elongate before division, with homogeneous, blue-green content, 1.7–2.5 µm in diameter.

Occurrence: Freshwater, epilithic, submersed in littoral of cold, mainly nordic lakes: Austria, N. Sweden.

## 21. **Gloeocapsa haematodes** Kützing 1849 (fig. **326**)

? *Gloeocapsa stegophila* (Itzigsohn) Rabenhorst 1863

Colonies microscopic, rarely macroscopic, small, nest-like, composed of (1) 2–8-celled, rarely up to 16–32-celled, spherical, ellipsoidal or polyhedral-rounded subcolonies, enveloped by narrow, reddish, indistinctly lamellate common, delimited mucilage, rarely diffluent at the margin. Mucilaginous envelopes around single cells within colonies narrow, not lamellate, 2–6 (8) µm in diameter, orange-red (*G. stegophila*), rusty red or dark (brownish) red (*G. haematodes*), usually not transparent. Cells small, blue-green, 1.6–3.2 µm in diameter. Resting cells with narrow, thin, firm sheath, coloured dark red.

Occurrence: In moorland, on wet peaty soils, rarely on wet granitic rocks (*G. haematodes*), or on wet old wood (*G. stegophila*); poorly understood species, recorded only from Europe (Czech Republic, Germany, Scandinavia, Switzerland, etc.).

---

GLOEOCAPSA

**Fig. 324.** *Gloeocapsa thermalis* (envelopes colourless, later reddish or brownish red): **a** – after Anagnostidis (1961), **b** – after Palik (1949).

**Fig. 325.** *Gloeocapsa tornensis* (envelopes red-brown): after Skuja (1964).

**Fig. 326.** *Gloeocapsa haematodes* (envelopes red): **a** – after Wille from Geitler (1932)), **b** – after Hollerbach from Kondrateva & al. (1984).

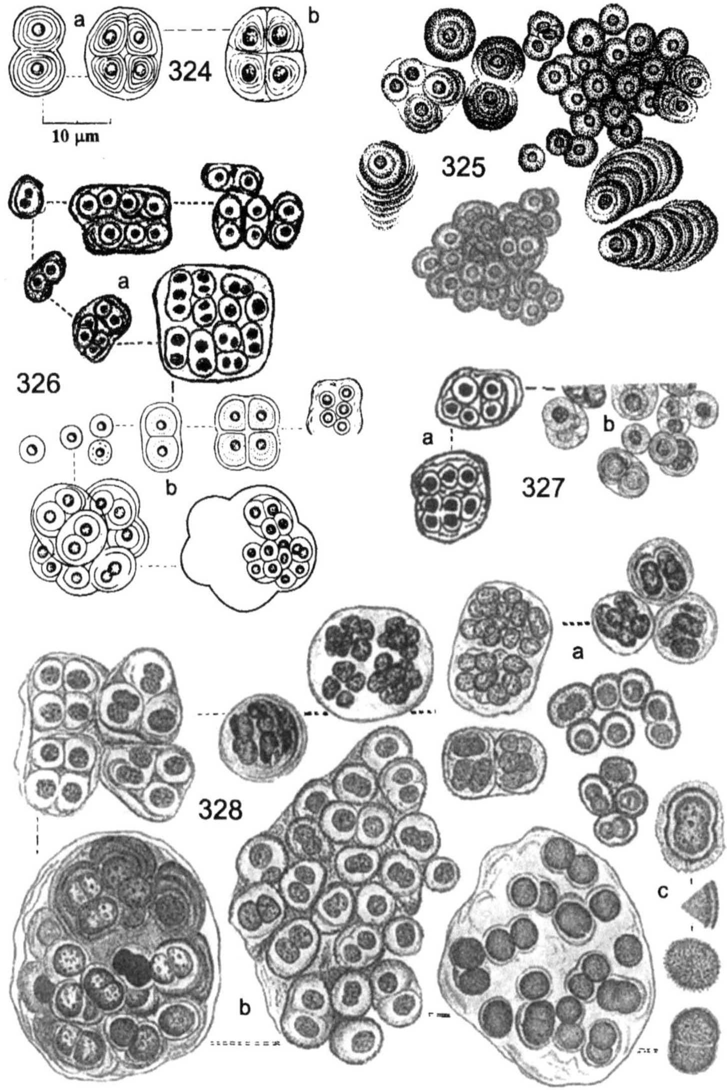


**Fig. 327.** *Gloeocapsa rupicola* (envelopes dark reddish brown): **a** – after Wille from Starmach (1966, redrawn from the type exsiccate of Kützing), **b** – after Bourrelly & Manguin (1952).
**Fig. 328.** *Gloeocapsa novacekii* (envelopes reddish): **a** – variation in colonies, **b** – old colonies, **c** – resting stages; after Nováček (1930, sub *G. sanguinea*).

## 22. Gloeocapsa rupicola Kützing 1849 (fig. 327)

? *Gloeocapsa sabulosa* (Meneghini) Richter in Wittrock et Nordstedt 1864; *G. quaternata* (Brébisson) Kützing 1848 sensu auct. post. (false *«quaternaria»*) (typus: sec. Van Den Hoek 1963 = subaerophytic green algae)

Colonies micro- or macroscopic, not very large, forming reddish, red brownish or red blackish gelatinous spots or nests on (usually) non-calcareous rocks, in fissures, among mosses, etc., composed from numerous irregular, 2 – more-celled subcolonies, containing a system of mucilaginous concentrical envelopes enclosing groups of cells with individual sheaths. Mucilaginous envelopes around single cells thin, firm, to 6 µm wide, indistinctly or distinctly lamellate, inside dark red or reddish brown, sometimes scarcely transparent, outside only reddish, to (peripheral envelopes) almost colourless, sometimes diffluent; envelopes never very wide. Cells often more or less densely packed. Cells blue-green, pale olive-green or yellowish, 2–3 (4.5) µm in diameter. *Aphanocapsa* stages and resting cells known with dark reddish, thin, firm sheaths.

Occurrence: Aerophytic, on wet, but sometimes periodically drying rocks, walls, from lowlands to mountains, often in deep river valleys; in central Europe distributed, other localities uncertain.

## 23. Gloeocapsa novacekii Komárek et Anagnostidis 1995 (fig. 328)

*Gloeocapsa sanguinea* sensu Nováček 1930

Colonies micro- or macroscopic, gelatinous, granular, rusty red (wet), dirty or blackish brown (dry), composed of subcolonies to 60µm in diameter. Mucilaginous envelopes wide, not or slightly striate, finely granular, colourless or reddish, extending up to 5 µm beyond cells. Cells spherical or wide oval, pale olive-green, blue-green or yellowish, (3)3.5–9(10) µm in diameter. Commonly forms *Aphanocapsa* stages, nanocytes and resting stages with firm, almost non-transparent, rusty red coloured envelopes.

Occurrence: Aerophytic on periodically wetted rocks from lowlands to mountains, in river valleys; described from serpentine rocks in Czech Republic (W. Moravia), probably has wider distribution (possibly identified as other species).

## 24. Gloeocapsa sanguinea (Agardh) Kützing 1843 (fig. 329)

*Gloeocapsa itzigsohnii* Bornet in Zopf 1882

Colonies micro- or macroscopic, mucilaginous, irregular, composed of numerous subcolonies, usually with several cells enveloped by a system of irregular oval, thin or more or less slightly widened envelopes. Mucilaginous sheaths firm, only outer ones slightly diffluent, not or simply lamellate, rarely with distinct lamellation, to 13–15 µm wide around solitary cells or few-celled (–8-celled) groups of cells, subcolonies to 150 µm (rarely more) in diameter; envelopes near the cells intensely red, outer slightly reddish to almost colourless, never much widened. Cells spherical, oval before division, pale blue-green or greyish blue, (3.5)4–6.5 µm in diameter. «*Aphanocapsa*»-stages, nanocyte production (nanocytes about 3 µm in diameter), resting cells with thin, firm, dark red-brown sheaths, to 10(11.5?) µm in diameter.

---

GLOEOCAPSA

**Fig. 329.** *Gloeocapsa sanguinea* (envelopes externally almost colourless, internally intensely red): **a** – solitary cells with envelopes, **b** – typical young colonies, **c** –

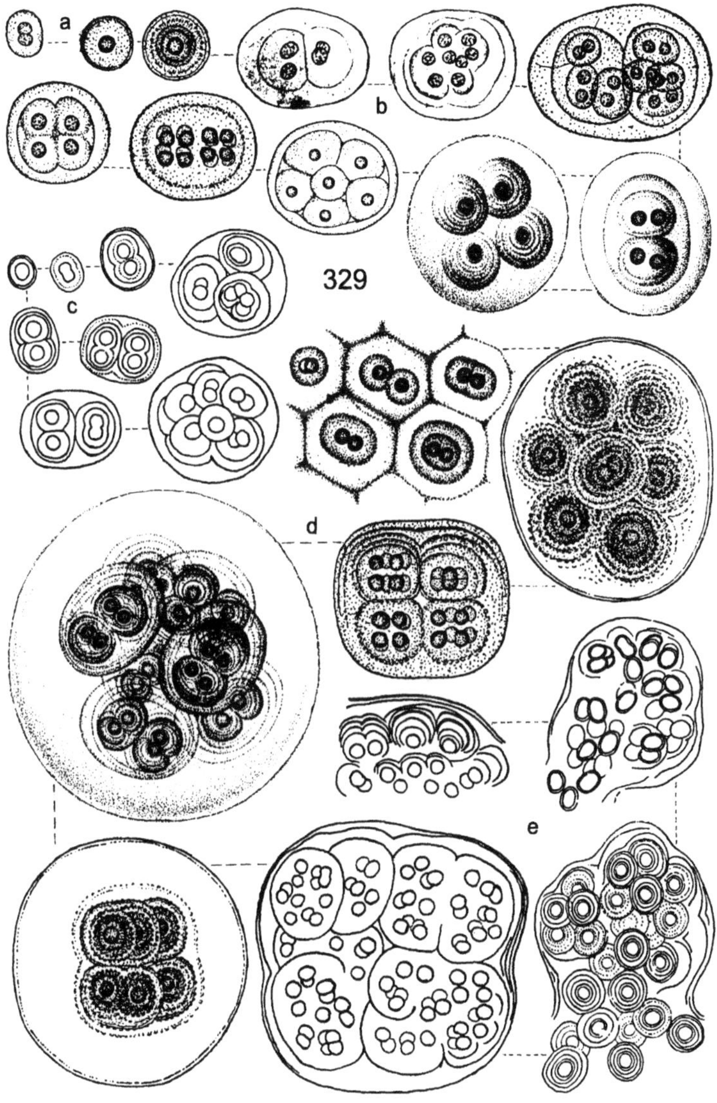


typical development of colonies during main part of life cycle, **d** – old colonies with intensely red inner envelopes, **e** – parts of old colonies with dormant cells and resting stages; after various authors, from Geitler (1932), Hollerbach & al. (1953), Starmach (1966), Skuja (1964) and Kondrateva & al. (1983).

Occurrence: Subaerophytic, on wet rocks (esp. on granite, sandstone, etc.), near waterfalls, particularly in high mountains; common in mountain systems of central Europe (Alps, Carpathians), but also in more humid valleys of S. European mountains and in northern regions (up to subpolar areas: Greenland, Svalbard, N. Scandinavia), sometimes lichenized (Jaag 1945, Büdel & Henssen 1988); probably present over the whole Eurasia, may be cosmopolitan in distribution, however, sometimes misinterpreted with other species.

### 25. **Gloeocapsa ralfsii** (Harvey) Kützing 1846 (fig. **330**)

*Palmella ralfsii* Harvey 1841 (false «*ralfsiana*»); *Gloeocapsa ralfsii* (Harvey) Lemmermann 1907

Colonies micro- or macroscopic, irregular, red, gelatinous, composed of numerous, more or less spherical subcolonies, only in atypical stages of irregular rounded outline. Subcolonies consisting of 2–64-celled clusters, enveloped by lamellate, intensely red envelopes, to 50 µm in diameter, enclosed in very wide, homogeneous, almost colourless or only pale reddish, more or less spherical, delimited sheath. Cells spherical or slightly elongate, pale blue-green, 4–8 µm in diameter. *Aphanocapsa* stages can occur or stages without wide envelopes (status «siccus»), resting cells up to 9 µm in diameter, with dark red, thin, firm, granular sheath. Nanocytes not known.
Occurrence: Subaerophytic, on non-calcareous wet rocks, known from various European mountains; common in granitic W. Carpathians (Slovakia – High Tatra Mts.), known also from Alps (Austria, Switzerland), Scandinavian mountains and Svalbard; distribution is not well known.

### 26. **Gloeocapsa reicheltii** Richter 1895 (fig. **331**)

Colonies micro- or macroscopic, gelatinous or crustaceous, composed of numerous subcolonies, forming dirty yellow-green or orange mass. Cells spherical, oval before division, pale blue-green, 2.5–3.5 µm in diameter. Mucilaginous envelopes around cells lamellate, thin, later widened, homogeneous, yellow-orange, limited or diffluent, slightly granular.
Occurrence: Aerophytic, described from walls of a stony bridge near Leipzig, Germany; in central Europe probably rare (mainly on sandstone). Not a well-known species.

### 27. **Gloeocapsa shuttleworthiana** Kützing 1845 (fig. **332**)

Colonies irregular, microscopic, joined into large, macroscopic, more or less orange aggregates, to 13 µm wide, composed of numerous subcolonies with one or more cells enveloped by non- lamellate or slightly lamellate sheaths; the inner

---

GLOEOCAPSA

**Fig. 330.** *Gloeocapsa ralfsii* (inner sheaths red, widened envelopes almost colourless, hyaline): **a** – from Geitler (1932, sub *G. ralfsiana*), **b** – after Cooke from Geitler (1925).
**Fig. 331.** *Gloeocapsa reicheltii* (yellow-orange sheaths): after Richter (1895).
**Fig. 332.** *Gloeocapsa shuttleworthiana* (yellow-orange to red-orange sheaths): after Wille from Geitler (1932, redrawn from the original exsiccate of Kützing).
**Fig. 333.** *Gloeocapsa fusco-lutea* (yellow-brown envelopes): after Wille from Geitler (1932, redrawn from the original exsiccate of Nägeli).

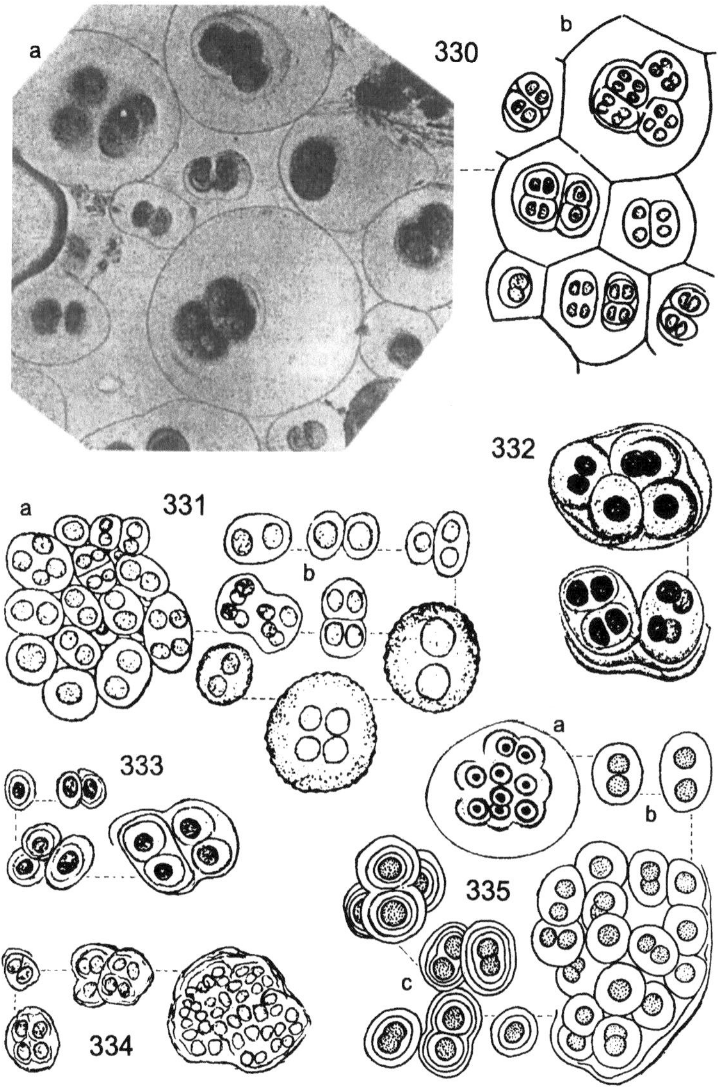


**Fig. 334.** *Gloeocapsa deusta* (yellow-brown envelopes): after Wille from Geitler (1932, redrawn from the original exsiccate of Meneghini).
**Fig. 335.** *Gloeocapsa kuetzingiana* (yellow envelopes): **a** – after Wille from Geitler (1932, redrawn from the original exsiccate of Nägeli), **b-c** – from Starmach (1966).

layers sometimes thin, particularly intensely coloured. Mucilaginous envelopes bright orange-yellow to orange-red, enveloping the subcolonies and forming groups of up to 65 or more µm in diameter, with outer layers less intensely coloured to almost colourless. Cells more or less spherical, pale blue-green, 6–8 µm in diameter.
Occurrence: Subaerophytic or aerophytic, on wet rocks and walls: common in humid areas of W. Europe (near Atlantic coast), in mountains of Ireland, Great Britain (e.g., in England in Lake District), but its distribution is not well known due to frequent confusion with other species.

### 28. **Gloeocapsa fusco-lutea** (Nägeli) Kützing 1849 (fig. **333**)

Colonies micro-, rarely macroscopically visible, irregular, composed of small, several-celled subcolonies which are up to 50 µm in diameter. Solitary or small groups of cells surrounded by 4.5–5.5 µm wide gelatinous layer. Mucilaginous envelopes more or less wide, usually slightly lamellate, delimited, yellow or yellow-brown. Cells spherical or slightly elongate, hemispherical after division, pale blue-green, 1.5–2 µm in diameter.
Occurrence: Aerophytic, on wet, non-calcareous rocks, not common; known from central and W. Europe, further distribution is unknown. Revision is needed.

### 29. **Gloeocapsa biformis** Ercegović 1925 (fig. **336**)

Colonies microscopic, rarely macroscopically visible, irregular, dirty yellow or brownish, composed of groups of small subcolonies. Mucilaginous envelopes colourless and hyaline or yellow, rarely yellow-brown, not or slightly lamellate. Cells spherical, pale blue-green, 0.8–3 µm in diameter. Nanocyte formation recorded.
Occurrence: Aerophytic, on wet limestone rocks, epilithic and chasmoendolithic (according to Nováček 1934 also on serpentine rocks in lowland regions, Czech Republic); known mainly from calcareous Alps, Carpathians and W. Balcanian mountains, also from N. Greece and Spain. Under the name *G. biformis* were designated numerous populations from various habitats and areas, but their identity is improbable. Revision of these specimens is desirable. *G. sabulosa* (Meneghini) Richter in Wittrock et Nordstedt 1884 is usually synonymized with *G. biformis*, but it grows usually in water biotopes (unpolluted streaming water) (fig. **336c**); comp. also *G. rupicola*.

### 30. **Gloeocapsa deusta** (Meneghini) Kützing 1849 sensu auct. (fig. **334**)

Colonies macroscopic, flat, granular, dirty yellow-brown, composed of many small, irregular rounded subcolonies. Subcolonies with 1–2–4– more cells, to 80 µm in diameter, with lamellate, not very wide, firm, delimited, yellow-brown or brown envelopes; outer layer sometimes intensely brown. Cells more or less spherical or slightly elongate, pale blue-green, 3–6 µm in diameter.

---

GLOEOCAPSA

**Fig. 336.** *Gloeocapsa biformis* (sheaths hyaline or yellowish – brownish): **a** – various stages in the life cycle after Nováček (1934) from Czech Republic, **b** –

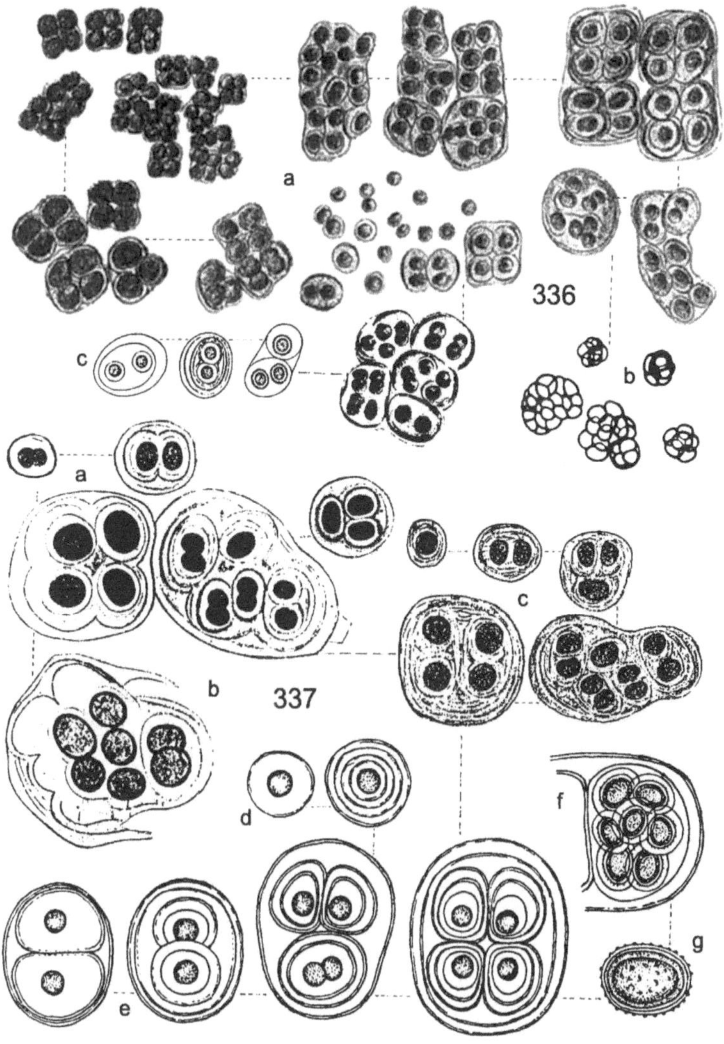


after Golubić (1967) from Venezuela (identity with European material is questionable – comp. *Cyanosarcina*), **c** – after Geitler (1932, redrawn from the Wittrock & Nordstedt's exsiccate nr. 698, sub *G. sabulosa* - recently synonymized with *G. biformis*).

**Fig. 337.** *Gloeocapsa rupestris* (sheaths yellow-brown): variability of colonies, **b,f-g** – dormant and resting stages; **a-b** – after Wille from Geitler (1932, redrawn from the original exsiccate of Kützing), **c** – after Frémy (1930), **d-g** – after Starmach (1966).

Occurrence: Marine, subaerophytic, epilithic on calcareous stones and rocks in the splash zone; Mediterranean coasts, known particularly from Adriatic Sea. Revision needed, probably different from the type (see *Entophysalis deusta*, p. 337).

### 31. Gloeocapsa kuetzingiana Nägeli 1849 (fig. 335)

Colonies micro- or macroscopic (sometimes covering several cm$^2$), irregular, mucilaginous, dirty brownish, composed of more or less spherical or ellipsoidal subcolonies (usually 2–4-celled with regularly distributed cells), enveloped by widened, outer firm, delimited, not or indistinctly lamellate mucilaginous sheaths, usually intensely yellow or yellow-brown coloured; outer layers sometimes less intensely yellow, even colourless. Cells spherical, blue-green, (2) 3–5 µm in diameter; the inner envelopes around single cells to 7.5 (10) µm in diameter. The dormant cells perhaps up to 12.5µm in diameter (confirmation is needed).
Occurrence: Subaerophytic, on wet, non-calcareous rocks in high mountains; common in Alps and Carpathians, known also from Scandinavian mountains, Italy, Spain etc., however several localities need confirmation; several records from subpolar regions; probably with wide distribution in corresponding biotopes.

### 32. Gloeocapsa rupestris Kützing 1845 (fig. 337)

*Gloeocapsa muralis* Kützing 1845 (sensu auct. post., typus sec. Van Den Hoek 1963 = ? *Chlorophyceae*)

Colonies micro- or macroscopic, irregular, gelatinous, reaching several cm$^2$, dirty greyish brown, composed of more or less spherical or oval subcolonies, 15–75 (200) µm in diameter. Subcolonies with (1) 2–4 – 16 (more) cells, surrounded by individual delimited, more or less wide, usually distinctly lamellate, pale up to intensely yellow or yellow-brown envelopes; outer sheaths usually pale, almost colourless. Cells dirty blue-green, spherical or slightly elongate after and before division, (5) 6–8 (11) µm in diameter. Resting cells spherical or oval with firm, narrow, dark brown, red-brown up to blackish brown sheaths (sometimes with granular surface and bluish, thin outer layer), to 15 (19) µm in diameter.
Occurrence: Subaerophytic, on periodically wetted rocks and walls from lowlands to mountains, on glasshouse walls; common in unpolluted regions, known from the whole Europe, N. America and probably widely distributed, but some identifications are not certain.

Notes: The inner taxonomy of the genus *Gloeocapsa* is unclear. The wide variation, morphological changes during vegetation cycles (status), formation of resting stages and variation in sheath colouration complicate correct identification. Numerous populations from different habitats and different regions are designated by the same name. On the other hand, they have been described many species, many of which are only stages of other species. Species surely exist in various biotopes that still are not taxonomically defined (see, e.g., Friedmann & al. 1967 from chasmolithic biotopes in Negev desert, Israel; Broady 1986, 1989 from Antarctica). Therefore, in the following review are mentioned only some of the taxa, the taxonomic position of which is unclear. The genus needs revision. Numerous species were transferred in other genera, mainly to *Aphanothece*, *Asterocapsa*, *Chondrocystis*, *Chroococcus*, *Gloeocapsopsis* and *Gloeothece*.

**Species outside Europe:**
- *Gloeocapsa acervata* Gardner 1927: Aerophytic on limestone in Porto Rico; envelopes hyaline, later yellow-brown.

- *Gloeocapsa africana* Cholnoky 1952: Problematic species, from Mozambique.
- *Gloeocapsa bahamensis* Collins 1920: In 4–24-celled colonies, cells 3–5 µm in diameter, envelopes wide, lamellose, browish or dark brown; in crusts on the border of a salt pond, Bahamas.
- *Gloeocapsa calcarea* Tilden 1898: From USA (Wisconsin); probably *Chroococcus* ?.
- *Gloeocapsa calcicola* Gardner 1927: From limestone rocks in Porto Rico; intensely lamellate hyaline envelopes.
- *Gloeocapsa cartilaginea* Gardner 1927: From red soils in Porto Rico, with hyaline envelopes.
- *Gloeocapsa incrustata* Chu 1944: From China (Szechwan).
- *Gloeocapsa minutula* Gardner 1927: On walls near thermal springs in China (Fukien).
- *Gloeocapsa montana* sensu El-Ayouty & Ayyad 1972: From Egypt, expanded strata; probably *Gloeocapsopsis*.
- *Gloeocapsa multisphaerica* Gardner 1927: From rocks (aerophytic) in China (Fukien).
- *Gloeocapsa ovalis* Gardner 1927: Aerophytic on limestone, described from Porto Rico; probably *Chroococcus* (cf. *C. ercegovicii*).
- *Gloeocapsa sparsa* Wood 1872: Aerophytic, described from rocks in USA (Pennsylvania).
- *Gloeocapsa sphaerica* Gardner 1927: From calcareous rocks in Porto Rico.
- *Gloeocapsa zostericola* Farlow 1882: Marine, described from Woods Hole, USA (Mass.).

**Unrevised European species:**

- *Gloeocapsa ambigua* Nägeli 1849
- *Gloeocapsa ampla* Vorce 1881
- *Gloeocapsa attingens* Schiller 1954
- *Gloeocapsa aurata* Stizenberger in Rabenhorst 1854 (on wet wood; drawing exists only from African samples by Frémy 1930; fig. **338**)
- *Gloeocapsa botryoides* Kützing 1843
- *Gloeocapsa conglomerata* Kützing 1846 (on wet soil among mosses; fig. **339**)
- *Gloeocapsa conspicua* Reinsch 1867
- *Gloeocapsa cryptococcoides* Kützing 1846
- *Gloeocapsa didyma* (Kützing) Kützing 1846
- *Gloeocapsa dirumpens* Beck-Mannagetta 1929
- *Gloeocapsa dubia* Wartmann in Rabenhorst 1861
- *Gloeocapsa endocodia* Vouk 1936
- *Gloeocapsa geminata* Kützing 1849
- *Gloeocapsa gigas* W. et G.S. West 1895 (described and known from tropical countries, the European localities are very probably wrong; fig. **340**)
- *Gloeocapsa holstii* Hieronymus 1895
- *Gloeocapsa ianthina* Nägeli in Kützing 1849
- *Gloeocapsa juliana* (Meneghini) Kützing in Rabenhorst 1869
- *Gloeocapsa livida* (Carmichael) Kützing 1846 (recorded from Spain)
- *Gloeocapsa magma* «(Brébisson) Hollerbach 1924» (evidently *Gloeocapsopsis* species, also in sense of Hollerbach 1924; however, in literature various *Gloeocapsa* species are designated by this name; fig. **341**)
- *Gloeocapsa mellea* (Meneghini) Kützing 1846
- *Gloeocapsa monococca* Kützing 1843
- *Gloeocapsa myxophila* Beck-Mannagetta 1931
- *Gloeocapsa nigra* (Meneghini) Grunow in Rabenhorst 1865
- *Gloeocapsa ocellata* Rabenhorst 1863

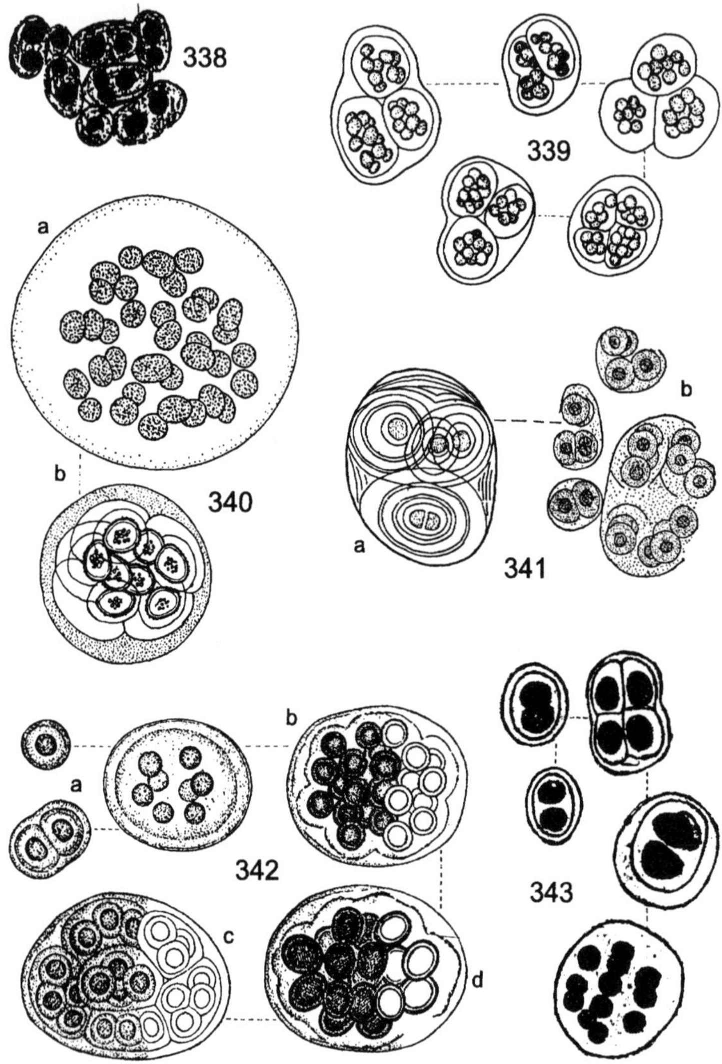


GLOEOCAPSA
Several species repeatedly identified in floristic lists, the taxonomic identity of which is unclear:

**Fig. 338.** *Gloeocapsa aurata*: after Frémy (1930) from Africa.

**Fig. 339.** *Gloeocapsa conglomerata*: after Kützing from Geitler (1925).

**Fig. 340.** *Gloeocapsa gigas*: after W. & G.S. West from Geitler (1932), Antillean

- *Gloeocapsa opaca* Nägeli 1849 (considered usually as var./f. of «*Gloeocapsa magma*» = *Gloeocapsopsis magma*, however, according to drawings of later authors, it belongs to the g. *Gloeocapsa*; fig. **342**)
- *Gloeocapsa opaca* var. *pellucida* Nägeli 1849
- *Gloeocapsa palea* Kützing 1843
- *Gloeocapsa paroliniana* Wille 1925
- *Gloeocapsa peniocystis* Kützing 1846 (Daily 1942 combined this species under the name «*Anacystis peniocystis* (Kützing) Drouet et Daily», which is probably a special species of the g. *Gloeothece*, described from USA.
- *Gloeocapsa punctata* (sensu Hegewald et al. 1975 from plankton in Hungary !; fig. **309e**)
- *Gloeocapsa purpurea* Kützing 1846
- *Gloeocapsa quaternata* (Brébisson) Kützing 1846 (see *G. rupicola*, p. 252)
- *Gloeocapsa rosea* Kützing 1846
- *Gloeocapsa rubicunda* Kützing 1843
- *Gloeocapsa sanguinolenta* Kützing 1843
- *Gloeocapsa saxicola* Wartmann in Rabenhorst 1859
- *Gloeocapsa scopulorum* Nägeli 1849
- *Gloeocapsa squamulosa* Brébisson in Kützing 1846 (marine)
- *Gloeocapsa stegophila* (Itzigsohn) Rabenhorst 1863 (possibly identical with *G. heamatodes*; fig. **343**)
- *Gloeocapsa stillicidiorum* Kützing 1843
- *Gloeocapsa thermalis* Kützing 1843
- *Gloeocapsa versicolor* Nägeli in Kützing 1849
- *Gloeocapsa* sp. (in Croasdale 1973 sub *Anacystis montana*; Ellesmere Isl., Canadian Arctic)
- *Gloeocapsa* sp. sensu Čado 1958 (probably *Gloeocapsopsis* sp.)
- Numerous *Gloeocapsa* species are known as phycobionts of lichens; the specificity of these taxa should be studied in future (comp., e.g., Büdel 1987, and others).

## 3. **Chondrocystis** Lemmermann 1899

Colonies free-living, more or less irregular or irregular-polygonal in outline, packet-like, gelatinous, laterally usually flattened or (in masses) slimy granular, later macroscopic, metaphytic or covering the substrate with a mucilaginous layer up to several cm thick, composed of numerous partial subcolonies, sometimes with precipitates (calcium carbonate particles). Subcolonies (small partial groups) surrounded by more or less cubic or polygonal flattened, widened, colourless, delimited slime, cells with individual thin, gelatinous, slightly stratified envelopes, usually following the spherical cell outline. Envelopes are partially thickened.

---

material corresponding rather to *Aphanothece* (subg. *Cyanogastrum*); **b** – resting stages (?).
**Fig. 341.** Two various *Gloeocapsa*-like concepts of «*G. magma*», differing from the original description: **a** – after Lemmermann from Geitler (1925), **b** – after G.M. Smith (1950).
**Fig. 342.** *Gloeocapsa opaca*: from Hollerbach & al. (1953, sub *G. magma* f. *opaca*), **d** – resting stages.
**Fig. 343.** *Gloeocapsa stegophila*: after Wille from Geitler (1932, redrawn from the original exsiccate of Itzigsohn).

Cell division not described, probably in 3 (more ?) planes in succeeding generations. Reproduction by small subcolonies after colony disintegration.
**Important references:** Geitler (1942); Kondrateva & al. (1984, sub *Chlorogloea sarcinoides*); Komárek & Anagnostidis (1995).

**Key to the species:**

1a Colonies compact, compound and confluent, hemispherical or subspherical; cells 1.5–3 µm in diameter; subaerophytic and in alpine streams . . . . . . . . . . . . . . . . . . . . . . . . . . . . . . . . . **1. C. dermochroa**

1b Colonies more or less packet-like, later irregular, amorphous; cells 3.5–6 µm in diameter; in shallow inland saline waters . . **2. C. sarcinoides**

**1. Chondrocystis dermochroa** (Nägeli) Komárek et Anagnostidis 1995 (fig. **344**)

*Gloeocapsa dermochroa* Nägeli 1849

Colonies microscopic, rarely macroscopic, more or less spherical, hemispherical or oval, composed of very densely packed, small subcolonies, which are confluent in the center of large colonies. Subcolonies spherical, subspherical to polygonal rounded, also composed of groups of clustered cells, enveloped more or less narrowly by firm, later diffluent, not or slightly lamellate yellow or yellow-brown sheaths, not much widened, to 8 or more µm in diameter. The whole, more or less spherical colony has the appearance of densely packed, small spherical clusters of cells; inside the colony, the envelopes disappear and colonies have an *Aphanocapsa*-like organization. Cells spherical, blue-green, 1.5–3 µm in diameter. Resting stages with firm, thin, dark yellow-brown sheaths.
Occurrence: Subaerophytic, usually in limestone areas, on rocks with periodically flowing water, in stony littoral splash zone of lakes, rarely in the water-level zone of streams; common in central Europe, further distribution not known (taxonomic identity of floristic data from different other regions is questionable).

**2. Chondrocystis sarcinoides** (Elenkin) Komárek et Anagnostidis 1995 (fig. **345**)

*Oncobyrsa sarcinoides* Elenkin 1923; *Chlorogloea sarcinoides* (Elenkin) Troickaja 1928

Young colonies microscopic, packet-like, sometimes slightly flattened, discoid, attached to the substrate, later macroscopic, gelatinous, irregular, more or less

CHONDROCYSTIS

scheme of
colony structure

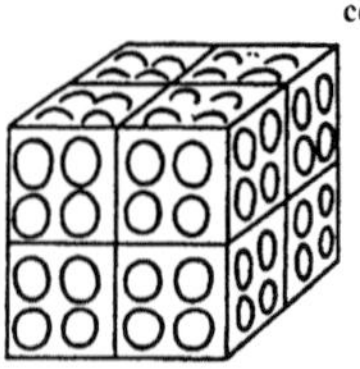

KOSINSKAJA
from KONDRATEVA & al. 1984

**Fig. 344.** *Chondrocystis dermochroa*: **a** – young colonies, **b–c** – parts of large colonies, **d** – old colonies; **a, b, d** – after Geitler (1925, sub *Gloeocapsa dermochroa*), **c** – after Starmach (1966, sub *Gloeocapsa dermochroa*).
**Fig. 345.** *Chondrocystis sarcinoides*: **a** – solitary cells, **b** – old colonies, **c** – packet-

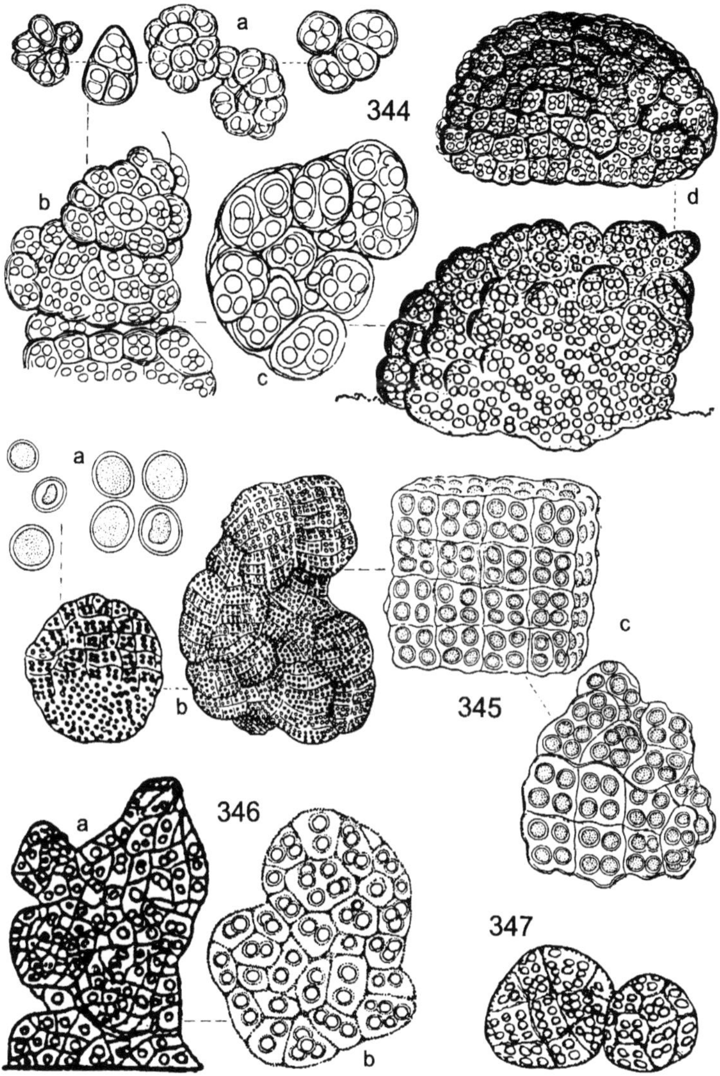


like colonies with typical tetradoid arrangement of cells; after Kosinskaja from Kondrateva & al. (1984, sub *Chlorogloea sarcinoides*).

**Fig. 346.** *Chondrocystis schauinslandii* (type species, described from Hawaii islands): two colonies; after Lemmermann from Geitler (1942).

**Fig. 347.** *Chondrocystis* sp.: after Wille from Geitler (1925, sub *Chroococcus rechingeri*, from Salomon islands).

amorphous or gelatinous-granular strata, blue-green, olive-green or brownish, to 2 cm high; sometimes disintegrating into subcolonies, 0.3–2 (4) mm in diameter. Mucilaginous envelopes colourless, around cells special thin slimy layers. Cells spherical, rarely slightly elongate, blue-green or yellow-green, 3.5–6 µm in diameter; individual envelope more or less 1–3 µm wide; cells arranged in the colony more or less in perpendicular rows (origin of sarcinoid packets).
Occurrence: In shallow inland saline localities, in littoral, benthic, on wet mud, sometimes small colonies secondarily free-floating; Ukraine, Russia, central Asia, E. Asia.

**Species outside Europe:**
- *Chondrocystis schauinslandii* Lemmermann 1899: Oceanic coastal lagoons (Hawaii Islands); type species (fig. **346**).
- *Chondrocystis bracei* Howe 1924: Marine coastal waters (Bahamas, Bermuda Islands).

**Unrevised species:**
- *Chroococcus rechingeri* Wille 1914: Cells oval, 1.8–2.5 µm in diameter; among mosses, Solomon Islands; (fig. **347**).

---

EUCAPSIS

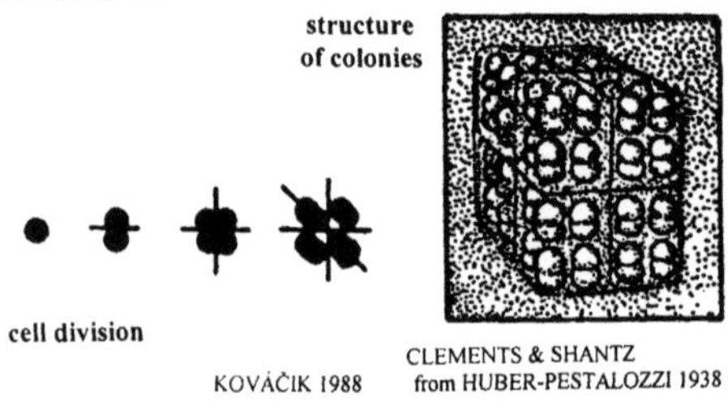


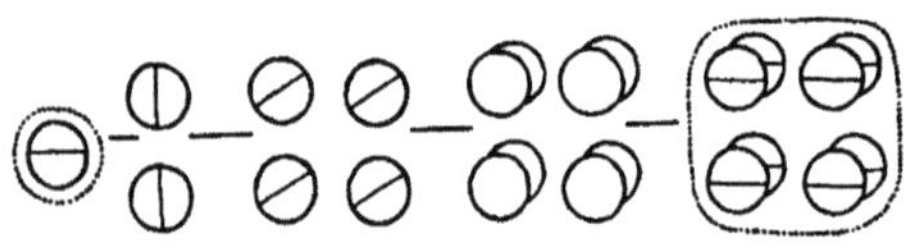

**Fig. 348.** *Eucapsis starmachii*: after Komárek & Hindák (1989).
**Fig. 349.** *Eucapsis parallelepipedon*: **a** – iconotype after Schmidle (1902, sub *Chroococcus parallelepipedon*) described from tropical Africa, **b** – after Komárek & Hindák (1989, sub *Eucapsis* cf. *parallelepipedon*) from Pannonian region.
**Fig. 350.** *Eucapsis minor*: **a** – after Skuja (1926, sub *E. alpina* var. *minor*), **b** – after Siemińska (1956, sub *E. alpina* var. *minor*), **c** – from Chu & al. ed. (1992, sub *E. alpina* var. *minor*), **d** – after Komárek & Hindák (1989).
**Fig. 351.** *Eucapsis terrestris*, European material from the volcanic island Santorini (Thera, Aegean Sea): after Anagnostidis & Economou-Amilli (1978).

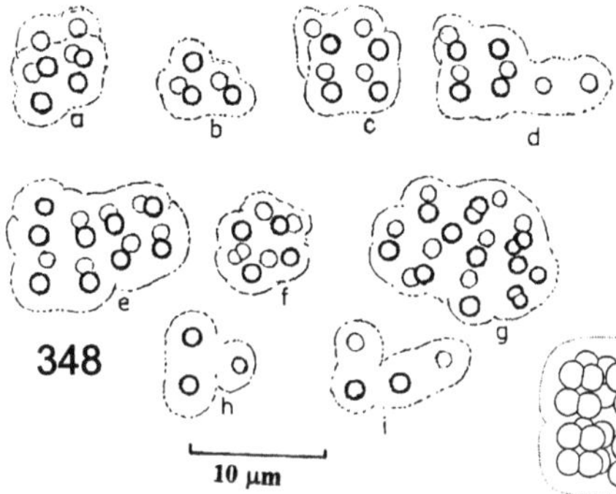


348

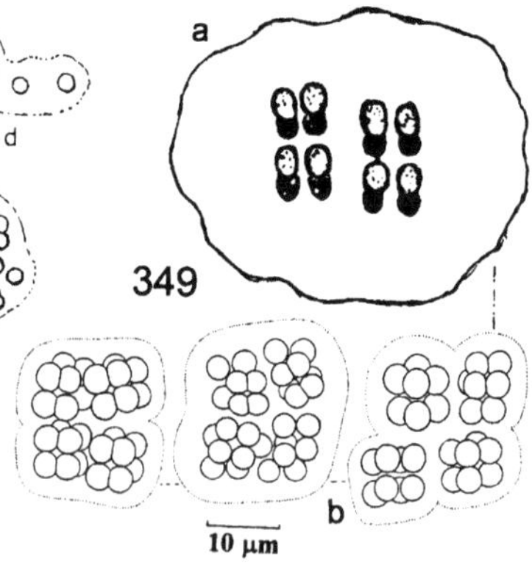


349

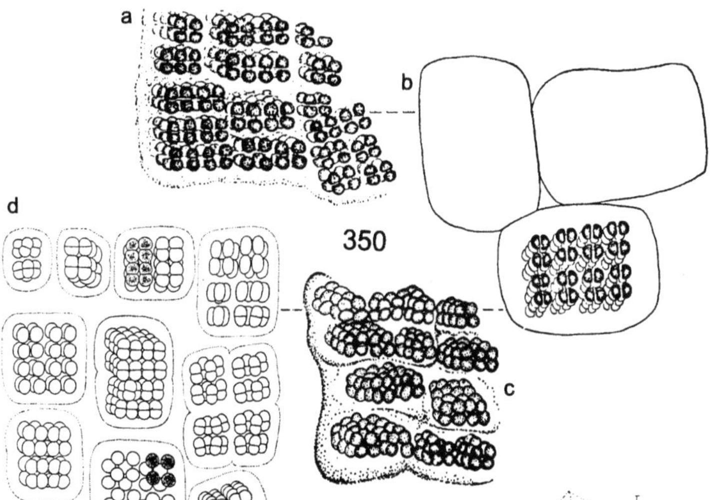


350

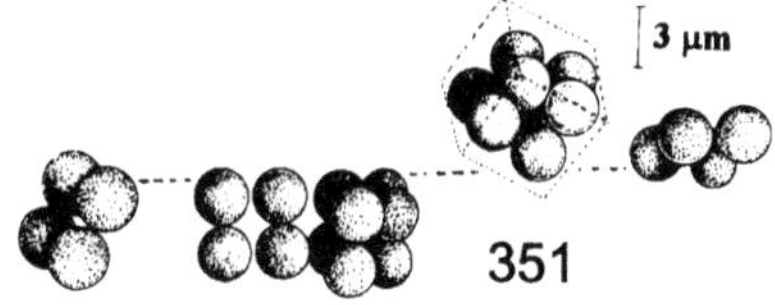


351

## 4. **Eucapsis** Clements et Shantz 1909

Colonies free-living, usually in metaphyton or rarely on soil, cubic, with cells arranged three-dimensionally in perpendicular rows; sometimes large colonies composed of subcolonies; colonies small, few-celled or multicellular with up to more than 128 cells; rows in old colonies sometimes a little disturbed. Mucilage colourless, hyaline, delimited, rarely diffluent (sometimes almost invisible); special gelatinous envelopes around cells are absent or indistinct, facultative. Cells spherical or widely oval before division.

Cells divide regularly in three perpendicular planes in successive generations without displacement of daughter cells (this type of division causes development of regular cubic colonies); cells grow into the original spherical form before next division. Reproduction by colony disintegration.

**Important references:** Geitler (1932, 1942, 1960); Hollerbach & al. (1953); Bourrelly (1961, 1970); Akiyama (1965); Starmach (1966); Anagnostidis & Economou-Amilli (1978); Compère (1981); Komárek & Anagnostidis (1986); Komárek & Hindák (1989).

**Key to the species:**

**1a** Cells less than 4 µm in diameter . . . . . . . . . . . . . . . . . . . . . . . . . . . . . . **2**
**1b** Cells usually more than 5 µm in diameter . . . . . . . . . . . . . . . . . . . . . . . . . **5**
**2a** Cells less than 2 (2.4) µm in diameter . . . . . . . . . . . . . . . **1. E. starmachii**
**2b** Cells 2–5 µm in diameter . . . . . . . . . . . . . . . . . . . . . . . . . . . . . . . . . . . . **3**
**3a** Cells pale blue-green, 2–4 (5) µm in diameter; more or less thermophilic species . . . . . . . . . . . . . . . . . . . . . . . . . . . . . . . . . . . . . . . . . . . . . **4**
**3b** Cells bright blue-green, usually 2–3.2 µm in diameter; psychrophilic species (moors, peaty bogs, etc.) . . . . . . . . . . . . . . . . . . . . . . . . . **3. E. minor**
**4a** Submersed in slightly alkaline swamps, usually in tropics and warm regions of temperate zones; cells 2–4 µm in diameter, colonies with 4–64 cells . . . . . . . . . . . . . . . . . . . . . . . . . . . . . . . . . . **2. E. parallelepipedon**
**4b** Terrestrial, known from volcanic soils; cells 2.5–5 µm in diameter, colonies with four to many cells . . . . . . . . . . . . . . . . . . . . . . . . **4. E. terrestris**
**5a** Cells bright blue-green or olive-green, sometimes slightly elongate, (3.5) 5–7.2 µm in diameter . . . . . . . . . . . . . . . . . . . . . . . . . . **5. E. alpina**
**5b** Cells greyish or pale blue-green, spherical, (5.2) 7–11 µm in diameter . . . . . . . . . . . . . . . . . . . . . . . . . . . . . . . . . . . . . . . . **6. E. carpatica**

### 1. **Eucapsis starmachii** Komárek et Hindák 1989 (fig. **348**)

*Chroococcus parallelepipedon* sensu Hansgirg 1892

Colonies small, delicate, somewhat irregular-cubic, usually with small number of cells, (1–3) 4–16 (32), which are more or less distant from one another (more than cell diameter), in short rows, sometimes with small irregularities; sometimes composed of several subcolonies. Mucilage colourless, diffluent. Cells spherical, 1–2.4 µm in diameter, with homogeneous, pale grey-blue content.

Occurrence: Freshwater, in acidic swamps, moorland and peaty bogs from lowlands to mountains in central Europe (Giant Mts., Šumava Mts., Carpathians), sporadic; probably over the whole European temperate zone.

### 2. **Eucapsis parallelepipedon** (Schmidle) Komárek et Hindák 1989 (fig. **349**)

*Chroococcus parallelepipedon* Schmidle 1902; *Eucapsis alpina* sensu auct.

Colonies more or less small, regular cubic, mainly with (4) 8–32 (64) more or less densely packed cells, enclosed within homogeneous, colourless, hyaline, sometimes diffluent mucilage. Cells spherical or slightly elongate, oval, pale blue-green, with homogeneous content, 2–4 µm in diameter to 6 × 3–4 µm before division. Occurrence: Freshwater, in tropical and probably also in warm areas of temperate zones, submerged in slightly alkaline swamps, pools, backwaters and lakes, often with water plants; there are records from whole of tropical Africa and from Iran (sometimes as «*E. alpina*»), those from S. and central Europe are not well documented, but they are probable.

### 3. **Eucapsis minor** (Skuja) Elenkin 1933 (fig. **350**)

*Eucapsis alpina* var. *minor* Skuja 1926

Colonies usually with large number of cells (32–64–128 and more), always regular cubic (slight disturbances in regularity occur only in colonies with large number of cells); sometimes composed of subcolonies. Cells spherical or hemispherical, before division slightly elongate, densely arranged in cubic colonies, bright blue-green, 2–3.2 (4) × 1.5–3.2 µm.
Occurrence: Freshwater, psychrophilic, cold stenotherm, metaphytic in moorland waters, peaty bogs, raised bogs, often in mountains, recorded also from winter plankton of clear lakes; temperate zones, but probably cosmopolitan in ecologically corresponding biotopes.

### 4. **Eucapsis terrestris** Akiyama 1965 (fig. **351**)

Colonies small, more or less regularly cubic, rarely slightly irregular, sometimes aggregated together, usually with (1) 4–16–64 (rarely to more than 128) densely packed cells (arranged in regular perpendicular rows). Mucilage delicate, diffluent, scarcely visible. Cells spherical, hemispherical or widely oval before division, pale blue-green with visible chromatoplasma, 2.5–3 (to 4.8 before division) µm in diameter.
Occurrence: Terrestrial, soils in volcanic ash regions; Greece (Thera Island – Mikra Kammeni), Japan (Okayama pref.).

### 5. **Eucapsis alpina** Clements et Shantz 1909 (fig. **352**)

*Merismopedia cubica* W. West 1912

Colonies regular cubic, usually small, with (2)8–16(32), rarely with numerous cells, mainly more or less grouped together, sometimes composed of small, 4–8-celled subcolonies, with colourless, unstructured, usually diffluent mucilage. Cells spherical or widely oval, pale or bright blue-green or olive-green, (3.5) 5–7.3 µm in diameter (before division to 6.7 × 5.2 µm). Varieties and forms were described with many cells and cell diameters 4.7–10.3 µm (separate species ?).
Occurrence: Freshwater, probably cold stenotherm, metaphytic in swamps, peaty bogs, clear lakes and moorland acidic waters; in temperate and subpolar zones, mainly in mountains, in Europe probably with boreo-alpine distribution; records from Africa are possibly correct if from high mountain biotopes, all others (from subaerophytic or thermal habitats, etc.) need confirmation.

### 6. **Eucapsis carpatica** Komárek et Hindák 1988 (fig. **353**)

Colonies exactly and regularly cubic, usually small, with 8–16 cells, but sometimes joining together and forming aggregates of subcolonies. Mucilage fine, colourless, distinct or diffluent. Cells spherical, only before division slightly widely oval, pale grey-blue, rarely pale blue-green, with sparsely granulate content, (5.2) 7–11 µm in diameter.

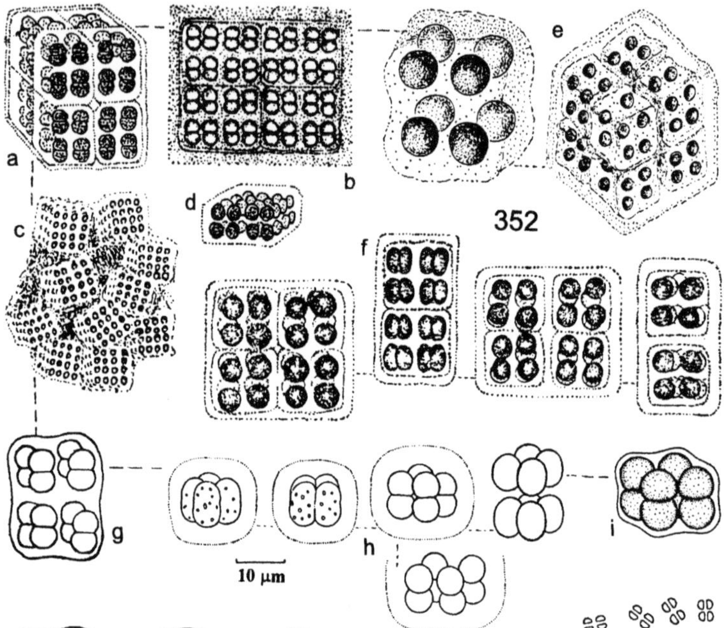


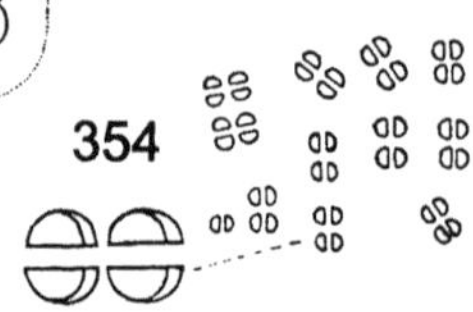

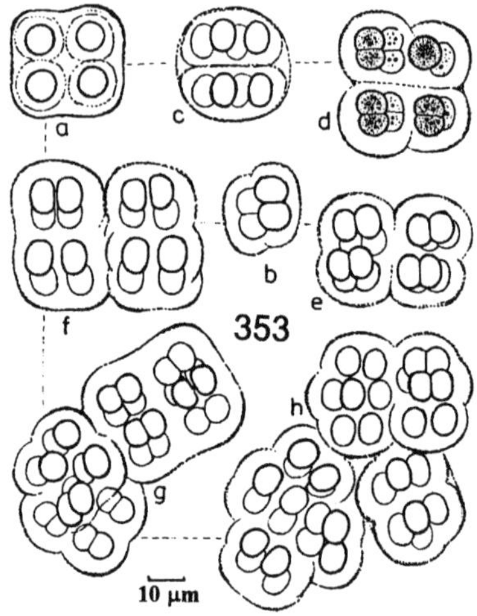


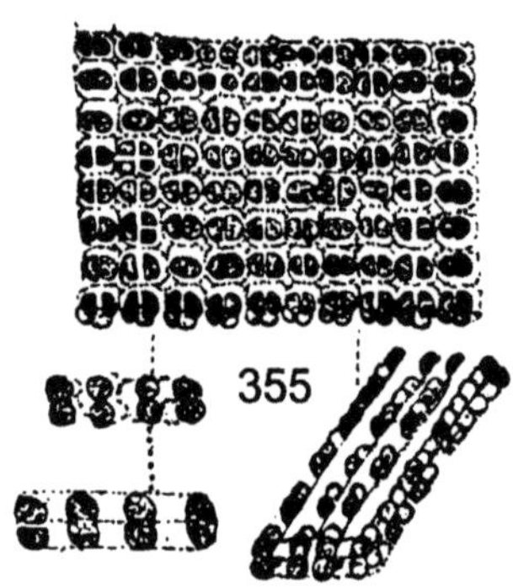

EUCAPSIS

**Fig. 352.** *Eucapsis alpina*: **a–f** – typical cubic colonies or compound colonies, corresponding with original description of Clements & Shantz, **a–b** – part of original drawings of Clements & Shantz from G.M. Smith (1933); **c–d** – after Huber-Pestalozzi (1938), **e** – after Bourrelly (1970), **f** – after Kosinskaja from

Occurrence: Freshwater, metaphyton in raised bogs, pools; Slovakia (W. Carpathians).

**Species outside Europe:**

- *Eucapsis himalayensis* M. Watanabe et Komárek 1994: Packet-like colonies with more or less distant cells from one another, cells blue-green, (3) 5–7.5 (10.5) × (3) 3.5–7.5 µm, mucilaginous colourless margin around cells 3–15 µm wide; described from central Himalaya region (Nepal, Sagarmatha Nat. Park).
- *Eucapsis minuta* Fritsch 1912: With small cells (1–1.5 µm) and large colonies, described from ice-waters of Antarctica (Fritsch 1912); record from Pakistan (Singh 1933; Desikachary 1959) obviously concerns another species; (fig. **354**).

**Unrevised taxa:**

- *Eucapsis alpina* var. *maior* Prescott 1965: With many-celled colonies (with up to several hundreds cells), described from Alaska, later homonym; (Prescott & Vinyard 1965).
- *Eucapsis alpina* var. *maior* Schwabe in Schwabe & Simonsen 1961: Cells 9–10 µm in diameter; epipsammic, from crater oasis in central Sahara (Libya).
- *Eucapsis alpina* f. *major* Poljanskij in Hollerbach et al. 1953: Colonies with 9.2–10.3 × 8.2 µm large cells, described from N. Russia.
- *Eucapsis alpina* f. *violacea* Thomasson 1952: With reddish cells, described from Sweden.
- *Eucapsis salina* Gonzalez-Guerrero 1948: Tabular, up to macroscopic colonies (to 8.4 mm in diameter, with sometimes over 1000 cells), in which cells grow in two layers, 4–5 mm in diameter, blue-green; planktic in marine waters near San Fernando, Spain (comp. *Microcrocis marina*, p. 188); (fig. **355**).
- *Eucapsis* sp. in Anagnostidis & Roussomoustakaki 1988: Taxonomically undefined species (microphoto) isolated from soils in mine deposits with high content of heavy metals, in Greece (Larymna, Lavrion).
- *Eucapsis* sp. (cf. *parallelepipedon*) in Komárek & Hindák 1988: Repeatedly found in backwaters and small ponds with submerged water plants in SW. Slovakia near Danube and Morava rivers.
- *Microcystis merismopedioides* Fritsch 1912: Spherical colonies with numerous cells (0.7 µm in diameter), arranged in perpendicular rows, distant from one another; described from South Orkneys, Antarctica.
- To the genus *Eucapsis* belongs probably also *Synechocystis* sp. sensu Castenholz & Waterbury (1989, Bergey's Manual): High GC-cluster, mainly with phycocyanin, facultatively motile cells, in one strain was found division in three planes.

---

Hollerbach & al. (1953); **g–i** – few-celled colonies, representing the initial stages or a special morphotype of *E. alpina* (separate species?), **g** – after Siemińska (1967), **h** – after Komárek & Hindák (1989), **i** – after Thompson (1938).

**Fig. 353.** *Eucapsis carpatica*: after Komárek & Hindák (1989).

**Fig. 354.** *Eucapsis minuta*: after Fritsch (1912).

**Fig. 355.** *Eucapsis salina*: after Gonzalez Guerrero (1946).

# 5. Family **Chroococcaceae** Nägeli 1849

*Gloeocapsaceae* Elenkin 1938 p.p. sine typo

Coccal, colonial; cells rarely solitary, mainly joined irregularly or more or less packet-like into mucilaginous, usually microscopic, few-celled, rarely many-celled, irregular-spherical colonies or groups, rarely in short rows, clustered densely together or (rarely) distant from one another. Colonies sometimes aggregated into macroscopic mats and layers. Modifications of cell arrangements in old colonies (resulting from the dominating type of division in later generations, e.g., shortly serially, radially, in cubic packets, etc.) are usable for generic definitions. Mucilage fine, diffluent or firm, stratified, gelatinous, mainly forming distinctly delimited envelopes or sheaths. Cells spherical, oval, hemispherical, rounded-polyhedral or irregular.
Cells divide successively in three or more different planes, later forming small, packet-like colonies, and reach the original shape before next division, but not into a spherical form; nanocytes lacking (taxonomic position of species in which nanocytes were mentioned is debatable).

**Key to the genera:**

**1a** Cell division more or less regular in first 2–3 generations, sometimes perpendicularly one to another («*Chroococcus*-like»), later irregularly; old colonies always with irregular arrangement of cells . . . . . . . . . . . . . . . . . . . . . . . . 2
**1b** Cells divide irregularly, but in later generations in one dominant way: this type of division results in a characteristic organization of cells in old colonies (in rows, radially arranged, regularly in cubic formations, etc.) . . . . . . . . 4
**2a** Mucilaginous envelopes colourless, usually slightly enlarged, sometimes not sharply delimited . . . . . . . . . . . . . . . . . . . . . . . . . . . . . **3. Chroococcus**
**2b** Mucilaginous envelopes (sheaths) coloured (at least in old stages), thin and firm, distinct, usually sharply delimited, very rarely slightly diffluent . . . 3
**3a** Colonies small, with one or several irregularly arranged, in outline more or less irregular-rounded cells; only in several species later sometimes more or even many cells accumulated in the center of a spherical colony; whole mucilaginous surface of colonies or a part thereof has granular consistency, covered by densely or sparsely dispersed small gelatinous protuberances or warts . . . . . . . . . . . . . . . . . . . . . . . . . . . . . . . . . . . . **1. Asterocapsa**
**3b** Colonies later many-celled, with densely packed, irregularly arranged, irregular-shaped cells, with either each cell or small groups of cells surrounded by individual, narrow, firm and usually distinct sheaths; mucilage smooth, but usually stratified and coloured . . . . . . . . . . . . . . . . . . **2. Gloeocapsopsis**
**4a** Colonies spheroid, without any mucilaginous stalks . . . . . . . . . . . . . . . 5
**4b** Cells localized at ends of thick, mucilaginous stalks . . . . . **8. Cyanostylon**
**5a** Colonies always microscopic, more or less oval or spherical, later with cells arranged randomly in two (rarely more) irregular, perpendicular rows; cells slightly distant from one another . . . . . . . . . . . . . . . . . **4. Cyanokybus**
**5b** Colonies microscopic, but later sometimes irregularly grouped together into amorphous, macroscopically visible conglomerations; cells packed together, in loose (rarely) or close contact with one another . . . . . . . . . . . . . . . . 6
**6a** Cells arranged in more or less regular, three-dimensional packets in old colonies; in somewhat perpendicular rows . . . . . . . . . . **5. Cyanosarcina**
**6b** Cells arranged more or less radially, cruciately or wreath-like in older colonies . . . . . . . . . . . . . . . . . . . . . . . . . . . . . . . . . . . . . . . . . . . 7
**7a** Cells arranged radially in developed colonies . . . . . . . . . . **6. Pseudocapsa**
**7b** Cells arranged more or less cruciately or wreath-like . . . **7. Nephrococcus**

## 1. **Asterocapsa** Chu 1952

Cells solitary, in few-celled groups or in more or less spherical, microscopic colonies, forming granular mats. Cells or their groups enveloped by distinct, delimited, usually firm sheaths, which are colourless or coloured by sheath pigments (reddish, yellow, etc.), and covered by irregularly, but evenly distributed, conical or rounded wart-like processes and forming granular or spiny surface of their colonies. Cells subspherical, oval or irregular in outline, sometimes slightly elongated or polygonal-rounded. Common envelopes thin or thick, with the surface covered by minute or stout, wart-like projections of different length.

Cells divide by irregular binary fission within their sheaths, in different planes in successive generations; the divided cells are liberated occasionally from splits in the firm envelopes. Within the genus were recognized three types of life cycle (probably on a subgenus or genus level): (i) Cells solitary or in small, 2–4- to few-celled colonies, enveloped only by firm sheath, which splits after cell division. (ii) Cells form many-celled clusters, enveloped by firm, more or less spherical and, on the surface, spiny sheaths; within colonies sometimes with nanocyte-like cell division; the large colonies gelatinize or disintegrate, and daughter cells are liberated with their own spiny sheaths. (iii) Irregular-shaped cells, enclosed in spiny sheaths, are localized in an amorphous gelatinous mass.

**Important references:** Gardner (1927a; sub *Anacystis* p.p.); Chu (1952); Chu & al. ed. (1991); Komárek (1993).

**All still described species are non-European (several unrevised):**

- *Asterocapsa atrata* Liang 1987: Described from China; (Chu & al. ed. 1991).
- *Asterocapsa badia* Komárek 1993: With dark, chocolate-brown sheaths, described from Himalayas; type (i) (fig. **357**).
- *Asterocapsa changbaishanensis* Wang 1985: Described from China; (Chu & al. ed. 1991).
- *Asterocapsa divina* Komárek 1993: Typical subaerophytic species of the above mentioned group (ii), described from limestone rocks in Mexico (San Luis Potosí); (fig. **359**).
- *Asterocapsa fujianica* Liang 1987: Described from China; (Chu & al. ed. 1991).
- *Asterocapsa gloeotheceformis* Chu 1952: Type species, described from China; (Chu 1952; Chu & al. ed. 1991); (fig. **356**).
- *Asterocapsa hyalina* Chu 1958 (with var. *rubra* Liang): Described from China (Szechwan); (Chu & al. ed. 1991).
- *Asterocapsa longipapilla* Liu 1985: Described from China: (Chu & al. ed. 1991).
- *Asterocapsa magnifica* (Gardner) Komárek 1993 [syn.: *Anacystis magnifica* Gardner 1927]: With hyaline sheaths, described from Porto Rico.
- *Asterocapsa masayuki-watanabei* Komárek 1993: With red sheaths, described from Himalayas; type (iii).
- *Asterocapsa pulchra* (Gardner) Komárek 1993 [syn.: *Anacystis pulchra* Gardner 1927]: With red sheaths, described from Porto Rico.
- *Asterocapsa purpurea* (Jao) Chu 1952 [syn.: *Anacystis ? purpurea* Jao 1944]: Described from China; (Chu & al. ed. 1991).
- *Asterocapsa rubra* Wang 1986: Described from China; (Chu & al. ed. 1991).
- *Asterocapsa sinica* Liang et Chen 1985: Described from China; (Chu & al. ed. 1991).
- *Asterocapsa trochiscioides* (Jao) Chu 1952 [syn.: *Anacystis ? trochiscioides* Jao 1944]: Described from China (Kwangsi, Szechwan); (Chu & al. ed. 1991); (fig. **358**).

356

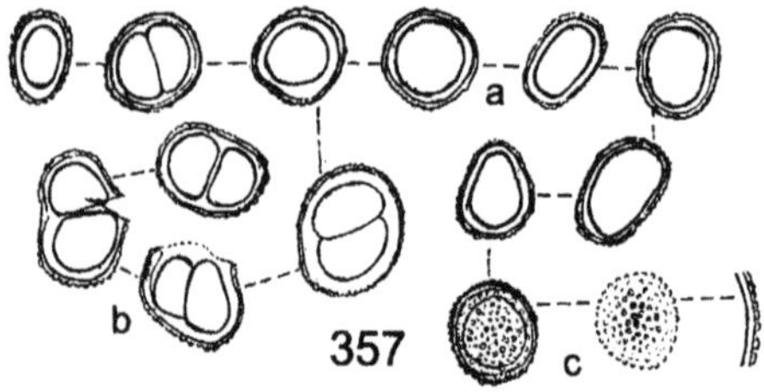
a
b
357
c

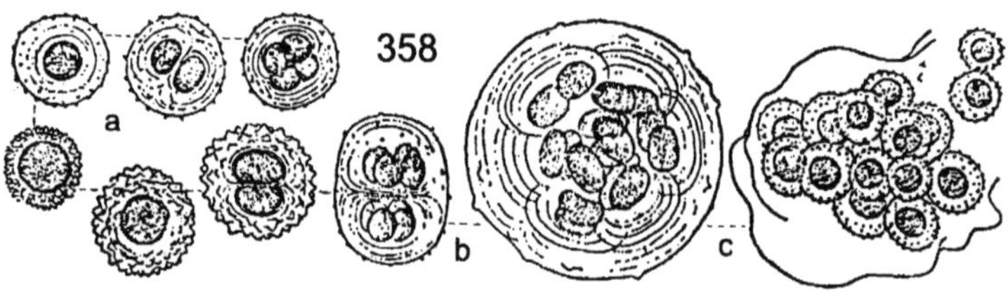
358
a
b
c

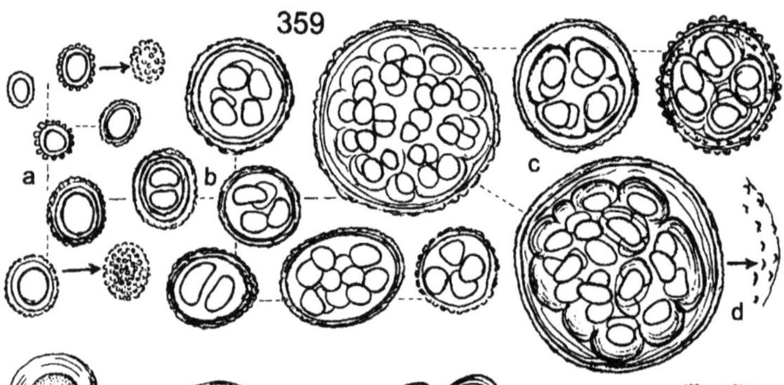
359
a
b
c
d

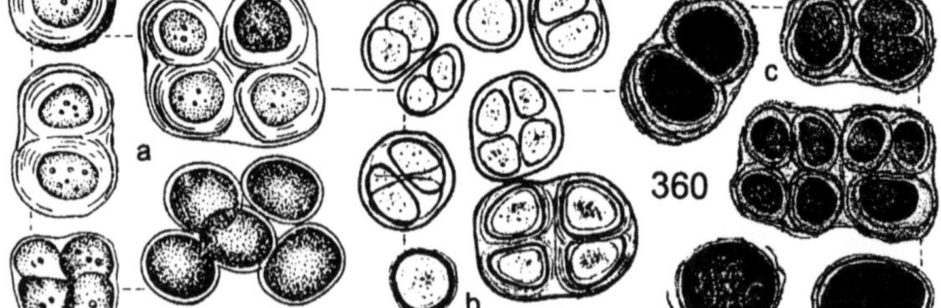
a
b
c
360

**Unrevised species:**
- *Anacystis amplivesiculata* Gardner 1927: Described from Porto Rico.
- *Chroococcidiopsis karnatakensis* Kamat 1962: Described from India.
- *Chroococcidiopsis spinosa* Kamat 1962: Described from India.
- *Gloeothece fusco-lutea* cf. var. *unilamellaris* Chu in Sant'Anna & al. 1995 from Atlantic supralittoral zone in Brazil (SP) is surely different from the Chu's type,

ASTEROCAPSA

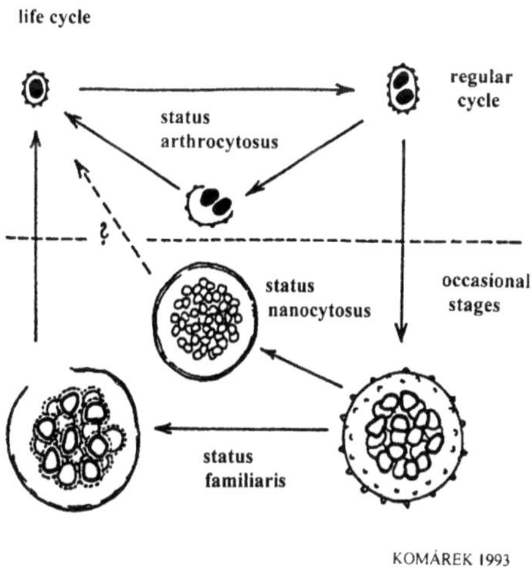


**Fig. 356.** *Asterocapsa gloeotheceformis*, type species from China; after Chu (1952).
**Fig. 357.** *Asterocapsa badia*, from Himalaya Mts.: species with simple life cycle and dark chocolate brown envelopes; after Komárek (1993).
Two species with full cycle:
**Fig. 358.** *Asterocapsa purpurea*, from China: after Jao (1944, sub *Anacystis? purpurea*).
**Fig. 359.** *Asterocapsa divina*, from Mexico: after Komárek (1993).

GLOEOCAPSOPSIS

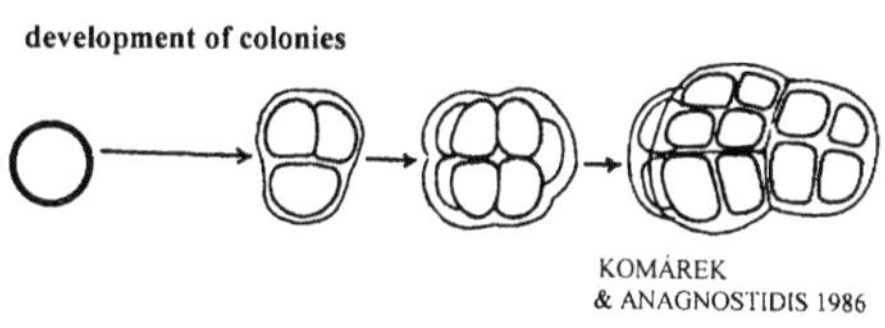


**Fig. 360.** *Gloeocapsopsis chroococcoides*: **a** – after Nováček from Hollerbach & al. (1953, sub *Gloeocapsa chroococcoides*), **b–c** – after Nováček (1934, sub *Gloeocapsa chroococcoides*).

and corresponds rather to the genus *Asterocapsa* according to the description of the Brazilian population.
– Several independent and taxonomically not yet defined species, corresponding to the genus *Asterocapsa*, were found on wet sandy soils (Ohtani & Kanda 1987, sub *Gloeocapsa* sp. 1–3) and wet rocks (Komárek in prep.) in Antarctic coastal regions (vicinity of Syowa Station, Arctowski Station, Fildes peninsula).

## 2. Gloeocapsopsis Geitler ex Komárek 1993

*Gloeocapsopsis* Geitler 1925 sine diagn.

Rarely solitary cells, mostly micro- or macroscopic, irregular, formless or granular colonies, composed of densely, irregularly aggregated cells or their small groups, surrounded by mucilaginous envelopes. Cells subspherical, more or less irregular-rounded in outline, sometimes slightly elongate (never spherical), enveloped by thin, narrow, clearly delimited, rarely slightly diffluent, sometimes feebly lamellate, often coloured sheaths (yellow, red, rusty red, etc.), usually following the cell outline.

Cell division irregularly in various planes in successive generations. Reproduction by liberation of divided and ensheathed cells from ruptured mother sheaths. Nanocyte production rarely observed in several species (?). Occasional presence of enlarged resting cells with thick, firm and usually intensely coloured envelopes.

**Important references:** Geitler (1925); Nováček (1929, 1934; sub *Gloeocapsa* p.p.); Komárek & Anagnostidis (1986); Komárek (1993).

**Key to the species:**

**1a** Mucilaginous envelopes (sheaths) colourless, violet or blackish, never yellowish, orange or red . . . . . . . . . . **2**
**1b** Envelopes yellow, orange or red (at least in part) . . . . . . . . . . **4**
**2a** Envelopes always colourless, cells intensely blue, 3–4 µm in diameter; on sandstone . . . . . . . . . . **1. G. cyanea**
**2b** Envelopes violet or blackish, cells over 4 µm in diameter . . . . . . . . . . **3**
**3a** Inner envelopes intensely violet, outer colourless, widened and diffluent; cells 5–9.2 µm in diameter . . . . . . . . . . **2. G. polyedrica**
**3b** Around cells only envelopes following the cell outline, colourless or blackish; cells 8–14 µm in diameter . . . . . . . . . . **3. G. chroococcoides**
**4a** Sheaths from yellowish to intensely yellow or yellow-brown . . . . . . . . . . **5**
**4b** Sheaths orange, rusty red or bloody red . . . . . . . . . . **6**
**5a** Sheaths colourless to yellowish, cells 4–8 µm in diameter, on marine coastal rocks . . . . . . . . . . **4. G. crepidinum**
**5b** Sheaths yellow to dark yellow-brown, cells 4.8–11 (20) µm in diameter, aerophytic on inland rocks . . . . . . . . . . **5. G. pleurocapsoides**
**6a** Sheaths rusty-red or orange-red, cells 3.5–11.5 (17.5) µm in diameter, aerophytic on serpentine, sandstone and volcanic rocks from lowlands to mountains . . . . . . . . . . **6. G. dvorakii**
**6b** Sheaths bloody red, cells 3–7 (18) µm in diameter, subaerophytic on wet rocks in high mountains . . . . . . . . . . **7. G. magma**

### 1. Gloeocapsopsis cyanea (Krieger) Komárek et Anagnostidis 1994 (fig. 366A)

*Gloeocapsa cyanea* Krieger 1944

Solitary cells or micro- to macroscopic, irregular, formless colonies, composed of aggregations of ensheathed cells or their small blackish blue-green groups, mostly

2–16- celled to many-celled. Cells in densely and irregularly or more or less packet-like arrangements, often compressed by mutual pressure, subspherical to spherical or more or less regular-rounded in outline or irregular. Cells bright blue-green, without envelopes 3–4 µm in diameter. Envelopes distinct, thin, narrow, following the cell outline, firm, colourless, not enlarged; inside layers scarcely visible, outside rough, sometimes slightly lamellate. Nanocytes and resting cells not observed.
Occurrence: Aerophytic, epilithic on damp or wet sandstone (sandy rubble) and on the entrance walls of caves (alkaline substrate), known only from the island Crete, Greece, probably wider distributed.

### 2. **Gloeocapsopsis polyedrica** (Ercegović) Komárek et Anagnostidis 1994

*Gloeocapsa alpina* (Nägeli) Brand var. *polyedrica* Ercegović 1925

Colonies amorphous, usually composed of few-celled (up to 14-celled) subcolonies, which are more or less spherical and up to 70 µm in diameter, later macroscopic, indefinite. Cells mostly obtuse polyedric, often oval, rarely spherical, 5–9.2 µm in diameter without envelope, 10–19 µm with envelope, blue-green. Common colonial mucilage enlarged, fine, structureless, colourless or slightly coloured, transparent; around cells narrow individual sheaths following the cell outline, distinct, delimited, homogeneous or two-layered, 2.5–5 µm wide, inside intensely violet to blackish and not transparent, outside slightly coloured to colourless.
Occurrence: Aerophytic, dispersed among other epilithic cyanoprokaryotes on limestone rocks; known only from Croatia (Samborski Ostrč, Plješevica).

### 3. **Gloeocapsopsis chroococcoides** (Nováček) Komárek 1993 (fig. **360**)

*Gloeocapsa chroococcoides* Nováček 1934

Colonies microscopic, composed of several subcolonies, pale green up to blackish blue, up to 200 µm in diameter, disintegrating into small groups of ensheathed cells. Cells more or less irregular-rounded, pale yellowish-green, with homogeneous content, 8–14 µm in diameter. Envelopes thin or more or less thick, following the cell outline, lamellate, colourless or blackish, never diffluent.
Occurrence: Aerophytic, on serpentine arid rocks: known only from Czech Republic (W. Moravia).

### 4. **Gloeocapsopsis crepidinum** (Thuret) Geitler ex Komárek 1993 (fig. **361**)

*Protococcus crepidinum* Thuret 1854; *Gloeocapsa crepidinum* (Thuret) Thuret in Bornet et Thuret 1876, not sensu Geitler 1932; *Pleurocapsa crepidinum* (Thuret) Ercegović 1930

Colonies large, up to macroscopic, gelatinous, irregular, dirty greenish. Cells irregular-spherical, (2.5) 4–8 µm in diameter, with thin, colourless and sometimes diffluent envelopes, which are yellowish or yellow-brown near the colonial surface, not lamellate; outer envelopes hyaline. Cell content pale blue-green or greyish, usually homogeneous.
Occurrence: Halophilic, epilithic, on stones and rocks or on wooden substrates, in the littoral and sprayed supralittoral of seas and inland saline lakes and swamps; cosmopolitan (?).

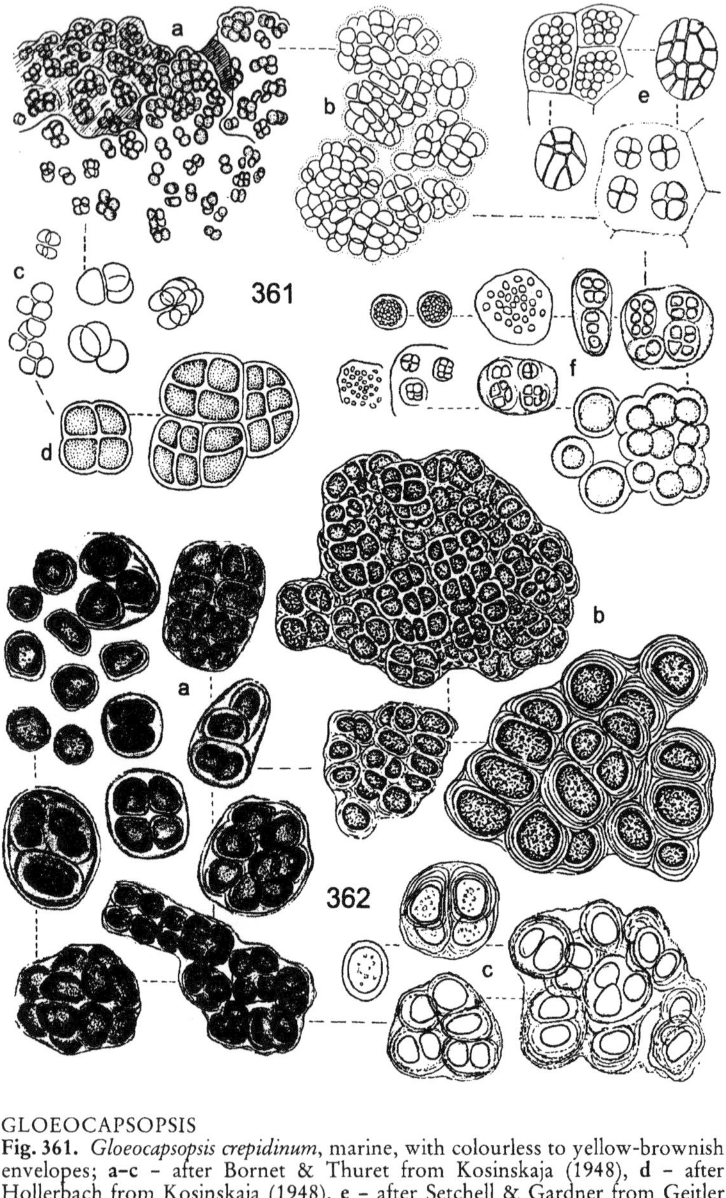


GLOEOCAPSOPSIS

**Fig. 361.** *Gloeocapsopsis crepidinum*, marine, with colourless to yellow-brownish envelopes; **a–c** – after Bornet & Thuret from Kosinskaja (1948), **d** – after Hollerbach from Kosinskaja (1948), **e** – after Setchell & Gardner from Geitler (19321), **f** – after Ercegović (1930), **e, f** – in part baeocyte formation; **a–e** – sub

**5. Gloeocapsopsis pleurocapsoides** (Nováček) Komárek et Anagnostidis 1986 (fig. **362**)

*Gloeocapsa pleurocapsoides* Nováček 1929

Colonies micro- or macroscopic, irregular, composed of aggregations of solitary, ensheathed cells or their groups, crustaceous, dark olive-green, brownish or blackish-brown, gelatinous (wet) or flat, squamulous (dry). Cells subspherical, more or less oval, ellipsoidal, irregularly rounded or of irregular outline, usually composed of subcolonies, with greyish blue-green, pale olive-green or blue-green, homogeneous or finely granular content, with differentiated chromato- and centroplasma, sometimes with small granules, (4)4.8–11(20) µm in diameter. Envelopes narrow, copying the cells, more or less lamellated, firm, not diffluent, yellowish or orange-brownish, inside more intensely, outer lightly coloured. Resting cells more or less spherical, irregular, yellow-brownish or brown, 5–11 µm in diameter.

Occurrence: Aerophytic on dry, temporary wetted rocks, rarely subaerophytic on wet rocks; known from Czech Republic (W. Moravia), several Asiatic localities (Burma – Mandalay, Himalayas – Khumbu region), and from Argentine (Tierra del Fuego).

**6. Gloeocapsopsis dvorakii** (Nováček) Komárek et Anagnostidis 1986 (fig. **363**)

*Gloeocapsa dvorakii* Nováček 1929

Solitary cells or groups of cells with simple, narrow (about 1–3 µm wide) gelatinous envelopes, usually aggregated into colonies. Colonies microscopic or later macroscopic, irregular, forming rusty, red-orange or dirty brownish-blackish spots on the stony substrate, crustaceous, thin, many-celled, to 140 µm in diameter. Cells irregularly spherical or oval, irregular and polygonal in outline with rounded ends or pear-shaped, pale blue-green with almost homogeneous content, sometimes with several scarcely distributed granules, 3.5–11.5 (17.7) µm in diameter. Envelopes dark rusty reddish, brownish or yellowish-brown (intensity of colouration very variable), sometimes indistinctly layered, never much widened, usually with yellow, very thin, delimited distinct sheath on the outer surface. The surface of envelopes is rarely slightly granular.

Occurrence: Aerophytic, xerotherm, on serpentine, volcanic and sandstone rocks, known from temperate zone, sometimes mixed with various other cyanoprokaryotic species; Czech Republic, Japan, probably more widely distributed and misinterpreted with various *Gloeocapsa* species.

---

*Gloeocapsa crepidinum*, **f** – sub *Pleurocapsa crepidinum*; however, populations **e** and **f** with baeocyte formation belong probably to a *Myxosarcina* species (see the identical fig. 556).

**Fig. 362.** *Gloeocapsopsis pleurocapsoides*, aerophytic, with yellow-brown envelopes: **a** – after Nováček (1934, from central Europe), **b** – after Skuja (1949, from Burma), **c** – after Komárek (1993, from Himalayas); **a–b** – sub *Gloeocapsa pleurocapsoides.*

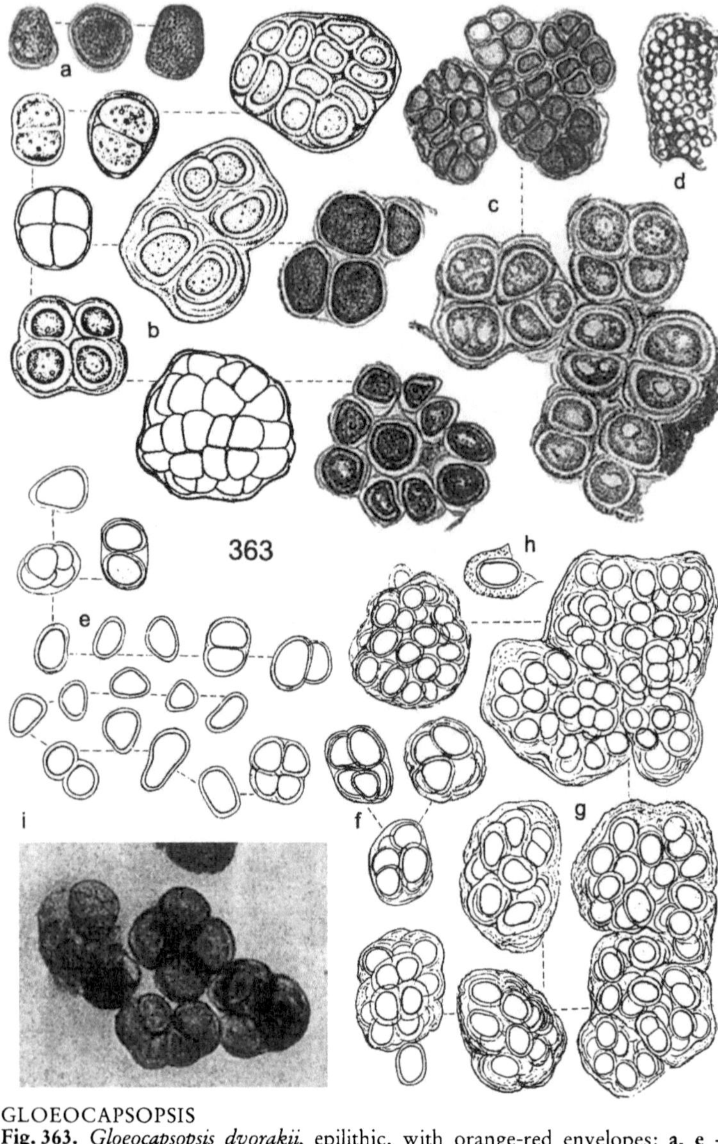


GLOEOCAPSOPSIS

**Fig. 363.** *Gloeocapsopsis dvorakii*, epilithic, with orange-red envelopes: **a, e** - solitary cells and initial stages, **b, f, i** - young colonies, **c, g** - colonies, **d** - nanocytic division, **h** - cell being liberated from the envelope; **a–d, i** - after Nováček (1929, populations from central Europe, sub *Gloeocapsa dvorakii*, partly redrawn), **e–h** - after Komárek (1993, populations from Japan).

**7. Gloeocapsopsis magma** (Brébisson) Komárek et Anagnostidis 1986 (fig. **364**)

*Protococcus magma* Brébisson in Brébisson et Godey 1836; *Pleurococcus magma* (Brébisson) Meneghini 1843; *Gloeocapsa magma* (Brébisson) Kützing 1846 sec. Hollerbach 1924; *Chroococcus simmeri* Schmidle in Simmer 1899; *Gloeocapsa magma* var. *simmeri* (Schmidle) Nováček ex Geitler 1932

Colonies more or less spherical or irregular, composed of numerous subspherical subcolonies, usually 30–60 µm in diameter, rarely larger, colourless, reddish or brownish, usually with a circa 1 µm thick, colourless, firm mucilaginous sheaths. Cells irregularly spherical, in outline polygonal-rounded, 3–7 (18) µm in diameter, with thin, 0.5–1.5 µm wide, coloured, individual envelopes and only sometimes also slightly widened, rarely indistinctly lamellate, envelope following the cell outline; never forming very wide, regularly lamellate envelopes. Cell content blue-green. In dried state, the envelope is thinner and intensely (dark) coloured, sometimes containing more than one cell in a single sheath. Nanocytous stages probably present. Enlarged cells without division were described as «resting cells» («Dauerzellen»), which, however, do not correspond to akinetes of filamentous types. Several described varieties [var. *pellucida* (Kützing) Nägeli and var. *opaca* (Nägeli) Kirchner ex Hansgirg] are probably only developmental stages (or morphological modifications or status).
Occurrence: Subaerophytic, on wet rocks, stony walls and roofs, usually in high mountains, in general not on alkaline substrates. Common in high mountains in Europe, possibly cosmopolitan in similar localities (also in Antarctica?). Records from warm regions or even from sea littoral are surely wrong.

**Unrevised species:**
- *Gloeocapsa* sp. sensu Čado 1958: Probably a *Gloeocapsopsis* species forming many-celled colonies (layers of 0.2 mm thick) on the limestone rocky shore of Ohrid lake (Montenegro); envelopes colourless or yellow-brown, cells 2.8–10.5 µm in diameter (various status were described); (fig. **365**).
- *Gloeocapsa* sp. sensu Guarrera & Ferrario (1979); (fig. **366**).
- Endolithic species (sub *Gloeocapsa* sp.) were depicted from hot desert Negev in Israel (Friedmann 1971, Golubić & al. 1981); comp. also *Chroococcus.*
- Epiphytic species, presented sub «*Gloeocapsa* sp. 3» from mosses, with cells 6–10 µm in diameter, reddish brown (content ?) and lamellate sheaths, from Antarctica (Ohtani & Kanda 1987).
- *Myxosarcina chroococcoides* sensu Bourrelly 1984: With cells 5–6 µm in diameter, colourless sheaths; probably a separate *Gloeocapsopsis* species; New Caledonia.

## 3. Chroococcus Nägeli 1849

Cells or groups of cells surrounded by mucilaginous envelopes, rarely solitary (initial stages), more commonly in microscopic, irregular or more or less spherical, gelatinous colonies, usually with a small cell number (2–16), rarely forming compound, many-celled agglomerations, never in rows or in regular arrangements. Cells form within multicellular colonies sometimes in packet-like, 2–8-celled groups. Mucilaginous envelopes colourless or yellowish, usually concentrically lamellate, often delimited, well-marked at the margin, only in certain species (subg. **Limnococcus**) irregularly disposed within homogeneous, amorphous, hyaline mucilaginous matrix; cells often surrounded by narrow individual envelopes, following the cell outline. Cells at first subspherical, widely oval or spherical, later hemispherical or in the form of a segment of the sphere. Cell

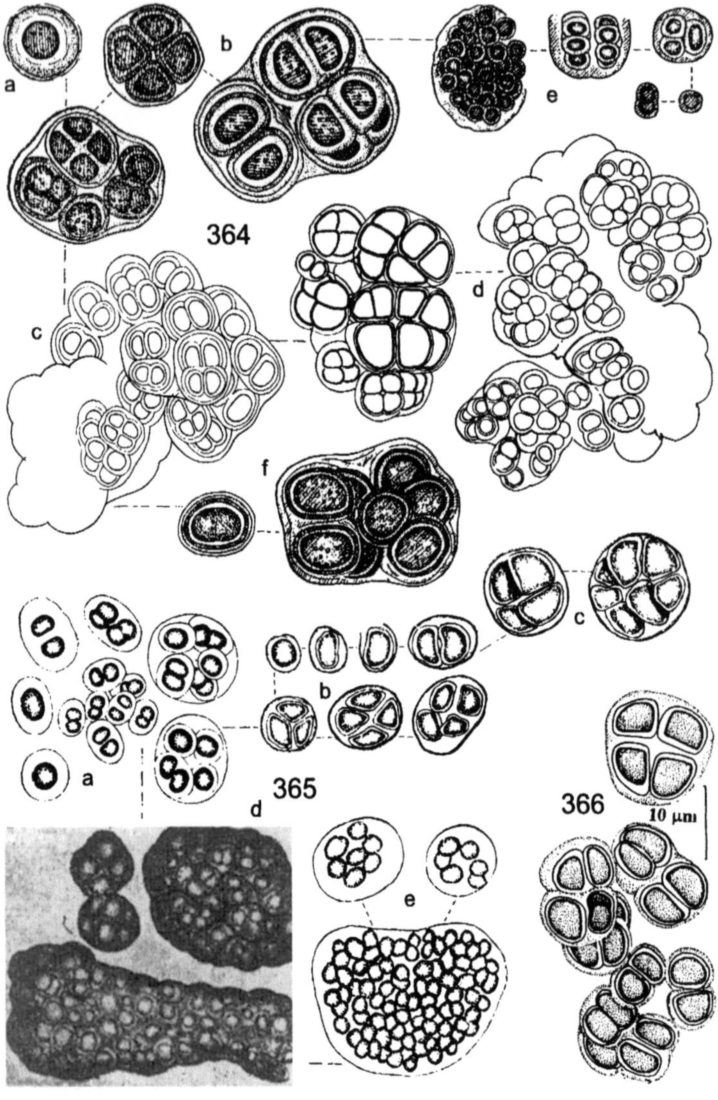


GLOEOCAPSOPSIS

**Fig. 364.** *Gloeocapsopsis magma*, subaerophytic in mountains, with dark red inner envelopes: **a** – solitary cell, **b** – young colonies, **c–d** – colonies, **e** – nanocyte formation, **f** – resting stages; **a–c, e–f** – after Hollerbach from different authors, **d** – after Geitler (1932).

content grey, grey-green, blue-green, olive-green, yellowish, orange, reddish or violet, homogeneous, granular, only in few planktic species with gas vesicles (visible aerotopes). Free-living, epiphytic or in periphyton, freshwater in ponds, lakes, moorland waters, rarely in thermal waters and aerophytic. In special (extreme) biotopes various other species undoubtedly exist.
Cell division in at least 3 planes in successive generations (usually perpendicular to one another), later sometimes irregular in old colonies. Cells divide by binary fission and do not reach the original shape before next division. Reproduction by cell division and by their liberation (or groups of cells) from colonies; rarely, nanocyte production.
**Important references:** Geitler (1932, 1942, 1960); Hollerbach in Elenkin (1938, sub *Gloeocapsa*); Skuja (1956, 1964); Komárek & Anagnostidis (1986); Kováčik (1988); Komárek & Novelo (1994).

**Key to the species:**

**1a** Mucilaginous envelopes enlarged, sometimes diffluent, not lamellate, in outline spherical or irregular; cells are in groups or separate after division, in colonies cells usually attain a spherical form before the next division, rarely the polygonal-rounded shape of cells remains (1. subg. **Limnococcus**) . . 2

Note: Diacritical features of the subg. *Limnococcus* (cells grow more or less into the original shape before next division) correspond rather to the family *Microcystaceae* (similarity with «*Aphanocapsa*-like» types, but with cells dividing in three planes in succeeding generations) and indicate a special genus within this family.

**1b** Mucilaginous envelopes follow the outline of cells more or less closely or that of cell groups, thickened, diffluent or delimited at the surface, sometimes lamellate; cells do not reach the original spherical form before the next division (2. subg. **Chroococcus**) . . . . . . . . . . . . . . . . . . . . . . . . . . 11
**2a** Cells with facultative aerotopes; planktic . . . . . . . . . . . . . . . . . . . . . . . . 3
**2b** Cells always without aerotopes . . . . . . . . . . . . . . . . . . . . . . . . . . . . 5
**3a** Cells up to 5 µm in diameter; old cells more or less spherical . . . . . . . . . 4
**3b** Cells 5–7 µm in diameter: cells remain polygonal rounded after division . . . . . . . . . . . . . . . . . . . . . . . . . . . . . . . . . . . . . . . **7. C. acumulatus**
**4a** Cells 1.5–2.7 µm in diameter . . . . . . . . . . . . . . . . . . . . **3. C. vacuolatus**
**4b** Cells 3–4 µm in diameter . . . . . . . . . . . . . . . . . . . . . **5. C. planctonicus**

---

**Fig. 365.** *Gloeocapsopsis* sp.: **e** – nanocyte formation; after Čado (1958, littoral of Ohrid lake, sub *Gloeocapsa* sp.).
**Fig. 366.** *Gloeocapsopsis* sp.: after Guarrera & Ferrario (1979; sub *Gloeocapsa* sp.).
**Fig. 366A.** *Gloeocapsopsis cyanea*, envelopes colourless: after Krieger (1944, sub *Gloeocapsa cyanea*).

**366A**

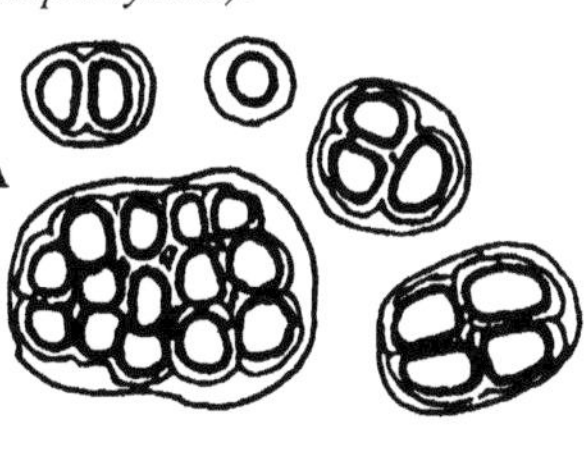

**5a** Cells remain in groups after division, irregularly distributed, not regularly distant from one another; in many-celled colonies; cells up to 3 µm in diameter . . . . . . . . . . **6**
**5b** Cells separate after division, usually all distinctly distant one from another, in colonies enveloped by enlarged, structureless mucilaginous envelopes; not in many-celled colonies; over (1.7) 2 µm in diameter . . . . . . . . . . **7**
**6a** Cells 0.7–1 µm in diameter; colonies usually with many irregularly clustered cells and/or forming groups of cells in a single colony **1. C. microscopicus**
**6b** Cells 1.4–2.3 µm in diameter; colonies with separate cell groups, which are usually distinctly distant from one another . . . . . . **2. C. aphanocapsoides**
**7a** Cells regularly arranged in more or less cubic formations, 5–9 µm in diameter; after division remaining irregular («rounded polygonal») in shape; mucilaginous envelopes lamellate . . . . . . . . . . **8. C. prescottii**
**7b** Cells arranged irregularly within colonies; usually attaining a spherical shape before next division; mucilaginous envelopes not lamellate . . . . . . . . . . **8**
**8a** Cells 1.7–4.5 µm in diameter . . . . . . . . . . **9**
**8b** Cells over (5) 6 µm in diameter . . . . . . . . . . **10**
**9a** Cells 1.7–3 µm in diameter . . . . . . . . . . **4. C. minimus**
**9b** Cells 3–4.5 µm in diameter . . . . . . . . . . **6. C. dispersus**
**10a** Cells 6.5–7 µm in diameter, usually without individual envelopes around cells . . . . . . . . . . **9. C. distans**
**10b** Cells 6–12 (22) µm in diameter; with narrow individual envelopes around cells (staining !) . . . . . . . . . . **10. C. limneticus**
**11a** Cells up to 10 µm in diameter, only in few species the largest cells reaching 15 µm . . . . . . . . . . **12**
**11b** Cells over 10 µm in diameter . . . . . . . . . . **26**
**12a** Thermal species . . . . . . . . . . **13**
**12b** Species living in other biotopes than thermal waters or thermal atmophytic localities . . . . . . . . . . **15**
**13a** Envelopes around cells very thin, colourless, cells 3–4 µm in diameter . . . . . . . . . . **13. C. subtilissimus**
**13b** Envelopes around cells somewhat wider, colourless, cells usually > 4 µm in diameter . . . . . . . . . . **14**
**14a** Cells more or less isodiametric, 3–8 µm in diameter, pale blue-green . . . . . . . . . . **17. C. membraninus**
**14b** Cells usually distinctly elongate, 6–16 × 6–10 µm, bright blue-green . . . . . . . . . . **11. C. globosus**
**15a** In aquatic localities, swamps with submerged vegetation, in metaphyton, littoral of lakes, etc. . . . . . . . . . . **16**
**15b** In different types of aerophytic or subaerophytic localities . . . . . . . . . . **19**
**16a** Cells 3–4 µm in diameter, in thin, diffluent envelopes, metaphytic in swampy biotopes . . . . . . . . . . **16. C. minor**
**16b** Cells > 4 µm in diameter . . . . . . . . . . **17**
**17a** Envelopes usually not distinctly delimited, not lamellate, enlarged, in metaphyton and often solitary, also in plankton . . . . . . . . . . **19. C. minutus**
**17b** Envelopes distinctly delimited, colourless, sometimes slightly or distinctly lamellate; littoral of lakes, swamps, ponds, channels, etc. . . . . . . . . . . **18**
**18a** Envelopes slightly lamellate, not very wide, cells 4–10 µm in diameter; common species . . . . . . . . . . **23. C. obliteratus**
**18b** Envelopes densely and clearly lamellate, 5.8–11 mm in diameter; rarely . . . . . . . . . . **24. C. schizodermaticus**
**19a** Envelopes colourless . . . . . . . . . . **20**
**19b** Envelopes in old colonies yellow or yellow-brown . . . . . . . . . . **25**
**20a** Kryosestic species, cells more or less 6 µm in diameter, in colourless, not lamellated envelopes . . . . . . . . . . **18. C. scherffelianus**

**20b** Aerophytic species, not in kryoseston . . . . . . . . . . . . **21**
**21a** Cells to 4 (rarely to 6.6) µm in diameter . . . . . . . . . . . . **22**
**21b** Cells more than (4) 5 µm in diameter . . . . . . . . . . . . **23**
**22a** Small, few-celled colonies in more or less delimited, indistinctly lamellate, colourless, thin envelopes; cells 2–4 µm in diameter . . . . . . . **14. C. varius**
**22b** Cells unified into large, confluent colonies with colourless, not lamellate envelopes; cells 2–5 (6.6) µm in diameter . . . . . . . . . . . **15. C. cohaerens**
**23a** Envelopes distinct, delimited and intensely and densely lamellate; cells up to 10 µm in diameter; aerophytic on limestone rocks . . . . . **26. C. ercegovicii**
**23b** Envelopes distinct, delimited, but without lamellation or only indistinctly lamellate . . . . . . . . . . . . **24**
**24a** Cells pale blue-green, 5–8 (13) µm in diameter, envelopes homogeneous, colourless, not lamellate . . . . . . . . . . . . **20. C. pallidus**
**24b** Cells blue-green, 4–9 µm in diameter, envelopes slightly lamellate . . . . . . . . . . . . **21. C. helveticus**
**25a** Envelopes thin, not lamellate, yellowish or brown when old, cells 5–6 (9) µm in diameter; on wet granitic (acid) rocks or on wet wood **22. C. montanus**
**25b** Envelopes thin, not or slightly lamellate, yellow-brown when old; cells 5–15 µm in diameter; on wet limestone rocks . . . . . . . . **25. C. lithophilus**
**26a** Aerophytic or subaerophytic species . . . . . . . . . . . . **27**
**26b** In various aquatic biotopes . . . . . . . . . . . . **30**
**27a** Envelopes not lamellate . . . . . . . . . . . . **28**
**27b** Envelopes distinctly lamellate . . . . . . . . . . . . **29**
**28a** Cells 13–15 µm in diameter, blue-green or yellowish, mainly on non-calcareous wet rocks . . . . . . . . . . . . **27. C. turicensis**
**28b** Cells 15–30 µm in diameter, blue-green, olive-green or slightly pinkish-violet; on wet limestone rocks . . . . . . . . . . . . **30. C. spelaeus**
**29a** Cells violet, lamellation distinct, but not very intense and not very dense . . . . . . . . . . . . **31. C. westii**
**29b** Cells blue-green or olive-green, lamellation very dense and distinct . . . . . . . . . . . . **32. C. tenax**
**30a** Thermal species, cells pale blue-green, up to 32 µm in diameter . . . . . . . . . . . . **29. C. thermalis**
**30b** In aquatic but not thermal biotopes . . . . . . . . . . . . **31**
**31a** Cells remain distinctly irregular after division, solitary or in groups, in enlarged, colourless, not or slightly lamellated, rounded envelopes, to 27 µm in diameter . . . . . . . . . . . . **12. C. quaternarius**
**31b** Cells divide regularly, after division remain grouped in typical chroococcoid colonies; envelopes not much wide, diffluent or delimited (and lamellate); cells (10) 15–35 (58) µm in diameter . . . . . . . . . . . . **32**
**32a** Envelopes not delimited and not lamellate, diffluent at the margin, usually narrow . . . . . . . . . . . . **33**
**32b** Envelopes always delimited and usually slightly or distinctly lamellate, more or less following the cell outline . . . . . . . . . . . . **34**
**33a** Cells olive-green to pinkish-violet, in salty swamps and ponds, halophilic, usually in brackish or salinic inland waters . . . . . . . . . **28. C. submarinus**
**33b** Cells grass-green or blue-green, in moorland waters, raised bogs, in lake littoral and swamps, mainly in mountains or northern regions . . . . . . . . . . . . **33. C. subnudus**
**34a** Cells to 32 µm in diameter, bright blue-green, in moorland waters and bogs, swamps and littoral of stagnant waters . . . . . . . . . . . . **34. C. turgidus**
**34b** Cells larger than 50 µm in diameter, dark blue-green, in swamps, mainly in northern regions . . . . . . . . . . . . **35. C. giganteus**

## 1. **Chroococcus microscopicus** Komárková-Legnerová et Cronberg 1994 (fig. **367**)

Colonies microscopic, free floating, cumulus-like, with numerous, densely packed groups of cells, irregular, to 100 µm in diameter. Mucilage colourless, delicate, more or less clearly delimited or diffluent, structureless. Cells spherical, 0.7–1 µm in diameter, with homogeneous, pale blue-green content, usually with small groups of 4–8 cells arranged regularly and surrounded by a sheath, which is easily recognizable after staining or in phase contrast.
Occurrence: Planktic, in mesotrophic lakes; known only from Sweden, but probably more widely distributed in northern lakes (found also in slightly brackish bays in Baltic Sea and in the Netherlands).

## 2. **Chroococcus aphanocapsoides** Skuja 1964 (fig. **368**)

Colonies microscopic, free floating, gelatinous, in outline oval, spherical or more or less irregular-rounded, usually 30–80 (rarely more, to 100) µm in diameter. Cells rarely solitary and evenly distributed, usually irregularly arranged in small aggregates (with 2–8 - more cells) within common, homogeneous, colourless, hyaline, structureless and delimited or diffluent mucilage, forming wide margin around cells. Cells spherical or hemispherical after division, or more or less rounded-angular, (1.4) 1.8–2.3 µm in diameter, without or with individual mucilaginous envelopes, with homogeneous, pale blue-green content, without aerotopes. *Aphanocapsa elachista* var. *planctonica* sensu Skuja (1964) belongs possibly to *C. aphanocapsoides*; (fig. **369**).
Occurrence: Planktic in oligotrophic or mesotrophic lakes; known only from Sweden, probably more widely distributed in corresponding biotopes.

## 3. **Chroococcus vacuolatus** Skuja 1939 (fig. **370**)

*Gloeocapsa vacuolata* (Skuja) Hollerbach et al. 1953

Colonies mucilaginous, microscopic, composed of 4–16–32 cells, rarely small groups of 2–3 cells, irregular in outline, with more or less densely arranged cells. Gelatinous envelopes colourless, not lamellate, diffluent at the margin; around

---

CHROOCOCCUS subg. LIMNOCOCCUS

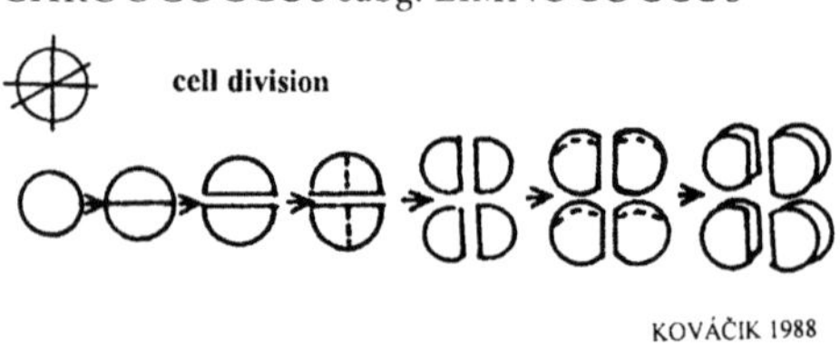


**Fig. 367.** *Chroococcus microscopicus*: **a** - colony, **b** - detail of cell arrangement; after Komárková-Legnerová & Cronberg (1994).
**Fig. 368.** *Chroococcus aphanocapsoides*: **a** - after Skuja (1956), **b** - after Komárková-Legnerová & Cronberg (1994).
**Fig. 369.** *Chroococcus* cf. *aphanocapsoides*?: after Skuja (1964, sub *Aphanocapsa elechista* var. *planctonica*), with the succesive cell division (nanocyte formation?).

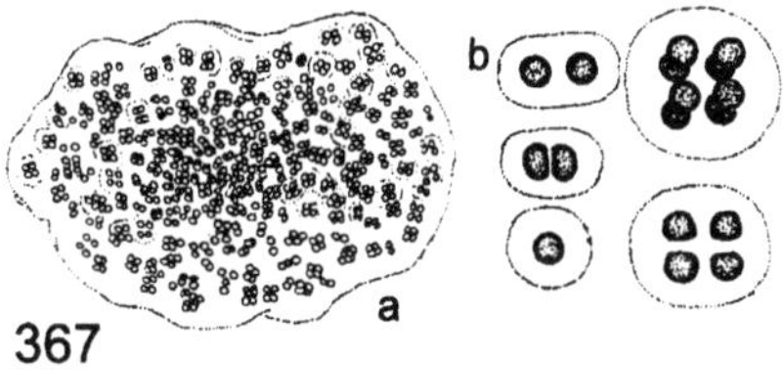


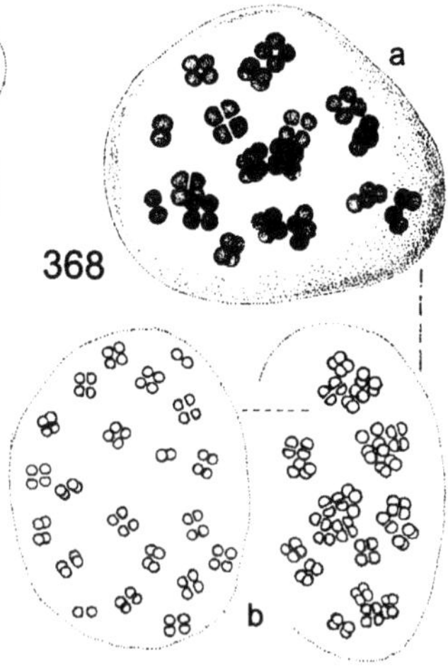


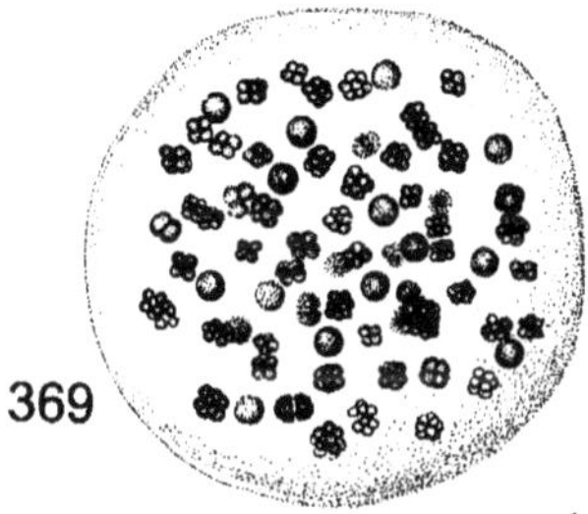


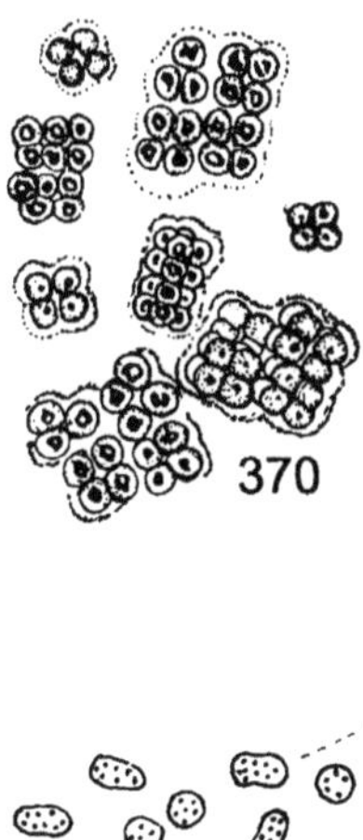


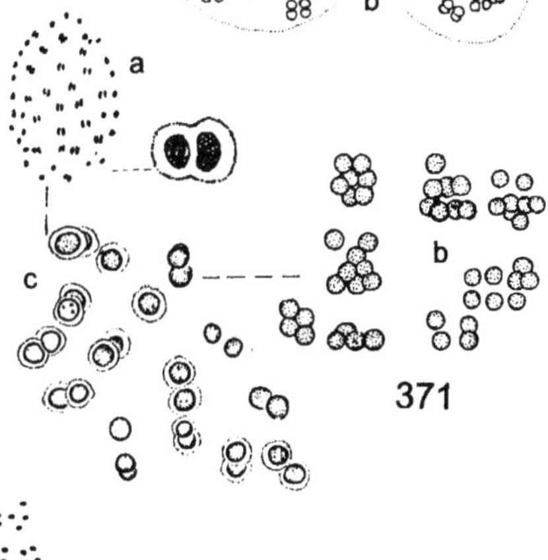


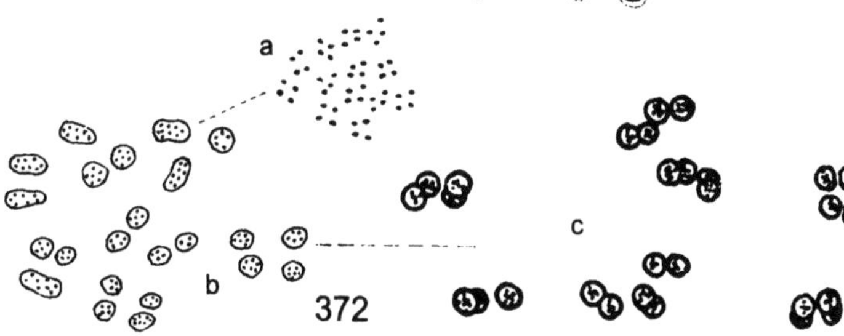


**Fig. 370.** *Chroococcus vacuolatus*, with aerotopes: after Skuja (1939).
**Fig. 371.** *Chroococcus minimus*: **a** – colony and detail of cells after Keissler from Huber-Pestalozzi (1938), **b** – after G.M. Smith (1920, sub *C. dispersus* var. *minor*), **c** – after Wojciechowski (1971).
**Fig. 372.** *Chroococcus planctonicus*, with aerotopes: **a** – colony, **b-c** – details of colony; **a-b** – after Bethge (1935), **c** – after Teiling (1942).

cells individual gelatinous colourless envelopes, sometimes indistinct, 1–1.5 µm wide. Cells spherical, hemispherical, 1.5–2.7 µm in diameter, pale blue-green or olive-green, at first (living in benthos) with few granules in cells, later (after transition into plankton) with (usually) one large, more or less central aerotope.
Occurrence: Benthic (epipelic), tychoplanktic or planktic, in small water bodies; described from Latvia, little known species of unknown distribution.

### 4. **Chroococcus minimus** (Keissler) Lemmermann 1904 (fig. **371**)

*Chroococcus minutus* var. *minimus* Keissler 1901; *C. dispersus* var. *minor* G.M. Smith 1920; *Gloeocapsa minima* (Keissler) Hollerbach in Elenkin 1938; ? *Microcystis chroococcoidea* var. *minor* Nygaard 1949; *Gloeocapsa minima* f. *smithii* Hollerbach et al. 1953

Gelatinous microscopic colonies free floating, spherical or irregular, with (2) – 8 – several cells, which are more or less regularly arranged or in small groups situated within colourless, homogeneous mucilage, diffluent at the margin; special envelopes around cells present or absent, but sometimes just recognizable (staining !), not lamellate. Cells spherical, after division hemispherical, pale blue-green or greyish blue-green, 1.7–3 µm in diameter (data about 8 µm are obviously wrong), without aerotopes.
Occurrence: Planktic, in oligotrophic or mesotrophic reservoirs and lakes, rarely among algae in metaphyton of stagnant waters; in temperate zones over the whole northern hemisphere, more common in northern regions, probably also in Argentine and Brazil; data from Africa (Nyassa lake), India and Srí Lanka evidently concern another species.

### 5. **Chroococcus planctonicus** Bethge 1935 (fig. **372**)

Colonies more or less irregular, gelatinous, free floating, with several or many cells, more or less flat, to 100 µm in diameter, with randomly arranged cells, which are sometimes in groups of 2–4 together. Colonial mucilage colourless, diffluent, usually almost invisible (staining !), special envelopes around cells lacking. Cells spherical or slightly elongate, later hemispherical or slightly irregular-rounded, pale yellowish, with aerotopes, 3–4 µm in diameter. Cell divisions usually in two perpendicular planes (resulting in flattened colonies), but sometimes in more planes.
Occurrence: In winter plankton of eutrophic ponds with muddy bottom and anaerobic layer ($H_2S$); described from Germany (near Potsdam) and recorded from Sweden, not well known.

### 6. **Chroococcus dispersus** (Keissler) Lemmermann 1904 (fig. **373**)

*Chroococcus limneticus* var. *subsalsus* Lemmermann 1901; *C. minor* var. *dispersus* Keissler 1902; *Gloeocapsa minor* f. *dispersa* (Keissler) Hollerbach in Elenkin 1938

Colonies free floating, microscopic, (1) 4–16 – (32) – (rarely more)-celled, more or less spherical or irregular in outline, with irregularly, sparsely or almost densely arranged cells; cells sometimes in groups (up to 16-celled), which are distant from

---

CHROOCOCCUS subg. LIMNOCOCCUS
**Fig. 373.** *Chroococcus dispersus*: **a** – after G.M.Smith (1920), **b** – after G.M.Smith (1920, sub *C. limneticus* var. *subsalsus*), **c** – after Komárková-Legnerová & Cronberg (1994), **d** – after Werner & Rosa (1992, from Brazil).

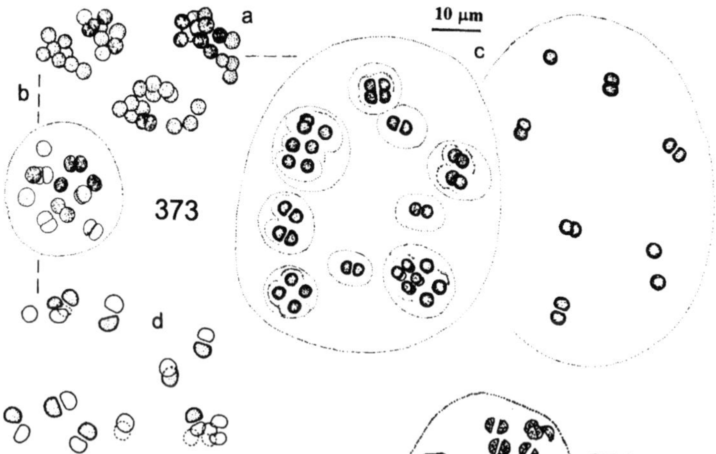


373

374

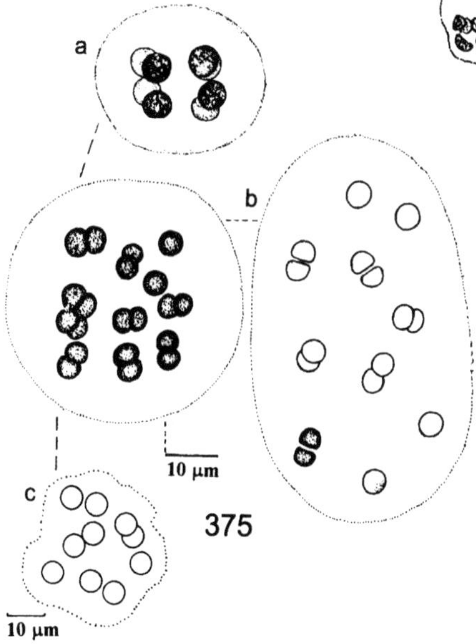


375

376

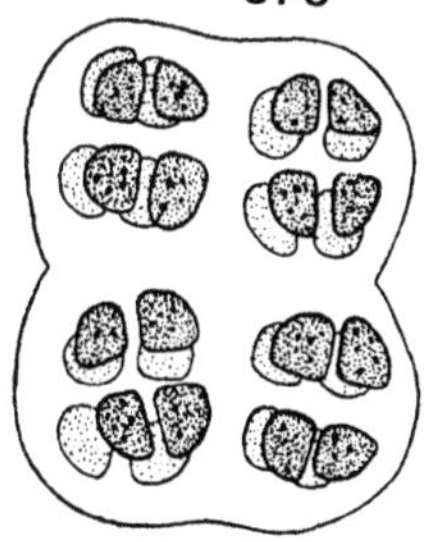

**Fig. 374.** *Chroococcus cumulatus*: after Bachmann (1921).
**Fig. 375.** *Chroococcus distans*: **a** – after G.M. Smith (1920, sub *C. limneticus* var. *distans*), **b** – after Komárková-Legnerová & Cronberg (1994), **c** – after Yacubson (1974, from Venezuela).
**Fig. 376.** *Chroococcus gomontii*, tropical species: after Nygaard (1926); comp. also *Cyanokybus venezuelae* (fig. 414).

one another, but in common mucilage. Colonial slime colourless, sometimes very wide around cells, at the margin distinct and clearly delimited or diffluent, rarely indistinct, homogeneous; around cells are sometimes homogeneous, 1–2 µm wide, usually hardly distinguishable and diffluent gelatinous envelopes. Cells spherical or hemispherical, pale grey-blue, pale blue-green, olive-green or bright blue-green, without gas vesicles, (3) 3.5–4.5 (5) µm in diameter; cell content homogeneous.
Occurrence: Planktic, in clear ponds and lakes, oligotrophic or mesotrophic reservoirs, rarely in slightly brackish waters; temperate zone, known from N. America and over the whole of Eurasia, southern to central Asia; sporadic data from rice-fields and other tropical biotopes (E. Africa, Brazil, Srí Lanka) need confirmation (comp. Werner & Rosa 1992; fig. **373d**). Data from aerophytic habitats are incorect.

### 7. **Chroococcus cumulatus** Bachmann 1921 (fig. **374**)

Colonies microscopic, free floating, amorphous, usually of irregular outline, with 8 or more sparsely and irregularly situated cells, or with cells aggregated in 4–8-celled groups, slightly distant from one another. Mucilage irregularly widened around cells, colourless, not striate, diffluent; special envelopes around cells lacking. Cells spherical, later hemispherical or in the shape of sector of a sphere, blue-green, with gas vesicles (visible aerotopes), 5–7 µm in diameter. Belongs probably into the subg. *Chroococcus.*
Occurrence: Little known species, described from small lakes in Greenland, but known also from large reservoirs in northern and central Europe.

### 8. **Chroococcus prescottii** Drouet et Daily 1942 (fig. **377–381**)

*C. gomontii* Nygaard sensu auctorum (European and Nordic populations ?); *Eucapsis alpina* sensu Siemińska 1967

Colonies microscopic, free floating, more or less subspherical or irregularly oval-cylindrical, with 4–16–32 cells arranged in cubic or sarcinoid clusters (sometimes with 4 closely packed cells), enclosed in widened, clearly delimited, sometimes rectangular, hyaline, often slightly lamellate envelopes. Cells spherical or hemispherical after division, with bright blue-green and finely granular content, 5–8 µm in diameter. Relationship with the genus *Eucapsis* need resolution (type of division?). The strain 9BG, designated in national culture collections as «*Chroococcus prescottii*» belongs to *Synechocystis fuscopigmentosa* Kováčik 1983 (p. 145).
Occurrence: Freshwater, tychoplanktic in clear acidic reservoirs, sometimes associated with typical *Chroococcus turgidus*; known from USA (Michigan, Wisconsin) and Argentine (Buenos Aires, upper Rio Limay, Tierra del Fuego); «European» planktic populations of the tropical *C. gomontii* are probably identical with this species (fig. **378**).

---

CHROOCOCCUS subg. LIMNOCOCCUS
**Fig. 377.** *Chroococcus prescottii*: **a** – cell division, **b-d** – colonies; **a-c** – after Drouet (1942, from USA), **d** – after Guarrera & al. (1968, from Argentine).
**Fig. 378.** *Chroococcus* cf. *prescottii*: after Nygaard (1977, from the lake Fures, Denmark, sub *C. limneticus*).

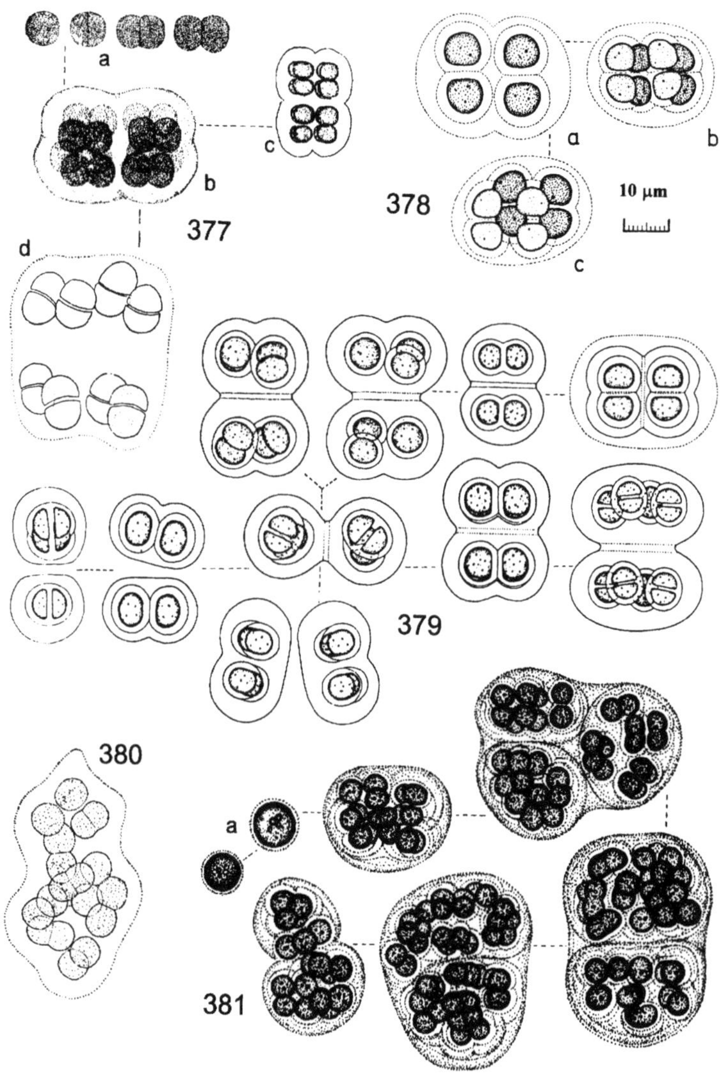


**Fig. 379.** *Chroococcus* cf. *prescottii*: after Nygaard (1984, from S. Greenland, sub *C. gomontii*; typical *C. gomontii* from tropical countries see on fig. 376).
**Fig. 380.** *Chroococcus* cf. *prescottii*: after Drouet (1942, sub *C. sonorensis*, from USA).
**Fig. 381.** *Chroococcus* cf. *prescottii*: after Skuja (1976, sub *Chroococcus* sp., from New Zealand).

**9. Chroococcus distans** (G.M. Smith) Komárková-Legnerová et Cronberg 1993 (fig. **375**)

?*Microcystis chroococcoidea* W. et G.S. West 1911; ?*Chroococcus microcystoideus* W. et G.S. West 1911; *Chroococcus limneticus* var. *distans* G.M. Smith 1920; *Gloeocapsa limnetica* var. *distans* (G.M. Smith) Hollerbach et al. 1953

Cells in microscopic, irregular or more or less spherical, mucilaginous colonies, free floating, usually with 4–32 cells, which are irregularly disposed within the slime, usually distant from one another, rarely in irregular groups. Colonial slime delicate, colourless, homogeneous, without any structure, at first distinct, delimited or later diffluent at the margin; special hyaline envelopes are lacking around cells. Cells spherical or widely oval (before division), after division hemispherical, with blue-green, usually homogeneous content, (4.5) 5.2–7.5 µm in diameter.
Occurrence: Planktic, in freshwater oligotrophic or mesotrophic lakes and reservoirs; northern areas of N. America and Eurasia, rarely in southern parts of temperate zone, in the North up to Novaja Zemlja. Sometimes occurs in slightly saline habitats. Data from tropical countries are also available (e.g. Yacubson 1974 from Venezuela; fig. **375c**).

**10. Chroococcus limneticus** Lemmermann 1898 (fig. **382**)

*Chroococcus limneticus* var. *carneus* (Chodat) Lemmermann 1904; *Gloeocapsa limnetica* (Lemmermann) Hollerbach in Elenkin 1938; *Anacystis limnetica* (Lemmermann) Drouet et Daily 1952; *Anacystis thermalis* f. *major* (Lagerheim) Drouet et Daily 1956 p.p.

Colonies with wide gelatinous slime, free floating, oval, rarely almost spherical or irregular and lobate, 4–40-celled, usually with cells slightly distant from one another, irregularly arranged or in indistinct 2–4-celled groups. Colonial mucilage colourless, delicate, structureless, homogeneous, clearly delimited or diffluent at the margin, sometimes scarcely visible, distant from the cells; around single cells are developed individual, colourless, not lamellate, narrow, sometimes diffluent and, in this case, indistinct envelopes. Cells spherical or hemispherical, greyish blue-green, bright blue-green, olive-green or yellowish, with finely granular protoplast, without aerotopes, 6–12.2–22 µm in diameter. Variable species, occurring in numerous local morphotypes and ecotypes, often of different cell-size (various varieties were described, e.g., typical var. **limneticus** with cells 6–12 µm in diameter, var. **elegans** G.M. Smith 1920 with cells 14–22 µm, etc.).
Occurrence: Planktic, in lakes and ponds, large reservoirs, swamps, secondarily in rivers; during the whole year, usually with maxima in summer and in the autumn, but never in large biomass. With cosmopolitan distribution, the exact variation, distribution and ecology not well known.

**11. Chroococcus globosus** (Elenkin) Hindák 1978 (fig. **383**)

*Aphanothece globosa* Elenkin 1938

Colonies 4-or 8-celled or many-celled, with solitary cells or in twos near to one another, sometimes forming more or less cubic packets, surrounded by mucilaginous envelopes. Envelopes colourless and structureless, forming wide, more or less oval, not lamellate margin around cells. Cells rounded oval, (6)8–16 × 6–10 µm, with blue-green or bright blue-green, homogeneous protoplast.
Occurrence: Thermophilic, in thermal mud of hot springs; known from Eurasian thermal springs: Russia (Kamčatka), Slovakia (Piešťany).

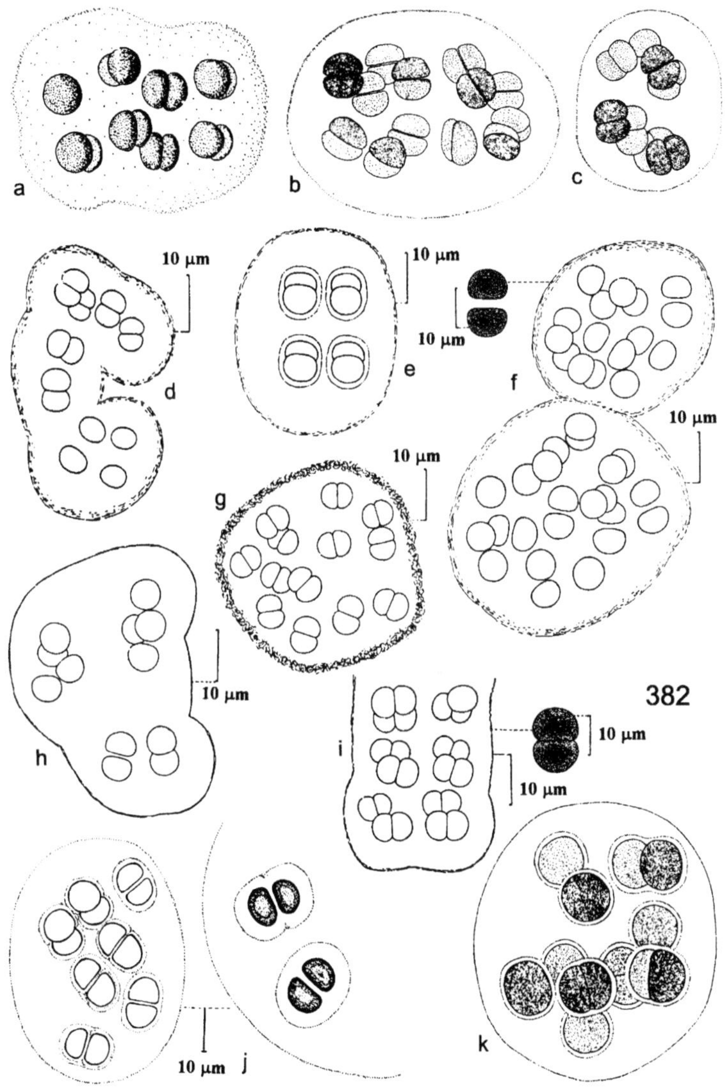


CHROOCOCCUS subg. LIMNOCOCCUS

**Fig. 382.** *Chroococcus limneticus*: variability of colonies; **a** – after Bourrelly (1970), **b** – after G.M.Smith (1920), **c** – after G.M.Smith (1920, sub *C. limneticus* var. *carneus*), **d-i** – after Komárek (1958), **j** – after Komárková-Legnerová & Cronberg (1994), **k** – after G.M.Smith (1920, sub *C.limneticus* var. *elegans*).

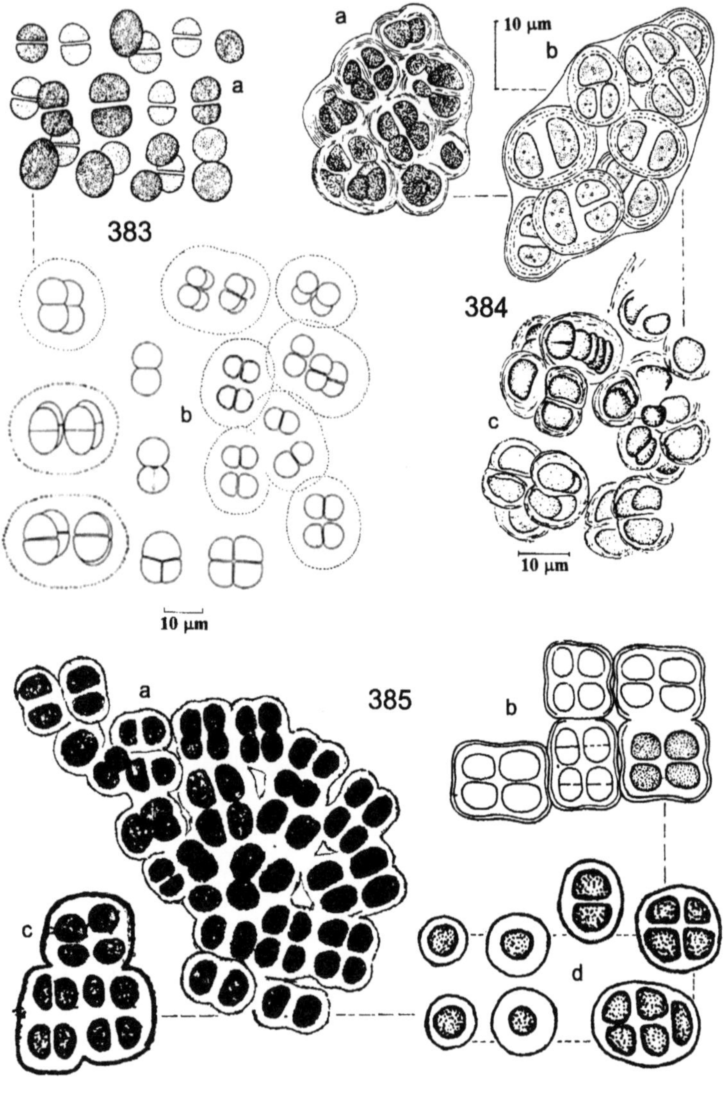

383
a
b
10 µm
384
a
b
c
10 µm
10 µm
385
a
b
c
d

**12. Chroococcus quaternarius** Zalessky 1926 (fig. **384**)

Colonies microscopic, forming nest-like thallus composed of small subcolonies with 2–4 - 8 (16) more or less clustered cells, irregularly or in slightly packet-like (more or less quadrangular) arrangements, enveloped by individual, finely striate, clearly delimited mucilaginous envelopes. Mucilage colourless. Cells more or less spherical, later usually hemispherical, irregular-rounded or with rounded angular (polygonal) outline (this shape is kept over the whole cell cycle), blue-green, to 27 µm in diameter.
Occurrence: Epipelic, in benthos of oligotrophic lakes; Israel, Poland (near Lublin), N. Russia, probably widely distributed in corresponding localities.

**13. Chroococcus subtilissimus** Skuja 1937 (fig. **386**)

Colonies composed of small, mostly 2–4-celled groups (rarely solitary cells), microscopic, closely aggregated to amorphous, mucilaginous or gelatinous, brown-rose, small, thin thallus, attached to the submerged rocky substrate; cells usually closely packed together, more or less in one plane, finally forming almost quadrate (crucially arranged), oval to nearly irregular in outline or very short uniseriate cylindrical 4-celled clusters, which are surrounded by individual, wide, diffluent, non-lamellate, colourless, scarcely recognizable envelopes. Cells spherical or hemispherical, sometimes even more or less discoid, pale blue-green or pinkish, 0.8–1 µm in diameter. Cell division sometimes mainly in one plane.
Occurrence: Thermal salty sulphur springs, benthic, epilithic on submerged not calcareous (flysh) rocks together with other cyanoprokaryotes, described and known only from Greece (Epirus).

**14. Chroococcus varius** A. Braun in Rabenhorst 1861 (fig. **387**)

Colonies microscopic or large, macroscopic, slimy or gelatinous, colourless, dirty olive-green or brownish, with solitary cells or in 2–4-celled groups, rarely in multicellular, irregular slimy clusters. Colonial slime wide, indistinctly lamellate

---

CHROOCOCCUS subg. CHROOCOCCUS

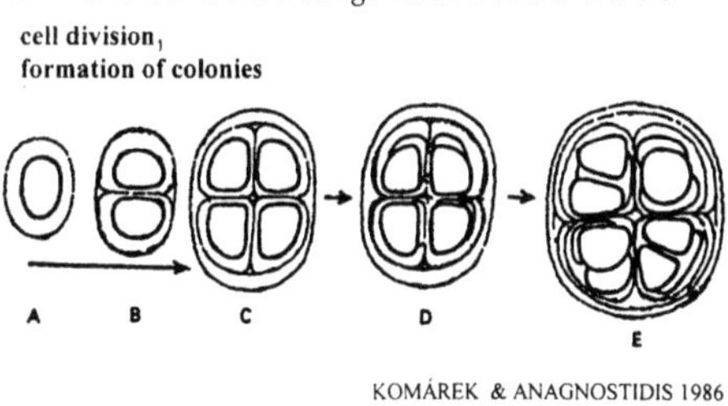


**Fig. 383.** *Chroococcus globosus*: **a** – after Elenkin (1938, sub *Aphanothece globosa*), **b** – after Hindák (1978, cultivated strain).
**Fig. 384.** *Chroococcus quaternarius*: **a** – after Zalessky (1926), **b** – after Dor (1967, from Israel), **c** – after Wojciechowski (1971, sub *Chroococcus* sp.).
**Fig. 385.** *Chroococcus cohaerens*: **a,c** – after Wille from Geitler (1932, redrawn from the type material), **b** – after Hollerbach from Starmach (1966), **d** – after Hollerbach from Starmach (1966, sub *Gloeocapsa bituminosa*).

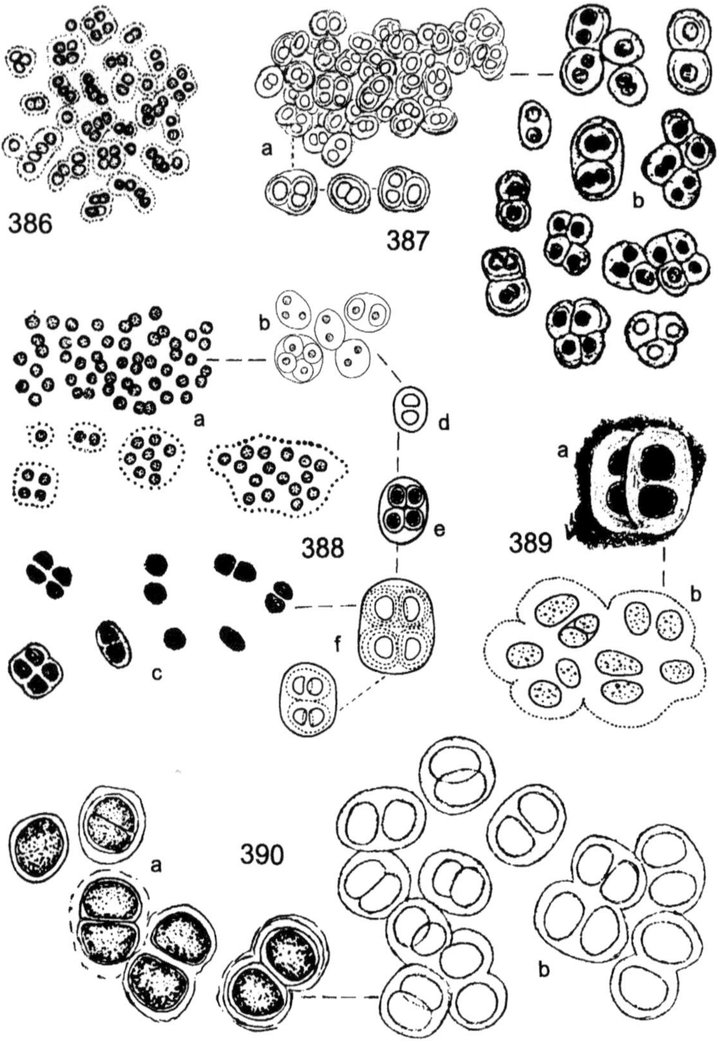


CHROOCOCCUS subg. CHROOCOCCUS
**Fig. 386.** *Chroococcus subtilissimus*: after Skuja (1937).
**Fig. 387.** *Chroococcus varius*: **a** – after M. Watanabe & Komárek (1994), **b** – after Wille from Geitler (1932, redrawn from type exsiccate).

or homogeneous, refractive, colourless or yellowish to orange-yellowish; cells and their small groups (2–4-celled) are surrounded by individual colourless, indistinctly and concentrically lamellate envelopes, to 8 µm in diameter. Cells spherical or hemispherical or in a form of a sector of a sphere, with pale blue-green, olive-green, rarely yellowish content, 2–4 µm in diameter. Cell division usually irregular, in different oblique planes.
Occurrence: Aerophytic, epilithic, on wet walls and rocks, in glasshouses; common in Europe, probably with cosmopolitan distribution (often misinterpreted with other species).

### 15. Chroococcus cohaerens (Brébisson) Nägeli 1849 (fig. 385)

*Chroococcus bituminosus* (Bory) Hansgirg 1892

Cells in 2–8-celled groups or forming micro- or macroscopic colonies, which are gelatinous, amorphous, dirty green or blackish blue-green, with irregularly or slightly fan-like arranged, usually numerous cells and groups of cells with individual, thin, unstratified mucilaginous and confluent, colourless envelopes; common mucilage colourless, not lamellate; cells with envelopes to 7 µm in diameter. Cells spherical or oval, later hemispherical or irregular, pale blue-green or olive-green, 2–5 (6.6) µm in diameter.
Occurrence: Subaerophytic, mainly in glasshouses, on wet rocks, walls and soils, data from swamps and saline waters probably concern another species; confirmed samples known only from temperate and subarctic zones and from mountains.

Note: **C. bituminosus** never forms compound colonies according to recent authors, and differs clearly from *C. cohaerens.* Colonies are always only (1)2–4(16)-celled, cells spherical or hemispherical, bright blue-green, 2–3.5 µm in diameter; *C. bituminosus* occurs on aerophytic localities, rocks in river valleys, from lowland to mountains; (fig. **385d**).

### 16. Chroococcus minor (Kützing) Nägeli 1849 (fig. 388)

Colonies microscopic, irregular, gelatinous, rarely larger, amorphous, dirty blue-green or olive-green, with irregularly arranged and more or less clustered cells; cells sometimes in 2–4-celled groups. Mucilage delicate, wide, diffluent, colour-

**387**

c

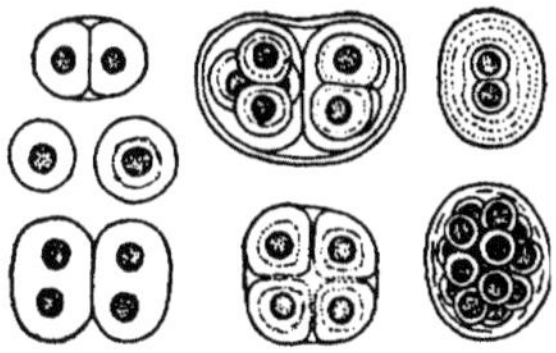

**Fig. 388.** *Chroococcus minor*: **a** – after Nägeli (1849), **b** – after W. & G. S. West from Geitler (1925), **c** – after Wille from Geitler (1932, redrawn from Kützing's type material), **d** – after M. Watanabe & Komárek (1988), **e** – after Palik (1949), **f** – after Werner & Rosa (1992, from Brazil).
**Fig. 389.** *Chroococcus scherffelianus*: **a** – after Kol (1928), **b** – after Lukavský (1993).
**Fig. 390.** *Chroococcus membraninus*: **a** – after Boye-Petersen (1946), **b** – after Wille from Geitler (1932).

less, not lamellate, sometimes scarcely visible. Cells spherical or hemispherical, during intense division irregular, 3–4 µm in diameter, pale blue-green, rarely with individual fine, thin, not lamellate, colourless, diffluent and usually scarcely visible envelopes.
Occurrence: On wet stones and wood near the water surface of streams and swamps, metaphytic in bogs, swamps and ponds with submerged vegetation, in rice-fields; in temperate zone and probably also tropical countries, sporadical, in some places frequent. Data from thermal springs (Fott 1971) probably concern another species. Numerous records in literature are debatable.

### 17. **Chroococcus membraninus** (Meneghini) Nägeli 1849 (fig. **390**)

Colonies (mats) micro- or macroscopic, mucilaginous, flat, leathery, violet-green to blackish. Common mucilage wide, not lamellate, colourless, with irregularly arranged numerous cells, or with groups of 2–4 cells in clusters. Cells spherical or hemispherical, with wide, colourless, structureless and more or less spherical individual envelopes, 3–8 µm in diameter, pale blue-green or blue-green, usually with finely granular content.
Occurrence: Near thermal waters, subaerophytic on wet mud, usually mixed with other cyanoprokaryotes; distribution not well known. *C. membraninus* var. *salinus* Ercegović 1932 was described from saline waters, but its taxonomic position is problematic. Branco & al. (1996) record this species from Brazilian mangroves, but the description and drawings do not enable to prove the identity with the type description.

### 18. **Chroococcus scherffelianus** Kol 1928 (fig. **389**)

Colonies microscopic with 2-4-(12), rarely more cells, irregularly distributed in unstratified or slightly lamellose, colourless, diffluent or finely delimited mucilage. Envelopes around cells indistinct, fine, structureless, sometimes invisible. Cells subspherical or irregular-rounded, blue-green, more or less 4–6 µm in diameter.
Occurrence: In summer kryoseston; known only from central Europe; Czech Republic (Šumava Mts.), Slovakia (High Tatra Mts.).

### 19. **Chroococcus minutus** (Kützing) Nägeli 1849 (fig. **391**)

*Protococcus minutus* Kützing 1843; *Chroococcus virescens* Hantzsch in Rabenhorst 1865; *Gloeocapsa minuta* (Kützing) Hollerbach in Elenkin 1938 p.p.

Cells solitary or in microscopic, few-celled colonies (usually with 2–8 cells), enclosed by a wide, more or less spherical or oval, homogeneous or slightly lamellate, delimited (rarely a little diffluent), sometimes with indistinct margin (staining !). Cells spherical or hemispherical after division, with pale blue-green or

---

CHROOCOCCUS subg. CHROOCOCCUS
**Fig. 391.** *Chroococcus minutus*, various morphotypes: **a-e** – original concept, all with the same morphology and ecology, **f-g** – morphotype with more distinct and stratified envelopes, **h-k** – planktic specimens, **l** – variability of cells and colonies from two strains isolated from ponds in Slovakia, **m-n** – tropical populations (**n** – with numerous cells without envelopes from Mexico), **o** – specimens from thermal springs in USA; **a** – after Wille from Geitler (1932, redrawn from the type exsiccate of Kützing), **b** – after Frémy (1929), c – after Pitschmann (1969), **d** – after Golubić (1967a), **e** – after Dor (1967), **f** – after Golubić (1967b), **g** – after

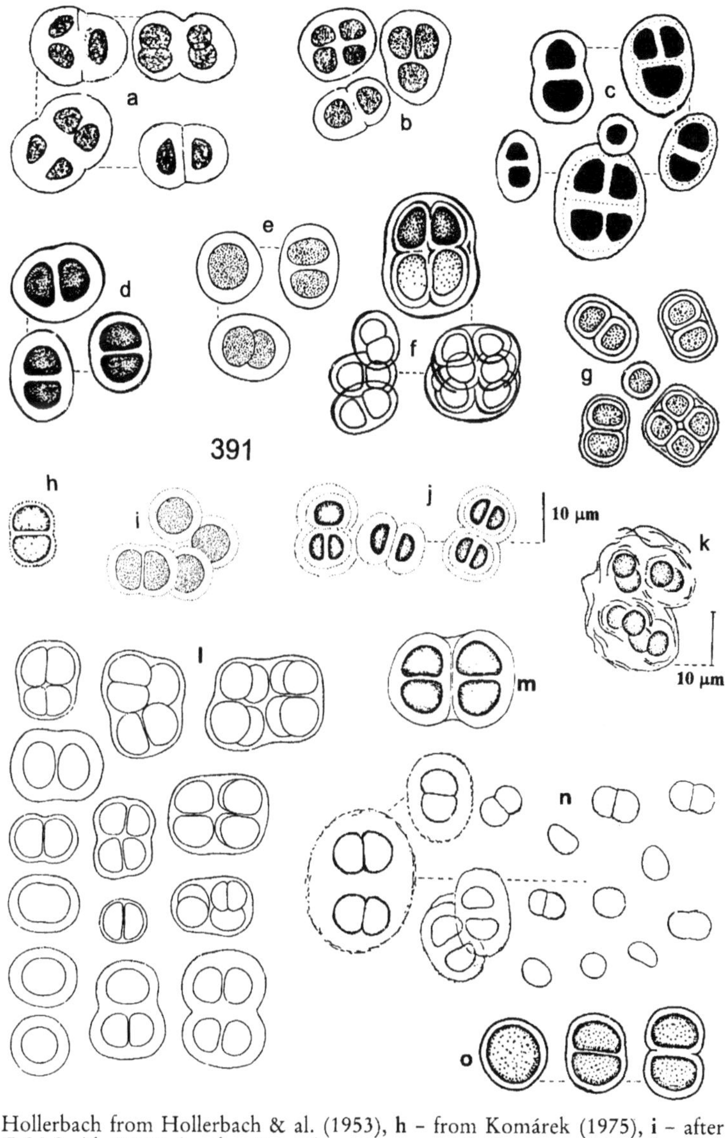


Hollerbach from Hollerbach & al. (1953), **h** - from Komárek (1975), **i** - after G.M.Smith (1920), **j** - after Komárková-Legnerová & Cronberg (1994), **k** - after Wojciechowski (1971), **l** - after Kováčik (1988), **m** - after Rino (1969), **n** - after Komárek & Novelo (1994), **o** - after Copeland (1936, sub *C. minutus* var. *thermalis*).

grey, homogeneous or granulate content, (4) 5–10 (12) µm in diameter. Polymorphic species with numerous morpho- and ecotypes.
Occurrence: Planktic, tychoplanktic or metaphytic in oligotrophic up to slightly eutrophic lakes, ponds, water reservoirs and swamps, often in littoral, but never in masses; found also in brackish waters (?); known mainly from temperate zones (N. America, Eurasia, Argentine), but probably with cosmopolitan distribution in various eco- and morphotypes; thermal (e.g., var. *thermalis;* Copeland 1936, Emoto & Hirose 1942, Palik 1949), aerophytic on bark of trees in W. Africa (Romo & al. 1995), or marine (e.g., var. *salinus*; Ardré 1969, de Halperin & Wenzel 1987) populations belong very probably to other species.

### 20. Chroococcus pallidus (Nägeli) Nägeli 1849 (fig. **396**)

Colonies usually microscopic, gelatinous, amorphous, colourless or yellowish, with irregularly disposed cells; cells arranged evenly in the common slime or in 2–4–8-celled groups. Common mucilage more or less clearly delimited, inside confluent, colourless, structureless. Around cells, special colourless, more or less rounded, not lamellate envelopes, 7.5–13 µm wide (with envelopes?). Cells spherical or hemispherical, pale blue-green, yellowish or yellow, (5) 6–11 (13) µm in diameter. Species not well known.
Occurrence: Aerophytic, on wet rocks and walls; described and known only from central Europe. All other records are unclear (USA – N. Carolina, Angola, Victoria lake, etc.).

### 21. Chroococcus helveticus Nägeli 1849 (fig. **392**)

*Gloeocapsa helvetica* (Nägeli) Starmach 1966

Solitary cells or few-celled, microscopic colonies, rarely in larger gelatinous groups with irregularly arranged cells, composed of 2–8-celled subcolonies. Around cells and cell groups thin, colourless, delimited, not lamellate or slightly lamellate sheaths. Cells spherical, hemispherical or irregular, 4–7.5 (9) µm in diameter, pale blue-green or yellowish green.
Occurrence: Subaerophytic on wet rocks, the data from swamps and lakes need confirmation (e.g., var. *consociato-dispersus* Elenkin 1923 with bright blue-green cells, 7.6–9.2 µm in diameter, described from one lake near Oloneck, Russia); probably cosmopolitan distribution.

### 22. Chroococcus montanus Hansgirg 1892 (fig. **393**)

Colonies micro- or macroscopic, mucilaginous, dark yellowish brown or blackish brown, with numerous, irregularly arranged cells, sometimes in 2–4-celled groups within mucilage. Around cells thin, not lamellate, yellow or yellow-brown envelopes. Cells spherical, hemispherical or green or yellow-brown. Species not well-known, it needs revision.
Occurrence: Aerophytic, on wet rocks and wood, stones, and wooden or iron tubes near springs and wells; described from central Europe, further distribution not known.

---

CHROOCOCCUS subg. CHROOCOCCUS
**Fig. 392.** *Chroococcus helveticus*: **a** – after Wille from Geitler (1932), **b-c** – after Golubić (1967a,b), **d** – after Hoffmann (1986).

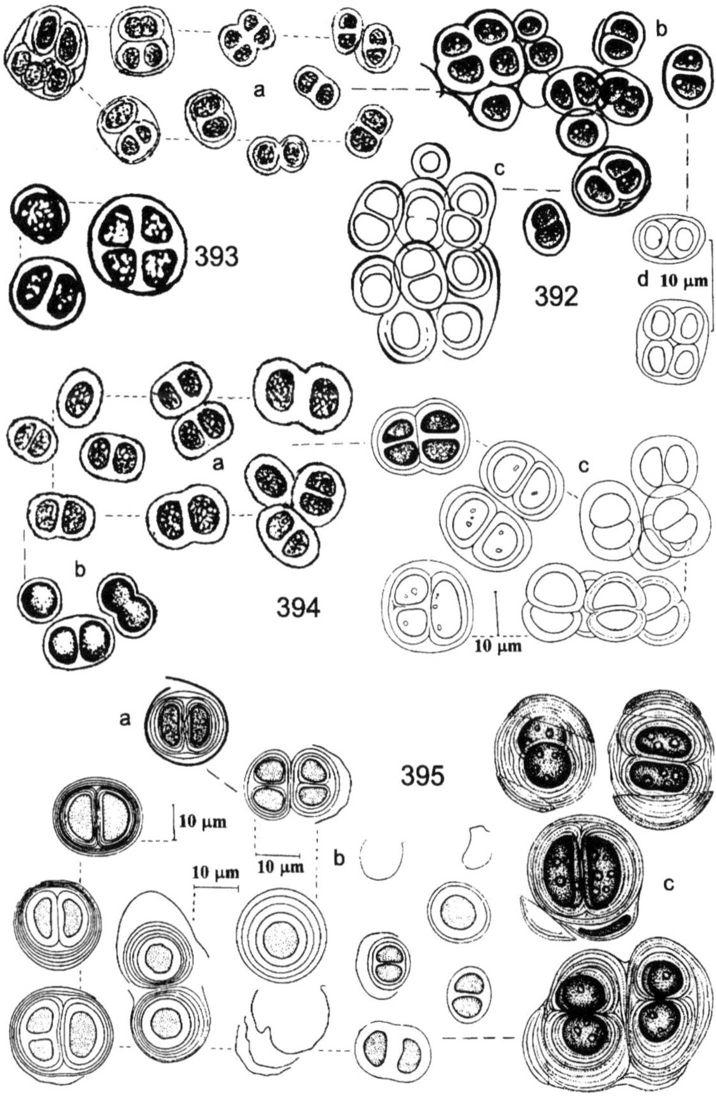


**Fig. 393.** *Chroococcus montanus*: after Frémy (1930).
**Fig. 394.** *Chroococcus obliteratus*: **a** – after Wille from Geitler (1932), **b** – after Palik (1960), **c** – after Komárek (1975, sub *Chroococcus minutus*).
**Fig. 395.** *Chroococcus schizodermaticus*: **a** – after West from Geitler (1932), **b** – after Geitler from Geitler & Ruttner (1935), **c** – after Skuja (1964).

## 23. **Chroococcus obliteratus** Richter 1886 (fig. **394**)

*Chroococcus minutus* sensu auct. post. (Komárek 1975)

Cells solitary or in 2–4(-8)-celled colonies, rarely with more cells. Common wide mucilage around colony lacking or very fine, colourless, diffluent; gelatinous envelopes around cells usually distinct, clearly delimited, colourless, more or less thin, following the outline of inner cells, 2–7 µm wide, usually not lamellate or slightly lamellate. Cells spherical, hemispherical or in the form of a segment of a sphere, pale blue-green, olive-green or yellowish, (4)6–10 µm in diameter.
Occurrence: Common species in shallow waters with submerged vegetation, in littoral of lakes and ponds, in swamps (sometimes also brackish), occasionally in plankton of shallow water bodies; common in north temperate zone, but not in abundance; records from tropical countries are not certain, the cosmopolitan distribution is possible. Often misinterpreted as other species (as «small *C. turgidus*»).

## 24. **Chroococcus schizodermaticus** W. et G.S. West 1892 (fig. **395**)

Colonies microscopic, usually only 2–4-celled, rarely with more cells, without common diffluent mucilage, to 42 µm in diameter. Envelopes around cells wide, delimited, yellow or yellow-brown, distinctly concentrically lamellate, with outer layers often enlarged. Cells spherical, hemispherical or irregular-rounded, blue-green, 5.8–11 µm in diameter.
Occurrence: Metaphytic in unpolluted swamps, in littoral of water bodies, among water plants; described from England, possibly cosmopolitan (Africa, Argentine, India, Mexico, N. America), but not frequent.

## 25. **Chroococcus lithophilus** Ercegović 1925 (fig. **397**)

Colonies micro- or macroscopic, amorphous, often mixed with other algae, with cells situated within firm, delimited, colourless or yellow-brown, not stratified mucilage, sometimes in 2–4-celled groups. Envelopes around cells firm, delimited, thin, not lamellate, colourless or yellowish brown, 7–17 µm wide. Cells spherical, after division hemispherical, blue-green, single or in pairs, 5–15 µm in diameter.
Occurrence: Aerophytic, epilithic or endolithic on wet, limestone rocks; known from Croatia, probably more distributed.

## 26. **Chroococcus ercegovicii** Komárek et Anagnostidis 1994 (fig.**398**)

*Chroococcus schizodermaticus* f. *pallida* Ercegović 1925

Colonies microscopic, usually (1)2–4(8)-celled, irregularly spherical, rarely forming macroscopic, gelatinous granular mats of small colonies. Cells and groups of cells surrounded by delimited, colourless or slightly yellowish, enlarged, strongly stratified envelopes (layered especially near outer margin). Cells oval, spherical, after divison rounded subspherical or hemispherical, sometimes

---

CHROOCOCCUS subg. CHROOCOCCUS
**Fig. 396.** *Chroococcus pallidus*: **a** – after Geitler (1932, redrawn from the type exsiccate of Nägeli), **b** – after Palik (1949).
**Fig. 397.** *Chroococcus lithophilus*: **a** – after Ercegović from Geitler (1932), **b** – after Palik (1938), **c** – after Gonzalves & Kamat (1958).

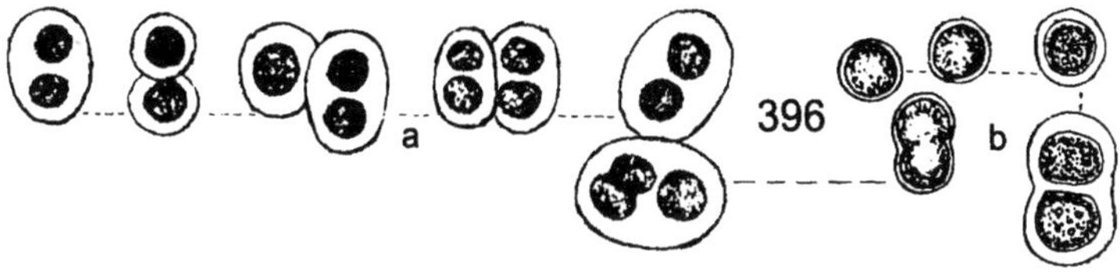


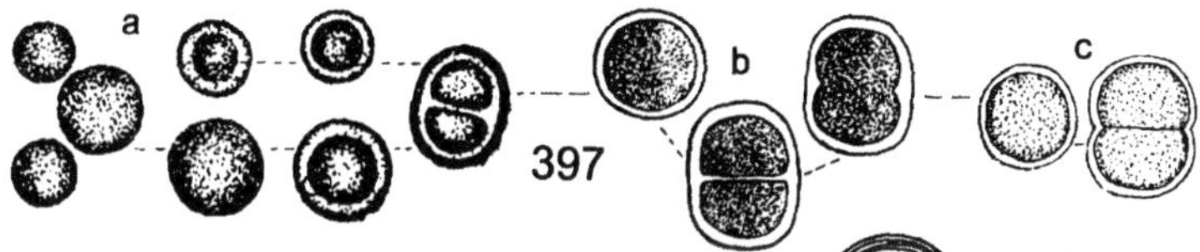


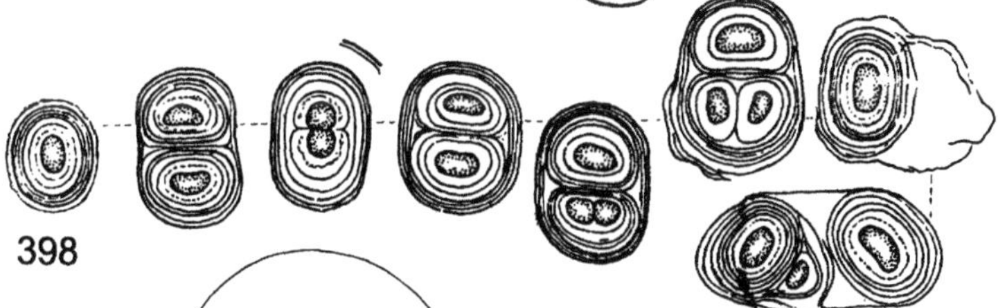


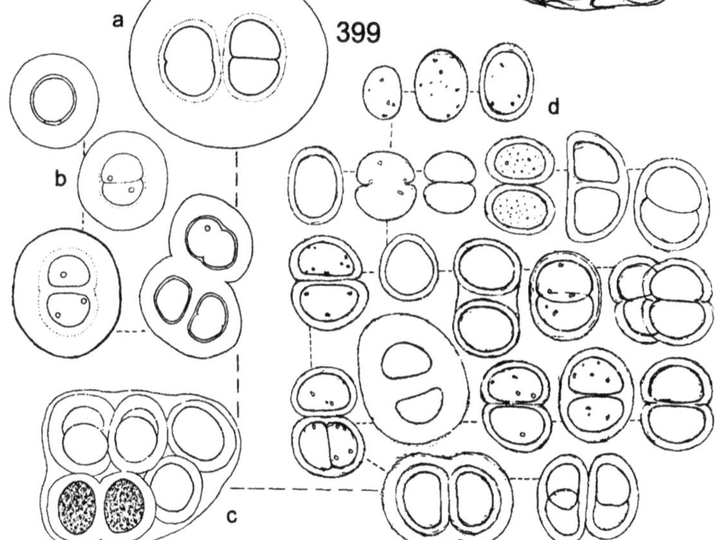


**Fig. 398.** *Chroococcus ercegovicii*: after Komárek & Anagnostidis (1995).
**Fig. 399.** *Chroococcus turicensis*: **a** – after Geitler (1932), **b** – after Starmach (1966), **c** – after Bourrelly & Manguin (1947), **d** – after M. Watanabe & Komárek (1994, from Himalayas).

slightly reniformly arcuate, bright or pale blue-green, sometimes olive-green, with more or less homogeneous content, 4.5–8 (11) × (2.2) 2.5–8 µm or 5.8–11 µm in diameter; 2-celled colonies to 28 µm in diameter.
Occurrence: Aerophytic, on limestone rocks; described from Croatia (Plitvica), rarely in calcareous Mediterranean and central European regions (e.g., Czech Republic), probably more distributed.

### 27. **Chroococcus turicensis** (Nägeli) Hansgirg 1892 (fig. **399**)

Microscopic, few-celled colonies, rarely large, gelatinous, pale yellow-orange; aggregates sometimes composed of subcolonies with 2–4 (-16) cells. Envelopes around cells to 19–34 µm in diameter, distinct, wide, not lamellate, colourless; the outermost layer sometimes extremely wide; special envelopes around cells lacking or present (in this case not lamellate, sometimes yellowish or yellow-brown). Cells pale blue-green, yellowish, olive-green or yellow-orange, slightly granular, 13–15 µm in diameter. During reproduction the gelatinous envelopes are sometimes completely diffluent.
Occurrence: Subaerophytic, on wet rocks and in springs, flushes, mainly in high mountains; rare over the temperate zone of Eurasia, up to subpolar regions, in high mountains also in tropical zone (e.g., in Himalayas).

### 28. **Chroococcus submarinus** (Hansgirg) Kováčik 1988 (fig. **400**)

*Chroococcus turgidus* var. *submarinus* Hansgirg 1889; *C. turgidus* var. *hookeri* Lagerheim 1883

Colonies micro- or macroscopic, amorphous, dirty greenish, greyish or pale violet, mucilaginous, with irregularly arranged cells; usually many-celled. Common colonial slime fine, diffluent, structureless; around cells fine, colourless, delimited or diffluent individual envelopes. Cells spherical, subspherical, hemispherical or irregular-rounded, olive-green, pinkish green or yellowish, (10) 15–25 (35) µm in diameter.
Occurrence: Halophilic, in coastal and inland swamps, metaphytic or among other algae, sometimes benthic in shallow waters, usually in saline or brackish waters; known from northern temperate zone, mainly from Europe: Croatia (Istria, Dalmatia), Italy, Romania (coast of Black Sea), Sweden; recorded also from the Pacific coast of N. America.

### 29. **Chroococcus thermalis** (Meneghini) Nägeli 1849 (fig. **401**)

*Chroococcus turgidus* sensu auct. (from thermal biotopes); *Pleurococcus thermalis* Meneghini 1837; *Protococcus thermalis* (Meneghini) Kützing 1846; *Chroococcus turgidus* var. *thermalis* (Meneghini) Rabenhorst ex Hansgirg 1892; ? *Chroococcus yellowstonensis* Copeland 1936

Colonies few-celled (usually 2–8-celled), microscopic, with typical chroococcoid arrangement of cells. Mucilaginous envelopes copying the cell outline, colourless,

---

CHROOCOCCUS subg. CHROOCOCCUS
**Fig. 400.** *Chroococcus submarinus*: **a** – after Setchell & Gardner (1919, sub *C. turgidus* var. *submarinus*), **b** – after Kováčik (1988).
**Fig. 401.** *Chroococcus thermalis*: **a** – after Palik (1949, sub *C. turgidus* var. *thermalis*), **b** – after Wenzel & Halperin (1991, sub *C. turgidus* from thermal springs), **c** –

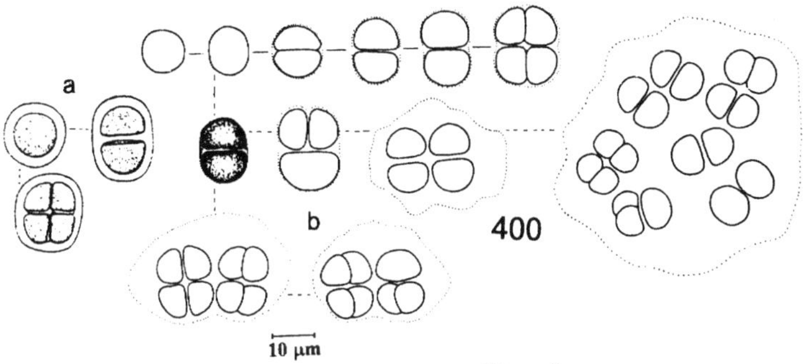


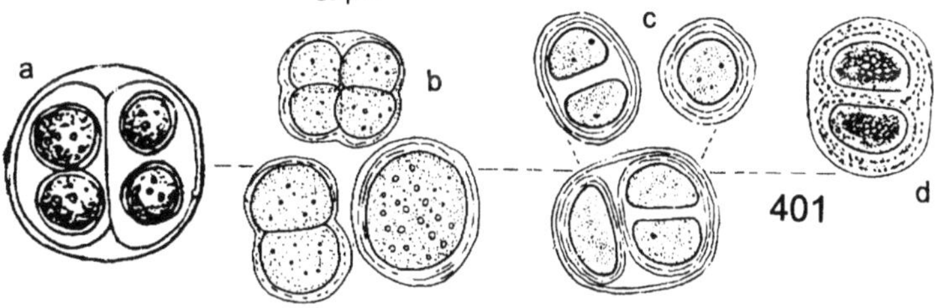


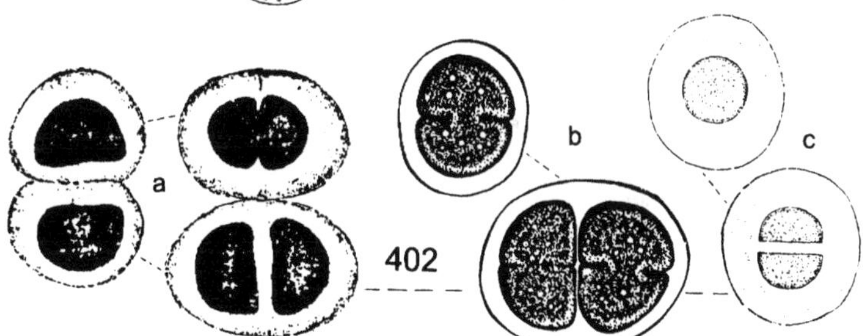


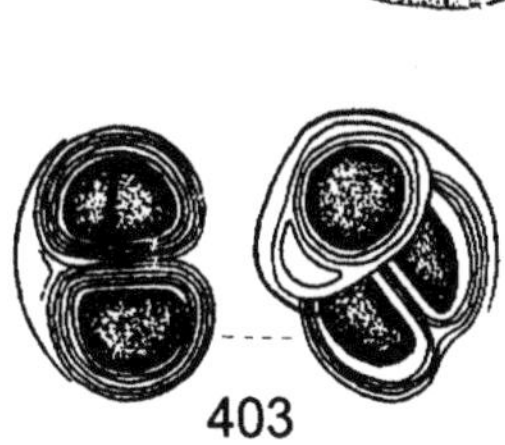


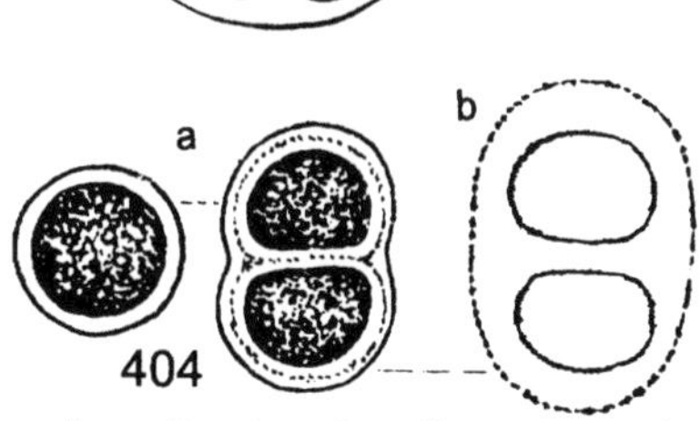


after Dor (1967, sub *C. turgidus* var. *thermalis*), **d** – *C. yellowstonensis* after Copeland (1936, possibly identical with *C. thermalis*).

**Fig. 402.** *Chroococcus spelaeus*: **a** – after Ercegović from Geitler (1932), **b** – after Golubić (1967), **c** – after Gonzalves & Kamat (1958, from India).

**Fig. 403.** *Chroococcus westii* (violet cells): after Boye-Petersen (1923).

**Fig. 404.** *Chroococcus subnudus*: **a** – after Hollerbach from Hollerbach & al. (1953, sub *Gloeocapsa turgida* var. *subnuda*), **b** – orig.

delimited, usually indistinctly or slightly lamellate or homogeneous, circa 4 µm wide. Cells with finely granular content or with homogeneous protoplasts with a single large granule, dull to bright blue-green, without sheath 5–10 (up to 32 ?) µm in diameter.
Occurrence: Probably widely distributed in thermal waters, pools, eventually also atmophytic in wet habitats up to 36.6 °C, among other cyanoprokaryotes; known from Europe, Japan and N. America; from thermal springs in USA a similar species was described (*C. yellowstonensis* Copeland) with distinctly lamellate sheaths, only from atmophytic habitats (up to 42 °C, pH 6.5–8.3). The specific identity of populations from various thermal localities should be reconsidered.

## 30. **Chroococcus spelaeus** Ercegović 1925 (fig. **402**)

Colonies micro- or macroscopic, formless, composed of subcolonies (usually with 2–4 cells), with diffluent common slime. Envelopes around cells thin, colourless, more or less distinct and delimited, to 27–40 µm wide, enveloping the cells by a circa 5–6 µm wide margin, not lamellate. Cells spherical or hemispherical after division, pale blue, pale violet or pale olive-green, (15) 16–24 (30) µm in diameter.
Occurrence: Aerophytic, among other algae on wet rocks (limestone ?); described from Croatia, distribution not known.

## 31. **Chroococcus westii** Boye-Petersen 1923 (fig. **403**)

*Chroococcus turgidus* var. *violaceus* W. West 1892; *C. turgidus* var. *subviolaceus* Wille 1913

Colonies microscopic, usually with (1) 2–4 cells, without wide mucilaginous slime. Envelopes around cells thick (up to 32 µm wide), delimited, colourless, distinctly concentrically lamellate. Cells spherical, hemispherical or rounded triangular, with violet, homogeneous or finely granular content, 13–27 µm in diameter.
Occurrence: Subaerophytic, on wet rocks, usually in mountains; rare in temperate zone (Austria, England, Germany, Norway, etc.), possibly also in subpolar and subtropical regions.

## 32. **Chroococcus tenax** (Kirchner) Hieronymus 1892 (fig. **405**)

*Chroococcus turgidus* var. *tenax* Kirchner 1878

Solitary cells, or colonies with 2–4 (-8–16) cells, microscopic, delimited. Mucilaginous envelopes thin, usually strongly lamellate, colourless and later yellowish brown, to 26 µm in diameter, usually copying the cell outline or groups of cells. Cells hemispherical or in the form of a segment of a sphere, blue-green or olive-green, 15–21 µm in diameter.
Occurrence: Subaerophytic, on wet rocks, in springs, secondarily metaphytic in clear, unpolluted shallow pools, frequent in high mountains; probably of worldwide distribution, but only in corresponding ecological environment. From saline waters in Spain (near Cádiz) was described var. *boeticus* Gonzalez Guerrero 1946, with smaller cells (7–17 µm in diameter) and other similar features to the type.

---

CHROOCOCCUS subg. CHROOCOCCUS
**Fig. 405.** *Chroococcus tenax*: **a** – after Geitler (1932, 1935), **b** – after Hollerbach

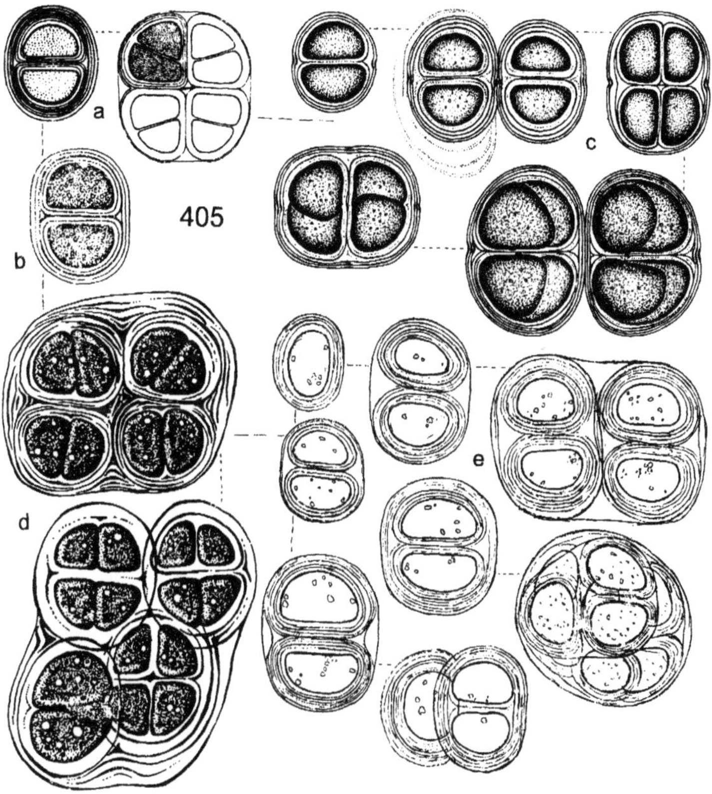


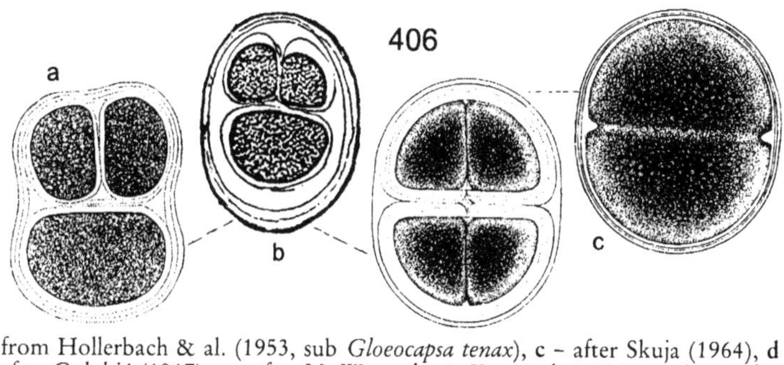


from Hollerbach & al. (1953, sub *Gloeocapsa tenax*), **c** - after Skuja (1964), **d** - after Golubić (1967), **e** - after M. Watanabe & Komárek (1994, specimens from Himalayas).

**Fig. 406.** *Chroococcus giganteus*: **a** - after G.M.Smith (1920), **b** - after Frémy (1930), **c** - after Skuja (1956); taxonomic identity of these specimens is uncertain.

**33. Chroococcus subnudus** (Hansgirg) Cronberg et Komárek 1994 (fig. 404)

*Chroococcus turgidus* var. *subnudus* Hansgirg 1892

Colonies microscopic, usually (1)2–4 (-8–32)-celled, without enveloping wide mucilage. Envelopes around cells thin, hyaline, not layered, slightly diffluent at the margin, but copying the cell outline, sometimes indistinct (staining !). Cells oval-spherical, hemispherical or subspherical-rounded, pale olive-green or intensely greenish (not dark blue-green), to 32 µm in diameter.
Occurrence: In moorland waters, raised bogs and swamps, on wet rocks, in lake littoral, usually in mountainous regions and in northern areas; probably mainly in temperate zone of the northern hemisphere, up to tundra regions.

**34. Chroococcus turgidus** (Kützing) Nägeli 1849 (fig. 407)

*Chroococcus dimidiatus* (Kützing) Nägeli 1849 = *Anacystis dimidiata* (Kützing) Drouet et Daily 1952 p.p.; *Gloeocapsa turgida* (Kützing) Hollerbach in Elenkin 1938; ? *Chroococcus turgidus* var. *maximus* Nygaard 1926 incl.

Colonies microscopic, usually 2–8 (-32)-celled, rarely solitary cells. Common mucilage around colonies lacking, however, older envelopes surround usually the whole colony; mucilaginous envelopes usually clearly delimited, firm, copying the cell outline or that of group of cells, colourless, generally striate (sometimes indistinctly), to 10 µm wide; individual cells with hyaline, lamellate sheaths. Cells spherical or widely oval, later (during successive fissions) hemispherical or in the form of segment of a sphere, intensely blue-green, rarely olive-green, yellowish or slightly pinkish (taxonomic forms ?), homogeneous or finely granular, without sheaths (6) 8–32 (45) µm in diameter.
Occurrence: Common species in detritus on wet rocks, in raised bogs, swamps, oligotrophic waters with a lot of submerse vegetation, in littoral of stagnant waters, ponds and lakes; prefers slightly acidic biotopes, particularly in moorland; common in temperate zones, but probably with cosmopolitan distribution (iden-

---

CHROOCOCCUS subg. CHROOCOCCUS

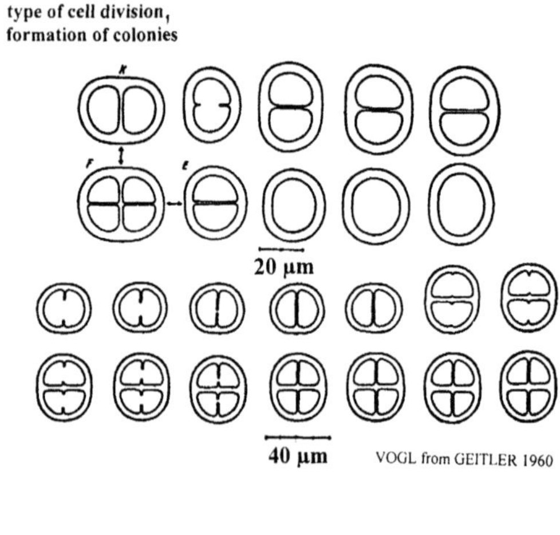

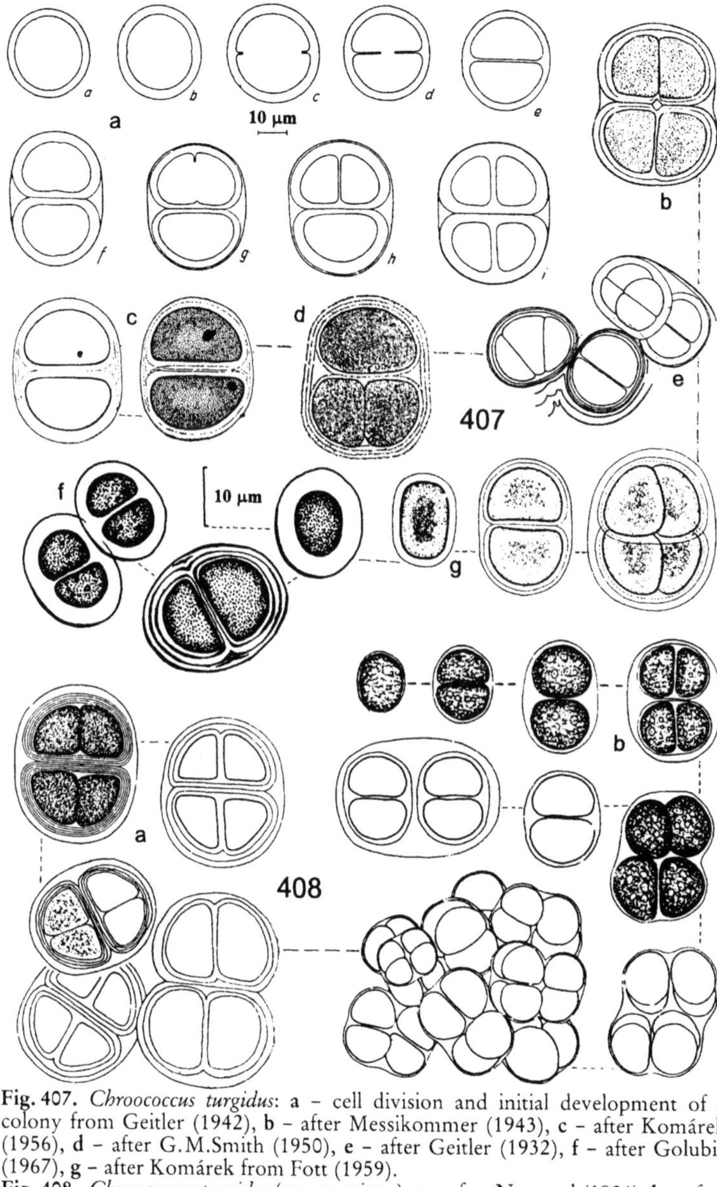


**Fig. 407.** *Chroococcus turgidus*: **a** – cell division and initial development of a colony from Geitler (1942), **b** – after Messikommer (1943), **c** – after Komárek (1956), **d** – after G.M.Smith (1950), **e** – after Geitler (1932), **f** – after Golubić (1967), **g** – after Komárek from Fott (1959).

**Fig. 408.** *Chroococcus turgidus* (var. *maximus*): **a** – after Nygaard (1926), **b** – after Kováčik (1988, from culture).

tity with tropical populations needs confirmation; fig. **409**). Data from thermal springs, endolithic or marine habitats etc. refer very probably to other species (often misinterpreted). Numerous varieties were described (e.g., fig. **408**), from which several evidently represent separate species.

### 35. Chroococcus giganteus W. West 1892 (fig. 406)

*Gloeocapsa gigantea* (W. West) Hollerbach et al. 1953

Colonies microscopic, usually 2-, rarely 4-celled, enveloped by hyaline, delimited envelopes. Mucilaginous envelopes thick, colourless, markedly lamellate; individual cells surrounded by conspicuous, distinctly lamellate, colourless sheaths. Cells spherical, oval or hemispherical (after division), bright blue-green, usually intensely and distinctly granulate, 54–58 (65) µm in diameter, with sheaths up to 80 µm in diameter.
Occurrence: In stagnant freshwaters, mainly in swamps, usually metaphytic among other algae and water plants, rarely tychoplanktic or planktic; in northern regions of the temperate zone, but recorded also from Spain, Argentine, etc.

**Species outside Europe:**
- *Chroococcus aeruginosus* Gardner 1927: On wood near thermal waters in Porto Rico.
- *Chroococcus bernardii* Van Oye 1921: Described from SE. Asia.
- *Chroococcus calcicola* Anand 1937: Aerophytic from India.
- *Chroococcus catenatus* Komárek et Novelo 1994: Planktic species (subg. *Limnococcus*) from E. African lakes.
- *Chroococcus constrictus* Gardner 1927: Freshwater, described from Porto Rico.
- *Chroococcus cronbergae* Komárek et Novelo 1994: Planktic species (subg. *Limnococcus*), described from S. Africa, probably pantropical.
- *Chroococcus cubicus* Gardner 1927: Aerophytic, from Porto Rico, probably *Cyanosarcina*.
- *Chroococcus deltoides* Komárek et Novelo 1994: Pantropical species, known from swamps and ponds; (fig. **412**).
- *Chroococcus endophyticus* Copeland 1936: Thermal waters in USA and Japan.
- *Chroococcus epiphyticus* Jao 1964: Epiphytic on *Cladophora*, described from China. (Another epiphytic «*Chroococcus dispersus* var. *minor* G.M. Smith ?» from *Oedogonium* was described by Emoto & Hirose 1942 from slightly thermal waters in Japan.)
- *Chroococcus goetzei* Schmidle 1902: Described from Africa.
- *Chroococcus gomontii* Nygaard 1926: Planktic, with wide, homogeneous mucilage and dice-like arrangement of cells, 7–9 µm in diameter; described from a

---

CHROOCOCCUS subg. CHROOCOCCUS

**409**

e

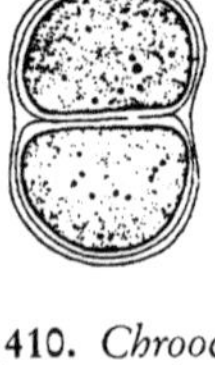

**Fig. 409.** *Chroococcus* cf. *turgidus* (tropical populations): **a** – after Schmidle (1902, from Africa), **b** – after Skuja (1949, from Burma, sub *C. turgidus* var. *maximus*), **c** – after Komárek (1989,from Cuba), **d** – after Welsh (1965, from South Africa, described as *Chroococcus pulcherrimus*), **e** – after Rino (1969, from Mozambique).

**Fig. 410.** *Chroococcus rufescens*, the type species of the genus *Chroococcus* from soil biotopes, but little known; however, the generic characters are evident. The

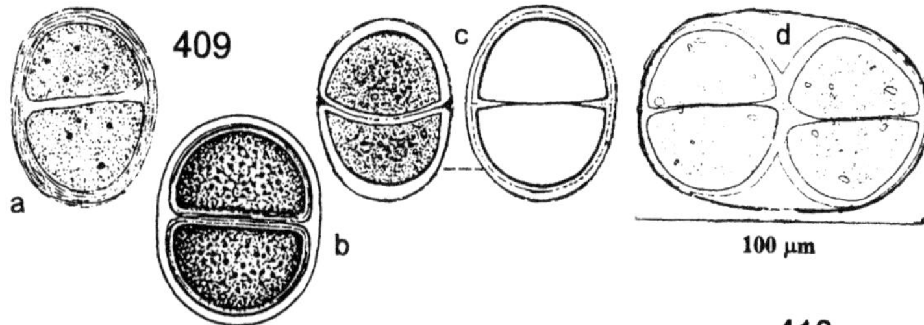


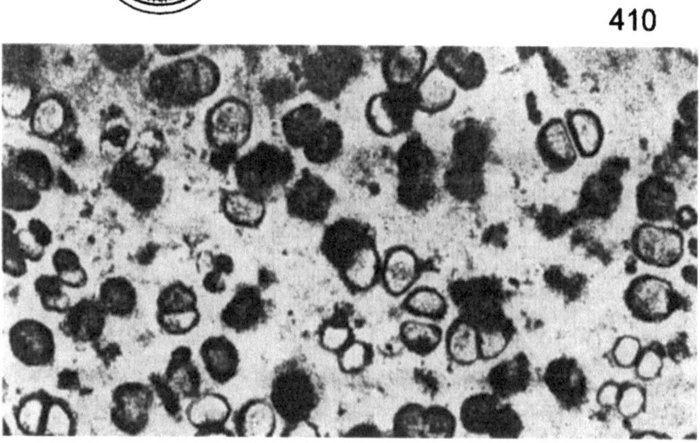


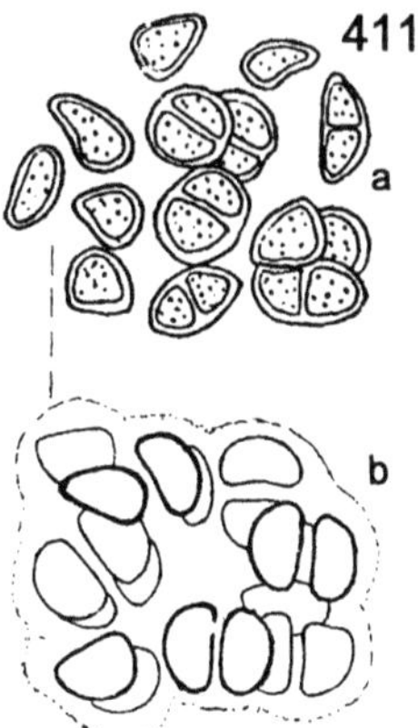


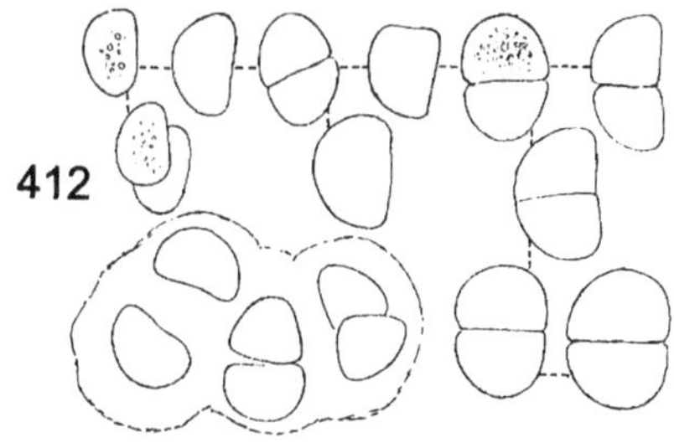

microphoto from Daily (1942) from USA is the only illustration of this species. Several common pantropical species:
**Fig. 411.** *Chroococcus polyedriformis*: **a** – after Schmidle (1902), **b** – after Komárek & Novelo (1994).
**Fig. 412.** *Chroococcus deltoides*: after Komárek & Novelo (1994).
**Fig. 413.** *Chroococcus mipitanensis*: after Skuja (1949, sub *Chroococcus minutus*).

lake on a coral island (SE. Asia); (fig. **376**). Possibly identical with *Cyanokybus venezuelae* (p. 311, fig. **414**). The European and nordic populations probably do not belong to the same type (see, e.g. *C. prescottii*, p. 288, fig. **377–381**).
– *Chroococcus heanogloios* Gardner 1927: Aerophytic, epiphytic species (on *Scytonema*) from Porto Rico.
– *Chroococcus indicus* Zeller 1873: Insufficiently known species, revision needed: in later concepts probably *Cyanosarcina* (e.g., Chu et al. ed. 1991).
– *Chroococcus irregularis* Huber 1929: Fine, planktic species (subgen. *Limnococcus*), described from S. Africa.
– *Chroococcus kerguelensis* Wille 1928: Described from Kerguelen Islands, but not found later (e.g., by Thérézien & Couté 1977).
– *Chroococcus mediocris* Gardner 1927: Aquatic, described from Porto Rico.
– *Chroococcus minutissimus* Gardner 1927: Epiphytic, described from waterfalls in Porto Rico.
– *Chroococcus mipitanensis* (Woloszyńska) Geitler 1925: Common and well defined pantropical species; (fig. **413**).
– *Chroococcus molischii* Schiller 1956: Irregular rounded cells, planktic and metaphytic in saline pools, Caribbean Islands.
– *Chroococcus multicoloratus* Wood 1872: Subaerophytic, described from wet rocks in USA (Philadelphia).
– *Chroococcus muralis* Gardner 1927: Aerophytic species from walls in Porto Rico.
– *Chroococcus mutisii* Gonzalez Guerrero 1947: Aggregates of cells or two-celled colonies; cells 14–18 µm in diameter; described from a river in Argentine.
– *Chroococcus polyedriformis* Schmidle 1902 [syn.: *C. multicellularis* (Chu) Chu 1952]: Probably pantropical species from swamps and ponds (India, China – Szechwan, Mexico, Tanganyika); (fig. **411**).
– *Chroococcus pulcherrimus* Welch 1965: From saline waters in S. Africa; morphologically similar to *C. turgidus*; (fig. **409d**).
– *Chroococcus purpureus* Snow 1903: Planktic species (subg. *Limnococcus*) described from lakes in N. America.
– *Chroococcus refractus* Wood 1872: Aerophytic, described from USA.
– *Chroococcus sarcinoides* Huber 1929: Species with fine cells (subg. *Limnococcus*), planktic in S. Africa.
– *Chroococcus sonorensis* Drouet & Daily 1942: Described from brackish water in Mexico (Baja California).
– *Chroococcus splendidus* Jao 1944: From freshwaters in China (Kwangsi).
– *Chroococcus subsphaericus* Gardner 1927: Aerophytic from walls near thermal springs in Porto Rico.
– *Chroococcus tenacoides* Copeland 1936: From thermal springs in USA.
– *Chroococcus verrucosus* Krieger 1931: Described from mountains (rocks about 1500 m above sea level, near Buea) in Cameroon.
– *Chroococcus* sp. sensu Skuja 1976: Probably from subg. *Limnococcus*, with spherical cells (5–9 µm in diameter), in mucilaginous, up to 32-celled clusters; described from a pond in New Zealand (South Island – Swampy Hill).

**Unrevised European species:**
– *Chroococcus alpinus* Schmidle 1900
– *Chroococcus atrochalybeus* Hansgirg 1892: Described from Croatia (Istria), halophilic.
– *Chroococcus aurantio-fuscus* (Kützing) Rabenhorst 1865
– *Chroococcus aurantius* (Agardh) Wille 1918
– *Chroococcus consociatus* Hariot 1891
– *Chroococcus crassus* (Kützing) Nägeli 1849

- *Chroococcus decoloratus* Migula 1906
- *Chroococcus fuscoviolaceus* Hansgirg 1890
- *Chroococcus glomeratus* (Meneghini) Forti 1907
- *Chroococcus macrococcus* (Kützing) Rabenhorst 1865; *C. macrococcus* var. *salinarum* Hansgirg 1892 was described from Croatia (Dalmatia).
- *Chroococcus rufescens* (Kützing) Nägeli 1849: Type-species of the genus. This species is not commonly known, but the generic identification is doubtless. Description is based on the data according to Daily (1942): Colonies to 64-celled (or more-celled), enveloped by homogeneous or slightly lamellose mucilage; cells olive-blue-green to yellowish or orange-brownish, 3–11 µm in diameter; on moist substrates, in soils or in shallow waters; Daily (1942) records numerous localities from USA, from central Europe (Czech Republic) are known similar populations from glasshouses and from soil; (fig. **410**).
- *Chroococcus smaragdinus* Hauck 1888: Marine, described from Italy (Izola near Trieste).
- *Chroococcus solitarius* Eichler 1896
- *Chroococcus violaceus* Rabenhorst 1865

## 4. **Cyanokybus** Schiller 1956

In colonies; colonies small, few-celled, microscopic, more or less spherical or subspherical, sometimes composed of subcolonies, with cells in short perpendicular rows, more or less distant from one another, sometimes three-dimensionally arranged within colonial slime. Mucilage fine, delimited or diffluent, colourless, forming wide margin around cells. Cells hemispherical, rectangular or cubic-rounded, without or with individual fine envelopes. Monotypic genus.
Cell division changes in three or more planes, later mostly in one plane (origin of short rows); cells later shift within colony and again divide in more planes. Reproduction by disintegration of colonies and by the liberation of solitary cells.
**Important references:** Schiller (1956); Komárek (1994).

**One species outside Europe:**
- *Cyanokybus venezuelae* Schiller 1956: In metaphyton and plankton of coastal swamps and pools on coral islands; known only from Caribbean district (fig. **414**).

## 5. **Cyanosarcina** Kováčik 1988

Colonies microscopic, packet-like or subspherical-irregular, more or less rounded, with densely aggregated cells, enveloped by thin, usually firm, colourless envelopes, adjacent to cell clusters; finally in dense, sarcinoid packets, forming agglomeration of macroscopic, irregular mats. Cells spherical, subspherical, oval, hemispherical or irregular-rounded, enveloped by thin mucilaginous sheaths, usually packed closely to one another.
Cell division irregular in three, later different planes; daughter cells do not reach the original form before next division; nanocytes (or baeocytes) always lacking. Reproduction by the liberation of cells or groups of cells from sheaths.
**Important references:** Anagnostidis & al. (1983; sub *Myxosarcina*); Komárek & Anagnostidis (1986); Kováčik (1988); Anagnostidis & Pantazidou (1991).

**Key to the species:**
1a Marine species, living on littoral calcareous substrates; cells 2–3 µm in diameter . . . . . . . . . . . . . . . . . . . . . . . . . . . . . . . . . . . . . . . **1. C. thalassia**

1b Aerophytic and freshwater species, cells over (2.5) 3 µm in diameter . . . **2**
2a Cells 2.5–6 µm in diameter . . . . . . . . . . . . . . . . . . . . . . . . . . . . . . **3**
2b Cells over 6 µm in diameter; in moorland waters . . . **5. C. chroococcoides**
3a Aerophytic (or freshwater ?) species; old colonies with more than 4 (8) cells . . . . . . . . . . . . . . . . . . . . . . . . . . . . . . . . . . . . . . . . . . . . . . . . . . . . . . . . **4**
3b Thermophilic species, in thermal mud; all colonies small, only 4 – (8)-celled, but sometimes aggregated together . . . . . . . . . . . . . . . . . **4. C. thermalis**
4a Epilithic and chasmoendolithic on calcareous substrates; cells olive-green, yellow-green or brownish, 2.5–5 µm in diameter . . . **2. C. parthenonensis**
4b Known only from cultures and glasshouses, living aerophytically or submersely, preferentially in pH < 7; cells bright blue-green, 3.5–5.4 µm in diameter . . . . . . . . . . . . . . . . . . . . . . . . . . . . . . **3. C. huebeliorum**

## 1. **Cyanosarcina thalassia** Anagnostidis et Pantazidou 1991 (fig. **418**)

Colonies microscopic, containing 2–32 cells, 12 × 12 µm in diameter, or composed of many-celled colonies, arranged in more or less cubic packet-like aggregates, with thin, colourless and structureless mucilaginous envelopes; thallus pale blue-green. Cells more or less spherical, after division hemispherical or subspherical, 2–3 µm in diameter, with blue-green, rarely olive-green, finely granulate protoplasts.
Occurrence: Marine, epilithic on marine calcareous substrates; described from the Greek coasts (Aegean Sea), recorded also from Brazil (SP).

## 2. **Cyanosarcina parthenonensis** Anagnostidis in Anagnostidis et Pantazidou 1991 (fig. **415**)

*Myxosarcina concinna* sensu Anagnostidis et al. 1983 p.p.

Thallus olive-green, consisting of 2–16-celled colonies, later compound and multicellular, in more or less cubic, dense, packet-like aggregates, 10–37 µm in

CYANOKYBUS

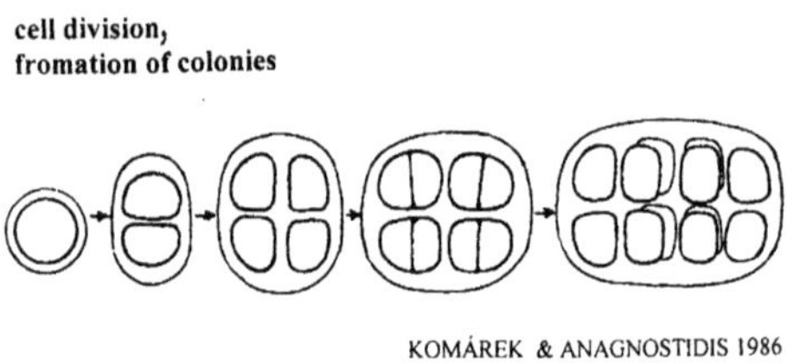


**Fig. 414.** *Cyanokybus venezuelae*: **a** – iconotype after Schiller 1956; **b** – young colony; **b-c** – after Komárek (1994).

CYANOSARCINA

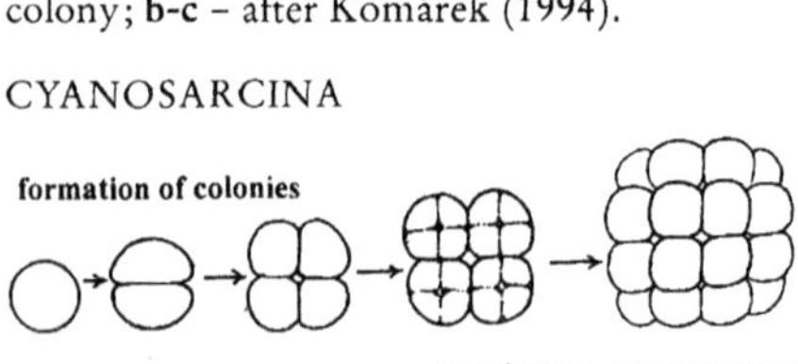

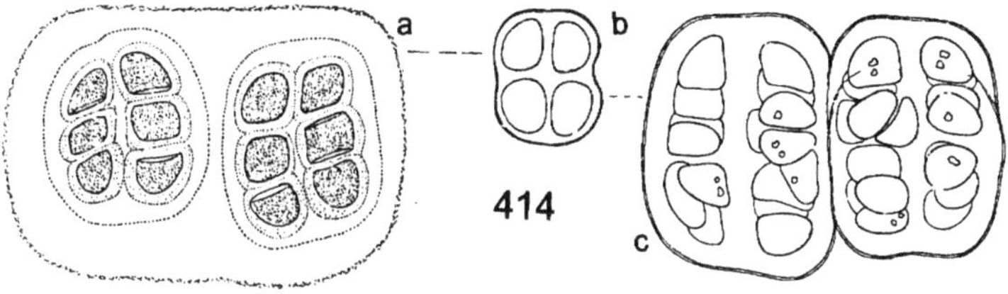


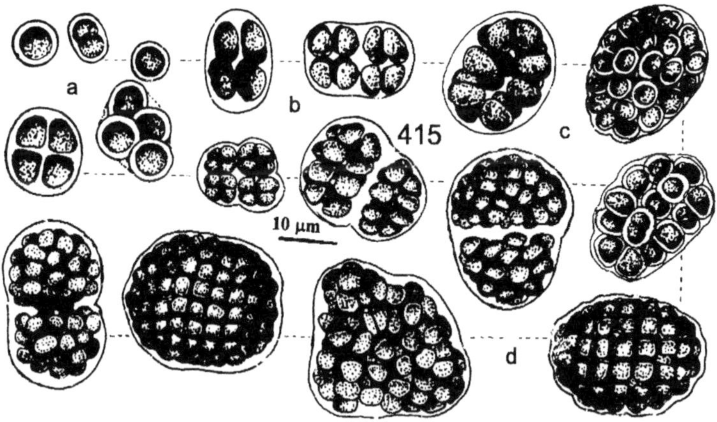


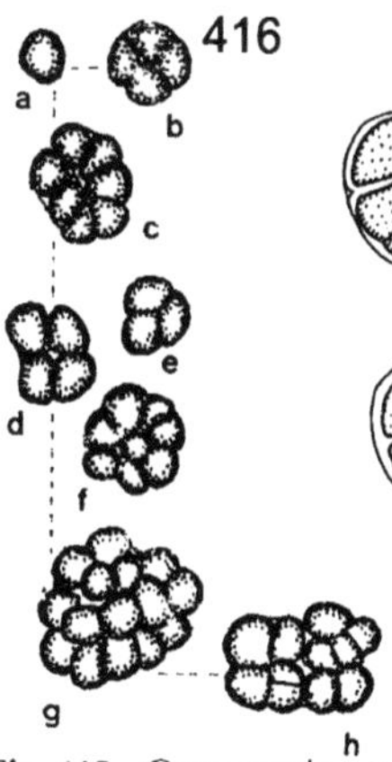


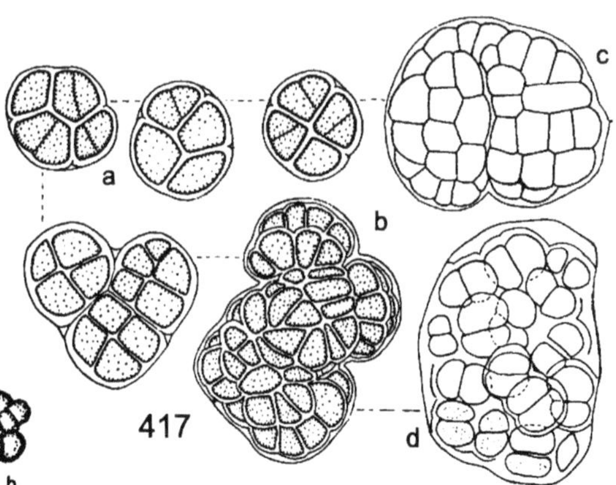


**Fig. 415.** *Cyanosarcina parthenonensis*: **a** – young colonies, **b-c** – colonies of various age; after Anagnostidis & al. (1983, sub *Myxosarcina concinna*).
**Fig. 416.** *Cyanosarcina huebeliorum*: after Komárek & Anagnostidis (1995).
**Fig. 417.** *Cyanosarcina chroococcoides*: **a** – young colonies, **b-c** – old colonies; **a-c** – after Geitler (1927, sub *Myxosarcina chroococcoides*), **d** – after Hoffmann (1986, sub *Myxosarcina chroococcoides*).

diameter, with thin, colourless and structureless, mucilaginous envelopes. Cells spherical, after division subglobose or subspherical, 2.5–4.5 (5) µm in diameter, with olive-green, yellow-green, rarely olive-brown, finely granulate protoplasts.
Occurrence: Aerophytic, chasmoendolithic on calcareous substrate; described and known only from marble statues and other substrates of the ancient Parthenon in Athens (Greece).

### 3. **Cyanosarcina huebeliorum** Komárek et Anagnostidis 1995 (fig. **416**)

Solitary cells or colonies; colonies microscopic, 2–4 – (32 – more ?)-celled, with irregularly or packet-like cell aggregates, sometimes packed together in a macroscopic, intensely blue-green, slightly granular thallus. Sheaths thin, firm, colourless, tightly enveloping cells or cell groups, without visible lamellation; sheaths sometimes split and then the inner vegetative cells can be liberated (without nanocyte formation !). Cells spherical, subspherical or irregular-rounded, bright blue-green, 3.5–5.4 µm in diameter, with more or less homogeneous content.
Occurrence: Described from a culture from the collection of Greifswald University, Germany (strain Greifswald-149), origin of the strain is uncertain (cultivated in liquid medium without high salt requirements), probably common in glasshouses (e.g., morphologically identical populations are known from glasshouses in Slovakia).

### 4. **Cyanosarcina thermalis** (Hindák) Kováčik 1988 (fig. **419**)

*Myxosarcina thermalis* Hindák 1978; *M. hindakii* Bourrelly 1985

Colonies small, composed of 4-celled packet-like aggregates (tetrads), to 15 µm in diameter, without visible mucilaginous envelopes (in light microscope), large, multicellular packets were never observed. Cells irregularly rounded, 3–6 µm in diameter, after division hemispherical, with blue-green, homogeneous protoplasts.
Occurrence: Thermophilic, in thermal mud; described from thermal springs in Piešt'any, Slovakia.

### 5. **Cyanosarcina chroococcoides** (Geitler) Kováčik 1988 (fig. **417**)

*Myxosarcina chroococcoides* Geitler 1927 (non sensu Starmach 1980, nec var. *tatrica* Starmach 1981 = *Myxosarcina*)

Colonies microscopic, more or less subspherical or irregular-rounded, composed of densely, irregularly aggregated cells of different size in sarcinoid, packet-like aggregates. Mucilaginous envelopes around colonies thin (about 1 µm thick), firm, colourless, tightly surrounding cell clusters, in old colonies sometimes diffluent. Cells irregular-rounded, 9–10 µm in diameter, olive-green or blue-green.
Occurrence: Metaphytic, in moorland waters and swamps; described from central Europe (Austria), probably with wide distribution over the whole temperate zone; «*Mixosarcinia chroococcoides*» from winter algal flora of ponds in Waldviertel, Austria (Wawrik 1978) is morphologically very different (fig. **423**); specimens from Luxembourg (Hoffmann 1986; fig. **417d**) have cells only 5 µm in diameter.

**Species outside Europe:**
– *Cyanosarcina burmensis* (Skuja) Kováčik 1988 [syn.: *Myxosarcina burmensis* Skuja 1949]: Cells 2–3 µm in diameter, planktic and metaphytic in shallow waters; known from Burma and India; (fig. **420**).

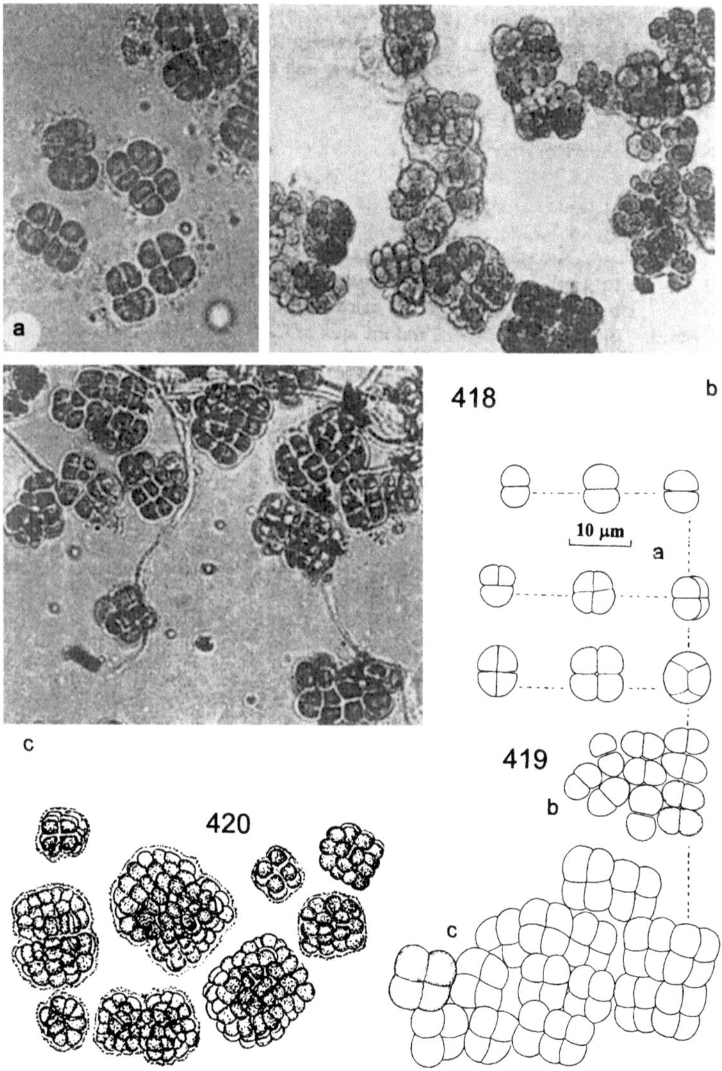


CYANOSARCINA

**Fig. 418.** *Cyanosarcina thalassia*: from Anagnostidis & Pantazidou (1991).

**Fig. 419.** *Cyanosarcina thermalis*: **a** – cell division, **b-c** – colonies; after Hindák (1978, sub *Myxosarcina thermalis*).

**Fig. 420.** *Cyanosarcina burmensis*: after Skuja (1949, sub *Myxosarcina burmensis*).

- *Cyanosarcina ferruginea* M. Watanabe et Komárek 1995: Cells 2–6.3 µm in diameter, envelopes yellow-brown (iron precipitates); metaphytic in cold katharobic waters of Himalayas, E. Nepal.
- *Cyanosarcina fontana* Kováčik 1988: From springs in central Asia, Armenia (Jerevan); (fig. **421**).
- *Cyanosarcina gelatinosa* (Emoto et Yoneda) Kováčik 1988 [syn.: *Myxosarcina gelatinosa* Emoto et Yoneda 1941]: Cells 2.5–6 µm in diameter; from thermal springs in Japan.
- *Cyanosarcina littoralis* (Schwabe) Kováčik 1988 [syn.: *Myxosarcina littoralis* Schwabe 1944]: Cells 4–6 µm in diameter; epilithic on granite rocks, thermal springs near sea shore (intertidal zone); known from Chile (Llancahue Isla).
- *Cyanosarcina regularis* (Geitler) Komárek et Anagnostidis 1995 [syn.: *Myxosarcina spectabilis* f. *regularis* Geitler in Geitler & Ruttner 1935]; Cells 3–6 µm in diameter; described from thermal springs in Indonesia.
- *Cyanosarcina spectabilis* (Geitler) Kováčik 1988 [syn.: *Myxosarcina spectabilis* Geitler 1933, non var. *decolorata* Varma et Mitra 1962]: Cells 6–8 µm in diameter; tropical species, described from thermal springs in Indonesia; (fig. **422**).

**Unrevised (probably *Cyanosarcina*) species:**
- *Endospora rubra* Gardner 1927: Aerophytic on rocks and wood, described from Porto Rico.
- *Microcystis splendens* Hollenberg 1939: Described from cultures of marine algae in S. California, USA.
- *Myxosarcina* sp. sensu Rippka et al. 1979: Known from cultures.
- *Oncobyrsa hispanica* M. Lewin 1888: Described from Spain, revision needed (fig. **424A**).

## 6. **Pseudocapsa** Ercegović 1925

Colonies spherical or irregularly packet-like, in aggregates, microscopic (aggregates often of macroscopic size), with radially or fan-like oriented, later irregularly, densely aggregated cells; colonies enveloped by thin, firm, smooth, colourless, sometimes striated sheaths; often composed of subcolonies. Cells irregular spherical, after division hemispherical, later irregular or in the form of

---

CYANOSARCINA
**Fig. 421.** *Cyanosarcina fontana*: **a** – cell division, **b** – colonies; after Kováčik (1988).
**Fig. 422.** *Cyanosarcina spectabilis*: **a** – young colonies, **b** – old colonies; after Geitler & Ruttner (1935, sub *Myxosarcina spectabilis*).
**Fig. 423.** *Cyanosarcina ?* sp.: after Wawrik (1978, sub «*Mixosarcinia chroococcoides*»).

PSEUDOCAPSA

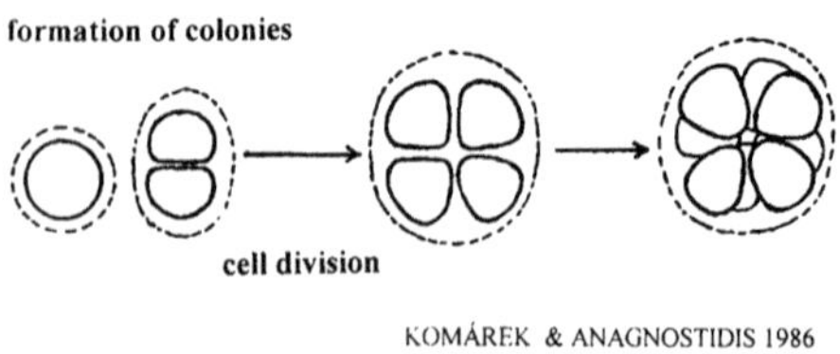

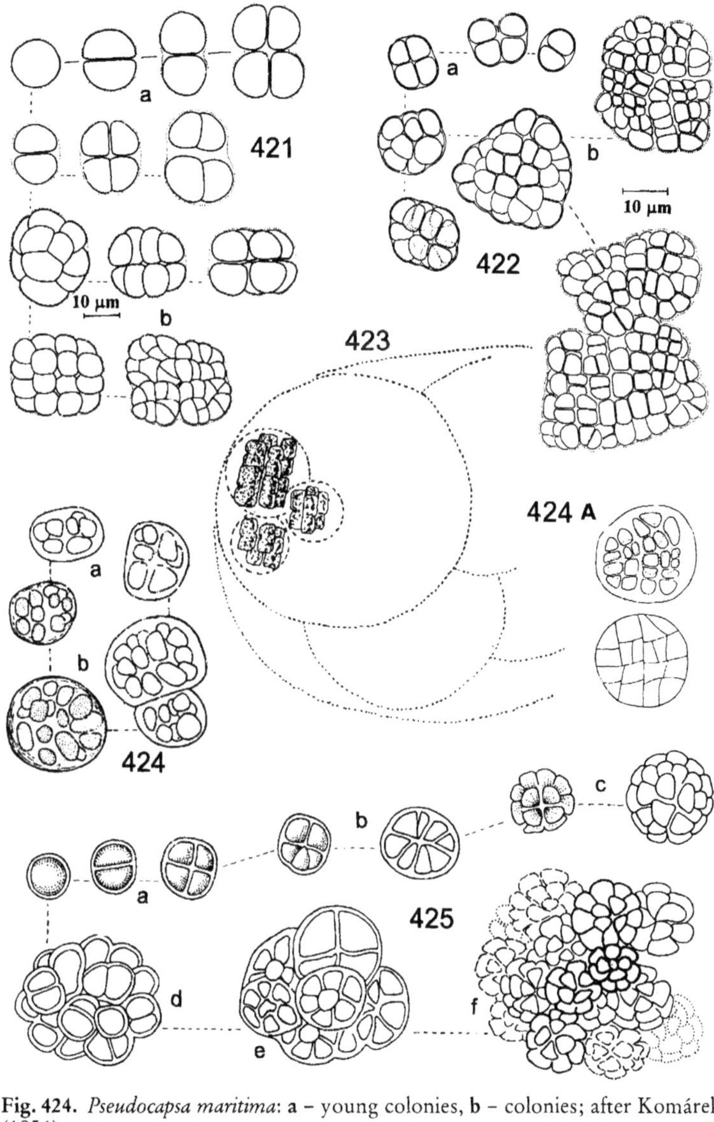


**Fig. 424.** *Pseudocapsa maritima*: **a** – young colonies, **b** – colonies; after Komárek (1956).
**Fig. 424A.** *Pseudocapsa ?* sp.: after M. Lewin (1888, sub *Oncobyrsa hispanica*, possibly *Cyanosarcina* or *Pseudocapsa* species).
**Fig. 425.** *Pseudocapsa sphaerica*: **a-c** – development of colonies, **d-f** – colonies; after Proškina-Lavrenko (1951, sub *Myxosarcina sphaerica*).

segments of a sphere, more or less radially oriented in the colonies; in old colonies the cells are densely and irregularly aggregated.
Cell division successively in three or more different planes (mainly radially in colonies); cells do not grow to the original, more or less spherical form before the next division. Nanocytes always lacking.
**Important references:** Ercegović (1925); Starmach (1936, 1966); Komárek & Anagnostidis (1986); Kováčik (1988); Abdelahad (1989); Anagnostidis & Pantazidou (1991).

**Key to the species:**
1a Cells less than 5 µm in diameter . . . . . . . . . . . . . . . . . . . . . . . . . . . . . . **2**
1b Cells 4–8 µm in diameter; metaphytic in water bodies with high content of mineral salts . . . . . . . . . . . . . . . . . . . . . . . . . . . . . . . . **3. P. sphaerica**
2a Cells 1–5 µm in diameter, from marine littoral . . . . . . . . . . **1. P. maritima**
2b Cells 2.5–4.5 µm in diameter; aerophytic species from limestone rocks . . . . . . . . . . . . . . . . . . . . . . . . . . . . . . . . . . . . . . . . . . **2. P. dubia**

## 1. Pseudocapsa maritima Komárek 1956 (fig. **424**)

Colonies more or less spherical, 6.5–12 µm in diameter, solitary or in aggregates and composed of 2–3 subcolonies. Mucilage firm, delimited, colourless, homogeneous or very slightly striate at the margin. Cells pale blue-green, with homogeneous content, spherical or polyhedral-rounded, 1–5 µm in diameter. Insufficiently described, needs revision (comp. with *Gloeocapsopsis crepidinum*).
Occurrence: Metaphytic (among detritus), together with other cyanoprokaryotes and algae on the stony marine littoral, known only from coast of Black Sea in S. Bulgaria, but recorded also from Atlantic seashore in Brazil (SP).

## 2. Pseudocapsa dubia Ercegović 1925 (fig. **426**)

Colonies spherical or irregularly spherical, sometimes slightly flattened or composed of several cells, forming 2–16-celled subcolonies, surrounded by thick, firm, colourless, delimited, structureless sheaths; sometimes in clusters reaching 150(170) µm in diameter. Mucilaginous envelopes thin, firm, yellowish to brownish or colourless, layered. Cells spherical or hemispherical, later irregular-rounded, in old colonies more or less radially arranged, without individual envelopes (exceptionally in reproduction stages), with yellowish or blue-green content, 2.5–4.5(–10) µm in diameter.

PSEUDOCAPSA

428

c

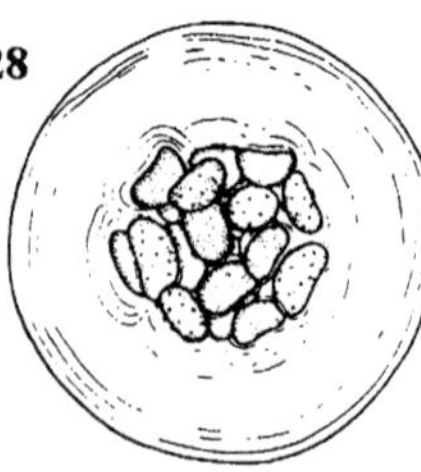

**Fig. 426.** *Pseudocapsa dubia*: **a** – development of colonies, **b-e** – variation in colonies, **f** – formation of resting cells; **a,d-f** – after Starmach (1966), **b** – after Ercegović (1925), **c** – after Palik (1938).
**Fig. 427.** *Pseudocapsa venkataramanii*: development of a colony on agar plate – variation and shape of colonies; after Kováčik (1988).

NEPHROCOCCUS
**Fig. 428.** *Nephrococcus confertus*: **a** – young colonies, **b-c** – colonies; after Li (1984).

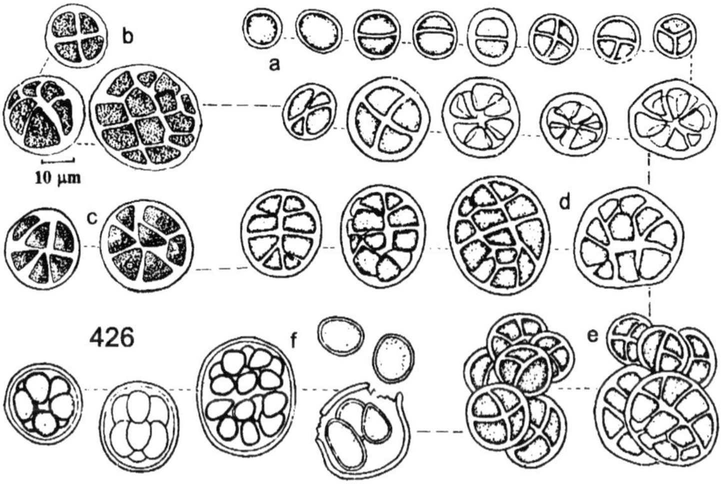


426

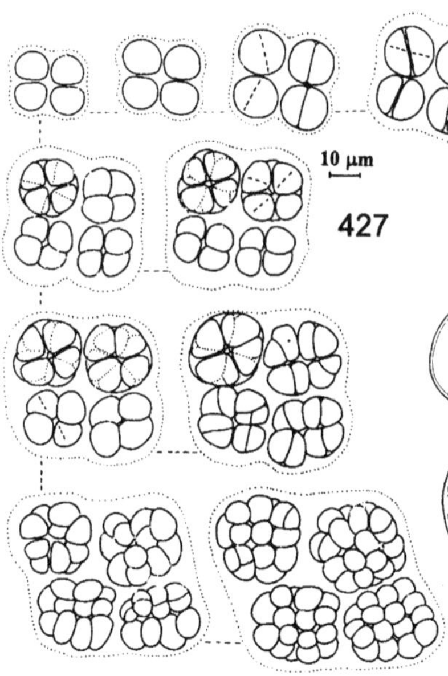


427

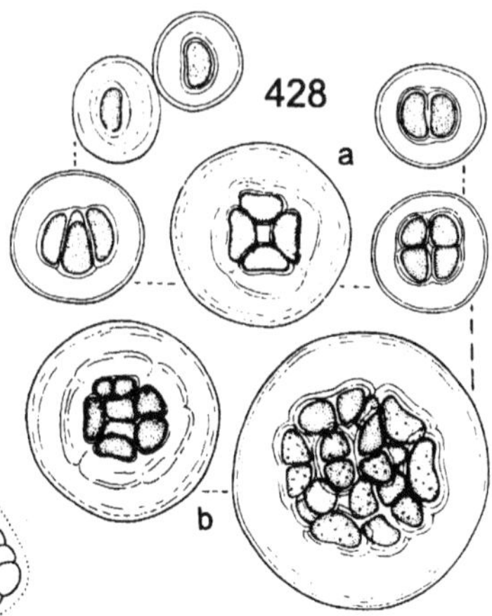


428

Occurrence: Aerophytic, epilithic on wet limestone rocks and shaded walls (entrances of caves); Croatia, Greece (island Crete), Italy, Poland (High Tatra Mts.).

**3. Pseudocapsa sphaerica** (Proškina-Lavrenko) Kováčik 1988 (fig. **425**)

*Myxosarcina sphaerica* Proškina-Lavrenko 1951

Colonies solitary or composed of several subcolonies, spherical, yellowish blue-green, to 150 µm in diameter, without visible gelatinous envelopes or with distinct, delimited, colourless envelopes. Cells (to 30–60 in a colony) arranged within colonial slime more or less radially (result of more or less radial division of cells), initially spherical, later hemispherical or in the form of a segment of a sphere, 4–8 µm in diameter, pale blue-green.

Occurrence: Metaphytic among algae in waters with high content of mineral salts, known only from the Ukraine (Sivas).

**Species outside Europe:**

– *Pseudocapsa venkataramanii* Kováčik 1988: Cells 7–15 µm in diameter, isolated from soil in India; (fig. **427**).

**Unrevised species:**

Probably other species belong to the genus *Pseudocapsa*, described under various names, e.g., *Gloeocapsa granosa* in Geitler 1925, *Myxosarcina chroococcoides* sensu Starmach 1989, and others.

## 7. Nephrococcus Li 1984

Colonies micro- to macroscopic, gelatinous, composed of more or less spherical subcolonies, forming a granular, slimy mass. Subcolonies (1) 2–4-many-celled, enveloped by mucilaginous, wide, stratified or unstratified, sometimes yellow-brownish coloured envelopes, distinctly delimited at the surface. Cells aggregated in the centre of subcolony, oval and slightly arcuate to kidney-shaped, or irregular-rounded after intense division, in an indistinct wreath-shaped arrangement; around cells sometimes indistinctly striated gelatinous envelopes.

Cells divide by binary fission, irregularly, mainly perpendicular to the longer cell axis (origin of the cell arrangement). Reproduction by disintegration of colonies.

**Important references:** Li (1984); M. Watanabe & Komárek (1988); Chu & al. ed. (1991).

**Species outside Europe:**

– *Nephrococcus confertus* Li 1984: Aerophytic, described from the bark of trees; S. China (fig. **428**).

– *Nephrococcus* sp. in M. Watanabe & Komárek 1988: Aerophytic, from wet rocks in Himalayas, slightly different from *N. confertus.*

## 8. Cyanostylon Geitler 1928

*Stilocapsa* Ley 1947

Colonies microscopic to macroscopic, mucilaginous, irregular, attached to the substrate or free living, composed of several irregularly or more or less radially arranged, thick and wavy gelatinous strands, which sometimes change into an amorphous mass, particularly in basal parts of a colony; cells are placed solitary or in small groups in terminal parts of strands, very rarely in short, few-celled rows. Gelatinous strands colourless, sometimes indistinctly or distinctly transversely lamellated. Cells spherical, after division hemispherical.

Cell division in two or more planes in successive generations, perpendicular to one another, later irregularly (successive binary fission); reproduction by solitary cells and by disintegration of colonies.
**Important references:** Geitler (1927, 1932, 1942); Skuja (1964); Hindák (1988b).

**Key to the species:**

1a Colonies free living and floating in lake plankton and metaphyton, in form of single strands or composed of 2 up to 10 fine, irregularly divaricated, not lamellated strands . . . . . . . . . . . . . . . . . . . . . . . . . . . . **2. C. plancticum**

1b Colonies joined to the substrate, with distinct, lamellated gelatinous strings **2**

2a Cells elongated, 3–12 × 3–6 µm . . . . . . . . . . . . . . . . . . . . **1. C. ovoideum**

2b Cells more or less spherical, 2.5–3 µm in diameter . . **3. C. microcystoides**

## 1. Cyanostylon ovoideum Skuja 1964 (fig. 429)

Colonies microscopic, shrub-like, partly creeping radially on the substrate, partly erected, composed of thick mucilaginous stalks, which are simple or (rarely) irregularly pseudodichotomously divaricated, with cells at the ends; stalks colourless, more or less transversely lamellated, 3.5–8 µm wide, up to 40 µm long. Cells oval, elongate-obovate or slightly pear-shaped, 3–12 × 3–6 µm, solitary or geminate after division, in stalks always terminally and radially localized, enveloped by thin layer of mucilage; cell content greyish olive-green, homogeneous, with distinct chromatoplasm; cell division probably only crosswise (number of planes ?).
Occurrence: Epilithic on stones among other algae, submersely in streams or in littoral of clear lakes; N. Sweden.

## 2. Cyanostylon plancticum Hindák 1988 (fig. 431)

Colonies microscopic, free floating, composed of simple or bifurcate to shrub-like, thick mucilaginous stalks with 2–8, rarely more cells; stalks slightly concentrically layered (not transversely striated), 10–45 µm in diameter, 80–200 µm long, pseudodichotomic ramified, hyaline, sometimes indistinct, with cells at the apical parts, in old colonies changing in irregular mucilaginous clusters with dispersed cells on the periphery. Cells solitary or in twos after division, shortly oval to hemispherical or slightly irregular, 10–13 (15) × 7–9 (11) µm, with bright blue-green content, without conspicuous inclusions. Disintegration of colonies by separation of individual stalks with cells.
Occurrence: Described from summer plankton of lake Volvi, N. Greece; probably known also from other localities.

## 3. Cyanostylon microcystoides Geitler 1928 (fig. 430)

Colonies microscopic, attached to the substrate, composed of more or less radially oriented, but irregularly waved mucilaginous stalks, which are colourless, slightly transversely lamellated; cells localized at the ends of stalks, solitary or in groups of 2–4 cells, rarely in short rows. Cells spherical or hemispherical after division, 2.5–3 µm in diameter, intensely blue-green with solitary granules and visible chromatoplasmic area, with own narrow envelopes; common mucilage fine, diffluent.
Occurrence: In littoral (*Schizothrix*-zone) of Alpine lakes and in waterfalls, in calcareous areas; recorded from Austria, Poland (High Tatra Mts.), N. Greece.

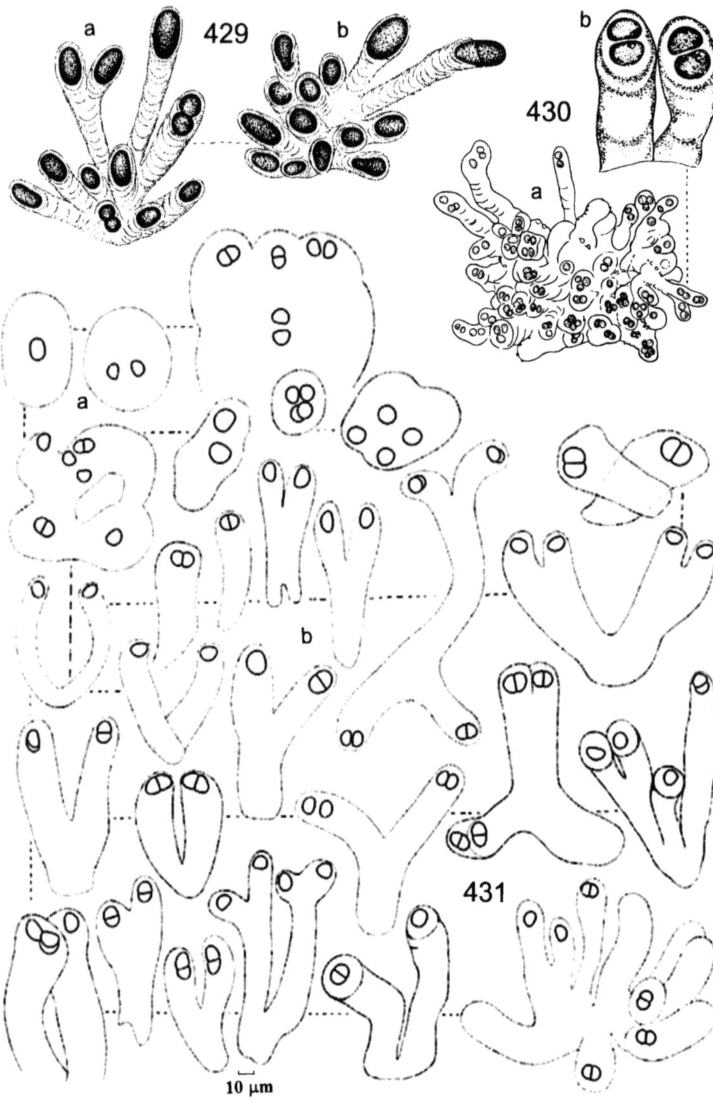
a
429
b
b
430
a
a
b
431
10 µm

**Species outside Europe:**
- *Cyanostylon gelatinosus* Azevedo et Sant'Anna 1994: Cells subspherical, 1.5–3.5 (4) µm in diameter, stalks hyaline, lamellated, subaerophytic from wet walls of an open cave near the sea shore, Brazil.
- *Cyanostylon giganteum* Liang 1987: Cells (2.5) 7.2–12.5 µm in diameter, described from China; (Chu et al. ed. 1991).
- *Cyanostylon purpureum* Chu et Chen 1985: Cells 0.2–1.5 µm in diameter, described from China; (Chu et al. ed. 1991).
- *Cyanostylon sinense* Chu 1952: Cells 2.5–9 µm in diameter, described from China (Szechwan); (Chu et al. ed. 1991).
- *Cyanostylon sinicum* (Ley) Bourrelly 1970 [syn.: *Stilocapsa sinica* Ley 1947]: Cells 2.5 µm in diameter, stipes up to 35 µm wide; aerophytic on wet rocks in China (N. Kwangtung).

**Not revised species:**
- *Stilocapsa fujianica* Liang 1983: Cells 5–9.5 µm in diameter, with numerous cells at the end of stalks; described from China.

Notes to the family *Chroococcaceae*: Within the family *Chroococcaceae* were described two genera, the taxonomic value of which is not solved and which consequently have to be revised:
**Anacystis** Meneghini 1837: The type species *A. marginata* was described from periphyton in springs in Italy; it is characterized by aggregates of oval cells within firm envelopes (this original concept resembles strikingly the genus *Cyanogastrum* Schiller 1956 = *Aphanothece* subg. *Cyanogastrum* - see p. 71). Later, this species was recorded several times in various taxonomic concepts, also as a planktic *Microcystis*.
The original Meneghini's data (particularly drawings) reminiscent the concept of periphytic or aerophytic colonial cyanoprokaryotes with elongate cells (without or with indistinct individual sheaths) and a special life cycle: cells divide repeatedly within the common firm, distinct and delimited gelatinous envelopes, later are liberated from the «mother» envelope by rupture, and are able to start a new cycle. This concept seems to be nearest to that of Gardner (1927), who described 19 *Anacystis* taxa from Porto Rico: several of them were transferred to the genus *Asterocapsa* (*A. magnifica*, *A. pulchra*), however, the taxonomic character of the other and the whole genus *Anacystis* corresponds to Meneghini's original concept

CYANOSTYLON

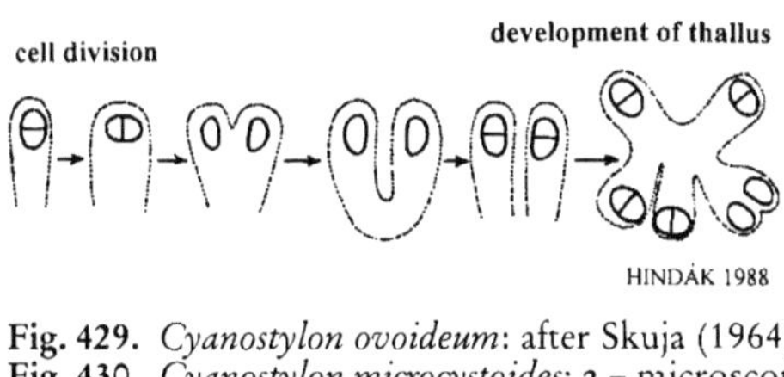


**Fig. 429.** *Cyanostylon ovoideum*: after Skuja (1964).
**Fig. 430.** *Cyanostylon microcystoides*: **a** – microscopic colony, **b** - detail of ends of stalks with divided cells; after Geitler (1928).
**Fig. 431.** *Cyanostylon planctıcum*: **a** – initial stages of colonies, **b** – variability of colonies; after Hindák (1988).

(*Anacystis nigropurpurea*, *A. nigroviolacea*, *A. compacta*, *A. distans*, *A. microsphaeria*, *A. amplivesiculata*, *A. gloeocapsoides*, *A. nidulans*, *A. minutissima*, *A. radiata*). «*A. compacta* Gardner» of Yamagishi (1992) from Taiwan (but from plankton), or *Aphanocapsa endolithica* Ercegović 1925 from Croatia (endolithic in limestone rocks in Plitvica region) are also similar to Gardner's concept. Daily (1942) included several species from the genera *Aphanothece* and *Gloeothece* into the g. *Anacystis*. The problematic aspects of this genus or subgenus (*Cyanogastrum* Schiller ?) will have to be solved in the future.
The generic name «*Anacystis*» is better known in modern literature in connection with the specific epithet «*nidulans*». The strain of Kratz & Allen (Bloom. 625 = PCC 6301 = ATCC 27144) which was designated as «*Anacystis nidulans*» by Drouet, is commonly used, and now is famous as an experimental cyanoprokaryotic model organism, but it evidently belongs to the genus *Synechococcus* (see *S. nidulans*, p. 120) and differs distinctly from *Aphanothece nidulans* Richter (p. 81–82). The taxon «*Anacystis nidulans* (Richter) Drouet et Daily 1952» (based on the basionym *Aphanothece nidulans* Richter 1884) was later included by Drouet & Daily (1956) in the synonymy of another species accepted by Drouet, «*Anacystis marina*», and the type species of the genus *Anacystis* (*A. marginata*) into the very broadly conceived *A. montana* which contains numerous species of *Gloeocapsa*, *Gloeothece*, *Chroococcus*, *Aphanocapsa* and *Microcystis*. Thus, the name «*Anacystis nidulans*» was not accepted as a valid species by Drouet & Daily (1956). However, this invalid name has remained in experimental papers as a monument.
**Endospora** Gardner 1927: The genus was described with 5 species from aerophytic and soil localities in Porto Rico. It has a similar life cycle to *Anacystis* (in Gardner's sense, see above), but the cell division and aggregation of cells in a colony are more like those of *Cyanosarcina* (particularly in the type species, *E. rubra*); also the gelatinous envelopes are tightly adherent to the cell cluster and do not form a wide margin around the cells. In Gardner's concept it is a very heterogeneous genus (several species resemble *Chroococcus* or *Chroococcidiopsis*, e.g., *E. bicoccus*); (comp. also *Asterocapsa*; p. 271–274).

## 6. Family **Entophysalidaceae** Geitler 1925

*Chlorogloeaceae* Geitler 1925 incl.

Colonies mucilaginous, polarized, more or less spherical, elongate or irregular with facultative pseudofilamentous cell orientation, with 1 (rarely) or many cell layers, sometimes consisting of pseudofilamentous rows of separate cells. Cells in young colonies spherical, subspherical, oval or irregular-rounded, in groups, usually distant from one another, not physiologically joined, but (particularly in marginal parts) in radial rows and sometimes with individual gelatinous envelopes. Mucilage firm, distinct, delimited, or fine. Thallus not differentiated in spite of a certain degree of polarity (different cell form in basal parts, radial rows of cells, etc.).
Cell division successive in various planes, or (in older stages or on the margin of colonies) in one plane, perpendicular to the radially oriented cells. Reproduction by solitary cells, monocytes, or by clusters of cells; nanocyte division lacking (?).

**Key to the subfamilies:**

1a All cells spherical, subspherical or irregular, but never polarized (only polarized colonies) . . . . . . . . . . . . . . . . . . . . . . **1. Entophysalidoideae**
1b Initial, young cells polarized, sometimes *Chamaesiphon*-like, later more or less spherical to irregular-rounded; colonies also polarized . . . . . . . . . . . . . . . . . . . . . . . . . . . . . . . **2. Siphononematoideae**

# 1. Subfamily **Entophysalidoideae** Komárek et Anagnostidis 1986

Without special polarized initial cells; colonies spherical or irregular, with cells arranged more or less in rows, sometimes polarized, irregularly or radially oriented (rows recognizable mainly in the marginal parts of colonies); firm gelatinous envelopes around colonies and individual envelopes around single cells usually produce a compact thallus. Cells initially mainly spherical or oval, later irregular-rounded.

**Key to the genera:**

**1a** Cells without distinct individual envelopes or with simple envelopes without stratification; colonies micro- or macroscopic . . . . . . . . . . . . . . . . . . . . 2
**1b** Cells always within stratified individual envelopes; colonies usually large, macroscopic . . . . . . . . . . . . . . . . . . . . . . . . . . . . . . . . . . . . 4
**2a** Thallus with iron precipitates . . . . . . . . . . . . . . . . . . . . . . **1. Paracapsa**
**2b** Thallus without iron precipitates . . . . . . . . . . . . . . . . . . . . . . . . . . 3
**3a** Thallus composed of «fascicles» of parallely arranged distinct rows of ensheathed cells (confluent individual envelopes around cells) **2. Lithocapsa**
**3b** Thallus irregular, cells in rows only in young colonies and at the colonial margin; cells without or with individual simple envelopes **3. Chlorogloea**
**4a** Colonies ± amorphous, at the margin slimy; the radial organization of cells is recognizable only within colony . . . . . . . . . . . . . . . . . . **4. Entophysalis**
**4b** Colonies firm, pseudodichotomously or irregularly ramified, with firmly delimited surface, polarized; cells organized irregularly, ± radially only at the colonial margin . . . . . . . . . . . . . . . . . . . . . . . . . . . . . . . . **5. Cyanoarbor**

Note: Interesting members of the *Chroococcaceae* or *Entophysalidaceae* were found as phycobionts of the lichens *Phyllisciella* and *Gonohymenia* (*Lichinaceae*), described by Henssen & Büdel (1984) and Büdel (1985) from central Europe (Czech Republic) and from the southern hemisphere (subantarctic Prince Edward Islands, Patagonia, New Zealand). Taxonomically not yet evaluated.

## 1. **Paracapsa** Naumann 1924

Thallus attached to the substrate, hard, leathery, more or less spherical or hemispherical, often with abundant ferric precipitations, to 0.5 cm in diameter. Cells arranged irregularly within structureless, irregular mucilage and with individual gelatinous envelopes, or in irregular rows, sometimes slightly radially oriented. Cells more or less spherical or irregular.
Cell division not described, but probably irregular and in various planes in successive generations. Little known, unrevised, monotypic genus; the ferric precipitates can hardly be used as the only generic differential feature, in other characters similar to *Chlorogloea*.
**Important references:** Naumann (1924); Geitler (1932, 1942); Bourrelly (1970); Komárek & Anagnostidis (1986).

### 1. **Paracapsa siderophila** Naumann 1924 (fig. **432**)

Colonies hemispherical, hard, tightly attached to the stony substrate. Cells with or without individual envelopes, in clusters or situated in irregular rows, 2.5–5 μm in diameter; gelatinous envelopes not stratified. Iron precipitates on the surface of the mucilage.

Occurrence: Freshwater, in stony littoral of lakes, growing on the stone surface, particularly in localities with high content of iron; S. Sweden. Little known species, needs revision; *Paracapsa* differs from the g. *Chlorogloea* only in the presence of ferric precipitates.

## 2. **Lithocapsa** Ercegović 1925

Colonies mucilaginous, microscopic, composed of distinct, parallely arranged and fasciculate rows of cells which are slightly distant from one another, situated within individual gelatinous, sometimes slightly lamellate, colourless envelopes, solitary or joined together and thus forming gelatinous «fascicles», which are sometimes attached by one end to the substrate; the mucilaginous envelopes surrounding cells are confluent laterally, but rows («mucilaginous fascicles») are not enveloped by any common sheath or mucilage (rows of cells or their «fascicles» are located more or less separately). Cells spherical or subspherical, after division hemispherical, probably with homogeneous content.
Cell division by binary fission in one or several (?) planes (sometimes probably in different directions in successive generations, but with one plane of division predominating). Reproduction not described, probably by clusters of cells enveloped by gelatinous sheaths. Little known genus, still of uncertain taxonomic status.
**Important references:** Ercegović (1925); Geitler (1932, 1942); Bourrelly (1970); Komárek & Anagnostidis (1986).

### 1. **Lithocapsa fasciculata** Ercegović 1925 (fig. **433**)

*Chlorogloea fasciculata* (Ercegović) Bourrelly 1970

Gelatinous colonies to 133 µm long and 95 µm wide. Cells more or less spherical, 1.4–3 µm in diameter, blue-green. Mucilaginous envelopes around cells colourless, indistinct, only rarely slightly striate.
Occurrence: Aerophytic, epilithic on limestone rocks; described from Croatia.

**Species outside Europe:**
- *Pleurocapsa fremyi* f. *minor* Bourrelly 1984: Gelatinous, radially and parallely arranged stalks, sometimes ramified, colourless, 2–3 µm wide, cells spherical, 1.5–2 µm in diameter, blue-green, colonies reddish; New Caledonia.
- *Pleurocapsa subgelatinosa* sensu Bourrelly 1984: Colonies composed of gelatinous, more or less parallely arranged stalks, inside with irregular rows of distant, more or less spherical cells, 4 µm in diameter; New Caledonia.

## 3. **Chlorogloea** Wille 1900

Colonies mucilaginous, multicellular, in outline more or less spherical, hemispherical, flattened or granular irregular with rough surface, often composed of subcolonies, micro-, later usually macroscopic, attached to the substrate (plants,

---

PARACAPSA
**Fig. 432.** *Paracapsa siderophila:* a – surface tangential section, **b-c** – parts of thallus with cells in rows; after Naumann (1924).

LITHOCAPSA
**Fig. 433.** *Lithocapsa fasciculata*: after Ercegović (1925).

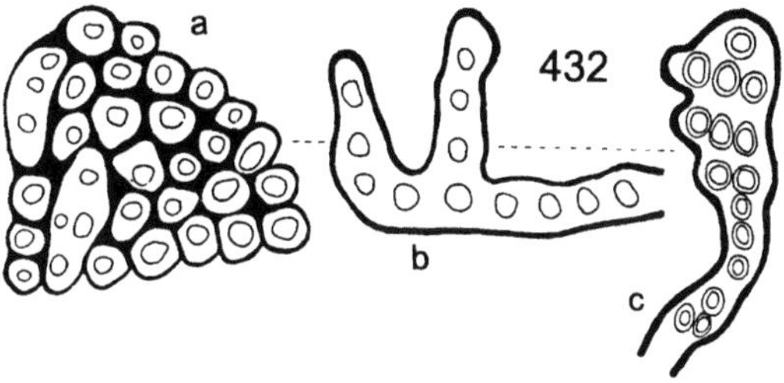

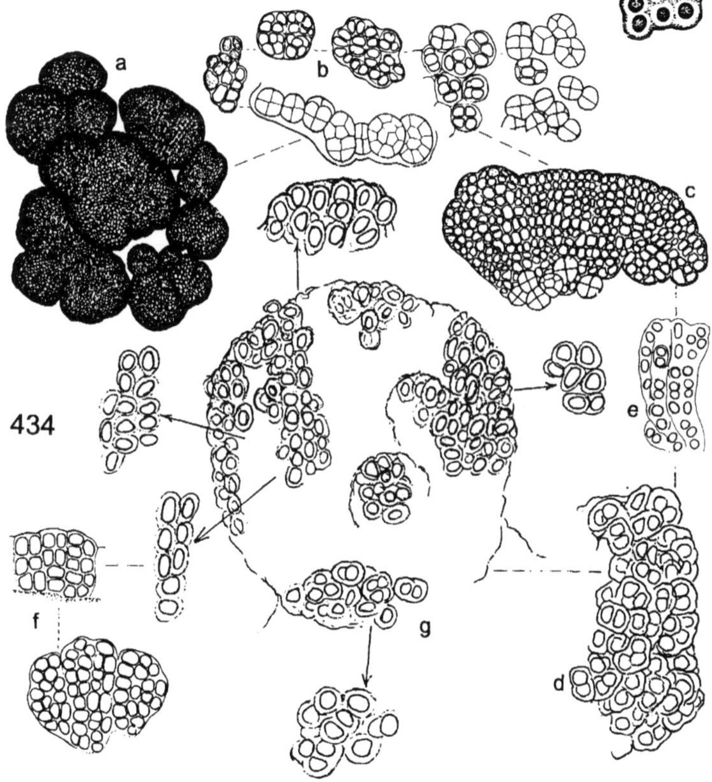


CHLOROGLOEA

**Fig. 434.** *Chlorogloea microcystoides*: **a** – cluster of colonies, **b** – young colonies and parts of colonies with dividing cells, **c-d** – marginal parts of old colonies, **e-f** – vertical sections of colonies, **g** – colony with detail of cells; **d**, and **g** – after Komárek & Montejano (1994), other from Geitler (1932).

stones) or free-living, epipelic or metaphytic, sometimes forming a large gelatinous mass composed of single microcolonies. Cells irregularly arranged in common mucilage, but sometimes forming (particularly in marginal parts) short, irregular rows of cells, usually radially oriented; the rows are sometimes oriented perpendicularly to the substrate. Cells spherical or oval, in old colonies sometimes rounded polygonal, without or with indistinct and diffluent, rarely distinct, delimited and firm individual mucilaginous envelopes, with blue-green, greyish, pale olive-green, yellowish or reddish, finely granular content.

Cell division in three perpendicular planes (less frequently irregularly), but occasionally repeatedly in one plane (origin of rows); cells more or less attain the original size (not always the original shape) before the next division. Nanocytes described in one species (*C. microcystoides*), but they arise from the successive division of cells without remaining in a common sheath (revision !). Reproduction by solitary cells or groups of cells, resulting from the fragmentation of colonies into several-celled mucilaginous clusters.

**Important references:** Geitler (1932); Frémy (1934); Geitler & Ruttner (1935); Skuja (1964); Watanabe & Komárek (1994); Komárek & Montejano (1994).

**Key to the species:**

**1a** Marine, epiphytic and epizoic in the littoral region; mucilage colourless, structureless, cells circa 1–2.5 µm in diameter ........ **1. C. tuberculosa**

**1b** Freshwater and aerophytic species ........................... 2

**2a** Cells spherical, 1.5–2.5 µm in diameter, reddish; benthic in clear (alpine) lakes; mucilage structureless, colourless .............. **2. C. purpurea**

**2b** Cells spherical, elongate or polygonal rounded, always over 2 µm in diameter, grey, blue-green or yellowish; in flowing or stagnant water or aerophytic ........................................ 3

**3a** Freshwater (rarely in supralittoral), in streams and in littoral of lakes, rarely subaerophytic on wet rocks ................................ 4

**3b** Aerophytic species from wet limestone rocks, growing in calcareous tuffs; cells usually enveloped by refractive individual oval or polygonal-rounded envelopes, which are 8.5–12 µm in diameter; mucilage colourless, in old colonies yellow-brown ........................ **5. C. novacekii**

**4a** Cells mainly in individual envelopes; epilithic in submersed biotopes ... **5**

**4b** Around cells very fine, hardly visible, diffluent individual envelopes; free-living (metaphytic) in benthos of nordic lakes (spheroidal/globose colonies) ........................................ **6. C. gentilis**

**5a** Cells 3–4 (6) µm in diameter, mucilage colourless; epilithic in clear, not calcareous streams ........................... **3. C. rivularis**

**5b** Cells 2–3.8 µm in diameter; mucilage colourless or yellow-brown (depending on insolation); epilithic in littoral to supralittoral zones, in calcareous streams and lakes ........................ **4. C. microcystoides**

Note: The genus *Chlorogloea* contains two distinguishable types of species: (i) with more or less small, microscopic colonies, composed mainly solely from irregular, radially or more or less parallely oriented short rows of cells, enveloped by colourless, diffluent slime (species **2**) and (ii) with large, sometimes macroscopically visible colonies, in which the arrangement of cells in rows is recognizable only in marginal parts of colonies (species **3–6**). Both types are characterized by slightly different life cycles.

## 1. **Chlorogloea tuberculosa** (Hansgirg) Wille 1900 (fig. **435**)

Colonies epiphytic, irregular, microscopic (usually not over 135 µm in diameter), only exceptionally macroscopic and in this case to several mm in diameter (?),

composed of numerous irregular, more or less short rows of cells, oriented radially from the colonial center or vertically, enveloped by fine, thin, homogeneous, colourless mucilage. Cells spherical, ellipsoidal, oval or slightly irregular-rounded, circa 1–2.5 × 1–1.5 µm, yellow-green, which divide usually in one plane, perpendicularly to the axis or to rows of cells.
Occurrence: Marine, epiphytic on various macroscopic algae and *Bryozoa*, probably widely distributed in littoral of temperate seas and oceans; Mediterranean Sea (Adria, Hellenic coasts), S. Norway, Faeroe Islands, recorded also from Pacific coast of N. America and Peru. Numerous incorrect identifications of this species in the literature evidently exist.

### 2. **Chlorogloea purpurea** Geitler 1928 (fig. **437**)

Colonies microscopic, mucilaginous, irregular, epilithic on stones and among other epilithic and metaphytic algae, with short, usually slightly radially or vertically arranged, sometimes indistinct rows of cells, distant from one another. Mucilage fine, structureless, homogeneous, colourless, confluent. Cells spherical or slightly elongate before division, sometimes subspherical (polygonal-rounded), 1.5–2.5 µm in diameter, pink, reddish or dirty violet, without individual envelopes (or with very fine sheaths enveloping groups of few cells).
Occurrence: Freshwater, described from benthos of deep stony alpine lakes; Austria. Later identified from other biotopes (in Indonesia, etc.), however, all these populations with reddish cells must be revised.

### 3. **Chlorogloea rivularis** (Hansgirg) Komárek et Anagnostidis 1995

*Pleurocapsa ? rivularis* Hansgirg 1886; *Cyanoderma (Myxoderma) rivulare* (Hansgirg) Hansgirg 1889; *Xenococcus rivularis* (Hansgirg) Geitler 1925

Colonies flat, thin, epilithic, bright blue-green, more or less circular in outline, microscopic, one-layered, or in form of small, flat spherical or dot-like spots, 1–2 (–4) mm in diameter, later sometimes confluent, larger and irregular, multi-layered. Cells 3–4 (–6) µm in diameter, usually densely packed together, spherical, ovoid or laterally flattened and polygonal-rounded, aggregated in a pseudoparenchymatous layer in old colonies, forming indistinct rows at the colonial margin; all cells in colonies of more or less uniform size and shape. Cells with pale blue-green or olive-green, homogeneous or finely granular content. Sheaths firm, thin, or slightly thickened, but distinct, colourless, or slightly yellow-brown at the margin.
Occurrence: Freshwater, epilithic, in clear and cold, usually mountain streams and waterfalls (usually not on a calcareous substrate, described from quartz substrates); known from numerous localities in central European mountains, rarely in rivers in lowlands (Czech Republic, probably also SE. Germany, N. Austria, etc.).

### 4. **Chlorogloea microcystoides** Geitler 1925 (fig. **434**)

Colonies mucilaginous, epilithic on the stony substrate, forming small, but macroscopically visible, granular or wart-like, dark green or brownish thallus, sometimes composed of hemispherical or flattened subcolonies. Cells without or with individual envelopes, which are fine, hardly visible or distinct; cells usually arranged in irregular rows (sometimes very densely), which are diversely divaricated and anastomozing within the colonial mucilage, sometimes in erect or radial positions. Cells spherical, ellipsoidal or polygonal-rounded, 2–3.8 µm in diameter, blue-green, olive-green or yellowish. Colonial mucilage colourless, rarely yellow-

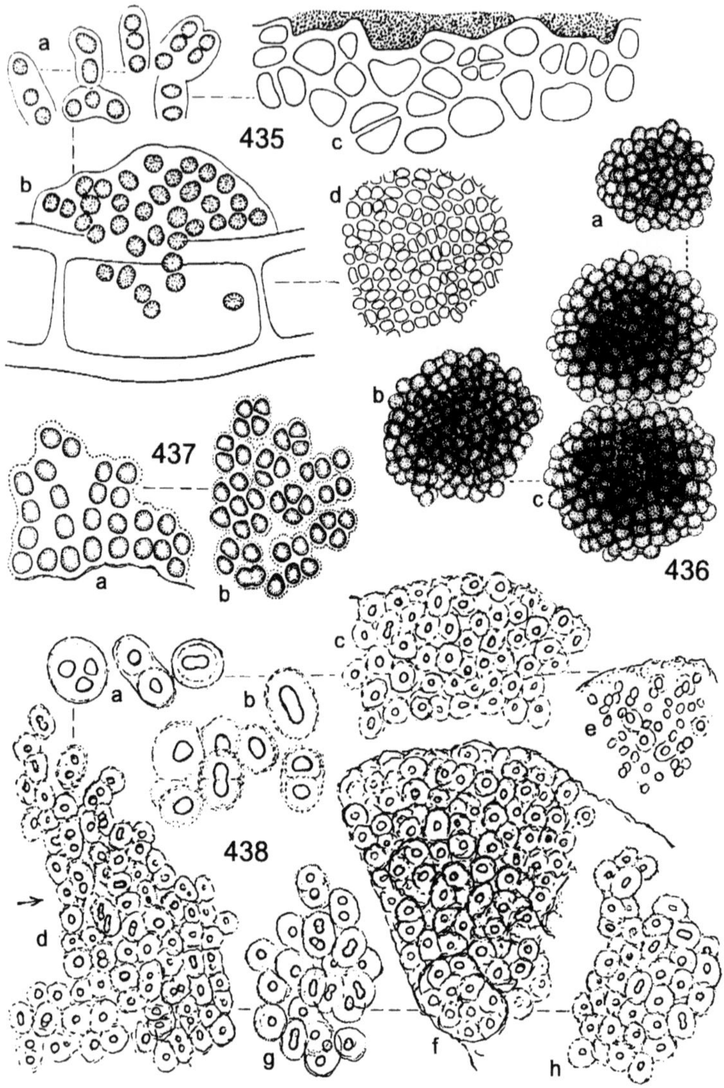


CHLOROGLOEA

**Fig. 435.** *Chlorogloea tuberculosa*: **a-b** – initial stages of colonies, **c** – colony from the surface of *Laminaria* (section), **d** – view on the colony from above; **a-c** – after Wille (1900), **d** – after Ardré (1969, sub *C. tuberculosa* f. *maior* = possibly a separate species).

ish or yellow-brown at the margin. Nanocytes have been described, originating from dividing larger «basal» cells (repeated cell division in different planes; such nanocytes do not remain in a common mucilaginous sheath).
Occurrence: Freshwater, epilithic in the shoreline zone in streams or in littoral of lakes, rarely subaerophytic on wet rocks; described from stony edges of wells and known from clear alpine streams and lakes in Austria; probably more widely distributed in corresponding localities over the whole of Europe (e.g., from the splash zone in cataracts in Greece - Macedonia) and, possibly, in the whole temperate zone; however, data from various other biotopes and from tropical and subtropical countries belong surely to other species (numerous *Chlorogloea*-populations identified as *C. microcystoides* must be especially reconsidered; e.g., from trunks of palms or from stony walls in Spain, from marine benthos in Spain, from thermal springs, from desert crusts in Israel, etc.).

## 5. Chlorogloea novacekii Komárek et Montejano 1994 (fig. 438)

Colonies mucilaginous, micro- up to macroscopic (few mm in diameter), small, more or less spherical to amorphous, yellowish or dirty grey-green or blackish, with cells densely and irregularly arranged, or in indistinct rows (indistinctly radially situated), distant from one another, surrounded by individual firm, but structureless and at the margin slightly diffluent and confluent, slightly refractive envelopes; mucilage colourless, in old parts yellowish-brown (depending on light intensity?). After division, the cells soon separate from one another and form their own, wide envelopes. Cells spherical, subspherical, 2–2.5 µm in diameter, or distinctly elongate (–7 µm long) before division; diameter of gelatinous envelopes around solitary cells = 8.5–12 µm. Cell division mainly by pinching.
Occurrence: Aerophytic, on wet limestone rocks, usually in shady localities, in caves, etc.; described from the central Bohemian karst (Czech Republic), probably more widely distributed.

## 6. Chlorogloea gentilis Skuja 1964 (fig. 436)

Colonies small, free-living, spherical, subspherical or slightly lobate, to 76 µm in diameter, with numerous, densely packed cells, indistinctly radially arranged in marginal parts, without visible gelatinous envelopes (probably very fine, diffluent). Cells spherical (globose) or subspherical, sometimes polygonal-rounded, dividing in three planes, 4–6 µm in diameter, with thin, homogeneous, scarcely visible individual envelopes; cell content olive-green or blue-green, homogeneous or with a few small granules.
Occurrence: Freshwater, benthic and tychoplanktic, together with other cyanoprokaryotes, chlorophytes and diatoms, described and known only from lakes in N. Sweden (Abisko).

**Species outside Europe:**
- *Chlorogloea aphanocapsoides* Geitler in Geitler et Ruttner 1935: Freshwater, with calcareous precipitates, described from Indonesia (Java).

---

**Fig. 436.** *Chlorogloea gentilis*: after Skuja (1964).
**Fig. 437.** *Chlorogloea purpurea*: from Geitler (1932).
**Fig. 438.** *Chlorogloea novacekii*: **a-b** – detail of cells with individual envelopes, **c-f** – marginal parts of old colonies, **g-h** – detail of cells from old colonies; after Komárek & Montejano (1994).

- *Chlorogloea cuauhtemocii* Komárek et Montejano 1994: Freshwater, epilithic and metaphytic; gelatinous colonies, in littoral of pools and ponds with water plants in limestone region, described and known from Mexico (Mexico City, Tolantongo, Yucatán).
- *Chlorogloea endophytica* Howe 1914 [syn.: *Myxohyella howei* Geitler 1925, *Entophysalis endophytica* (Howe) Drouet et Daily 1948]: Marine, described from the coasts of Peru; (fig. **443**).
- *Chlorogloea epiphytica* Komárek et Montejano 1994: Freshwater, epiphytic species; gelatinous colonies reaching macroscopic size, in streams in calcareous regions, described from central Mexico; (fig. **442**).
- *Chlorogloea gessneri* Schiller 1956: Described from salty waters in Islas de los Aves (Venezuela); needs revision; (fig. **439**).
- *Chlorogloea lithogenes* Komárek et Montejano 1994: Dominant travertine forming species in streams and rivers of the limestone region of San Luis Potosí, Mexico; (fig. **441**).
- *Chlorogloea major* Geitler 1933 [syn.: *Chlorogloea minor* Geitler in Geitler et Ruttner 1935]: Freshwater, epilithic, described from Indonesia (Java).
- *Chlorogloea minutissima* (Geitler) Komárek et Anagnostidis 1995 [syn.: *Chlorogloea purpurea* var. *minutissima* Geitler 1933]: Freshwater, epilithic, with reddish cells, described from Indonesia (Java); from *C. purpurea* differs morphologically and in ecology; with this species is probably identical *C. purpurea* sensu Chu & al. ed. (1991); (fig. **440**).
- *Chlorogloea pharaonis* Komárek et Montejano 1994: Known from hypersaline desert lakes in Egypt.
- *Chlorogloea regularis* Setchell et Gardner 1924: Marine, described from Baja California (Mexico), known also from SW. Pacific Atolls (Gilbert Island).
- *Chlorogloea simplex* M. Watanabe et Komárek 1994: Freshwater, described from Himalayas (Kathmandu Valley).
- *Chlorogloea sinensis* Chu 1952: Freshwater, described from China.

**Unrevised species:**
- *Chlorogloea microcystoides* sensu Gonzalez Guerrero 1946: Polymorphic colonies to 200 µm in diameter, yellow, blue-green or brownish, with firmly delimited envelopes, aggregated in masses; cells 3–6 µm in diameter, rounded or polygonal-rounded, mainly arranged in rows within colonies; aerophytic species, growing on trunks of palms and *Dracaena* on the sea shore near Cádiz, Spain; evidently a separate species.
- *Chlorogloea tuberculosa* f. *maior* Ardré 1969: Colonies small, verrucose, to 1 mm in diameter, with rounded-polygonal or slightly oval cells, 3.5–5 (8) × 2–3.5 µm. Marine, growing epiphytically on *Jania longifurca* near Sagres, Portugal. Evidently a different morphotype from the typical *C. tuberculosa* (structure of colonies, size of cells), possibly at the species level; (fig. **435d**).

---

CHLOROGLOEA
Examples of characteristic species unknown in Europe:
**Fig. 439.** *Chlorogloea gessneri*, from Caribbean island near Venezuela: **a** -part of a colony, **b** – detail of a small colony; after Schiller (1956).
**Fig. 440.** *Chlorogloea minutissima*, from Indonesia and China: **a-b** – initial colonies from side (**a**) and from above (**b**), **c-f** – colonies of different age; **a-b** – after Geitler from Geitler & Ruttner (1935), **c-f** – from Chu & al. ed. (1991, sub *C. purpurea*, probably identical with *C. minutissima*).
**Fig. 441.** *Chlorogloea lithogenes*, travertine forming species from central Mexico:

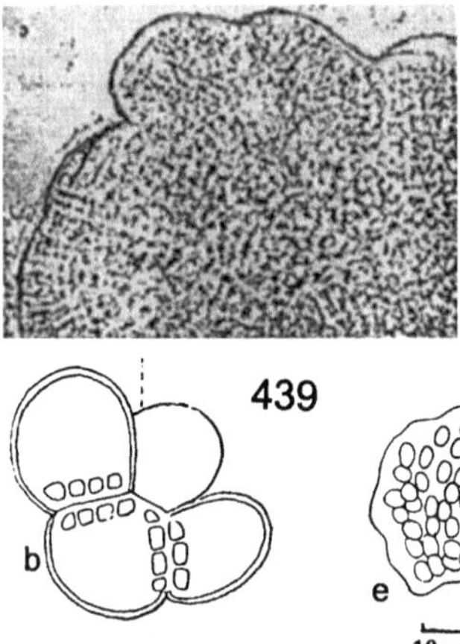


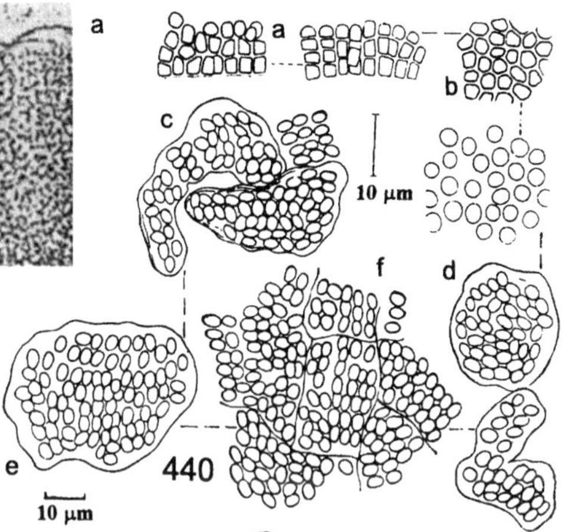


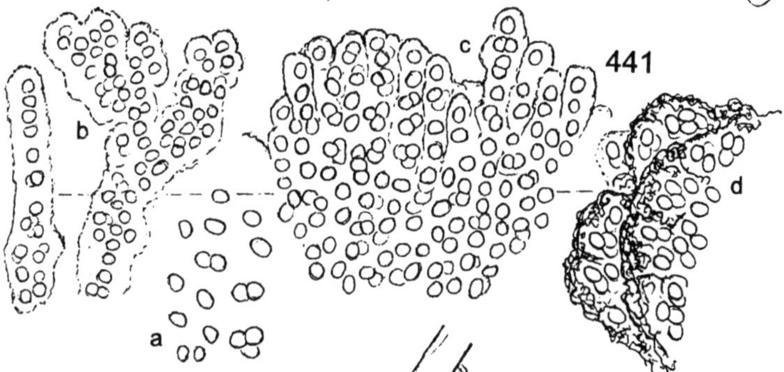


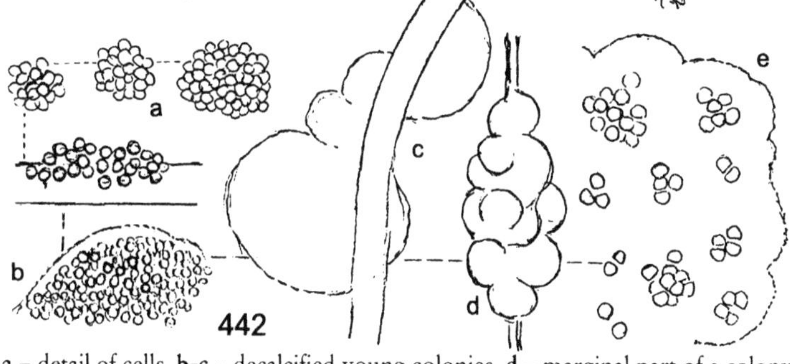


a – detail of cells, **b-c** – decalcified young colonies, **d** – marginal part of a colony; after Komárek & Montejano (1994).

**Fig. 442.** *Chlorogloea epiphytica*, from Mexico: **a** – initial colonies, **b** – marginal part of a colony, **c** – young colonies on filamentous alga, **d** – old macroscopic colony, **e** – detail of disintegrating colony with initial stages of colonies; after Komárek & Montejano (1994).

– The genus *Chlorogloea* contains probably other species, described under different names, particularly from tropical or polar habitats (*Anacystis anomala* Gardner 1927, *Placoma willei* Gardner 1927, *Vanhoeffenia antarctica* Wille 1923, *Polycystis clarionensis* Setchell et Gardner 1937, etc.).

## 4. Entophysalis Kützing 1843

Colonies gelatinous, irregular, granular, microscopic to macroscopic and covering the substrate, sometimes (especially when young) polarized, attached to the substrate, later composed of subcolonies forming aggregates, which are sometimes radially or serially disposed within the thallus. Cells or their groups disposed in colonies more or less in radial or erect rows, enveloped by their own, usually layered and distinct, delimited mucilaginous envelopes; cells spherical to irregular, often of variable size in a colony, with pale blue-green, olive-green or yellowish, finely granular content.
Cells divide in three or more planes in successive generations and reach approximately the original size before the next division. The rows of cells originate probably by the prevalence of cell division in one plane. Reproduction mainly by small colonies or gelatinous groups of cells.
**Important references:** Geitler (1932, 1942); Drouet & Daily (1948, 1952); Komárek & Anagnostidis (1986).

**Key to the species:**
**1a** Aerophytic or freshwater inland species . . . . . . . . . . . . . . . . . . . . . . . . **2**
**1b** Marine coastal species . . . . . . . . . . . . . . . . . . . . . . . . . . . . . . . . . . . . . . **3**
**2a** Aerophytic on periodically dried and wetted rocks; sheaths at the outer margin dark violet or blackish; cells 5–9 µm in diameter; central Europe . . . . . . . . . . . . . . . . . . . . . . . . . . . . . . . . . . . . . . . **1. E. atroviolacea**
**2b** Freshwater, on stones in streams; sheaths colourless; cells slightly elongate, 4–6 × 2–5 µm; in Europe known from the Mediterranean region . . . . . . . . . . . . . . . . . . . . . . . . . . . . . . . . . . . . . . . . **2. E. cornuana**
**3a** Sheaths colourless or yellowish at the margin; cells without distinctly lamellated individual sheaths . . . . . . . . . . . . . . . . . . . . . . . . . . . . . . . . . . **4**
**3b** Sheaths distinctly yellow (at least near the outer margin) or becoming brown; all cells with stratified envelopes . . . . . . . . . . . . . . . . . . . . . . . . **5**
**4a** Cells circa 6–8 µm in diameter; on limestone coastal rocks, known only from warm seas . . . . . . . . . . . . . . . . . . . . . . . . . . . . . . . . . . . **3. E. maior**
**4b** Cells smaller than 3 µm; epiphytic species on marine algae . . **4. E. conferta**
**5a** Thallus not composed of clearly delimited spherical subcolonies; cells 2–5 µm in diameter; Atlantic coasts, North Sea, W. Baltic Sea **5. E. granulosa**
**5b** Thallus composed of subcolonies, enveloped by firm sheaths; cells 3–6 µm in diameter; limestone Mediterranean coasts . . . . . . . . . . . . . . . **6. E. deusta**

### 1. Entophysalis atroviolacea Nováček 1932 (fig. 444)

Colonies in the form of a gelatinous, thin, black layer, in dry state of a thin, fine, paper-like mat; colonies and subcolonies composed of short, irregular rows of cells, all of which are surrounded by firm, stratified (*Gloeocapsa*-like) envelopes; common envelopes also around subcolonies and colonies. Young colonies polarized, attached to the substrate by basal, stalk-like sheaths. Cells more or less spherical, subspherical or oval, 5–9 µm in diameter, pale blue-green. Cellular envelopes colourless in the colonial center, in the marginal parts blackish violet, sometimes even becoming opaque, granular. Resting stages one- or two-celled, with spherical, ellipsoidal or polygonal-rounded cells, with black-violet, slightly

granular sheath, enveloped by colourless, smooth envelope, 7–13 µm in diameter.
Occurrence: Typical aerophyte, occurs on insolated or shady, periodically drying rocks (serpentine, quartz, etc.) with low humidity; described from rocks in deep valleys of rivers in Czech Republic (W. Moravia) in the central Europe.

### 2. **Entophysalis cornuana** Sauvageau 1892 (fig. **445**)

*Radaisia cornuana* (Sauvageau) Sauvageau 1895

Thallus gelatinous-crusty, bright blue-green, composed of parallely (radially) arranged rows of sheathed cells (usually 60–120 µm long), oriented usually perpendicularly to the substrate. All cells enveloped by their own thin, colourless sheaths. Cells subspherical or polygonal-rounded, sometimes slightly elongate, 4–6 × 2–5 µm; cells in colonies often of diverse size and shape. Without nanocytes (therefore not *Radaisia*, nevertheless, comp. with *Cyanodermatium*; p. 351–354).
Occurrence: Freshwater, on stones in flowing water; known from France and Algeria; recorded also from Gabon (Frémy 1930).

### 3. **Entophysalis maior** Ercegović 1932 (fig. **446**)

Thallus gelatinous-crusty, to 1 mm thick, colourless to brownish, with divaricate and radially oriented rows of cells, later more or less irregularly disposed. Cells spherical or subspherical, of varied size, 6–8 µm in diameter. Gelatinous envelopes around cells indistinct, but common mucilage surrounding the cell rows is sharply delimited, colourless, hyaline, or later slightly pale yellowish-brownish at the margin, hardly visible, overlapping the cells by the 3 µm wide margin.
Occurrence: Marine, epilithic on rocks, mixed with other algae, sometimes dominating in mats in salty seapages or in intertidal zones of tropical lagoons; known from Croatia (Dalmatia, limestone Adria coast) and Greece (coast of Aegean Sea, Attiki Peninsula); recorded also from SW. Pacific reefs (Nauru Islands) and other islands in tropical and subtropical zones (however, some of these populations should be revised).

### 4. **Entophysalis conferta** (Kützing) Drouet et Daily 1948

*Chlorogloea conferta* (Kützing) Setchell et Gardner 1918 in Gardner 1918

Colonies large, gelatinous, granular, irregular, yellowish or green-brown, composed of numerous subcolonies; cells in young colonies in radial rows, later irregularly arranged, subspherical, polygonal-rounded or sometimes slightly elongate, in old colonies with densely packed cells, 0.8–1.2 µm in diameter, pale blue-green with visible chromatoplasma. Common mucilage dense, copious, slightly stratified, colourless or pale yellow.
Occurrence: Marine, epiphytic on algae in the coastal zone; described from Germany, but recorded also from Pacific coast of North America (California), Atlantic coast of Brazil (SP) and probably widely distributed in the littoral zones of the whole temperate world. Needs revision, the generic identification needs confirmation (*Chlorogloea* ?).

### 5. **Entophysalis granulosa** Kützing 1843 (fig. **447**)

Thallus large, macroscopic, gelatinous or crusty, to 1 mm thick, brown-blackish, composed of irregular and sometimes indistinct rows of ensheathed cells. Cells spherical or subspherical, to irregular rounded, 2–5 µm in diameter, with pale

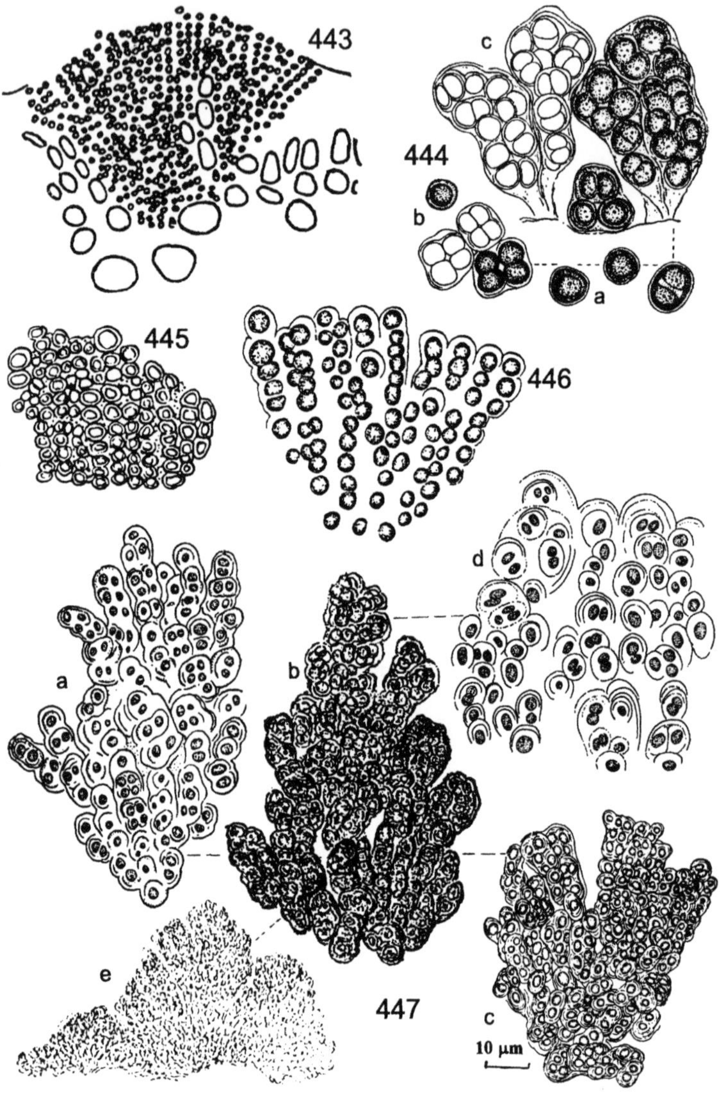


CHLOROGLOEA

**Fig. 443.** *Chlorogloea endophytica*, marginal endophytic species from algae (*Leptocladia peruviana*) from Pacific coasts of Peru: after Howe from Geitler (1932).

blue-green content. Enveloping *Gloeocapsa*-like sheaths are stratified, often there are groups of cells enveloped by a firm, common sheath; envelopes around cells colourless, yellow or brown.
Occurrence: Marine, widely distributed in the shoreline (splash) zone on stones and rocky sea coasts, participating substantially in the formation of a gelatinous «black band» immediately above the surf zone; on all European Atlantic coasts, North Sea, western parts of Baltic Sea, Mediterranean Sea; reported also from Chile and SW. Pacific coasts (Lord Howe Islands).

### 6. **Entophysalis deusta** (Meneghini) Drouet et Daily 1948

*Coccochloris deusta* Meneghini 1841; *Gloeocapsa deusta* (Meneghini) Kützing 1849

Thallus flat, mucilaginous, granular, composed of numerous, more or less rounded subcolonies, with sheathed cells situated more or less in rows within the colonies; subcolonies with 2 – more cells, up to 80 µm in diameter, enveloped by firm sheaths. Cells subspherical or rounded-polygonal, 3–6 µm in diameter, pale blue-green. Mucilaginous envelopes colourless to brown, lamellate, with distinct outer brown layer.
Occurrence: Marine, on rocks and stones in the splash zone, mainly on calcareous coasts; described from the Mediterranean (Italy), known also from Adriatic and Aegean coasts and from Portugal.

Notes: *Entophysalis lemaniae* (Agardh) Drouet et Daily 1956 [syn.: *Nostoc lemaniae* Agardh 1824; *Oncobyrsa lemaniae* (Agardh) Gomont 1896; *Hydrococcus lemaniae* (Agardh) Bornet in Gomont 1896 pro syn.; *Oncobyrsa brebissonii* Meneghini 1843] was very widely conceived by Drouet & Daily (1956), but originally described by Agardh (1824) in a substantially more restricted sense, from the western part of Baltic Sea (Sweden, Scanian coast); in this concept it should be revised, possibly it is a distinct species (comp. Gomont 1896), which was sometimes misinterpreted with «marine *Hydrococcus/Oncobyrsa rivularis*». The probably synonymous *Oncobyrsa brebissonii* Meneghini was described from France (Falaise). The original drawing according to the Bornet's exsiccate (designated «*Oncobyrsa lemaniae*, fixe sur *Lemanea*») was presented by Bourrelly (1970, fig. 88:4), which corresponds, however, more to *Cyanodermatium*. Thallus gelatinous, attached to the substrate (epiphytic on *Lemanea;* freshwater ?), composed of radially and densely, more or less parallely arranged, divaricate pseudofilaments, consisting of elongate cells; apical cells near the surface are shortened (monocytes?). – Within the genus *Entophysalis* exists a group of species, characterized by distinctly radially arranged cells gradually enlarging towards the marginal parts of colonies (*E. maior, E. marginalis,* or gonidial *Entophysalis* sp. in Büdel 1985 isolated from lichens collected in Czech Republic).

---

ENTOPHYSALIS
**Fig. 444.** *Entophysalis atroviolacea*: **a** – solitary cells (resting cells), **b** – small groups of cells representing young colonies, **c** – colonies; redrawn from Nováček (1930).
**Fig. 445.** *Entophysalis cornuana*: after Sauvageau (1892).
**Fig. 446.** *Entophysalis maior*: after Ercegović (1932).
**Fig. 447.** *Entophysalis granulosa*: **a-c** – colonies from various (distant) localities, **d** – detail of a colony, **e** – habitus of a colony; **a** – after Kosinskaja (1948), **b** and **e** – after Bornet from Geitler (1932), **c** – from Chu & al. ed. (1991, from China), **d** – after Bourrelly (1970).

**Species outside Europe** (some of them need revision !):
- *Entophysalis atrata* Tavera et Komárek 1996: Stromatolite-forming species in littoral of volcanic, slightly salinic and alkaline lake Alchichica, central Mexico, with violet-blackish sheaths; morphologically similar type with reddish sheaths from the same locality can be only a different morphotype of this species.
- *Entophysalis collinsii* de Toni 1934 [syn.: *Entophysalis violacea* Collins 1920]: Marine, on rocks at low water mark, Bahama islands.
- *Entophysalis cryptarum* (Farlow) Drouet 1942: Marine, coasts of Central America, partic. of Caribbean Sea (Bermuda, Puerto Rico, Jamaica, Guatemala, Venezuela).
- *Entophysalis lithophila* Tavera et Komárek 1996: From inland salinic, volcanic lake Alchichica, Mexico; stromatolite forming species with yellow-brown sheaths.
- *Entophysalis magnoliae* Farlow 1881: Marine, USA (Mass.).
- *Entophysalis marginalis* Hollenberg 1939: Marine, described from USA, S. California.
- *Entophysalis perrieri* Frémy 1930: Rocks in Madagascar.
- *Entophysalis robusta* Chu 1952: Freshwater, described from China.
- *Entophysalis rubra* M. Watanabe et Komárek 1994: Freshwater, described from Himalayas.
- *Entophysalis samoensis* Wille 1913: Rocks, in Samoa Islands.
- *Entophysalis sinensis* Chu 1944: Freshwater, described from China.
- *Entophysalis violacea* Weber van Bosse 1913: Marine, described from Indonesia.
- *Entophysalis willei* Gardner 1927 [syn.: *Entophysalis violacea* Gardner 1927]: Rocks in Porto Rico.
- *Entophysalis zonata* Gardner 1927: Rocks in China.
- *Pleurocapsa entophysaloides* Setchell et Gardner in Gardner 1918: Marine, from California, USA; probably *Entophysalis*-species.

**Unrevised European species:**
- *Entophysalis brebissonii* (Meneghini) Drouet et Daily 1948: Marine, described from Croatia (Istria – Adriatic Sea).
- *Entophysalis crustacea* (J. Agardh) Drouet et Daily 1952: Freshwater, described from Germany.
- *Entophysalis elongata* (Wille) Drouet et Daily 1952: Freshwater, described from Arctica (Russia – Novaja Zemlja).
- *Entophysalis papillosa* (Kützing) Drouet et Daily 1948: Freshwater, described from Germany (Harz Mts.).

## 5. **Cyanoarbor** Wang 1989

Colonies gelatinous, polarized, firm, ± «erect», ramified, macroscopically visible when old; ramification is pseudodichotomous or irregular. Attached to the substrate. Cells in colonies arranged irregularly or indistinctly radially. Cells irregular-rounded, more or less spherical, hemispherical, ellipsoidal or angulate-rounded, sometimes grouped 2–4 together in colonies; cells in marginal parts or at the apex of the colony sometimes indistinctly elongated. Around cells narrow, distinct envelopes. Colonial mucilage distinct, firm on the surface, confluent within colony.
Cells divide evidently in various planes in successive generations and grow more or less into the original size before the next division. Reproduction by fragmentation of thallus in small groups of cells or in solitary cells. – Monotypic genus.
**Important references:** Wang (1989).

**Species outside Europe:**
- *Cyanoarbor rupestris* Wang 1989: Thallus up to 10 mm high, cells 5.5–10 µm in diameter, usually 2–4 together in own envelopes; this single (type) species is described and known only from one locality, from moist rocky substrate or mixed with other algae among mosses near the Jingbo Lake, Heilongjiang province, China.

# 2. Subfamily **Siphononematoideae** (Geitler) Komárek et Anagnostidis 1986

Initial cells polarized, usually elongated, sometimes *Chamaesiphon*-like (= unicellular status), later intensely dividing in their apical part and then subspherical or irregular-rounded, mainly in the upper part of many-celled colonies; colonies more or less polarized, multilayered, spherical or elongate, gelatinous, with more or less radially oriented cells in rows; mucilage firm, distinct, delimited, cells often with individual envelopes, sometimes striate, in old colonies reproduction by monocytes. Monotypic subfamily.

## 1. **Siphononema** Geitler 1925

Thallus (colonies) polarized, attached by one end to the substrate, micro- to macroscopic. Initial cells solitary, cylindrical or subspherical to irregularly and narrowly club-shaped, attached by the basal end to the substrate, straight or slightly curved, with colourless and/or later yellowish pseudovagina, later dividing intensely in the upper part and forming pseudofilamentous to irregular, large colonies of aggregated cells; cells are oriented more or less in rows within thallus, or (in old parts) irregularly. Cells sometimes produce their individual mucilaginous, colourless envelopes, which are lamellate in old parts of thallus and yellowish-brown or orange coloured (in one species).
Cells divide irregularly, usually in three or various planes in successive generations (development of «stigonematoid» thallus), but in apical parts or in sides of thallus with predominating crosswise fission (development of rows of cells). Reproduction by solitary cells or groups of cells with their own gelatinous envelopes; solitary cells (motile monocytes, planocytes) escape from the splitting envelopes at the marginal and apical parts of thallus (after which remain the cup-shaped remnants of sheaths). Resting stages (cells enveloped by firm sheaths) were also described.
**Important references:** Geitler (1925, 1932, 1942); Hindák (1978).

**Key to the species:**
1a Cells in colonies mainly spherical; envelopes not stratified, colourless; in thermal springs . . . . . . . . . . . . . . . . . . . . . . . . . . . . **1. S. thermophilum**
1b Cells in colonies subspherical, oval or irregular-rounded; envelopes more or less stratified, in old colonies intensely yellow-brown; in mountain limestone streams . . . . . . . . . . . . . . . . . . . . . . . . . . . . . . . . **2. S. polonicum**

### 1. **Siphononema thermophilum** Hindák 1978 (fig. **448**)

Colonies dirty greyish blue-green, variously shaped, usually elongate, sometimes to 500 µm long (rarely longer) and ramified, bunchy, sausage-like, irregular or irregularly globose, surrounded by a firm and distinctly delimited gelatinous sheath; older colonies slightly diffluent at the margin; colonies reproduce by fragmentation. Cells more or less spherical, (2) 3–4 (5) µm in diameter, very

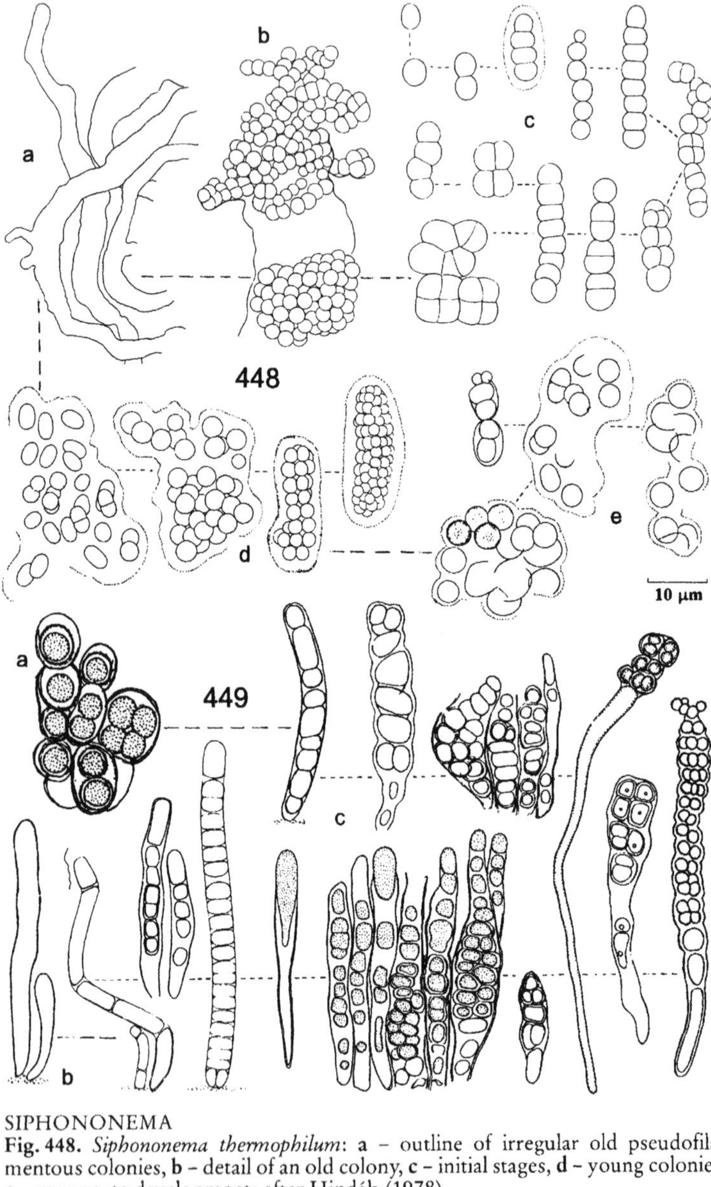


SIPHONONEMA

**Fig. 448.** *Siphononema thermophilum*: **a** – outline of irregular old pseudofilamentous colonies, **b** – detail of an old colony, **c** – initial stages, **d** – young colonies, **e** – monocyte development; after Hindák (1978).

**Fig. 449.** *Siphononema polonicum*: **a** – after Geitler (1932, redrawn from the

densely packed, sometimes arranged in short filamentous or packet-like formations; cell content blue-green, homogeneous, with fine granules. Cells divide normally in two or more planes; reproduction sporadically by monocytes liberated from their own sheaths after division. Old cells, before division and before monocyte formation, to 8 µm in diameter.
Occurrence: In periphyton of small open basins with thermal water (49.5 °C); described from W. Slovakia (Piešt'any); possibly identical is *S. polonicum* sensu Anagnostidis & Zehnder 1964 from thermal springs (40 °C) St. Lorenz (Switzerland).

**2. Siphononema polonicum** (Raciborski) Geitler 1925 (fig. **449–451**)
*Pleurocapsa polonica* Raciborski 1910
Colonies orange, rusty red, red-brown, composed of clusters of cells with distinct envelopes arranged more or less in pseudofilamentous, uni- or multiseriate rows, 15–30 (later up to 54) µm wide and up to 300 µm long, but forming also smaller and irregular groups of cells, enveloped by mucilaginous sheaths. Thallus is more or less polarized, attached by one end to the substrate, with different morphological modifications (developmental stages); basal (and sometimes also apical) cells are subspherical, elongate cylindrical, irregularly club-shaped, irregularly polygonal-rounded, etc., to 90 µm long and 4–7.5 µm wide. Cells within thallus spherical, oval or irregular-rounded, surrounded by their own thin or thick, lamellate and usually yellow-brown envelopes, tightly gathered together, sometimes in short rows, (2.5) 3.5–7 µm in diameter. Resting stages were described: cells more or less 10–11 × 7–8 µm, enveloped by thick sheath with yellow inner and brown outer layer.
Occurrence: Freshwater, epilithic in mountain streams on calcareous substrate; Austria (Alps), Poland (High Tatra Mts.). Other localities must be revised; numerous misidentifications (*Pleurocapsa polonica,* separated in Geitler 1932 from *S. polonicum,* is based on the same basionym and type and can not be classified as a separate species; fig. **449a**).

# 7. Family **Hydrococcaceae** Kützing 1843

Thallus is more or less radially differentiated and/or polarized, with pseudofilamentous rows or layers of cells of variable kind, or irregularly covering the substrate (in form of nematoparenchymatous or pseudoparenchymatous strata), forming more or less flat, one- to several-layered, hemispherical or irregular-spherical colonies of cells (discs, spheres, hemispherical to irregular clusters, etc.); radial structure of colonies is recognizable particularly in marginal parts of the thallus. Colonies are usually sessile on different substrates. Cells are irregular-spherical, polygonal-rounded, more or less irregular or irregularly elongate. Mucilaginous sheaths are firm, rarely fine, narrow.
Cell division by irregular binary fission in various planes, or (in distinct developmental stages) mainly perpendicularly to the long axis of elongate cells (and of pseudofilaments). Without nanocytic cell division.

---

Raciborski's original exsiccate of «*Pleurocapsa polonicum*», which is, hovewer, nomenclatoric type and basionym of *Siphononema polonicum*), **b** – outline of young attached colonies, **c** – various types of initial stages and young colonies; **b-c** – after Geitler (1925).

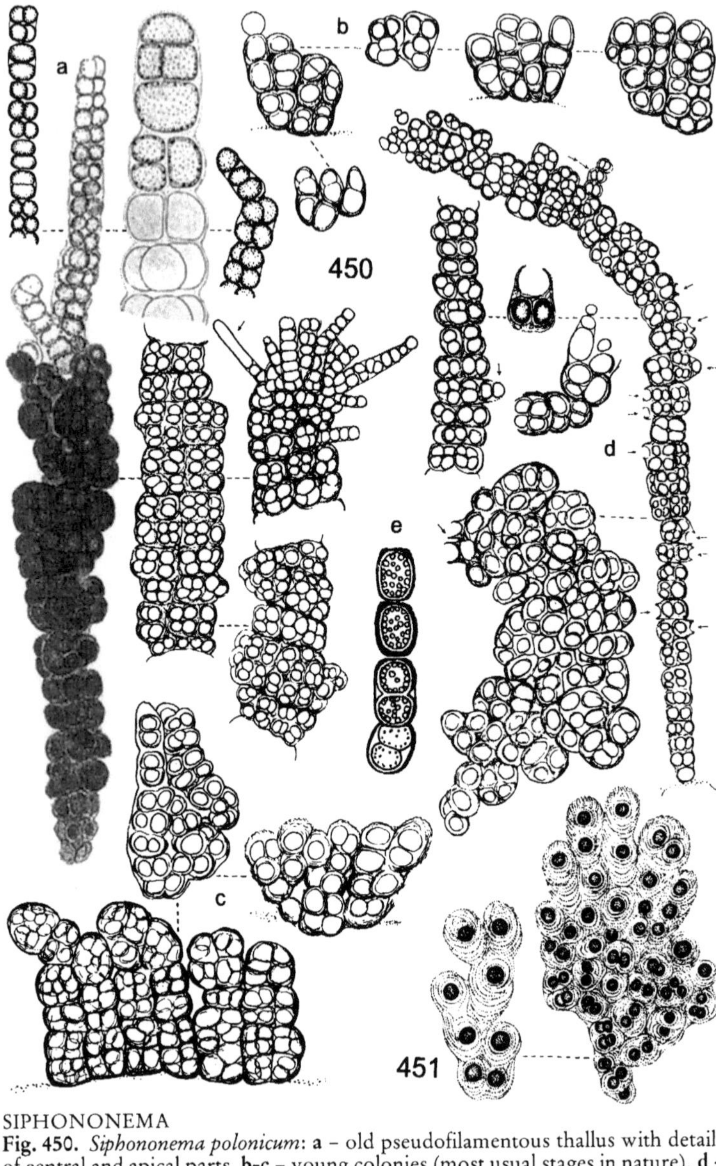


SIPHONONEMA

**Fig. 450.** *Siphononema polonicum*: **a** – old pseudofilamentous thallus with details of central and apical parts, **b-c** – young colonies (most usual stages in nature), **d** – parts of thallus with monocyte development, **e** – resting cells; after Geitler (1925, 1932).

**Key to the genera:**

**1a** Thallus (at least partly) endolithic; envelopes (sheaths) fine, hyaline, more or less wide, sometimes diffluent; cells distinctly separated (distant) from one another . . . . . . . . . . 2
**1b** Thallus not endolithic (epilithic, epiphytic, etc.); enveloping sheaths thin, firm; spaces between cells lacking, or, if present, narrow . . . . . . . . . . 3
**2a** Epilithic part of thallus very reduced, endolithic pseudofilaments few, distinct rows of elongate cells . . . . . . . . . . **1. Tryponema**
**2b** Epilithic part of thallus well developed, cells in endolithic pseudofilaments few, distant from one another . . . . . . . . . . **2. Hormathonema**
**3a** Thallus in a form of a thick layer on the substrate or with clustered cells with vertically or radially growing cells in marginal parts, sometimes composed of vertically growing pseudofilaments . . . . . . . . . . 4
**3b** Thallus in principle flat, creeping on the substrate with radially growing cells, forming sometimes radial rows of cells at the colonial margin; only later in the center of old, flat, discoid colonies cells also grow more or less vertically, forming multilayered strata . . . . . . . . . . 6
**4a** Irregular or spherical, mucilaginous colonies with all-sided growth (spatially) or with cells arranged in peripheral layers . . . . . . . . . . 5
**4b** Thallus forming hemispherical colonies or layers on the substrate, composed of vertically, parallely and densely growing, usually short, sometimes divaricate pseudofilaments . . . . . . . . . . **5. Cyanodermatium**
**5a** Irregular colonies with multilateral, spatial growth, on the periphery sometimes forming short pseudofilaments . . . . . . . . . . **3. Onkonema**
**5b** Hemispherical or irregular-spherical, gelatinous colonies, attached to the substrate, with irregular-rounded cells in the peripheral layer; colonies «hollow» inside . . . . . . . . . . **4. Placoma**
**6a** Flat colonies more or less small, microscopic, usually one-layered with short, uniseriate or few-seriate radial rows of cells at the margin; mainly freshwater species . . . . . . . . . . **6. Hydrococcus**
**6b** Flat colonies microscopic, later large, with radial creeping multiseriate and sometimes divaricate pseudofilaments; mainly marine species **7. Myxohyella**

## 1. **Tryponema** Ercegović 1929

Thallus (= pseudofilaments) endolithic, growing usually perpendicularly from the surface into the substrate and more or less parallely to one another, composed of isolated, irregularly laterally divaricate, uniseriate pseudofilaments. Cells are irregularly elongate, oval, cylindrical, sometimes several times longer than wide, at the ends of pseudofilaments slightly club-shaped widened. Sheaths wide, homogeneous or stratified, colourless, distinct, delimited.
Cells evidently divide crosswise, perpendicularly or obliquely, before divarication of pseudofilaments also laterally (the cells form a lateral bud in its apical part, which later separates from the mother cell by fission). Reproduction not known. Nanocytes never observed. Unrevised genus, which contains two species, only one known from Europe.

---

**Fig. 451.** «*Siphonema*» *polonicum* status pleurocapsoides sensu Skuja 1964 from Sweden (probably a separate species of *Cyanostylon*).

Note: *Tryponema* resembles the filamentous (stigonematalean) type. However, the trichal structure was not yet proved and no special type of cells (e.g., heterocytes) or hormogonia were observed.

**Important references:** Ercegović (1929a); Geitler (1932, 1942); Frémy (1934); Kosinskaja (1948); Bourrelly (1970).

## 1. **Tryponema endolithicum** Ercegović 1929 (fig. **452**)

Pseudofilaments 10–15 µm wide and 200–450 µm long, cells up to 80 × 4–8 µm. Sheaths are wide, distinct, delimited, homogeneous, colourless, 2–3 µm thick. Cells sometimes almost isodiametric or elongate and to more than 10-times longer than wide, irregularly cylindrical, at the ends sometimes slightly club-shaped. Cell content pale blue-green, heavily «vacuolized» (artifact after decalcification ?).
Occurrence: Marine, in littoral zone of the limestone coasts, endolithic (to 1 mm depth); described from Adriatic Sea (coast of Croatia – Dalmatia). The species (and the whole genus) is not well known and needs further investigation.

**Species outside Europe:**
- *Tryponema indicum* Thomas et Gonzalvez 1965: Freshwater, from thermal waters in India, needs revision.
- With the genus *Tryponema* corresponds also *Rhodostichus fluminensis* (Fritsch) Bourrelly sensu Bourrelly 1984, described from flowing waters in New Caledonia: gelatinous stalks with rows of small, blue-green cells, irregular in outline, 2–2.5 µm in diameter.

**Unrevised species:**
- *Entophysalis deusta* «status tryponematoides» sensu Anagnostidis 1968: Marine, euendolithic, reported from Greece (Aegean Sea coasts); probably *Tryponema*-species.

## 2. **Hormathonema** Ercegović 1929

Thallus mainly epilithic, with irregularly or ± in rows clustered cells, or pseudofilamentous. Epilithic pseudofilaments are composed of the irregular uni- or multiseriate rows of irregular groups of more or less spherical cells, which are surrounded by their own mucilaginous, sometimes widened and slightly lamellate envelopes; sometimes develop endolithic (boring) pseudofilaments, which grow directly into the limestone substrate and are composed of uniseriate rows of cells which are usually distant from one another, and separated by thick, mucilaginous, usually transversely lamellate and layered sheaths; all pseudofilaments are envel-

---

TRYPONEMA
**Fig. 452.** *Tryponema endolithicum*: after Ercegović (1929).

HORMATHONEMA
**Fig. 453.** *Hormathonema sphaericum*: after Ercegović (1932).
**Fig. 454.** *Hormathonema epilithicum*: after Ercegović (1932).
**Fig. 455.** *Hormathonema paulocellulare*: **a** – top view, **b-d** – short endolithic pseudofilaments, **e** – monocyte formation; after Ercegović (1932).
**Fig. 456.** *Hormathonema longicellulare*: epilithic and endolithic parts of thallus; after Ercegović (1932).

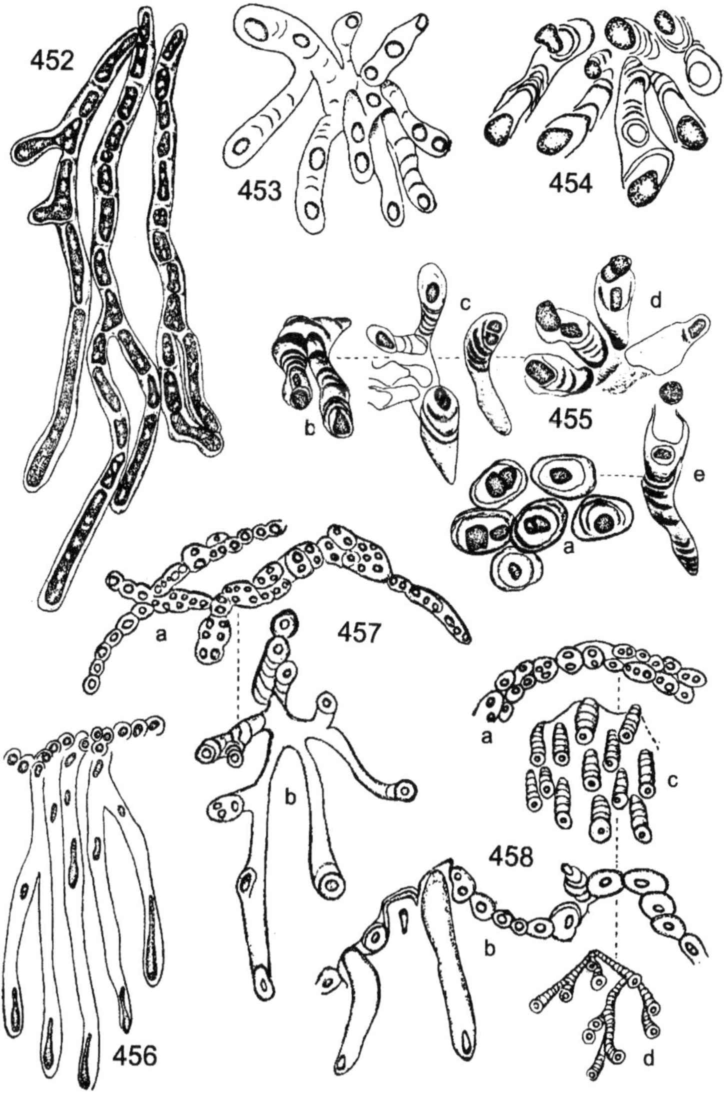


**Fig. 457.** *Hormathonema luteo-brunneum*: **a** – epilithic thallus, **b** – endolithic pseudofilaments; after Ercegović (1930).

**Fig. 458.** *Hormathonema violaceo-nigrum*: **a** – epilithic thallus, **b** – epilithic thallus with initial endolithic pseudofilaments, **c-d** – endolithic pseudofilaments; after Ercegović (1930).

oped by lamellate, gelatinous, wide and delimited sheaths, sometimes coloured, yellowish-brown or violet, sometimes laterally (or pseudodichotomously) divaricate («branched»). «Branches» are lateral and arise after the lengthwise division of cells in pseudofilaments; endolithic pseudofilaments sometimes consisting only of one (–2) terminal cells. Cells spherical, irregular or polygonal, slightly elongate in euendolithic pseudofilaments, usually wider at their ends.
Cells divide irregularly in various planes, in pseudofilaments with predominant crosswise binary fission. After division, they separate soon from one another and produce their own sheaths. Reproduction by solitary cells (monocytes) which are liberated from the mucilaginous sheaths in the upper part of pseudofilaments, or by small gelatinous clusters of several cells. Baeocytes not observed (even not in culture).
**Important references:** Ercegović (1929, 1930, 1932); Geitler (1932, 1942); Frémy (1934); Kosinskaja (1948); Le Campion-Alsumard & Golubić (1985b); Pantazidou (1991 – diss.); Le Campion-Alsumard, Golubić & Pantazidou (1996).

Notes: The genus *Hormathonema* was originally described with only one (type) species *H. paulocellulare* Ercegović 1929 (later corrected with slightly modified descriptions; Ercegović 1930, 1932). However, this first species corresponds to the previously described genus *Solentia* Ercegović 1927 (endolithic, with baeocytic reproduction), and the generic name *Hormathonema* was later commonly used for another type of mostly epilithic cyanoprokaryotes, from which *H. violaceo-nigrum* and *H. luteo-brunneum* are the most characteristic members. The species *H. paulocellulare* was transferred therefore into the g. *Solentia* by Le Campion- Alsumard & Golubić (1985b) and Le Campion-Alsumard & al. (1996) (see p. 457), and the generic name *Hormathonema* was proposed for conservation in the later sense, with the type species *H. violaceo-nigrum*.

**Key to the species:**

**1a** Sheaths always colourless, cells subspherical, oval or ovoid . . . . . . . . **1. H. sphaericum**
**1b** Sheaths at least in surface layers of old colonies coloured . . . . . . . . **2**
**2a** Sheaths of old colonies yellowish or yellow-brown . . . . . . . . **3**
**2b** Sheaths of old colonies more or less violet or dark violet . . . . . . . . **4**
**3a** Epilithic pseudofilaments to 20 µm wide, cells in endolithic parts short, to 9 µm long . . . . . . . . **4. H. luteo-brunneum**
**3b** Epilithic pseudofilaments to 8 µm wide, cells in endolithic parts very long, to 40 µm . . . . . . . . **3. H. longicellulare**
**4a** Cells over (4) 5 µm wide, distant from one another, few in epilithic pseudofilaments . . . . . . . . **2. H. epilithicum**
**4b** Cells 2–5 µm wide, usually numerous in epilithic pseudofilaments . . . . . . . . **5. H. violaceo-nigrum**

## 1. Hormathonema sphaericum Ercegović 1932 (fig. 453)

Thallus microscopic, among other algae. Pseudofilaments not morphologically different in epilithic and endolithic habitats, to 15 µm wide, with cells distant from one another in epilithic as well as in endolithic parts. Cells spherical, subspherical, oval or ovoid, 6–9 µm in diameter, only a few in each pseudofilament, with blue-green content. Sheaths distinct, delimited, colourless, «one-sided» lamellate.
Occurrence: Marine, on wet coastal limestone rocks, in the lower zone of lithophytic algae; known only from the Adriatic Sea (Croatia – Dalmatia). Pantazidou (1991) described from Greece *H.* cf. *sphaericum* with pseudofilaments up to 75 µm long, with coloured sheaths, 17–26 µm (at the basis) and 26–41 µm

(at the end) wide and cells (4.2) 7.2–9.6 (12.1) µm in diameter, growing only euendolithic; evidently a separate species.

## 2. **Hormathonema epilithicum** Ercegović 1932 (fig. **454**)

Thallus with limestone precipitates, greyish, composed of pseudofilaments to 130 µm long and 40 µm wide. Endolithic pseudofilaments not described in detail. Cells elongate, irregular club-shaped, only a few in each pseudofilament, 9–26 × 9–18 µm, slightly club-shaped at the ends, with blue-green content. Sheaths blackish-violet, distinct, delimited, firm, «one-sided» lamellate, never diffluent. Monocytes also occur in this species after drawings. According to Frémy (1934) and Golubić & al. (1975), this species contains morphotypes with transitions to *H. paulocellulare*, therefore it is distinguishable from this species only with difficulty.
Occurrence: Marine, on littoral limestone rocks, in the upper zone of lithophytic algae; known from Mediterranean Sea (Croatia – Dalmatia, S. coast of France), reported also from Aldabra (coral) Atoll (Indian Ocean).

## 3. **Hormathonema longicellulare** Ercegović 1932 (fig. **456**)

Epilithic thallus yellow-brown, composed of irregularly creeping pseudofilaments up to 8 µm in diameter, with more or less spherical or subspherical cells, 3–5 µm wide, surrounded by their own gelatinous envelopes. Euendolithic pseudofilaments to 16 µm wide and 250 µm long, with hyaline gelatinous sheaths, without distinct lamellation, with only few, 4–6 (8) uniseriate cells. Cells diversiform at their bases (spherical, subspherical, shortened to elongate cylindrical, isodiametric), 2–4 × 2–3 µm; intercalary cells isodiametric or elongate cylindrical, rarely ellipsoid, 4–7 × 2.3–2.5 µm, with bluish to blue-green, finely to coarsely granular content. Apical cells wider and elongate, pear-shaped, to 6 µm wide and 40 µm long; gelatinous sheaths around cells colourless or yellow-brown. Nanocytes not observed even in culture. Pantazidou (1991) found only euendolithic pseudofilaments.
Occurrence: Marine, in supralittoral down to intertidal zone of limestone coasts (and on carbonate rubble), in small depressions and rock-pools, wetted by the waves; described from Adriatic Sea (Croatia – Dalmatia), known also from Aegean Sea (Greece – Attiki Peninsula); reported also from SW. Pacific atolls (Ellice Islands).

## 4. **Hormathonema luteo-brunneum** Ercegović 1930 (fig. **457**)

Thallus yellow-brown, with epilithic and euendolithic parts. Epilithic part composed of spheroid to cylindroid many-celled aggregations (clusters), 8–52 × 4–30 µm (or more), which consist of 2–4 – 8-celled, more or less densely arranged subcolonies enveloped by gelatinous, yellow-brown, non-lamellated, more or less thin sheaths; cells bluish to blue-green, spherical, subspherical to ellipsoid, symmetrically divided in two planes perpendicular to one another, (1.5) 2–3 µm in diameter, usually enveloped by individual thick, not lamellate envelopes. Irregular short, sometimes moniliform pseudofilaments, grow out from epilithic cell clusters to 20 µm wide, with cells 2–9 × 2–5 µm, which are densely arranged and usually surrounded by their own envelopes. From these parts grow more or less perpendicularly long gelatinous processes into the substrate, gradually developing into euendolithic, rarely branched, few-celled (2–4) pseudofilaments, in which the cells are sparsely distributed, uniseriate, regularly localized within the ends of mucilaginous strands. Euendolithic pseudofilaments 7–9 µm wide, usually

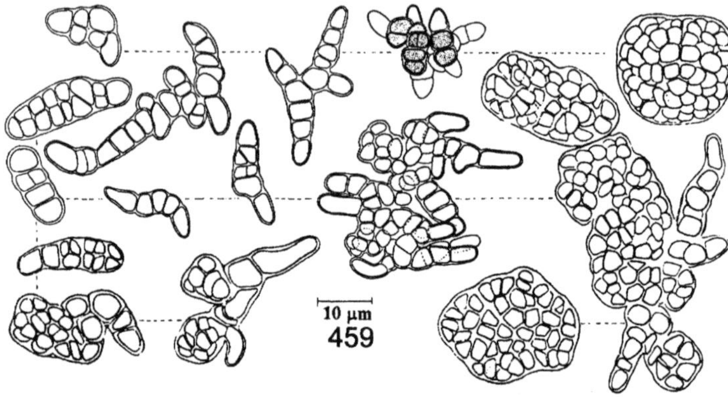


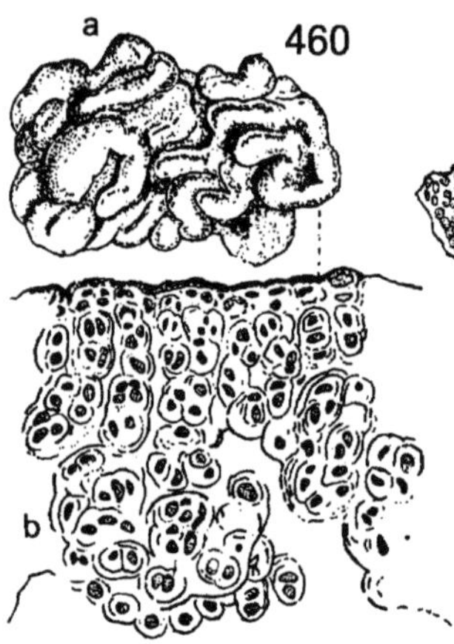


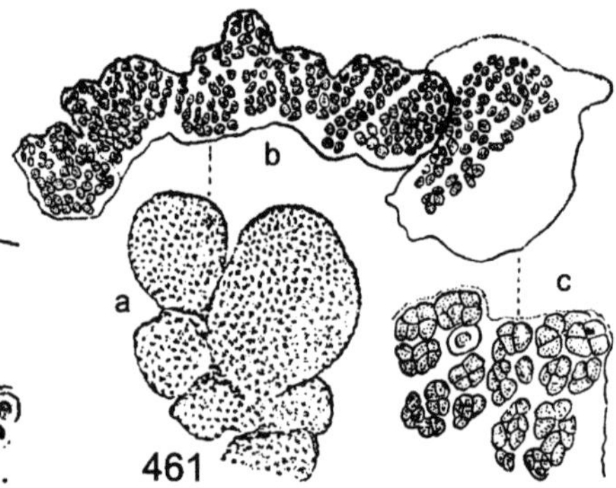


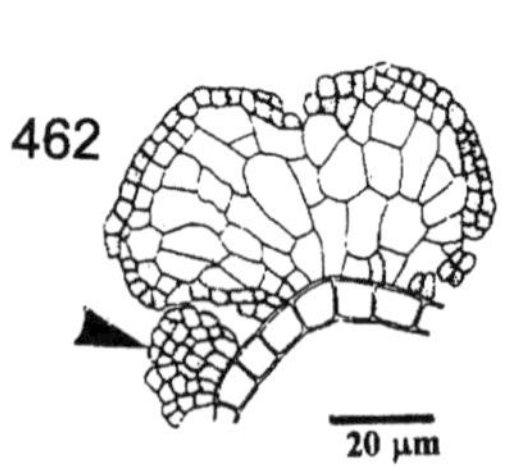

ONKONEMA

**Fig. 459.** *Onkonema compactum*, thermophilic species from tropical thermal springs; after Geitler from Geitler & Ruttner (1935).

12–19 μm long, with yellow-brown (scytonemin), rarely colourless (inside layers) sheaths; sheaths are gelatinous (rarely firm) and one-sided (excentric) stratified. Cells bluish to blue-green, subspherical to cylindrical or irregularly oval, with finely granular content, asymmetrically horizontally divided, 2.5–4 × 2–2.5 μm. Apical cells cylindrical, irregularly oval or pear-shaped, longer than the other cells, 4–9 μm long, 3–4 μm wide. Nanocytes not observed, even not in culture.
Occurrence: Marine, on wet rocks in the splash zone of limestone coasts (frequently together with *H. violaceo-nigrum, H. paulocellulare,* etc.) boring also within mollusc shells, calcareous skeletons of invertebrates, calcified rhodophytes (*Corallina, Melobesia*) and chlorophytes (*Halimeda*); described from the Adriatic Sea (Croatia – Dalmatia), known also from other similar localities in the Mediterranean Sea (France – near Marseille), widely distributed along the continental and island coasts of the Aegean Sea (Greece); known also from Sinai Peninsula, Florida, Bermuda and Bahama Islands, Australia, probably more widely distributed.

## 5. Hormathonema violaceo-nigrum Ercegović 1930 (fig. **458**)

Thallus brownish black to blue-black, forming irregular spherical spots on stony substrate, 1–2 mm in diameter, composed of many-celled epilithic and few-celled euendolithic parts. Epilithic part with regular or irregular pseudofilaments (to 30 μm wide), or in form of subspherical or irregular cell aggregations, 20–56 × 10–18 μm, which consist of clusters with numerous (2–16) cells. Cells spherical, subspherical or slightly elongate, (1) – 2.5–3 × (1) – 2–3 μm, surrounded by their own, blue, blackish to dark violet (gloeocapsin), gelatinous, to 3 μm thick, concentrically (with 1–2 layers) lamellate envelopes. From this thallus euendolithic pseudofilaments (up to 55 μm long, and 4–6 – (10) μm wide) grow into the substrate, with thin colourless (inside layer) to black-blue (outside layer) or intensely one-sided concentrically stratified sheaths; within sheaths solitary or up to 6, more or less spherical to cylindrical or ellipsoidal, transversely dividing cells, (1) 2.5–3.5 (8) × (1) 2–3 (5) μm, located mainly towards the ends of pseudofilaments. Apical cells of pseudofilaments (which arise from the first asymmetrical cell division) are long cylindrical to pear-shaped, to 8 μm long (longer than the others). Cells in both epilithic and euendolithic parts bluish to blue-green, finely granular. Blackish violet gelatinous envelopes turn reddish under acidic environment.
Occurrence: Marine, on wet limestone rocks, in the tidal zone (frequently together with *H. luteo-brunneum* etc.); known from the Mediterranean Sea (Croatia – Dalmatia, S. coast of France, along the continental and islandic Aegean Sea coasts of Greece); known also from Egypt (Suez canal), Indian Ocean (Aldabra) and W. Atlantic coasts (Florida, Bermuda), probably more widely distributed.

---

PLACOMA

**Fig. 460.** *Placoma vesiculosum*: **a** – macroscopic colony, **b** – detail of marginal part of thallus; after Bornet & Thuret (1876).

**Fig. 461.** *Placoma violaceum*, marine species from California: **a** – colonies, **b-c** – parts of thallus; from Gardner (1918).

**Fig. 462.** *Placoma regulare*, freshwater species from mountain streams in New Zealand: after Broady & Ingerfeld (1991).

**Fig. 463.** *Placoma adriaticum*, marine species from Adriatic Sea: after Hauck (1885, sub *Oncobyrsa adriatica*).

**Unrevised species:**
- *Hormathonema onychoides* Gonzalez-Guerrero 1951: Thallus composed of hyaline, not stratified (?), gelatinous strands, widening towards the ends, in which are located solitary apical, hemispherical cells, 1–3 × 2–4 µm, with homogeneous blue-green content; reproduction not mentioned. Endolithic species described from calcareous coastal rocks near Alicante, Spain; classification into the genus *Hormathonema* is uncertain.
- Chu & Wu (1984) mention two species with provisory names from the Xisha Islands, China (*H. sinensis, H. dichotoma*).
- *Entophysalis deusta* «status hormathonematoides» sensu Anagnostidis 1968: reported from the Aegean Sea coasts, Greece; evidently a *Hormathonema* species.
- *Hormathonema*-like cyanoprokaryote was isolated from the lichen *Phyllisciella marionensis* by Henssen & Büdel (1984) and Büdel (1985), from the subantarctic Marion Island.

## 3. **Onkonema** Geitler 1933

Cells in pseudofilamentous colonies; thallus microscopic, irregular, periphytic, creeping over the substrate, with cells arranged into flat or three-dimensional groups (chroococcacean stages) or in short, irregular rows, enveloped by thin, firm, but confluent sheaths, or into pseudofilaments with irregular divarications; rows of cells uni- to two- or multiseriate, mainly with apical growth (e.g., with prevailing division of «terminal» cells). Cells irregular, polygonal-rounded, in rows and especially at their ends elongate or more or less cylindrical, slight narrowed and rounded.
Cells divide irregularly in different planes, in rows mostly perpendicular to the growth direction; sometimes successive division in several smaller cells (monocytes ?, nanocytes ?), which are liberated from thallus and serve to the reproduction. Monotypic genus.
**Important references:** Geitler (1933, 1942); Geitler & Ruttner (1935).

**Species outside Europe:**
- *Onkonema compactum* Geitler 1933: Periphytic (epiphytic on other algae and cyanoprokaryotes) in warm and thermal springs in tropical countries (Indonesia – Java, Sumatra; Zambia); (fig. **459**).

## 4. **Placoma** Schousboe ex Bornet et Thuret 1876

Cells in more or less globular colonies, located near the surface of gelatinous, irregular spheres; thallus is small or expanded, mucilaginous, more or less lobate or vesiculose with wavy or smooth surface, usually «hollow», with cells situated in one or several layers in the outer part of the colony, irregularly or radially arranged in irregular rows or groups, or in flattened layers. Gelatinous envelopes colourless, yellow-brownish or pinkish coloured. Cells without or with individual envelopes, which are sometimes layered, more or less spherical, irregular in outline or slightly elongate, with homogeneous, blue-green, olive-green, yellow-green or pinkish, sometimes slightly granular content.
Cell division irregular, in various planes, predominantly perpendicular to the surface of a colony. Reproduction by solitary liberated cells or by mucilaginous clusters of cells. Freshwater (usually from running waters) as well as marine species (rocky littorals of seas and oceans); only one species known from Europe.

**Important references:** Geitler (1932, 1942); Frémy (1934); Broady & Ingerfeld (1991).

### 1. Placoma vesiculosum Schousboe in Bornet et Thuret (fig. 460)

Thallus gelatinous, macroscopic, irregularly spherically lobate, composed usually of hemispherical subcolonies; vesiculose formations of thallus, wavy and smooth on the surface, inside more or less hollow, irregularly covering the substrate. Cells in vesicles concentrated mainly in the peripheral and subperipheral layers; mucilage usually yellow-brown, inside the colony diffluent, fine and colourless. Cells more or less irregular or polygonal-rounded, blue-green, about 4 µm in diameter.
Occurrence: Marine, on coastal rocks; known from the Atlantic European coasts, W. coast of France (Biarritz), Portugal (Lagos) and Spain (Aguitania Bay, Cádiz), NW. Africa (Tangier) and from the Mediterranean Sea (Algeria, France, Greece, Morocco).

**Species outside Europe:**
- *Placoma africanum* Wille 1903: Cells 2–3 µm in diameter; freshwater, benthic, epipelic, Kachembe River, S. Africa.
- *Placoma regulare* Broady et Ingerfeld 1991: Macroscopic, reddish hemispherical colonies, cells to 30 µm in diameter, thylakoids peripherally arranged in cells; freshwater, epilithic on rocks and epiphytic on mosses in mountain streams, New Zealand; (fig. **462**).
- *Placoma violaceum* Setchell et Gardner in Gardner 1918: Cells 3.5–4 µm in diameter, blue-green; coastal marine habitats, described from W. coast of USA; (fig. **461**).
- *Placoma willei* Gardner 1927: Cells 2–2.4 µm in diameter; known from wet rocks and tree trunks in Porto Rico and China.
- *Placoma* sp. in Skuja 1949: Freshwater, attached to sand grains and bricks; near Rangoon, Burma.
- *Oncobyrsa rivularis* sensu Frémy 1930: Freshwater, spherical colonies, often aggregated to form a violet mass, epilithic in streams in Gabon; it seems to be a *Placoma* species according to drawings.

**Unrevised European species:**
- *Placoma adriaticum* (Hauck) Umezaki et M. Watanabe 1994 [syn.: *Oncobyrsa adriatica* Hauck 1885]: Cells (2) 4–10 µm in diameter, rounded or irregular-rounded or slightly elongate; marine species, epiphytic on *Gelidium* or epilithic in splash zone; needs taxonomic revision; (fig. **463**).
- *Placoma fluviatile* (Agardh) Umezaki et M. Watanabe 1994 [syn.: *Oncobyrsa fluviatilis* Agardh 1827]: Freshwater, needs confirmation.
- *Placoma micrococcum* (Hansgirg) Umezaki et M. Watanabe 1994 [syn.: *Oncobyrsa adriatica* var. *micrococca* Hansgirg 1893]: Marine species, needs confirmation.
- *Pleurocapsa fluviatilis* var. *subsalsa* Hansgirg 1890 [syn.: *Oncobyrsa fluviatilis* var. *subsalsa* (Hansgirg) Hansgirg ex Forti 1907]: Marine, described from Croatia (Dalmatia), needs taxonomic revision.

## 5. Cyanodermatium Geitler 1933

Gelatinous, flattened, irregularly hemispherical, microscopic or macroscopic colonies or crusty layers on a substrate, composed of more or less parallely arranged pseudofilamentous rows of cells, oriented usually perpendicular to the

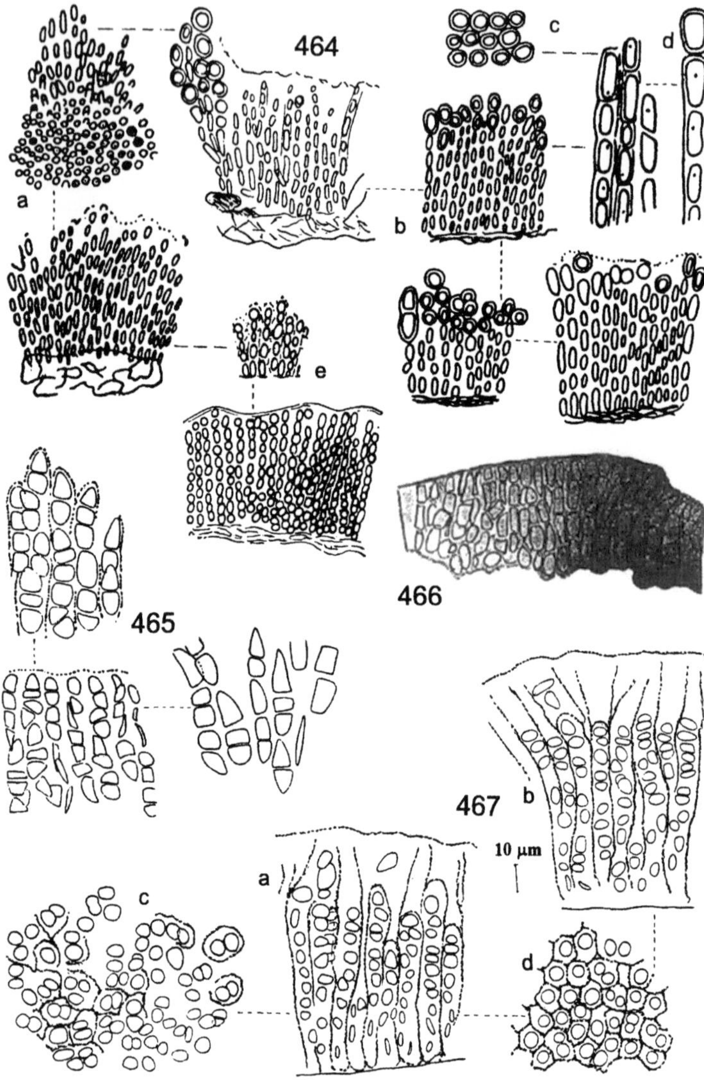


CYANODERMATIUM

**Fig. 464.** *Cyanodermatium fluminense*, with ensheathed apical reproductive cells: **a-d** – after Fritsch (1929), **e** – from Chu & al. ed. (1992).

Species from Indonesia, in the original sense of Geitler:

**Fig. 465.** *Cyanodermatium gelatinosum*: after Geitler from Geitler & Ruttner (1935).

substrate, sometimes irregularly divaricate; at first growth is radial, later sometimes flat strata arise, pseudoparenchymatous from above. Cells irregularly polygonal-rounded, sometimes slightly elongate (and/or shortened in upper colonial parts), enveloped by fine sheaths, which usually coalesce. Upper surface of colonies usually delimited and firm, mucilage in some species intensely coloured.
Cells divide irregularly in various planes, but evidently predominantly crosswise to the vertical axes (= origin of erect pseudofilaments). Nanocytes never observed, reproduction by solitary cells liberated from the terminal parts of a thallus (monocytes). In several species were found special, often ensheathed spherical apical cells, differentiated from terminal cells and serving for distribution (special genus ?). Little known genus, probably more widely distributed; in Europe only one well known species.
**Important references:** Geitler (1933, 1942); Geitler & Ruttner (1935).

## 1. Cyanodermatium fluminense (Fritsch) Komárek et Anagnostidis 1995 (fig. 464)

*Pseudoncobyrsa fluminensis* Fritsch 1929; *Rhodostichus fluminensis* (Fritsch) Bourrelly 1970

Colonies flat, microscopic, about 70 µm thick, attached to the substrate, composed of numerous erect, parallely oriented rows (pseudofilaments) of cells, in apical parts distinctly widened, mucilaginous envelopes colourless or yellow. Cells irregular elongate-rounded to cylindrical or rounded-isodiametric, without individual envelopes, pale blue-green, (2) 3.5–5 × ± 2–3.5 µm. In the apical parts of vertical pseudofilaments differentiate terminal, distinctly widened, solitary (or in short rows) spherical or slightly elongate reproductive cells (monocytes), 2.8–4 µm in diameter, with distinct yellow or brown sheath. According to Geitler (1932, 1942) and Bourrelly (1970, 1985) near to *Rhodostichus.*
Occurrence: Freshwater, on stones in rivers; England.

**Species outside Europe:**
- *Cyanodermatium gelatinosum* Geitler 1933: Freshwater, epilithic on rocks and stones (forming crusts), described from lakes in Indonesia (S. Sumatra); (fig. **465**).
- *Cyanodermatium gigas* (Geitler) Komárek et Anagnostidis 1995 [syn.: *Radaisia gigas* Geitler 1933]: Cells grey-blue, 5–8 (12) × 3–4 (6) µm, sometimes with individual envelopes, thallus crusty to 120 µm thick, colourless sheaths; freshwater, epilithic on stones and tuffs, described from lakes and streams in Indonesia (Sumatra); (fig. **467**).
- *Cyanodermatium violaceum* Geitler 1933: Freshwater, epilithic crusts, described from littoral of Ranau Lake in Indonesia (S. Sumatra); (fig. **466**).

**Unrevised species:**
- A *Cyanodermatium*-like species with intense orange coloured sheaths (Lindström in litt.) occurs in the periphyton of clear lakes in S. Norway.

---

**Fig. 466.** *Cyanodermatium violaceum*: after Geitler from Geitler & Ruttner (1935).
**Fig. 467.** *Cyanodermatium gigas*: **a-b** – vertical section of thallus, **c-d** – top view; after Geitler from Geitler & Ruttner (1935; sub *Radaisia gigas*).

– *Entophysalis chlorophora* Gardner 1927: Described from streams in Porto Rico; probably a *Cyanodermatium* species.
– *Entophysalis lemaniae* sensu Gomont (see Bourrelly 1970): Corresponds better with the diagnosis of the g. *Cyanodermatium*.
– *Entophysalis* sp. in Sant'Anna & al. 1995: Marine, from the Brazilian coast, corresponds rather to the genus *Cyanodermatium*.
– *Radaisia violacea* Frémy 1930: Forming crustose layers up to 80 µm thick, composed of erect, parallely arranged pseudofilaments, with violet cells 2–3 µm in diameter, mucilage colourless or yellowish, never forming nanocytes; described from flowing water in central Africa (Gabon); similar type known from Guadeloupe (Bourrelly & Manguin 1952). Bourrelly (1984), who earlier joined the genus *Radaisia* with *Pleurocapsa*, renamed this species *Pleurocapsa fremyi* with respect to *Pleurocapsa violacea* Weber van Bosse 1926, and described a new f. *minor* of this species from New Caledonia. Both taxa (*R. violacea* Frémy and its f. *minor* Bourrelly) correspond better to the g. *Cyanodermatium* (in which, however, the epitheton «*violacea* Frémy» has priority against *Cyanodermatium violaceum* Geitler 1933).

## 6. **Hydrococcus** Kützing 1833, nomen cons.

*Oncobyrsa* Meneghini 1842

Cells grouped initially into unilayered, pseudoparenchymatous, nemato- or blastoparenchymatous colonies, more or less circular in outline, covering the substrate (stones, plants), from which radially oriented rows of tightly packed cells grow on the margin; old colonies remain monolayered, or erect pseudofilaments (?) grow in the center, which are more or less parallel to one another, or form aggregates of clustered cells; pseudofilaments enveloped by thin, colourless sheaths; sheaths laterally confluent. The cells in old colonies form flattened layers up to hemispherical formations, or thin or thick, on the surface granular, black-green, black-brown or violet macroscopic mats. In old parts of the thallus are the cells or their groups (usually with 4–8 cells) more distant from one another, enveloped by their individual mucilaginous sheaths, sometimes even forming sarcinoid packets. Cells are spherical, oval, irregular or elongate at the ends of pseudofilaments, pale blue-green or olive-green, with more or less homogeneous content.
Cell division in various planes in successive cell cycles, but the crosswise binary fission prevails in marginal parts or in surface layers of colonies. Typical nanocytes not observed. Reproduction by solitary cells or by small clusters of cells. Not a well known genus, almost all the species must be revised; data about *Hydrococcus* from saline waters and mineral lakes are still uncertain.
**Important references:** Geitler (1932, 1942).

**Key to the species:**

**1a** Cells usually 1.2–2.5 (–4) µm in diameter, colonies always microscopic . . . . . . . . . . . . . . . . . . . . . . . . . . . . . . . . . . . . . . . . . . **1. H. cesatii**

**1b** Cells (2.5) 3–6 µm in diameter, old colonies to 2.5 mm in diameter . . . . . . . . . . . . . . . . . . . . . . . . . . . . . . . . . . . . . . . . . **2. H. rivularis**

### 1. **Hydrococcus cesatii** Rabenhorst 1860 (fig. **468**)

*Oncobyrsa cesatiana* (Rabenhorst) Rabenhorst 1865

Young colonies flat, more or less circular in outline, usually nemato-, rarely blastoparenchymatic, with usually elongate cells (particularly in marginal parts), to 4 × ± 2 µm; marginal creeping rows of cells usually grow radially, terminal cells

are sometimes arcuate. The cells grow later in the colonial center also spatially (in more layers), forming sometimes short, vertical, simple and parallely arranged rows, and so producing a flattened-rounded (to hemispherical) crustose layer (but usually only of microscopic size). Cells in upper layers (in vertical rows) usually isodiametric, 1.2–2.5 µm in diameter. Cells blue-green, olive-green, grey-green or dirty violet. Mucilaginous sheaths firm, thin, colourless.
Occurrence: Freshwater, epiphytic on filamentous algae (*Lemanea, Cladophora,* etc.), mosses (particularly on *Fontinalis*), in clear mountain running or stagnant waters, mainly on calcareous substrates; described from the Alps, rarely occurring in other European mountains; numerous data, particularly outside Europe, have to be revised.

### 2. **Hydrococcus rivularis** Kützing 1833 (fig. **469**)

*Oncobyrsa rivularis* (Kützing) Meneghini 1846; *Entophysalis rivularis* (Kützing) Drouet 1943

Initial flat colonies more or less circular in outline, usually nematoparenchymatic, rarely blastoparenchymatic, with cells in radially oriented rows; cells irregular-rounded in outline, elongate and arcuate only in marginal parts. Cells are densely arranged in old colonies, where they divide intensely also in the central colonial parts and form vertical (radial) rows (sometimes pseudodichotomously or pseudotetrachotomously divaricate) of more or less isodiametric, polygonal-rounded cells; this process results in formation of hemispherical, firm, microscopic colonies (to 2.5 mm in diameter), which are brown-green, black-brown or violet. Cells blue-green, dirty olive-green or dirty violet, (2.5) 3–6 µm in diameter, rarely slightly elongate. Mucilaginous envelopes more or less fine, colourless; around cells or their groups sometimes special individual gelatinous sheaths (in old parts of colonies). Within colonies sometimes carbonate precipitates.
Occurrence: Freshwater, on stones, algae, mosses, water plants and wooden substrates in rapidly flowing, clear, cold streams, rivers and waterfalls, mainly in mountains, probably with worldwide distribution (revisions of localities are necessary). Records from marine biotopes (e.g., England, Portugal), thermal springs (e.g., Hungary) and from tropical regions (fig. **470**) concern evidently other species.

**Unrevised species:**
- *Hydrococcus marinus* Grunow 1861 [syn.: *Oncobyrsa marina* (Grunow) Rabenhorst 1865; *Placoma marinum* Bornet et Thuret 1892]: Gelatinous, more or less spherical colonies, even reaching macroscopic size (to 2 mm in diameter), green, with more or less irregular cells 6–7.5 × 5 µm, with individual envelopes; epiphytic on algae in the Mediterranean Sea, described from Greece.
- *Hydrococcus ulvaceus* Kützing 1843: Thallus spherical-vesiculose, macroscopic, olive-green or green; cells spherical, 2–3 µm in diameter, with distinct envelopes, blue-green; freshwater, epiphytic on mosses in streams, Croatia and Italy; judging by the description not *Hydrococcus* (*Placoma ?, Cyanodermatium ?*).

## 7. **Myxohyella** Geitler 1925

Thallus composed of a flat pseudoparenchymatous monolayer, from which arise radially, distinctly multiseriate pseudofilaments, growing over the substrate, composed of one (at the apex) or more nematoparenchymatous rows of cells; pseudofilaments irregularly, repeatedly, laterally divaricate, with distinct «apical» growth, usually with fine, thin, gelatinous, confluent, colourless and not layered, homogeneous sheaths; radial pseudofilaments not confluent, often separated

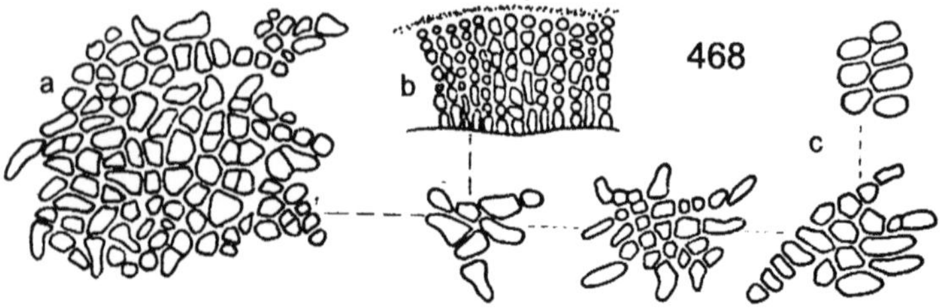


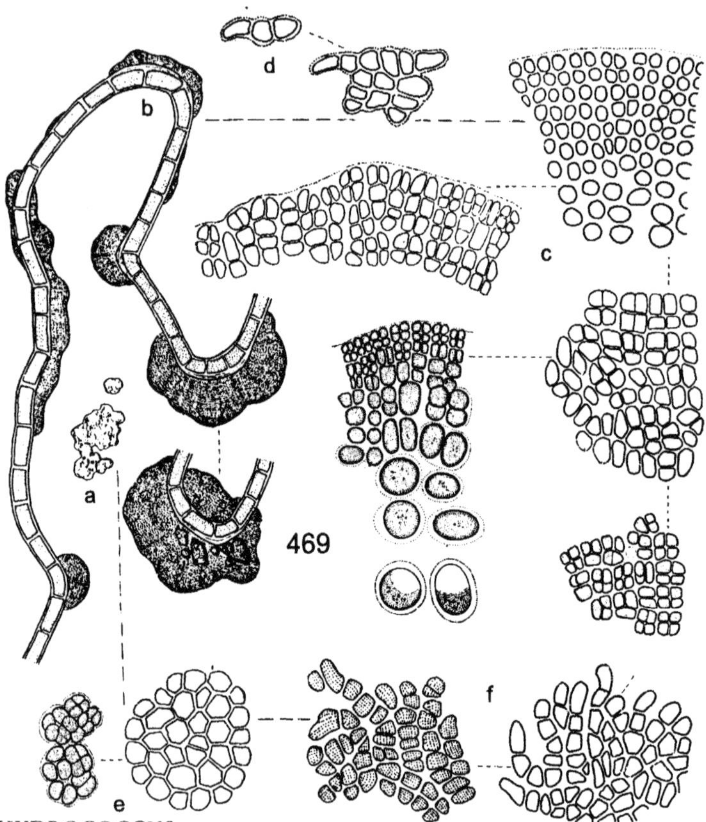


HYDROCOCCUS

**Fig. 468.** *Hydrococcus cesatii*: **a** – colony from the top view, **b** – vertical section through the colony, **c** – young (initial) colonies; from Geitler (1932).

**Fig. 469.** *Hydrococcus rivularis*: **a** – habitat of old macroscopic colonies, **b** – colonies on filament of a green alga, **c** – vertical section through colonies, **d-e** – young colonies from the top view, **f** – marginal parts of characteristical colonies (top view); from Geitler (1932, 1960).

from one another. Old parts of pseudofilaments change into a pseudoparenchymatous layer, but later sometimes with clustered cells. Cells rounded-polygonal or irregular in outline, at the ends of pseudofilaments sometimes elongate. All species epiphytic or partly endophytic, forming flat creeping colonies on the surface of marine algae.
Cells divide in various planes in successive generations, in apical parts of pseudofilaments with mainly transverse binary fission. Nanocytes never observed (*Myxohyella*-like species with nanocyte production belong to other genera). Up to now only marine species are known, growing epiphytically and/or endophytically on red and brown algae and other cyanoprokaryotes. Little known genus, from which only one species has been found in Europe.
**Important references:** Gardner (1918, sub *Chlorogloea*); Setchell & Gardner (1919, sub *Chlorogloea, Hyella*); Geitler (1925, 1932, 1942); Kosinskaja (1948); Feldmann (1956).

## 1. Myxohyella lutea (Setchell et Gardner) Geitler 1925 (fig. **471**)

*Chlorogloea lutea* Setchell et Gardner in Gardner 1918

Colonies extremely variable in shape and size, initially spreading over the surface of the host, forming cushion-like (to 100 µm thick) cell clusters; cells are arranged initially irregularly or more or less in radial rows; later, certain cells from the lower side of the colony penetrate by transverse divisions into the host, forming crooked, branched pseudofilaments increasing in length by apical growth; cells of these pseudofilaments divide in various planes, producing masses of cells, encroaching on one another, and finally coalescing into a solid mass in the central part of the colony. Cells angular, nearly isodiametric, 0.9–1.5 µm in diameter; apical cells penetrating the host to 4 µm long; cell content pale yellow-green. Sheaths colourless.
Occurrence: Marine, epiphytic on *Iridaea* in the lower littoral zone, described from California, known also from Aegean Sea in Greece.

**Species outside Europe:**
- *Myxohyella infestans* (Howe) Geitler 1925 [syn.: *Hyella infestans* Howe 1914]: Marine, endophytic in *Leptocladia*, described from the Pacific coast of Peru; (fig. **473**).
- *Myxohyella papuana* L. Hoffmann 1992: Marine coasts, described from Papua New Guinea, recorded also from the Atlantic sea shore in Brazil (SP).
- *Myxohyella socialis* (Setchell et Gardner) Geitler 1925 [syn.: *Hyella socialis* Setchell et Gardner in Gardner 1918]: Marine, endophytic in algae and cyanoprokaryotes, described from California, known also from coasts of Sea of Japan; (fig. **472**).

**Unrevised species:**
- *Myxohyella cavanillesiana* Gonzalez Guerrero 1946: Dense aggregates and pseudofilamentous formations 10–17 µm wide, with spherical cells 2–5(7) µm in diameter; the author describes «gonidangia» (20–30 µm in diameter) with numerous, 1–3 µm large baeocytes («endospores»); endophytic in submerged leaves of vascular plants in saline waters of the river Avillo near San Fernando, Spain; the taxonomic position of this species (even the baeocytic division) is unclear.
- *Hyella intermedia* Varma et Mitra 1966: Described from India; habitually corresponds more to the g. *Myxohyella* (without baeocyte formation).
- Several *Myxohyella* types (designated as *Myxohyella* sp. div.) were found on the Atlantic coasts in Brazil (SP); (Sant'Anna & al. 1995, Branco & al. 1996).

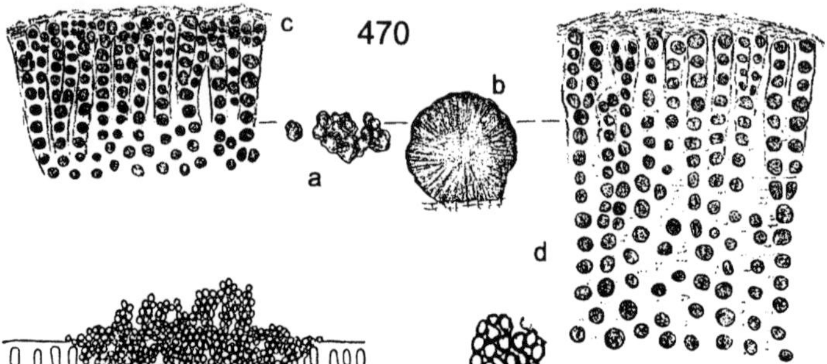


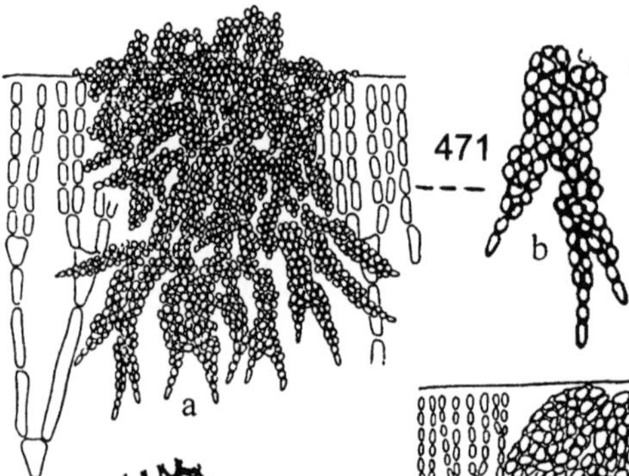


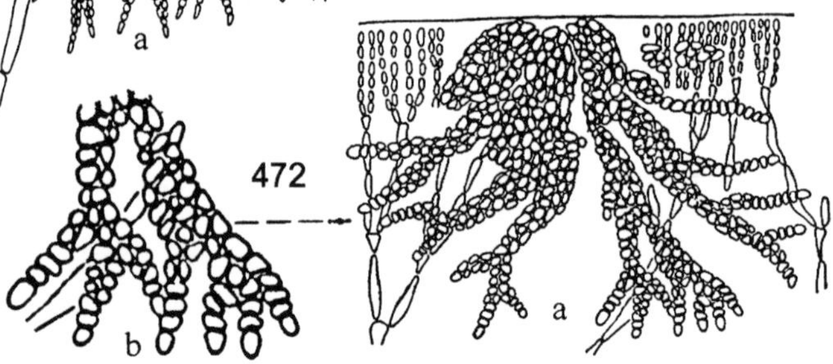


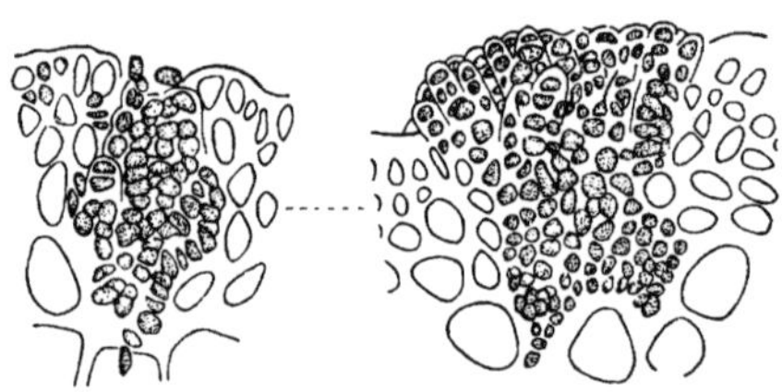


HYDROCOCCUS

**Fig. 470.** *Hydrococcus rivularis* sensu Frémy, from Africa: **a** – macroscopic colonies, **b** – section through the single colony, **c-d** – radial section through the colony; after Frémy (1930, sub *Oncobyrsa rivularis*, possibly different from *H. rivularis* described from Austria, comp. Fig. 469).

# 8. Family **Chamaesiphonaceae** Borzì 1882

Cells sessile, heteropolar, solitary, in groups or forming layers on the substrate, sometimes in dendroid colonies or in polarized pseudofilaments (with one row of cells). Cells or their groups with mucilaginous, multilayered, mainly firm sheaths (pseudovaginae), which are rarely fine or reduced to the mucilaginous basal mucous cluster.
Cell division in principle asymmetric, successively or almost simultaneously in one plane perpendicular to the long (vertical) axis (with the exception of basal cells), or (in unicellular types) asymmetrical in one or more directions in the upper part of the cell (rarely in the middle); in pseudofilaments only the terminal cell divides or the separated «daughter» cells are able to divide repeatedly (*Stichosiphon*), sometimes in a rapid sequence. Terminal divided cells (exocytes) separate often from the basal component of the mother cell and germinate in new vegetative cells; sometimes a new cell grows from the remaining basal part of the mother cell.

**Key to the genera:**

1a Pseudovagina almost lacking or reduced to a small mucilaginous, colourless basal gelatinous cluster or short cup-shaped formation; cells narrow cylindrical; exocytes separate at the cell apex, solitary or in rows . . . . . . . . . . . 2

1b Pseudovagina cup-shaped or cylindrical, enveloping major part or the whole cell; cells cylindrical, obovoid, club-shaped, oval or almost spherical; exocytes separate from the cell apex as single cells, as rows of cells, or spatially (three-dimensional) . . . . . . . . . . . . . . . . . . . . . . . . . . . . . . . . . . 3

2a Old cells cylindrical, attached to the substrate by a scarcely visible, small, colourless gelatinous cluster of slime; exocytes cylindrical, separate singly (after unequal or almost equal cell division, division plane usually displaced to the upper part of a cell) . . . . . . . . . . . . . . . . . . . . **1. Geitleribactron**

2b Cells cylindrical, rarely divaricate («branched», Y-formed), attached to the substrate directly by their base (sometimes slightly enlarged), by a small gelatinous pad or by a cup-shaped, colourless, fine gelatinous pseudovagina; exocytes small, differentiate in rows in the upper part of the long, cylindrical cells . . . . . . . . . . . . . . . . . . . . . . . . . . . . . . . . **2. Cyanophanon**

3a Old cells cylindrical or elongate-obovoid, at the apex narrow-pointed and ending in elongate, more or less long, hyaline hair-like gelatinous strand (= sheath material), which disappears after exocyte differentiation . . . . . . . . . . . . . . . . . . . . . . . . . . . . . . . . . . . . . . . **3. Clastidium**

3b Cells (sheaths) not ending in gelatinous strands . . . . . . . . . . . . . . . . . . 4

4a Cells mostly cylindrical, changing almost completely into a long row of exocytes in the upper part; exocytes divide repeatedly forming more or less long pseudofilamentous formations . . . . . . . . . . . . . . . **4. Stichosiphon**

4b Cells rarely cylindrical, usually oval, ovoid, obovoid or club-shaped; exocytes differentiate singly, in short rows or spatialy in clusters from the upper part of heteropolar cells . . . . . . . . . . . . . . . . . . . . . . . . . . . . . . . . . 5

5a Exocytes differentiate singly or in short rows in the apical part of heteropolar,

---

MYXOHYELLA

**Fig. 471.** *Myxohyella lutea*: **b** – detail of a thallus; from Gardner (1918).
**Fig. 472.** *Myxohyella socialis*: **b** – detail of a thallus; from Gardner (1918).
**Fig. 473.** *Myxohyella infestans*: after Howe (1914) from Kosinskaja (1948).

sessile cells, only exceptionally forming groups at the ends of cells; cells solitary, in groups or forming layered thallus . . . . . . . . **5. Chamaesiphon**

5b Sessile, elongate, club-shaped mother cells first divide transversely (horizontally; first division, preliminary to exocyte formation), then the upper part divides spatially into several exocytes (transition to nanocyte formation); cells solitary or in groups . . . . . . . . . . . . . . . . . . . . . . . **6. Chamaecalyx**

## 1. Geitleribactron Komárek 1975

Heteropolar cells joined to the substrate by one end, solitary or in irregularly parallel or stellate groups. Cells ovoid-oval or shortly club-shaped when young, later cylindrical, rod-shaped, rounded at the apex, slightly and shortly attenuate or rounded at the base, attached to the substrate by fine mucilaginous pad, scarcely visible in light microscope (staining !); without sheaths. Cell content homogeneous, grey-blue, blue-green or olive-green, rarely slightly granulate, sometimes with differentiated chromato- and centroplasmic areas (peripheral arrangement of thylakoids).

Cell division transverse, more or less in the middle or in the upper part of a cell (asymmetrically displaced towards the apex), rarely two divisions occur almost simultaneously. The liberated, usually elongate apical daughter cell becomes attached to the substrate by either end.

**Important references:** Geitler (1966, 1970, sub *Chamaesiphon*; 1975); Golubić (1967b, sub *Cyanophanon*); Komárek (1975b); Preisig (1979); Hällfors & Munsterhjelm (1982); Komárek & Anagnostidis (1986).

**Key to the species:**

1a Cylindrical cells with slightly flattened base, usually to 7 × 1.4–1.6 µm, exceptionally to 19.5 µm long . . . . . . . . . . . . . . . . . . . **1. G. subaequale**

1b Cylindrical cells with slightly and shortly attenuated base, usually larger than 5 × 1.3–2.7 µm . . . . . . . . . . . . . . . . . . . . . . . . . . . **2. G. periphyticum**

### 1. Geitleribactron subaequale (Geitler) Komárek 1975 (fig. 474)

*Chamaesiphon subaequalis* Geitler 1970

Cells solitary, sessile, perpendicularly oriented to the substrate, parallely organized into one blue-green layer. Cells straight or slightly curved, cylindrical, rounded at the apical end and slightly flattened at the base, with more or less homogeneous protoplast with clearly recognizable chromatoplasma, pale grey-green or intensely blue-green, 4–5 (19.5) × 1.4–1.6 (1.8) µm. Cell division by transverse fission almost in the middle of cells or slightly displaced to the upper end, rarely to the basal end.

Occurrence: Periphytic, on submerged water plants (roots, stems), filamentous algae (*Cladophora, Oedogonium*) and detritus in oligotrophic and oligohaline water bodies, lakes and backwaters; known only from Austria (lakes Neusiedler See, Piburger See, and backwaters near Danube).

### 2. Geitleribactron periphyticum Komárek 1975 (fig. 475, Tab. 2 B/d)

*Cyanophanon* sp. in Schwabe 1964, sec. Golubić 1967

Cells solitary or forming periphytic, dense groups, layers or star-like colonies on the substrate; mass development causes an intense blue-green colour of the substrate. Cells straight or slightly arcuate, cylindrical, usually shortly and slightly attenuate at the base with colourless mucilaginous pads, rounded at the ends, with solitary granules, blue-green, more or less homogeneous content, to 25 (50) ×

1.3–2.7 µm. Reproduction by 1- (2) elongate, cylindrical exocytes, separating more or less in the upper part of a mother cell.
Occurrence: Freshwater, periphytic on mollusc shells, stones, leaves of water plants and mosses, filamentous algae, detritus, etc., in littoral of oligotrophic lakes and ponds; Czech Republic (S. Bohemia), Germany (Schwarzwald – Titisee), Switzerland (near Zürich), Finland (lower course of the river Svartan).

**Species outside Europe:**
- *Geitleribactron crassum* Gold-Morgan et al. 1996: With cells 14.5–22(30) × 3.5–6 µm, epiphytic on filamentous algae in running waters in central Mexico.

## 2. **Cyanophanon** Geitler 1956

Heteropolar cells solitary or in groups, elongate, attached to the substrate by one end, long cylindrical when old, sometimes irregularly laterally or Y-shaped divaricate, at the base slightly widened-rounded and sometimes with short, cup-shaped, simple sheath (pseudovagina). Cell content pale blue-green, olive-green or light reddish, homogeneous or with solitary granules, usually with visible, separate chromatoplasma (lateral position of thylakoids).
Reproduction by simultaneous or successive (from the apex) differentiation of exocytes in the upper part of the cylindrical cells, rarely along the whole cell; cell division by transverse fission. Exocytes separate one by one, or a great part of a cell is transformed into a row of exocytes; they are liberated terminally from the mother cell end, become attached by their long sides to the substrate, change their polarity and grow into a new cell from an upper side (fig. **476**).
**Important references:** Geitler (1955, 1957, 1960, 1966); Friedmann (1964); Skuja (1964); Starmach (1966); Bourrelly (1970); Kann (1978); Schanz (1983); Komárek et al. (1985); Komárek & Anagnostidis (1986).

**Key to the species:**
1a Cells to 8 (30) µm long, exocytes in rows with up to 5 (8) daughter cells . . . . . . . . . . . . . . . . . . . . . . . . . . . . . . . . . . . . . . . . . . . . . **1. C. minus**
1b Cells to 90 (190) µm long, exocytes in rows with up to 8–16 (> 40) daughter cells . . . . . . . . . . . . . . . . . . . . . . . . . . . . . . . . . . . . . . **2. C. mirabile**

### 1. **Cyanophanon minus** Geitler 1966 (fig. **476,** Tab. **2 B/d**)

Cells solitary or in small groups or clusters, very near to or slightly distant from one another, more or less cylindrical, slightly widened and rounded both at the base and at the apex, rarely straight, usually irregularly curved or arcuate. Reproductive stages composed of 2–5-(8)-celled (rarely more-celled) uniserial pseudofilaments (fission into exocytes); cells are attached to the substrate by their widened basal ends, rarely with a visible pseudovagina or mucilaginous pad; cells rarely irregularly Y-shaped divaricate; cell dimensions 1.5–8-(30) × (1)-1.4–1.8(3) µm, widened base 2–3.5 µm broad. Young cells spherical, oval or more or less cylindrical, straight or curved, asymmetric, sometimes variable (club-shaped, cordiform, sigmoid, Y-, T- or V-shaped). Cell content grey-pinkish or olive-green, with distinctly visible chromatoplasmic region (peripheral thylakoids), homogeneous, without granules. Cell division in the upper part of a cell, exocytes of variable length, usually 3.5–5 µm long. Sometimes fine mucilaginous, colourless, diffluent, 1–4 µm wide envelopes around cells (staining !).
Occurrence: Freshwater, epiphytic on *Cladophora* and *Oedogonium*, exocytes rarely free floating, in cold stagnant pools and ponds; Austria (Lunz), SW. Slovakia.

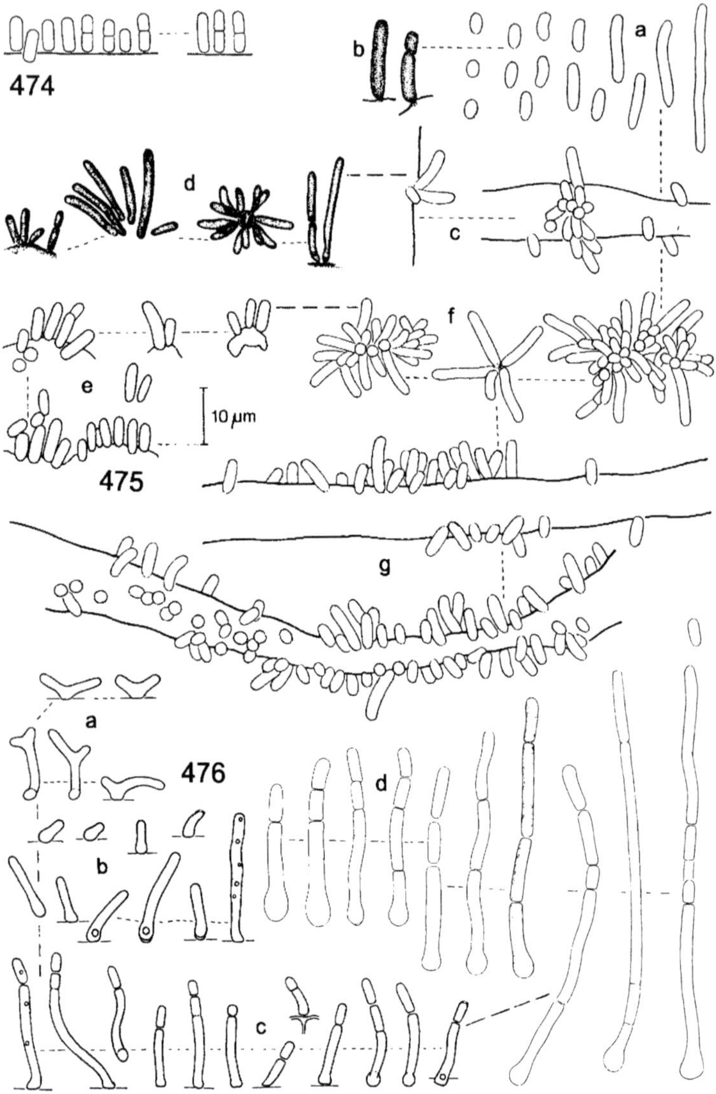
474
a
b
d
c
f
e
10 µm
475
g
a
476
b
d
c

**2. Cyanophanon mirabile** Geitler 1955 (fig. **477–478**)

Cells solitary or in groups, elongate, cylindrical, arcuate or irregularly sinuose, sometimes more or less pseudodichotomously divaricate, at the base slightly widened, widely rounded, usually with a small pad and a short, cup- to vase-shaped, thin, colourless, simple pseudovagina, at the apex rounded and later in the upper part (rarely along the whole length) dividing into a row of cylindrical to

GEITLERIBACTRON

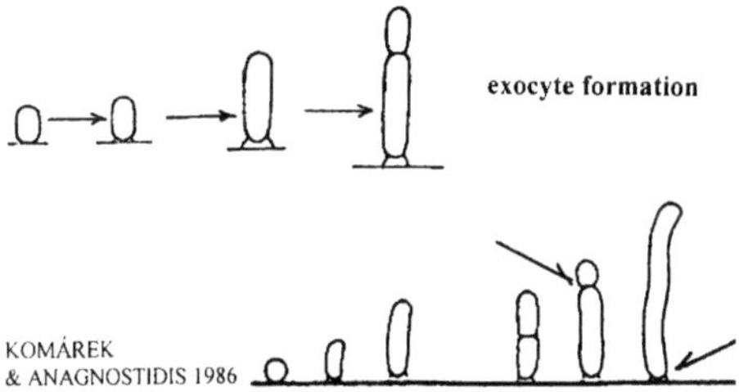


**Fig. 474.** *Geitleribactron subaequale*: after Geitler (1970, sub *Chamaesiphon subaequalis*).
**Fig. 475.** *Geitleribactron periphyticum*: **a** – solitary cells from culture, **b** – detail of cells, **c-e** – small attached groups of cells, **f** – groups of cells attached to detritus, **g** – parts of filaments with *Geitleribactron* cells; **d** – after Golubić (1967, sub *Cyanophanon* sp.), **e** – after Hällfors & Munsterhjelm (1982), other after Komárek (1975).

CYANOPHANON

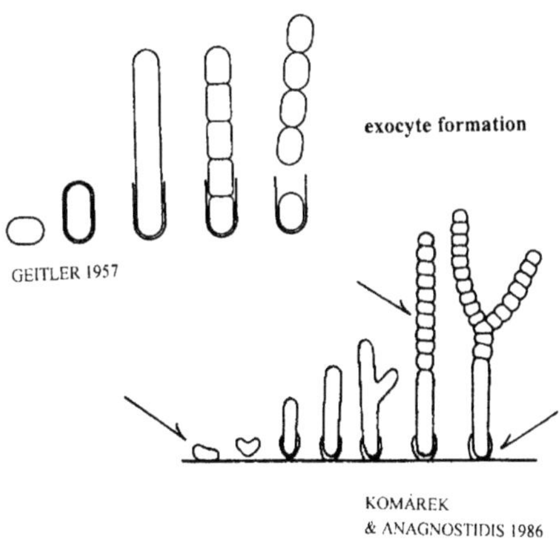


**Fig. 476.** *Cyanophanon minus*: **a** – young divaricated cells, **b** – young simple cells, **c-d** – old cells with exocyte formation; **a-c** – after Geitler (1966), **d** – after Komárek & al. (1985).

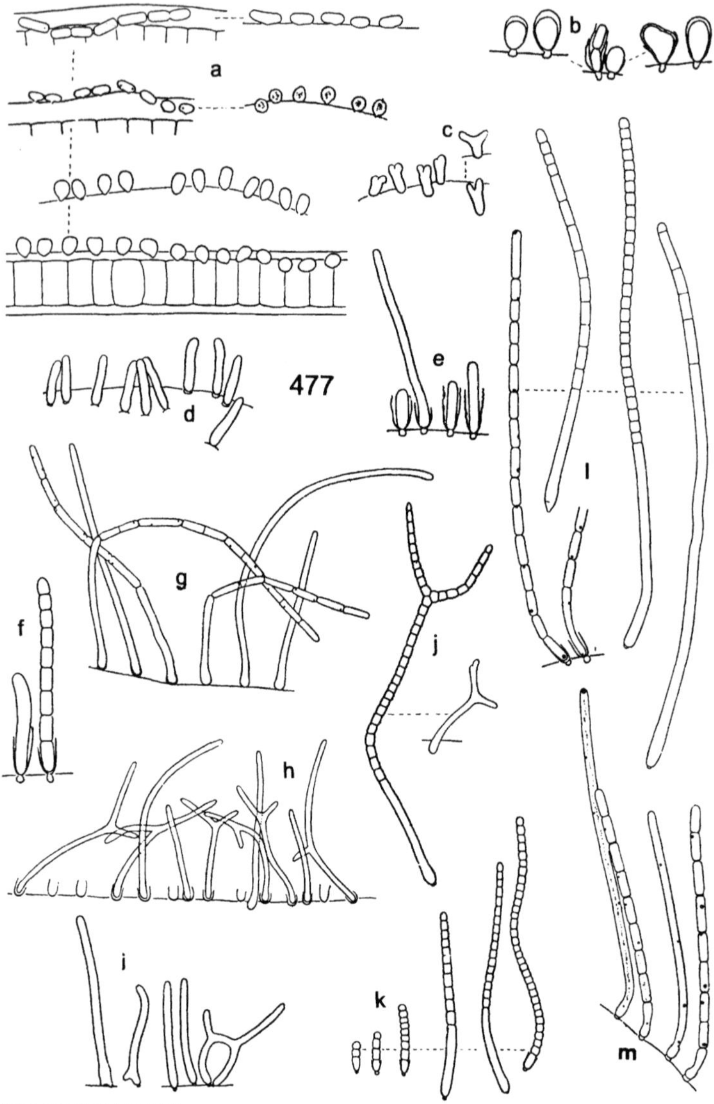


CYANOPHANON

**Fig. 477.** *Cyanophanon mirabile*: **a** – germinating exocytes attached to filamentous algae, **b** – resting cells, **c** – germinating divaricated cells, **d** – germinating simple cells, **e-f** – germinating resting cells (one cell dividing into exocytes), **g-m** – various stages of old cells (several divaricated, several with exocyte differentiation); after Geitler from Starmach (1966).

barrel-shaped exocytes; cells pale olive-green, grey-blue or pinkish, homogeneous, sometimes with visible, distinct chromatoplasma, solitary granules, to 90(190) × 1–2.6(2.8) µm. Exocytes shortly cylindrical, usually 3.5–5.5 × 1–2 µm, arise in a row of to 8–16 cells (rarely up to more than 40 exocytes); the row of exocytes separates sometimes as a unit and attaches to the substrate as a whole (in this case each exocyte germinates laterally into a new cell).
Occurrence: Freshwater, epiphytically on filamentous algae, on rocks with flowing water, streams and waterfalls with clear, katharobic water, usually in mountains; Austria, England, Norway, Portugal, Sweden, Japan, etc.

## 3. **Clastidium** Kirchner 1880

Cells solitary or in groups, heteropolar, attached to the substrate by their basal ends, elongate, ovoid, ellipsoidal, pear-shaped to cylindrical, widely or narrowly rounded at their bases, more or less narrowed to the apex and extended into a fine gelatinous hair; cells with homogeneous content. Mucilaginous sheaths (pseudovagina) thin, firm, sometimes slightly and indistinctly lamellate.
Cell division by successive transverse division in the apical part or (very rarely ?) along the whole length of the cell (exocyte formation). Exocytes (motile) more or less spherical or slightly elongate, separate from the apical cell end solitary or in a cluster, and attach again to the substrate. The basal remnant of the mother cell develops often into a new mother cell.
**Important references:** Geitler (1932, 1942, 1975, 1977); Starmach (1966); Golubić (1967); Economou-Amilli & Anagnostidis (1981); Hällfors & Munsterhjelm (1982); Li (1984); Komárek & Anagnostidis (1986).

**Key to the species:**

**1a** Cells club-shaped or cylindrical, with length:width ratio (without «hairs», without exocytes) > 2:1, usually narrowed towards the base; sometimes the longer part of a cell is transformed into exocytes . . . . . . . . . . . . . . . . . . **2**
**1b** Cells ovoid, with length:width ratio (without «hairs» and exocytes) always < 2:1, widest near the base; exocytes separate only from the upper end of mother cells . . . . . . . . . . . . . . . . . . . . . . . . . . . . . . . . **1. C. rivulare**
**2a** Cells short, more or less cylindrical, with slightly narrowed bases, 2–5 × 1–1.5 µm; apical hairs short, to 4 µm long; Arctic species . **2. C. cylindricum**
**2b** Cells ellipsoidal or slightly club-shaped, narrowed towards the base, 8–15 (38) × 2–4 µm; apical hair usually longer than 4 µm (up to 75 µm long); in temperate zones, usually in mountains . . . . . . . . . . . . . . **3. C. setigerum**

### 1. **Clastidium rivulare** (Hansgirg) Hansgirg 1892 (fig. **479**)

*Clastidium setigerum* var. *rivulare* Hansgirg 1890

Cells usually in dense gelatinous colonies forming thin, yellowish or brownish spots on stones, joined to the substrate by their widened bases; cells shortly ovoid to oval or slightly pear-shaped, never more than twice as long as wide, pale yellowish or grey-green, –6 × 2–4(6) µm; uniserial rows of more or less barrel-shaped exocytes to 45 µm long; mucilaginous hair 1–6x longer than the cell, towards the basis slightly thickened (to 0.5 µm thick). Exocytes differentiate solitarily or in rows of up to 8 (rarely more), only in the upper cell section. Rarely the exocytes join to the end of pseudovagina and small shrub-like colonies arise. Mucilaginous sheath thin, firm, later clearly visible and usually slightly lamellate, colourless.
Occurrence: Freshwater, epilithic, in fast flowing clear water, waterfalls and

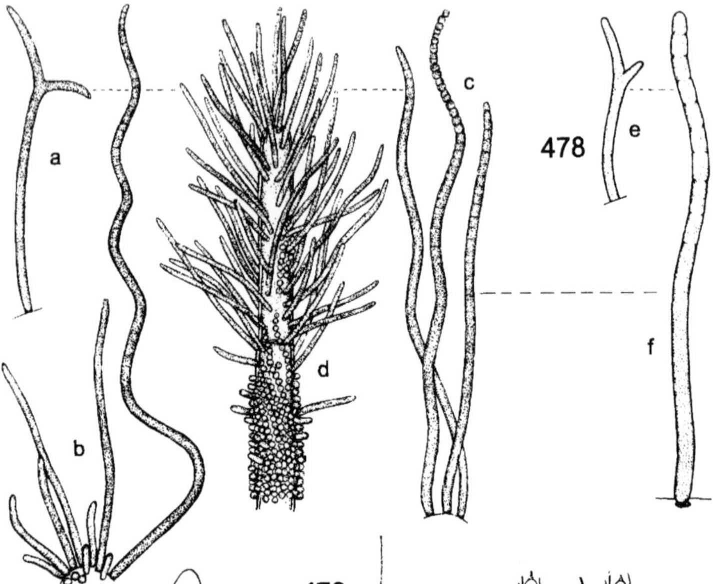


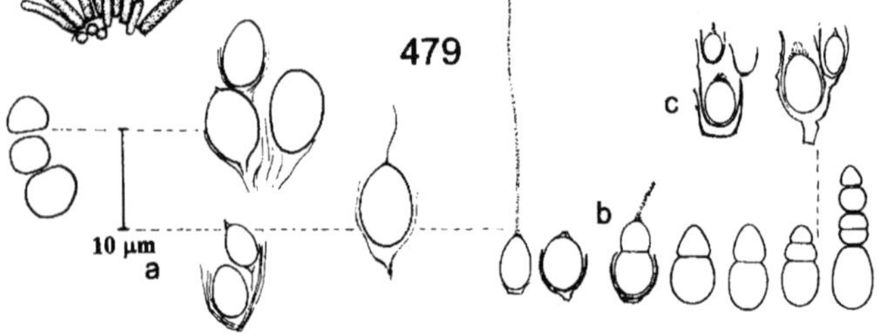


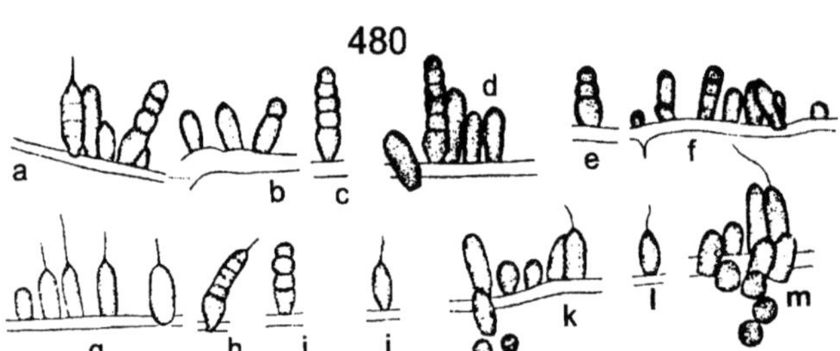


CYANOPHANON

**Fig. 478.** *Cyanophanon mirabile*: **a-d** – after Skuja (1964) from Sweden, **e-f** – after Watanabe & al. (1984) from Japan.

benthic in tarns in mountains; Austria (Alps), Bulgaria, Croatia (Istria), Czech Republic (Šumava Mts.), Greece (Macedonia), Slovakia (High Tatra Mts.).

## 2. Clastidium cylindricum Whelden 1947 (fig. 480)

Cells solitary or in groups, cylindrical, rarely slightly ellipsoid or club-shaped and then indistinctly conically narrowed towards the base, with hemispherical or roundly conical apex, with homogeneous, blue-green content; apical hair very fine, short, usually 2–4 µm long, rarely a little longer, sheaths not described. Exocytes differentiate at the cell apex after the hair disappears, (1) 3–6, spherical, 1–1.5 µm in diameter; they elongate after attachment to the substrate to 2–5 µm before further exocyte differentiation takes place; the width of exocytes remains unchanged during this process.
Occurrence: Freshwater, epiphytic on floating filamentous algae (mainly *Oedogonium*); described from an Arctic lake in Canada (N. Baffin Land), probably also in Svalbard (not confirmed).

## 3. Clastidium setigerum Kirchner 1880 (fig. 481)

Cells solitary or in multicellular groups, attached by bases to the substrate (stones, filamentous algae, always in submersed biotopes), ellipsoidal or slightly irregularly club-shaped (widest between the middle and the apex), to almost cylindrical, narrowed to both ends, with blue-green, olive-green or yellowish content, rarely with solitary granules, usually without visible sheaths (very fine, staining !), 8–15(38) × 2–4 µm; mucilaginous hair to 52(75) µm long, slightly widened at the base. Rows of exocytes uniserial, in the upper part of a cell or (rarely) down to the base of the mother cell.
Occurrence: Freshwater, epiphytic (on algae and water plants) and epilithic in clear stagnant or slow flowing waters (benthic in lakes, streams and springs), on wet rocks, in small artificial channels, etc., frequent in high mountains in temperate countries: Austria, Bulgaria, Czech Republic, Finland, Greece, Poland (High Tatra Mts.), Slovakia, Sweden, etc., in USA (including Alaska) and in Argentine; Japanese populations with cells 4–10 × 2–4 –5 µm differ probably from European specimens (comp. Watanabe & al. 1984).

CLASTIDIUM

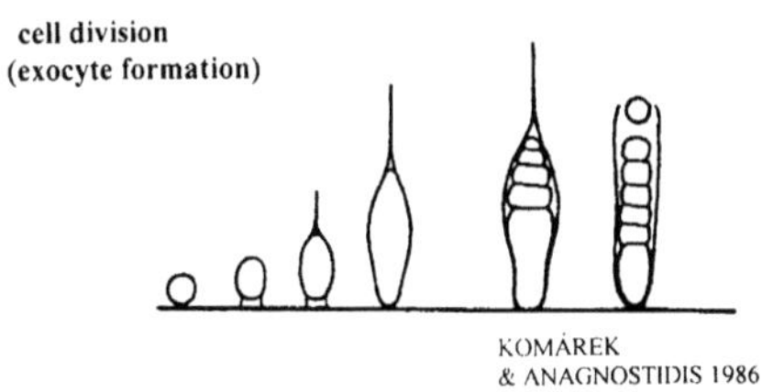


**Fig. 479.** *Clastidium rivulare*: **a** – after Kann (1978), **b-c** – after Geitler (1932).
**Fig. 480.** *Clastidium cylindricum*: after Whelden (1947).

**Species outside Europe:**
- *Clastidium sicyoideum* Li 1984: Described from China.
- *Clastidium nepalense* M. Watanabe et Komárek 1998: Epiphytic on algae and cyanoprokaryotes in cold katharobic water in the Himalayas (E. Nepal).

## 4. Stichosiphon Geitler 1932

Cells solitary or in groups, more or less short cylindrical or club-shaped, heteropolar, elongate, attached by one end to the substrate usually by a mucilaginous pad, very soon (during the cell development) dividing transversely and successively into uniseriate, cylindrical, straight or curved, pseudofilamentous rows of numerous exocytes (sometimes very long), enveloped by firm, colourless sheath (pseudovagina), which later opens at the apex; exocytes remain for long time in pseudofilamentous formations. Cell content pale blue-green, greyish blue, olive-green, sometimes with visible keritomy (position of thylakoids in fascicles throughout the whole cell) and solitary granules.
Cell division by repeated binary fission, transverse to the longer (vertical) axis into a row of more or less short cylindrical or discoidal exocytes, liberating at the apex from the opened sheath (pseudovagina); separated exocytes in pseudofilaments are able to divide repeatedly. The exocytes are attached to the substrate by one end (of the elongated exocyte) and grow into a new cell along the original exocyte long axis. (According to special cytology and biology probably a special cyanoprokaryotic type.)
**Important references:** Geitler (1932); Skuja (1949); Drouet & Daily (1952); Kann (1972, sub *Chamaesiphon* p.p.); Sarma (1981); Hindák (1983); Komárek & al. (1985); Komárek & Anagnostidis (1986); Komárek (1989); Montejano & al. (1996).

CLASTIDIUM
**Fig. 481.** *Clastidium setigerum*: vegetative attached cells, several with exocyte formation; **a** – after Kirchner (1880), **b** – after Golubić (1967), **c-d** and **h** – after Prichod'kova from Kondrateva & al. (1984), **e** – after Geitler from Kondrateva & al. (1984), **f** – from Starmach (1966), **g** – after Hällfors & Munsterhjelm (1982).

STICHOSIPHON

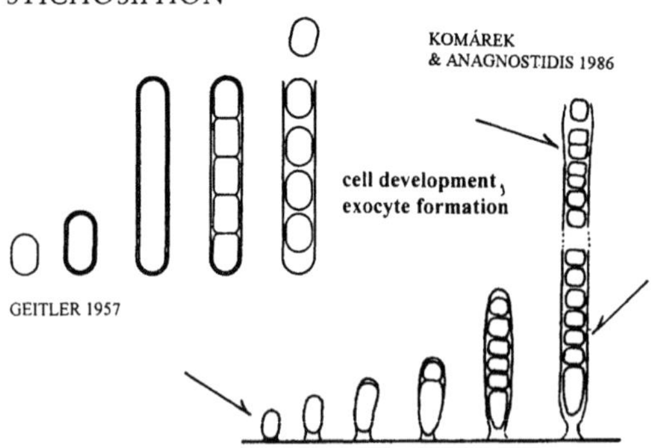

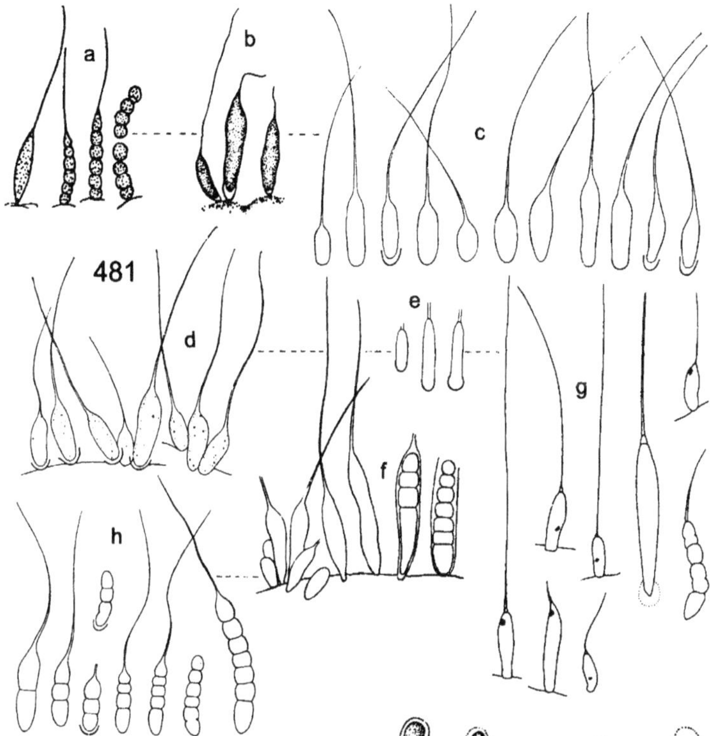


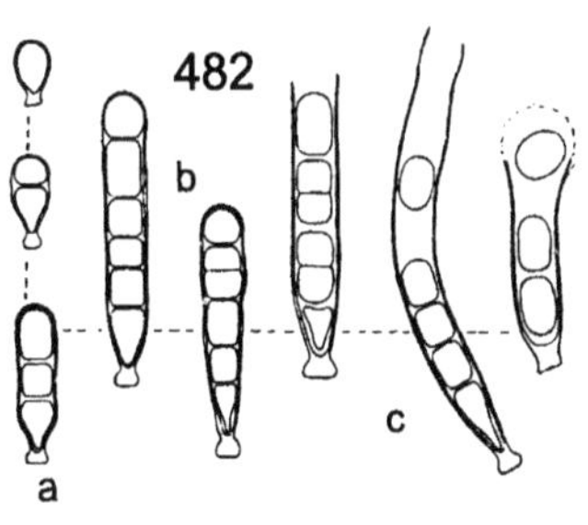


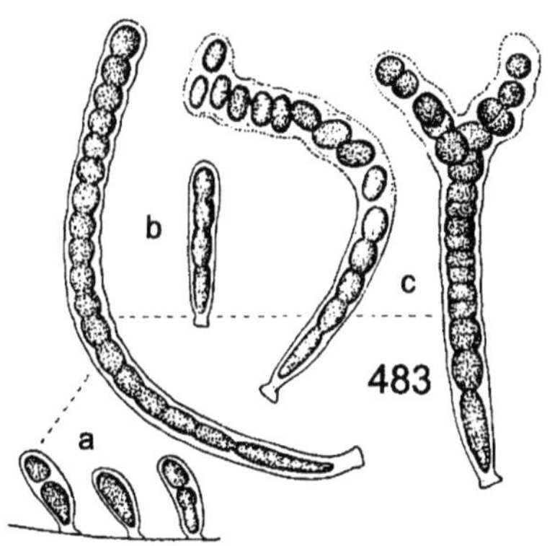


**Fig. 482.** *Stichosiphon regularis*: **a-b** – specimens of various age with pseudofilamentous rows of exocytes, **c** – liberation of exocytes; from Geitler (1932).
**Fig. 483.** *Stichosiphon skujae*: **a** – young cells with first exocytes, **b** – specimens of various age with differentiated exocytes, **c** – specimens with liberating exocytes; after Skuja (1949).

**Key to the species:**

**1a** Cells rounded at the base; pseudofilaments of exocytes more or less creeping on the substrate, forming colonial clusters; exocytes narrower than the basal cell, more or less spherical or shorter than wide; epilithic in mountain streams . . . . . . . . . . . . . . . . . . . . . . . . . . **3. S. pseudopolymorphus**

**1b** Cells narrowed towards the base, attached by the base to the substrate; pseudofilaments of exocytes oriented more or less perpendicularly to the substrate; exocytes more or less of the same width as the basal cell, more or less cylindrical; epiphytic on filamentous algae in tropical waters, thermal basins and warm water aquaria . . . . . . . . . . . . . . . . . . . . . . . . . . . . . **2**

**2a** Pseudofilaments short, up to 40-(60) µm long, with up to 6(12) exocytes, more or less 5–7 µm wide . . . . . . . . . . . . . . . . . . . . . . . . **1. S. regularis**

**2b** Pseudofilaments to 50 (400) µm long, with numerous exocytes, in lower parts 5.8–9 µm wide . . . . . . . . . . . . . . . . . . . . . . . . . . . . **2. S. willei**

## 1. Stichosiphon regularis Geitler 1932 (fig. 482)

Individuals (cells or pseudofilamentous rows of exocytes within sheaths) solitary or in groups, straight or slightly curved, more or less parallely oriented, joined to the substrate by a distinct pad, with cylindrical or very slightly club-shaped sheaths enveloping short rows of up to 6 (rarely up to 12) exocytes, 30–40(60) × 5–7 (apex) µm; exocytes of variable length, successively escaping from the sheath; the apical exocyte widely rounded and sometimes slightly widened. Sheath (pseudovagina) firm, distinct, colourless, closed and widely rounded at the apex, later open (gelatinizing of apical part), slightly funnel-shaped and indistinctly layered at the apex after opening; basal cells and pseudovaginae narrowed to the basal pad. Cell content pale blue-green, more or less homogeneous.
Occurrence: Freshwater tropical species, on submersed algae and plants, in clear water bodies; originally described from Indonesia (E. Java), known also from Mexico (Yucatan), in Europe rarely found in tropical aquaria.

## 2. Stichosiphon willei (Gardner) Komárek et Anagnostidis 1995 (fig. 485–486, Tab. 2 D/k)

*Chamaesiphon willei* Gardner 1927; *Stichosiphon indicus* Rao 1953; *Chamaesiphon/Stichosiphon sansibaricus* sensu auct. post. (Desikachary 1959, fig. 32: 9–13; Hindák 1983; Komárek 1989; Komárek & al. 1985)

Individuals (young cells or pseudofilamentous uniseriate rows of exocytes in sheaths) more or less solitary or in loose groups, club-shaped when young, later cylindrical along the whole length of the exocyte row, rarely straight, usually slightly curved, joined to the substrate by indistinct or distinct gelatinous pads (discs); rows with a few (2) to numerous exocytes, to 200(–450) × 5.8–9 µm; sheaths (pseudovaginae) distinct, colourless, sometimes slightly elongate and narrowed to the substrate. Exocytes cylindrical or widely oval, embedded in a fine, diffluent mucilaginous layer (to 2 µm wide; this envelope is lacking in sessile pseudofilaments), more or less isodiametrical or slightly shorter or longer (to 2 times) than broad. Cell content grey-blue or blue-green, homogeneous or slightly keritomized (arrangement of thylakoids in fascicles), mainly with several small granules and large (to 5 µm in diameter) irregular bodies.
Occurrence: Freshwater, probably pantropical species, epiphytic on filamentous algae in clear pools, littoral of lakes, etc.; known from India and central America (Antilles: Cuba, Porto Rico), in Europe occurs secondarily in warm aquaria and in pools with thermal water and tropical vegetation (e.g., in Slovakia – Piešt'any).

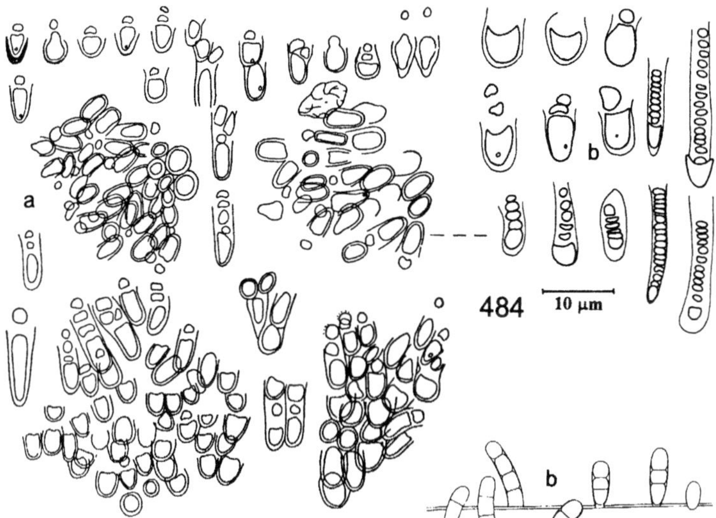


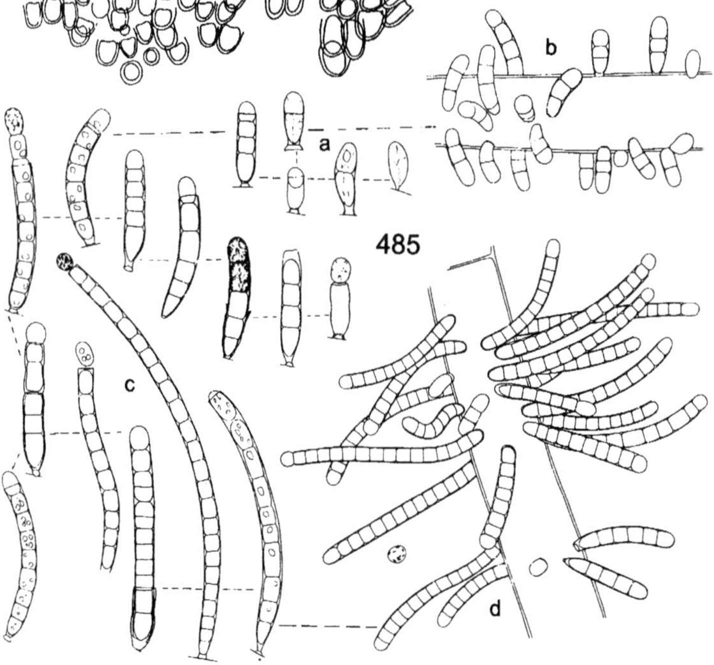


STICHOSIPHON

**Fig. 484.** *Stichosiphon pseudopolymorphus*: **a** – after Fritsch (1929), **b** – after Kann (1972).

**Fig. 485.** *Stichosiphon willei*, from basins with thermal water in Piešt'any, Slovakia: **a-b** – young specimens with primary exocytes, **c-d** – characteristic specimens with differentiated exocytes; after Hindák (1983).

Morphologically similar type was found in small reservoirs in eastern Spain, epizootic on *Daphnia* (Cambra & al. 1989, fig. 5A,B).

**3. Stichosiphon pseudopolymorphus** (Fritsch) Komárek 1989 (fig. **484**)

*Chamaesiphon pseudopolymorphus* Fritsch 1929

Solitary cells with rows (usually short) of exocytes among other algae or in colonies in the form of irregular, macroscopically visible, brown or reddish brown spots on stones; cells with pseudofilamentous rows of exocytes more or less creeping on the substrate. Basal cells more or less irregular-ellipsoidal, usually slightly elongate, rounded et the base, on the apical end flattened or almost concave (after exocyte division), with pale blue-green, yellow-brown or reddish cell content, often with a distinct central or lateral granulum, (4)8–10 × (2.5)6–8 µm. Pseudovagina distinct, firm, sometimes thickened, cup-shaped, yellowish or yellow-brown. Exocytes few, or differentiated into «many-celled» (to 17 or more ?), long, uniseriate rows (they are able to divide repeatedly after separation from the mother cell); cells narrower than the basal mother cell, spherical, hemispherical or flattened (to almost discoidal).

Occurrence: Freshwater, epilithic in clear streams and in waterfalls, usually in mountains; Austria (Alps), Bulgaria, England, Greece (Macedonia).

**Species outside Europe:**

- *Stichosiphon filamentosus* (Ghose) Geitler 1932 [syn.: *Chamaesiphon filamentosus* Ghose 1924; *Stichosiphon indicus* sensu auct. post. (Gupta 1957, Gonzalves & Kamat 1958); *S. sansibaricus* sensu Bourrelly 1970]: Tropical species; (Montejano & al. 1996).
- *Stichosiphon gardneri* Komárek 1989: With elongate, cylindrical exocytes, to 4-times longer than wide, described from Cuba (epiphytic in freshwater swamps).
- *Stichosiphon himalayensis* Jao et Zhu 1974: Variable exocytes, telescopic sheaths, described from Tibet.
- *Stichosiphon mangle* Branco et al. 1994: With aggregates of exocytes at the end of sheaths, which are sometimes telescopic; described from mangrove biotopes in Brazil, known also from Mozambique; (Silva 1991, Branco & al. 1996).
- *Stichosiphon sansibaricus* (Hieronymus) Drouet et Daily 1956 [syn.: *Chamaesiphon sansibaricus* Hieronymus in Engler 1895; *Stichosiphon regularis* sensu Whelden 1941 and G.M.Smith 1950 ]: With pseudofilamentous uniserial rows (to 200 µm ? long) of 2–25 (rarely more ?) exocytes, which are slightly continually widened from the base towards the apex; the apical exocyte widely rounded or hemispherical; known from freshwater swamps, probably with pantropical distribution (India, E. Africa, USA – Fla, Mexico); (Desikachary 1959).
- *Stichosiphon sansibaricus* var. *marinus* L. Hoffmann 1992: From marine biotopes in Papua New Guinea, probably a special species.
- *Stichosiphon* cf. *sansibaricus* in Sant'Anna & al. 1995: Marine, with pseudofilaments to 84µm long and 3–4µm wide, with a widened base; recorded from the «mesolittoral zone», epiphytic on *Caulerpa fastigiata*; probably a separate species (distinct morphology and ecology).
- *Stichosiphon skujae* Montejano et al. 1996 [syn.: *S. indicus* sensu Skuja 1949]: With spatially aggregated, more or less oval or barrel-shaped exocytes at the end of sheaths, known from freshwater localities in Burma and India; (fig. **483**).

## 5. Chamaesiphon A.Braun et Grunow in Rabenhorst 1865, nom.cons.

*Aristella* Kützing 1834; *Brachythrix* A.Braun in Rabenhorst 1865 pro syn.; *Sphaerogonium* Rostafinski 1883; *Godlewskia* Janczewski 1883 incl.; *Chamaesiphonopsis* Fritsch 1929 incl.

Cells heteropolar, slightly or distinctly elongate, attached by their bases to the substrate (stones, plants), to the margin of sheaths of their mother cells or radially and more or less parallely arranged in gelatinous colonies; solitary cells, or micro- to macroscopic colonies, which are shrub-like, more or less spherical (with radially arranged cells in common mucilage) or flat (composed of a few layers of densely and parallely clustered cells). Cells spherical, later growing in club-shaped, cylindrical, pear-shaped, oval or ellipsoidal form; always enveloped by a sheath (pseudovagina), which is thin, colourless or slightly lamellate and yellowish or brownish, opening at the apex after cell division and exocyte separation, at the base usually narrowed and sometimes ending in an adhering pad (disc, sometimes with a very short stipe); the cells are sometimes slightly withdrawn from the base of a sheath; in some species the sheaths are diffluent forming a gelatinous mass, in which daughter cells remain enclosed. Cell content homogeneous, sometimes with distinguishable chromatoplasma (thylakoids are wavy, situated principially parietally but in several modifications, sometimes partly protruding into the centroplasma), in several species with prominent, solitary granules (cyanophycin, carboxysomes); cell content usually pale blue-green, yellowish, olive-green, greyish, pinkish or reddish-violet, sometimes almost colourless (the colour changes in numerous species in one and the same colony. Cell division always asymmetric, crosswise near the apex; the separated daughter cells (exocytes) are liberated from opened sheaths or are attached to the empty pseudovagina (origin of shrub-like or layered colonies) or remain in the common colonial mucilage (origin of gelatinous colonies with radially oriented cells). Reproduction by the liberation of exocytes from mother cells or from colonies.

Notes: Several species (*Chamaesiphon subglobosus, C. polonicus)* are cultured as monospecific strains. In this case they grow in irregular groups, not forming characteristic colonies, *C. subglobosus* sometimes without sheaths (fig. **487**). The heteropolarity is manifested in culture usually only by asymmetric cell division.
**Important references:** Geitler (1925, 1932); Fritsch (1929); Geitler & Ruttner (1935); Behre (1961); Kann (1966b, 1972, 1978); Starmach (1966, 1969); Backhaus (1967a, 1968c); Golubić (1967a, 1967b); Bourrelly (1970, 1985); Waterbury & Stanier (1977); Gromov & Mamkaeva (1980); Hällfors & Munsterhjelm (1982); Komárek & Ludvík (1982); Wujek & Gretz (1984); Komárek & Anagnostidis (1986); Komárek (1989); Cantonati & al. (1996).

**Key to the species:**

**1a** Cells solitary, or more or less parallely arranged in groups or in isolated shrub-like formations, but individually attached to the substrate, not forming coherent or layered colonies (not with exocytes remaining in the colonial mucilage) (subg. 1: **Chamaesiphon**) . . . . . . . . . . . . . . . . . . . . . . . . . . . **2**

**1b** Cells in colonies, forming coherent shrub-like formations, grouped in rows or in layered flat aggregations (with exocytes attached to the margin of mother pseudovaginae or remaining in colonial mucilage) . . . . . . . . . . **12**

**2a** Cells more or less oval, ellipsoidal to shortly cylindrical with widely rounded ends, minute, straight, maximally 3-times longer than broad . . . . . . . . . **3**

**2b** Cells cylindrical or club-shaped, straight or slightly to intensely curved; old cells always more than 3-times as long as wide . . . . . . . . . . . . . . . . . . . **4**

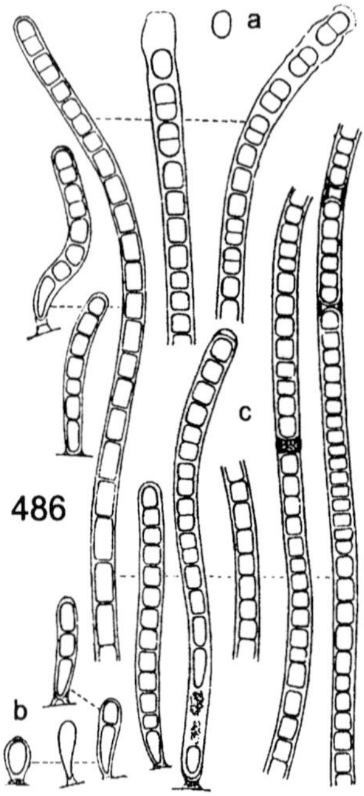


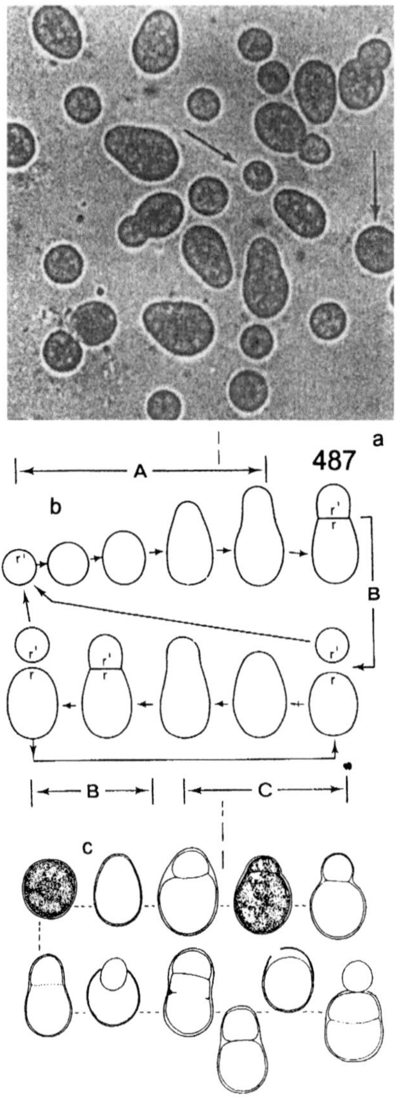


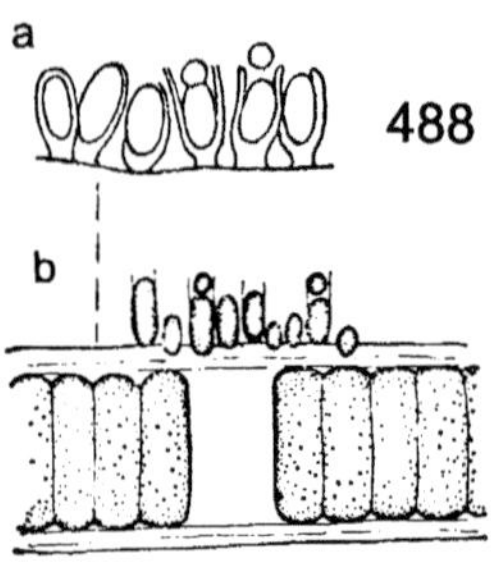


STICHOSIPHON

**Fig. 486.** *Stichosiphon willei*, from tropical Cuba: **a** – liberated exocyte, **b** – young specimens with first exocytes, **c** – characteristic specimens with rows of differentiated exocytes; after Komárek (1989).

**3a** Cells –4 × 1.5–1.8 µm, with thick, distinct pseudovaginae . **1. C. minimus**
**3b** Cells –8 × 1.5–5 µm, with fine, thin, indistinct pseudovaginae . **2. C. minutus**
**4a** Cells straight or very slightly arcuate, to 30 (35) µm long, cylindrical or slightly narrowed towards the base . . . . . . . . . . . . . . . . . . . . . . . . . . . **5**
**4b** Cells usually more or less curved, narrowed towards the base, old cells often longer than 30 µm . . . . . . . . . . . . . . . . . . . . . . . . . . . . . . . . . . . . . . **8**
**5a** Cells straight, more or less rod-shaped, (1)4.5–9 × 1–2 µm, usually parallely and densely attached to the substrate . . . . . . . . . . . . . . . **3. C. investiens**
**5b** Cells straight or slightly curved, sometimes slightly narrowed towards the base, near the apex wider than 2 µm . . . . . . . . . . . . . . . . . . . . . . . . . . **6**
**6a** Cells always solitary, exocytes never remain and develop on the margin of sheaths of mother cells . . . . . . . . . . . . . . . . . . . . . . . . . . . . . . . . . . . . . **7**
**6b** Exocytes often remain on the apical edge of the sheath (pseudovagina) of mother cells and later form solitary shrub-like formation; cells club-shaped, 14–35 × 3–6.5 µm; on green filamentous algae . . . . . . . . **11. C. carpaticus**
**7a** Cells straight, cylindrical or slightly widened at the apex, at the base with distinct short stipe and basal pad (disc), 8–22.6 × 3.5–6.5 µm; exocytes 1(2), slightly flattened . . . . . . . . . . . . . . . . . . . . . . . . . . **4. C. amethystinus**
**7b** Cells straight or slightly curved, cylindrical or sometimes with conically

CHAMAESIPHON

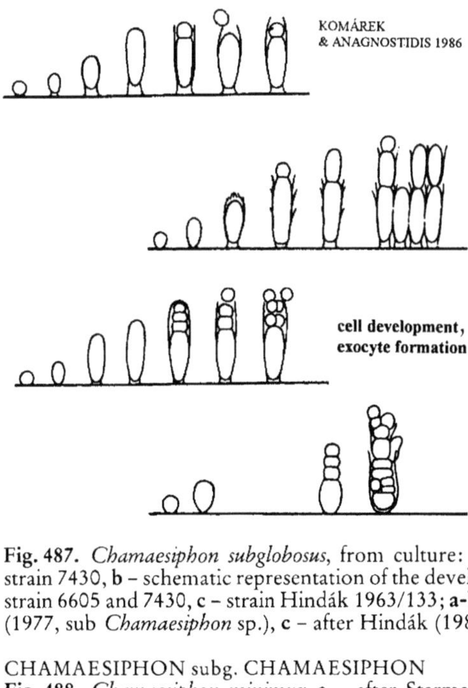


**Fig. 487.** *Chamaesiphon subglobosus*, from culture: **a** – light micrograph of the strain 7430, **b** – schematic representation of the developmental cycle derived from strain 6605 and 7430, **c** – strain Hindák 1963/133; **a-b** – after Waterbury & Stanier (1977, sub *Chamaesiphon* sp.), **c** – after Hindák (1983, in litt.).

CHAMAESIPHON subg. CHAMAESIPHON
**Fig. 488.** *Chamaesiphon minimus*: **a** – after Starmach from Geitler (1932), **b** – from Chu & al. ed. (1991, from China).

narrowed base, without distinct basal pad, 7–13 × 1.5–5 µm; exocytes 1–3, usually spherical . . . . . . . . . . . . . . . . . . . . . . . . . . . . . . . . **5. C. incrustans**

8a Cells epizoic on crustaceans (rarely attached to floating detritus particles) in plankton, club-shaped, 17–45 × 2–3 µm; exocytes cylindrical, 4–6 × 1.5–2 µm . . . . . . . . . . . . . . . . . . . . . . . . . . . . . . **7. C. cylindrosporus**

8b Cells epiphytic or epilithic in periphyton; exocytes more or less spherical . 9

9a Sheaths brown, incrusted with ferric precipitates; cells –61 × 3–7.5 µm . . . . . . . . . . . . . . . . . . . . . . . . . . . . . . . . . . . . . . . . . **8. C. sideriphilus**

9b Sheaths usually colourless, not incrusted with ferric precipitates . . . . . . 10

10a Cells very narrow, up to 40 (100) µm long, continually widened towards the apex, at the end 1–2.5(3) µm wide . . . . . . . . . . . . . . . . **9. C. rostafinskii**

10b Cells near the apex more than 3 µm wide . . . . . . . . . . . . . . . . . . . . . . 11

11a Cells club-shaped, distinctly attenuated towards the base, usually a little withdrawn from the base of the pseudovagina (with «hyaline» base), slightly or distinctly curved, 15–200 × 3–13 (apex) µm, without basal adherence pad (disc) . . . . . . . . . . . . . . . . . . . . . . . . . . . . . . **6. C. confervicolus**

11b Cells almost cylindrical or slightly narrowed towards the base, more or less straight, slightly curved or sinuous, up to184 × 5.5–8.5 (apex) µm, usually with well developed adherence pad (disc) . . . . . . . . . . . . . . **10. C. longus**

12a Exocytes differentiate solitarily from basal cells and soon separate or remain for a short time attached to the margin of sheaths; colonies in the form of a unilayer with more or less densely, parallely oriented cells (subg. 2: **Chamaesiphonopsis**); cylindrical cells very densely and parallely arranged in one layer, in lower part with colourless sheaths, in upper part the sheaths disappear, cells 10.5–14 × 3.5–5 µm . . . . . . . . . . . . . . **12. C. britannicus**

12b Exocytes attach to the margin of opened sheaths, or remain in colonial mucilage, forming sometimes rows, layers, or multilayered strata on the substrate; colonies in the form of small shrub-like groupings, or of large (in nature), layered colonies, often macroscopically visible (spots on the substrate) (subg. 3: **Godlewskia**) . . . . . . . . . . . . . . . . . . . . . . . . . . . 13

13a Colonies compact shrub-like, cells bottle-shaped, rear-shaped, oval or irregular ovoid, widest in the lower third, 7.1–12 (16) × 3.3–6.3 µm, endogloeic on *Batrachospermum* . . . . . . . . . . . . . . . . . . . . . . . **13. C. aggregatus**

13b Colonies layered or with irregularly aggregated cells, not shrub-like . . . 14

14a Cell length:width ratio less than 2:1; cells oval, widely oval or ovoid, sometimes arranged radially in rows in colonies . . . . . . . . . . . . . . . 15

14b Cell length:width ratio (old cells) always over 2:1; cells cylindrical, long oval, pear-shaped or club-shaped, usually in firm, lamellate, coloured sheaths and layered (sometimes crustose) colonies with densely and parallely arranged cells . . . . . . . . . . . . . . . . . . . . . . . . . . . . . . . . . . . . . . . . . 19

15a Cells very small, 1–2 × 1–1.5 µm, with blue-violet or blackish sheaths . . . . . . . . . . . . . . . . . . . . . . . . . . . . . . . . . . . . . . . . . **14. C. niger**

15b Cell length (old cells) always over 2 µm . . . . . . . . . . . . . . . . . . . . . 16

16a Cells to 9 µm long, sheaths colourless or only slightly yellowish or orange, rarely absent . . . . . . . . . . . . . . . . . . . . . . . . . . . . . . . . . . . . . . 17

16b Cells 4–15 × 3–9 µm, parts of sheaths intensely rusty yellow-orange or brownish; large rusty spots on stones, mainly in streams and lake littoral . . . . . . . . . . . . . . . . . . . . . . . . . . . . . . . . . . . . . . **18. C. polonicus**

17a Sheaths firm, not intensely gelatinizing (only slightly superficially); cells in colonies partly densely clustered in more or less radial rows, enveloped by firm (sometimes yellowish) sheaths; apical cells with open sheaths, usually with fine mucilaginous terminal «caps», dark olive-green, blue-green or slightly pinkish coloured, 3–6 µm . . . . . . . . . . . . . . **17. C. polymorphus**

17b Sheaths fine or firm, more or less gelatinizing in various ways; cells in colonies usually slightly distant from one another (in common fine mucilage); cells without apical fine mucilaginous «cap», usually with very pale content, 1.8–4 µm wide . . . . . . . . . . . . . . . . . . . . . . . . . . . . . . . . . . **18**

**18a** Colonies with many cells, oriented radially and more or less regularly in radial rows within fine, diffluent, colourless or slightly orange common mucilage; cells 2.5–5 × 1.8–3.6 µm . . . . . . . . . . . . . **15. C. oncobyrsoides**

**18b** Colonies small (sometimes only solitary cells or irregular groups of cells), with few radially oriented cells in short irregular rows; mucilage always colourless, diffluent (in culture sometimes lacking, cells then are solitary); cells 1.9–8.6 × 1.9–4 µm . . . . . . . . . . . . . . . . . . . . . . . **16. C. subglobosus**

**19a** Cell shape variable, cells irregularly oval, ovoid or pear-shaped, with reddish-violet content, 6–18 × 4–7.2 µm . . . . . . . . . . . . . . **19. C. starmachii**

**19b** Cell shape in colonies more constant, cylindrical-oval or club-shaped, cell content always pale . . . . . . . . . . . . . . . . . . . . . . . . . . . . . . . . . . . **20**

**20a** Cells more or less oval or cylindrical with rounded ends, 5–22 × 2.5–7 µm; sheaths U-shaped, dark brown or blackish . . . . . . . . . . . . **20. C. fuscus**

**20b** Cells more or less club-shaped, narrowing to the base, 4–30 × 2.5–9.6 µm; sheaths V-shaped, yellow-brown . . . . . . . . . . . . . . . . . . **21. C. geitleri**

## 1. Chamaesiphon minimus Schmidle 1902 (fig. 488)

*Chamaesiphon africanus* var. *minimus* (Schmidle) Lemmermann 1910

Cells solitary or densely parallely aggregated, individually attached to the substrate, oval, ellipsoidal to shortly cylindrical with rounded ends, blue-green, with very distinct, firm, more or less thick, colourless pseudovagina, with a short basal stipe (and pad ?), circa 3.5–4 × 1.5–1.8 µm. Possibly belongs to the subg. *Chamaesiphonopsis.*

Occurrence: Freshwater, epiphytic on *Chantransia* in running waters; tropical species (Africa, Asia), described from Cameroon, later recorded also in Poland (probably a different species; in concepts of several later authors possibly initial stages of another species).

## 2. Chamaesiphon minutus (Rostafinski) Lemmermann 1910 (fig. 489)

*Sphaerogonium minutum* Rostafinski 1883; *Chamaesiphon minutus* var. *maior* Geitler 1932

Cells solitary or in groups, all attached separately to the substrate (epiphytic), ovoid, oval, slightly pear-shaped to shortly cylindrical with rounded ends, pale blue-green or olive-green, usually with solitary granules, 3.5–6(8) × 1.5–3(5) µm. Pseudovagina thin, colourless, usually indistinct, rarely visible without staining. Exocytes divide solitarily from the apical cell end.

Occurrence: Freshwater, cold-stenotherm, epiphytic on algae, mosses, filamentous cyanoprokaryotes (*Tolypothrix*) in clear stagnant waters (usually in mountains, in glasshouses, etc.), pools, springs; Austria, Czech Republic, Italy (Alpine regions), Poland, Slovakia, found also in mountains of Argentine and Mexico; records from tropical countries (e.g., var. *gracilior* Geitler 1933) and from Japan (Watanabe & al. 1984; with cells 3–4 × 1.5 µm) belong probably to other species (different morphology and ecology).

Note: Hindák (1995) found morphologically almost identical endogloeic species, attached to the cells or near the cells in the colonial mucilage of *Snowella lacustris* in a fishpond near Bratislava, Slovakia. Cells were cylindrical oval, with a conspicuously thickened basal part of pseudovagina, 3–9 × 3–3.5 µm. From the apical part separate solitary spherical exocytes.

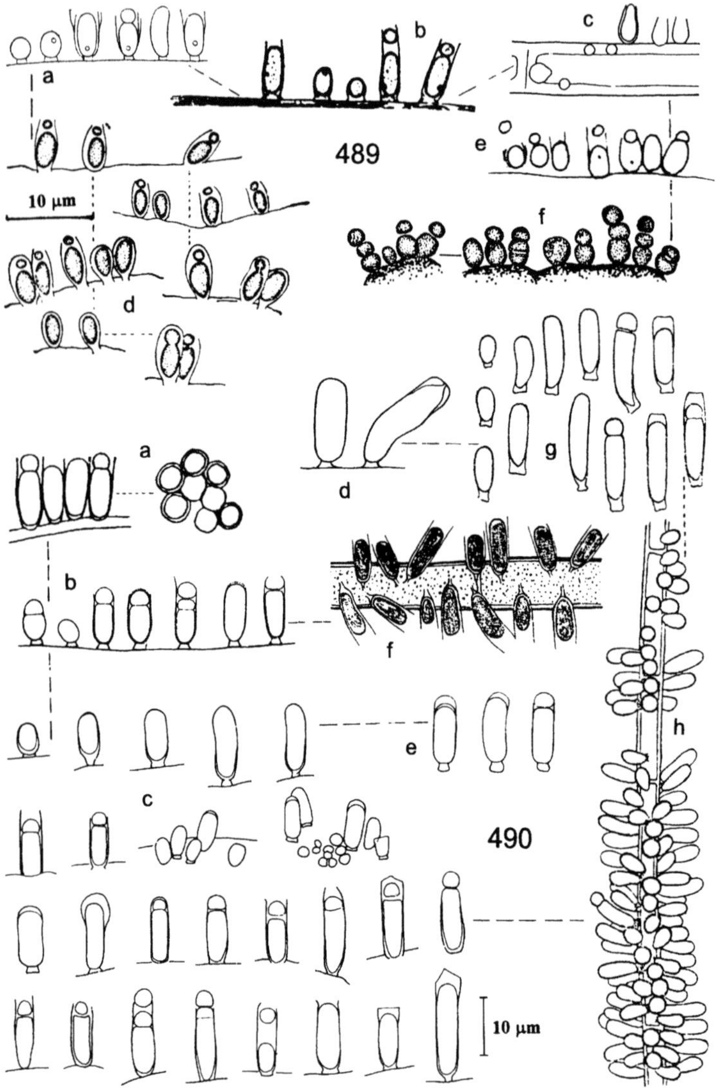


CHAMAESIPHON subg. CHAMAESIPHON

**Fig. 489.** *Chamaesiphon minutus*: **a** – after Geitler (1932, from Austria), **b** – after Copeland (1936, from U.S.A.), **c** – after M. Watanabe & Komárek (1988, from Himalayas), **d** – from Chu & al. ed. (1991, from China), **e** – after Kann (1978,

### 3. Chamaesiphon investiens Skuja 1964 (fig. 491)

Cells usually densely parallely aggregated, individually attached (perpendicularly) to the substrate, forming almost a continuous layer of rod-shaped cells, covering the surface of the host plant. Cells elongate, cylindrical or slightly ovoid, rarely slightly arcuate, rounded or bluntly pointed at the ends, at the base slightly narrowed, sometimes with short stipes and mucilaginous, hyaline pads, with homogeneous, pale blue-green (var. **investiens**) or pinkish violet (var. **roseus** Skuja 1964) content; cells (1)4.5–9 × 1–2 µm. Pseudovagina very fine, colourless, usually invisible (staining !); the presence of a pseudovagina still is uncertain (not yet recorded; if lacking, the species probably belongs to *Geitleribactron*). Exocytes usually separate solitarily, (rarely 2, rounded or slightly rounded-polygonal). *C. investiens* belongs possibly to the subg. *Chamaesiphonopsis.*
Occurrence: Freshwater, epiphytic on algae, waterplants and submersed mosses, in small clear streams; Austria , Italy (Alps), Sweden.

### 4. Chamaesiphon amethystinus (Rostafinski) Lemmermann 1910 (fig. 490)

*Sphaerogonium amethystinum* Rostafinski 1883

Cells solitary or arranged parallel to one another in dense groups, attached individually, but often in masses, covering the substrate, cylindrical or slightly widened towards the ends, widely rounded at the apex (sometimes with «thickened» apical sheath), blunt at the base, straight or rarely very slightly curved, with distinct, gelatinous pad (or adhering discs combined sometimes with short stipes, or at least with distinct separation of the basal part of pseudovagina by a constriction); cell content pale blue-green or grey-blue and lightly pinkish, brownish or violet (all transitions of colour change in cells from the same colony), finely and regularly granular; cell dimensions (with pseudovagina) 8–16.5(22.6) × (2.5)3.5–4.5(6.5) µm (very rarely, longer cells). Pseudovagina distinct, colourless. Exocytes in 1-(2) at the apex, usually slightly flattened, 2–3 × 2.5–4 µm. There are two well defined taxonomic forms: f. **amethystinus** – cells (with pseudovagina) 7–16.5(20) × (2.5)3.5–4.5 (6.5) µm, and f. **africanus** (Schmidle) Komárek et Anagnostidis 1986 [syn.: *Chamaesiphon africanus* Schmidle 1902] – cells (with pseudovagina) –18–22.6 × 4–5(6) µm, known from tropical countries.
Occurrence: Freshwater, epiphytic, mainly on *Chantransia* and filamentous cyanoprokaryotes in clear streams in temperate zone (f. *amethystinus;* Finland, Italy -Alps, Poland, but also in Mexico), or in tropical countries (f. *africanus*; usually in limestone areas, central Africa, Argentine, Cuba, Indonesia, Mexico); records from marine habitats (e.g., from Brazil, Sant'Anna & al. 1995) concern probably another species.

---

from Austria), **f** – after Golubić (1967, from Germany); taxonomic identity of all populations is not evident.
**Fig. 490.** *Chamaesiphon amethystinus:* **a** – from Starmach (1966), **b** – after Geitler from Geitler & Ruttner (1935, possibly f. *africanus*), **c** – after Hällfors & Munsterhjelm (1982), **d** – after Schmidle from Geitler (1932, f. *africanus*), **e** – after Geitler from Geitler & Ruttner (1935), **f** – after Frémy (1930, f. *africanus*), **g–h** – after Komárek (1989, f. *africanus*).

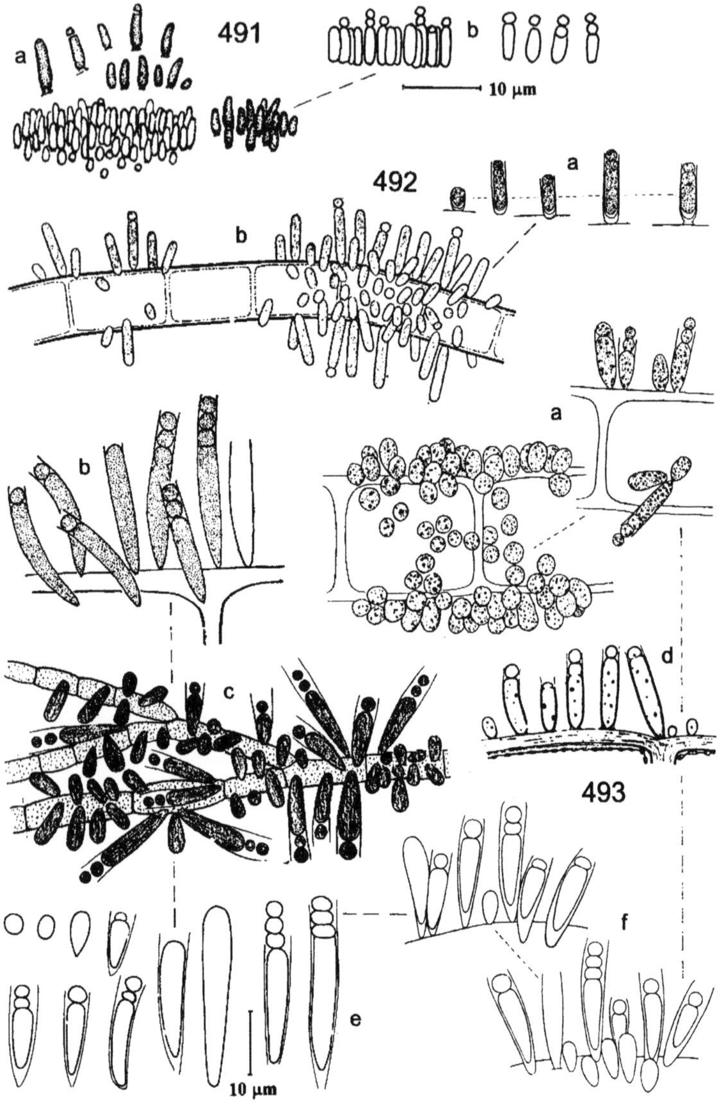


CHAMAESIPHON subg. CHAMAESIPHON

**Fig. 491.** *Chamaesiphon investiens*: **a** – after Skuja (1964), **b** – after Kann (1972, var. *roseus*).

**Fig. 492.** *Chamaesiphon incrustans* f. *cylindricus*: **a** – after Boye-Petersen from Geitler (1932), **b** – after Komárek in Fott (1959).

**5. Chamaesiphon incrustans** Grunow in Rabenhorst 1865 (fig. **492–493**)

*Sphaerogonium incrustans* (Grunow) Rostafinski 1883

Cells solitary or in sparse groups, individually attached to the substrate, more or less cylindrical, widely rounded at the base and always straight (f. **cylindricus** (Boye-Petersen) Komárek et Anagnostidis 1986, fig. **492** - syn.: *Chamaesiphon cylindricus* Boye-Petersen 1923 - with 1(2) exocytes), or slightly club-shaped, slightly conically narrowed towards the base, straight or slightly curved (f. **incrustans**, fig. **493** - with 1–3 exocytes), sometimes with small adhering gelatinous discs, rarely slightly withdrawn from the sheath base, blue-green, grey-green, rarely olive-green or pinkish (at shaded localities), with more or less homogeneous content, without granules in the protoplast; cell dimensions (with pseudovagina) 7–20(26.5–30? - old specimens) × (1–3 at the base) 1.5–5.2(5.5 with sheath) µm (at the apex); dimensions of forms: –20(26.5–30?) × (1 at the base) 2.5–5.5 µm (f. *incrustans*), –13.2 × 2–2.5 µm (f. *cylindricus*). Pseudovagina firm, thin, colourless, distinct, but not lamellate. Exocytes more or less spherical, slightly flattened, differentiate in the apical part of the cell, (1.5)2–3(4.5) µm in diameter.

Occurrence: Freshwater, epiphytic (on green and red algae, vascular plants) in stagnant and flowing unpolluted waters, rarely epilithic, particularly in mountains; probably cosmopolitan, in temperate to subpolar regions generally distributed, f. *cylindricus* known from Czech Republic, Germany, Greece (Macedonia), Iceland and Spain (Pyrenean Mts.).

**6. Chamaesiphon confervicolus** A.Braun in Rabenhorst 1865 (fig. **495–499**)

*Brachythrix confervicola* A.Braun in Rabenhorst 1865; *Chamaesiphon curvatus* Nordstedt 1878; *C. torulosus* Borzì 1882; *C. confervicolus* var. *curvatus* Borzì ex Hansgirg 1883

Cells solitary or in groups, sometimes parallel to one another, more or less densely aggregated, attached to the substrate, more or less club-shaped or almost cylindrical, slightly (rarely intensely) curved, rounded at the apex, clearly narrowed to the bluntly pointed base (without discs), usually withdrawn from the base of the pseudovagina, pale blue-green, olive-green or grey-green, rarely slightly pinkish-violet, with more or less homogeneous content; cell dimensions (with pseudovagina) (4)15–70(200) × (1)3–4 (base)-8.5(13) µm. Pseudovagina distinct, thin, colourless, sometimes slightly widened at the base in old cells or attenuated to a short stipe-like base. Exocytes differentiate in 1–3-many, in one row or (rarely) spatially aggregated at the apical end (before opening of pseudovagina; «Stauchung»), spherical or slightly elongate, (2)2.7–3.5(5.2) µm in diameter; in var. *elongatus* the cells sometimes divide transversely before the exocytes differentiate. Three varieties are distinguishable by their dimensions: var. **confervicolus** (syn.: *Chamaesiphon schiedermayeri* Grunow 1865; *C. confervicola* var. *schiedermayeri* (Grunow) Borzi ex Hansgirg 1892; *C. incrustans* f. *asiaticus* Wille 1922; *C. incrustans* var. *elongatus* Starmach 1929; fig. **495, 497**) with cells –40(68) × (1)3–8.5(10) µm; var. **angustus** (Behre) Komárek et Anagnostidis 1986 (syn.: *Chamaesiphon incrustans* var. *angustus* Behre 1956; *C. gracilis* Rabenhorst 1865; C.

---

**Fig. 493.** *Chamaesiphon incrustans* f. *incrustans:* a – after West from Geitler (1932), **b** – after G.M.Smith (1950), **c** – after Frémy (1930), **d** – after Copeland (1936), **e** – after Kann (1972), **f** – after Prichod'kova from Kondrateva & al. (1984).

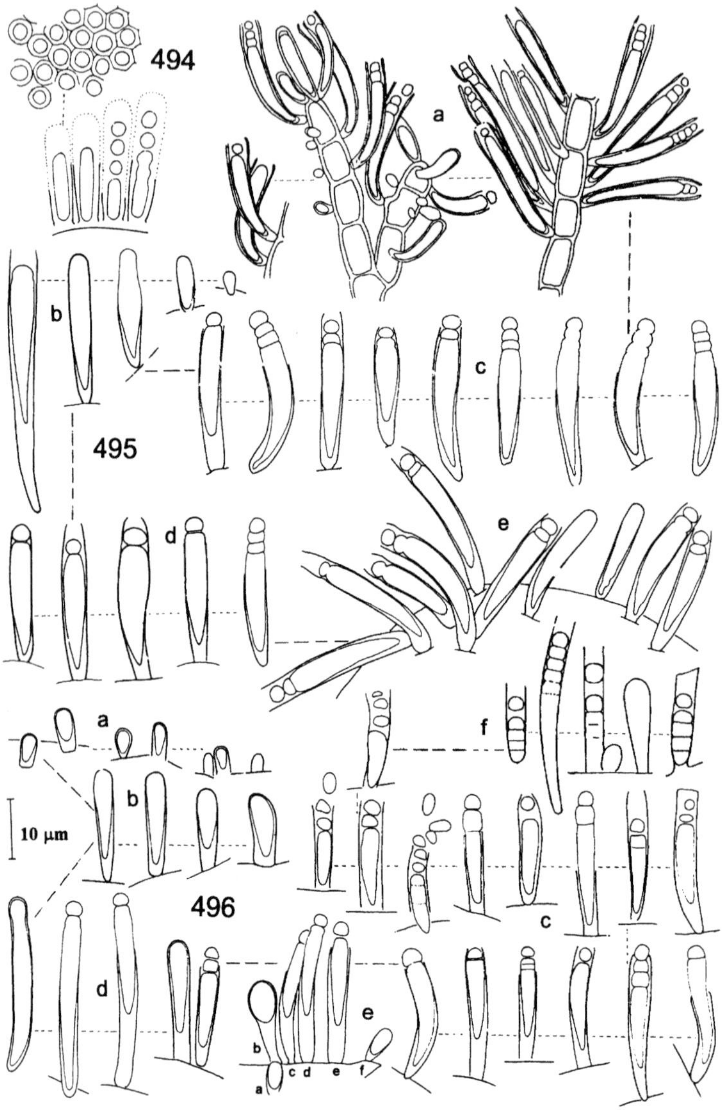


CHAMAESIPHON subg. CHAMAESIPHON

**Fig. 494.** *Chamaesiphon* sp.: after Godward (1937, sub *Chamaesiphon cylindricus* var. *ampla*).

**Fig. 495.** *Chamaesiphon confervicolus* var. *confervicolus*: **a** – from Starmach (1966,

*prescottii* Kullberg 1971; fig. **496**) with cells 10–29(40) × (1)2.8–3.6(5.5) µm; var. **elongatus** (Nordstedt) Kann 1972 (syn.: *Chamaesiphon curvatus* f. *elongatus* Nordstedt 1878; fig. **498**) with cells -60(160–200) × (1.5–2)3–7(9–13?) µm.
Occurrence: Freshwater (rarely in slightly brackish water), epiphytic on filamentous algae (*Cladophora, Chantransia*, and others), mosses and water plants, rarely epilithic, in stagnant or flowing unpolluted waters, often in limestone areas; probably cosmopolitan, distributed over the whole temperate zones (incl. high mountains in central and E. Asia), in central Mexico, Brazil, Argentine, etc., nevertheless not very common; records from tropical regions (central Africa) are less frequent and some of them should be taxonomically revised.

## 7. **Chamaesiphon cylindrosporus** Skuja 1948 (fig. **500**)

Cells solitary or in small groups, individually attached by one end to the substrate, elongate club-shaped to almost cylindrical, straight or slightly arcuate, rounded at the apex, narrowed towards the base and sometimes forming a short stipe, 17–45 × 2–3 µm, with homogeneous, dark blue-green or olive-green protoplast. Sheaths (pseudovaginae) thin, firm, colourless. Exocytes 1–2, cylindrical, rounded in the apical part, truncate at the proximal end, 4–6 × 1.5–2 µm.
Occurrence: Epizoic, on planktic crustaceans (*Harpacticidae*) in littoral zone of lakes, probably also on floating detritus; Sweden, perhaps found also in central Europe (Slovakia, Switzerland) and Greece (Trichonis lake).

## 8. **Chamaesiphon sideriphilus** Starmach 1929 (fig. **501**)

Cells solitary, narrow club-shaped to cylindrical, clearly narrowed towards the base, attached to the substrate by the narrow end, usually slightly curved, with pale yellow-green or yellow-brown, homogeneous cell content; cell dimensions (with pseudovagina) 12.5–61 × 3–5.8(7.5) µm. Pseudovagina thick, distinct, sometimes slightly lamellate, colourless when young, later yellow or brown by ferric precipitates and granular on the surface. Exocytes differentiate solitarily, rarely in twos, at the cell apex.
Occurrence: Epiphytic on filamentous algae (*Cladophora, Oedogonium, Vaucheria*, etc.) in shallow stagnant waters; Poland (north slopes of W. Carpathians, W. Beskids, near Cracow, etc.).

## 9. **Chamaesiphon rostafinskii** Hansgirg 1887 (fig. **503**)

*Chamaesiphon gracilis* f. *maior* Magnus et Wille 1882; *Sphaerogonium gracile* Rostafinski 1883; ? *Chamaesiphon gracilis* f. *elongatus* Wille 1884; *C. incrustans* f. *longissimus* Wille 1922; *C. macer* Geitler 1925

Cells solitary or parallel to one another in groups, individually attached to the substrate, epiphytic, distinctly elongate and narrow, slightly and narrowly club-shaped, straight or arcuate, rarely cylindrical in the apical part, pale olive-green, yellowish pink, reddish or to almost colourless, without distinct granules,

---

sub *Chamaesiphon incrustans* var. *elongatus*), **b-e** – from Hällfors & Munsterhjelm (1982, sub *Chamaesiphon incrustans* var. *elongatus*).
**Fig. 496.** *Chamaesiphon confervicolus* var. *angustus*: **a** – initial stages, **b** – young cells, **c-e** – cells with differentiated exocytes, **f** – various developmental stages; **a-e** – from Hällfors & Munsterhjelm (1982, sub *Chamaesiphon incrustans* var. *angustus*), **f** – after Gomont from Geitler (1932, sub *Chamaesiphon gracilis*).

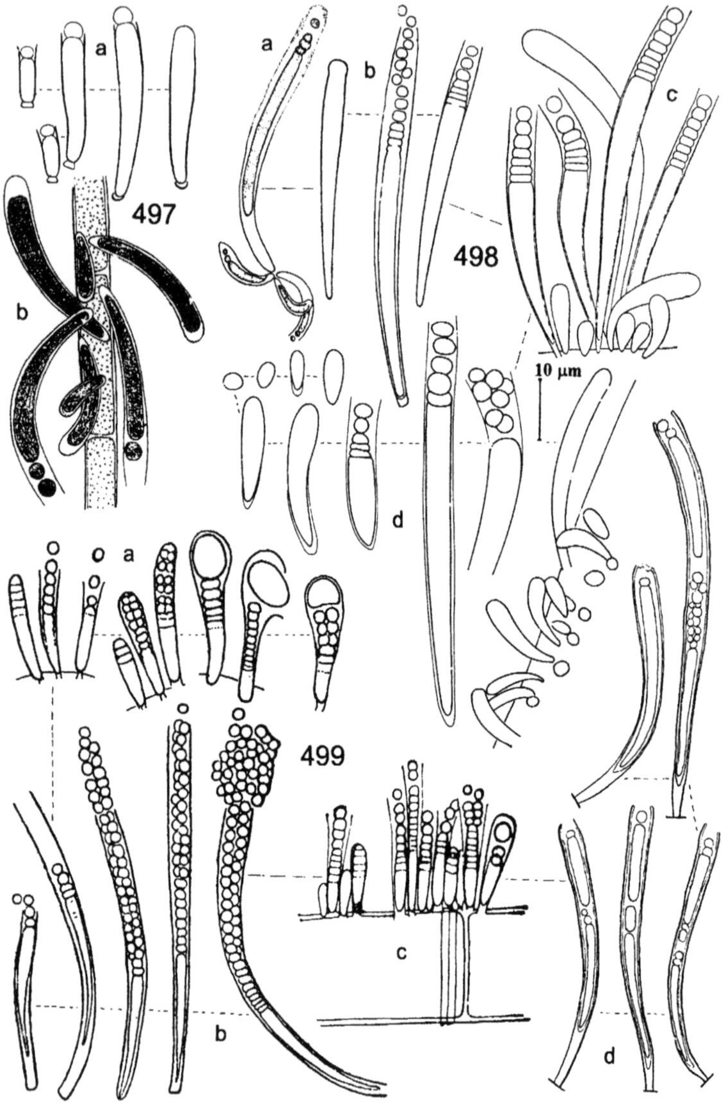


CHAMAESIPHON subg. CHAMAESIPHON

**Fig. 497.** *Chamaesiphon confervicolus* var. *confervicolus*: **a** – after Geitler from Kondrateva & al. (1984, sub *Chamaesiphon curvatus*), **b** – from Frémy (1930, sub *Chamaesiphon curvatus*).

**Fig. 498.** *Chamaesiphon confervicolus* var. *elongatus*: **a** – after Nordstedt from

(4)10–60(100) × 1–2.5(3) µm. Pseudovagina very thin, firm, colourless, not layered, but usually distinct; cells usually withdrawn from the base of the pseudovagina. Exocytes 1–4(9) in a short irregular row at the cell apex.
Occurrence: Freshwater, epiphytic on other cyanoprokaryotes and mainly filamentous algae in katharobic water, in streams, waterfalls, rarely in stagnant waters (littoral splash zone of clear alpine lakes), usually in calcareous areas; high mountains in temperate zone (Austria – Alps, Poland – Carpathians, Scandinavia, central Asia – Pamir), known also from Arctica (Greenland, Novaja Zemlja), maybe cosmopolitan (Geitler & Ruttner 1935: Indonesia; Magnus & Wille 1882: near Bangkok), but with very disjunctive area of distribution (different ecotypes ?).

## 10. **Chamaesiphon longus** Hällfors et Munsterhjelm 1982 (fig. **502**)

Cells solitary or in groups, very rarely forming shrub-like colonies, club-shaped, later long cylindrical, broadly rounded at the apex, slightly narrowed toward the base, usually with distinct adhering disc to 6.5 µm in diameter, straight or slightly and irregularly bent or sinuose, usually slightly withdrawn from the base of the pseudovagina, pale blue-green with numerous solitary granules; cell dimensions (with pseudovagina) (21.5)59–184 × 5.5–8.5 µm. Pseudovagina distinct, thin, firm, colourless. Exocytes in rows, usually numerous (4–10), 4–6 µm in diameter.
Occurrence: Freshwater, epiphytic on water plants (filamentous algae, mosses, vascular plants), usually in rivers; known only from S. Finland.

## 11. **Chamaesiphon carpaticus** Starmach 1929 (fig. **504**)

Cells joined to form solitary, shrub-like colonies, with daughter cells attached to the margin of mother cell pseudovagina (with 2–3 generations); colonies to 80 µm long. Cells narrow club-shaped, straight or slightly curved, rounded at the ends and narrowed to the base, slightly withdrawn from the base of pseudovagina, pale blue-green, greyish or pale pink-violet (in shady localities), 14–35 × 3–4(6.5) µm. Sheaths of basal cells to 10 µm thick; pseudovagina thick, colourless, slightly telescopic, lamellate around old cells. Exocytes arise from the cell apexes in (1)2–6, forming irregular rows.
Occurrence: Freshwater, epilithic or epiphytic on algae (*Cladophora*) in flowing, not very polluted water, mainly in streams and backwaters: Poland, Slovakia (W. Carpathians, Beskid Mts., Tatra Mts.).

---

Geitler (1932, sub *Chamaesiphon curvatus*), **b** – after Geitler (1932, sub *Chamaesiphon curvatus*), **c** – after Prichod'kova from Kondrateva & al. (1984, sub *Chamaesiphon curvatus* f. *curvata*), **d** – after Kann (1972, sub *Chamaesiphon confervicola*).
**Fig. 499.** *Chamaesiphon confervicolus* occurs in various morphotypes, with unusual exocyte production. The identity of all these morphotypes is not evident. Several examples: **a** – after Borzì from Geitler (1932, sub *Chamaesiphon confervicolus*), **b** – after Širšov from Kondrateva & al. (1984, sub *Chamaesiphon curvatus* f. *polysporinus*), **c** – from Chu & al. ed. (1991, sub *Chamaesiphon confervicolus*), **d** – after Starmach from Geitler (1932, sub *Chamaesiphon curvatus*).

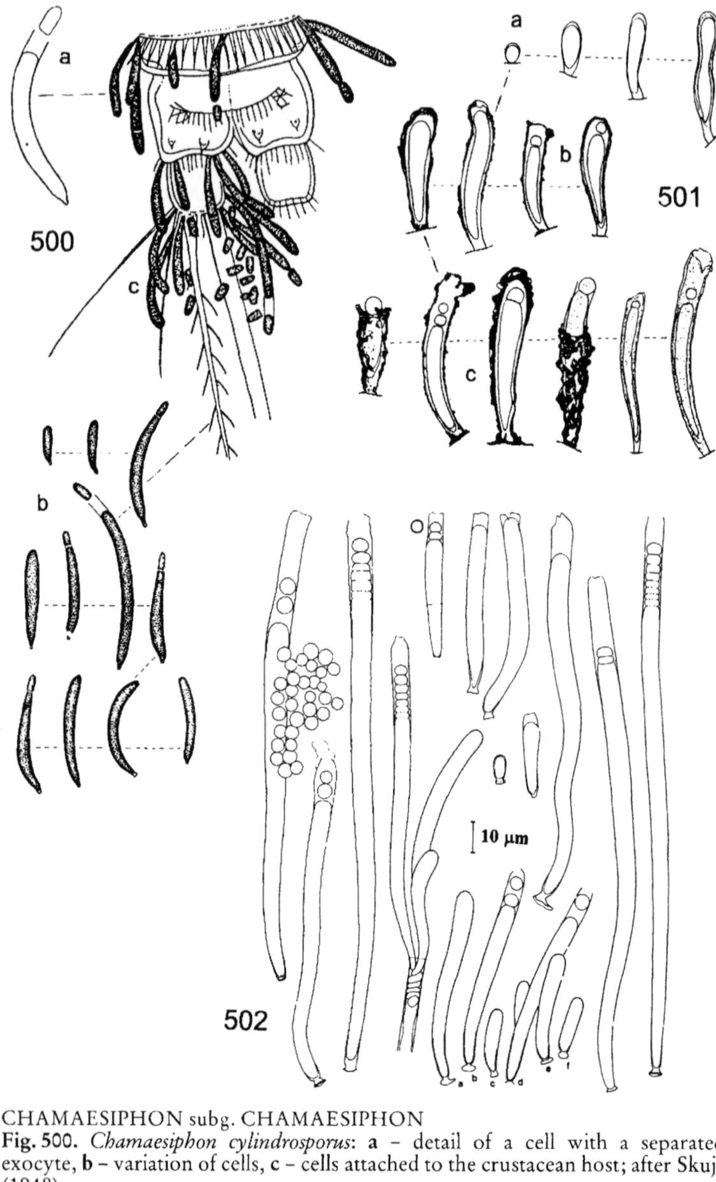


CHAMAESIPHON subg. CHAMAESIPHON

**Fig. 500.** *Chamaesiphon cylindrosporus*: **a** – detail of a cell with a separated exocyte, **b** – variation of cells, **c** – cells attached to the crustacean host; after Skuja (1948).

**Fig. 501.** *Chamaesiphon sideriphilus*: **a** – young cells, **b-c** – cells with ferric precipitates and with exocyte formation; after Starmach (1929).

**Fig. 502.** *Chamaesiphon longus*: after Hällfors & Munsterhjelm (1982).

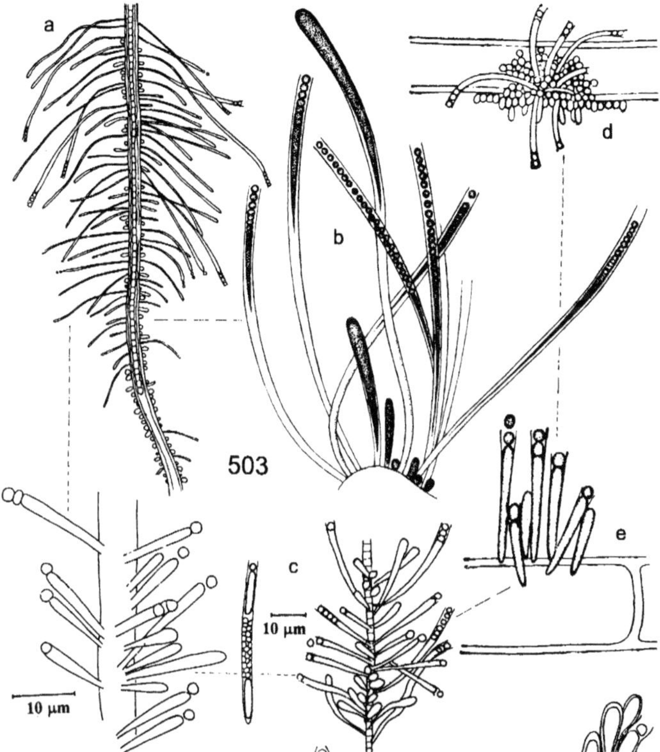


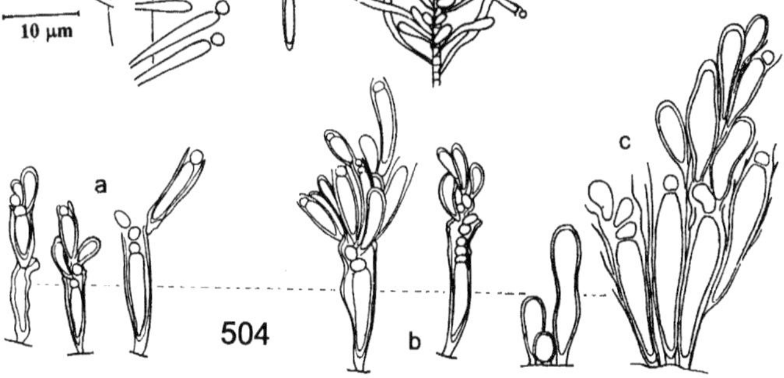


CHAMAESIPHON subg. CHAMAESIPHON

**Fig. 503.** *Chamaesiphon rostafinskii*: **a** – cells attached to the filament of the cyanoprokaryote *Schizothrix tinctoria*, **b-e** – cells from distant areas; **a** – from Geitler (1932, sub *Chamaesiphon macer*), **b** – after Skuja (1964, sub *Chamaesiphon curvatus* var. *elongatus*), **c** – after Kann (1972), **d** – from Chu & al. ed. (1991, sub *Chamaesiphon macer*), **e** – from Chu & al. ed. (1991, sub *Chamaesiphon rostafinskii*). – *Ch. macer* is synonymized with *Ch. rostafinskii*, but the identity of all described populations of both these species need confirmation (comp. figs. **a-b** and **c-e**).

**Fig. 504.** *Chamaesiphon carpaticus*: after Starmach (1929, 1966).

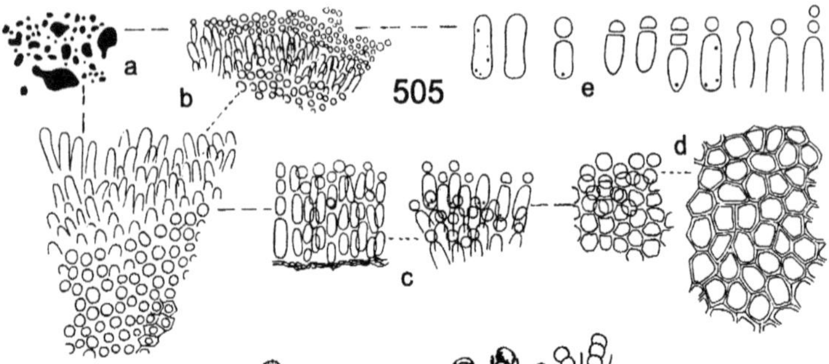


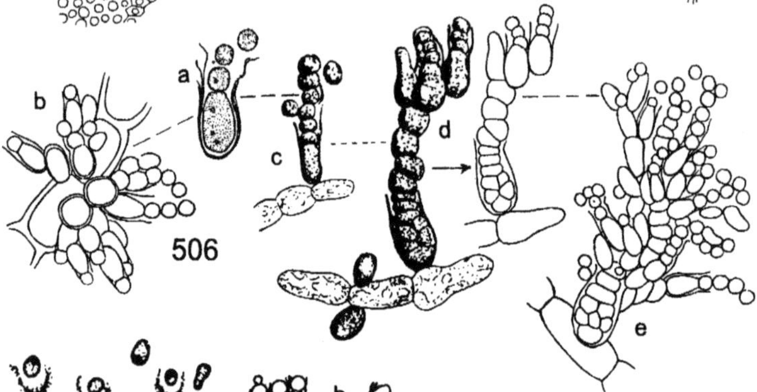


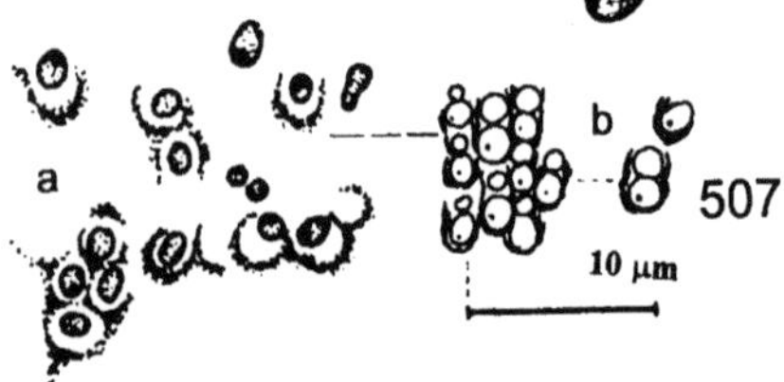


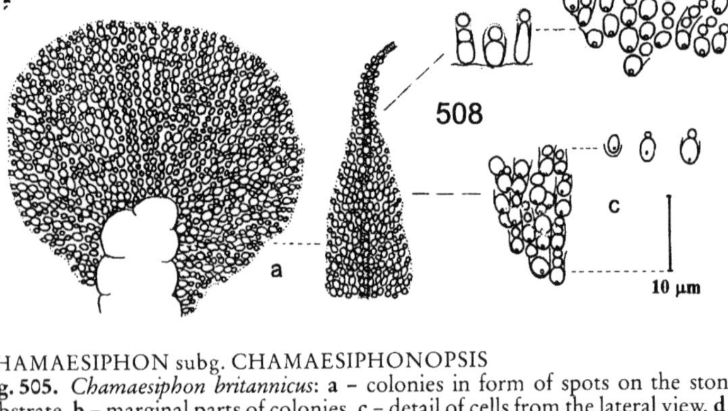


CHAMAESIPHON subg. CHAMAESIPHONOPSIS

**Fig. 505.** *Chamaesiphon britannicus*: **a** – colonies in form of spots on the stony substrate, **b** – marginal parts of colonies, **c** – detail of cells from the lateral view, **d** – detail of colonies from the top view, **e** – detail of cells; after Fritsch from Geitler (1932).

**12. Chamaesiphon britannicus** (Fritsch) Komárek et Anagnostidis 1995 (fig. **505**)

*Oncobyrsa britannica* Fritsch 1927 pro synon.; *Xenococcus britannicus* Fritsch in West & Fritsch 1927; *Chamaesiphonopsis regularis* Fritsch 1929; *Chamaesiphon regularis* (Fritsch) Geitler 1929

Colonies composed of 1–3(4) layers of very densely packed cells, which are arranged more or less parallel to one another and perpendicular to the substrate, forming several $mm^2$ large brown spots on stones. Basal cells densely packed or clustered, more or less polygonal in outline in cross-section, in the upper part more randomly arranged, elongated, without visible pseudovagina; cells ovoid, ellipsoidal, cylindrical or pear-shaped, sometimes slightly narrowed in the middle, widely rounded at both ends, pale blue-green, rarely reddish, 10.5–14 × 3.5–4.5(5) µm. Sheaths faintly visible, fine, colourless, diffluent, in upper colonial parts wholly lacking. Exocytes separate in 1–2 from the ends of apical part of cells, 3.5–4.5 µm in diameter.
Occurrence: Freshwater, in flowing and stagnant clear waters (splash zone), epiphytic on filamentous cyanoprokaryotes; England, Germany.

**13. Chamaesiphon aggregatus** (Janczewski) Geitler 1925 (fig. **506**)

*Godlewskia aggregata* Janczewski 1884

Cells aggregated in isolated, sessile, irregular, shrub-like colonies with daughter cells attached to the margin of mother pseudovagina. Cells bottle-shaped, oval or ovoid, widest in lower third, straight or slightly arcuate, attached to the substrate by their wide, rounded base, enveloped by thin, colourless sheath, which is later (usually in lower part of colonies) thickened and brownish, funnel-shaped and widened at the apex and gelatinizing in the upper part of colony, 7.1–12(9–15 with exocytes) × 3.3–6.3 µm; cell content olive-green, grey-green, blue-green or greyish violet. From young cells separate 1–2 exocytes, arranged serially, in old cells (in lower old parts of colonies) sometimes more exocytes are formed, which divide also lengthwise or irregularly; exocytes usually 3 µm in diameter.
Occurrence: Freshwater, endogloeic and epiphytic in mucilage of *Batrachospermum*; known from streams on northern slopes of W. Carpathians, S. Poland.

**14. Chamaesiphon niger** (Starmach) Golubić 1967 (fig. **507**)

*Chroococcus niger* Starmach 1934

Colonies microscopic, forming dark violet or blackish groups of ensheathed cells, to 50 µm in diameter, sometimes grouped into agglomerations enclosed in common mucilage, with more or less radially oriented cells in short, irregular rows. Cells spherical, ellipsoidal or oval, rarely pear-shaped, sometimes 2–4 near to one another, with pale blue-green content, sometimes with 1 small, solitary, usually excentrically localized granules, 1–2 × 1–1.5 µm. Pseudovagina firm, layered,

---

CHAMAESIPHON subg. GODLEWSKIA
**Fig. 506.** *Chamaesiphon aggregatus*: **a-c** – young cells with exocyte formation, **d-e** – characteristic colonies; after Janczewski and Starmach from various authors (comp. Starmach 1966).
**Fig. 507.** *Chamaesiphon niger*: **a** – after Golubić (1967), **b** – after Kann (1972).
**Fig. 508.** *Chamaesiphon oncobyrsoides*: **a** – epiphytic colonies, **b-c** – details of colonies and cells; **a-b** – from Geitler (1932), **c** – after Kann (1972).

more or less thick, sometimes gelatinizing on the surface, always pigmented, blue or violet; pigmentation sometimes weak in the center of colonies, or more intense at the outer side of a cell. Exocytes relatively large, 1–2 remaining attached to the apex of a mother cell.
Occurrence: Subaerophytic on wet rocks and epilithic on stones in water-level zone of mountain lakes, rarely in alpine streams; known from Poland (High Tatra Mts.), Austria (Alps), Slovenia and Croatia.

### 15. **Chamaesiphon oncobyrsoides** Geitler 1925 (fig. **508**)

Colonies small, irregular spherical or hemispherical, mucilaginous, almost colourless or pale orange, composed of radially and more or less parallely arranged rows of many cells. Cells oval (sometimes almost spherical), ellipsoidal or (rarely) even cylindrical, rarely narrowed at the apex, pale pink, dirty violet, pale blue-green, yellowish, sometimes with pale orange mucilage, often with solitary granules in protoplasts, 2.5–5 × 1.8–3(3.6) µm. Pseudovagina fine, colourless, sometimes finely lamellate, usually gelatinizing and diffluent in colourless and structureless common mucilage. 1–3(4) exocytes separate at the apical cell end, usually remain connected with mother cell and in the mucilage within the colony.
Occurrence: Freshwater, epiphytic on filamentous algae and cyanoprokaryotes (and among other algae), on wet rocks (in dripping water) and in clear streams, often in limestone mountains, but also in katharobic and oligotrophic water in streams of non-calcareous regions, rare; known only from central European mountains (Alps, Böhmerwald, Šumava Mts.).

### 16. **Chamaesiphon subglobosus** (Rostafinski) Lemmermann 1910 (fig. **487, 509,** Tab. **2 C/g**)

*Sphaerogonium subglobosum* Rostafinski 1883

Cells solitary or aggregated in small, irregular colonies, which are later 3- or more-layered, or with several cells in irregular clusters. Cells spherical, ovoid to oval, pale greyish, pale blue-green or slightly pinkish, usually with one or several prominent granules, mainly parietally arranged thylakoids (visible chromatoplasmic region), sometimes combined with groups of coiled thylakoids in centroplasma, (1.9)3–6(8.6) × (1.9)2.1–3.5(5) µm. Sheaths (pseudovaginae) colourless, indistinct or later (in old colonies) distinct, sometimes layered and slightly incrusted (with lime ?) or diffluent (sometimes completely lacking in cultures). 1–2 exocytes differentiate at the cell apex, growing during division («budding»; Waterbury & Stanier 1977), spherical or slightly flattened, 0.9–2.2 µm long. Several strains are known from culture collections.
Occurrence: Freshwater, epilithic (also in detritus) and epiphytic (on algae, mosses) in stagnant and running oligotrophic to mesotrophic, unpolluted waters, lake littoral, waterfalls, streams; known mainly from central Europe, from lowlands to mountains, less in northern and southern subpolar regions; data from tropical countries are uncertain (e.g., epiphytic in Gabon).

### 17. **Chamaesiphon polymorphus** Geitler 1925 (fig. **510,** Tab. **2 A/c**)

Colonies usually small, in the form of thin, gelatinous, dark green, brownish green or dirty violet (dried), irregular clusters, rarely forming macroscopically visible spots among other algae on stony substrate; composed of more or less radially arranged, short, irregular rows of cells enveloped by mucilage. Cells oval-ellipsoidal, pear-shaped to almost cylindrical with widely rounded ends, sometimes slightly polygonal (in colonies with tightly packed cells), variable, mainly

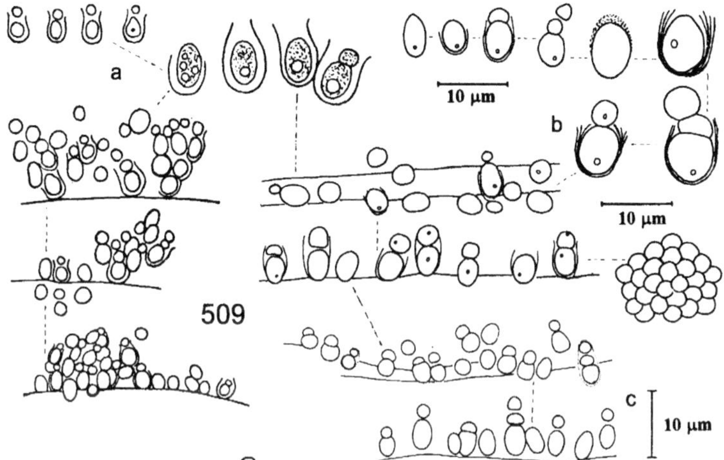


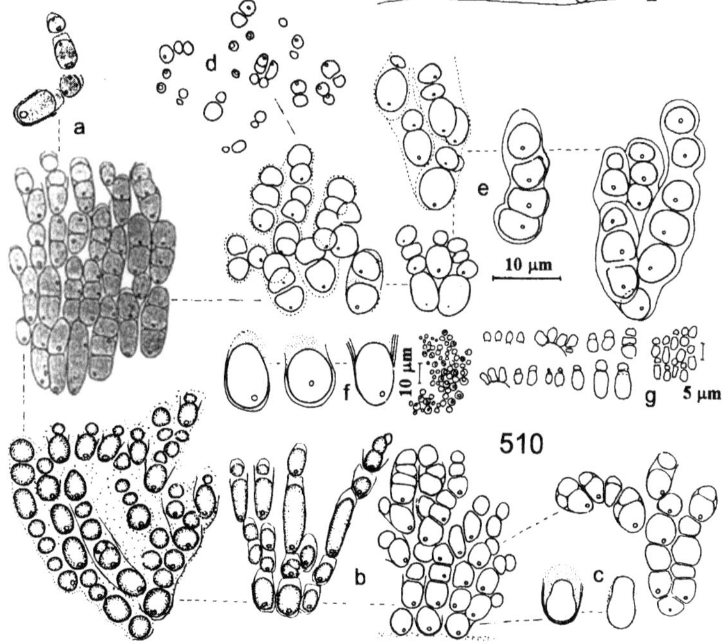


CHAMAESIPHON subg. GODLEWSKIA

**Fig. 509.** *Chamaesiphon subglobosus*: **a** – after Starmach from Geitler (1932), **b** – after Kann (1972), **c** – after Hällfors & Munsterhjelm (1982).

**Fig. 510.** *Chamaesiphon polymorphus*: **a-c** – from Geitler (1932), **d-g** – after Kann (1972).

with dark olive-green content, rarely blue-green, dirty violet, grey-green or yellowish, usually with a typical gelatinous, colourless and fine cap at the apex and 1–2 distinct, small granules within the protoplast, 3–9 × 3–6 µm. Pseudovagina thin, firm, colourless, outer part sometimes gelatinizing, in old colonies yellowish, very rarely brown. Exocytes 1–4 at the cell ends, in a short row or later displaced from the radial direction, usually remaining in the common colonial mucilage and forming irregular, radial rows within colony.
Occurrence: Freshwater, on stones in clear mountain, non- calcareous streams and in the splash zone of the littoral of tarns; in mountains of central Europe, Alps (Austria, Germany), Carpathians (Czech Republic, Poland, Slovakia); Indonesia (?).

### 18. **Chamaesiphon polonicus** (Rostafinski) Hansgirg 1892 (fig. **511**, Tab. **2 A/b**)

*Sphaerogonium polonicum* Rostafinski 1883

Colonies in the form of large, flat, thin, rusty or yellow-orange to brownish, irregular spots on submersed stones, sometimes several $dm^2$, composed of 1(2) or more compact layers of clustered cells; cells are sometimes arranged in irregular rows. Cells almost spherical, later ellipsoidal or oval, longer than wide, rarely pear-shaped, ovoid or shortly cylindrical, with pale content, yellowish, pale olive-green or greyish green, usually with 1 or more, more or less large prominent granules, (4)6–10(15) × 3–6(9) µm. Pseudovagina thick, firm, layered, frayed at the apical end, later yellow-orange, rusty brown, to dark brown, not gelatinizing. Exocytes 1–2(4), solitary or in very short irregular rows, remaining attached to the mother cell sheath, usually 2.6–4 µm in diameter.
Occurrence: Freshwater, in oligo- or stenotherm, katharobic habitats, forming large rusty spots on stones in running water (clear streams) from lowlands to high mountains, rarely in stony littoral of clear lakes, also in periodically dried localities; in central Europe widely distributed, probably present all over the temperate zone.

### 19. **Chamaesiphon starmachii** Kann 1972 (fig. **512**)

*Chamaesiphon fuscoviolaceus* Starmach 1969, non *C. fuscoviolaceus* (Hansgirg) Margalef 1952

Colonies in form of dark brown and slightly reddish, distinctly delimited spots on stony substrates, mainly up to 1 cm in diameter, irregular in outline, composed of several (2–12) layers of perpendicularly oriented, elongate cells. Cells ellipsoidal, ovoid or pear-shaped, rarely cylindrical, slightly narrowed towards the top, sometimes slightly arcuate, with pink or reddish-violet content (in lower layers very pale), usually with 1–3 prominent granules in the basal part, 6–18 × 4–7.2 µm. Pseudovagina cup-shaped, thin and colourless around young cells, later thick and brownish, frayed, more or less lamellate, gelatinizing in the apical part. Exocytes spherical or more or less rounded-polygonal, remain adherent in 1–3 at the apical cell end.

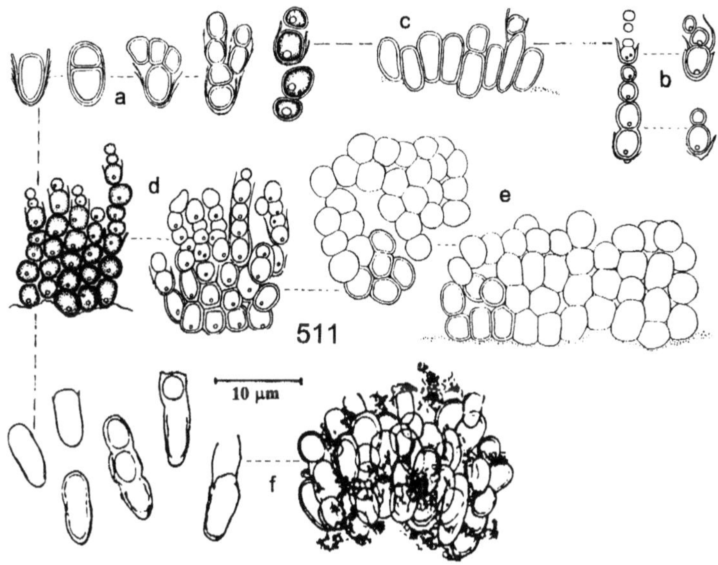


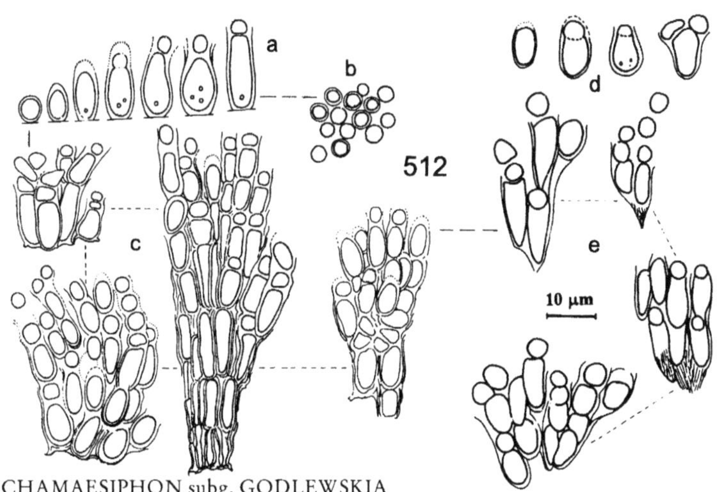


CHAMAESIPHON subg. GODLEWSKIA

**Fig. 511.** *Chamaesiphon polonicus*: **a-c** - detail of cells and young colonies, **d-e** – colonies, **f** – solitary cells and a colony; **a-e** – after Geitler from various authors, **f** – after Komárek & Kalina (1965).

**Fig. 512.** *Chamaesiphon starmachii*: **a** – solitary cells, **b** – young cells from a top view, **c** – colonies, **d** – solitary cells and differentiation of exocytes, **e** – young colonies; **a-c** – after Starmach (1969, sub *Chamaesiphon fuscoviolaceus*), **d-e** – after Kann (1972).

Occurrence: Freshwater, on stones in cold running water in high mountain streams and in the splash zone of stony littoral of lakes and tarns; W. Carpathians (S. Poland, Slovakia), Alps (Austria).

### 20. **Chamaesiphon fuscus** (Rostafinski) Hansgirg 1888 (fig. **513**)

*Sphaerogonium fuscum* Rostafinski 1883; *Chamaesiphon ferrugineus* Fritsch 1929

Colonies in the form of shrub-like («*Dinobryon*-like») formations or composed of 1–2(3) layers of densely and parallely (rarely irregularly) arranged elongate cells, perpendicularly oriented to the substrate, forming dark rusty brown to blackish irregular spots on the substrate, several $mm^2$ or $cm^2$ in extent. Cells more or less irregular club-shaped, rarely almost cylindrical or oval, rounded at both ends, with pale blue-green, olive-yellow, reddish or almost colourless cell content, without granules, 5–13(22) × (2.5)3–7 µm. Pseudovagina distinct, thin, colourless when young, later brown (Fe-precipitates), elongate cup-shaped, U-shaped, slightly lamellate, sometimes with slightly widened opening at the apical end and very short basal narrowed stipe; old sheaths usually perpendicularly lamellate and frayed. Exocytes usually 1–2, as wide as the cell, (2)2.4–4.5 µm in diameter.
Occurrence: Freshwater, on granitic (non-calcareous) stony substrates in clear running waters, mainly in mountain streams; Austria (Alps), Czech Republic, England, Finland, Poland (High Tatra Mts.), Slovakia, etc.

### 21. **Chamaesiphon geitleri** Luther 1954 (fig. **514**)

*Chamaesiphon fuscus* (Rostafinski) Hansgirg sensu Geitler 1925

Colonies in the form of dark brown or blackish-brown, distinct, delimited irregular spots on the substrate, composed of densely, parallely arranged cells, oriented perpendicularly to the substrate, usually shrub-like or in two or more (–5) layers, to several $mm^2$ in diameter. Cells oval when young, later elongate, narrow club-shaped or almost cylindrical, widely rounded at the apex, narrowed towards the base, yellowish, pale olive-green, reddish violet, without prominent granules, (4)5–20(30) × 2.5–6.5(9.6) µm. Pseudovagina thick, lamellate, funnel-shaped and frayed, more or less V-shaped, yellow-brown, dirty brown or brown-blackish; young sheaths thin, colourless, 3–5-times longer than wide. Exocytes 1–2, rarely more in a short, simple row, narrower than the mother cell.
Occurrence: Freshwater, oligo- to stenotherm, in katharobic waters, on stones in rapid mountain (alpine) streams and waterfalls, rarely in the splash zone of the stony littoral of alpine tarns, always in submersed localities; Austria (Alps), Greece (Macedonia).

**Species outside Europe:**
- *Chamaesiphon fallax* Geitler 1933: Subaerophytic, epiphytic species on *Nostoc*-colonies, subg. *Godlewskia,* described from Indonesia; (fig. **518**). Similar type was found at the margin of concrete basin in Mexico City (*C.* cf. *fallax* Gold-Morgan & al. 1996).

---

CHAMAESIPHON subg. GODLEWSKIA
**Fig. 513.** *Chamaesiphon fuscus:* a,c,e – detail of cells and groups of cells, **b,f,g** – top view on the colony, **d, h** – solitary cells and groups of cells, **i** – macroscopic spots

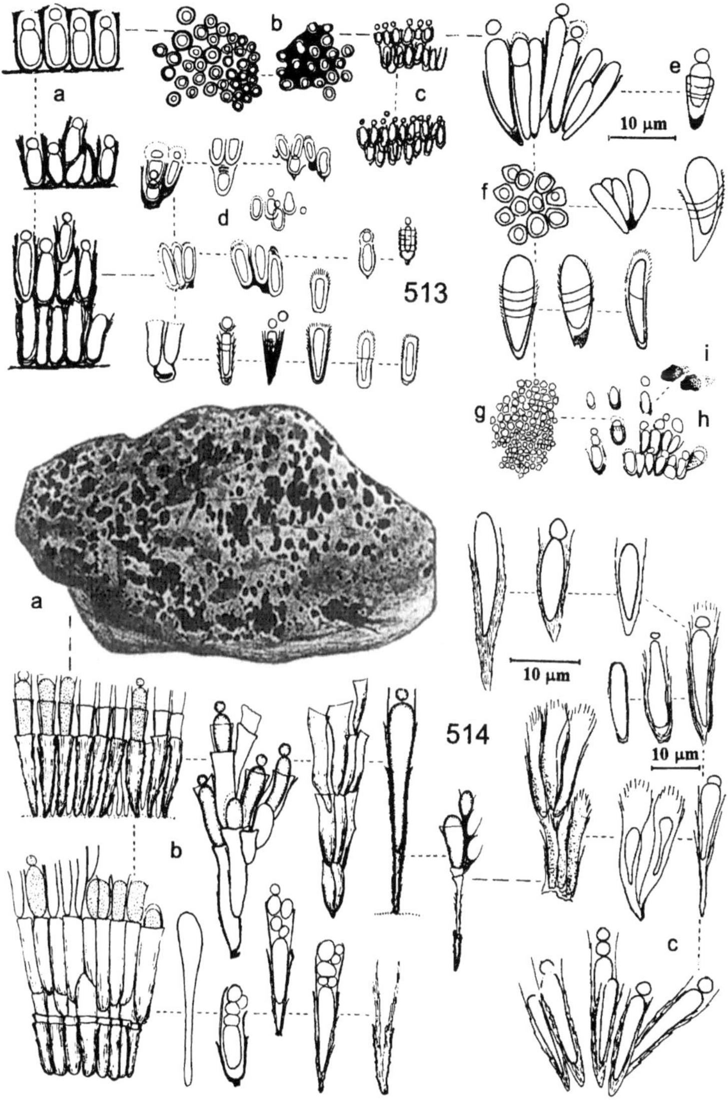


on stony substrate in running water; **a-d** – after Fritsch from Geitler (1932), **e-i** – after Kann (1972).

**Fig. 514.** *Chamaesiphon geitleri*: **a** – stone from a mountain stream (Austrian Alps) with spots of *C. geitleri*, **b-c** – detail of cells and colonies; **a-b** – from Geitler (1932), **c** – after Kann (1972).

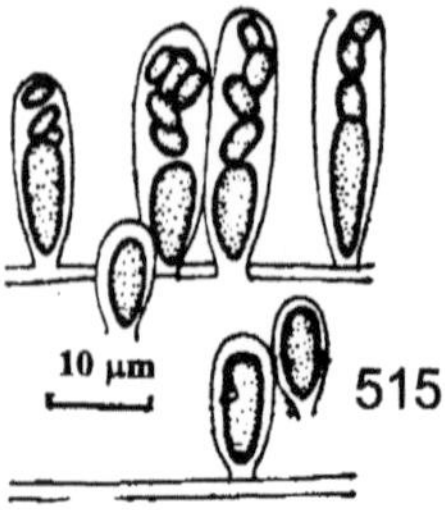


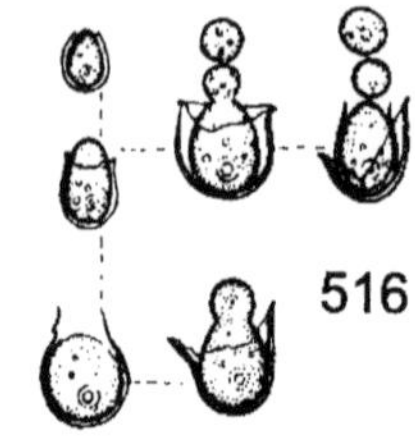


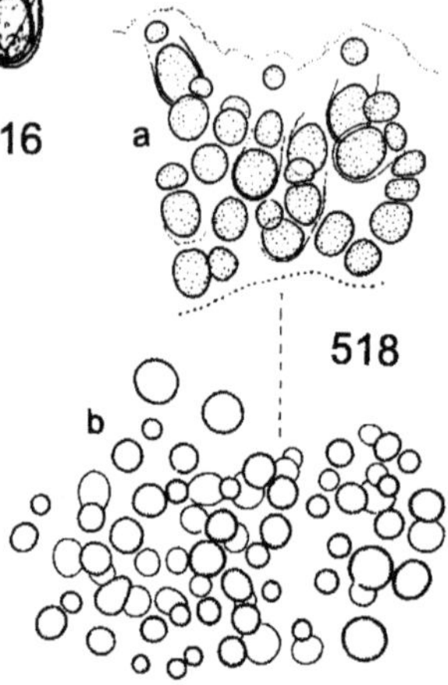


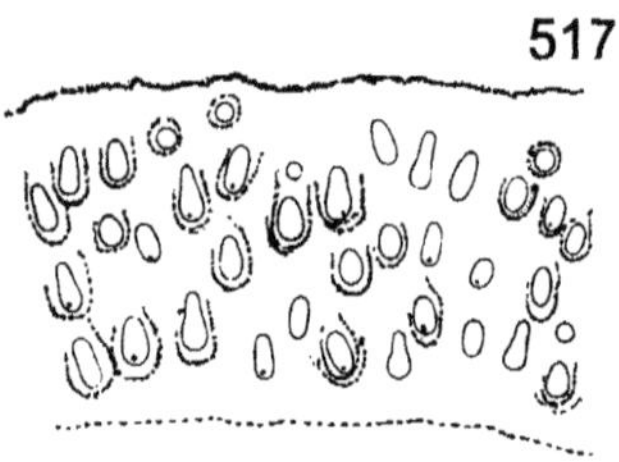


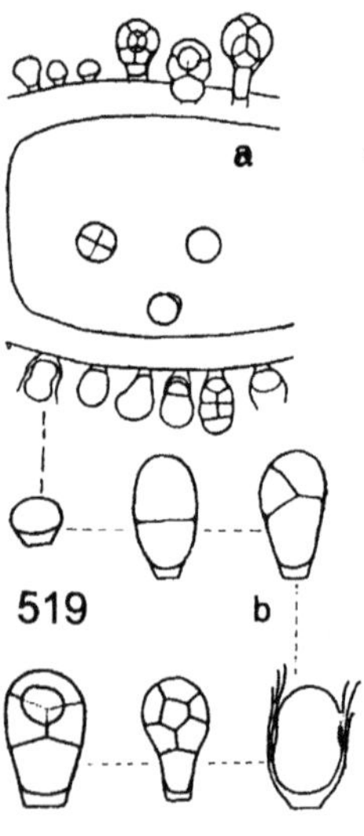


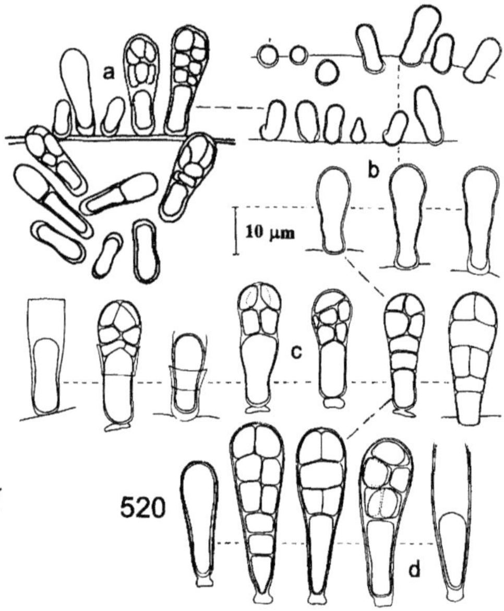


CHAMAESIPHON
Examples of *Chamaesiphon*-species from China and Indonesia:
**Fig. 515.** *Chamaesiphon* sp., from subg. *Chamaesiphon*: from Chu & al. edit. (1991, sub *Dermocarpa clavata*); from China.

- *Chamaesiphon halophilus* Tavera et Komárek 1996: With reddish cells, from the subg. *Chamaesiphon*, epiphytic on filamentous algae, described from a slightly salinic volcanic lake in central Mexico.
- *Chamaesiphon jaoi* Hällfors 1982 [syn.: *C. clavatus* Jao 1939, non *C. clavatus* Setchell et Gardner 1930 = *Chamaecalyx*]: Epiphytic and submersed species, described from China; (Chu & al. ed. 1991).
- *Chamaesiphon major* (Geitler) Komárek et Anagnostidis 1986 [syn.: *C. subglobosus* var. *major* Geitler 1933]: Described from Indonesia.
- *Chamaesiphon mollis* Geitler 1933: From tropical, clear springs, subg. *Godlewskia*, described from Indonesia; (fig. **517**).
- *Chamaesiphon portoricensis* Gardner 1927: Described from Porto Rico, similar to *C. incrustans*, or *C. minutus* (the similar species from central Mexico was recorded under this name; Gold-Morgan & al. 1996); revision is needed.
- *Chamaesiphon tibeticus* Jao et al. 1974: Small species from streams in Tibet; (fig. **516**).
- *Dermocarpa clavata* Geitler sensu Chu & al. ed. 1991: Evidently a separate *Chamaesiphon*-species with shortly oval or cylindrical exocytes, known from China; (fig. **515**).

**Unrevised species:**

- *Chamaesiphon cylindricus* var. *ampla* Godward 1937: Epiphytic on *Isoetes* and *Nitella*, described from the lake Windermere, England; possibly a separate species; (fig. **494**).
- *Chamaesiphon duran-milleri* Gonzalez Guerrero 1946: Cells 30–50 × 15–17 μm, exocytes 10–11 × 8–9 μm; epiphytic on algae in saline waters near Cádiz, Spain; resembles *C. confervicolus*, but is thicker and has a different ecology; not recorded since this first description.

---

**Fig. 516.** *Chamaesiphon tibeticus*, from subg. *Chamaesiphon*: from Chu & al. edit. (1991); from China.
**Fig. 517.** *Chamaesiphon mollis*, from subg. *Godlewskia*: after Geitler from Geitler & Ruttner (1935); from Indonesia.
**Fig. 518.** *Chamaesiphon fallax*, from subg. *Godlewskia*: **a** – lateral view, **b** – top view; after Geitler from Geitler & Ruttner (1935); from Indonesia.

CHAMAECALYX

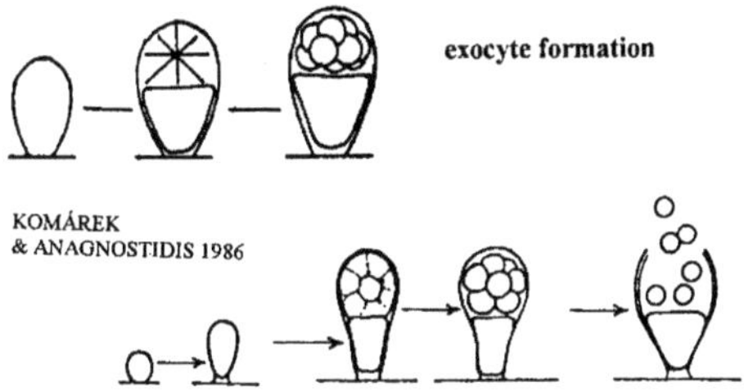


**Fig. 519.** *Chamaecalyx chamaesiphonoides*: from Geitler (1932).
**Fig. 520.** *Chamaecalyx swirenkoi*: **a** – group of cells attached to the filamentous alga, **b** – young cells, **c-d** – cells with differentiating exocytes; **a** – after Širšov (1929), **b-c** – after Hällfors & Munsterhjelm (1982), **d** – from Geitler (1932).

– *Chamaesiphon fuscoviolaceus* (Hansgirg) Margalef 1952: Described from Spain, not satisfactorily defined.
- *Chamaesiphon hyalinus* Scherffel 1907: Epiphytic on Diatoms; described from Hungary, probably a bacterium.
- *Chamaesiphon marinus* Wille et Rosenvinge 1885: Marine, described from Arctic coasts (Novaja Zemlja), never confirmed, problematic species.
– Species, which were classified into the genus *Xenococcus*, grow in monolayers of uniform small cells, divide possibly only by exocytic separation of apical daughter cells and never produce baeocytes, belong probably to the genus *Chamaesiphon* (subg *Chamaesiphonopsis*) (*X. gracilis* Lemmermann 1898, *X. laysanensis* Lemmermann 1905, *X. minimus* Geitler 1922, *X. laminariae* Feldmann 1953); see p. 428–429.
– *Dermocarpa rosea* var. *major* (Hansgirg) Hansgirg ex Forti 1907: Marine, described from Italy, belongs possibly to the genus *Chamaesiphon*.
– *Sphaenosiphon minimus* Reinsch 1874: Marine, epiphytic on seaweeds, described from Italy, problematic species.

## 6. **Chamaecalyx** Komárek et Anagnostidis 1986

Cells heteropolar, slightly elongate, more or less club-shaped, attached to the substrate individually by their narrower ends, solitary or in groups, always enveloped by a thin (rarely a little thickened), firm, colourless sheath (pseudovagina); joined to the substrate by the narrow base or by a pad (more or less discoid), which is a part of a sheath. Pseudovagina opens at the apex before the exocyte liberation. Cell content homogeneous (thylakoids are regularly distributed throughout the cell), usually pale blue-green, olive-green or reddish. Several marine and a few freshwater species are known.
Cell division starts by the transverse (horizontal) binary fission, in which the first division plane is perpendicular to the longer (vertical) axis of the polarized cell; later the upper (distal) part of the original mother cell divides spatially, successively or almost simultaneously, while the basal (proximal) part remains undivided. This division type results in more small daughter cells (exocytes) differentiated within the upper part of the mother sheath. After their liberation from the sheath (pseudovagina) the basal part is able to grow into the form of the original mother cell again (or divides later into two or four cells). This process is sometimes repeated. Rarely complete cell division occurs. Reproduction by liberated exocytes, which attach again to the substrate. In a few isolated strains, the cells do not grow (in agar cultures or in liquid medium) attached in the typical position, but the type of heteropolar division is retained.
**Important references:** Geitler (1932, 1967, sub *Dermocarpa*); Feldmann & Feldmann (1953, sub *Dermocarpella*); Ardré (1960, 1969, sub *Dermocarpella*); Starmach (1966, sub *Dermocarpa*); Bourrelly (1970, 1985, sub *Dermocarpa*); Waterbury & Stanier (1978, sub *Dermocarpella*); Hällfors & Munsterhjelm (1982, sub *Cyanocystis*); Hindák (1983, sub *Dermocarpa*); Komárek & Anagnostidis (1986).

**Key to the species:**

**1a** Freshwater; maximally with 10 exocytes . . . . . . . . . . . . . . . . . . . . . . . . **2**
**1b** Marine; numerous, usually up to 20 exocytes . . . . . . . . . . . . . . . . . . . . **3**
**2a** Cells maximally 11 × –7 µm; apical part more or less spherical, distinctly separated from the narrow basal part . . . . . . . . **1. C. chamaesiphonoides**
**2b** Cells up to 30(40) × –10.5 µm; cells club-shaped . . . . . . . **2. C. swirenkoi**
**3a** Old cells over 10 µm wide (in widest part) . . . . . . . . . . . . . . . . . . . . . . **4**
**3b** Old cells 3.5–10 µm wide; cell content reddish . . . . . . . **3. C. algarvensis**
**4a** Cell content olive-green, blue-green or brownish . . . . . . . **4. C. leibleiniae**
**4b** Cell content red-violet . . . . . . . . . . . . . . . . . . . . . . . . . . . **5. C. pedicellatus**

**1. Chamaecalyx chamaesiphonoides** (Geitler) Komárek et Anagnostidis 1986 (fig. **510**)

*Dermocarpa chamaesiphonoides* Geitler 1922

Cells heteropolar, attached to the substrate by the narrower part, solitary, rarely in groups, obovoid or slightly club-shaped, widely rounded to hemispherical in the apical part, attenuate (stalk-like) at the base, with thin, colourless sheath, a little withdrawn from the sheath base, which is bluntly widened; sheath a little thickened (easily seen by light microscopy) in old cells and after repeated exocyte differentiation with telescopic layers. Cell content greyish blue-green or olive-green. Exocytes (6–8) differentiate successively from the upper, spherical part of the cell. Cell dimensions -11 × -7 µm.

Occurrence: Freshwater, epiphytic on filamentous algae (*Cladophora*) and vascular plants (*Lemna*, etc.); described and known only from basins with cultivated water plants in Austria (Wien); Indonesia (?).

**2. Chamaecalyx swirenkoi** (Širšov) Komárek et Anagnostidis 1986 (fig. **520**)

*Dermocarpa swirenkoi* Širšov 1929; *Dermocarpa clavata* Geitler 1932; *D. clavata* var. *aquae-dulcis* Geitler 1932; ? *Dermocarpella clavata* (Geitler) J. & G. Feldmann 1953; *Cyanocystis swirenkoi* (Širšov) Hällfors et Munsterhjelm 1982; *Dermocarpa clavata* (Setchell et Gardner) Geitler f. *provinciae* Bourrelly 1950 (sensu Yamagishi in Yamagishi & Akiyama ed. 1997)

Cells solitary, club-shaped, straight or very slightly arcuate, widely rounded at the apex, with tightly enveloping firm, thin, colourless sheath (pseudovagina), slightly withdrawn from the base; sheath forms at the base a very short stipe ending in a cushion-shaped enlargement (by which the cell is attached to the substrate). Cell content greyish blue-green or dirty violet. Exocytes (6–10) differentiate successively from the upper (widened) part of the cell, rarely from the whole mother cell, the basal remnant of the mother cell can grow repeatedly; in old mother cells a thin telescopic pseudovagina arises. The apical part of pseudovagina gelatinizes prior to exocyte liberation. Cell dimensions -20–30(40) × -8–10.5 (apical part) µm. Liberated exocytes are usually rounded-conical (in optical section rounded-triangular) and are attached to the substrate probably by the widened end (3–5 µm in diameter).

Occurrence: Freshwater, epiphytic on filamentous algae, cyanoprokaryotes, mosses and vascular water plants in clear, stagnant waters (lakes), usually in limestone areas; probably cosmopolitan, but with a very discontinuous area of distribution (Austria, S. Finland, Indonesia - Java, Japan, Mexico, Slovakia, Ukraine). The marine «*Dermocarpa clavata* Geitler f. *marina*» is probably confuse name, concerning *Chamaecalyx clavatus.*

**3. Chamaecalyx algarvensis** (Ardré) Komárek et Anagnostidis 1995 (fig. **521**)

*Dermocarpella algarvensis* Ardré 1969

Cells heteropolar, attached to the substrate by the narrow end, elongate ellipsoid, obovoid or pyriform, widely rounded at the apex, slightly narrowed towards the base, solitary or in groups loosely arranged. Sheath (pseudovagina) thin, colourless, not lamellate, at the base slightly widened. Cells 15–17 × 3.5–10 µm; cell content reddish (fixed material), homogeneous, finely granular. Exocytes numerous, differentiate successively from the upper (widened) part of the cell, the basal

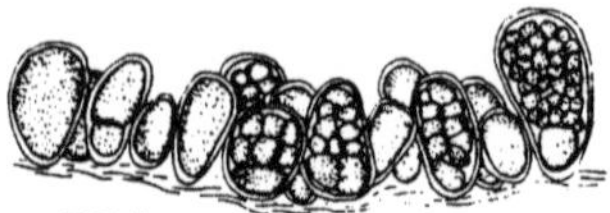

521

522

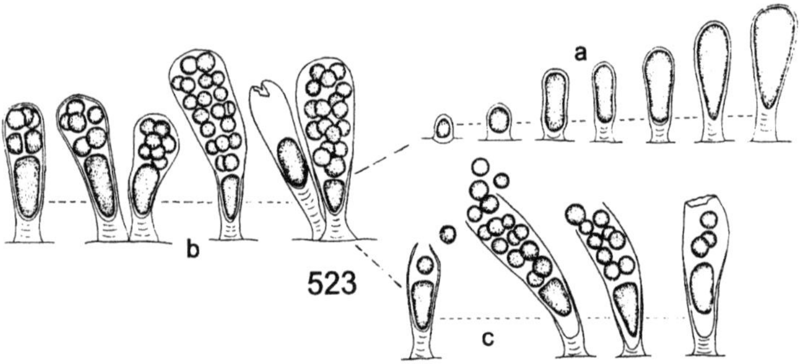


523

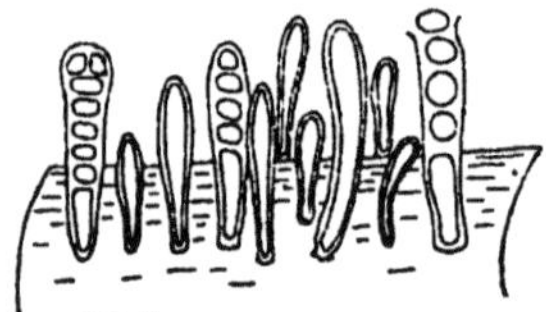

524

525

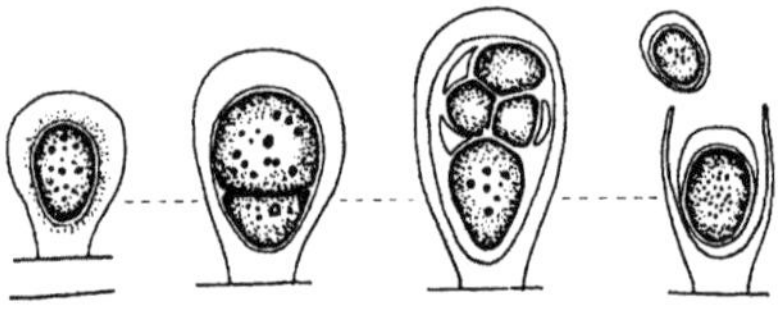

526

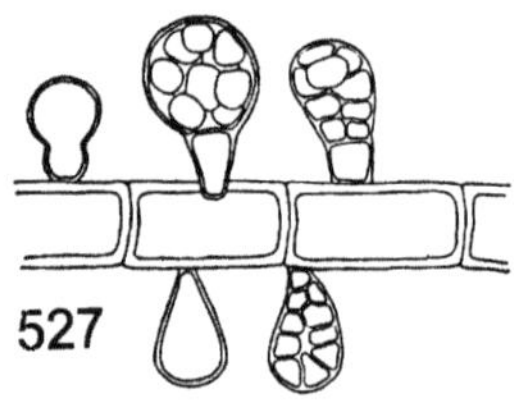

527

remnant remaining undivided; liberated exocytes spherical, 1.5–2 µm in diameter.
Occurrence: Marine, epiphytic on *Fucus*, *Gigartina*, described and known only from the coasts of Portugal (Parede, Carrapateira).

**4. Chamaecalyx leibleiniae** (Reinsch) Komárek et Anagnostidis 1986 (fig. **522**)

*Sphaenosiphon leibleiniae* Reinsch 1874; *Dermocarpa leibleiniae* (Reinsch) Bornet et Thuret 1880; *Dermocarpella leibleiniae* (Reinsch) J. et G. Feldmann 1953

Cells heteropolar, attached to the substrate by the narrow end, slightly elongate,

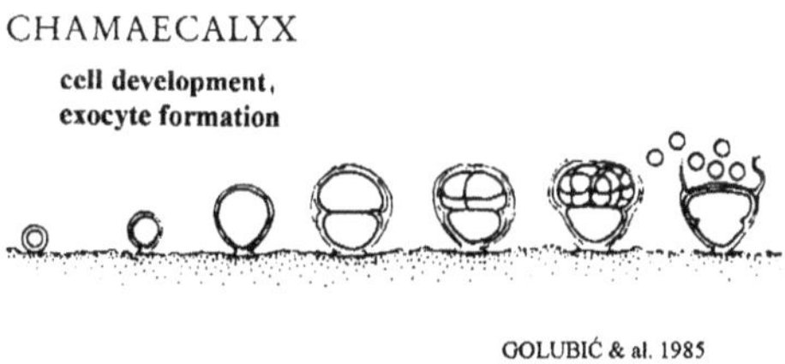


**Fig. 521.** *Chamaecalyx algarvensis*: after Ardré (1969, sub *Dermocarpella algarvensis*).
**Fig. 522.** *Chamaecalyx leibleiniae*: **a** – after Wille from Kosinskaja (1948), **b-c** – after Bornet from Kosinskaja (1948, sub *Dermocarpa leibleinii*).

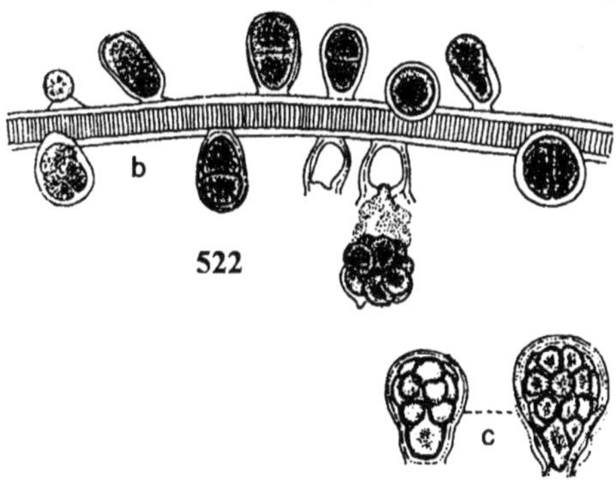


**Fig. 523.** *Chamaecalyx pedicellatus*: **a** – young cells, **b-c** – cells with differentiating exocytes; after J. & G. Feldmann (1953, sub *Dermocarpella stipitata*).
**Fig. 524.** *Chamaecalyx clavatus*: after Setchell & Gardner from Kosinskaja (1948, sub *Dermocarpa swirenkoi* f. *marina*).
**Fig. 525.** *Chamaecalyx fucicolus*: after Setchell & Gardner from Geitler (1932, sub *Dermocarpa fucicola*).
**Fig. 526.** *Chamaecalyx incrassatus*: after Lemmermann from Geitler (1932, sub *Dermocarpa incrassata*).
**Fig. 527.** *Chamaecalyx suffultus*: after Setchell & Gardner from Geitler (1932, sub *Dermocarpa suffulta*).

obovoid or obpyriform, widely rounded at the apex, slightly narrowed towards the base, solitary or in groups. Sheath (pseudovagina) thick, colourless, lamellate, at the base widened into a massive adherence disc. Cell content blue-green, olive-green or brownish. Exocytes (12–20) differentiate from the apical part of the cell or from the whole mother cell. Cells 8–20 µm wide in the upper part and usually ~1.5(2)x longer than wide. Exocytes liberate from the mother sheath together in a gelatinous cluster and separate later; liberated exocytes are more or less spherical.
Occurrence: Marine, epiphytic on various algae and filamentous cyanoprokaryotes along Mediterranean and Atlantic coasts (both hemispheres), found also in Pacific and Indian Oceans, and Sea of Japan; sporadically.

**5. Chamaecalyx pedicellatus** (J. et G. Feldmann) Komárek et Anagnostidis 1986 (fig. **523**)

*Dermocarpella pedicellata* J. et G. Feldmann 1953 (false *Dermocarpella stipitata* in J. et G. Feldmann 1953, fig. 3)

Cells solitary, with colourless sheath, attached to the substrate by a thickened pad at the base. Old cells obovoid or club-shaped, to 40 µm long and 12–13 µm wide. Cell content homogeneous, finely granulate, red-violet (fixed material). Cells dividing into exocytes are to 55 µm long and 15–18 µm wide; first cell division perpendicular to the vertical axis (horizontal); numerous (up to 20) spherical exocytes, 5–6 µm in diameter, develop from the upper part of the mother cell and liberate after the rupture of a sheath; lower part undivided or rarely dividing after liberation of exocytes. Lower part sometimes grows again into a vegetative cell within the original sheath.
Occurrence: Marine, epiphytic on red and green algae, described from the coasts in Red Sea, known also from the Atlantic coasts in Portugal (Arrabida).

**Species outside Europe:**
- *Chamaecalyx calyculatus* Gold-Morgan et al. 1996: Freshwater, epiphytic on *Audouinella* in streaming water in limestone areas in San Luis Potosí, Mexico; cells 15–16 × 4–5 µm, with characteristic thickened basal part of the sheath.
- *Chamaecalyx clavatus* (Setchell et Gardner) Komárek et Anagnostidis 1986 [syn.: *Chamaesiphon clavatus* Setchell et Gardner 1930 = *Dermocarpa clavata* (Setchell et Gardner) Geitler 1932 = *Dermocarpa swirenkoi* f. *marina* Kosinskaja 1948 = *Cyanocystis clavata* (Setchell et Gardner) Hällfors 1982]: Marine species, epiphytic on algae and cyanoprokaryotes, known from Baja California (Mexico) and tropical islands (Guadeloupe, New Guinea – Papua, Philippines, W. Samoa); (fig. **524**).
- *Chamaecalyx fucicolus* (Saunders) Komárek et Anagnostidis 1986 [syn.: *Dermocarpa fucicola* Saunders 1901]: Marine, Pacific coasts of N. America; cells 40–60 × 22–30 µm, exocytes 4–5 µm in diameter; (fig. **525**).
- *Chamaecalyx incrassatus* (Lemmermann) Komárek et Anagnostidis 1986 [syn.: *Dermocarpella incrassata* Lemmermann 1900 = *Dermocarpa incrassata* (Lemmermann) Geitler 1925]: Epiphytic, in brackish waters, described from Chatham Island; cells 20.5–27 × 11–16 µm, with 8 exocytes; (fig. **526**).
- *Chamaecalyx papuanus* L. Hoffmann 1992: Marine, epiphytic on *Lyngbya semiplena*, described from Papua New Guinea.
- *Chamaecalyx suffultus* (Setchell et Gardner) Komárek et Anagnostidis 1986: Marine, epiphytic on algae, described from USA (California); cells 17–20 × 10–14 µm, ovoid, pear-shaped or slightly stipitate; exocytes (8–12) 4–6 µm in diameter; (fig. **527**).

**Unrevised species:**
- Several marine species, described under various *Dermocarpa* and *Dermocarpella* taxa, e.g. *Dermocarpa leibleiniae* var. *pelagica* Wille 1904, *Dermocarpa solitaria* Collins et Harvey 1917, *Dermocarpella* sp. in Waterbury et Stanier 1978 sec. Rippka et al. 1979, etc. belong probably into the genus *Chamaecalyx*.
- *Dermocarpella hemisphaerica* sensu J. et G. Feldmann 1953, sec. Ardré1969: Epiphytic on marine algae (Algeria, France, Portugal, USA - California ?), morphologically similar to the freshwater *Dermocarpella hemisphaerica* Lemmermann, described from clear lakes in Chatham Island. It differs in ecology and probably mainly by the type of cell division: in the original Lemmermann's species the whole protoplast divides into baeocytes, while in the marine species the lower part remains undivided in a sheath (*Chamaecalyx* type of cell division). (See p. 417, figs. **546–547**).

# 9. Family **Dermocarpellaceae** Ginsburg-Ardré ex Christensen 1980

Cells solitary or in irregular groups, usually attached to the substrate, exceptionally free-living, in colonies irregularly arranged, or parallel to one another, and perpendicular to the substrate. Cells spherical and attached by side to the substrate (*Stanieria*), or elongate and polarized, attached by one cell end to the substrate. Cells are enveloped by firm, usually not very thick sheaths, which are colourless, homogeneous or slightly striated.
Cells divide completely only by multiple fission into numerous baeocytes; cell differentiation is more or less simultaneous or in rapid sequence, in different planes. Baeocytes are liberated from mucilaginous envelopes by their gelatinization or by rupture of the sheath. Mother cells reach a size several times larger than the baeocytes before the next division.

Note: The exclusive type of multiple fission is the diacritical feature of the family. *Stanieria* with not polarized cells belongs therefore in this family as well as genera with polarized cells.

**Key to the genera:**

**1a** Cells more or less spherical, not polarized, usually in irregular groups, attached to the substrate (and one to another) by any side . . . **1. Stanieria**
**1b** Cells spherical, more or less hemispherical or elongate, oval, obovoid, club-shaped or pear-shaped, usually parallel to one another in flat or hemispherical sessile colonies, attached to the substrate by one, usually narrowed end or by the flat, widened base . . . . . . . . . . . . . . . . . . . . . . . . . . . 2
**2a** First division plane (start of multiple fission) is vertical, more or less perpendicular to the base (substrate) . . . . . . . . . . . . . . . . . . . . . . **2. Cyanocystis**
**2b** First division plane or planes are horizontal to the substrate (with respect to the polarized cell) . . . . . . . . . . . . . . . . . . . . . . . . . **3. Dermocarpella**

## 1. **Stanieria** Komárek et Anagnostidis 1986

Cells solitary or in groups, more or less spherical, usually attached to the substrate by any side (cells are not polarized), secondarily (and in cultures) in free or attached irregular, usually microscopic aggregates. Cells in a colony of variable size, with homogeneous, pale blue-grey, blue-green, yellowish, olive-green, reddish to dark red or violet content. Sheaths (vaginae) thin, firm, mucilaginous, colourless, forming a narrow hyaline layer («wall») around solitary cells.

Cell division only by multiple fission, more or less simultaneously or in rapid sequence in various directions, resulting in numerous spherical, motile baeocytes (planocytes), which liberate from the split or gelatinized sheaths. Small baeocytes reach the original size (several times larger than baeocytes) before the next multiple fission. Reproduction only by baeocytes.
**Important references:** sub *Chroococcidiopsis* – Komárek & Hindák (1975), Komárek & al. (1975); sub *Dermocarpa* and *Xenococcus* – Lindstedt (1943), Stanier & Cohen-Bazire (1977), Waterbury & Stanier (1978), Rippka & al. (1979), Stanier & al. (1981), Waterbury (1989); sub *Stanieria* – Komárek & Anagnostidis (1986), Anagnostidis & Pantazidou (1991).

**Key to the species:**

**1a** Old cells 5–7 µm in diameter; cell content mainly red ...... **1. S. suecica**
**1b** Old cells 6–25 µm in diameter; cell content pale blue-green, bluish or red-violet ........................................ **2**
**2a** Cells 7–25 µm, baeocytes up to 64 or 128, 1.5–2.5 (4) µm in diameter ........................................ **2. S. sublitoralis**
**2b** Cells 6–16 µm, baeocytes up to 32, 1.5–2 (3) µm in diameter **3. S. sphaerica**

## 1. **Stanieria suecica** (Kylin) Komárek et Anagnostidis 1995

*Dermocarpa suecica* Kylin 1937

Cells solitary or densely aggregated in groups, finally forming an almost closed (delimited) layer attached to an animal substrate; in masses or in culture grows in form of red, red-violet or blue-violet mats (chromatically adaptable pigment ratio; Kylin 1937). Cells spherical or angular-rounded (in tight clusters), without any attaching pad, just before the multiple fission 5–7 µm in diameter, with red, red-violet or rarely bluish-violet content. Cells with thin, firm, colourless sheaths. Reproduction by spherical baeocytes, 4–8 (16) per cell, after a successive division of the whole protoplast.
Occurrence: Marine, epizoic on various animals (*Membranipora, Flustra, Alcyonium,* Ascidia), often together with *S. sublitoralis* in sublittoral zone (in a depth of 10–20 m); described and known only from west coasts in Sweden, repeatedly recorded.

## 2. **Stanieria sublitoralis** (Lindstedt) Anagnostidis et Pantazidou 1991

*Dermocarpa sublitoralis* Lindstedt 1943

Cells solitary, spherical, attached to the substrate, 7–12(18) µm in diameter, to 25 µm just before the multiple fission, sometimes in groups, with bluish to reddish, dark red or violet and more or less homogeneous content. Reproduction by up to 64 or 128 spherical baeocytes, which are 1.5–2.5(4) µm in diameter; liberation of baeocytes by rupture of a sheath.
Occurrence: Marine, epilithic on limestone substrate, less commonly epiphytic or on calcareous shells of animals (also on non-calcareous coasts), first described from the sublittoral zone (from a depth 1-20 m), forming red spots on animal shells, algae and live animals near west coast of Sweden (Baltic Sea), locally widely distributed in the littoral zone of marine coasts of Greece (Aegean Sea).

## 3. **Stanieria sphaerica** (Setchell et Gardner) Anagnostidis et Pantazidou 1991 (fig. **528**) .

*Dermocarpa sphaerica* Setchell et Gardner 1918; *Cyanocystis sphaerica* (Setchell et Gardner) Komárek et Anagnostidis 1986

Cells spherical, solitary or in loose groups, attached to the substrate (epiphytic), 6–10–16 (19.5) µm in diameter, with bluish or pale blue-green, more or less homogeneous content. Sheaths narrow and homogeneous, colourless, but distinct, forming a narrow layer around cells. Reproduction by up to 32 spherical baeocytes arising by simultaneous cell division, which are 1.5–2–3 (3.4) µm in diameter; liberation of baeocytes by gelatinization of sheaths.
Occurrence: Marine, epiphytic on various filamentous cyanoprokaryotes and algae, probably only on calcareous rocks of the littoral zone; described from USA (California), known also from Croatia (Senj), France, Greece (Aegean Sea coasts) and Spain (near Cádiz); recorded also from Atlantic coast of Brazil (SP), from Chile and NW. coasts of Bering Sea.

**Species outside Europe:**
- *Stanieria cyanosphaera* (Komárek et Hindák) Komárek et Anagnostidis 1986 [syn.: *Chroococcidiopsis cyanosphaera* Komárek et Hindák 1975]: Spherical cells to 30 (40) µm in diameter, with greyish blue-green, homogeneous content, in irregular aggregates, from alkaline swamps and lake littoral in Cuba. Waterbury & Stanier (1978) designated two strains of this species as «*Dermocarpa*». Type species of the genus; (fig. **529,** Tab. **2 D/j**).

**Unrevised species:**
- *Dermocarpa* sp., strain Lewin 1965/Camb. 1416–1: Marine, with reddish content; (Stanier & Cohen-Bazire 1977; Waterbury & Stanier 1978; Rippka & al. 1979; Komárek & Anagnostidis 1986). Waterbury & Stanier (1978) studied three other marine strains, probably taxonomically identical. Strains of this type appear in collections under numbers: UTEX 1635, CCAP 1416/1, CCMEE 527, PCC 7301, ACMM 424, ATCC 29367; (sometimes incorrectly sub «*Dermocarpa violacea*»); (Tab. **2 C/i**).
- *Dermocarpa* sp.: Marine; (Rich 1938).
- *Dermocarpa sphaerica* sensu Rao 1940: Freshwater, cells 16.5(23.1) µm in diameter, epiphytic on *Pithophora*, Delhi, India; (Desikachary 1959).
- *Dermocarpa simulans* Setchell et Gardner 1937: Cells solitary, spherical, free floating among *Rhizoclonium* filaments, to 35–38 µm in diameter, blue-green, baeocytes 2–2.5 µm in diameter; marine, Galapagos Islands; probably a *Stanieria* species. Very similar (with smaller cells) is *Dermocarpa sphaerica* var. *galapagensis* Setchell et Gardner 1937, growing in the same biotope on the surface of colonies of *Aphanocapsa salinarum*.
- *Stanieria* sp., strain Kusel 1983/01025 (Wien, Austria): Freshwater European species, isolated from littoral of the Neusiedler Lake, Austria.

## 2. **Cyanocystis** Borzì 1882

? *Sphaenosiphon* Reinsch 1874; *Krkia* Pevalek 1929; *Dermocarpa* Crouan p.p. excl. typo (incl. sensu Waterbury & Stanier 1978 p.p.)

Cells solitary or in flattened or hemispherical groups (parallely or radially arranged), attached to the substrate; cells heteropolar, usually slightly or distinctly elongate, widely oval, obovoid, club-shaped or pear-shaped, rarely almost spherical or hemispherical, attached to the substrate more or less perpendicularly by the narrowed end and sometimes by discoidally widened sheath, or by widened base, at the apex rounded; in cultures only irregular groups of cells. Cells in colonies of variable size, with homogeneous, pale blue-green, olive-green or violet content. Sheaths (pseudovaginae) thin, firm, colourless.
Cell division only by rapid successive multiple fission, resulting in numerous, spherical baeocytes, which are liberated from the sheath (pseudovagina) through a

rupture at the cell apex; cell division in various planes, but the first plane is always «vertical», from the apex to the base. Baeocytes grow to the mother cell size before the next division (i.e. several times larger than baeocytes). Reproduction only by non-motile baeocytes, which attach to the substrate.
**Important references:** Geitler (1932 – sub *Dermocarpa*, 1967, 1975); Feldmann & Feldmann (1953); Ginsburg-Ardré (1966); Ardré (1969); Bourrelly (1970, 1985); Christensen (1980); Kaas (1985); Komárek & Anagnostidis (1986); Hua & al. (1989).

**Key to the species:**

1a All cells spherical, hemispherical or conical-rounded, not longer than wide, at the base widened and flattened . . . . . . . . . . . . . . . . . . . . . . . . . . . . **2**
1b Old cells elongate, longer than wide, oval, ellipsoidal, pear-shaped, etc. . **3**
2a Cells 5–7 µm in diameter, pale blue-green; marine, epiphytic on cyanoprokaryotes . . . . . . . . . . . . . . . . . . . . . . . . . . . . . . . . . . . . . **1. C. minima**
2b Cells 10–13 × 18–21 µm, flattened, bright blue-green, marine, attached to thallous algae and to sand particles . . . . . . . . . . . . . . . . **2. C. hemispherica**
3a Cells usually more or less ellipsoid or oval (or rarely indistinctly pear-shaped), not distinctly narrowed towards the base (not clearly pear-shaped or club-shaped) . . . . . . . . . . . . . . . . . . . . . . . . . . . . . . . . . . . . . . **4**
3b Cells clearly elongate obovoid, club-shaped or pear-shaped, attached to the substrate by narrowed base . . . . . . . . . . . . . . . . . . . . . . . . . . . . . **5**
4a Cells more or less ellipsoidal to indistinctly pear-shaped, to 9 µm long; freshwater, in small clear streams . . . . . . . . . . . . . . . . . . **3. C. tatrensis**
4b Cells more or less oval, to 16 µm long; freshwater, in streams . . . . . . . . . . . . . . . . . . . . . . . . . . . . . . . . . . . . . . . . . . . . . **4. C. versicolor**
5a Colonies free floating, spherical with radially arranged cells; cells 7–8 × 0.7–1.8 µm; freshwater in Mediterranean coastal areas . . . . . **5. C. croatica**
5b Colonies sessile, hemispherical or flat; over 13 µm long; rarely freshwater, mainly marine species . . . . . . . . . . . . . . . . . . . . . . . . . . . . . . . . . . . . **6**
6a Freshwater, on mosses in alpine streams; cells blue-green, 13–17 × 6.5–8.5 µm . . . . . . . . . . . . . . . . . . . . . . . . . . . . . . . . . **6. C. aquae-dulcis**
6b Marine, on seaweeds in coastal waters; cells blue-green, olive-green, reddish or violet . . . . . . . . . . . . . . . . . . . . . . . . . . . . . . . . . . . . . . . . . . . . **7**
7a Cells olive-green, reddish or violet, 13–31 µm long . . . . . . . . . . . . . . . . **8**
7b Cells blue-green, 60–90 × –30 µm . . . . . . . . . . . . . . . . . . . **9. C. farlowii**
8a Cells club-shaped or pear-shaped, olive-green or violet, 13–25 × 9.5–17 µm . . . . . . . . . . . . . . . . . . . . . . . . . . . . . . . . . . . . . . . . . . . . **7. C. olivacea**
8b Cells obovoid, clavate or club-shaped, pinkish-red or violet, 16–31 × 8–17(28?) µm . . . . . . . . . . . . . . . . . . . . . . . . . . . . . . . **8. C. violacea**

---

STANIERIA

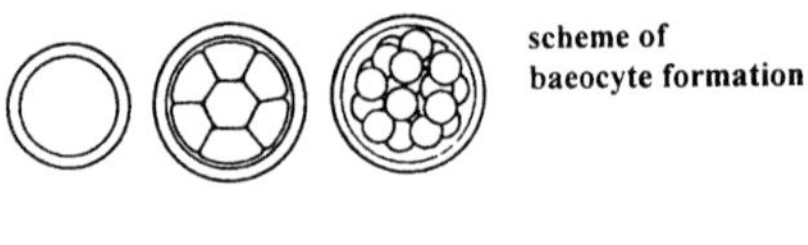


**Fig. 528.** *Stanieria sphaerica*: after Setchell & Gardner from Geitler (1932, sub *Dermocarpa sphaerica*).

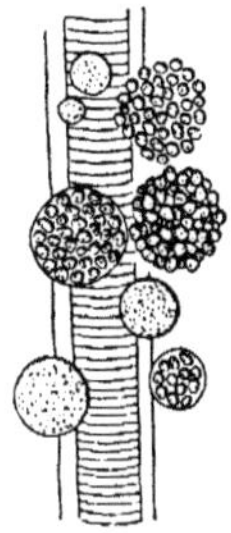

528

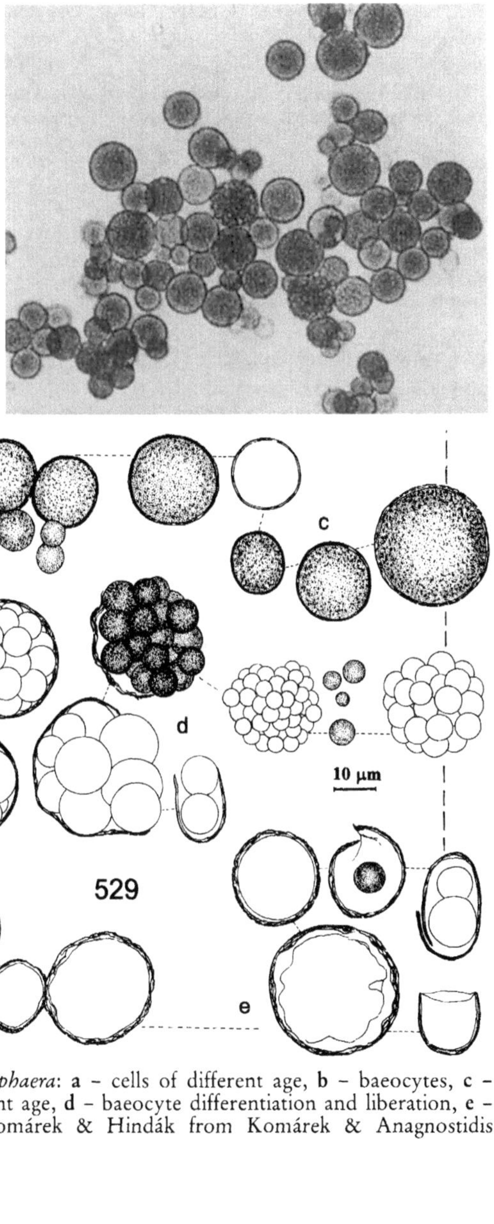


**Fig. 529.** *Stanieria cyanosphaera*: **a** – cells of different age, **b** – baeocytes, **c** – vegetative cells of different age, **d** – baeocyte differentiation and liberation, **e** – empty sheaths; after Komárek & Hindák from Komárek & Anagnostidis (1986).

**1. Cyanocystis minima** (Geitler) Komárek et Anagnostidis 1986 (fig. **530**)

*Dermocarpa minima* Geitler 1932

Cells epiphytic, solitary or in small tight groups (in this case with slightly flattened sides), almost spherical, at the base slightly flattened and usually with wide, very short and indistinctly recognizable pad; cells 5–7 µm in diameter. Sheaths very thin, but firm, colourless. Cell content pale blue-green, grey or yellowish. Baeocytes 8–12 (rarely –16), arise after successive division; liberation of baeocytes through the apical opening in the sheath.

Occurrence: Marine, epiphytic on filamentous cyanoprokaryotes (*Oscillatoria*, *Lyngbya*) and green algae (*Cladophora*); described from Croatia (Dalmatia – Trpanj), known from Greece (Attiki Peninsula); reported also from SW. Pacific islands.

**2. Cyanocystis hemisphaerica** (Setchell et Gardner) Kaas 1985 (fig. **531**)

*Dermocarpa hemisphaerica* Setchell et Gardner in Gardner 1918

Cells solitary, hemispherical or pyramidal-rounded, attached to the substrate by a widened, flat base; at the base 18–21 µm wide, 10–13 µm high; cell content homogeneous, bright blue-green. Sheaths thin, but distinct, not striate, firm, colourless. Cell division starts in vertical planes, later the whole content divides into numerous baeocytes, 0.8–1.2 µm in diameter; baeocytes are liberated through a small opening at the apex.

Occurrence: Marine, epiphytic on algae (e.g., *Rhodochorton*) or sand grains; described from Californian coasts and known also from Chile, SW. Pacific islands;

CYANOCYSTIS

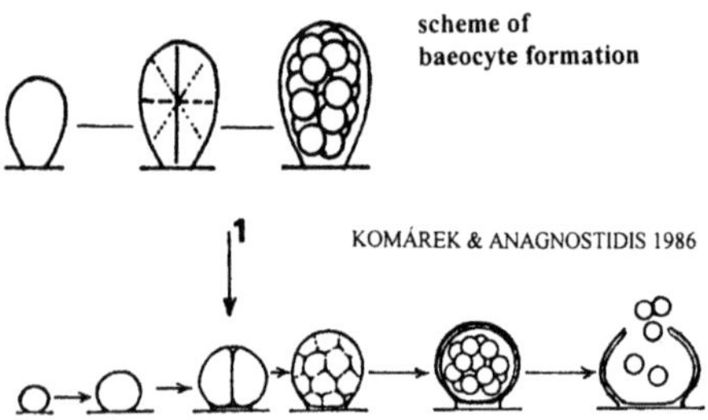


**Fig. 530.** *Cyanocystis minima*: after Geitler (1932, sub *Dermocarpa minima*).
**Fig. 531.** *Cyanocystis hemisphaerica*: after Setchell & Gardner from Kosinskaja (1948, sub *Dermocarpa hemisphaerica*).
**Fig. 532.** *Cyanocystis tatrensis*: **a** – groups of sessile cells, **b** – top view on cells, **c** – baeocyte production, **d** – group of cells on a filamentous alga; **a-c** – from Starmach (1966, sub *Dermocarpa aquae-dulcis* var. *tatrensis*), **d** – from Chu & al. edit. (1991, sub *Dermocarpa aquae-dulcis* var. *tatrensis*).
**Fig. 533.** *Cyanocystis versicolor*: **a** – after Borzì from Geitler (1932), **b** – after Proškina-Lavrenko from Kondrateva & al. (1984, sub *Dermocarpa versicolor*), **c** – from Chu & al. edit. (1991, sub *Dermocarpa versicolor*).
**Fig. 534.** *Cyanocystis croatica*: after Pevalek from Bourrelly (1970).

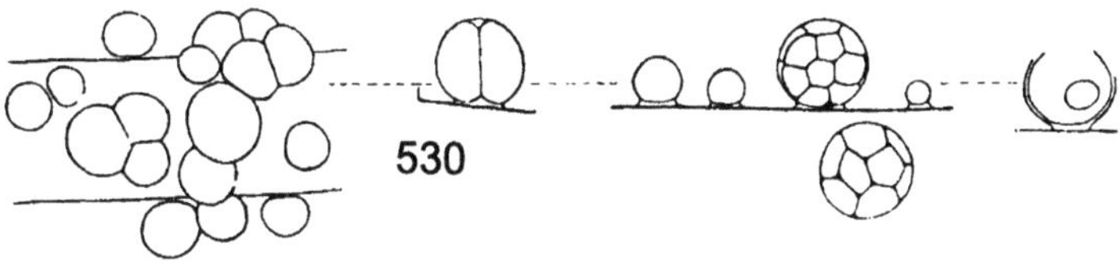


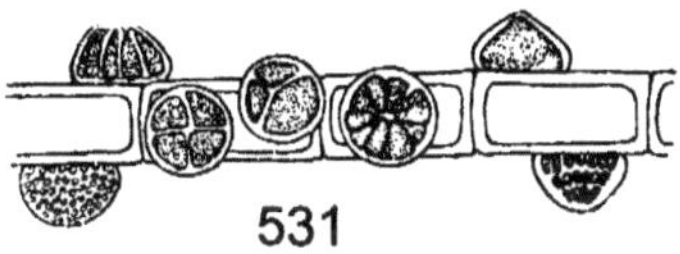


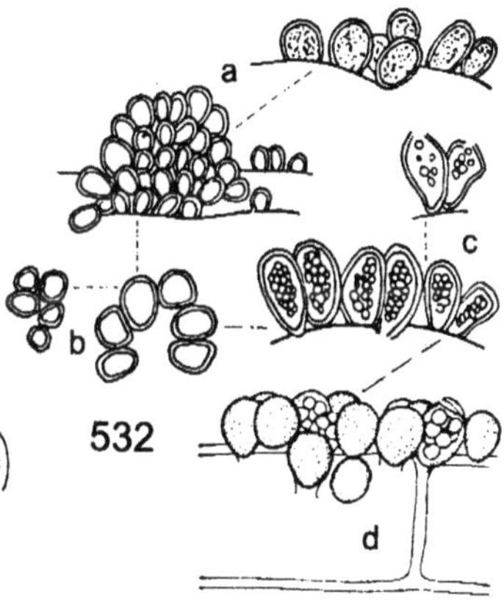


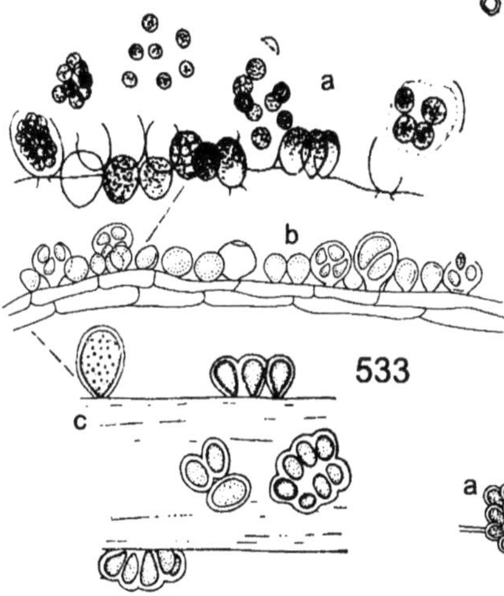


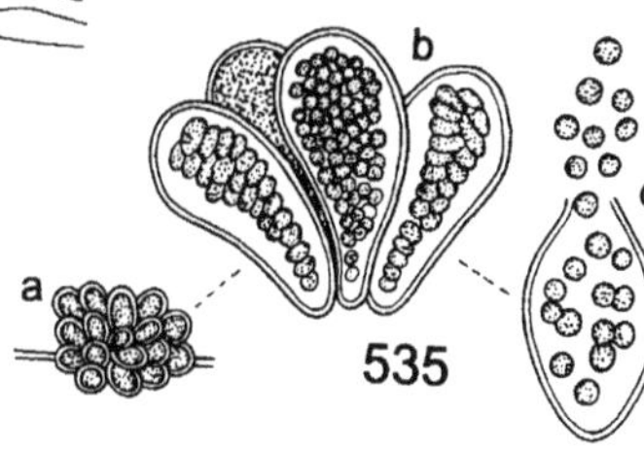


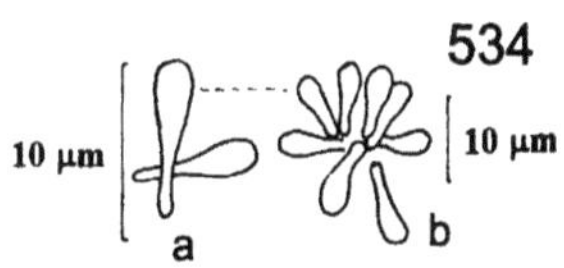


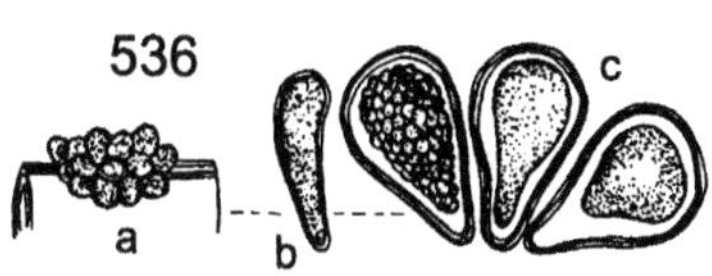


**Fig. 535.** *Cyanocystis aquae-dulcis*: **a** – group of young cells, **b** – differentiation of baeocytes, **c** – liberation of baeocytes; after Reinsch from Starmach (1966, sub *Dermocarpa aquae-dulcis*).

**Fig. 536.** *Cyanocystis olivacea*: **a** – group of young cells, **b** – young cell, **c** – old cells and baeocyte differentiation; after Reinsch from Geitler (1932, sub *Dermocarpa olivacea*).

in Europe found in Baltic Sea (Denmark, Wadden Sea; Kaas 1985) and in Spain (near Cádiz; Gonzalez Guerrero 1946).

**3. Cyanocystis tatrensis** (Starmach) Komárek et Anagnostidis 1995 (fig. **532**)

*Dermocarpa aquae-dulcis* var. *tatrensis* Starmach 1927

Colonies microscopic, forming flat layers on the substrate, sometimes with cells distant from one another, rarely more or less hemispherical, with irregularly aggregated, elongate and polarized cells, attached by the narrowed end to the substrate. Cells slightly pear-shaped, oval, obovoid or ellipsoid, rounded at the apex, 6.2–9 × 3.6–5.3 µm. Sheath distinct, thin or thick, colourless. Baeocytes small, spherical, liberated through an apical opening.
Occurrence: Freshwater, epilithic and epiphytic (*Cladophora*), in clear, not polluted, mainly mountain streams and rivers; known only from W. Carpathians (Poland – High Tatra Mts.).

**4. Cyanocystis versicolor** Borzì 1882 (fig. **533**)

*Dermocarpa versicolor* (Borzì) Geitler 1925

Cells in free or more or less close groups on the substrate, heteropolar, slightly elongate, attached by the narrow end to the substrate, spherical to oval, to 10(16?) µm long, blue, pale blue-green or violet, with thin sheath, sometimes with indistinct hyaline, discoid pad on the base. Baeocytes 4–16, spherical, liberated through the wide apical opening, occasionally into a mucilaginous vesicle, which later disappears.
Occurrence: Freshwater, epiphytic on filamentous green algae (*Cladophora, Oedogonium*) and cyanoprokaryotes, in flowing, not polluted waters; sporadic in south and central Europe. Needs revision; the population described by Frémy (1930) from Gabon with cells up to 16 µm long, and the halophilic var. *subsalsa* Proškina-Lavrenko 1950 from Russia, belong probably to other species.

**5. Cyanocystis croatica** (Pevalek) Komárek et Anagnostidis 1995 (fig. **534**)

*Krkia croatica* Pevalek 1929

Cells rarely solitary, usually in small, spherical, gelatinous, free floating groups with 2–24 cells, which are rounded when young, later pear-shaped or club-shaped, 7–8 × 0.7–1.8 µm, with bright blue-green content. Reproduction by baeocytes; multiple fission initiates by lengthwise division. Poorly known and unrevised species.
Occurrence: Freshwater (?; free floating in a lake?); described from the island Krk, Croatia (Dalmatia).

**6. Cyanocystis aquae-dulcis** (Reinsch) Kann 1978 (fig. **535**)

*Sphaenosiphon aquae-dulcis* Reinsch 1874; *Dermocarpa aquae-dulcis* (Reinsch) Geitler 1925

Colonies small, microscopic, flat or hemispherical, sessile, rarely solitary cells or few-celled, irregular groups. Cells ellipsoid to pear-shaped, densely and parallely arranged, heteropolar, attached by the narrower end to the substrate, widely rounded at the apex, flattened at sides, blue-green, 13–17 × 6.5–8.5 µm, with distinct, slightly widened sheath. Numerous spherical baeocytes (their dermocarpelloid origin is not clear!), liberate through an apical opening.

Occurrence: Freshwater, in katharobic alpine streams, usually in high mountain regions, rare; Austria, Croatia (Dalmatia?), Switzerland.

### 7. **Cyanocystis olivacea** (Reinsch) Komárek et Anagnostidis 1986 (fig. **536**)

*Sphaenosiphon olivaceus* Reinsch 1874; *Dermocarpa olivacea* (Reinsch) Tilden 1910

Cells in small, microscopic, more or less hemispherical or flat, sessile groups, composed of parallely and radially arranged cells. Cells elongate, heteropolar, spherical when young, later wide pear-shaped or club-shaped, narrowed towards the attached base, even stalk-like, 13–25 × 9.5–17 µm, with olive-green or violet (?) content. Sheaths thick, lamellate, colourless. Baeocytes numerous, spherical.
Occurrence: Marine, epiphytic on various algae and epilithic (?); known from Atlantic coasts of England and Spain, Mediterranean Sea (Spain, Adria coasts), outside Europe from Canada, Labrador and Brazil (SP); the species and localities should be checked.

### 8. **Cyanocystis violacea** (Crouan) Komárek et Anagnostidis 1986 (fig. **537**)

*Dermocarpa violacea* Crouan 1878

Cells rarely (when young) solitary, usually in groups, forming a layer on the substrate, more or less densely, parallely arranged; cells spherical when young, later polarized, elongate, obovoid, clavate or club-shaped, attached by the narrower end (by «fibrous adhesive substance») to the substrate, widely rounded at both ends, 16.3–30.7 × 8–17(28) µm, with pinkish-red, violet or yellow-brown (?) cell content, sometimes with recognizable centro- and chromatoplasmic regions (thylakoids parietally arranged). Sheaths thin, firm, colourless, to 1.2 µm thick, a little thicker towards the base, sometimes with discoid thickenings at the bases. Baeocytes numerous (12–220), arising from rapid, simultaneous multiple fission, 2.4–3.2 µm in diameter, liberated by rupture of the sheath at the apex.
Occurrence: Marine, epiphytic on algae; known from Atlantic coasts (England, Faeroe Islands, France, Norway), Baltic Sea (Sweden) and Mediterranean coasts (Adriatic Sea, Aegean Sea). Hua & al. (1989) studied a characteristic strain of this species (CCMEE 407), and described its life cycle and fine structure in detail.

### 9. **Cyanocystis farlowii** (Börgesen) Komárek et Anagnostidis 1986 (fig. **538**)

*Dermocarpa farlowii* Börgesen 1903

Cells in hemispherical or flattened colonies, densely radially or more or less parallely arranged, elongate, attached by one end to the substrate, cylindrical or more or less club-shaped, widely rounded at the apex and slightly narrowed towards the base, with blue-green content, 60–90 × –30 µm. Sheaths slightly thickened, not lamellate, firm, colourless. Baeocytes numerous, spherical, more or less 2.5 µm in diameter.
Occurrence: Marine, epiphytic, on algae (*Polysiphonia*) in coastal areas; from Europe known only from northern parts of the Atlantic and North Sea (Faeroe Islands), but also from N. Pacific (Japan).

**Species outside Europe:**
- *Cyanocystis flahaultii* (Sauvageau) Komárek et Anagnostidis 1995 [syn.: *Dermocarpa flahaultii* Sauvageau 1892, *Chamaecalyx flahaultii* (Sauvageau) Komárek et

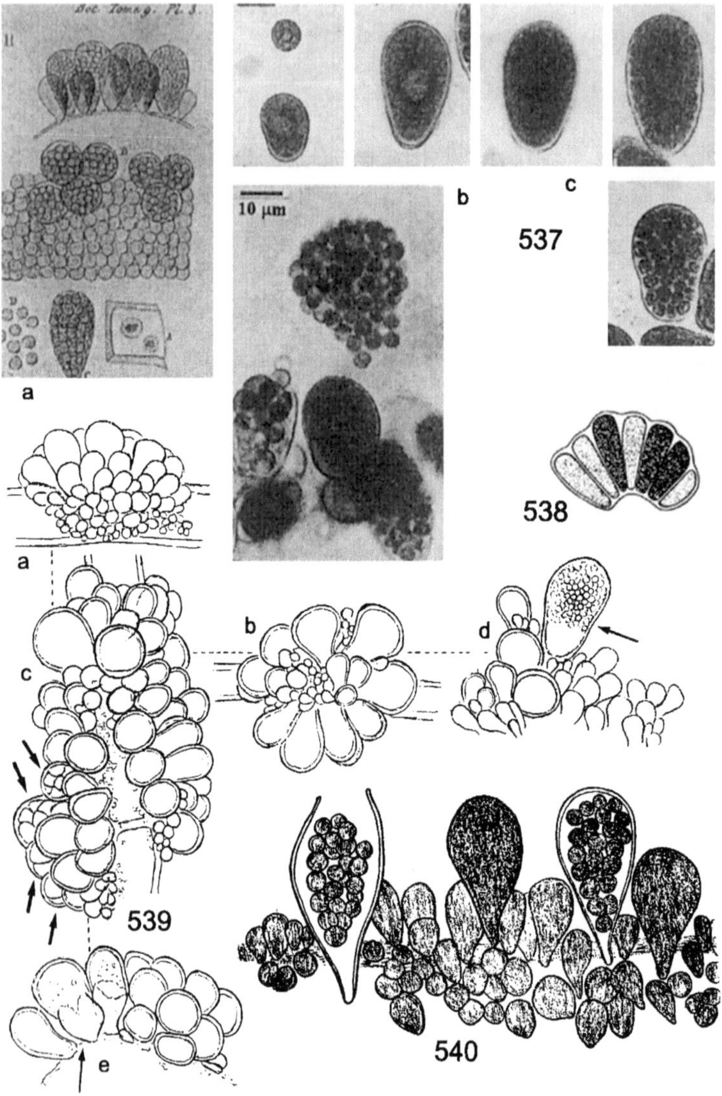


CYANOCYSTIS

**Fig. 537.** *Cyanocystis violacea*: **a,b** – cells of different age, several with baeocyte division, **c** – development (life cycle) of cells; **a** – after Crouan from Hua & al. (1989), **b,c** – from Hua & al. (1989).

Anagnostidis 1986]: Cells oval or almost spherical, 6–18 μm in diameter; freshwater, epilithic, found on stones in one creek in Algeria and epiphytic in cascades in Gabon; (Frémy 1930).
- *Cyanocystis mexicana* Montejano et al. 1993: Cells pear-shaped, in hemispherical colonies, to 50 μm long and to 39 μm broad, pale blue-green; freshwater, epiphytic on filamentous algae in central America (Mexico); (fig. **539**).
- *Cyanocystis pacifica* (Setchell et Gardner) Komárek et Anagnostidis 1986 [syn.: *Dermocarpa pacifica* Setchell et Gardner in Gardner 1918]: Cells more or less spherical with blue-green content, 20–45 μm in diameter, baeocytes 2 μm in diameter; marine, epiphytic, known from Californian coast.
- *Cyanocystis plectonematis* (Frémy) Komárek et Anagnostidis 1986 [syn.: *Dermocarpa plectonematis* Frémy 1930]: Cells spherical, later club-shaped or pear-shaped, 35–45 × 15–20 μm, baeocytes 3.5–4 μm in diameter, violet; epiphytic on *Plectonema wollei*, described from Gabon; (fig. **540**).
- *Cyanocystis pseudoxenococcoides* (Bourrelly) Bourrelly 1970 [syn.: *Dermocarpa pseudoxenococcoides* Bourrelly in Bourrelly & Manguin 1952]: Cells pear-shaped, violet, more or less 38 × 25 μm; freshwater, epilithic, Guadeloupe; (fig. **541**).
- *Cyanocystis sphaeroidea* (Setchell et Gardner) Komárek et Anagnostidis 1986 [syn.: *Dermocarpa sphaeroidea* Setchell et Gardner in Gardner 1918]: Cells more or less spherical with blue-green content, 18–25 μm in diameter; marine, described from Californian coast, known also from Brazil, Chile and SW. Pacific islands.
- *Cyanocystis valiae-allorgei* (Bourrelly) Bourrelly 1970 [syn.: *Dermocarpa valiae-allorgei* Bourrelly in Bourrelly & Manguin 1952]: Cells irregular rounded, blue-green, to 18–20 μm in diameter; epiphytic on algae, thermal waters, Guadeloupe; (fig. **542**).
- *Cyanocystis xenococcoides* (Geitler) Compère 1985 [syn.: *Dermocarpa xenococcoides* Geitler 1933]: Cells spherical, irregular-spherical or pyriform when old, circa 26 μm in diameter, later up to 50 × 30 μm, with red-violet to brown-violet content (rarely grey-bluish), with 64–129 baeocytes (the f. *minor* Geitler 1935 has cells only 10–17 rarely to 20 μm long); freshwater, epiphytic on various filamentous cyanoprokaryotes, algae and on mosses, described from Indonesia (Java, Sumatra). Probably identical with *C. plectonematis* (Geitler & Ruttner 1935); (fig. **543**).

**Unrevised species:**
- *Cyanocystis itzigsohniana* Rabenhorst 1866 pro synon.
- *Cyanocystis parva* Conrad 1914 [syn.: *Dermocarpa parva* (Conrad) Geitler 1925]: Cells spherical or almost spherical, 3–4 μm in diameter; freshwater,

---

**Fig. 538.** *Cyanocystis farlowii*: group of cells, several with baeocyte division; after Börgesen from Geitler (1932, sub *Dermocarpa farlowii*).

Characteristic species outside Europe:
**Fig. 539.** *Cyanocystis mexicana*, freshwater, from central America: **a,b** – young cells, **c** – group of cells, several cells with initiating baeocyte division (arrows), **d** – group of cells, one with differentiated baeocytes (arrow), **e** – group of cells and two empty sheaths (pseudovaginas); after Montejano & al. (1993).
**Fig. 540.** *Cyanocystis plectonematis*, from Gabon, W. Africa: group of cells of various ages, oldest cells with baeocyte production; after Frémy (1930, sub *Dermocarpa plectonematis*).

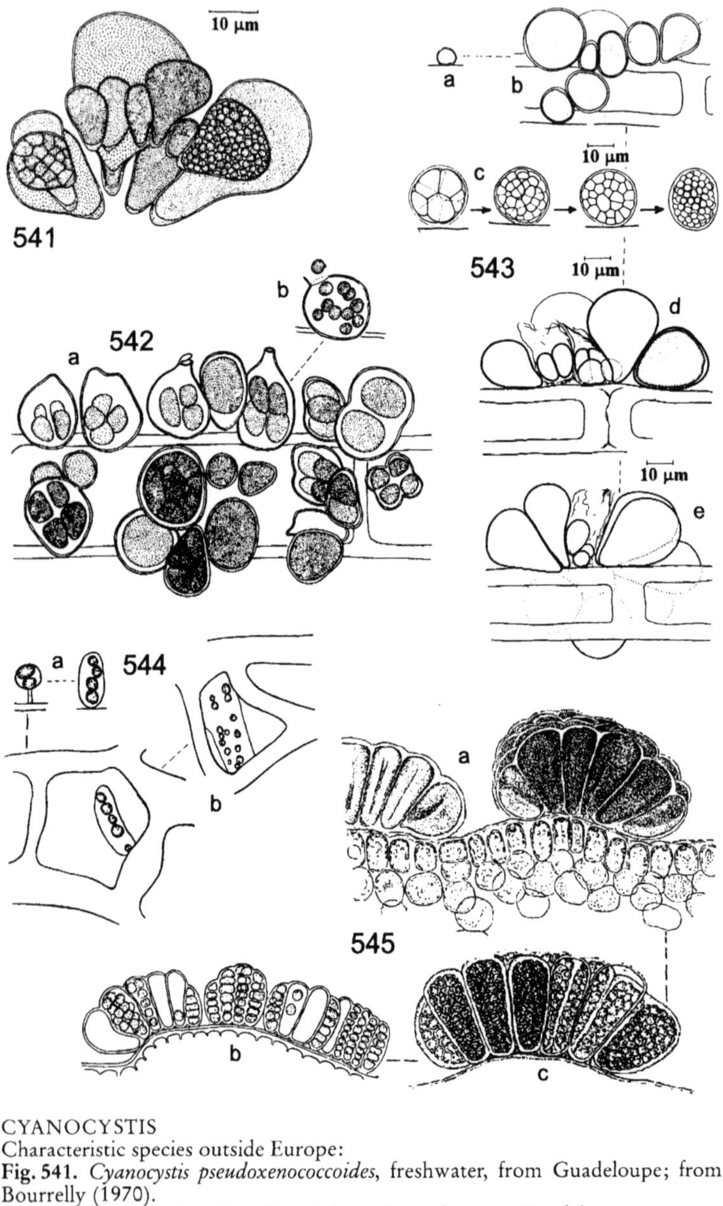


CYANOCYSTIS

Characteristic species outside Europe:

**Fig. 541.** *Cyanocystis pseudoxenococcoides*, freshwater, from Guadeloupe; from Bourrelly (1970).

**Fig. 542.** *Cyanocystis valiae-allorgei*, from thermal waters, Guadeloupe; a – group

epiphytic on *Cladophora*, Germany; reproduction (?); (see also Chu & al. ed. 1991).
- *Dermocarpa biscayensis* Sauvageau 1895: Cells narrow club-shaped, parallely arranged, 25–30 × 4–6 µm; marine, epiphytic on algae, described from France, recorded also by Ardré (1969) from Portugal; reproduction (?). According to Ardré (1969) synonymous with *Dermocarpa* (= *Dermocarpella*) *prasina*.
- *Dermocarpa cavanillesiana* Gonzalez Guerrero 1946: Thallus hemispherical, cells more or less pyriform, 8–24 × 8–20 µm, blue-green (?); epiphytic on *Lola lubrica*, estuary of the river Avillo, Spain; reproduction (?).
- *Dermocarpa cuspidata* (Reinsch) Geitler 1932: Thallus hemispherical, cells narrow fusiform, rounded at the base, 8–12 × 1.5–3 µm, blue-green; marine, epiphytic on *Sargassum*, Adriatic Sea; reproduction (?); (Frémy 1934).
- *Dermocarpa depressa* W. et G. S. West 1897: Cells «irregularly discoid», 2.7–3.8 × 5–10 µm, producing 8 baeocytes, 1.8 µm in diameter; described from an aquarium with brackish water, Angola.
- *Dermocarpa elliptica* Skvorcov 1946.
- *Dermocarpa enteromorphae* Anand 1937: Marine, described from England; (*Xenococcus* ?).
- *Dermocarpa fossae* Fjerdingstad 1950: Cells epiphytic, greenish, spherical to ovoid, 7.5 × 6 µm, with first horizontal division plane (*Dermocarpella* ?); freshwater, in ditches with stagnant water and rich in $H_2S$, Denmark; according to Drouet & Daily (1956) *Characium*, comp. also with *Chamaecalyx*.
- *Dermocarpa gaditana* Gonzalez 1947: Cells solitary, pear-shaped, with bluish, homogeneous content, 16–43 (50) × 10–20µm, first division plane is vertical, baeocytes 3–6 µm in diameter; partly endolithic in shells of living gasteropods, described from the river Arillo near San Fernando (Cádiz), Spain.

---

of cells attached to a filamentous alga, several with baeocyte formation, **b** – liberation of baeocytes from the sheath; from Bourrelly (1970).
**Fig. 543.** *Cyanocystis xenococcoides*, freshwater, from Indonesia: **a** – young cell, **b** – vegetative cells, **c** – baeocyte differentiation, **d-e** – groups of cells of various ages and empty sheaths (pseudovaginae); after Geitler from Geitler & Ruttner (1935, sub *Dermocarpa xenococcoides*).

DERMOCARPELLA

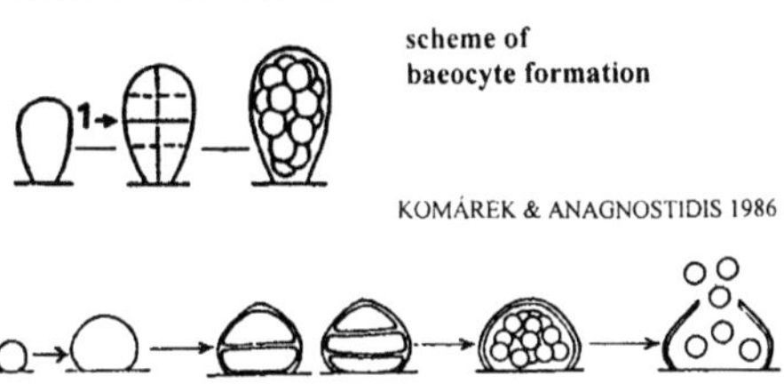


**Fig. 544.** *Dermocarpella sphagnicola:* a – baeocyte formation, **b** – groups of baeocytes within intercellular spaces in leaves of *Sphagnum*; after Maillefort from Geitler (1932).
**Fig. 545.** *Dermocarpella prasina*: **a** – groups of cells attached to the host alga, **b** – differentiation of baeocytes, **c** – cells differentiated into baeocytes; after Bornet & Flahault from various authors.

- *Dermocarpa gardneriana* Drouet 1942: Violet strata, cells blue-green to violet, 3–10(40) µm in diameter, reproduction (?); on rocks and shells in freshwater, described from San Francisco, California, known from southern states of the USA.
- *Dermocarpa hollenbergii* Drouet 1942: Cells blue-green or violet, 3–15(20) µm wide, freshwater, epiphytic, described from springs and ponds in Mojave desert, California, USA; reproduction (?); (*Xenococcus* ?).
- *Dermocarpa marchantae* Setchell et Gardner 1924: Marine, described from Baja California, Mexico.
- *Dermocarpa minuta* Drouet 1942: Marine, type from California, USA.
- *Dermocarpa reinschii* Setchell et Gardner 1924: Marine, described from Baja California (Tortuga Island), Mexico.
- *Dermocarpa rosea* (Reinsch) Batters 1889: Cells ovoid elliptical, 4–5 µm wide, pink; marine (England, Adriatic Sea, Red Sea, Labrador); poorly described, reproduction (?); (Frémy 1934).
- *Dermocarpa setchellii* Drouet 1942: Violet strata, epilithic, described from hot springs, reproduction (?); California, USA.
- *Dermocarpa smaragdina* (Reinsch) Tilden 1910: Cells pyriform or club-shaped, narrowing to the base, 16.5–33.5 × 8.5–11 µm, deep bluish-green; marine, epiphytic on algae, Canada, New Foundland, reported also from Adriatic Sea; reproduction (?); data from Africa concern probably another species; (Frémy 1934).
- *Dermocarpa solheimii* Drouet 1942: Cells blue-green, 2–4(5) µm wide, epiphytic, freshwater, described from Wyoming, USA.
- *Dermocarpaærediformis* (Reinsch) Geitler 1932: Cells pyriform, 8.5–42.5 × 5.5–8.5 µm, olive-green; reproduction (?); marine, Adriatic Sea; (Frémy 1934).
- *Dermocarpa strangulata* Sauvageau 1895: Cells rounded cylindrical, 15–30 × 4–24 µm, blue-green to violet; marine, epiphytic on *Sargassum*, described from France; reproduction (?); (Frémy 1934).
- *Dermocarpa vickersiae* Collins 1911: Described from Barbados; reproduction (?).
- *Dermocarpa xenococcoides* f. *minor* Geitler in Geitler & Ruttner 1935: Cells more or less spherical, 10–14 (20) µm in diameter, with red-violet to red content, baeocytes up to 32, circa 3 µm in diameter; freshwater, epiphytic, known from Indonesia (Java); (Geitler & Ruttner 1935).

## 3. **Dermocarpella** Lemmermann 1907

Cells solitary or aggregated into groups, polarized, attached by one end to the substrate (by narrower, almost stalk-like, or widened, flattened basal part), hemispherical, oval or club-shaped, with thin or thick, sometimes lamellate, colourless sheaths.

Reproduction by successive multiple fission of cells into baeocytes; first cell division always horizontal, perpendicular to the long («vertical») axis, later in various planes. Baeocytes numerous, usually several times smaller than the original mother cell.

**Important references:** Geitler (1932, 1942, 1967); Feldmann & Feldmann (1953); Ardré (1969); Komárek & Anagnostidis (1986).

**Key to the species:**

1a Freshwater, in intercellular spaces of *Sphagnum* in Alpine regions; cells solitary, spherical to cylindrical, 6–20 × 3–6 µm ....... **1. D. sphagnicola**

1b Marine, epiphytic on algae; cells in hemispherical colonies, cylindrical to club-shaped, 15–30 × 4–24 µm ...................... **2. D. prasina**

**1. Dermocarpella sphagnicola** (Maillefort) Komárek et Anagnostidis 1986 (fig. **544**)

*Chamaesiphon sphagnicola* Maillefort 1907; *Dermocarpa sphagnicola* (Maillefort) Geitler 1925

Solitary cells, spherical when young, later cylindrical, attached to the substrate by short hyaline stalk, 6–12 × 3–6 µm. Cells divide crosswise («horizontally»), successively into 2 or more cells, arranged in rows, later spatially. Sheaths colourless, thin, later gelatinizing and thickened. Baeocytes 0.7–1 µm in diameter.

Occurrence: Freshwater, endophytic, in intercellular spaces in leaves of *Sphagnum*; Switzerland (in alpine zone to about 1300 m), data from Hungary should be checked.

**2. Dermocarpella prasina** (Reinsch) Komárek et Anagnostidis 1995 (fig. **545**)

*Sphaenosiphon prasinus* Reinsch 1874; *Dermocarpa prasina* (Reinsch) Bornet et Thuret 1880, ? var. *incrustans* (Reinsch) Holmes et Batters ex Forti 1907 incl.; *Xenococcus prasinus* (Reinsch) Ardré 1961; *Cyanocystis prasina* (Reinsch) Komárek et Anagnostidis 1986

Cells elongate, cylindrical, obovoid or slightly club-shaped, very densely packed in hemispherical clusters or layers, joined by one end to the substrate, parallely or radially arranged, with flattened sides, widely rounded at the apex; cells 15–30 × 4–24 µm, with blue-green, olive-green or brownish content. Sheaths thin, firm, colourless. Cell division successive, first in several horizontal planes, later spatially into numerous baeocytes; in thin cells sometimes only one row of baeocytes arises (transition to exocytes); baeocytes liberated by gelatinization of sheaths. Life cycle not well known (also binary fission and in this case *Xenococcus*?).

Occurrence: Marine, epiphytic on various algae; known from the North Sea, Baltic Sea, Adriatic Sea, Aegean Sea, from the Atlantic coasts of Europe and N. and S. America, recorded also from the coast of Peru, Sea of Japan (Russia), Philippines and SW. Pacific islands, probably has a worldwide distribution.

**Species outside Europe:**

- *Dermocarpella hemisphaerica* (Lemmermann) Lemmermann 1907: Cells hemispherical with widened base, 12–27 µm in diameter, with blue-green content, baeocytes 0.8–1 µm in diameter; freshwater, epiphytic on algae in clear lakes, Chatham island, New Zealand; (fig. **546**).
- *Dermocarpella hemisphaerica* sensu J. et G. Feldmann (1953) and Ardré (1969): Cells 25–35 µm in diameter, attached to various small seaweeds, from marine coasts of western Europe and western Mediterranean Sea; morphologically very similar to *D. hemisphaerica*, however, found in marine biotopes and probably with different reproduction; the taxonomic identity with the freshwater species from New Zealand needs confirmation; (fig. **547**).
- *Dermocarpella protea* (Setchell et Gardner) Komárek et Anagnostidis 1995 [syn.: *Dermocarpa protea* Setchell et Gardner in Gardner 1918; *Cyanocystis protea* (Setchell et Gardner) Komárek et Anagnostidis 1986]: Cells club-shaped, in hemispherical colonies, 40–120 × 6–40 µm, pale blue-green, baeocytes 3–3.5 µm in diameter; marine, epiphytic on algae, described from Pacific coast of N. America.

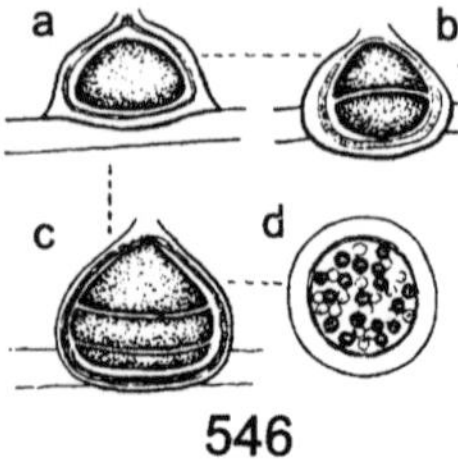


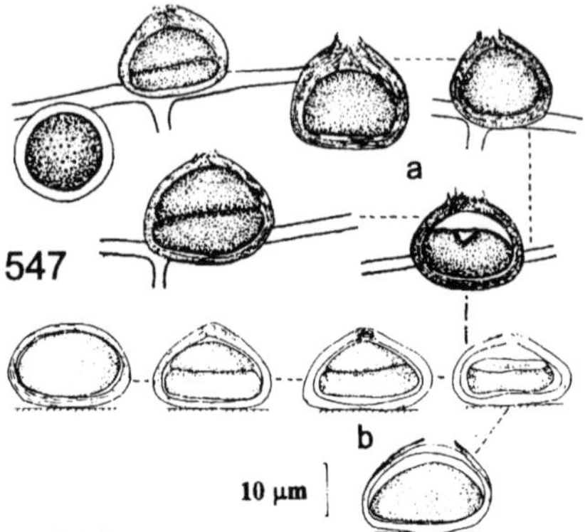


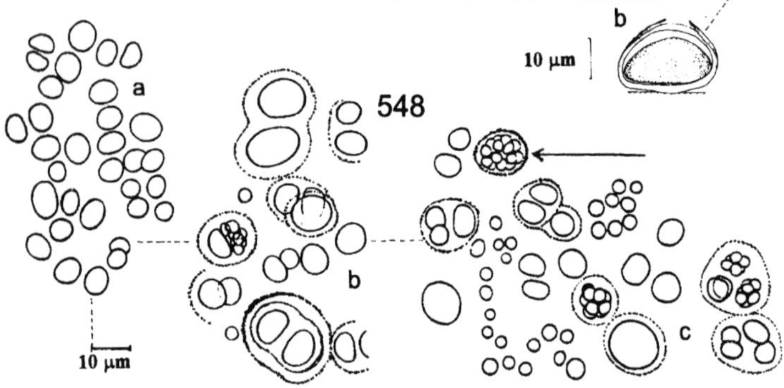


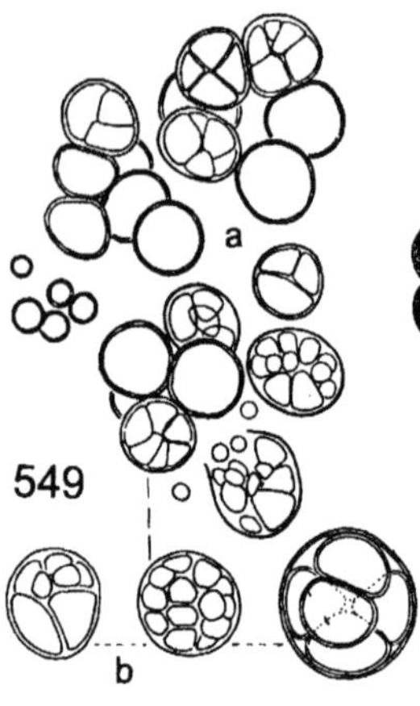


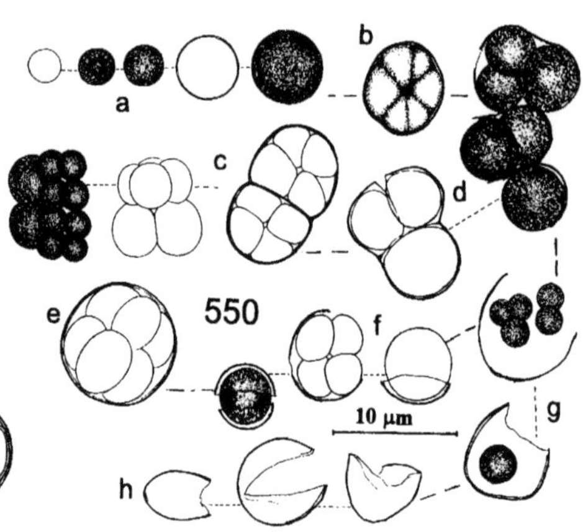


DERMOCARPELLA

**Fig. 546.** *Dermocarpella hemisphaerica*, type species from freshwater lakes in Chatham island, New Zealand: **a** – vegetative cells, **b-d** – differentiation of baeocytes; after Lemmermann from Geitler (1932).

**Fig. 547.** *Dermocarpella* cf. *hemisphaerica*, marine populations from western Med-

# 10. Family Xenococcaceae Ercegović 1932

Cells spherical or irregular rounded to rounded-polygonal, solitary or in irregular, packet-like colonies, or forming flat layers on the substrate; young cells and cells in layers polarized, attached to the substrate. Sheaths enveloping cells and groups of cells colourless, diffluent or firm, usually narrow, less frequently slightly widened and lamellate.

---

iterranean region and SW. Europe: **a** – after J. et G. Feldmann (1953), **b** – after Ardré (1969).

CHROOCOCCIDIUM

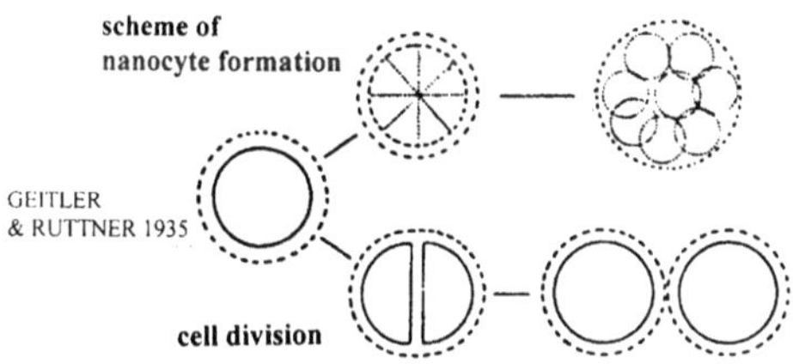


**Fig. 548.** *Chroococcidium gelatinosum* from tropical volcanic lakes: **a** – top view, **b-c** – cells of various ages and with nanocyte formations (arrow); after Geitler from Geitler & Ruttner (1935).

CHROOCOCCIDIOPSIS

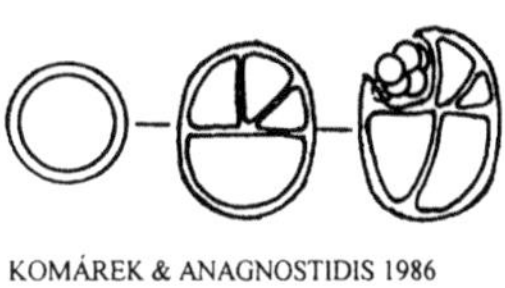


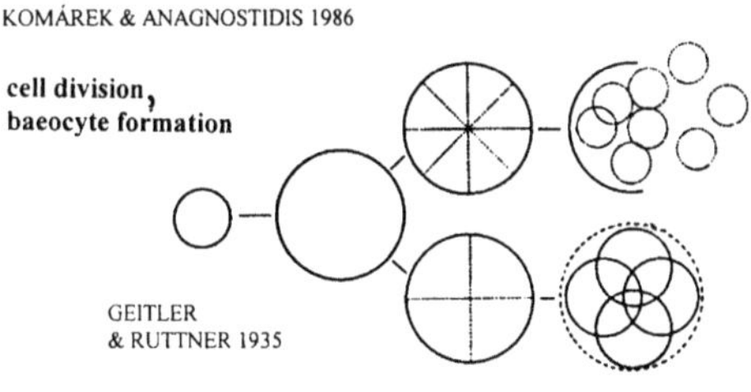


**Fig. 549.** *Chroococcidiopsis thermalis*: **a** – group of cells of various ages, partly with baeocyte differentiation, **b** – various stages of baeocyte formation; after Geitler from Geitler & Ruttner (1935).
**Fig. 550.** *Chroococcidiopsis thermalis*: **a** – growth of cells, **b-f** – various stages of baeocyte differentiation, **g** – liberation of baeocytes, **h** – empty sheaths; after Komárek & Hindák (1975), from various cultivated strains.

Cell division irregular, in various planes, successive (repeated binary fission), sometimes perpendicular to the substrate; facultatively (in older colonies and aggregates) special cells (often terminal) divide rapidly and successively into baeocytes or nanocytes. Reproduction by liberated cells, disintegration of colonies, and by baeocytes or by nanocytes.

**Key to the genera:**

**1a** Ensheathed cells, colonies or groups of cells not polarized, free-living (without distinct polarity when attached to the substrate or in periphyton); cells more or less spherical or later irregular-rounded . . . . . . . . . . . . . . . . . . **2**

**1b** Solitary cells or colonies always attached to the substrate, usually with distinct polarity, in one or more layers or in radially or packet-like arranged rows; cells irregular-rounded, often elongate and widely rounded at the apex . . . . . . . . . . . . . . . . . . . . . . . . . . . . . . . . . . . . . . . . . . . . . . . . **4**

**2a** Mucilaginous envelopes always fine, thin or wide, diffluent; free-living gelatinous groups of cells; reproduction by binary fission and nanocytes . . . . . . . . . . . . . . . . . . . . . . . . . . . . . . . . . . . . **1. Chroococcidium**

**2b** Mucilaginous envelopes firm, thin, delimited; reproduction usually by binary fission combined with baeocyte production . . . . . . . . . . . . . . . . . . **3**

**3a** Cells a short distance from one another, irregular-rounded, forming more or less spherical colonies . . . . . . . . . . . . . . . . . . . . . . . **2. Chroococcidiopsis**

**3b** Cells in dense packet-like groups, more or less irregular-polygonal, in subspherical, compound colonies . . . . . . . . . . . . . . . . . . . **3. Myxosarcina**

**4a** Young cells solitary, later more or less aggregated into one layer on the substrate; all cells of very similar size; when rows arise, they are short, indistinct; any cell in a colony may produce baeocytes . . . . **4. Xenococcus**

**4b** Cells in older colonies agglomerated spatially, sometimes in several layers, forming irregular clusters or more or less hemispherical colonies . . . . . . **5**

**5a** Cells in short irregular but distinct pseudofilamentous rows, later sometimes growing radially and spatially from the colonial center; terminal cells usually distinctly enlarged, widely rounded at the apex; baeocytes arise usually from terminal cells . . . . . . . . . . . . . . . . . . . . . . . . . . . **5. Chroococcopsis**

**5b** Cells forming two- to several-layered, more or less hemispherical colonies, attached to the substrate (usually epiphytic on algae and plants); all cells approximately the same size . . . . . . . . . . . . . . . . . . . . . . . **6. Xenotholos**

## 1. **Chroococcidium** Geitler 1933

Mucilaginous aggregates of cells on a substrate, with irregularly organized cells of diverse size, enveloped by common amorphous, diffluent, colourless slime; sometimes fine, diffluent individual envelopes around cells and around cell groups. Cells spherical, subspherical, irregular-rounded, ovoid or oval, with homogeneous cell content.

Cell division irregular, in various planes, daughter cells usually do not reach the original size and shape before the next division; occasionally intense nanocytic division.

**Important references:** Geitler (1933, 1942); Geitler & Ruttner (1935); Komárek & Anagnostidis (1986).

**Species outside Europe:**

– *Chroococcidium gelatinosum* Geitler 1933 [syn.: *Chroococcopsis gelatinosa* (Geitler) Bourrelly 1970]: Colonies dark brown-green, flat, with not very densely disposed cells, sometimes 2–4 (several) in small subgroups, around cells sometimes very finely delimited or diffluent individual envelopes. Cells in colonies, of varied size, to 18 µm in diameter, blue-green or yellow-green. Nanocytes 8–16 (rarely to 32), circa 3 µm in diameter. Freshwater, periphytic (epiphytic) on submersed plants or on limestone rocks; known from volcanic lakes in Indonesia (S. Sumatra) and in central Mexico; (fig. **548**).

## 2. **Chroococcidiopsis** Geitler 1933

Cells solitary or in more or less spherical or irregular groups, with thin, firm, colourless sheaths, usually periphytic on a variety of aerophytic or submersed substrates. Cells spherical, hemispherical or irregular-rounded, enveloped by sheaths. Without pseudofilamentous stages. Thylakoids distributed usually irregularly throughout the protoplast. Cell wall S-layer of special ribbon-like type found in two strains (Büdel & Rhiel 1985).
Cells divide irregularly (in various planes), successively by binary fission and retain usually their irregular shape during the cell cycle; mother cells divide often into daughter cells of different shape and size. Some cells later change into baeocytes (or only some of the cells in a colony); baeocytes few or many, of different number in various species. Waterbury & Stanier (1978) have never found motile baeocytes in *Chroococcidiopsis*, but this phenomenon does appear temporarily, being dependent on environmental factors (e.g., Andersen, in litt.). Divided cells or single cells sometimes are liberated from irregularly split mucilaginous envelopes (sheaths). Sheaths sometimes gelatinize. Different types of cell division and baeocyte production (large variability) were described by Geitler & Ruttner (1935, fig.12) and Friedmann (1961, fig. 25).
**Important references:** Ercegović (1932, sub *Pleurocapsa*); Geitler (1933); Geitler & Ruttner (1935); Friedmann (1961, 1971, 1980); Komárek (1972); Komárek & Hindák (1975); Komárek & al. (1975); Hindák (1978); Waterbury & Stanier (1978); Büdel & Henssen (1983); Friedmann & Ocampo-Friedmann (1984); Büdel (1985); Büdel & Rhiel (1985); Komárek & Anagnostidis (1986); Dor & al. (1991).

**Key to the species:**
1a Nanocytes 2–3.5(4) µm, cells to 10(18) µm in diameter; in thermal springs . . . . . . . . . . . . . . . . . . . . . . . . . . . . . . . . . . . . . . . . **1. C. thermalis**
1b Nanocytes circa 0.5 µm, cells 2–4 µm in diameter; chasmoendo-lithic in the limestone supralittoral on Mediterranean coasts . . . . . . . **2. C. fissurarum**

### 1. **Chroococcidiopsis thermalis** Geitler 1933 (fig. **549, 550**, Tab. **2 C/h**)

Cells, groups of cells and more or less spherical or irregular colonies surrounded by envelopes in microscopic or macroscopically visible thin mats on various substrates; cells spherical or in groups and then slightly polygonal, in colonies usually of varied size. Mucilaginous envelopes (sheaths) thin, but distinct, firm, colourless in young baeocytes to 1.5 µm wide, later well recognizable. Cells grow to 9–10 (when differentiated in baeocytes up to 20) µm in diameter (according to Hindák 1978 to 35 µm); after irregular division of cells (perpendicular, radial, etc.) beaocyte production proceeds occasionally (with 8–16–32 baeocytes); baeocytes (1.5)2–3.5(4) µm in diameter, spherical to polygonal-spherical; cell content homogeneous or slightly granular, blue-green.
Occurrence: Aquatic, in periphyton on stones in mineral and thermal springs, rarely in heliothermal pools; known mainly from tropical countries (Cuba, Indonesia, Mexico), less frequently from the temperate zone, in the Europe found in thermal springs in Greece (Euboea Island) and in Slovakia (Piešt'any).

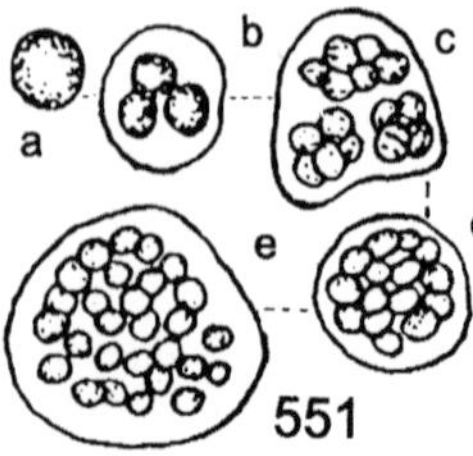


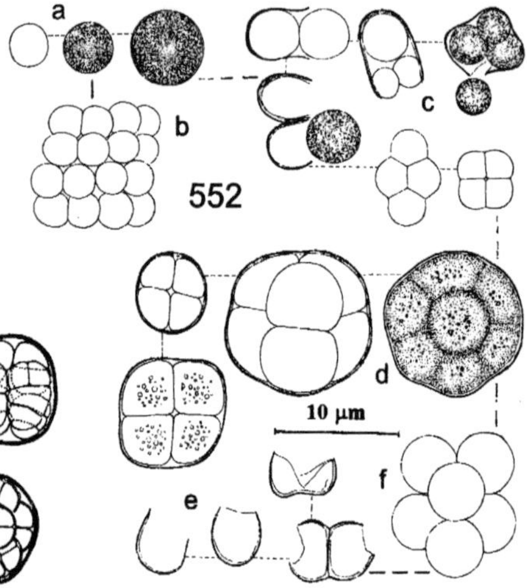


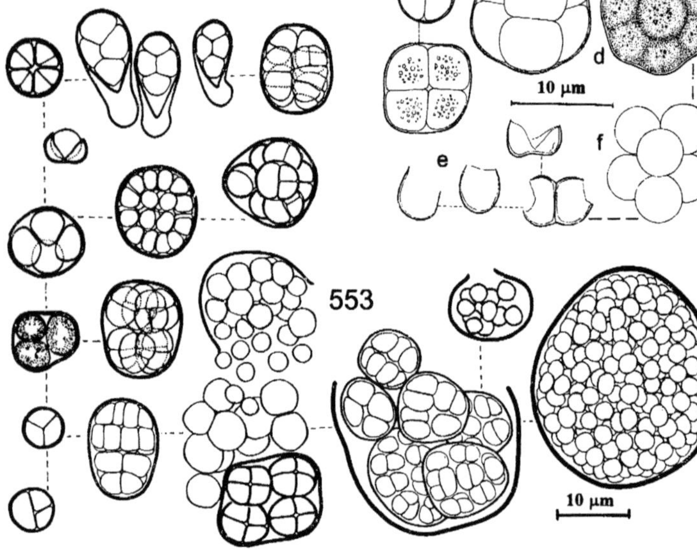


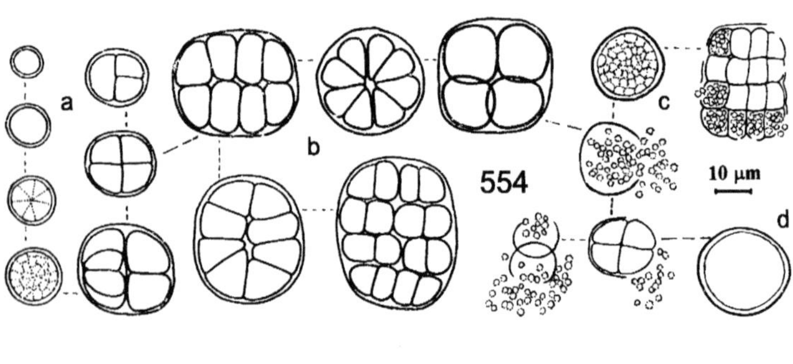


CHROOCOCCIDIOPSIS

**Fig. 551.** *Chroococcidiopsis fissurarum:* a – vegetative spherical cell, **b-e** – various stages of baeocytic cell division; after Ercegović (1932, sub *Pleurocapsa fissurarum*).

**2. Chroococcidiopsis fissurarum** (Ercegović) Komárek et Anagnostidis 1995 (fig. **551**)

*Pleurocapsa fissurarum* Ercegović 1932

Cells solitary, in groups or in large aggregates. Colonies form thin, flat, blue-green spots on the substrate. Mucilaginous sheaths thin, clearly delimited, colourless, not striate. Vegetative cells spherical, subspherical or later rounded-polygonal (in agglomerations), to 2 µm, before division to 4 µm in diameter; more or less spherical colonies with enveloping sheaths sometimes to 40 µm in diameter. Cell content blue-green. Baeocytes numerous (?), 0.5 µm in diameter or smaller.

Occurrence: Chasmoendolithic in supralittoral of calcareous rocks (marine splash-zone, probably up to the high supralittoral with at least a small influence of «salty» winds from the sea); coasts of Mediterranean Sea (Croatia – Dalmatia).

**Species outside Europe:**

- *Chroococcidiopsis bourrellyana* Compère 1997: Described from brackish waters in the lower part of rivers in Senegal.
- *Chroococcidiopsis codiicola* Beljakova 1989: Marine, epiphytic on *Codium fragile*, Russia, Sea of Japan (Popov island).
- *Chroococcidiopsis cubana* Komárek et Hindák 1975: Described from periphyton of pools and drainage ditches, and more or less alkaline swamps and the littoral of shallow water bodies, often near sea coasts; known from Cuba, Brazil and Florida (USA); (fig. **552**).
- *Chroococcidiopsis doonensis* R.B.Singh 1968: Aerophilic (?), described from surface of stones in dried paddy field in Uttar Pradesh, India; possibly *Cyanosarcina*.
- *Chroococcidiopsis indica* Desikachary 1959: Aquatic, described from laboratory culture in India (Madras).
- *Chroococcidiopsis kashayi* Friedmann 1961: Epilithic (aerophytic) on rocky substrates in caves, described from Israel; (fig. **553**). Recently recorded also from Italy (Abdelahad 1985) and Spain (Asencio & Aboal 1996); the taxonomic identity of all these populations should be revised.
- *Chroococcidiopsis mysorensis* Tiwari 1972: Free floating in enriched cultures from paddy field soils in India; (fig. **554**).
- *Chroococcidiopsis polansiana* Andersen, in prep.: Isolated from oceanic water near Galapagos islands.
- *Chroococcidiopsis supralittoralis* Dor et al. 1991: Freshwater, described from supralittoral (splash zone) of lake Kinneret, Israel.
- *Chroococcidiopsis umbratilis* Dor et al. 1991: Subaerophytic, from drying, shaded stone wall, Ein Karem, Jerusalem, Israel.
- *Chroococcidiopsis versatilis* Dor et al. 1991: Halophilic, from sublittoral of hypersaline solar pond in Dead Sea area, Israel.

---

Examples of characteristical species described from localities outside Europe:

**Fig. 552.** *Chroococcidiopsis cubana*, from alkaline swamps in Central America: **a** – cells of various ages, **b** and **f** – groups of liberated baeocytes, **c-d** – various stages of cell division (baeocyte formation), **e** – empty sheaths; after Komárek & Hindák (1975).

**Fig. 553.** *Chroococcidiopsis kashayi*, described from Israel: various stages of cell division and baeocyte differentiation; after Friedmann (1961).

**Fig. 554.** *Chroococcidiopsis mysorensis*, from India: **a** – cell division, **b** – cell division within sheaths, empty sheath; after Tiwari (1972).

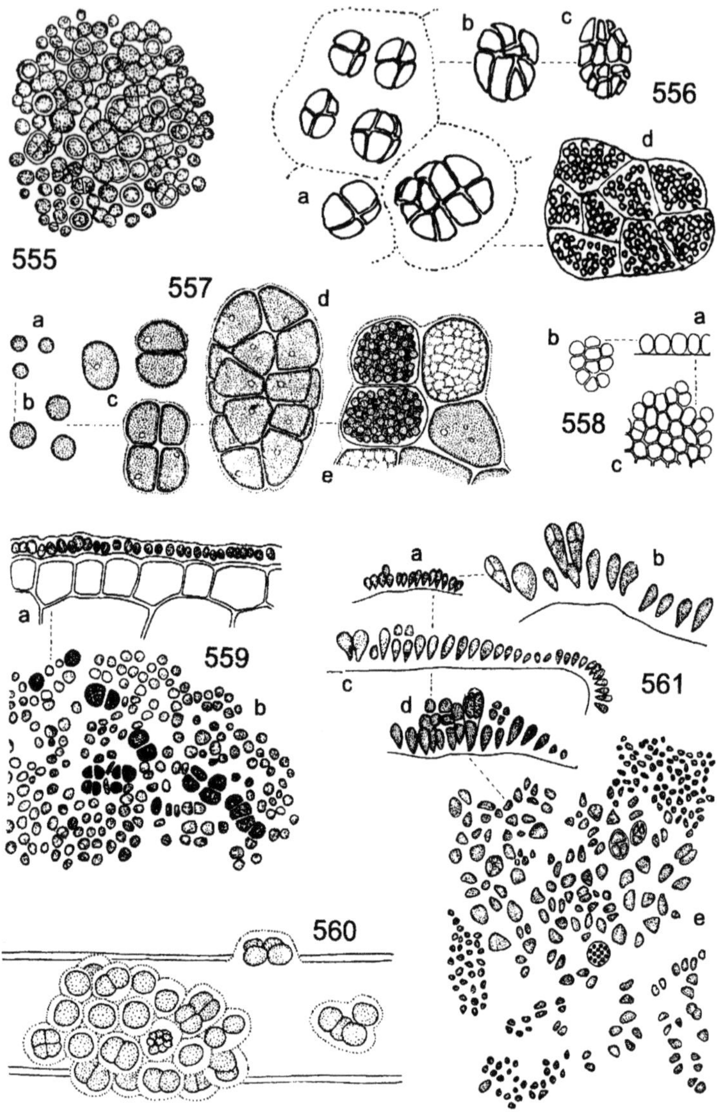


CHROOCOCCIDIOPSIS

Unrevised European species (comp. also Fig. 201a-b and 361f):

**Fig. 555.** *Chroococcidiopsis* sp., marine species from France: after Frémy from Kosinskaja (1948, sub *Aphanocapsa roberti-lamyi*).

- *Chroococcidiopsis* sp., ecologically similar to previous populations, was recorded from desert crusts in Dead Sea Valley, Israel (Dor & Danin 1996).
- *Chroococcidiopsis* sp. in Friedmann & al. 1967; Friedmann 1971, 1980; Friedmann & Ocampo-Friedmann 1984 (5 strains): Endolithic species from hot deserts worldwide.
- *Chroococcidiopsis* sp. in Broady 1981; Friedmann & al. 1988: Endolithic species from arid Antarctica.
- *Chroococcidiopsis* sp. div. in Wessels & Büdel 1989 and Weber & al. 1996: Cryptoendolithic, not identified species were published from semidesert ecosystems of S. Africa (Transvaal, Northern Province).
- *Chroococcidiopsis* sp. div. (comp. also *Myxosarcina*) in Büdel 1985, 1987; Büdel & Henssen 1983: Different phycobiontic strains isolated from lichens in Austria, Czech Republic, Spain, Australia and Mexico; members of this genus are cited as phycobionts of various genera of the family *Lichinaceae* (comp. also Henssen & Büdel 1984), often from arid and semiarid regions.
- Several species, described by Gardner (1927) from Porto Rico (*Anacystis compacta, A. gloeocapsoides, A. microsphaeria, A. minutissima, Endospora bicoccus,* and others) and by Schwabe (1944) as *Coccopedia andina* (cells 2.5–4 µm, baeocytes 1.8–2 µm in diameter, epiphytic, from thermal springs in Chile), as *Gloeocapsa* sp. «Nak» (blue and red forms, from thermal water in Japan – Nakabusa) and as *Synechocystis* sp. (from marine littoral near Tsingtao, China), belong possibly also to the genus *Chroococcidiopsis.*

**Unrevised European species:**
- *Chroococcidiopsis* sp. in Komárek & Hindák 1975: Described from psammon of Black Sea beaches, Roumania.

---

MYXOSARCINA

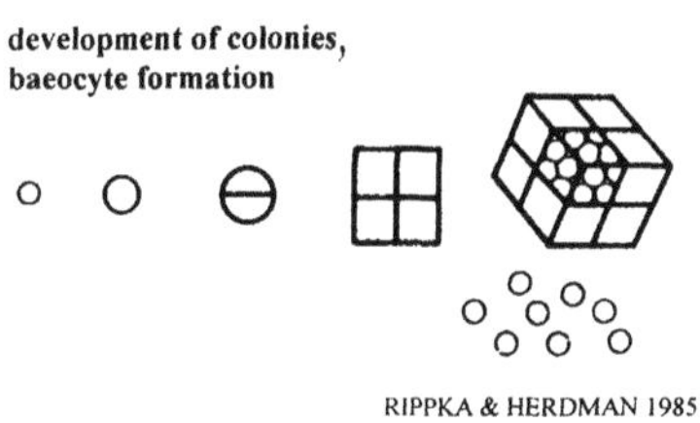


**Fig. 556.** *Myxosarcina gloeocapsoides*: **a-c** – colonies, **d** – cells divided into baeocytes; after Setchell & Gardner in Gardner (1918, sub *Pleurocapsa gloeocapsoides*).
**Fig. 557.** *Myxosarcina amethystina*: **a-c** – cell development and division, **d** – colony, **e** – baeocyte differentiation; after Copeland from G.M.Smith (1950).

XENOCOCCUS
**Fig. 558.** *Xenococcus minimus*: **a** – side view, **b-c** – top views; after Geitler (1932).
**Fig. 559.** *Xenococcus laminariae*: **a** – side view, **b** – top view; after Feldmann (1953).
**Fig. 560.** *Xenococcus pallidus*: from G.M.Smith (1950).
**Fig. 561.** *Xenococcus elenkinii*: **a-d** – side view, **e** – top view; after Pohribniak from Kondrateva & al. (1984).

– *Chroococcidiopsis* sp. div. in Büdel & Henssen 1983; Henssen & Büdel 1984; Büdel 1985: Various phycobiontic strains isolated from lichens in Austria, Czech Republic and Spain (Canary Islands).
– *Chroococcidiopsis* sp. in Anagnostidis & Roussomoustakaki 1991: From heliothermal saltworks in Greece (Messolongion).
– *Aphanocapsa biformis* sensu Geitler 1932 and Anagnostidis 1961: From various biotopes (Geitler) and thermal springs in Greece (atmophytic); (fig. **201b**).
– *Aphanocapsa (Microcystis) roberti-lamyi* Frémy 1934: Macroscopic colonies, cells reddish-violet, with thick sheaths, described from marine water, France; (fig. **555**).
– *Pleurocapsa crepidinum* Ercegović 1930 [false «(Thuret) Ercegović 1930», sensu Ercegović (= *Chroococcus crepidinum* Hansgirg 1887 = *Pleurocapsa ercegovicii* de Toni 1934), not sensu Thuret; nec *Pleurocapsa crepidinum* Collins 1901]: Cells spherical or irregular-rounded, 3–5 µm in diameter, aggregated into colonies; envelopes colourless or yellow-brown, slightly lamellate; cells dividing into baeocytes, to 11 µm, baeocytes about 1 µm in diameter; marine, described from coast of Croatia – Dalmatia. Ercegović's new Latin diagnoses not in original sense of Thuret; possibly *Chroococcidiopsis* species; (fig. **361f**).
– Two morphologically distinguishable types (differing in size of baeocytes and vegetative cells) were studied by Waterbury & Stanier (1978) among 8 strains isolated from freshwater or terrestrial biotopes, without taxonomic evaluation.

## 3. **Myxosarcina** Printz 1921

Cells in small groups (subcolonies) with up to 32 cells, later in packet-like, cubic-rounded or irregular, sometimes slightly flattened or blastoparenchymatous colonies, forming granular mats on the substrate; cells densely packed. Colonies free-living among other algae or freely on the substrate. Mucilaginous envelopes (sheaths) thin or very slightly widened, distinct, firm, colourless or yellow-brown. Cells rounded polygonal or rounded-irregular, in young colonies in more or less short perpendicular rows, later irregularly disposed.
Cells divide irregularly by binary fission in three or more planes, in older colonies often obliquely, forming groups of cells. In old colonies some cells occasionally (mainly in the margin) change into temporarily motile baeocytes (planocytes). Reproduction by disintegration of colonies and by baeocytes.
**Important references:** Geitler (1932, 1942); Waterbury & Stanier (1978); Anagnostidis & al. (1983); Komárek & Anagnostidis (1986).

Note: The genera *Chroococcidiopsis* and *Myxosarcina* are distinguishable only by motility of baeocytes according to Waterbury & Stanier (1978). This feature is, however, problematic; the traditional characters distinguish well both these taxa on the generic level.

**Key to the species:**

| | | |
|---|---|---|
| 1a | Freshwater, subaerophytic in high mountains ........... | **1. M. tatrica** |
| 1b | Marine, on calcareous coastal rocks ............ | **2. M. gloeocapsoides** |

### **1. Myxosarcina tatrica** (Starmach) Komárek et Anagnostidis 1995

*Myxosarcina chroococcoides* f. *tatrica* Starmach 1981

Cells in small sarcinoid groups (subcolonies) with up to 32 cells and up to 25 µm in diameter, packet-like, enveloped by thin, firm, hyaline, rarely yellow-brown sheath, 1–1.5 µm wide. Cells with dark or grey blue-green, homogeneous content, 4.5–6.5 µm in diameter; baeocytes develop rarely, 4–8 in number.

Occurrence: Subaerophytic, on moistened rocks among other algae; described from Poland (High Tatra Mts.).

**2. Myxosarcina gloeocapsoides** (Setchell et Gardner) Komárek & Anagnostidis 1995 (fig. **361e, 556**)

*Pleurocapsa gloeocapsoides* Setchell et Gardner in Gardner 1918; ? *Gloeocapsa crepidinum* Collins et al. sensu Setchell & Gardner in Gardner 1918, not Thuret 1876 (comp. *Gloeocapsopsis crepidinum,* fig. **361e**)

Colonies packet-like, glistening, gelatinous, with wide mucilaginous envelopes, aggregated in gelatinous mass, 1–2 mm thick. Cells globose when solitary, angular from mutual pressure, 4–8 µm in diameter, with a conspicuous brownish sheath; cell content homogeneous, pale blue-green. Cells grouped into 2- to 8-celled packets, surrounded by an ample, homogeneous, hyaline, soft envelope. Cell division regularly in three planes and into 2–4 (up to numerous) baeocytes which arise from unchanged vegetative cells, 2.5–3.5 µm in diameter.
Occurrence: Marine, described from a salt marsh in California, USA, known also from coastal rocks in Mediterranean Sea (Dalmatia).

**Species outside Europe:**
- *Myxosarcina amethystina* Copeland 1936: Described from thermal springs in Yellowstone National Park, USA; probably more widely distributed (comp. Vasishta 1969); (fig. **557**).
- *Myxosarcina cloacarum* Saxena et al. 1971: Grows among other algae in periphyton on bricks in sewage water, India.
- *Myxosarcina concinna* Printz 1921: Type species, described as aerophytic from tree trunks in S. Africa; recorded also from other areas (and from other biotopes, e.g., from Antarctica), but such populations need taxonomic revision.
- *Myxosarcina decolorata* (Varma et Mitra) Komárek et Anagnostidis 1995 [syn.: *Myxosarcina spectabilis* var. *decolorata* Varma et Mitra 1962]: Aerophytic species from concrete floor and moist bricks, described from India.
- *Myxosarcina rubra* (Gardner) Bourrelly 1970 [syn.: *Endospora rubra* Gardner 1927]: Probably atmophytic on rocks and old wood near springs and lake shores, known from Porto Rico; baeocytes unknown, possibly *Cyanosarcina* (see p. 311–316).
- *Myxosarcina salina* Frémy in de Toni 1940: Subspherical colonies to 60 µm in diameter, cells 5–7 µm in diameter, green or blue-green, with homogeneous content, baeocytes 2–2.5 µm in diameter; described from saline waters in Djibouti.
- *Myxosarcina spectabilis* sensu Vasishta 1963: Typical *M. spectabilis*, which was described from thermal springs in Indonesia, does not divide into baeocytes, however, Vasishta designated by this name an aerophytic population from tree trunks in India, reproducing by baeocytes; evidently another species.
- *Myxosarcina thermalis* Thomas et Gonzalves 1965: Described from thermal springs in India.
- Several species described by Gardner (1927) from Porto Rico, e.g., *Endospora mellea, E. nigra, E. olivacea* belong probably also into the genus *Myxosarcina*.
- Under «*Myxosarcina* sp.» were described different, taxonomically not validly published species, e.g., in Ley (1948), Rippka & al. (1979), Broady (1981) and others. *Myxosarcina* sp. is mentioned also as the possible phycobiont of the lichen genera *Lichinella, Peccania* and *Peltula,* often from arid and semiarid regions (Büdel 1985, 1987).

**Unrevised species:**
- *Myxosarcina minuta* Vogl 1955.
- *Pleurocapsa salevensis* Chodat 1909: Cells grey-green or violet, subspherical, 15–16 µm in diameter when old, reproduction by binary fission (aggregates or packets of cells) or by baeocytes (about 1 µm in diameter), isolated from rocks, associated with *Trentepohlia umbrina*, near Salève, Switzerland; probably a species of *Myxosarcina* or *Chroococcidiopsis*.

## 4. **Xenococcus** Thuret in Bornet et Thuret 1880

Cells attached to the substrate, solitary or in monolayered groups, sometimes forming blastoparenchymatous colonies or a crustose or flat thin thallus with densely arranged cells; young stages characterized usually by discontinuous groups of sessile cells, sometimes with spaces between one another. Cells more or less polarized, sometimes slightly or distinctly elongate along the vertical axis (this polarity is particularly recognizable in solitary cells). Around cells usually firm, less frequently gelatinous sheaths. Cells in old colonies are sometimes organized into short, horizontal, irregular rows. All cells more or less the same size, at the colonial margin usually rounded or elongate.
Cells divide irregularly in two or more different planes, but usually perpendicular to the substrate; divided cells separate from one another or remain closed in a common sheath. Occasionally in solitary cells occurs successive baeocytic division into numerous baeocytes, which are liberated from the enveloping sheaths by their splitting or gelatinization. The baeocytes differentiate from the whole mother cell, or the cell first divides into two parts by more or less horizontal binary fission and only one (upper) part divides into baeocytes.
**Important references:** Gardner (1918, 1927); Geitler (1932); Frémy (1934); Geitler & Ruttner (1935); Kondrateva & al. (1984); Komárek & Anagnostidis (1986); Montejano & al. (1993); Gold-Morgan & al. (1994).

**Key to the species:**

1a Cells 1–4(6) µm in diameter; baeocytes not known; marine or freshwater species . . . . . . . . . . . . . . . . . . . . . . . . . . . . . . . . . . . . . . . . 2
1b Cells (1–3 = baeocytes)-5–11(–23) µm in diameter; baeocytes occasionally; marine or saline species . . . . . . . . . . . . . . . . . . . . . . . . . . . . . 4

Note: In species 1–3 (**2a-3b**) with small cells baeocyte formation is not known, which should be one of the generic diagnostic features of *Xenococcus*. They belong, therefore, very probably to another genus (e.g., *Chamaesiphon* subg. *Chamaesiphonopsis*, or *Mantellum*).

2a Marine, epiphytic on *Laminaria*; cells 1.5–3 (5) µm in diameter . . . . . . . . . . . . . . . . . . . . . . . . . . . . . . . . . . . . . . . . **2. X. laminariae**
2b Freshwater . . . . . . . . . . . . . . . . . . . . . . . . . . . . . . . . . . . . . . . . 3
3a Cells spherical, 2.2 µm in diameter or widely oval, 1.5–2.2 × 1.6 µm . . . . . . . . . . . . . . . . . . . . . . . . . . . . . . . . . . . . . . . . **1. X. minimus**
3b Cells 3–5.5 µm in diameter (or long) . . . . . . . . . . . . . . . . **3. X. gracilis**
4a Cells 3–5 µm in diameter . . . . . . . . . . . . . . . . . . . . . . . . **4. X. pallidus**
4b Cells (not baeocytes) 4–11–23 µm in diameter . . . . . . . . . . . . . . . . . . 5
5a Cells 4–11 µm in diameter . . . . . . . . . . . . . . . . . . . . . . . . . . . . . . 6
5b Old cells usually over 11 µm in diameter (or long; 2–23 × 2–14 µm) . . . . 7
6a Cells more or less flattened or more or less spherical, 4–9 µm in diameter . . . . . . . . . . . . . . . . . . . . . . . . . . . . . . . . . . . . . . . . **5. X. schousboei**
6b Cells more or less drop-like or more or less club-shaped, 2–11 × (1)2–5 µm . . . . . . . . . . . . . . . . . . . . . . . . . . . . . . . . . . . . . . . . **6. X. elenkinii**

7a Old cells mainly pyriform, 12–20 × 10–15 µm, usually radially arranged, epiphytic on algae . . . . . . . . . . . . . . . . . . . . . . . . . . . . . . **7. X. pyriformis**
7b Old cells oval or club-shaped, 2–23 × 2–20 µm, arranged in a pseudoparenchymatous layer, epizoic on shells of *Gasteropodum* . . . **8. X. gaditanus**

## 1. **Xenococcus minimus** Geitler 1922 (fig. **558**)

Cells spherical or widely oval (in this case attached to the substrate by one end), sometimes very slightly narrowed towards the base, usually parallel and very densely packed together (blastoparenchymatically), always in one layer on the substrate; cells in colonies flattened at sides and polygonal from above; colonies usually discoid, with circular or later irregular outline. Cells to 2.2 × ± 1.6 µm, blue-green, olive-green or greyish green.
Occurrence: Epiphytic on filamentous algae (*Cladophora*) and on *Lemna minor*, mosses etc.; described from stagnant freshwaters in Austria (near Vienna). Reported also from saline waters (Spain) and from tropical countries (Indonesia); all these localities should be checked.

## 2. **Xenococcus laminariae** Feldmann 1953 (fig. **559**)

Cells in one-layered, flat colonies, which form reddish, macroscopically visible spots on the substrate (*Laminaria*); cells small, spherical, hemispherical, irregular or ovoid, with homogeneous, pinkish content, 1.5–3 (5) µm in diameter, irregularly arranged on the substrate.
Occurrence: Marine, epiphytic on thallus of *Laminaria hyperborea*; France (Atlantic coast near Roscoff).

## 3. **Xenococcus gracilis** Lemmermann 1898

Cells spherical or oval, aggregated in an one-layered colony on the substrate, forming discoid, circular or polygonal flat colonies, surrounded by thin, fine, colourless, gelatinous, homogeneous envelope, blue-green, 3–5.5 × 1.5–3 µm. Possibly initial stages of another species.
Occurrence: Epiphytic on filamentous algae (*Tribonema, Cladophora*, etc.) in channels and depressions with water, pools, drainage waters and small, shallow water bodies; central Europe, Latvia, recorded also from Argentine and Pacific islands, sometimes in marine water (?) (the identity of populations from diverse ecological niches and distant localities needs revision).

## 4. **Xenococcus pallidus** (Hansgirg) Komárek et Anagnostidis 1995 (fig. **560**)

*Xenococcus schousboei* var. *pallidus* Hansgirg 1889

Cells more or less spherical or rounded flattened, usually densely and irregularly aggregated on the surface of other filamentous algae, in one-layered, irregular colonies, pale blue-green or yellow-green, 3–5 µm in diameter.
Occurrence: Marine, epiphytic on seaweeds; along coast of Adriatic Sea (Croatia – Dalmatia), Aegean Sea (Greece – Macedonia), North Sea (Norway – Trondhjemfjord) and N. America (USA – Massachusetts).

**5. Xenococcus schousboei** Thuret 1880 (fig. **562**)

*Coleonema arenifera* Schousboe 1880 pro syn.; *Dermocarpa schousboei* (Thuret) Bornet in Batters 1889

Cells more or less spherical or flattened, usually dense massed in one-layered, irregular groups or rows, within colony ± polygonal in outline (from above); colonies crustose, flat, attached to the substrate (marine algae), very irregular. Cells blue-green, 4–9 µm in diameter, all more or less of the same size. Baeocyte formation occasional and rare, in marginal cells.
Occurrence: Marine littoral, epiphytic on various filamentous cyanoprokaryotes and algae; common on the W. European Atlantic coasts, less frequently in Baltic Sea, Mediterranean region and Black Sea; recorded also from Japan, SW. Pacific coasts (Ellice Islands), Africa and coasts of both Americas.

**6. Xenococcus elenkinii** Pohribniak 1937 (fig. **561**)

Cells solitary, in sparsely dispersed groups or forming one-layered, brownish violet or reddish-violet colonies on the substrate. Young cells almost spherical, later oval, drop-like or pear-shaped, attached by narrower end to the substrate, sometimes densely packed and parallel to one another, polyhedral in outline from above, distinctly changing in size during the cell cycle, 3–11 × (1)2–5 µm. Baeocytes arise successively.
Occurrence: Marine, usually in coastal bays, epiphytic on filamentous algae, in the upper littoral zone; Ukraine, coast of Black Sea.

**7. Xenococcus pyriformis** Setchell et Gardner in Gardner 1918 (fig. **563**)

*Dermocarpa pyriformis* (Setchell et Gardner) Kaas 1985

Cells in small groups and colonies, flattened and in outline from above circular or rounded-polygonal, later heteropolar, slightly elongate, obovoid to pear-shaped or subspherical, bright blue-green, (10) 12–20 × 10–15 µm; in colonies more or less radially, fan-like or star-like arranged, joined to the substrate by the narrower end. Sheaths distinct, thin, colourless. Baeocytes arise from single cells after successive rapid divisions, numerous, 2.5–3.5 µm in diameter.
Occurrence: Marine, epiphytic on filamentous and thalloid algae; described from Pacific coast of N. America (Oregon), later recorded from various localities on

XENOCOCCUS

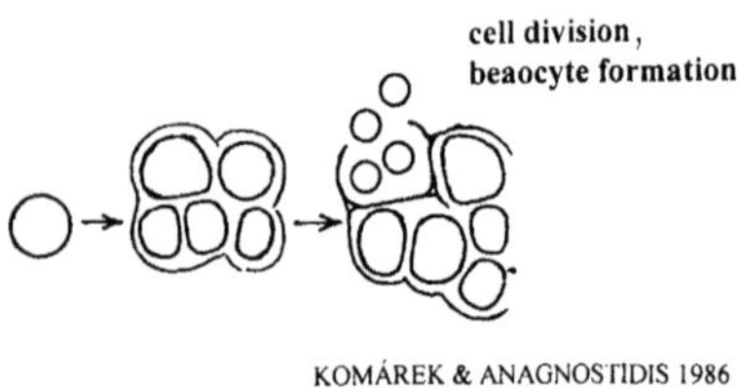


**Fig. 562.** *Xenococcus schousboei*: **a** – after Bornet from Geitler (1942), **b** – after Bornet from Geitler (1932), **c** – after Komárek (1956), **d** – after Kosinskaja (1948).
**Fig. 563.** *Xenococcus pyriformis*: after Setchell & Gardner from Kosinskaja (1948).

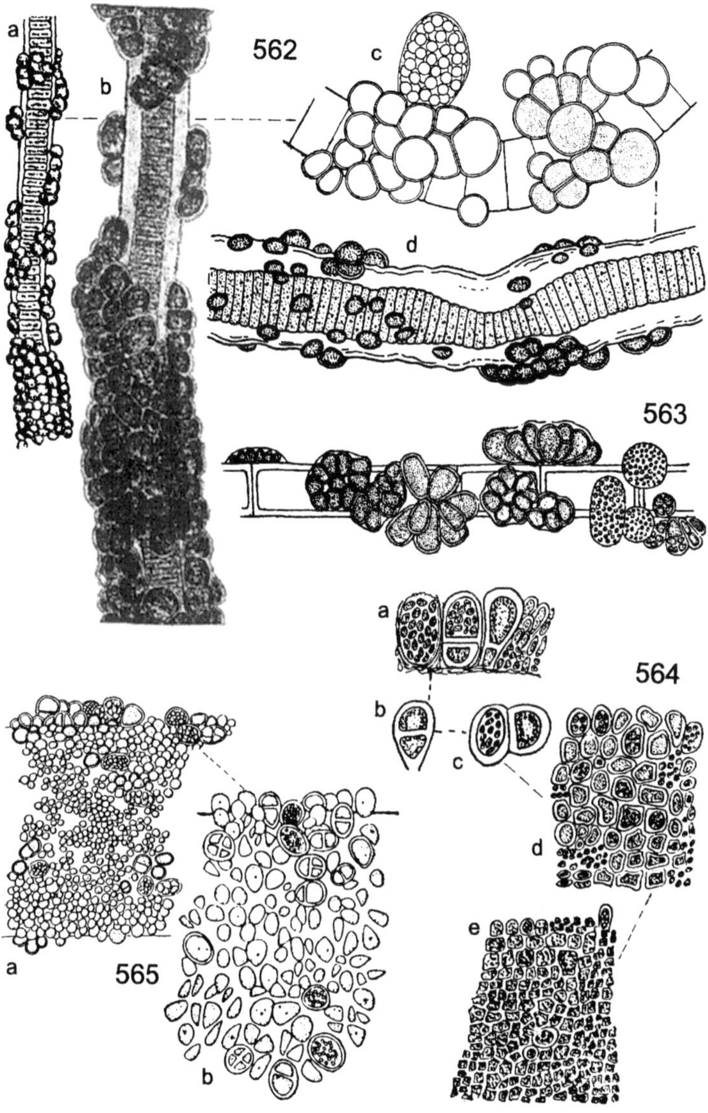


**Fig. 564.** *Xenococcus gaditanus*: **a-c** – detail of cells dividing into baeocytes, **d-e** – top views; after Gonzalez-Guerrero (1946).
**Fig. 565.** *Xenococcus cladophorae:* a – after Setchell & Gardner from Geitler (1932), **b** – after Umezaki (1956).

coasts of the whole northern hemisphere and from Chile, in Europe from Black Sea (Ukraine) and Baltic Sea (Kaas 1985); the identity of European records with American populations is not certain.

## 8. **Xenococcus gaditanus** Gonzalez Guerrero 1947 (fig. **564**)

Cells in pseudoparenchymatous monolayers on the substrate, irregular rounded, slightly vertically elongate when old, heteropolar, oval or club-shaped, 2–23 × 2–14 (20) µm, arranged irregularly in the colony. Sheaths thick, hyaline. Baeocytes arise from old elongate cells, from the whole mother cell or only from the upper part of the horizontally divided mother cell, spherical.
Occurrence: In saline biotopes near the sea coast, epizoic on shells of *Gasteropodum*; known from Spain (near Cádiz).

**Species outside Europe:**
- *Xenococcus alpinus* Emoto et Yoneda 1942: Epiphytic, one-layered colony, cells spherical or oval, 3–7 µm in diameter, with blue-green content, baeocytes ?; described from thermal springs in Japan.
- *Xenococcus angulatus* Setchell et Gardner 1937: Marine, endophytic, growing in the wall layers of *Callymenia angustata*, California; cells angular in outline, sometimes slightly elongate; baeocyte production not observed.
- *Xenococcus bicudoi* Montejano et al. 1993: Freshwater, epiphytic on filamentous algae in streams, known from tropical and subtropical regions in S. and central America.
– *Xenococcus candelariae* Tavera et Komárek 1996: Described from inland slightly salinic, volcanic lake in central Mexico, with reddish cells, epiphytic on filamentous algae.
- *Xenococcus chaetomorphae* Setchell et Gardner in Gardner 1918: Marine, epiphytic on algae, described from California; cells pale blue-green, polygonal spherical or distinctly elongate, to 45 µm long.
- *Xenococcus cladophorae* (Tilden) Setchell et Gardner 1918: Marine, epiphytic, from coast of British Colombia, later found in Antilles, Chile, India, Japan, Russia (Sea of Japan), with some doubt also recorded for Greece; cells pale blue-green, 8–22 µm in diameter; (fig. **565**).
- *Xenococcus deformans* Setchell et Gardner 1924: Marine, epiphytic on *Gelidium*, described from N. America (Baja California); cells bright blue-green, 9–14 µm in diameter.
- *Xenococcus endophyticus* Setchell et Gardner 1937: Marine, endophytic in walls of seaweeds, Galapagos Islands; cells more or less lenticular, in groups of 4–12, cells blue-green, 14–20 × 8–12 µm, dividing in two planes (!), baeocyte formation not observed.
- *Xenococcus gilkeyae* Setchell et Gardner in Gardner 1918 [syn.: *Dermocarpella gilkeyae* (Setchell et Gardner) Komárek et Anagnostidis 1986]: Cells spherical, solitary, 4–9 µm in diameter, pale blue-green, with binary fission and *Chamaecalyx*-like baeocytic division; marine, described from Alaska, recorded also from coasts of Kamchatka (Russia); (fig. **567**).
– *Xenococcus lamellosus* Gold-Morgan et al. 1994: Epiphytic on cyanoprokaryotes and algae from flowing water, described from central Mexico; (fig. **568**).
- *Xenococcus laysanensis* Lemmermann 1905: Marine, similar to *Chamaesiphon* (subg. *Chamaesiphonopsis*), described from Chatham Island; belongs to small species without baeocytes; (fig. **566**).
- *Xenococcus lyngbyeae* Jao 1939: Marine, described from China, maybe *Cyanocystis.*
- *Xenococcus pulcher* Hollenberg 1939: Marine, epiphytic, described from Cal-

ifornia; cells with reddish content, to 25 (30) µm, baeocytes 2–2.5 µm in diameter.
– *Xenococcus rivularis* sensu Frémy 1930: Epiphytic on *Plectonema wollei* in rivers in Gabon; ecologically and morphologically distinctly different from the European epilithic *X. rivularis* (cells 1.5–2.5 µm in diameter); (fig. **569**).
- *Xenococcus violaceus* Anand 1937: Marine, epiphytic on seaweeds, described from India, colonies violet, cells subspherical, 8–15 µm in diameter, with pale red content, baeocytes 1.5–2 µm in diameter.
- *Xenococcus willei* Gardner 1927: Freshwater, epiphytic on filamentous algae in springs and streams, known from India, Japan, Mexico and Porto Rico; (fig. **570**).
– *Xenococcus yellowstonensis* Copeland 1936: Freshwater, thermophilic, among other cyanoprokaryotes in thermal springs (51–62 °C) in Yellowstone, USA.
- *Xenococcus yonedae* Umezaki et M. Watanabe 1994 [syn.: *Xenococcus violaceus* Yoneda 1942]: Freshwater, epiphytic on *Chaetomorpha* in streams, violet cell content; Japan (Kyoto – Botanical garden).

**Unrevised species:**
– *Dermocarpa enteromorphae* Anand 1937: From India, maybe *Cyanocystis* (?).
– *Dermocarpa flahaultii* sensu Starmach 1928: Described from Poland, marginal cells of irregular colonies sometimes divide into numerous small, spherical baeocytes (comp. also *Pleurocapsa*; Geitler 1932, p. 404).
– *Dermocarpa violacea* Crouan sensu J. et G. Feldmann 1953: Epiphytic on *Laminaria* near Roscoff, France; probably a separate *Xenococcus* or *Pleurocapsa* species; (fig. **623**).
– *Entophysalis conferta* sensu Umezaki 1961: From marine biotopes in Japan; it belongs to *Xenococcus* according to drawings; (fig. **572**).
– *Xenococcus acervatus* Setchell et Gardner in Gardner 1918: Cells epiphytic, dividing in two planes perpendicular to the substrate, colonies at first one-layered, later heaped up; cells angular, spherical or pear-shaped, (2.1)3–6 µm in diameter (up to 8.2 µm long), protoplasts homogeneous, pale blue-green; baeocytes unknown. Marine, epiphytic on algae and cyanoprokaryotes in salt marsh pools, described from San Francisco Bay, California, later recorded from Mediterranean Sea (Banyuls), Turkey (Hekin Island), Canary Islands, Antilles, Brazil, Srí Lanka, Japan. Generic classification unclear (*Xenotholos* ?). The freshwater var. *dispersus* Emoto et Hirose 1942 evidently represents another species (? *Chamaesiphon* subg. *Chamaesiphonopsis*); (fig. **571**).
– *Xenococcus cladophorae* (Tilden) W. et G.S. West forma sensu Krieger 1944: Cells 4–8 µm, baeocytes 1–1.2 µm in diameter, epiphytic on freshwater *Cladophora*; described from Greece (Athos island).
– *Xenococcus concharum* Hansgirg 1899: Marine species, described from Adriatic Sea; insufficiently described, never recorded again nor revised.
– *Xenococcus irregularis* Gruia 1966: Aggregations of polyhedral cells, circa 5.2 × 4.8 µm in diameter, baeocytes unknown; described from soils in Roumania (Bucegi Mts.).
– *Xenococcus kerneri* sensu Emoto et Hirose 1942 and *X. schousboei* f. *thermalis* Emoto et Hirose 1942 from thermal waters in Japan represent probably separate species (substantial morphological and ecological differences); (fig. **573**).
– *Xenococcus rivularis* (Hansgirg) Geitler 1925: Freshwater, probably belongs to the g. *Chamaesiphon*, subg. *Chamaesiphonopsis*.
– Three *Xenococcus* strains of marine origin without taxonomic identification were studied by Waterbury & Stanier (1978).

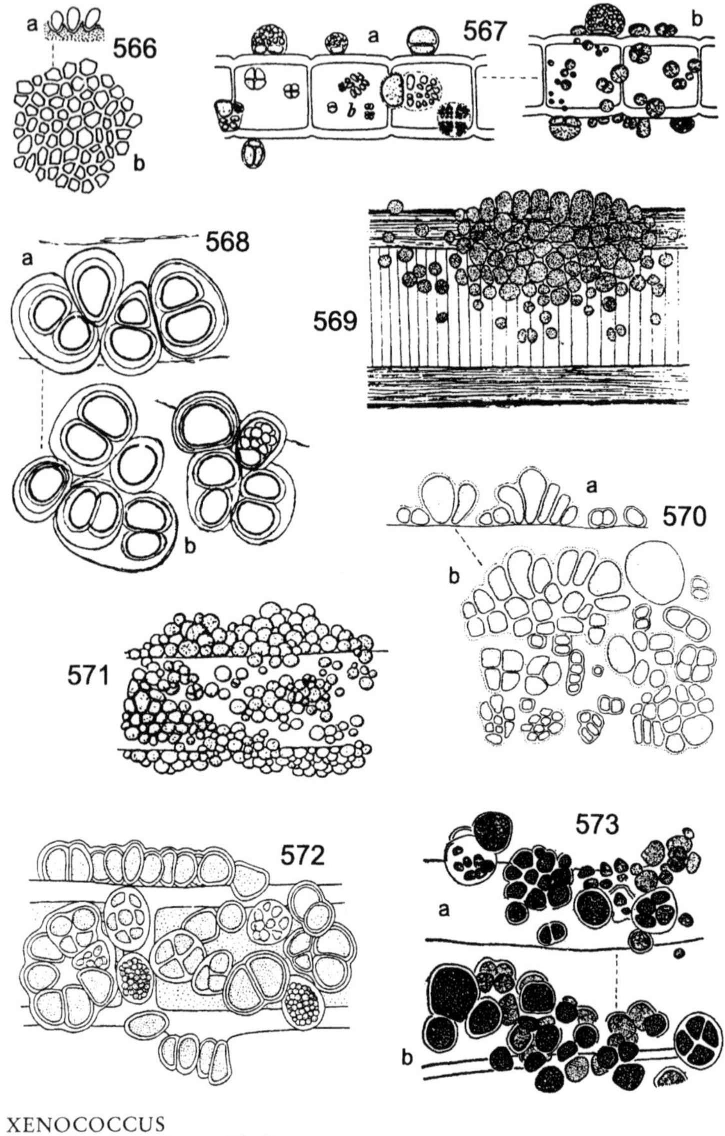


XENOCOCCUS

Examples of species outside Europe:

**Fig. 566.** *Xenococcus laysanensis*, small species without baeocyte formation (? *Chamaesiphon* subg. *Chamaesiphonopsis*):**a** – side view, **b** – top view; after Lemmermann from Geitler (1932).

## 5. Chroococcopsis Geitler 1925

Cells solitary or in compact groups, composed of ensheathed cells or of small, few-celled clusters, surrounded by common firm envelope; cells sometimes arranged in short irregular and indistinct rows, in larger colonies radially, with distinctly larger cells at the colonial margin. Sheaths firm, thin or slightly thickened, colourless, occasionally layered. Cells spherical, subspherical, oval or irregular, sometimes indistinctly club-shaped, pyriform and polarized (attached to the substrate), in older colonies radially situated; different size of cells in a single colony is a characteristic feature.

Cells divide by binary fission into two daughter cells, or later (particularly enlarged cells at the colonial margin) by successive multiple fission in three or more planes, producing numerous baeocytes. Transitions exist between both mentioned ways of cell division. Liberation of cells from common sheath usually by gelatinizing of envelopes.

**Important references:** Geitler (1925, 1932, 1980); Geitler & Ruttner (1935); Kondrateva & al. (1984).

**Key to the species:**

1a Vegetative cells 2.5–6.5 μm in diameter, marginal cells before baeocyte formation to 11.5 μm in diameter; epiphytic species ....... **1. C. epiphytica**

1b Vegetative cells (4)6–10(20) μm in diameter, in marginal parts (particularly before the baeocyte formation) to 36 μm in diameter ............... 2

2a Almost all cells slightly elongate and more and less obovate, 10–17 × 8.5–10 μm ............................. **2. C. chroococcoides**

2b Vegetative cells usually spherical, subspherical, polygonal-rounded, only towards the margin slightly elongate and widely rounded, 4–25–36 μm in diameter ........................................... 3

3a Cells to 10 (–25, dividing into baeocytes) μm in diameter, with blue-green content ..................................... **3. C. fluviatilis**

3b Cells (undivided) to 36 μm in diameter, bright blue-green, dirty olive-green or violet .................................. **4. C. gigantea**

---

**Fig. 567.** *Xenococcus gilkeyae*: after Setchell & Gardner from Geitler (1932).
**Fig. 568.** *Xenococcus lamellosus*, freshwater species from central Mexico; after Gold-Morgan & al. (1994).
**Fig. 569.** *Xenococcus rivularis* sensu Frémy, epiphytic, from streams in central Africa; after Frémy (1930).
**Fig. 570.** *Xenococcus willei*, freshwater, epiphytic, from Porto Rico: **a** – side view, **b** – top view; after Gardner (1927).

Unrevised species:
**Fig. 571.** *Xenococcus acervatus*, marine species: after Setchell & Gardner from Geitler (1932).
**Fig. 572.** *Xenococcus* sp., marine species from Japan: after Umezaki (1961, sub *Entophysalis conferta*).
**Fig. 573.** *Xenococcus* sp., thermophilic species from Japan: after Emoto & Hirose (1942, sub *Xenococcus schousboei* f. *thermalis*).

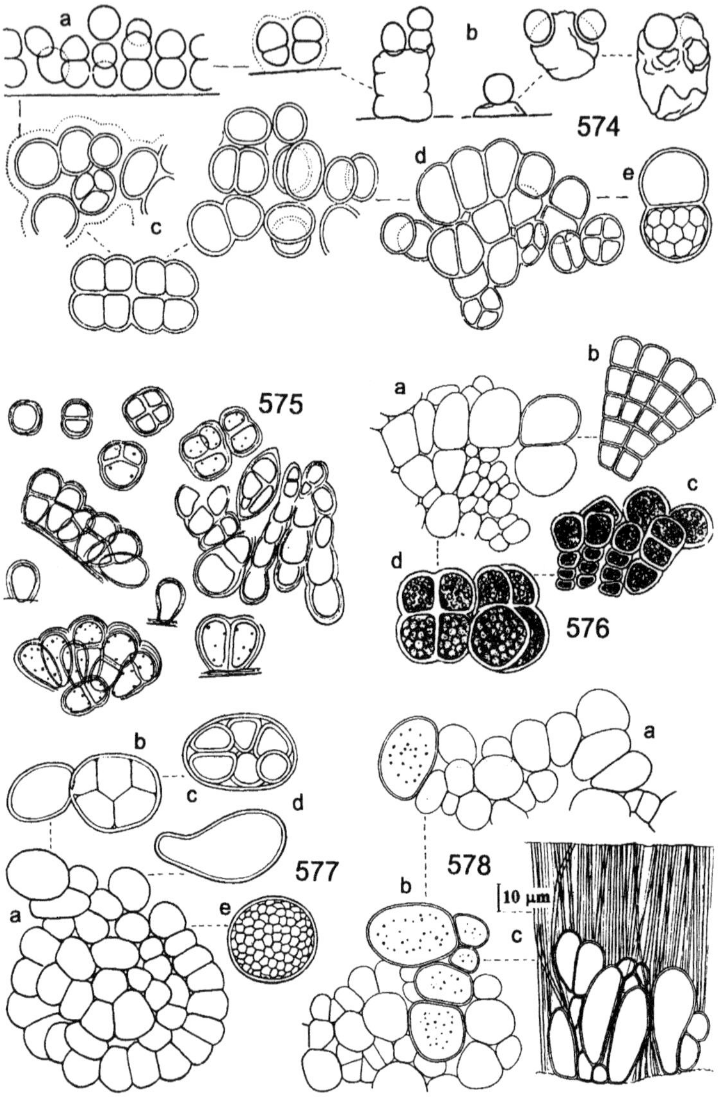


CHROOCOCCOPSIS

**Fig. 574.** *Chroococcopsis epiphytica*: after Geitler (1980).

**Fig. 575.** *Chroococcopsis chroococcoides:* a – after Fritsch from Geitler (1932, sub *Xenococcus chroococcoides*).

**Fig. 576.** *Chroococcopsis fluviatilis*: **a** – after Geitler (1932), **b** – after Hansgirg from

## 1. **Chroococcopsis epiphytica** Geitler 1980 (fig. **574**)

Cells solitary or in irregular groups, sparsely or densely aggregated, sometimes forming clusters of 2–4–8 or more cells, or chroococcoid conglomerations; old colonies more or less flat, 1–3-layered, with cells arranged sometimes in irregular rows. Cells more or less spherical or subspherical, in older colonies sometimes flattened laterally, blue-green, (2.5)4–6.5 (11.5) µm in diameter. Sheaths (envelopes) colourless, firm, thin, enveloping solitary cells or groups of cells, diffluent at the margin. Baeocytes sometimes arise from marginal enlarged cells, in number up to 32, 1.6–1.8 µm in diameter.
Occurrence: Freshwater, epiphytic on *Rhizoclonium* in stagnant waters; Austria.

## 2. **Chroococcopsis chroococcoides** (Fritsch) Komárek et Anagnostidis 1995 (fig. **575**)

*Xenococcus chroococcoides* Fritsch 1929

Cells solitary, spherical or slightly club-shaped or pear- shaped, attached to the substrate, later forming groups of 2–4 or more cells, or short rows and (later) two-layered, flat, microscopic colonies. Sheaths thick, layered, colourless. Cells blue-green, 10–17 × 8.5–10 µm.
Occurrence: Freshwater, in small streams and springs, epiphytic on green algae (*Cladophora*) and higher plants; England, Ukraine, recorded also from central and northern Asia.

## 3. **Chroococcopsis fluviatilis** (Lagerheim) Komárek et Anagnostidis 1995 (fig. **576**)

*Pleurocapsa fluviatilis* Lagerheim 1888; *Xenococcus fluviatilis* (Lagerheim) Geitler 1925

Cells rarely solitary, more or less spherical or rounded-polygonal, later usually in dense irregular groups and flattened at sides, usually aggregated in small, flat, hemispherical or wart-like colonies which are hard, later sometimes crustose, blackish, blue or dark brown; cells in colonies sometimes in short (mainly radially oriented) rows. Sheaths thick, colourless. Cells with blue-green content, 4–10 (20) µm in diameter, larger and rounded at the colonial margin. Baeocytes arise from spherical or slightly elongate cells, to 25 µm in diameter, at the end of «rows»; number of baeocytes 16–32, baeocytes spherical, 2–3 µm in diameter. Probably there are several morphotypes.
Occurrence: In small, clear streams and rivers, or littoral of lakes, on stones, mosses, vascular plants, rarely on wooden substrates, most frequently in mountains; mainly in European northern areas, recorded from Belorussia, Russia (N. Caucasus, Volha river, W. Siberia), and also from USA (another morphotype (?).

---

Geitler (1932), **c-d** – after Lagerheim from Geitler (1942); (all sub *Pleurocapsa fluviatilis*).
**Fig. 577.** *Chroococcopsis gigantea*: **a** – colony, **b-e** – cells, partly dividing into baeocytes; after Geitler (1925).
**Fig. 578.** *Chroococcopsis gigantea*, population from Indonesia; after Geitler from Geitler & Ruttner (1935).

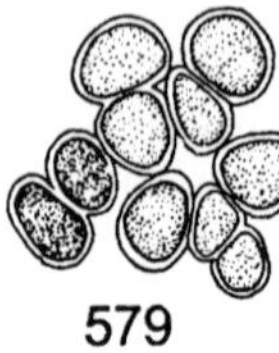


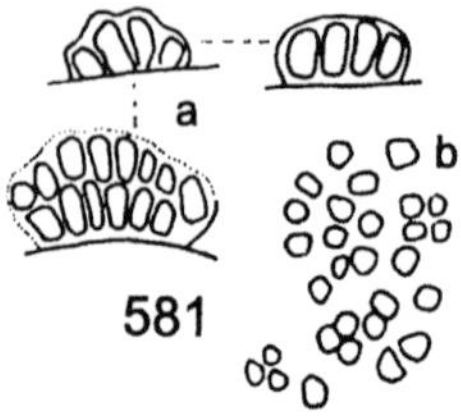


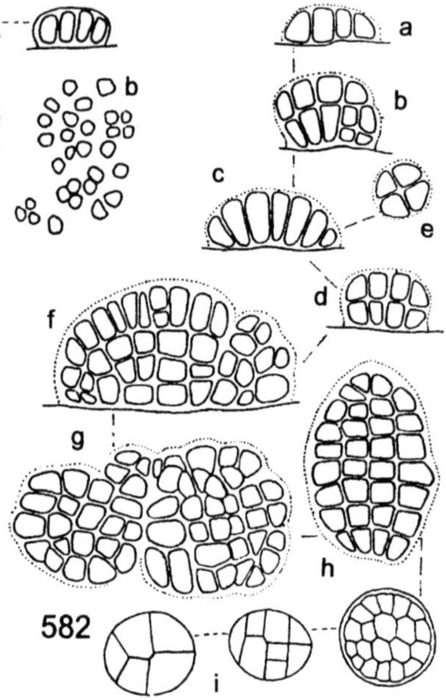


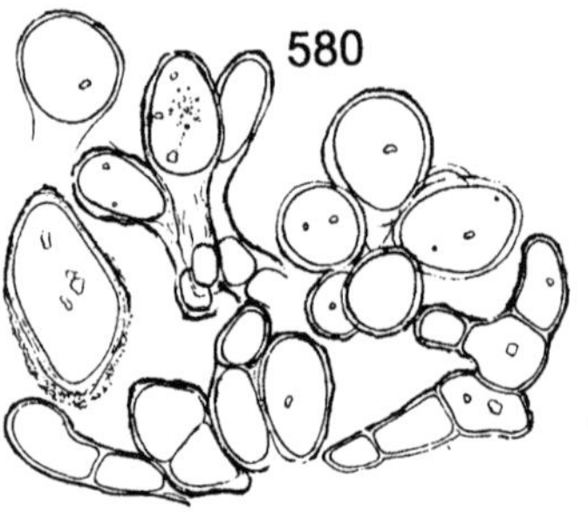


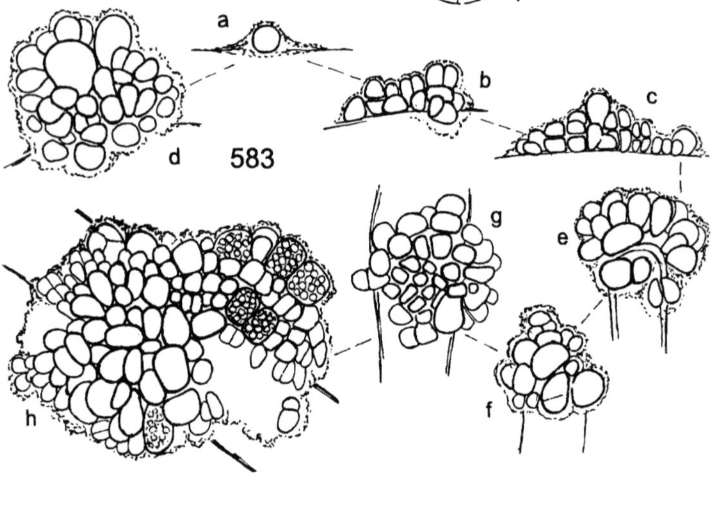


CHROOCOCCOPSIS
Unrevised species:

**Fig. 579.** *Chroococcopsis amethystea*: after Rosenvinge (1893, sub *Pleurocapsa amethystea*).

**Fig. 580.** *Chroococcopsis*? sp., from Cuba: after Komárek (1985, sub *Hyella* sp.).

### 4. Chroococcopsis gigantea Geitler 1925 (fig. **577, 578**)

Cells solitary or in irregular groups, densely packed. Sheaths thick, refractive, usually two-layered or sometimes lamellate. Cells spherical or elliptical, rarely irregular or rounded-polygonal in outline, 7–36 µm in diameter, bright blue-green, dirty olive-green or dark violet, rarely yellowish, with thick, colourless, firm sheath. Baeocytes in large number (–64 ?), 1.5–2.5 µm in diameter.
Occurrence: Freshwater, benthic on stones and bottom of shallow stagnant and flowing waters, on shells of molluscs, rarely epiphytic; Austria (Wien, Danube), Latvia, Ukraine, probably more widely distributed; records from tropical biotopes (Indonesia) need confirmation, from marine coasts (e.g. Chile) evidently concern other species. Similar to *C. fluviatilis*, possibly identical.

**Unrevised species:**
- *Chroococcopsis amethystea* (Rosenvinge) Geitler 1925 [syn.: *Pleurocapsa amythystea* Rosenvinge 1893]: Marine coasts, epiphytic, described from Greenland, known also from England, Iceland (var. *schmidtii* Collins 1911), France and Norway (Trondhjemfjord); with violet cells, insufficiently described (fig. **579**).
- Probably more species exist in tropical regions (see, e.g., «*Hyella* sp. population Rio Maximo» sensu Komárek 1985 from Cuba; fig. **580**).

## 6. Xenotholos Gold-Morgan et al. 1994

Colonial, thallus attached to the substrate (usually growing submersed, epiphytic on algae and waterplants, rarely epilithic); sessile, solitary cells occur only in the beginning of the life cycle, later, after repeated cell division, arise (1) 2-layered or multilayered, thick, usually rounded, hemispherical or irregularly hemispherical thalli, in which the cells are irregularly aggregated or organized more or less into short radial rows; colonies enveloped by thin, firm, colourless sheaths. Cells hemispherical or polygonal-rounded, sometimes shortly but distinctly elongate, occasionally with their own thin firm sheaths.
Cells divide by binary fission in various planes forming spatially layered aggregations. Occasionally some marginal or surface cells divide into baeocytes. Reproduction by baeocytes which escape from split sheaths, and by solitary cells (ensheathed ?).

XENOTHOLOS

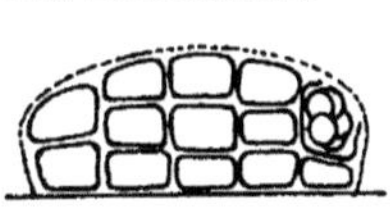


**Fig. 581.** *Xenotholos starmachii*: **a** – side views, **b** – top view; after Starmach from Geitler (1932, sub *Xenococcus minimus* var. *starmachii*).
**Fig. 582.** *Xenotholos kerneri*: **a-d** – side views of young colonies, **e** – top view of initial dividing cell, **f** – side view of a large colony, **g-h** – top views of large colonies, **i** – top view of a cell dividing into baeocytes; from Geitler (1932).
**Fig. 583.** *Xenotholos huastecanus*, freshwater, epiphytic species from Mexico: **a** – young cell, **b-f** – side view of young colonies, **g** – top view of a young colony, **h** – old colony with cells dividing into baeocytes; after Gold-Morgan & al. (1994).

**Important references:** Geitler (1932, sub *Xenococcus*); Gold-Morgan & al. (1994).

**Key to the species:**

1a Colonies 1–2-(3)-layered, cells to 1.6 µm in diameter . . . . **1. X. starmachii**

1b Old colonies several-layered, more or less hemispherical, cells to 10 µm in diameter . . . . . . . . . . . . . . . . . . . . . . . . . . . . . . . . **2. X. kerneri**

## 1. Xenotholos starmachii (Geitler) Gold-Morgan et al. 1994 (fig. **581**)

*Xenococcus minimus* var. *starmachii* Geitler 1932

Cells slightly but distinctly elongate, oval, ovoid, solitary or in loose groups, initially bipolar, joined by one narrow end to the substrate, later sometimes parallel and very densely packed, forming one- or two-layered, slightly hemispherical (globose) colonies, with pale blue-green content, –1.6 × 0.9–1.4 µm, surrounded by thin, firm, colourless envelope.

Occurrence: Freshwater, epiphytic on filaments of green algae; known from Poland (High Tatra Mts.). The baeocyte division is not known, the generic identification has to be revised.

## 2. Xenotholos kerneri (Hansgirg) Gold-Morgan et al. 1994 (fig. **582**)

*Xenococcus kerneri* Hansgirg 1893; *Dermocarpa kerneri* (Hansgirg) Bourrelly 1970

Cells rarely solitary (initial stages), later in groups of 4 or 8 cells and soon forming one-layered, later two or more layered (up to 6-layered) blastoparenchymatous to hemispherical or crustose, confluent spots on the substrate with marginal growth, occasionally in short, irregular rows within colony; cells divide commonly both vertically and horizontally to the substrate (origin of several layers in colonies), later are densely packed, polygonal, at the margin with the outside rounded shape; all cells of more or less similar size. Envelopes thick, hyaline or sometimes lamellate, confluent in colonies, usually colourless, rarely yellowish. Cells –10 × 3.5–6 µm, blue-green. Baeocytes arise occasionally from single marginal or superficial cells, rarely from isolated cells, by successive division, up to 32 per mother cell, circa 3 µm in diameter.

Occurrence: Freshwater, epiphytic on mosses and algae (*Cladophora, Lemanea,* etc.) and epilithic on stones in clear, mountain streams, waterfalls and rivers, locally frequent; probably in all European mountains, but distribution and ecology still not well known. There exist records of this species from tropical countries (Indonesia, Mexico, Mozambique), from Antarctica, N. America, New Zealand and even from marine biotopes (Bering Sea); the identity of all these populations is improbable.

**Species outside Europe:**

– *Xenotholos amplus* Gold-Morgan et al. 1994: Epiphytic on algae, known from central (mountainous) Mexico.

– *Xenotholos caeruleus* Gold-Morgan et al. 1994: Epiphytic on algae, known from central (mountainous) Mexico.

– *Xenotholos huastecanus* Gold-Morgan et al. 1994: Type-species of the genus; epiphytic on algae in streams with high carbonate content, described from San Luis Potosí, Mexico; (fig. **583**).

– *Xenotholos luteo-violaceus* M. Watanabe et Komárek 1999: With yellow-brown sheaths, epiphytic on cyanoprokaryotes, algae and mosses in cold katharobic waters in Himalaya Mts., E. Nepal.

Notes to the *Xenococcaceae*: The insufficiently described genus **Epilithia** Ercegović 1932 is usually classified also into this family: Thallus epilithic, pulverous, thin, irregular, with densely packed cells, sometimes gloeocapsoid, later with cells arranged in very short, pseudodichotomously divaricate, parallel and densely situated rows, oriented perpendicularly to the substrate. Cells heteropolar, rounded at the base, towards the apex ellipsoidally or almost cylindrically elongate, more or less enclosed in sheaths. Sheaths more or less of the same thickness, often open at the apex. – Cell division irregular and in various planes, occasionally the basal cells divide into baeocytes; cells dividing into baeocytes are of the same form and size as the vegetative cells. – The single species, *E. adriatica* Ercegović 1932, is marine, described from coastal rocks in Croatia (Dalmatia), and later recorded from France, but still taxonomically unclear (fig. **584**).

## 11. Family Hyellaceae Borzì 1914

Thallus with physiologically unconnected cells (without pores, not forming true filaments), but distinctly diversified, more or less polarized, composed of usually heteropolar, simple or divaricate (sometimes much shortened) rows of cells, or forming pseudofilamentous colonies, on one end (rarely irregularly) with cells dividing into nanocytes. Pseudofilaments creeping on the substrate, later growing vertically and so erect from the substrate, or endolithic, boring into limestone rocks, shells of molluscs, etc.; growth of pseudofilaments often is apical. Cells of varying form, usually elongate, irregular rounded, more or less cylindrical, in apical parts usually widened, pear-shaped or club-shaped. Sheaths mainly firm, often lamellate. In the thallus cell dimorphism often occurs (morphologically distinct apical cells).

Cell division by binary fission in one plane perpendicular to the pseudofilament axis, facultatively in other planes; rapid, simultaneous or successive division into baeocytes occurs facultatively in terminal or basal (rarely also in intercalar) cells; baeocytes are liberated from firm envelopes after their gelatinization or rupture. Baeocyte arise in vertically growing pseudofilaments usually from terminal cells, in endolithic (boring) and endophytic species from oldest «basal» cells near the surface of bored stones or host plants.

**Key to the subfamilies:**

1a Pseudofilaments very short, composed of only a few cells, the apical cells greatly widened and dividing by crosswise and/or lengthwise binary fission . . . . . . . . . . . . . . . . . . . . . . . . . . . . . . . . . . . . . . . . **1. Podocapsoideae**

1b Pseudofilaments distinctly composed of rows of cells **2. Hyelloideae** (p. 447)

## 1. Subfamily Podocapsoideae (Ercegović) Komárek et Anagnostidis 1986

Pseudofilaments reduced, very short, heteropolar, with cells more or less ovoid, pyriform, rarely spherical or irregular, epi- or endolithic; cells divide by lengthwise or cross-wise binary fission (in distal parts usually lengthwise); the proximal cells (in relation to the substrate surface) divide intensely into baeocytes; sheaths lamellate.

**Key to the genera:**

1a Short pseudofilaments epilithic, growing more or less vertically (erect) from the substrate; baeocytes develop from «basal» or apical cells . . . . . . . . . **2**

584

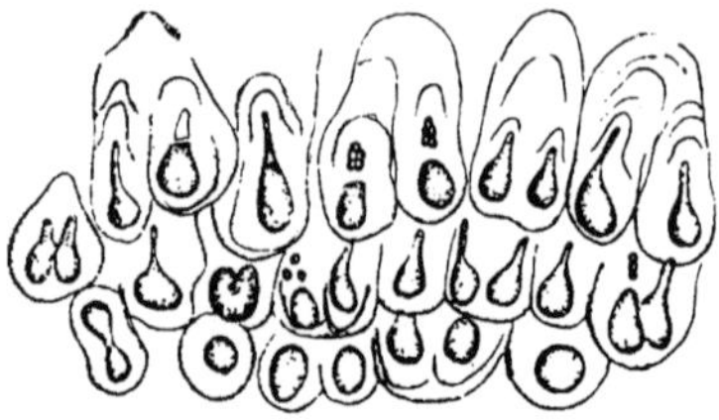
585

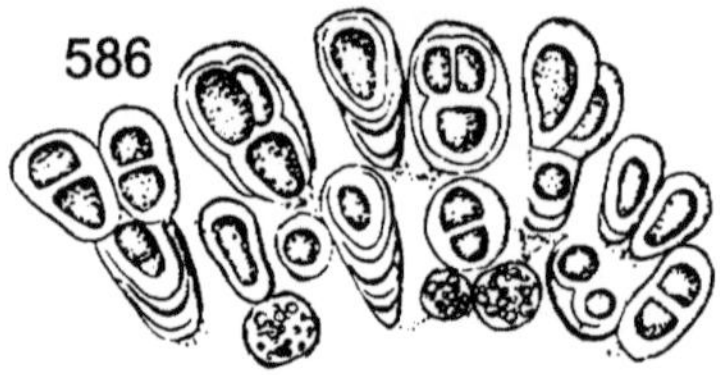
586

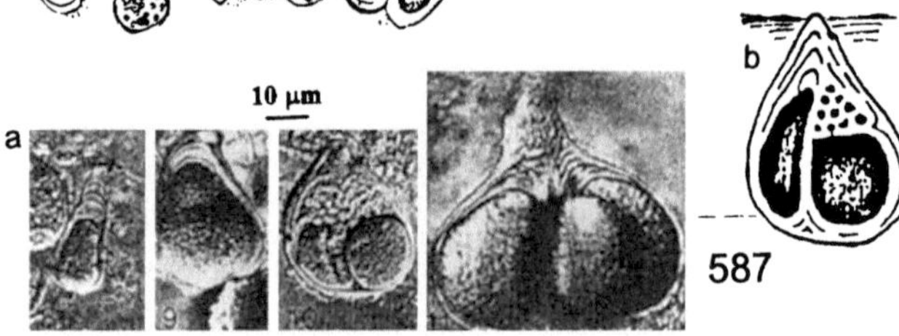

587

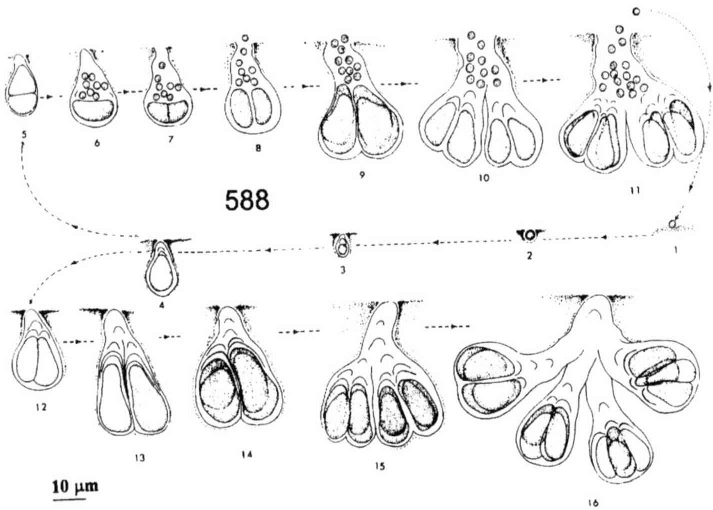

588

EPILITHIA
**Fig. 584.** *Epilithia adriatica*: after Ercegović (1932).

ERCEGOVICIA
**Fig. 585.** *Ercegovicia litoralis*: after Ercegović (1931).

**1b** Short pseudofilaments endolithic, terminating in a very distinct, large, pear-shaped apical cell; other cells (near the surface) are distinctly smaller, later dividing into baeocytes (= opposite structure to **1a**) . . . . . **3. Cyanosaccus**

**2a** Basal (wide) cells form spherical clusters, from which short pseudofilaments grow vertically; sheaths without calcium-carbonate precipitates; largest cells at the bases of pseudofilaments, the few superior cells distinctly smaller . . . . . . . . . . . . . . . . . . . . . . . . . . . . . . . . . . . . . . . . . . . **1. Ercegovicia**

**2b** Pseudofilaments without basal clusters of cells; sheaths with calcium-carbonate precipitates; largest cells at apices of pseudofilaments . . . **2. Podocapsa**

## 1. **Ercegovicia** de Toni 1936

*Brachynema* Ercegović 1931

Thallus epilithic, in the form of small, microscopic, slightly globular or formless colonies, consisting of basal groups of irregular-rounded, ellipsoidal or pear-shaped cells, which are irregularly arranged and mainly enveloped by stratified sheaths (gloeocapsoid stages); later very short, only few-celled pseudofilaments are formed, which grow parallel to one another and vertically from basal clusters, and are pseudodichotomously or pseudotetrachotomously diversified; sheaths around cells within pseudofilaments, or enveloping also the whole row of cells (pseudofilaments), sometimes stratified.
Cells in basal parts divide in various planes, apical cells sometimes divide into baeocytes. Still a problematic and not revised monotypic genus.
**Important references:** Ercegović (1931, sub *Brachynema*); Frémy (1934, sub *Brachynema*); de Toni (1936); Geitler (1942); Kosinskaja (1948).

### 1. **Ercegovicia litoralis** (Ercegović) de Toni 1936 (fig. **585**)

*Brachynema litorale* Ercegović 1931

Basal cells circa 4–6 µm in diameter, to 10 µm long (if elongate), with yellow-green or blue-green, homogeneous or granular content. Pseudofilaments composed of 2–4 (very rarely more) cells, to 25 µm long, narrowed to the apical (terminal) end.

---

PODOCAPSA
**Fig. 586.** *Podocapsa pedicellata*: after Ercegović (1931).

CYANOSACCUS

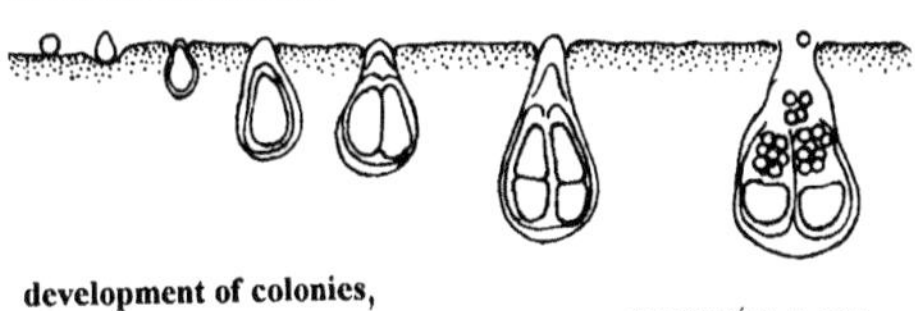


**Fig. 587.** *Cyanosaccus piriformis*: **a** – development of thallus, cell division and baeocyte formation, **b** – scheme of endolithic thallus; from Lukas & Golubić (1981).
**Fig. 588.** *Cyanosaccus aegeus*: development of an endolithic thallus, cell division, baeocyte formation and variation; after Anagnostidis & Pantazidou (1985).

Sheaths around cells lamellate, firm, yellow-brown. Baeocytes arise from apical cells in large number, 1–2 µm in diameter.
Occurrence: Marine, in littoral on calcareous rocks in splash zone; known only from Croatia (coast of Adriatic Sea in Dalmatia). Needs confirmation and revision.

## 2. **Podocapsa** Ercegović 1931

Thallus epilithic, granular, thin, irregularly covering the substrate, composed of more or less erect or aggregated, more or less parallel or irregular rows of several separate cells (up to 4–6), which are polarized, narrowed towards the basal end and wider in the upper part. «Terminal» cells irregular spherical, ellipsoidal, pear-shaped, drop-like or club-shaped; the short rows of cells sometimes pseudodichotomously divaricate, enveloped by thick, firm, layered and coloured sheaths. Cells divide in several planes, but with predominant transverse division, perpendicular to the long axis of the row of cells (origin of short pseudofilaments) or lengthwise; the divarication of rows is initiated by the lengthwise cell division (with respect to the axis of the pseudofilament). The smaller or narrower «basal» cells sometimes divide into baeocytes. Reproduction by baeocytes and probably also by small groups of cells, surrounded by gelatinous envelopes. Little known monotypic genus, which needs revision.
**Important references:** Ercegović (1931); Geitler (1932, 1942); Frémy (1934); Kosinskaja (1948); Anagnostidis & Pantazidou (1985, 1988a).

### 1. **Podocapsa pedicellata** Ercegović 1931 (fig. **586**)

Colonies mixed with other algae, epilithic, incrusted with calcium carbonate. Cells in pseudofilaments few, usually 1–4(6), pale blue-green with more or less homogeneous content, 8–22 × 1–16 µm, in basal parts sometimes narrowed and elongate («stipitate» = club-shaped form). Sheaths to 22 µm wide, colourless or yellow-brown, near the base diffluent, in the upper part firm, lamellate and more intensely coloured. Baeocytes develop from basal cells, usually 2–6 from one mother cell, 2–3 µm in diameter.
Occurrence: Marine, in the surf (splash) zone of the calcareous rocky littoral of the Mediterranean coasts in Croatia (Dalmatia).

## 3. **Cyanosaccus** Lukas et Golubić 1981

Thallus endolithic, penetrating into carbonate substrates, such as calcareous, marine littoral stones, rocks, ooid grains and mollusc shells, microscopic (to 85 µm long), sometimes simply finger-like branched, composed of polarized groups of cells (= shortened heteropolar pseudofilaments). Cells solitary, in pairs or in groups (clusters) with up to 12 cells, usually more or less ovoid, pear-shaped or club-shaped, sometimes pear-shaped with remarkably elongate pedicell, on both ends rounded, enclosed within ends of mucilaginous solitary, pseudodichotomous or finger-like branches of thallus, orientated by their narrowed ends to the surface of the upper substrate; the whole colony appears as a finger-like fascicle of pear-shaped cells immersed in the substrate. The gelatinous envelopes are firm, colourless or yellowish to yellow-brown, layered, narrowed and elongate toward the substrate surface, and provide the connection with the outer environment by a small opening. The divarication of the thallus is caused by the lengthwise division of endolithic cells and cell clusters, and by the consequent directional production of gelatinous stalks. Cell content of variable colour (greyish, bluish, olive-green, blue-green, red, pink), finely granular, without conspicuous inclusions.

Cell division at first horizontal, i.e. perpendicular to the vertical axis, later in further directions, by binary fission; the upper part divides after separation of the upper-outer part of a cell simultaneously by multiple fission or by successive (up to 3) binary fissions into many (to 32 or more) small, rounded baeocytes (to 5 µm in diameter), which escape from the sheaths into the environment, become attached to the limestone and penetrate again into the substrate. The «apical» large cells divide usually only lengthwise by binary fission, rarely both lengthwise and crosswise; (fig. **587**).
**Important references:** Lukas & Golubić (1981); Anagnostidis & Pantazidou (1985, 1988a); Pantazidou (1991 – diss.).

**Key to the species:**

1a Thallus finger-like radially divaricating into 2–4 pseudobranches; cells –27 × –14.5 µm, without separated basal (superficial) cells (pedicells) . **1. C. aegeus**
1b Thallus terminated by a pear-shaped formation, with only two nearly pseudodichotomously divaricate pseudobranches; basal pedicells sometimes remain near surface of bored substrate . . . . . . . . . . . . . . . . . . . . . . . . 2
2a Sheaths contain 1–12 cells each, cells –16.6 × –12 µm, pedicells present . . . . . . . . . . . . . . . . . . . . . . . . . . . . . . . . . . . . . . . . . . **2. C. atticus**
2b Sheaths contain 1–2 (4) cells each, cells –52 × –32 µm, pedicells rarely present . . . . . . . . . . . . . . . . . . . . . . . . . . . . . . . . . . . . . . . **3. C. piriformis**

## 1. Cyanosaccus aegeus Anagnostidis et Pantazidou 1985 (fig. **588**)

Thallus euendolithic, greyish to olive-green, boring into marine carbonate substrates, composed of 1–4 (6) pear-shaped clusters of short gelatinous finger-like (radially) divaricate sheaths, to 52 µm wide in the apical part (–28 µm in the basal part), containing several cells each, orientated by their widely rounded wider ends into the substrate (usually obliquely from the surface). The divarication is caused by separation of cell clusters through the production of gelatinous stalks after lengthwise cellular division. The number of cells in each envelope (branch) ranges from 1–4, rarely up to 8 (in culture only 4–8-celled clusters observed). Sheaths are firm, gelatinous, colourless, rarely yellowish, layered (usually with 1–2 layers) and stalked. Cells bluish, blue-green or olive-green, rarely blue-pinkish, usually pear-shaped, ellipsoid, rarely ovoid, spherical or subspherical, with finely granular content, 4.7–27 × 3.6–14.5 µm. In the upper part (near the stone surface) the cells differentiate by horizontal division, and later simultaneously by multiple fission divide into small spherical (to 32, rarely more) baeocytes, 1.5–2.5 µm in diameter. The large pear-shaped cells divide obliquely or longitudinally by binary fission. Pedicell lacking.
Occurrence: Marine, boring into carbonate rocks and mollusc shells (*Littorina*) in the upper intertidal zone (0.5–1 m above mean low water), usually in a community containing other boring cyanoprokaryotes; described from eastern Mediterranean Sea (Greece – Aegean Sea, Thira Island), known also from Attiki Peninsula and Paros Island; recorded recently from coasts of S. Africa.

## 2. Cyanosaccus atticus Anagnostidis et Pantazidou 1988 (fig. **589**)

Thallus euendolithic, greyish, bluish, yellowish to yellow-brown, boring into marine carbonate substrates, composed of 1–2 pear-shaped, ellipsoid to ovoid clusters of short, not or slightly dichotomously divaricate sheaths, –41 × –30 µm, containing few pear-shaped cells each (branch), orientated by their widely rounded wider ends into the substrate. The branching is caused by separation of

589

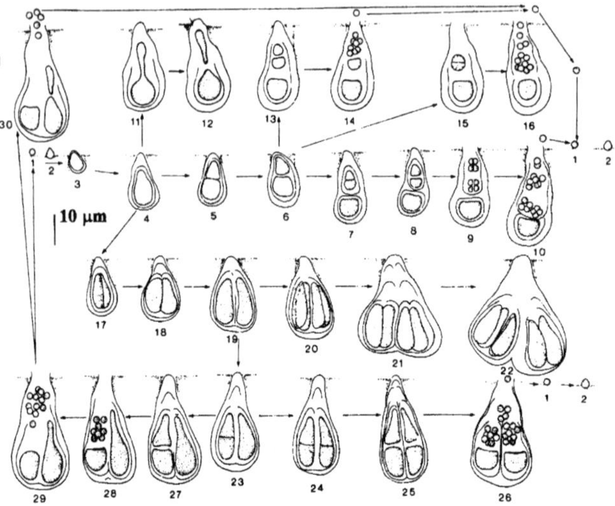


590

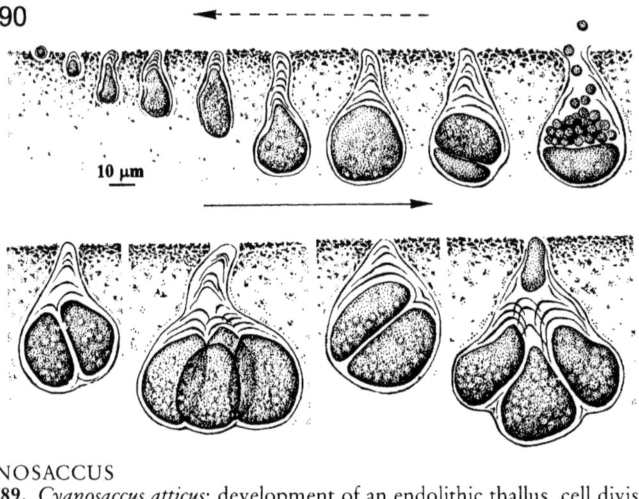


CYANOSACCUS

**Fig. 589.** *Cyanosaccus atticus*: development of an endolithic thallus, cell division, baeocyte formation and variation; after Anagnostidis & Pantazidou (1988).

**Fig. 590.** *Cyanosaccus piriformis*: development of an endolithic thallus, cell division, baeocyte liberation and variation; after Lukas & Golubić (1981).

cell clusters through directional production of gelatinous stalks after lengthwise cellular division. The number of cells in each envelop (branch) ranges from 1–2, occasionally 4 (in culture mostly 4–12-celled clusters). Sheaths firm, gelatinous, colourless to yellow-brown, layered and stalked. Cells blue-green or greyish, rarely pink, polymorphic, polarized, usually pear-shaped or club-shaped, flask-shaped, cylindrical, ovoid, ellipsoid, rarely subspherical or truncate at the proximal end, with finely granular content, 4.5–16.6 × 3.5–12 µm (in culture 5–9 × 4–7 µm, more or less without characteristic polarity); remarkably elongate narrow cells (pedicells, 3–7 × 1–3 µm) remain frequently near the surface of the substrate. In the upper part (closest to the substrate surface) the cells differentiate by transverse division, and later divide either simultaneously by multiple fission or by successive (up to 3) binary fissions into small spherical (to 32 per cell) baeocytes, 1.5–2.3 µm (in culture to 3 µm) in diameter. The large pear-shaped cells divide transversely, obliquely (?), or longitudinally by binary fission.
Occurrence: Marine, boring into carbonate rocks and mollusc shells in the intertidal zone (1.2 m above to 0.3 m below mean low water level), usually in a community of other boring cyanoprokaryotes; described from eastern Mediterranean Sea (Greece – Aegean Sea, Attiki Peninsula, known also from the islands Aegina, Paros and Sifnos); probably more widely distributed.

**3. Cyanosaccus piriformis** Lukas et Golubić 1981 (fig. **587, 590**)

Thallus euendolithic, blue-green to yellow-brown, pear-shaped, –85 × –70 µm, boring into marine carbonate substrates, composed of a short gelatinous pear-shaped, not divaricate sheath, containing usually 1–2, rarely 4 cells, orientated by their widely rounded wider ends into the substrate. Sheaths gelatinous, firm, colourless, stalked and layered (especially in the stalk). Cells bluish, blue-green to red, spherical, ovoid or pear-shaped when solitary, hemispherical and mutually flattened after cell division or when densely packed, with finely granular content, –53 × –43 µm. In the upper part (closest to the substrate surface) the cells differentiate by transverse division, and later divide by multiple fission into 32–64 spherical baeocytes, circa 4 µm in diameter. The large pear-shaped cells divide transversely, obliquely or longitudinally by binary fission.
Occurrence: Marine, boring into mollusc shells along the E. Florida continental margin, in ooid grains in Bahama Islands and in carbonate rocks, Bermuda Islands, from above 0.5 m down to 75 m; later found (with somewhat smaller dimensions) also in calcareous marine substrates in N. Aegean Sea, Greece (Thasos Island); reported also from Arabian/Persian Sea, S. Africa and SW. Pacific coasts (Ellice Islands).

## 2. Subfamily **Hyelloideae**

*Pleurocapsaceae* Geitler 1925 incl.

Pseudofilamentous, heteropolar thallus; cells more or less irregular, dividing mainly perpendicularly to the long axis of the pseudofilament; pseudofilaments simple, uniseriate or multiseriate and divaricate. Cells, dividing into baeocytes, situated on the proximal or distal end of pseudofilaments.

**Key to the genera:**

**1a** Pseudofilaments grow from the basal flat, discoidal colony or are erect, perpendicular to the substrate; the cells dividing into baeocytes develop in the apical parts of pseudofilaments . . . . . . . . . . . . . . . . . . . . . . . . . . **2**

**1b** Pseudofilaments (sometimes only rows of cells within mucilaginous strands) creeping over the substrate or growing into the substrate (endophytically or endolithically); cells dividing into baeocytes develop from the «oldest» («basal»), lateral or from the apical cells . . . . . . . . . . . . . . . . . . . . . . . . 4

2a Erect pseudofilaments parallel and unified into a gelatinous mass (their sheaths are confluent), so forming an one-layered thallus . . . . **1. Radaisia**

2b Erect pseudofilaments separate from one another . . . . . . . . . . . . . . . . . 3

3a Pseudofilaments never divaricate, of the same width from the base to the apex, apical cells not widened . . . . . . . . . . . . . . . . . . . . **2. Pascherinema**

3b Pseudofilaments simple or divaricate, with widened apical cells, from which baeocytes differentiate . . . . . . . . . . . . . . . . . . . . . . . . . . . **3. Cyanoderma**

4a Pseudofilaments composed of uniseriate or multiseriate rows of cells, which are distinctly distant from one another, enclosed within mucilaginous, sometimes stratified strands, creeping on the substrate and later boring into the substrate . . . . . . . . . . . . . . . . . . . . . . . . . . . . . . . . . . . . . . . . . . . . 5

4b Pseudofilaments with «joined» cells, without distinct spaces between them, but cells sometimes form irregular clusters; creeping on or boring into the substrate . . . . . . . . . . . . . . . . . . . . . . . . . . . . . . . . . . . . . . . . . . . . 6

5a Boring pseudofilaments uniseriate, baeocytes arise from terminal, lateral or the oldest cells in thallus . . . . . . . . . . . . . . . . . . . . . . . . . . . **4. Solentia**

5b Boring pseudofilaments usually multiseriate, baeocytes arise from oldest, more or less epilithic parts of thallus . . . . . . . . . . . . . . . . . . **5. Dalmatella**

6a Pseudofilaments irregular, more or less short, often only in the form of irregular groups of cells, only at the ends (on the colonial margin) pseudodichotomously divaricate, creeping on the substrate; cells forming baeocytes develop irregularly in old parts of thallus . . . . . . . . . . . . . . **6. Pleurocapsa**

6b Pseudofilaments distinctly uni- or multiseriate, simple or divaricate, creeping on the substrate and later usually growing into the substrate (endolithic pseudofilaments mainly uniseriate); cells dividing into baeocytes always develop from the initial (oldest) cells near the substrate surface . . **7. Hyella**

## 1. **Radaisia** Sauvageau 1895

Thallus at first in the form of a flat, discoidal or irregular, nemato- or blastoparenchymatous colony (these stages rarely lacking), attached to the substrate, from which more or less erect, short, parallely arranged, straight or slightly curved, simple or sometimes pseudodichotomously divaricate, uniseriate or multiseriate rows of cells (pseudofilaments) arise. Mucilaginous sheaths around pseudofilaments confluent, forming later homogeneous mucilage; old colonies in the form of a narrow, flat, gelatinous or crustose layer with parallely arranged pseudofilaments, oriented perpendicularly to the substrate. Cells more or less irregular-polyhedral-rounded, at the ends of pseudofilaments enlarged; the apical cells, dividing later into baeocytes, are therefore larger than vegetative cells and develop terminally or subterminally, on the top of pseudofilaments, i.e., near the surface of flat colonies.

Cells divide within thallus irregularly, in pseudofilaments predominantly crosswise in one plane in successive generations. Baeocytes differentiate by simultaneous or successive cell divisions of apical cells. Reproduction by baeocytes, many per cell, and later liberated from the colony. From about ten described species, several are satisfactorily characterized, however, only one (the type species) is known from Europe.

**Important references:** Sauvageau (1895); Gardner (1918, 1927); Setchell & Gardner (1919); Geitler (1932, 1942); Frémy (1934).

### 1. **Radaisia gomontiana** Sauvageau 1895 (fig. **591**)

Colonies small, dot-shaped, epiphytic, flattened to hemispherical, violet. Pseudofilaments 50–100 µm long or more. Cells more or less isodiametric, rounded, 4.5–7 µm in diameter, with violet content. Cells dividing into baeocytes are terminal or (rarely) intercalary, enlarged, oval, ovoid or elongate and rounded, 8–10 µm wide and to 46 µm long. Baeocytes arise by successive cell division, and are 0.8–0.9 µm in diameter.
Occurrence: Marine, on *Fucus* in Atlantic; known from the W. coast of France (Biarritz) and Portugal (Viana).

**Species outside Europe:**
- *Radaisia clavata* Setchell et Gardner in Gardner 1918: Thallus to 100 µm high, pseudofilaments 70–100 µm long, cells 4–9 µm in diameter, baeocytes 1–1.5 µm in diameter; marine, epiphytic, described from California, USA.
- *Radaisia confluens* Gardner 1927: Thallus 50–60 µm high, pseudofilaments with 8–13 cells, cells 2.5–3.2 µm in diameter, with purplish violet content; freshwater, on wood and stones in fountains, brooks and streams, Porto Rico.
- *Radaisia epiphytica* Setchell et Gardner in Gardner 1918: Thallus 50–60 µm high, pseudofilaments composed of groups of angular cells, 3–5 µm, baeocytes 1.8–2.4 µm in diameter; marine, epiphytic, from the coast of California, USA; (fig. **592**).
- *Radaisia gardneri* Komárek et Anagnostidis 1995 [syn.: *Pleurocapsa epiphytica* Gardner 1927 = *Radaisia epiphytica* (Gardner) Gardner 1932]: Thallus 12–20 µm high, pseudofilaments few-celled, short, with cells 2.8–3.2 (6) µm, baeocytes 1–1.5 µm in diameter; freshwater, epiphytic, in stagnant water, Porto Rico; (fig. **593**).
- *Radaisia pusilla* Weber van Bosse 1926: Thallus to ± 40 µm in diameter, pseudofilaments 8–20µm long, cells 3.6–4 µm wide; marine, found to a depth of about 50 m, epiphytic on *Chaetomorpha aerea*, USA (Florida, Key Island).
- *Radaisia willei* Gardner 1927: Thallus 30–40 µm high, cells 3.5–5 µm in diameter, shorter than wide, purplish violet; freshwater, epilithic, described from stones in brooks, Porto Rico.
- *Pleurocapsa subgelatinosa* Geitler 1933: Freshwater, epiphytic on cyanoprokaryotes, algae and mosses; colonies composed of parallel arranged, to tetrachotomous divaricate pseudofilaments, enveloped by fine mucilage, in upper part with larger cells dividing into baeocytes (without typical pleurocapsoid divarication); described from Indonesia; (fig. **594**).

**Unrevised species:**
- *Pleurocapsa entophysaloides* Setchell et Gardner in Gardner 1918: Gelatinous colonies with more or less radially arranged rows of cells, to 200 µm in diameter, with mucilaginous, yellowish brown envelopes; cells more or less spherical or angular, 4–10 µm in diameter; vegetative cells sometimes divide into baeocytes, 3–3.5 µm in diameter; marine, described from the upper littoral belt, USA (California), and found also on the Atlantic coast of Brazil (RJ). The description corresponds better to *Radaisia* (comp. generic diacritical features).

## 2. **Pascherinema** de Toni 1936

*Endonema* Pascher 1928

Pseudofilaments of «connected» cells solitary, uniseriate, more or less cylindrical, heteropolar, more or less erect, without wide mucilaginous envelopes, attached by one end (with modified basal, usually ovoid cell with narrowed basal stipe) to the substrate. Cells in pseudofilaments more or less of the same size and shape,

rounded or barrel-shaped (causing the moniliform appearance of pseudofilaments), only the basal cell is modified and the terminal cell(s) slightly enlarged and usually elongate-rounded, but not exceeding distinctly the pseudofilament width. Around cells narrow, firm sheaths, visible only during cell division or after baeocyte liberation.
Cell division transversely to the pseudofilament axis, or (mainly from enlarged terminal cells) successively into few or several (2–16) oval or spherical, motile baeocytes. Reproduction by means of baeocytes. Little known genus, never found again for about 70 years, needs revision.
**Important references:** Pascher (1929); Geitler (1932, 1942); de Toni (1936); Villeret (1953); Starmach (1966); Bourrelly (1966, 1970, 1985).

**Key to the species:**
**1a** Cells obovoid, 2–3 µm wide . . . . . . . . . . . . . . . . . . . . . . . . . **1. P. gracile**
**1b** Cells barrel-shaped, 6–8 µm wide . . . . . . . . . . . . . . . . . . **2. P. moniliforme**

## 1. Pascherinema gracile (Pascher) de Toni 1936 (fig. 595)

*Endonema gracile* Pascher 1929

Pseudofilaments straight, later slightly wavy, sometimes with more than 20 cells; cells pale blue-green, obovoid, narrowed towards the pseudofilament base, 2–3 µm wide.
Occurrence: Freshwater, epiphytic on various filamentous cyanoprokaryotes (e.g., *Hapalosiphon*) in clear, more or less acidic swamps; described and still known only from one locality in N. Bohemia (Czech Republic).

## 2. Pascherinema moniliforme (Pascher) de Toni 1936 (fig. 596)

*Endonema moniliforme* Pascher 1929

Pseudofilaments slightly wavy, with up to 20 or more cells; cells very pale blue-green, barrel-shaped, 6–8 µm wide.
Occurrence: Freshwater, in moorland and acidic swamps, epiphytic on leaves of *Sphagnum* and remains of plants; found only a few times in N. and NW. Bohemia (Czech Republic).

---

RADAISIA

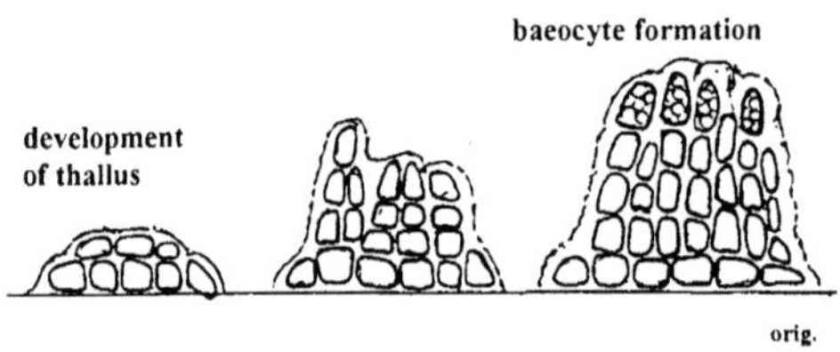


**Fig. 591.** *Radaisia gomontiana*: vertical section through the colony with apical large cells, dividing into baeocytes; after Sauvageau from Kosinskaja (1948).
**Fig. 592.** *Radaisia epiphytica*, marine epiphytic species from California: **a** – basal part of a thallus, **b** – thallus with terminal cells dividing into baeocytes; after Setchell & Gardner from Kosinskaja (1948).

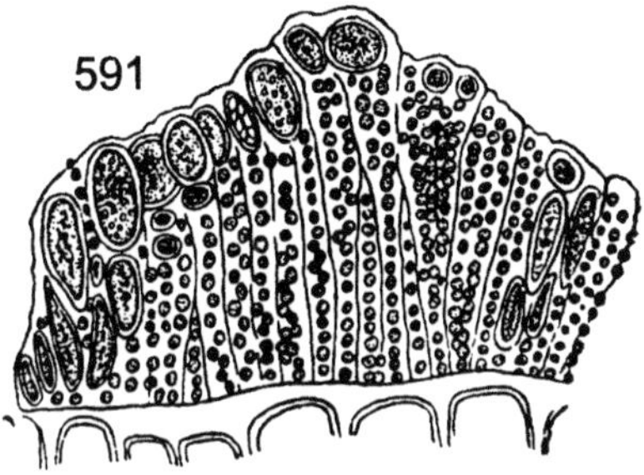


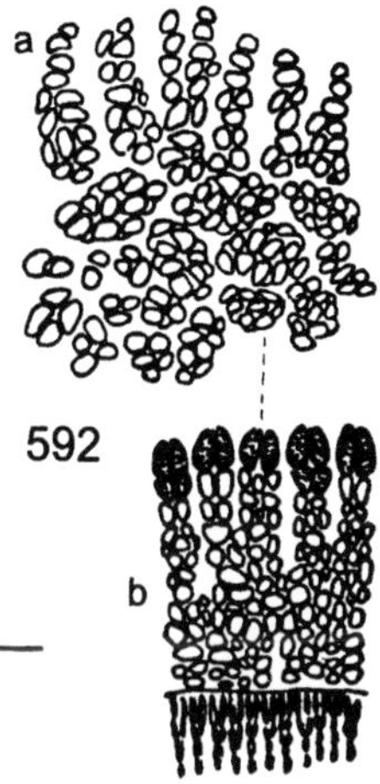


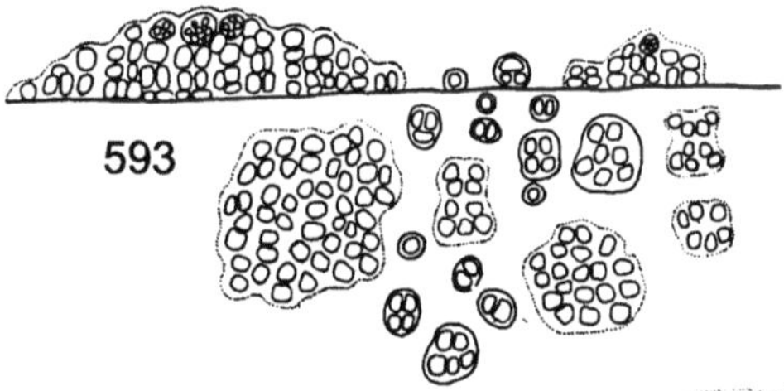


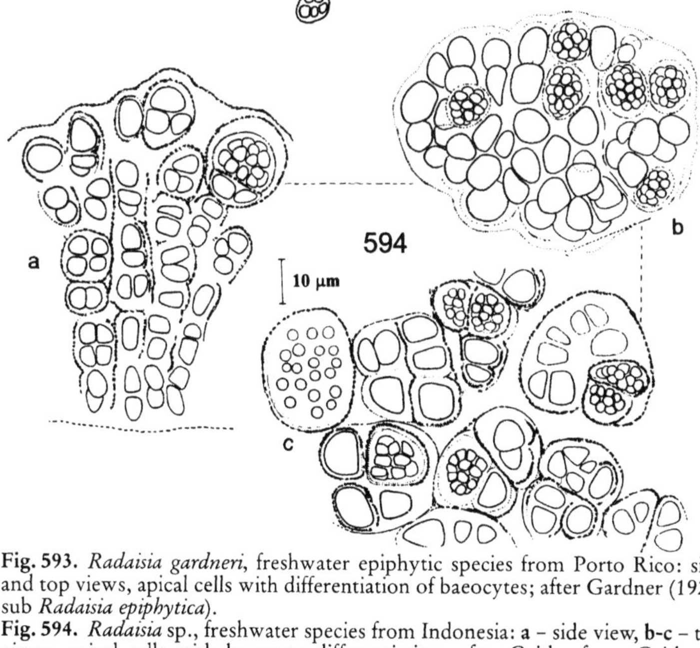


**Fig. 593.** *Radaisia gardneri*, freshwater epiphytic species from Porto Rico: side and top views, apical cells with differentiation of baeocytes; after Gardner (1927, sub *Radaisia epiphytica*).
**Fig. 594.** *Radaisia* sp., freshwater species from Indonesia: **a** – side view, **b-c** – top views, apical cells with baeocyte differentiation; after Geitler from Geitler & Ruttner (1935, sub *Pleurocapsa subgelatinosa*).

## 3. **Cyanoderma** Weber van Bosse 1887

*Nematoradaisia* Geitler 1925; *Radaisiella* Geitler 1925; *Geitleriella* de Toni 1936

The initial thallus forms the basal, usually discoid in outline, flat, blastoparenchymatous cell clusters (flat colonies), attached to the substrate, circular in outline when young, with irregularly and densely aggregated cells, or composed of short, radially creeping, pseudodichotomously divided rows of cells (these basal parts can be variously modified or almost completely reduced); from this initial thallus (colonies) the erect, simple or divaricate pseudofilaments arise, which are short, straight or slightly curved, parallel to one another, enveloped by firm sheaths, which are not confluent. Cells blue-green, irregular, in pseudofilaments sometimes elongate or isodiametric, widened at the ends, more or less spherical; cells dividing into baeocytes develop terminally and usually are larger than vegetative cells.

Cells in the basal discoid part and in pseudofilaments divide by binary fission, irregularly in various planes (particularly in the basal parts of a thallus), later (in basal or vertical pseudofilaments) crosswise, perpendicular to the pseudofilament axis. Baeocytes develop usually simultaneously from the apical spherical cells. Reproduction by liberation of baeocytes.

---

PASCHERINEMA

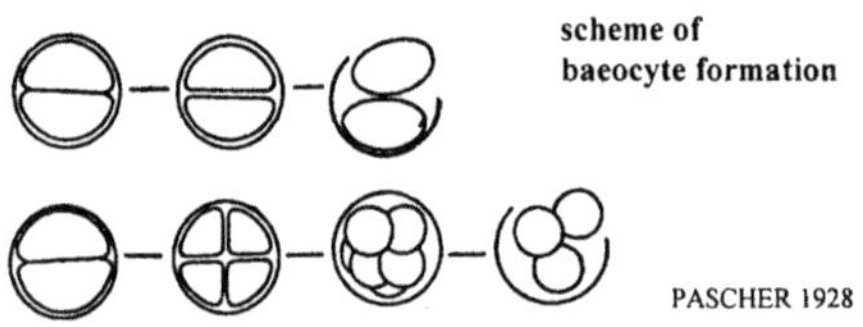


**Fig. 595.** *Pascherinema gracile*: after Pascher (1929, sub *Endonema gracile*).
**Fig. 596.** *Pascherinema moniliforme*: **a** – baeocyte differentiation in the apical cells, **b** – separation of baeocytes from the apical parts of pseudofilaments, **c** – short pseudofilaments, **d** – detail of the base of pseudofilament, **e** – two pseudofilaments; after Pascher (1929, sub *Endonema moniliforme*).

CYANODERMA

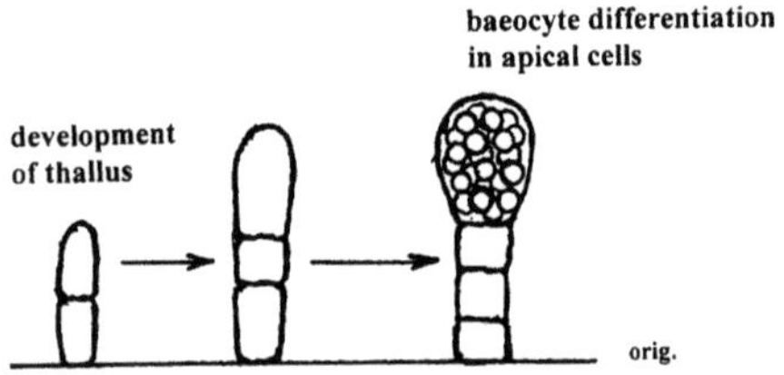


**Fig. 597.** *Cyanoderma bradypii*, from hairs of the three-toed sloth *Bradypus tridactylus*: **a** – group of pseudofilaments with apical cells dividing into baeocytes, **b** – groups of pseudofilaments (arrows) on the hair (transverse section); after Weber van Bosse (1887, sub *Cyanoderma bradypodis*).
**Fig. 598.** *Cyanoderma laminariae*, marine, epiphytic, described from California: **a**

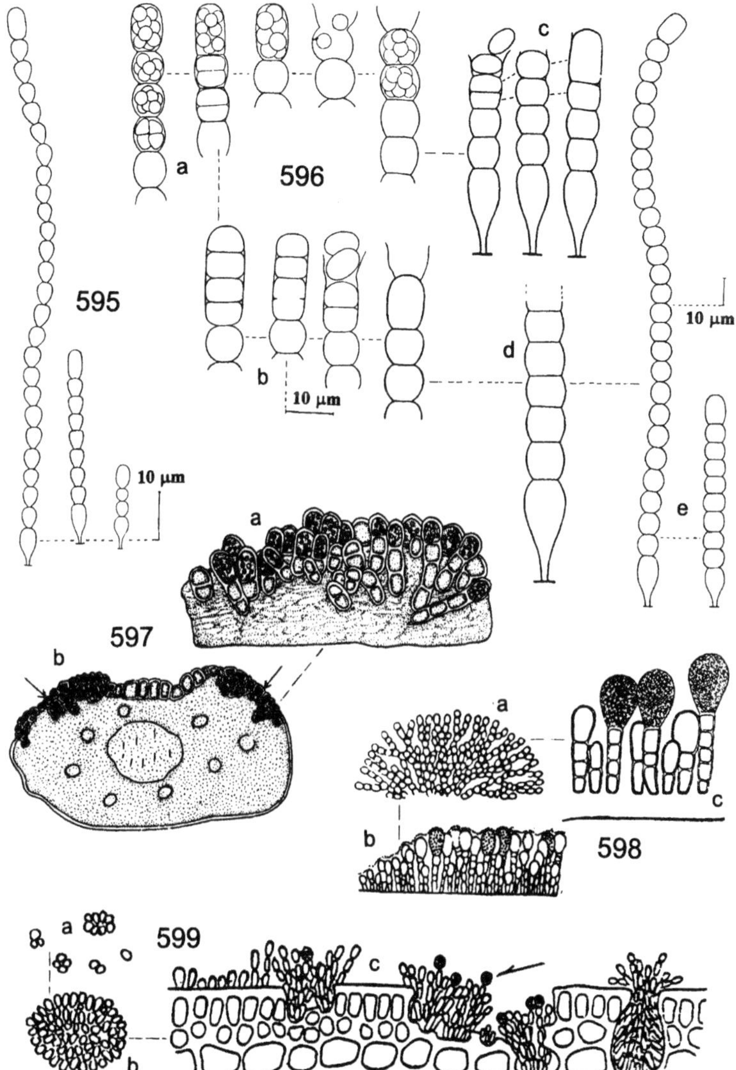


– basal part of thallus from top view, **b-c** – erect pseudofilaments with apical cells dividing into baeocytes; after Setchell & Gardner from Kosinskaja (1948, sub *Nematoradaisia laminariae*).

**Fig. 599.** *Cyanoderma subimmersa*, marine, epiphytic, from California: **a-b** – basal parts of thallus of various ages from top view, **c** – erect pseudofilaments from side; after Setchell & Gardner from Geitler (1932, sub *Radaisiella subimmersa*).

Notes: *Cyanoderma* does not differ in generic features from the later described *Nematoradaisia* according to the original diagnoses. Bourrelly (1954a) found in the original material of *Cyanoderma* another type which was a Rhodophyte (Bourrelly 1985, pl. 43: 3–6) and included this genus in this group of eukaryotic algae. However, Thompson (1972) supported the cyanoprokaryotic character of the original *Cyanoderma* (*Pleurocapsa*-like; see also Bourrelly 1985, p. 538). Therefore, we use this name in the original concept of Weber van Bosse (1887).
**Important references:** Weber van Bosse (1887); Gardner (1918, sub *Radaisia*); Setchell & Gardner (1919, sub *Radaisia*); Geitler (1925, 1932, 1942, sub *Radaisiella, Nematoradaisia, Geitleriella*); Bourrelly (1954a, 1970, 1985); Thompson (1972).

**Only species outside Europe:**
- *Cyanoderma bradypii* (Kuhn in Welcker) Weber van Bosse 1887 [syn.: *Pleurococcus bradypii* Kuhn in Welcker 1866, *Cyanoderma bradypodis* (Kuhn in Welcker) Weber van Bosse 1887 – orthogr. mut.]: Tropical (S. America), aerophytic, epizoic on hairs of *Bradypus*; (fig. **597**).
- *Cyanoderma laminariae* (Setchell et Gardner) Komárek et Anagnostidis 1995 [syn.: *Radaisia laminariae* Setchell et Gardner in Gardner 1918, *Nematoradaisia laminariae* (Setchell et Gardner) Geitler 1925]: Marine, described from California, USA; (fig. **598**).
- *Cyanoderma lineare* (Setchell et Gardner) Komárek et Anagnostidis 1995 [syn.: *Hyella linearis* Setchell et Gardner in Gardner 1918, *Myxohyella linearis* (Setchell et Gardner) Geitler 1925]: Marine, endophytic in *Prionotis* on the Pacific coast of N. America.
- *Cyanoderma subimmersum* (Setchell et Gardner) Komárek et Anagnostidis 1995 [syn.: *Radaisia subimmersa* Setchell et Gardner in Gardner 1918, *Radaisiella subimmersa* (Setchell et Gardner) Geitler 1925, *Geitleriella subimmersa* (Setchell et Gardner) de Toni 1936]: Marine, epiphytic on Rhodophytes, described from the Pacific coast of California, USA; (fig. **599**).

**Unrevised species:**
- *Nematoradaisia gasteropodum* Gonzalez Guerrero 1947: Colonies to 120 µm in diameter, consisting of basal pseudoparenchymatous clusters of cells, from which grow erect pseudofilamentous, dichotomously divaricate, to 70 µm long rows of cells; cells 2–8 × 2–5 µm, baeocytic division not described; boring into shells of living *Gasteropodum* in the river Avillo, Spain (near Cádiz). Generically surely different from *Cyanoderma* (*Nematoradaisia*).

## 4. **Solentia** Ercegović 1927

Thallus composed of epilithic (mainly petrified), and endolithic, coiled and/or curred pseudofilaments with apical and/or intercalary growth. Epilithic part (if developed) spreads over the surface of limestone substrates and is composed of simple, short uniseriate or multiseriate rows of spherical or slightly elongate, irregularly disposed cells, intensely calcified; endolithic pseudofilaments are common, long, enclosed within widened mucilaginous stalks, within which is a uniseriate row of solitary, usually elongate cells, which are distinctly distant one from another; at the ends of pseudofilaments one enlarged, oval, obovoid, club-shaped or irregular terminal cell is developed; cells within pseudofilaments are distant from one another, separated by a transversely layered mucilage. All pseudofilaments are enveloped by wide, lamellate, gelatinous sheath, sometimes coloured, pseudodichotomously or laterally and repeatedly divaricate («branched»), sometimes widened at the apex (the cells are sometimes wider in

the apical parts of branches than near their bases); lateral branches develop after the lengthwise division of apical cells (or cells in their apical parts) in endolithic pseudofilaments and grow perpendicularly or obliquely to the stony surface. Cells are of irregular shape, more or less spherical or elongate, with more or less homogeneous, pale blue-green content.
Cell division in various planes, apical or subapical (rarely lateral or basal) cells (endolithic, near the surface of rocks and stones) sometimes divide into several to numerous baeocytes. Reproduction by baeocytes and small clusters of cells enveloped by mucilage (?).
**Important references:** Ercegović (1927, 1929, 1930, 1932); Geitler (1932, 1942); LeCampion-Alsumard & Golubić (1985); Pantazidou (1991 - diss.); LeCampion-Alsumard, Golubić & Pantazidou (1996).

Note: According to the revision of LeCampion-Alsumard & al. (1996), the genus *Solentia* is euendolithic, characterized by the mainly endolithic, wide gelatinous stalks, in which the few cells are situated at the ends or sparsely along the stalk, distant one far from another. Reproduction by baeocyte formation.

**Key to the species:**

**1a** Endolithic pseudofilaments 3–9 µm wide, with usually several to numerous cells, which are sometimes only 2 µm distant from one another; cells 2–30 × 2–6 µm . . . . . . . . . . **1. S. intricata**
**1b** Endolithic pseudofilaments usually (especially at the apex) more than 10 µm wide (to 60 µm), mainly with only few cells, about 10 or more than 10 µm distant from one another . . . . . . . . . . **2**
**2a** Terminal cells very long and narrow, 15–90 × 7–16 µm, thallus greyish-green . . . . . . . . . . **2. S. achromatica**
**2b** Terminal cells shorter, to 45 µm long . . . . . . . . . . **3**
**3a** Endolithic pseudofilaments 3–25 µm wide (wider towards ends) . . . . . . . . . . **3. S. paulocellulare**
**3b** Endolithic pseudofilaments up to 40–45µm wide . . . . . . . . . . **4**
**4a** Thallus pale blue-green, pseudofilaments to 40 µm wide, sheaths colourless, cells –45 × 3–12 µm . . . . . . . . . . **4. S. stratosa**
**4b** Thallus greyish, pseudofilaments to 60 µm wide, sheaths colourless to blackish-violet, cells –40 × 5–20 µm . . . . . . . . . . **5. S. foveolarum**

## 1. **Solentia intricata** Ercegović 1927 (fig. **600**)

Thallus mainly endolithic, pale blue-green. Pseudofilaments 3–9 µm wide, with firm, colourless, more or less distinctly lamellate sheaths, with cells to 30 µm distant from one another, but usually numerous along the pseudofilament. Terminal and subapical cells usually several times longer than wide, cells in the middle of pseudofilaments shorter, only 2 µm distant from one another. Cells 2–30 × (2)3–4(6) µm. Baeocytes arise from apical cells (to 8 per mother cell), 2–3 µm in diameter.
Occurrence: Marine, boring into shells of molluscs and limestone substrates in the sea coast splash zone, sometimes occurring down to 3 mm within the substrate; known mainly from Croatia (Dalmatia); reported also from Aldabra Islands.

## 2. **Solentia achromatica** Ercegović 1932 (fig. **601**)

Thallus mainly euendolithic, greyish-green, bluish to blue-green. Pseudofilaments 9–50 µm wide, to 320 µm long, with firm, colourless, distinct transversely (excentrically) layered sheaths, with cells very distant from one another, only few

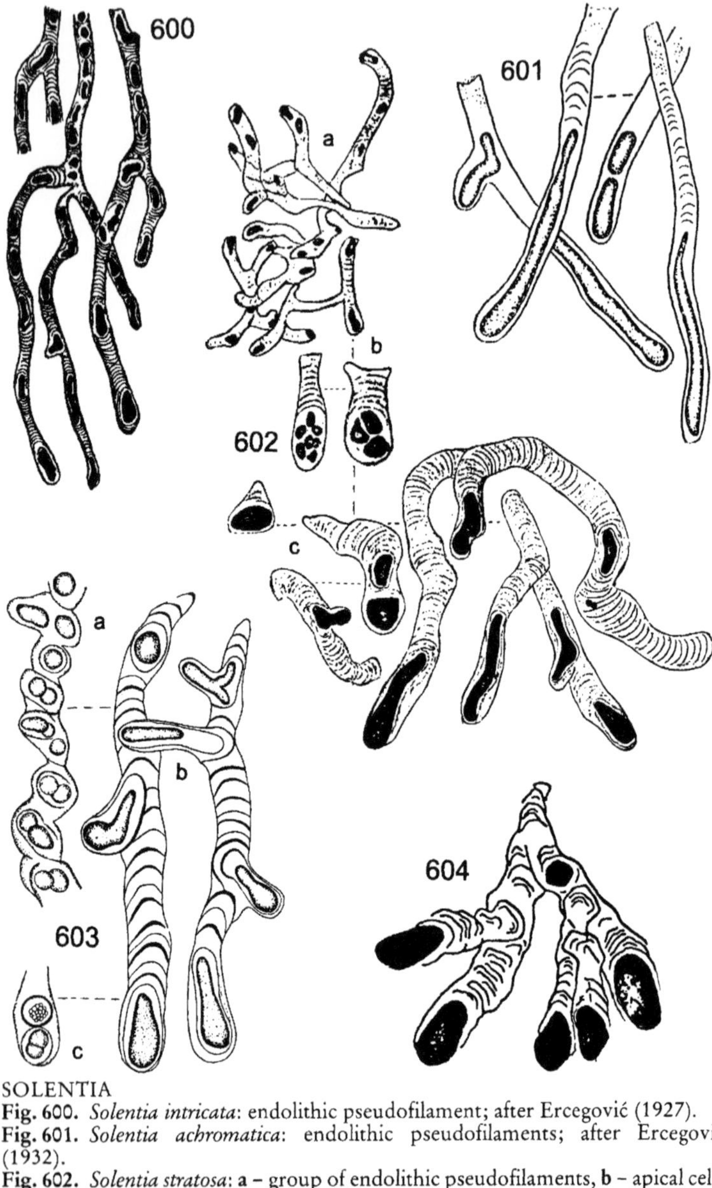


SOLENTIA

**Fig. 600.** *Solentia intricata*: endolithic pseudofilament; after Ercegović (1927).

**Fig. 601.** *Solentia achromatica*: endolithic pseudofilaments; after Ercegović (1932).

**Fig. 602.** *Solentia stratosa*: **a** – group of endolithic pseudofilaments, **b** – apical cells dividing into baeocytes, **c** – detail of endolithic pseudofilaments of various ages; after Ercegović (1927).

(1–4, rarely more) in each sheath; intercalary pseudobranches arise from lateral projections at the end of cells. Cells cylindroid, (8?-15)38–65(90) × (5) 7–16 μm, greenish, blue-green, bluish or yellow-green, with finely granular content. Apical cells remarkably elongate, with widened end, rarely cylindrical, 45–180 × 5–11 μm. Reproduction by cell division in the upper part of pseudofilaments. Baeocyte production not yet observed, even in culture.
Occurrence: Marine, in rock pools in the upper part of the splash zone on limestone coasts; known from Adriatic Sea in Croatia (Dalmatia) and Aegean Sea in Greece (Attiki Peninsula).

### 3. Solentia paulocellulare(Ercegović) LeCampion-Alsumard et Golubić 1985

*Hormathonema paulocellulare* Ercegović 1929 (fig. **455**)

Epilithic thallus greyish to blackish blue, forming spots, 4–12 mm in diameter, composed of pseudofilaments growing on the substrate (surface of stones) among other algae, later becoming euendolithic, boring perpendicularly into the substrate. Euendolithic pseudofilaments circa (14) 40–120 μm long (to 75 μm in culture), at the base (upper part) 3–14 μm wide, at the end (deeper part) 9–25 μm (or wider); intercalary pseudobranches scarce, arising from lateral protrusion of the upper part of the cell. Cells in one pseudofilament 2–6, rarely solitary, «uniseriate», clearly distant from one another, spherical, subspherical to oval and 5–9 × 2.5–9 μm at the base, ellipsoidal to long cylindrical, 4–17 × 4–10 μm in the middle, bluish, blue-green, rarely grey-green, with finely or coarsely granular content. Apical cell pear-shaped or obtuse conical, distinctly longer than the other cells, 6.7–28 × 4–12 μm; this apical cell arises after the first asymmetrical cell division. Thylakoids dispersed, irregularly arranged at the margin of cytoplasm (other cell inclusions are arranged in the same way as in other not euendolithic cyanoprokaryotes). Sheaths gelatinous, firm, 1–5 μm thick (4–22 μm between the cells), never diffluent, lamellate («one-sided» excentric production of slime layers), blue-black, violet or blackish violet. Monocytes 7–10 μm in diameter, spherical or subspherical surrounded by their own thin envelopes. Solitary cells with individual envelopes (about 2.5 μm in diameter, but not baeocytes) appear in the upper part of thallus. Life cycle see in Pantazidou (1991).
Occurrence: Marine, on rocks and stones in splash and intertidal zone; especially on flat surfaces and extending into shallow cavities, boring also into mollusc shells (but not into calcified *Rhodophytes* and *Chlorophytes*); described from Croatia (Adriatic coast in Dalmatia), but known from the whole Mediterranean Sea (France – near Marseille, widely distributed along the continental and islandic Aegean Sea coasts in Greece); known also from W. Atlantic (Bermuda and Bahama Islands) and Pacific coasts (Chile, Sea of Japan), probably more widely distributed (cosmopolitan ?).

### 4. Solentia stratosa Ercegović 1927 (fig. 602)

Thallus mainly euendolithic, pale blue-green to blue-black, with (epi- and) endolithic pseudofilaments. Euendolithic psudofilaments 7–40 μm wide, to

---

**Fig. 603.** *Solentia foveolarum*: **a** – epilithic pseudofilament, **b** – endolithic pseudofilaments, **c** – apical cells dividing into baeocytes; after Ercegović from Kosinskaja (1948).
**Fig. 604.** *Solentia sanguinea*: after Al-Thukair & Golubić (1991).

200 µm long, with firm, clearly stratified, rarely diffluent, delimited sheaths, bluish-black in the outer part, colourless in the inner part, and with cells distinctly distant (10–300 µm) from one another; in one (euendolithic) pseudofilament usually only 1–6 cells, often several times longer than wide. Cells (4)10–25(50) × (3)6–9(12) µm, blue-green or yellow-green, with finely granular content. Baeocytes (5–10 per cell) arise from terminal or intercalary, 4–10 µm wide cells, and themselves are 2–6 µm in diameter. (Life cycle described in Pantazidou 1991 – diss.).
Occurrence: Marine, boring into calcareous rocks and stones down to 1 mm of depth, in the littoral splash zone of limestone coasts; described from Croatia (Dalmatia), known from numerous localities in Greece (Aegean Sea), reported also from Red Sea (Sinai) and Indian Ocean (Aldabra Islands).

### 5. Solentia foveolarum Ercegović 1930 (fig. 603)

Thallus greyish, bluish to blue-black, with epi- and euendolithic pseudofilaments. Epilithic part of thallus composed of more or less spherical or irregular-rounded cells with their own envelopes, euendolithic filaments to 60 µm wide and 170 µm long (in culture to 245 µm), with colourless to blackish violet, firm, rarely diffluent, gelatinous, lamellate sheaths with transverse excentric layers, containing 1–4 (rarely more) cells in one row (uniseriate), to 10–400 µm distant from one another. Pseudobranches intercalary arising from lateral cell protrusions. Cells bluish to blue-green, ellipsoid to cylindrical, –40 × 5–20 µm, with finely granular content. Apical cell cylindrical, rarely with widened apex, longer than the other cells, to 15–33(to 45) µm long. Baeocytes to 32 per cell, produced by successive or simultaneous cell divisions in the upper part of the pseudofilaments, 2–4 µm in diameter. (Life cycle described in Pantazidou 1991 – diss.).
Occurrence: Marine, boring into limestones substrates and mollusc shells, in rock pools in the upper part of the splash zone and in the intertidal zone; widely distributed in Mediterranean Sea, Croatia (Dalmatia), France (Marseille), Greece (Aegean Sea, Ionian Sea coasts and islands), known also from Red Sea (Sinai) and Atlantic coasts (Florida, Bermuda, Bahamas).

**Species outside Europe:**
- *Solentia sanguinea* Golubić et al. 1996: Marine, euendolithic, from Arabian Gulf and the Bahama Bank, with reddish cells; (fig. **604**).

**Unrevised species:**
- *Entophysalis deusta* «status solentioides» in Anagnostidis 1968: Marine, euendolithic, penetrating into calcareous rocks, recorded from the Aegean Sea coasts in Greece; probably a *Solentia* species.
- Chu & Wu (1984) report two «lime-boring», provisionally designated *Solentia* species from Xisha islands, China (*Solentia sinica, S. xishaensis).*

## 5. Dalmatella Ercegović 1929

Thallus at first epilithic, more or less thick, gelatinous, forming amorphous mass with irregularly disposed cells (sometiems ensheathed), in groups or forming irregular rows of cells or pseudofilaments; later endolithic. Endolithic pseudofilaments grow parallely to one another and perpendicularly into the calcareous substrate; they are thick, enveloped by wide, distinct, clearly delimited, colourless and more or less homogeneous gelatinous envelopes. Endolithic pseudofilaments are composed of one or more rows of cells and are sometimes pseudodichotomously divaricate. Cells more or less spherical or irregular-rounded, in epilithic

parts of thallus smaller, subspherical, oval or polygonal, rarely elongate, but more spherical than in endolithic pseudofilaments, where they are usually ellipsoidal or cylindrical, clearly distant from one another. End cells sometimes wider than cells in pseudofilaments. Gelatinous envelopes in epilithic parts are usually more delimited, coloured and lamellate, than in endolithic parts, where they are colourless, more or less homogeneous. Cells pale blue-green.
Cell division irregular in three or more planes, in endolithic pseudofilaments mostly transverse, but sometimes lengthwise to the axis of pseudofilaments. In pseudofilaments division occurs in both intercalary and apical cells. Sometimes there is division into baeocytes in epilithic pseudofilaments by successive cell division. Reproduction by solitary cells, baeocytes, or by small gelatinous clusters of cells. The existence of the genus was recently confirmed by Pantazidou (1991), but all the species need revision.
**Important references:** Ercegović (1929, 1932); Geitler (1932, 1942); Frémy (1934); Kosinskaja (1948); LeCampion-Alsumard (1979 – diss.); Pantazidou (1991 – diss.).

**Key to the species:**

**1a** Epilithic thallus violet, cells in endolithic pseudofilaments with more or less small distances between them; terminal cells distinctly widened and larger than the others (endolithic cells 8–18 × 5–15 µm, terminal cells 12–25 × 10–20 µm) . . . . . . . . . . . . . . . . . . . . . . . . . . . . . . . . **1. D. violacea**

**1b** Epilithic thallus yellow-brown, cells in endolithic pseudofilaments usually clearly separate from one another; terminal cells in endolithic pseudofilaments not distinctly larger than other cells . . . . . . . . . . . . . . . . . . . . . . 2

**2a** Epilithic thallus encrusted by calcium carbonate, cells in endolithic pseudofilaments distant from one another, and also (at least in the upper parts) with yellow-brown individual sheaths . . . . . . . . . . . . . . . . . **5. D. polyformis**

**2b** Epilithic thallus not encrusted by calcium carbonate, cells in endolithic pseudofilaments with colourless or with yellow-brown envelopes . . . . . 3

**3a** Epilithic thallus composed largely of ensheathed cells with layered or special individual envelopes; endolithic pseudofilaments to 200 µm long (?), with cells 4–15 µm wide . . . . . . . . . . . . . . . . . . . . . . . . . . . . . . . . . . . . 4

**3b** Epilithic thallus composed of pseudofilaments, in which cells with distinctly coloured envelopes are arranged in rows; endolithic pseudofilaments to 500 µm long, with cells 4–15 × 3–7 µm . . . . . . . . . . . . . . . **4. D. buaensis**

**4a** Endolithic pseudofilaments to 100 µm long, sometimes pseudodichotomously divaricate, never anastomozing, with cells to 23 × 15 µm . . . . . . . . . . . . . . . . . . . . . . . . . . . . . . . . . . . . . . . . . **2. D. littoralis**

**4b** Endolithic pseudofilaments usually longer than 100 µm, sometimes anastomozing, with cells 6–12 × 4–10 µm . . . . . . . . . . . . . . . . . . **3. D. anomala**

## 1. Dalmatella violacea Ercegović 1932 (fig. **605**)

Epilithic thallus very thin, dark violet, composed of pseudofilaments up to 40 µm wide (rarely wider); cells spherical, 4–7 µm in diameter. Endolithic pseudofilaments to 30 µm wide, to 150 µm long, with two or several rows of cells; cells spherical or oval, 8–18 × 5–15 µm, sometimes flattened; apical cells often widened, 12–25 × 10–20 µm. Cell content olive-green. Envelopes colourless or yellow-brown. The violet colour of thallus seems to be strange, if the sheaths are yellow-brown and cells olive-green.
Occurrence: Marine, on coastal limestone rocks; known only from Adriatic Sea (Croatia – Dalmatia).

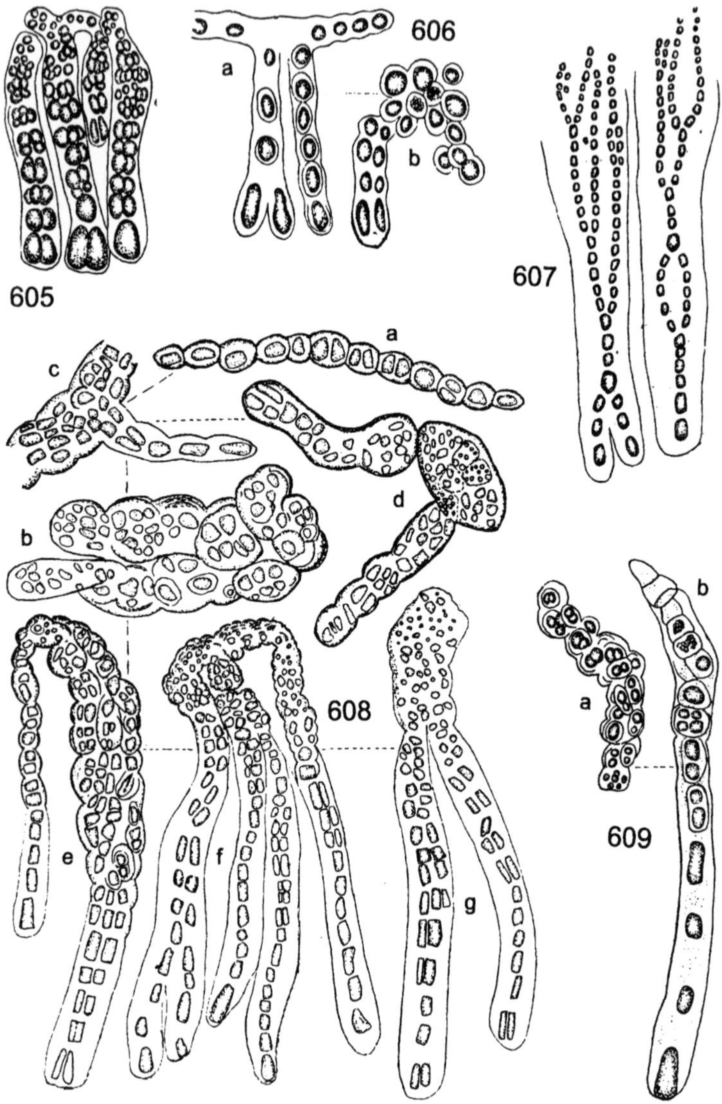


DALMATELLA

**Fig. 605.** *Dalmatella violacea:* after Ercegović (1932).

**Fig. 606.** *Dalmatella littoralis*: epilithic and endolithic pseudofilaments, two epilithic cells dividing into baeocytes; after Ercegović (1932).

## 2. **Dalmatella littoralis** Ercegović 1932 (fig. **606**)

Epilithic thallus yellow-brown, irregular, small (microscopic), thin, in the form of groups of ensheathed spherical cells, sometimes arranged in short rows, enveloped by a common sheath; usually among other epilithic microalgae. Cells, from which develop nanocytes, 9–12 µm in diameter, nanocytes numerous, around 1 µm in diameter. Endolithic pseudofilaments to 20 µm wide and to 100 µm long, with one or two rows of cells, rarely pseudodichotomously divaricate. Cells to 23 × 15 µm, usually with their individual, more or less thick and sometimes lamellate and yellow-brown envelopes, or colourless and homogeneous.
Occurrence: Marine, on limestone rocks in the upper lithophytic part of splash zone; known only from the Adriatic coast, Croatia (Dalmatia).

## 3. **Dalmatella anomala** Ercegović 1932 (fig. **607**)

Epilithic thallus composed of irregularly arranged pseudofilaments or groups of cells on a stony surface, with gloeocapsoid envelopes around the cells. Endolithic pseudofilaments at first uniseriate, later two- or multiseriate, sometimes anastomozing, with cells near or distant from one another, in special sheaths, which envelope cells or groups (rows) of cells. Endolithic pseudofilaments with ellipsoid, barrel-shaped or almost cylindrical cells, close to one another or slightly distant (clearly separate from one another). Cells 6–12 × 4–10 µm.
Occurrence: Marine, on limestone rocks in the middle part of the lithophytic band of the splash zone; known only from Adriatic Sea in Croatia (Dalmatia).

## 4. **Dalmatella buaensis** Ercegović 1929 (fig. **608**)

Epilithic thallus yellow-brown, gelatinous, to 1/2 mm thick, composed of 6–10 (or more) µm wide pseudofilaments, which are 2–3-multiseriate; cells are 2–8 × 2–6 µm. Endolithic pseudofilaments 12–30 µm wide, to 500 µm long, one- or two-, rarely more-seriate, with cells 4–15 × 3–7 µm. Cells in epilithic parts often with special yellow-brown envelopes; the common sheaths are also yellow-brown, lamellate, 6–12 µm wide; sheaths of endolithic pseudofilaments homogeneous, colourless, thin. Cell content blue-green.
Occurrence: Marine, on rocks on limestone coasts, in the splash zone; Croatia (Dalmatia, continental and island shores), France (near Marseille), Greece (Saronikos Gulf), known also from Aldabra Atoll Islands (Indian Ocean).

## 5. **Dalmatella polyformis** Ercegović 1932 (fig. **609**)

Epilithic thallus yellow-brown, granular, encrusted by calcium carbonate, to 0.5 mm thick, composed of pseudofilaments up to 50 µm wide; cells more or less spherical or ellipsoidal, to 8 µm in diameter, irregularly arranged in multicellular rows, each with its own, yellow-brown, thick and lamellate envelope. Endolithic

---

**Fig. 607.** *Dalmatella anomala*: endolithic pseudofilaments; after Ercegović (1932).
**Fig. 608.** *Dalmatella buaensis*: **a-b** – epilithic pseudofilaments, **c-d** – parts of epilithic thallus (a few cells dividing into nanocytes) with short endolithic pseudofilaments, **e-g** – well developed endolithic pseudofilaments; after Ercegović (1929).
**Fig. 609.** *Dalmatella polyformis*: **a** – part of epilithic thallus, **b** – endolithic pseudofilament with «basal» cells dividing into baeocytes; after Ercegović (1932).

pseudofilaments to 40 µm wide and to 500 µm long, with cells to 40 × 25 µm, arranged in one or more rows and often distinctly distant from one another (with spaces of up to 60 µm between them). Cells in the basal («upper») parts of pseudofilaments also with special yellow-brown envelopes, but more homogeneous and colourless in terminal parts. Cell content homogeneous or granular, greenish or yellowish blue-green.
Occurrence: Marine, on rocks on limestone shores, in the middle part of the lithophytic splash zone; known only from Croatia (Dalmatia).

**Unrevised species:**
- *Entophysalis deusta* «status dalmatelloides» in Anagnostidis 1968: Marine, euendolithic, penetrating into calcareous rocks and mollusc shells, recorded from the Aegean Sea coasts in Greece; probably a *Dalmatella* species.
- *Dalmatella* sp. in Pantazidou 1991: Special type of *Dalmatella* characterized by mainly euendolithic multiseriate pseudofilaments to 300 µm long and 57 µm wide, with sheaths outside gold-yellow to yellow-brown, inside colourless, without lamellation; cells diversiform in different parts of thallus, 3.5–31 µm in

PLEUROCAPSA

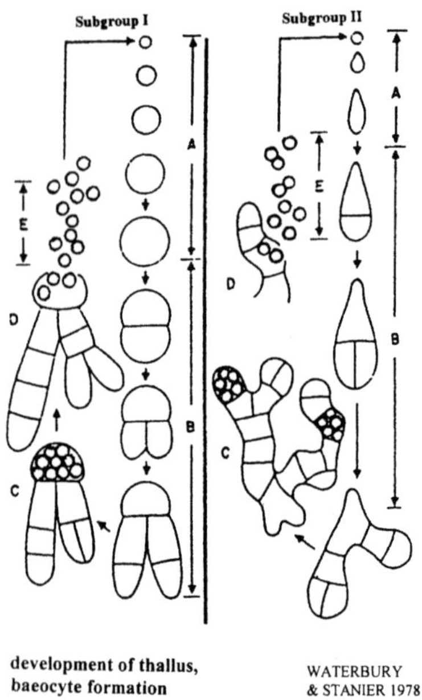


**Fig. 610.** Typical development of thallus in *Pleurocapsa*: *Pleurocapsa* sp., from Waterbury & Stanier (1978).
**Fig. 611.** *Pleurocapsa concharum*, from Indonesia: after Geitler from Geitler & Ruttner (1935).

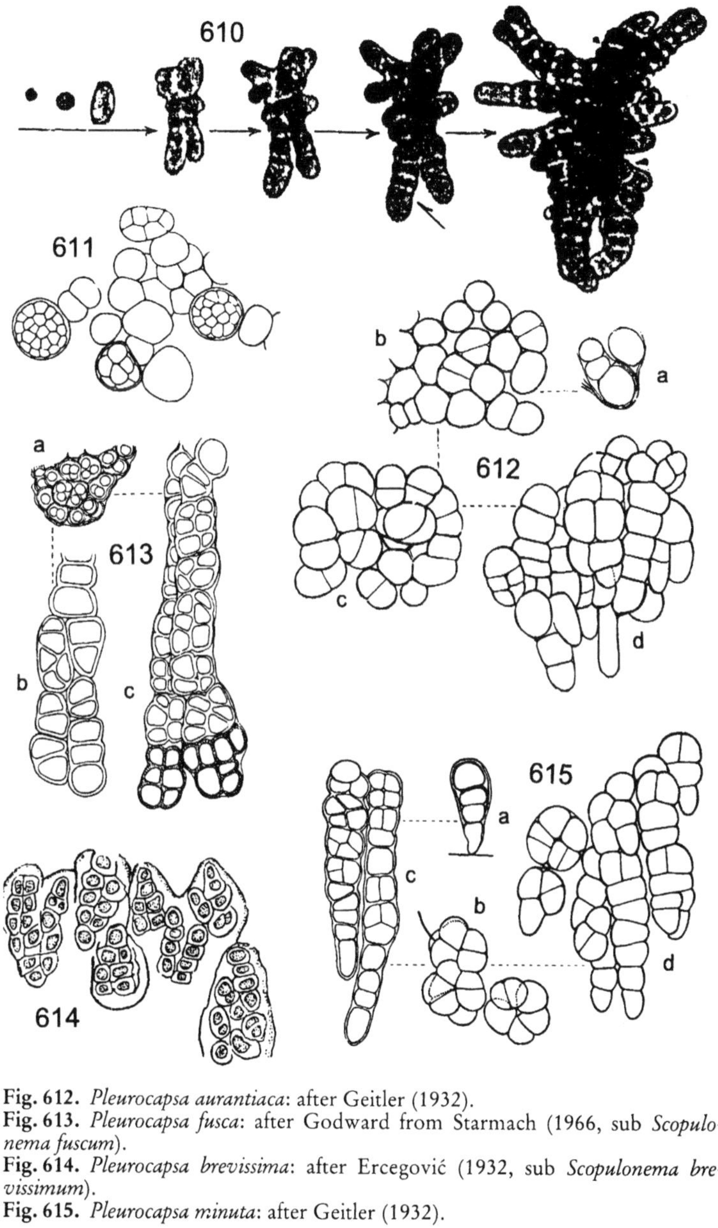


**Fig. 612.** *Pleurocapsa aurantiaca*: after Geitler (1932).
**Fig. 613.** *Pleurocapsa fusca*: after Godward from Starmach (1966, sub *Scopulonema fuscum*).
**Fig. 614.** *Pleurocapsa brevissima*: after Ercegović (1932, sub *Scopulonema brevissimum*).
**Fig. 615.** *Pleurocapsa minuta*: after Geitler (1932).

diameter; reproduction by liberation of cells from the upper part of pseudofilaments. Found in Greece (marine coasts of Aegean Sea).

## 6. **Pleurocapsa** Thuret in Hauck 1885

*Scopulonema* Ercegović 1930

Thallus creeping on the substrate (mainly on stones), crustose, in a thin layer or in three-dimensional clusters, composed of irregular groups of cells (chroococcacean stages), from which irregular or radial rows of cells grow, sometimes irregularly and pseudodichotomously branched; in several species the pseudofilaments are partly endolithic. Rows of cells uni- or multiseriate, enveloped by more or less thin, firm, sometimes lamellate and yellowish-brownish coloured sheaths which are laterally confluent. Cells irregular, of variable size, sometimes (in pseudofilaments) slightly elongate, with homogeneous or slightly granular content, sometimes with several prominent granules, blue-green, pale blue-green or pinkish.

Cells divide irregularly by binary fission in various planes, in pseudofilaments predominantly crosswise. The lengthwise division of apical cells in pseudofilaments resulting in pseudodichotomous divarication is a typical feature of the genus (fig. **610**). Enlarged cells, which arise in different parts of thallus, sometimes divide into baeocytes. Reproduction by baeocytes.

**Important references:** Geitler (1932, 1942); Ercegović (1930, sub *Scopulonema*); Frémy (1934); Geitler & Ruttner (1935); Starmach (1936); Kosinskaja (1948); Bourrelly (1970, 1972, 1984); Komárek (1972); Waterbury & Stanier (1978); Rippka & al. (1979); Kondrateva & al. (1988); Waterbury (1989).

**Key to the species:**

**1a** Freshwater species, usually in mountain streams and lakes . . . . . . . . . . . **2**
**1b** Marine species, in littoral (splash or tide) zone . . . . . . . . . . . . . . . . . . . . **5**
**2a** Sheaths colourless . . . . . . . . . . . . . . . . . . . . . . . . . . . . . . . . . . **3**
**2b** Sheaths coloured . . . . . . . . . . . . . . . . . . . . . . . . . . . . . . . . . . **4**
**3a** Cells 3–9(–12) µm wide, baeocytes 0.8–3 µm in diameter . . . . **1. P. minor**
**3b** Cells 4–17(–20) µm wide, baeocytes 3–4 µm in diameter . **2. P. concharum**
**4a** Sheaths intensely yellow-orange up to orange-red or orange- brown . . . . . . . . . . . . . . . . . . . . . . . . . . . . . . . . **3. P. aurantiaca**
**4b** Sheaths inside colourless, outside yellowish to dark olive- green or blackish-brown . . . . . . . . . . . . . . . . . . . . . . . . . . . . . . . . **4. P. fusca**
**5a** Cells 3–9 µm in diameter, of the uniform size; sheaths coloured . . . . . . . **6**
**5b** Cells 5–20(25) µm in diameter, usually of variable size in one and the same thallus; sheaths colourless . . . . . . . . . . . . . . . . . . . . . . . . . . . . **9**
**6a** Pseudofilaments to 20 µm wide, cells 3–5(–7) µm wide; sheaths dark brown or blackish brown . . . . . . . . . . . . . . . . . . . . . . . . . . . . . . . **7**
**6b** Pseudofilaments to 40–50 µm wide, cells 4–9 µm wide; sheaths yellow-brown . . . . . . . . . . . . . . . . . . . . . . . . . . . . . . . . . . . . . **8**
**7a** All cells more or less of the same size and shape, 3–5 µm in diameter . . . . . . . . . . . . . . . . . . . . . . . . . . . . . . . . **5. P. brevissima**
**7b** Basal cells more or less elongate, cells 3–7 µm wide . . . . . . . . **6. P. minuta**
**8a** Cells 4–8 µm wide, pseudofilaments to 80 µm long; cell content pale blue-green . . . . . . . . . . . . . . . . . . . . . . . . . . . . . . . . . . **7. P. mucosa**
**8b** Cells 5–9 µm wide, pseudofilaments to 150 µm long; cell content blue-green, olive-green, yellowish or pale violet . . . . . . . . . . . . . . **8. P. hansgirgiana**
**9a** Cells yellowish, yellow-green, yellow-reddish or dirty violet **9. P. fuliginosa**
**9b** Cells pale blue-green . . . . . . . . . . . . . . . . . . . . . . . . . . . **10. P. crepidinum**

Notes: Waterbury & Stanier (1978) studied twelve *Pleurocapsa* strains (eleven of marine origin, one moderate thermophile from a hot spring) and divided them into two subgroups according to morphological changes during the life cycle and reproduction strategy (see also Waterbury 1989). However, the «subgroup I» corresponds rather to the life cycle of *Hyella*. According to their description, at least three traditional species from the subgroup I (including the mentioned type from hot springs) and one from the subgroup II are recognizable. Unfortunately, they did not define their strains (and species) taxonomically.

## 1. **Pleurocapsa minor** Hansgirg 1891 (fig. **616, 618**)

*Scopulonema minus* (Hansgirg) Geitler 1942

Colonies composed of irregular, creeping or slightly erect, short pseudofilaments, sometimes boring into the limestone substrate, forming flat aggregations or apparently pseudoparenchymatous layers; well-developed colonies composed of short, irregular, simple or sometimes short and irregularly pseudodichotomously or laterally divaricate pseudofilaments, united laterally by the confluence of their gelatinous sheaths; pseudofilaments uni- or several-seriate, 3–9 μm wide, sometimes with morphologically different ends. Cells of various shapes, barrel-shaped, irregularly polygonal-rounded, sometimes asymmetrical, 3–12 μm in diameter, at the ends sometimes elongate and slightly arcuate, to 3-times longer than wide, with blue-green, rarely olive-green, brownish or light grey-violet content, sometimes with their own thin sheaths. Sheaths firm, distinct, thin, colourless, rarely yellowish, sometimes thicker and gelatinizing. All cells in pseudofilaments (rarely only terminal cells) able to divide into baeocytes, 8–32 in number, 0.8–1.5(3) μm in diameter. Sometimes stages with irregular or packet-like clusters of cells (or groups with 4–8 cells) are present. Variable species.
Occurrence: Freshwater, commonly on calcareous stones, rocks or toofs and on shells in mountain, usually katharobic streams, waterfalls and rivers, or in the stony littoral of clear lakes; known from mountain areas throughout Europe, but possibly a heterogeneous species (in cultures several different types occur); taxonomic identity of records from Gabon and Indonesia (numerous localities) should be revised (fig. **618**).

## 2. **Pleurocapsa concharum** Hansgirg 1891 (fig. **611, 617**)

Colonies composed of aggregates of short creeping pseudofilaments, growing on shells and stony (limestone?) substrates, sometimes pseudodichotomously divaricate, uni- or multiseriate. Cells of different shape and size, barrel-shaped or irregular-rounded, 4–17(–20) μm in diameter, with pale blue-green or olive-green content. Sheaths colourless. Both intercalary and terminal cells able to divide into baeocytes, which are 3–4 μm in diameter. Not a well-known species.
Occurrence: Freshwater, on stones (probably only calcareous) and shells in unpolluted streams and rivers; central Europe, data from tropical countries should be revised; however, good drawings in literature are only from tropical countries (fig. **611**), and the life cycle was described by Varma (1968) from Allahabad, India (fig. **617**); identity of all these populatins should be confirmed.

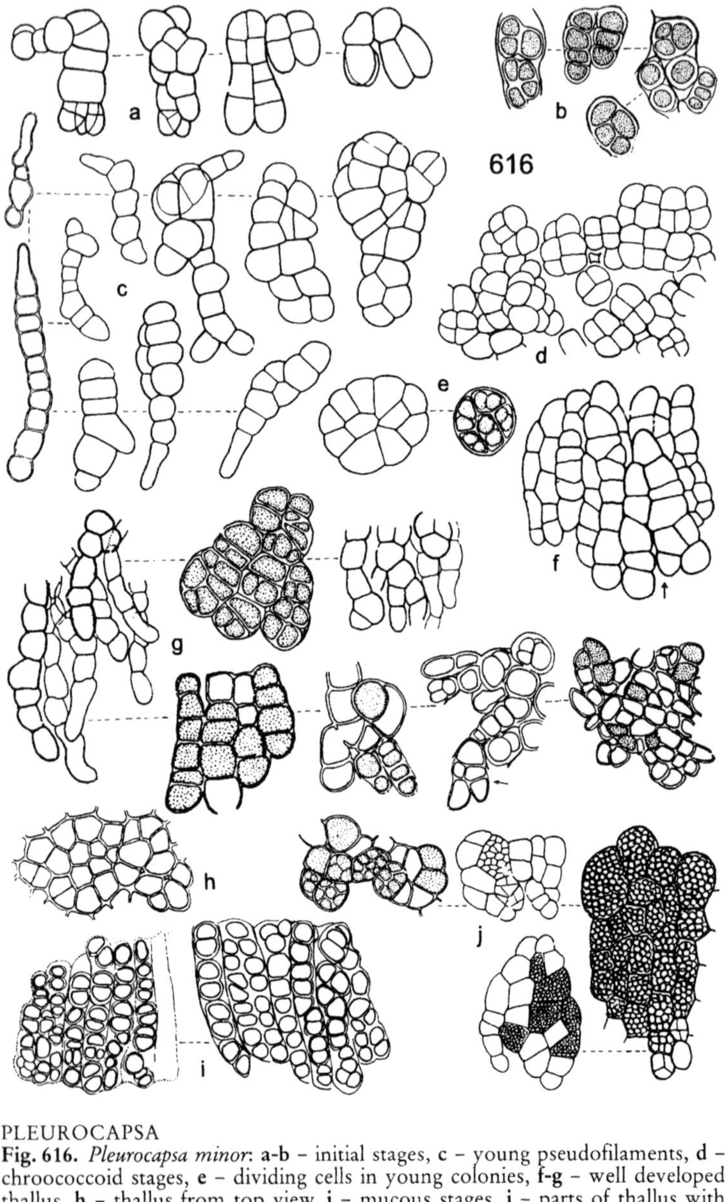


PLEUROCAPSA

**Fig. 616.** *Pleurocapsa minor*: **a-b** – initial stages, **c** – young pseudofilaments, **d** – chroococcoid stages, **e** – dividing cells in young colonies, **f-g** – well developed thallus, **h** – thallus from top view, **i** – mucous stages, **j** – parts of thallus with baeocyte production; from Geitler (1932).

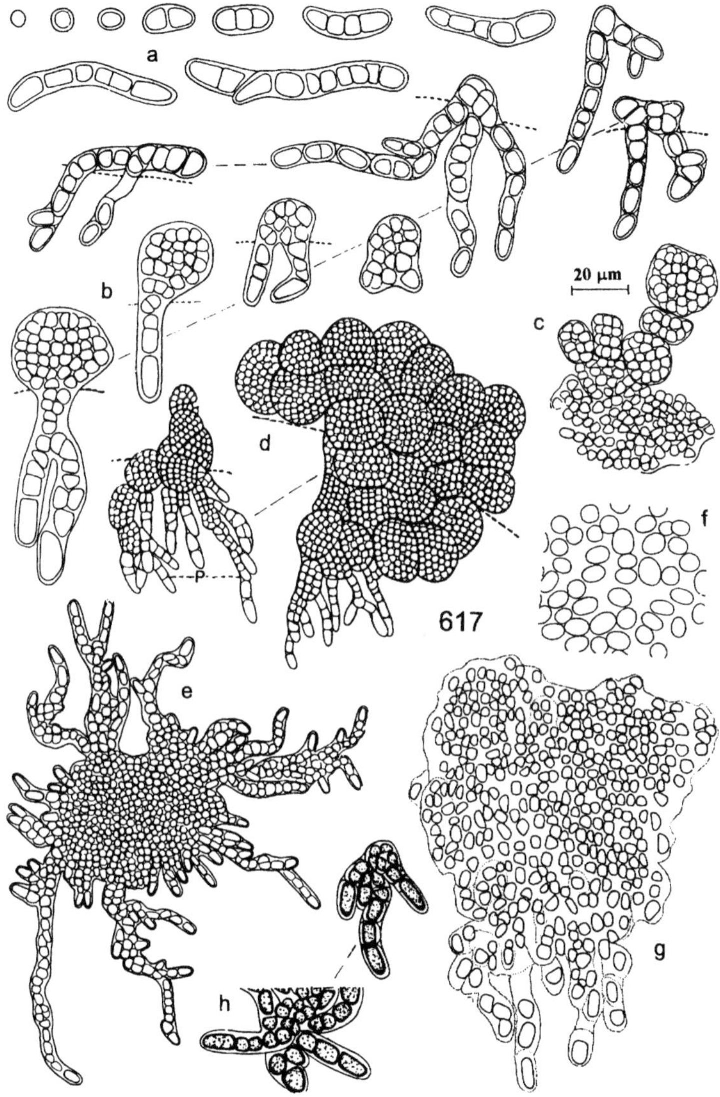


PLEUROCAPSA

**Fig. 617.** *Pleurocapsa concharum* sensu Varma, from India: **a** – initial stages of epilithic thallus, **b-g** – development of chroococcoid epilithic thallus with endolithic pseudofilaments, **h** – detail of young pseudofilaments; after Varma (1968).

**3. Pleurocapsa aurantiaca** Geitler 1932 (fig. **612**)

*Pleurocapsa polonica* Raciborski 1910 sensu auct. [type = basionym of *Siphononema polonicum* (Raciborski) Geitler 1925]; *Scopulonema polonicum* (Raciborski) Geitler 1942 sensu auct.; *Scopulonema aurantiacum* (Geitler) Geitler 1942

Colonies flat, covering stony substrates, 30–80 μm thick, forming reddish brown or yellow-brown to dark blackish-brown crustose spots on stones, composed of densely packed, irregular and sometimes pseudodichotomously divaricate, uni- or more-seriate pseudofilaments, which are separated one from another, or united laterally by confluent sheaths. Cells more or less spherical, barrel-shaped or polygonal-rounded, in packet-like (often 4 together) aggregates or in rows, circa 4 to 6–13 μm wide, sometimes shorter than wide (in pseudofilaments), or (in basal parts) to 3-times longer than wide. Sheaths firm, thin or thick, sometimes irregularly thickened, orange-yellow to reddish-brown, rarely purely yellow or brown; only the inner sheath layer sometimes colourless. Cell content pale blue-green. Occasional division of single cells into baeocytes (often in terminal parts of pseudofilaments).
Occurrence: Freshwater, epilithic on permanently wetted rocks or submersed in clear streams and rivers and in littoral of clear lakes in high mountains; known from Alps and western Carpathians (High Tatra Mts.).

**4. Pleurocapsa fusca** Godward 1937 (fig. **613**)

*Scopulonema fuscum* (Godward) Starmach 1966

Colonies forming flattened spots on stony substrates, to 100 μm thick, dark yellow-reddish or dirty orange; young stages composed of irregular clusters of cells, surrounded (solitary or in groups) by lamellate envelopes, later growing in irregular, uni- or several-seriate pseudofilaments, tightly joined together, sometimes pseudodichotomously divaricate. Cells of varied shape, mainly polygonal-rounded, 2.5–4.5 μm wide, with pale blue-green content. Sheaths distinct, firm, not very thick, inside colourless, outside yellowish, dark olive-green or blackish-brown. Marginal cells sometimes divide into baeocytes, usually 4 per cell and about 2 μm in diameter. Palmelloid stages with packet-like arranged cells occur commonly, arising by gelatinization of sheaths.
Occurrence: Freshwater, on stones, in the littoral (splash) zone of unpolluted lakes; England (Lake District), Poland (High Tatra Mts.).

**5. Pleurocapsa brevissima** (Ercegović) Komárek et Anagnostidis 1995 (fig. **614**)

*Scopulonema brevissimum* Ercegović 1932

Colonies epilithic, forming distinct gelatinous layer on the substrate, dark brown or black, composed of short pseudofilaments, which are confluent by their sides (sheaths). Pseudofilaments short, more or less uniseriate, occasionally pseudodichotomously divaricate, to 50 μm long and about 20 μm wide. Cells spherical, subspherical or polygonal-rounded, 3–5 μm wide, with their own envelopes. Sheaths thick, firm, distinct, delimited, dark brown or blackish, but sometimes, seen from outside, with a colourless outer layer. Baeocytic division not yet observed.
Occurrence: Marine, on littoral rocks in the upper lithophytic zone; known from several localities in Croatia (Dalmatia – Adriatic coast) and France (Mediterranean coast).

## 6. **Pleurocapsa minuta** Geitler 1932 (fig. **615**)

Colonies small, mixed with other algae, or in the form of brown or blackish spots on stones and shells, composed of more or less erect, parallel, tightly arranged but not always confluent pseudofilaments. Pseudofilaments in basal parts uniserial, in upper parts 2- or several-seriate. Basal cells a little elongate (longer than wide) and narrowed towards the substrate, other cells usually shorter than wide, polygonal-rounded, (2.4)4–7 µm wide, with pale blue-green, greyish blue or grey-green content, apical cells again slightly elongated. Sheaths thin, firm, lamellate, in basal parts colourless, in upper parts sometimes yellow to dark brown. Chroococcoid stages are common. The form of pseudofilaments is evidently pleurocapsoid, but the baeocytic division is still unknown (comp. with *Hydrococcus*). Not very common species.
Occurrence: Marine, in splash zone or submersed on limestone and various other rocks, stones and on snails and *Balanus*-shells, exceptionally epiphytic on algae (*Cladophora*, on the basis of *Isactis*, etc.); described from Adriatic Sea (Croatia – Dalmatia: port of Trpanj), later found on S. coast of France and commonly on shells and stones of the western coasts of Baltic Sea (Germany, Sweden); reported also from the Sea of Japan.

## 7. **Pleurocapsa mucosa** (Ercegović) Komárek et Anagnostidis 1995 (fig. **619**)

*Scopulonema mucosum* Ercegović 1932

Colonies epilithic, widened on the substrate, forming layers up to 0.4 mm high, composed of aggregations of pseudofilaments. Psudofilaments to 50 µm wide and 80 µm long, with densely packed cells, shortly pseudodichotomously divaricate. Cells barrel-shaped or irregular-rounded, 4–8 µm wide, more or less all of similar size, with pale blue-green content. Sheaths firm, distinct, delimited, inside sometimes colourless and slimy, outside yellow-brown. Baeocytic division was found in various cells, both solitary and in groups; baeocytes about 1 µm in diameter.
Occurrence: Marine, on limestone rocks, in the upper part of the splash lithophytic zone; known only from Adriatic Sea in Croatia (Dalmatia) and Aegean Sea in Greece (Euboea).

## 8. **Pleurocapsa hansgirgiana** (Ercegović) Komárek et Anagnostidis 1995 (fig. **620**)

*Scopulonema hansgirgianum* Ercegović 1930

Colonies epilithic, flattened, spreading over the substrate, forming layers up to 0.5 mm thick, brown or black, composed of aggregations of pseudofilaments. Pseudofilaments to 40 µm wide and 150 µm long, more or less creeping over the substrate, with cells grouped irregularly or in distinct rows and arranged in (1)2-more-seriate formations, in terminal parts pseudodichotomously or laterally divaricate; in basal parts sometimes there is an irregular chroococcoid mass. Cells of varied shape, usually polygonal-rounded, in basal parts sometimes smaller, in rows rarely slightly elongate, 5–9 µm wide. Sheaths thin or slightly thickened, clearly delimited, firm, not distinctly lamellate, yellow-brown; no special envelopes around cells, only sometimes around groups of cells. Cell content blue-green, yellowish or violet (f. *rosea* Ercegović 1932; this form also has shorter pseudofilaments). Apical cells conical-rounded or widely rounded. Baeocytic division in any cell, often in «basal» parts of pseudofilaments; baeocytes 1–2 µm in diameter.
Occurrence: Marine, on littoral rocks in the splash zone, in the lithophytic zone,

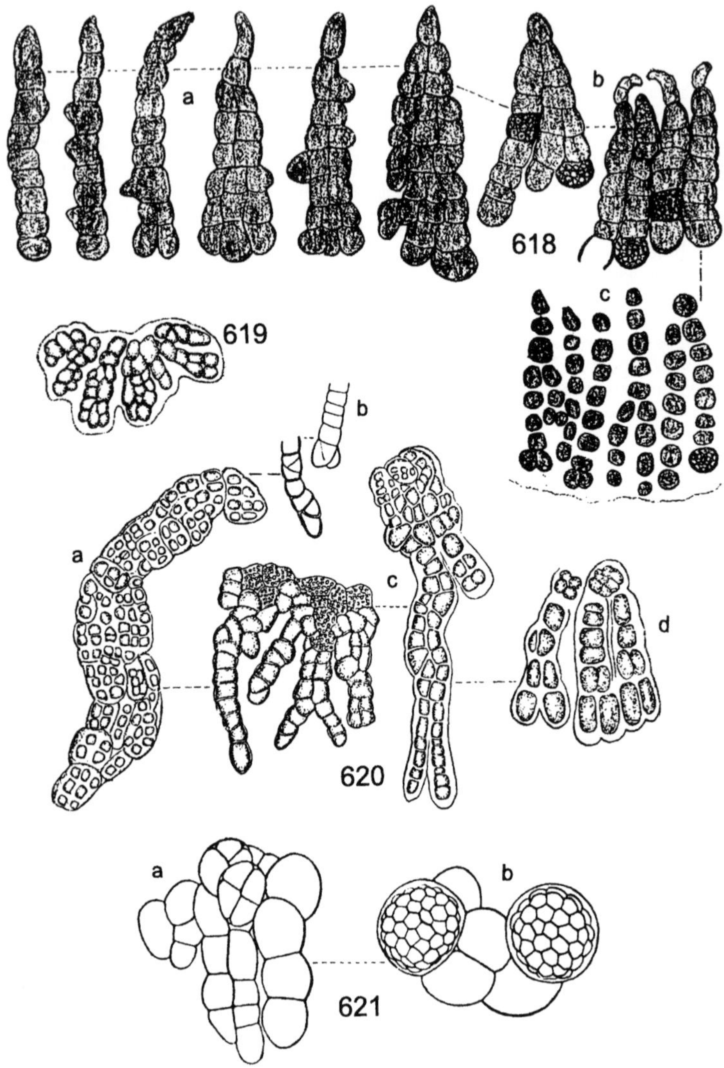


PLEUROCAPSA

**Fig. 618.** *Pleurocapsa minor* sensu Frémy, from Africa: **a** – development of thallus, **b** – old thallus with cells dividing into baeocytes, **c** – mucous stage (status mucosus) with cells dividing into baeocytes; after Frémy (1930).

**Fig. 619.** *Pleurocapsa mucosa*: after Ercegović (1932, sub *Scopulonema mucosum*).

**Fig. 620.** *Pleurocapsa hansgirgiana*: **a** – epilithic part of thallus, **b** – division of terminal cells before divarication (pseudobranching), **c** – endolithic parts of thallus

usually in shady places (partic. f. *rosea*); known from limestone coasts of Mediterranean Sea (Algeria, Croatia – Dalmatia, France, Greece – Lesbos island, Italy, Libya, Tunisia).

### 9. **Pleurocapsa fuliginosa** Hauck 1885 (fig. **622**)

Colonies irregular, to 100 μm in diameter, thin or/and crustose, forming aggregations of cells, groups of cells and pseudofilaments on the substrate, blackish; cell agglomerations irregular, with cells of varied shape and size, pseudofilaments short, irregular. Cells (3) 5–20 μm in diameter, irregular or polygonal-rounded, solitary or in 2-, 4- or more-celled aggregates, enveloped by their own, usually slightly lamellate, 50–100 μm thick, colourless sheaths; cell content homogeneous, pale blue-green, olive-green, brownish or slightly violet. Baeocytes probably arise from randomly localized cells (which are up to 25 μm in diameter), 1.5–3 μm in diameter.
Occurrence: Marine, in splash zone in stony littoral; described from Adriatic Sea (Italy), known also from Aegean Sea, Baltic Sea, Black Sea and North Sea (Norway – in Trondhjemfjord), from Atlantic coasts (England and N. America) and SW. Pacific coasts (Nauru and Lord Howe Islands); the distribution should be revised.

### 10. **Pleurocapsa crepidinum** Collins 1901 (fig. **621**)

Cells spherical or polygonal-rounded, 5–15 μm in diameter, with pale blue-green or greyish blue content, with colourless, thin sheaths; cells sometimes aggregated into groups or short irregular pseudofilaments. In marginal parts sometimes larger cells develop (to 25 μm in diameter) which divide into numerous baeocytes, about 2 μm in diameter.
Occurrence: Marine, described from coasts of N. America (on rocks, shells of *Balanus* and epiphytic on algae); recorded also from France (on algae, near Banyuls).

**Species outside Europe:**
- *Pleurocapsa africana* Gayral et Mazancourt 1960: Marine, described from African coast; it does not correspond to the g. *Pleurocapsa* morphologically.
- *Pleurocapsa brasiliensis* Schmidle 1901.
- *Pleurocapsa magna* Weber van Bosse 1913: Cells 18–36 μm in diameter, with pale violet content; marine, described from Indonesia.
- *Pleurocapsa mayori* Setchell 1924: Subspherical colonies of aggregated cells, which are about 2 μm in diameter; according to description and original drawing rather a *Chlorogloea*, *Cyanosarcina*, or *Myxosarcina* species; marine, collected from the inner surface of the *Tridacna* shells, about 7 m above high-tide level, Rose Atoll, Samoa Islands.
- *Pleurocapsa muralis* Lagerheim in Wittrock et al. 1893: Aerophytic, described from Ecuador (Quito).

---

with «basal» cells dividing into baeocytes, **d** – f. *rosea*; **a-c** – after Ercegović (1930, sub *Scopulonema hansgirgianum*), **d** – after Ercegović from Kosinskaja (1948, sub *Scopulonema hansgirgianum* f. *rosea*).
**Fig. 621.** *Pleurocapsa crepidinium*: **a** – part of thallus, **b** – basal cells, two with baeocyte division; after Geitler (1932, redrawn from the original exsiccate of Collins).

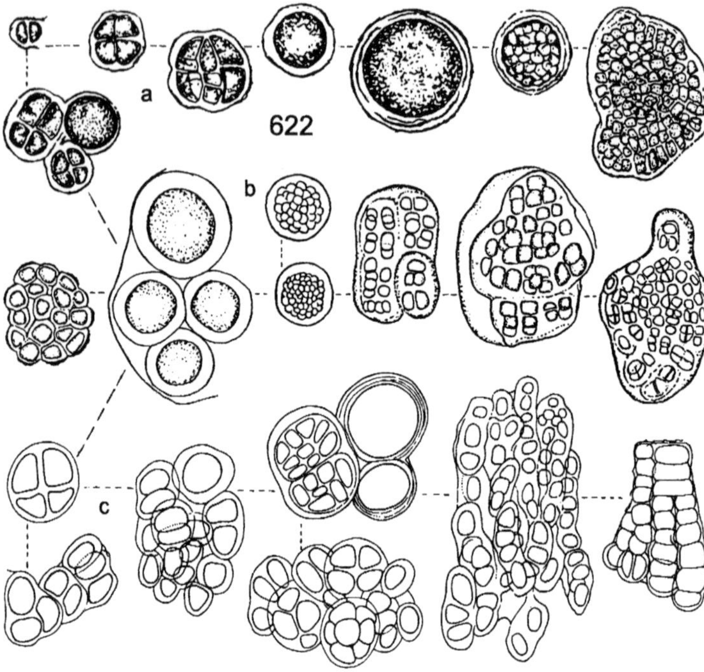


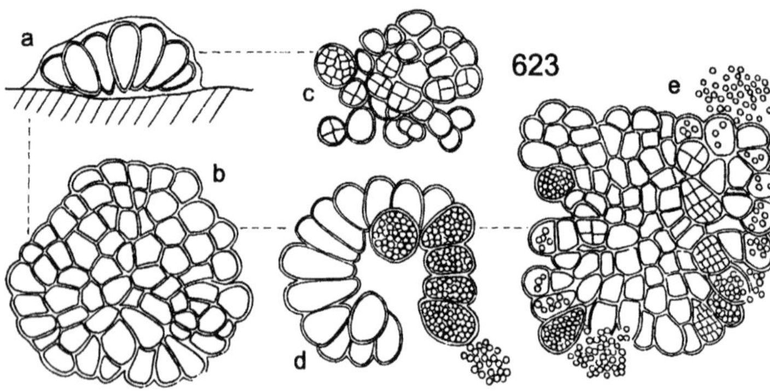


PLEUROCAPSA

**Fig. 622.** *Pleurocapsa fuliginosa*: typical cell division, variation in form of thallus, baeocytic division; **a** – after Hauck from Kosinskaja (1948), **b** – after Ercegović (1930), **c** – after Komárek (1972, redrawn from the original exsiccate of Hauck).
**Fig. 623.** *Dermocarpa violacea* Crouan sensu J. et G. Feldmann, taxonomically not definitely solved (comp. *Xenococcus* and *Pleurocapsa*): **a** – side view of a young

- *Pleurocapsa parenchymatica* Geitler 1935: Freshwater, described from Indonesia; it does not correspond precisely with the generic diagnosis.
- *Pleurocapsa usteriana* Schmidle 1904: Freshwater, epiphytic, described from a waterfall in Indonesia (Java); probably a *Xenococcus* species; (Geitler 1932).
- *Pleurocapsa violacea* Weber van Bosse 1925: Cells violet or yellow-green, divide three-dimensionally, ± 2 µm wide; marine, epizoic, on *Amphibia tenella*, Kei Island, Pacific; probably a *Cyanocystis* or *Xenococcus* species.

**Unrevised species:**
- *Pleurocapsa deeveyi* Drouet 1942: Planktic (!?) in brackish water, described from Texas, USA; without drawing or photo; description unclear.
- *Pleurocapsa rivularis* Hansgirg 1893: Freshwater, unsufficiently described species from running water.
- *Pleurocapsa varia* (A.Braun) Drouet et Daily in Daily 1942: Unclear concept, without baeocyte production, recorded from numerous localities in USA.
- *Dermocarpa violacea* Crouan sensu J.& G. Feldmann 1953: Marine, taxonomically unclear, described from France; probably a separate *Xenococcus* or *Pleurocapsa* species; (fig. **623**).

## 7. **Hyella** Bornet et Flahault 1888

Thallus highly differentiated, composed of irregular, more or less ramified (rarely simple) pseudofilaments, creeping on calcareous substrates (usually limestone or shells of molluscs, rarely on other stones or on plants), but mainly euendolithic (or endophytic) growing into the substrate (boring types). Pseudofilaments with one or more rows of cells, sometimes in older parts are formed irregular groups of cells, often nematoparenchymatously arranged; pseudofilaments usually laterally (rarely pseudodichotomously) divaricate. Thus, thallus is differentiated into (i) initial, epilithic or superficially endolithic part, from which grow (ii) pseudofilaments into the substrate (sometimes with branching); apical parts of endolithic pseudofilaments (i.e., deepest parts in the substrate) are usually morphologically different from initial pseudofilaments. Mucilaginous sheaths are firm, thick, usually layered, rarely gelatinized, commonly visible also between cells. Cells more or less spherical, subspherical, oval or rounded polygonal, of diverse size, sometimes distinctly elongate, cylindrical, ovoid, at the ends of pseudofilaments oval or club-shaped (rarely hemispherical). Thylakoids irregularly arranged (*H. balani, H. caespitosa*).
Cells divide by binary fission, irregularly in various planes, in pseudofilaments crosswise (perpendicularly to the long axis of pseudofilament), before branching also lengthwise or obliquely. Oldest cells near the substrate surface often change into baeocytes, which arise after rapid successive or almost simultaneous repeated cell division (multiple fission). Reproduction by baeocytes, rarely by liberation of solitary small cells (?).
**Important references:** Bornet & Flahault (1888); Gardner (1918); Setchell & Gardner (1919); Geitler (1932, 1942); Ercegović (1932); Frémy (1934); Elenkin (1938); Kosinskaja (1948); LeCampion-Alsumard (1969, 1970, 1972, 1979-diss., 1991); Lukas & Golubić (1983); Lukas & Hoffman (1984); LeCampion-Alsumard & Golubić (1985a); Anagnostidis & Pantazidou (1988); Al-Thukair & Golubić (1991); Pantazidou (1991-diss.); Al-Thukair (1992).

---

colony, **b** – top view of a young colony, **c-e** – colonies of various age with cells dividing into baeocytes; after Feldmann & Feldmann (1953).

**Key to the species:**

**1a** Freshwater or inland subaerophytic species . . . . . . . . . . . . . . . . . . . . . . **2**
**1b** Marine epi/endolithic, epi/endozoic (on shells), rarely epi/endophytic species . . . . . . . . . . . . . . . . . . . . . . . . . . . . . . . . . . . . . . . . . . . . **6**
**2a** Cells in pseudofilaments 2.5–3 µm wide; known from soil cultures . . . . . . . . . . . . . . . . . . . . . . . . . . . . . . . . . . . . . . . . . **1. H. terrestris**
**2b** Cells in pseudofilaments over 3 µm wide; species from aquatic biotopes (springs, streams) . . . . . . . . . . . . . . . . . . . . . . . . . . . . . . . . . . . . **3**
**3a** Cell content reddish violet . . . . . . . . . . . . . . . . . . . . . . . . . . **4. H. jurana**
**3b** Cell content grey-green, pale green or bright green . . . . . . . . . . . . . . . . **4**
**4a** Pseudofilaments more or less of the same width, more or less long, with rare branching; cells in pseudofilaments –10x longer than wide; cell content grey or pale blue-green . . . . . . . . . . . . . . . . . . . . . . . . . **10. H. kalligrammos**
**4b** Pseudofilaments of variable width, with frequent branching; cells in pseudofilaments 2–3x longer than wide; cell content blue-green or bright blue-green . . . . . . . . . . . . . . . . . . . . . . . . . . . . . . . . . . . . . . . . . . . . **5**
**5a** Cells 5–10 µm wide, terminal cells elongate, 3–4x longer than wide, not widened; cell content blue-green . . . . . . . . . . . . . . . . . . . **5. H. fontana**
**5b** Cells 7–30 µm wide, terminal cells sometimes distinctly enlarged and widely rounded, globose, sometimes over 30 µm in diameter; cell content bright blue-green . . . . . . . . . . . . . . . . . . . . . . . . . . . . . . . . **6. H. maxima**
**6a** Endophytic, subcuticular in marine seaweeds . . . . . . . . . . . . . . . . . . . **7**
**6b** Species growing into calcium carbonate substrates (limestone rocks, shells of molluscs), usually euendolithic (boring into the substrate) . . . . . . . . . . . **8**
**7a** Intercellular species in Rhodophytes (*Dilsea*); colonies to 1.5 mm in diameter, composed of numerous parallel rows of cells; cell content reddish . . . . . . . . . . . . . . . . . . . . . . . . . . . . . . . . . . . . . . . . . . **2. H. dilseae**
**7b** Endophytic in Phaeophytes; pseudofilaments short, few-celled; cell content blue-green . . . . . . . . . . . . . . . . . . . . . . . . . . . . . . **3. H. endophytica**
**8a** Pseudofilaments of variable width, more or less short (usually less than 100 µm long), frequently branched; cells 2–3x longer than wide, cells dividing into baeocytes 3–5x > than vegetative cells . . . . . . . . . . . . . . . . . . **9**
**8b** Pseudofilaments more or less of the same width, more or less long (in old colonies usually over 100 µm long), little branched; cells –10x longer than wide, cells dividing into baeocytes slightly bigger than vegetative cells . . **11**
**9a** Cell content blue-green; euendolithic pseudofilaments uni- or multiseriate, to over 30 µm wide; pseudobranches usually relatively short, sometimes do not differ from main pseudofilaments . . . . . . . . . . . . . . . . . . . . . . . **10**
**9b** Cell content grey reddish; euendolithic pseudofilaments mainly uni- or biseriate, to 19 µm wide, with distinct pseudobranches . . . . . **9. H. pyxis**
**10a** Cells 4–8 µm in diameter; pseudofilaments in outer part (i.e., near the surface of stones) always multiseriate, to 32 µm wide, pseudobranches lateral, usually short . . . . . . . . . . . . . . . . . . . . . . . . . . . **7. H. balani**
**10b** Cells 4.4–28.5 µm in diameter; pseudofilaments uniseriate or few-seriate, irregularly radially arranged, irregularly waved, to 37.5 µm wide, pseudobranches not distinctly differentiated from the main pseudofilaments . . . . . . . . . . . . . . . . . . . . . . . . . . . . . . . . . . . . . . . . . . . **8. H. gigas**
**11a** Cells in pseudofilaments near to one another, 4–10 µm wide, –60 µm long . . . . . . . . . . . . . . . . . . . . . . . . . . . . . . . . . . . . . . **11. H. caespitosa**
**11b** Cells in pseudofilaments distinctly separate from one another, 2–8 µm wide, –40 µm long . . . . . . . . . . . . . . . . . . . . . . . . . . . . . . . . . . . . . . . . . **12**
**12a** Cells –40 × 2–5 µm, pseudofilaments to 400 µm long . . . . . **12. H. tenuior**
**12b** Cells –30 × 2–8 µm, pseudofilaments to 30 µm long . . . . **13. H. dalmatica**

Notes: Two groups of species, differing in the method of thallus differentiation (LeCampion-Alsumard & Golubić 1985) were recognized within the genus *Hyella*. These differences result from the different strategies of cell division and branching patterns, and from the succeeding life cycles. Anagnostidis & Pantazidou (1988) proposed also two subgenera under the provisional names «*Megahyella*» and «*Microhyella*», based on cell division, branching pattern, mode of reproduction and life cycles. Both models express two patterns of thallus development, characteristic for (i) *H. balani* or *H. fontana* (and more or less species **1–8**), and (ii) *H. caespitosa* (and more or less for species **9–12**). The third modification is probably represented by endophytic species (*H. dilseae, H. endophytica, H. seriata*).

## 1. **Hyella terrestris** Chodat 1921

Thallus free living (soil species?). Cells in pseudofilaments cylindrical, 7–9 × 2.5–3 µm or more or less globose (barrel-shaped?) and up to 6 µm wide; cell content dirty green or violet. Sheaths thin, colourless, to 1 µm thick. Unsufficiently described and unrevised species (divariacation, baeocytic division, etc.? ). Without documentation.
Occurrence: Described from soil cultures from a coniferous forest; Switzerland.

## 2. **Hyella dilseae** (Feldmann) Komárek & Anagnostidis 1995 (fig. **624**)

*Myxohyella dilseae* Feldmann 1956

Colonies small, endophytic, growing near the surface of thallus of a rhodophyte host, irregularly rounded, 1–1.5 mm in diameter and to 500 µm high (long), dark purple violet, located in intercellular spaces. Cells more or less irregular-rounded, 5–6 µm in diameter, with homogeneous, finely granular content, arranged in more or less parallel or irregular pluriseriate rows, growing from the surface into the host plant. Cells which divide into baeocytes, 7–8 µm in diameter, arise only from oldest vegetative cells near the host surface; baeocytes small, rounded, 1–1.5 µm in diameter.
Occurrence: Marine, epiphytic and mainly endophytic in *Dilsea carnosa* and other red algae; Atlantic coast near Roscoff, France.

## 3. **Hyella endophytica** Börgesen 1903 (fig. **625**)

*Myxohyella endophytica* (Börgesen) Geitler 1925

Thallus small, microscopic, few-celled, endophytic, composed of short, uniseriate, pseudodichotomously divaricate, 40–80 µm wide pseudofilaments, growing perpendicularly from the host plant surface into the phylloid. Sheaths thin, firm, colourless. Cells more or less of the same size, more or less cylindrical rounded, shorter than wide, isodiametric or longer than wide (particularly towards the ends), 3–5.4 µm wide; cell content blue-green. Cell division by binary fission, cells in the proximal part (near the plant surface), which are usually longer and wider than the other vegetative cells (4–10 µm in diameter), divide sometimes into numerous baeocytes.
Occurrence: Marine, endophytic under the cuticula of phylloids of seaweeds (*Chondrus crispus, Laminaria hyperborea*); found and described from the coast of Faeroe islands, not revised later.

## 4. **Hyella jurana** Chodat 1898 (fig. **626**)

*Hyella fontana* var. *jurana* (Chodat) Nadson 1900

Thallus epi- and endolithic, reddish. Pseudofilaments irregularly wavy and laterally or pseudodichotomously (acc. to drawings) branched, uniseriate, but in proximal parts forming multiseriate irregular clusters of cells. Sheaths thin, firm, probably colourless. Divarication (branchings) by the lateral outgrowth of intercalary cells (from the middle of cell sides), from which arise the typical «T-like» cells; according to original drawings also pseudodichotomous divarications after the longitudial division of an apical cell. Cells longer than wide, sometimes with irregularly wavy margins, only in proximal parts (in the clusters) irregular or polygonal-rounded; apical cells elongate, often club-shaped; cell content probably red. Cells divide by binary fission, reproduction by baeocytes; cells from any part of the thallus can divide into baeocytes; these cells are larger than vegetative cells before division, often T-shaped or of very irregular («bizarre») form. Chroococcoid stages are also known.The species is not well described (cell dimensions, baeocytes, etc.).

Occurrence: Freshwater, growing on limestone substrates (stones) in stagnant water; described from France (Jura Mts.).

## 5. **Hyella fontana** Huber et Jadin 1892 (fig. **628–630**)

Thallus (colonies) euendolithic, apparently pseudoparenchymatous, flat, grey-green or grey-brown, composed of densely packed pseudofilaments, usually radiating from the point of penetration into the substrate. Pseudofilaments bore into carbonate substrates, perpendicularly or obliquely, more or less branched,

HYELLA

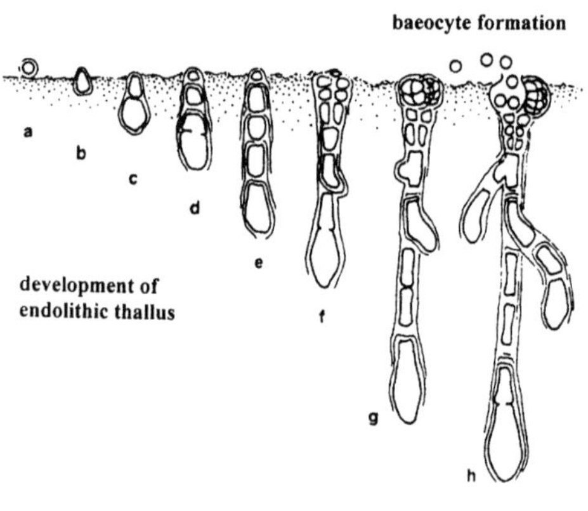


GOLUBIĆ & al. 1985

**Fig. 624.** *Hyella dilseae*, marine, endophytic: **a** – top view of the surface of thallus, **b** – vertical section through the margin of the host (*Dilsea carnosa*) with penetrating pseudofilaments, **c** – detail of pseudofilaments with the superficial cells dividing into baeocytes; after Feldmann (1956, sub *Myxohyella dilseae*).

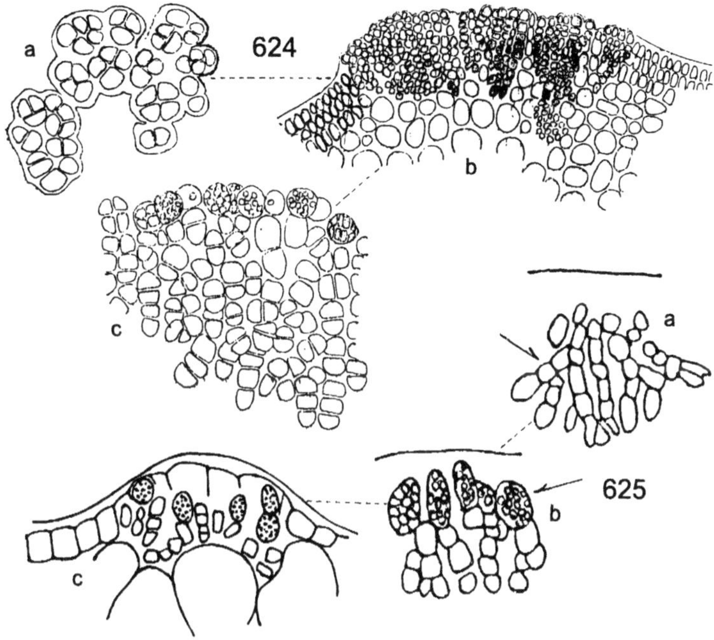


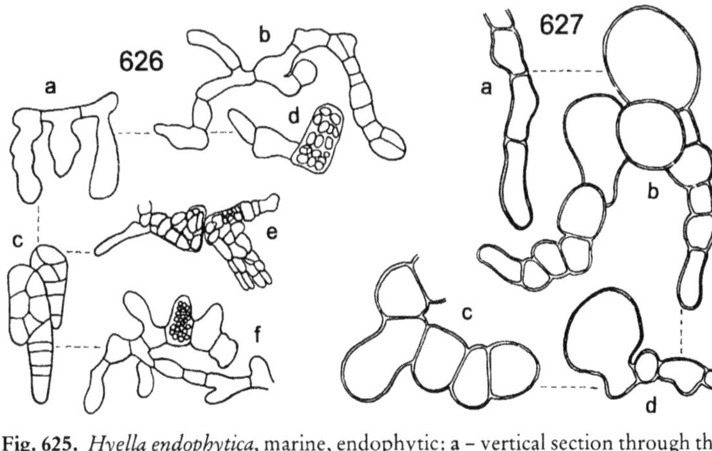


**Fig. 625.** *Hyella endophytica*, marine, endophytic: **a** – vertical section through the marginal part of a phaeophycean host with penetrating pseudofilaments, **b-c** – «basal» cells (near the surface) dividing into baeocytes; after Börgesen (1903).
**Fig. 626.** *Hyella jurana*, freshwater, epilithic (limestone); after Chodat (1898).
**Fig. 627.** *Hyella maxima*, freshwater, epi- and endolithic (limestone): from Geitler (1932, sub *H. fontana* var. *maxima*).

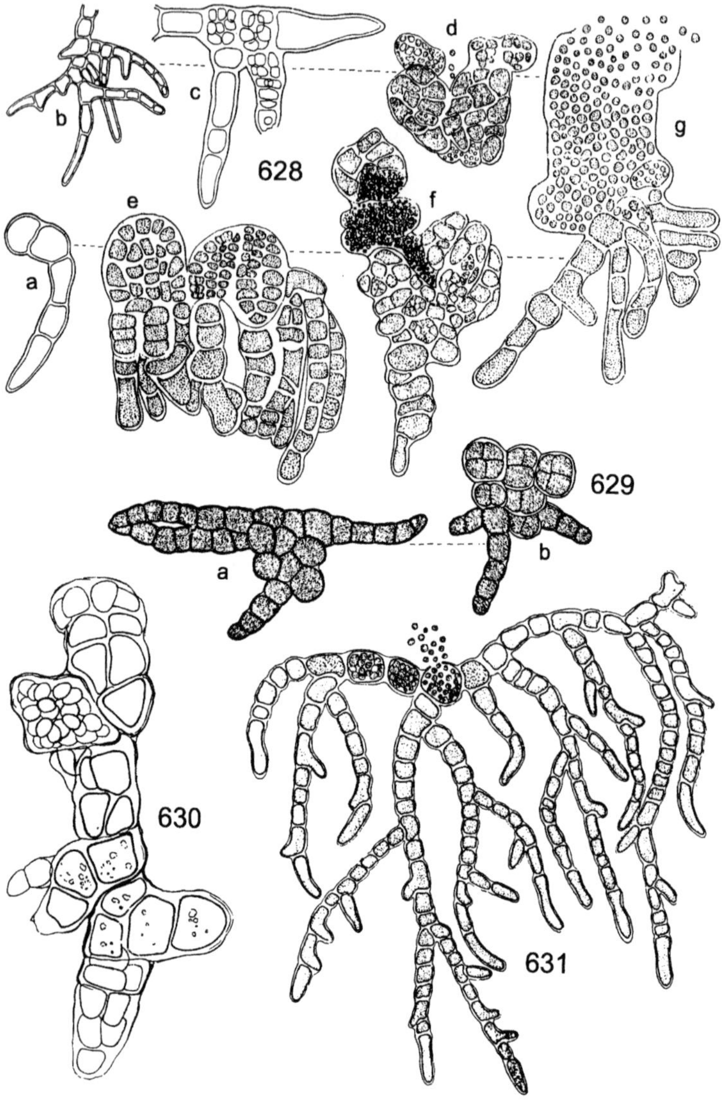


HYELLA

**Fig. 628.** *Hyella fontana*, freshwater, mainly endolithic (limestone): **a-b** – development of thallus, **c-g** – thallus with «basal» cells dividing into baeocytes; after Huber & Jadin (1892), partly from Geitler (1932).

4.5–7 μm wide; proximal part of pseudofilaments 1-, 2- or multiseriate, consisting of coccoid cells arranged sometimes in three-dimensional clusters; distal parts usually uniseriate, composed of cylindrical cells. Gelatinous sheaths firm, thin, colourless, homogeneous. Branching pattern of two types: (i) initiated at the distal end of an intercalary cell by a change of growth into a lateral direction, or (ii) initiated by a lateral protrusion of an intercalary cell («lateral branching»). Proximal cells are spherical, subspherical, piriform or angular, 7–11.2–17 × (4.6)5–9.2(12) μm, distal cells cylindrical (isodiametric or a little shorter or longer than wide), long cylindrical, rarely subspherical, 3–6.8–12 × 4–6 μm, terminal cells 3-4-times longer than wide, 6–10.3(18) × 3–5.5 μm, sometimes conically rounded or curved. Cell content pale blue-green or greyish, finely granular, rarely with conspicuous inclusions. Reproduction by baeocytes produced by successive division of proximal cells; sometimes part of a mother cell remains undivided; mother cells, which are able to divide into baeocytes are usually larger and more rounded than vegetative cells, 9–18.6–23 × 6–12.2(18) μm; baeocytes are spherical or angular-rounded, usually densely packed, in number to 32 or rarely even more, 1.2–1.6 μm in diameter. Chroococcoid stages were recorded.
Occurrence: Freshwater, boring into calcareous substrates (stones, rocks, shells) in clear, running water, more or less common; Austria, Croatia, Greece, Lithuania, Poland, Russia, Slovakia, Ukraine, outside Europe reported from USA and Cuba; records from Gabon (Frémy 1930) are related more to specimens from Greece (Anagnostidis & Pantazidou 1988) than to the original figures of Huber & Jadin (fig. **631**).

### 6. Hyella maxima (Geitler) Anagnostidis et Pantazidou 1988 (fig. 627)

*Hyella fontana* var. *maxima* Geitler 1928

Thallus euendolithic through almost the whole life cycle, dark blue-green to olive-green, boring into carbonate substrates, composed of divaricate pseudofilaments, which are more or less straight and penetrate perpendicularly into the substrate; pseudofilaments usually not densely aggregated. Pseudofilaments (5)8–11 μm wide, consisting of proximal coccoid and distal cylindrical cells. Gelatinous sheaths firm, colourless or occasionally bluish, homogeneous, not layered. Branching of three types: (i) initiated laterally from large proximal (superficial) globose cells (initiators of pseudofilamentous growth), (ii) initiated at the distal end of an intercalary cell by change of growth into a lateral direction, or (iii) initiated by lateral protrusion of an intercalary cell near its middle or (more frequently) at its distal end (sometimes forming «T-cells»); in all these cases the branching intercalary cell occasionally take on the role of the apical cell of a new branch (pseudodichotomous branching was not observed). Cells spherical, widely oval, subspherical or angular-rounded, 10–22.8–31 × (9)-17.8–23 μm in proximal parts, cylindrical (to long cylindrical) or slightly irregular, (9)-13.6–18 × 4.5–6.2–8(10) μm in distal parts, and irregular-rounded and distinctly elongate (to 3-times longer than wide), 22–53.4–120 × (3.5)-5.6–8 μm at the apical ends. Cell content blue-green to olive-green, or intensely blue-green, finely granular, usually

---

**Fig. 629.** *Hyella fontana*; a – young thallus, **b** – pseudofilaments penetrating into the substrate, basal cells dividing into baeocytes; after G.M.Smith (1950).
**Fig. 630.** *Hyella* cf. *fontana*, from the small calcareous stream in Cuba, epilithic; after Komárek (1985, sub *Hyella* sp.).
**Fig. 631.** *Hyella* sp., from stream in Gabon, Africa; after Frémy (1930, sub *Hyella fontana* - probably a separate type).

with conspicuous inclusions. Reproduction by baeocytes produced by multiple fission or successive binary fission of one (rarely more) proximal, large (spherical) cell in the same pseudofilament (sometimes a part of the mother cell remains undivided). Cells dividing into baeocytes measure 15–45 × 14–26 μm, baeocytes are spherical or angular-rounded, up to 64 per cell, densely packed, (1.5)2.4 μm in diameter. Chroococcoid clusters not observed.
Occurrence: Freshwater, epi- and mainly euendolithic, submersed in calcareous streams and in lake littoral (to 15 m depth), boring into carbonate substrates (rocks, stones); Austria, Greece, Montenegro.

## 7. **Hyella balani** Lehmann 1903 (fig. **632**)

Thallus brown to yellow-brown, only slightly epilithic, mainly euendolithic, composed of proximal, irregularly arranged groups of cells, from which grow

HYELLA

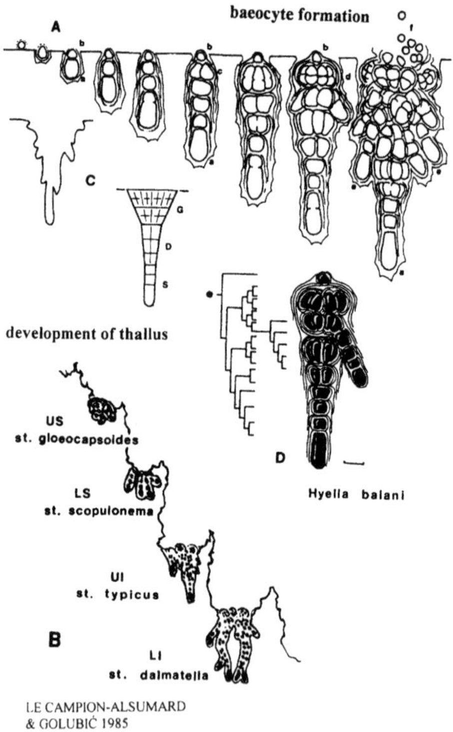


**Fig. 632.** *Hyella balani*, marine, endolithic: **a** - development of thallus, **b** - cell division into baeocytes, **c** - various stages of thallus; **a-b** - after Lehmann from Geitler (1933), **c** - from Le Campion-Alsumard & Golubić (1985).

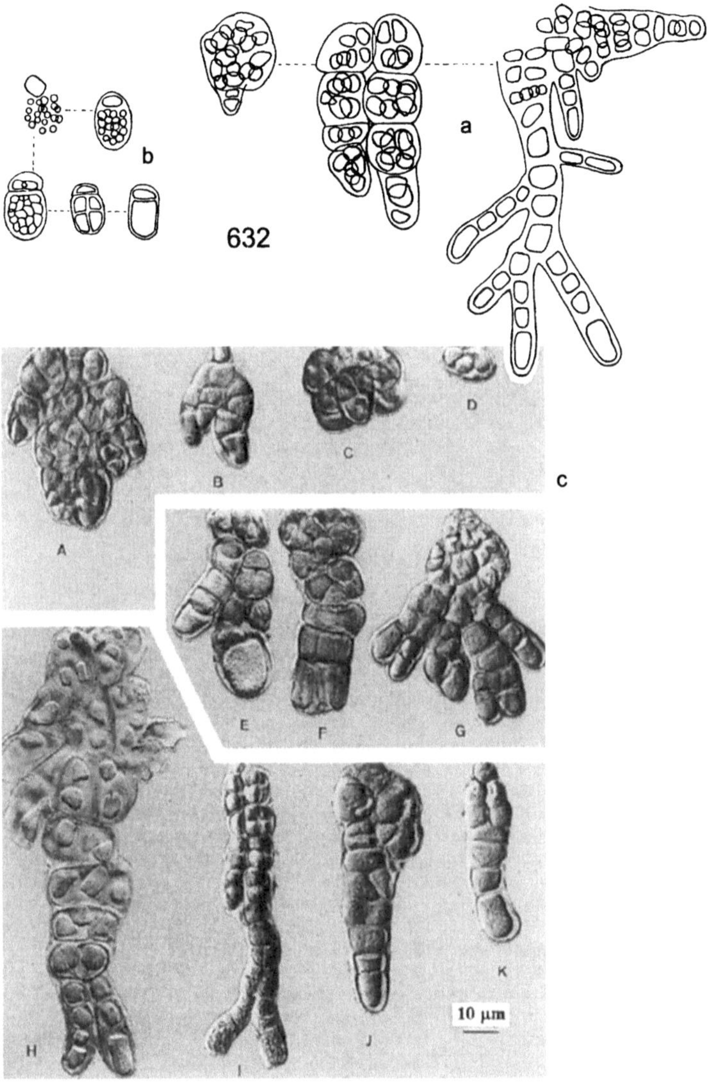
a
b
632
A
B
C
D
c
E
F
G
H
I
J
K
10 μm

simple, at the basis multiseriate, but later uniseriate, irregularly divaricate pseudofilaments, boring perpendicularly into carbonate substrates; the first cell division is always transverse. Gelatinous sheaths firm, colourless, lamellate. Lateral branching originates from external cells of chroococcoid clusters; these cells protrude out of a cluster and assume the role of a new apical cell. Pseudodichotomous branching results from the longitudinal division of the apical cell. Cells in proximal parts irregularly polygonal-rounded, sometimes in groups, which are surrounded by individual envelopes; in pseudofilaments cells are more or less cylindrical, usually longer than wide, 4–8 µm wide; terminal cells elongate and widely rounded, to 20 µm long. Cell content blue-green, thylakoids irregularly arranged throughout the cell. Reproduction by baeocytes, which develop from proximal cells (near the surface), which are usually ellipsoidal, to 13 µm long and circa 7–8 µm wide; baeocytes differentiate by successive division, many per cell, more or less spherical, 1.5–4 µm in diameter; part of the mother cell sometimes remains undivided. Also known are developmental gloeocapsoid, chroococcoid, and other stages (with irregularly arranged cells surrounded by a common mucilaginous envelope). (Life cycle described in LeCampion & Golubić 1985 and Pantazidou 1991-diss.).
Occurrence: Marine, on limestone coasts on rocks and stones, or (in the non-calcareous littoral) on shells of molluscs (*Balanus, Patella, Littorina*, etc.); mainly in Mediterranean Sea, but known also from Denmark, Germany, Norway and outside Europe from Red Sea, Sea of Japan, Bering Sea, Indian Ocean (Aldabra Atoll, S. Africa) an SW. Pacific coasts (New Hebrides, Nauru and Ellice Islands).

### 8. **Hyella gigas** Lukas et Golubić 1983 (fig. **633**)

Thallus microscopic, euendolithic, blue-green. Pseudofilaments more or less short, much ramified, penetrating into the carbonate substrate in short «fascicles», with 6–9(10) cells, (22)-100 (and more) µm long, 6.5–37.5 µm wide (in European localities 10–25 µm wide); usually uniseriate, rarely biseriate (mainly in the proximal part). Mucilaginous sheaths thin or more or less widened, to 5 µm thick, lamellate (up to 3-layered), colourless or yellow-brownish; special envelopes develop sometimes around cells. Branching (divarication of pseudofilaments) lateral. Proximal («basal») cells more or less spherical, subspherical or polygonal-rounded, rarely isodiametric cylindrical, (7)9–32 µm (in Europe circa 7–16 µm) in diameter, intercalary (central) cells more or less isodiametric, 4.5–11.7–26 × 4.5–9.2–28.5 µm (in Europe 6–13 × 7–16 µm), apical cells 6.5–14.5–28.5 × 5.5–9.4–22 µm. Cell content bluish to blue-green, homogeneous to granulate or with conspicuous inclusions, rarely brown-yellowish. Reproduction by baeocytes, which arise from proximal cells, usually about 32 (rarely more) per cell; they are spherical or angular, 4–8 µm (in Europe 2–3(4) µm) in diameter.
Occurrence: Marine, euendolithic on limestone rocks and stones, boring into carbonate substrates, less frequently on shells of molluscs (*Patella*), in lower sublittoral down to 25(75?) m depth; described from the coasts of USA (E. Florida) and Bahama Islands, in Europe known from the Mediterranean Sea (France – near Marseille, Greece – Attiki Peninsula, Paros Island). Both Medi-

---

HYELLA
**Fig. 633.** *Hyella gigas*, marine, endolithic (limestone): **a** – developmental stages, **b-c** – colonies from decalcified substrates; from Lukas & Golubić (1983), **a** – from Pantazidou (1991, after unpublished photomicrographs of K.Lukas).

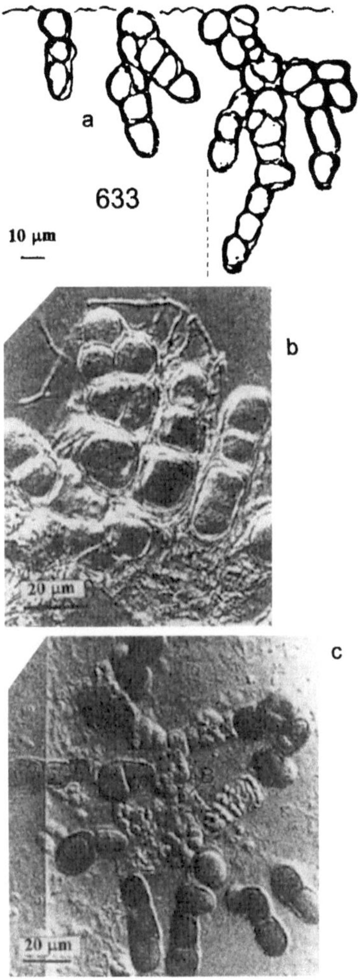


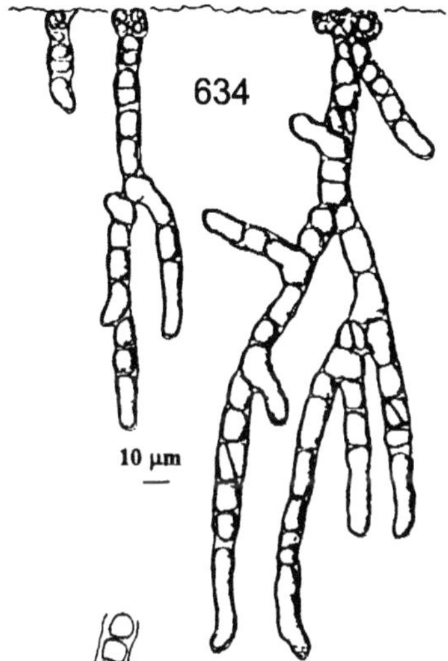


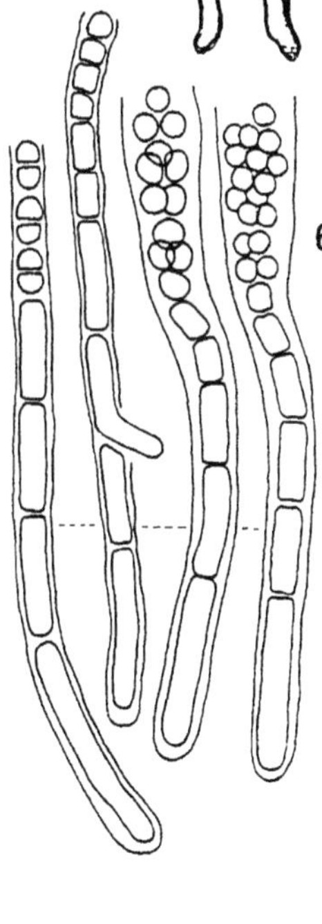


**Fig. 634.** *Hyella pyxis*, marine, endolithic (limestone): developmental stages; from Pantazidou (1991).
**Fig. 635.** *Hyella kalligrammos*, freshwater, mainly epilithic (limestone): after Anagnostidis & Pantazidou (1988).

terranean descriptions (LeCampion-Alsumard 1979-diss.; Pantazidou 1991-diss.) differ a little from the original diagnosis, mainly in cell and nanocyte dimensions and the colour of cell protoplasts.

### 9. **Hyella pyxis** Lukas et Hoffman 1984 (fig. **634**)

Thallus euendolithic, bluish, blue-green to brown reddish, composed of branched pseudofilaments boring into calcareous substrates. Pseudofilaments radiating from point of entry into substrate (penetration), to 350 µm long (in Europe 100–220 µm), 5.5–19 µm wide (in Europe 6.5–15 µm in the proximal part, 6.5–8.8 µm in the intercalary and apical parts), uniseriate, biseriate to multiseriate, mainly in proximal part. Gelatinous sheaths firm, thin, colourless, not layered, 0.5–1.2 µm thick. Pseudofilaments laterally divaricate; branches frequent, initiated by lateral projections of an intercalary or distal cell or by reorientation of growth in a lateral direction (alteration of cell division plane). Cells 2–52 × 2.7–21 µm, blue-green, in Europe in proximal part of diverse shape (subspherical, shortened to isodiametric cylindrical or of irregular form), 2–8 × 4–5 (7) µm; widened cells dividing into baeocytes 7–21 × 7–18 µm; intercalary cells ellipsoid, shorther or longer than wide, or isodiametrically cylindrical, 3–14 × 3–5.2 µm. Apical cells cylindrical, ellipsoid to pear-shaped, 6–55 × 4–14 µm (in Europe 9–30 × 4–5 µm). Reproduction by spherical to angular baeocytes (to 32 per cell), arising from the proximal parts of pseudofilaments by multiple fission (successive cell division in more than one plane), 3–7 µm in diameter (in Europe 1–2 µm in diameter).
Occurrence: Marine, boring into bivalve and gastropod mollusc shells and carbonate rocks; described from the coasts of the continental margin of E. Florida (lower intertidal zone, down to 50 m), known also from Bermuda and S. African coasts; in Europe known from the Aegean Sea (Greece – Attiki Peninsula). The Aegean description (Pantazidou 1991-diss.) differs from the original diagnosis, mainly in cell and baeocyte dimensions.

### 10. **Hyella kalligrammos** Anagnostidis et Pantazidou 1988 (fig. **635**)

Thallus euendolithic, greyish to dark grey, boring into carbonate substrates. Penetrating pseudofilaments straight or slightly curved, growing perpendicularly into the substrate, (2.5)4–5(6) µm wide; proximal part of pseudofilaments 1-, 2- or (rarely) multiseriate, distal parts always uniseriate. Gelatinous sheaths firm, colourless, not layered. Pseudofilaments laterally divaricate; branches are initiated from a lateral protrusion of a vegetative cell. Cells cylindrical, isodiametric or slightly shorter than wide in proximal parts, 1.5-(4) × 2–4 µm, intercalary cells cylindrical, 4–17 × 2.5–4 µm, apical cells longer than wide, 17–47 × 2.5–3.5(4.5) µm; cell content pale blue-green to dark grey, finely granular. Reproduction by equidimensional cells, liberated from the proximal parts of pseudofilaments and (rarely) by baeocytes, which arise from proximal cells by multiple fission or by successive binary fission; baeocytes spherical or subspherical, circa 1.5 µm in diameter.
Occurrence: Freshwater, submersed in clear streams in limestone areas, in carbonate pebbles and pieces of calcareous rocks; described from Greece (the authors describe also a form that morphologically is slightly different from the type). Probably widely distributed in warmer calcareous regions (e.g., surely in Mexico – San Luis Potosí).

**11. Hyella caespitosa** Bornet et Flahault 1889 (fig. **636**)

Thallus (epi- and) euendolithic, on the surface (macroscopically) dot-shaped, or in the form of small spots or small flattened discs, 1–2 mm in diameter (later sometimes confluent), yellow-green or yellowish brown, gelatinous on the surface. Primary, superficial pseudofilaments more or less horizontally creeping just under the stone surface, wavy and coiled, uniseriate or (rarely) multiseriate; later dividing into irregular groups of cells or into clusters of chroococcoid cell groups. Deeply penetrating pseudofilaments elongate, more-seriate at the basis, later uniseriate, laterally divaricate, towards the apical ends with very long cells; cells 4–10 µm wide, at the pseudofilamentous ends to 60 µm long; cell content yellow-green or blue-green (red-violet in depths of 10–15 m). Thylakoids contorted, densely distributed throughout the cell. Sheaths more or less thick, in superficial parts often lamellate, colourless, sometimes gelatinizing at the margin. Reproduction by solitary cells, liberated from the surface cell clusters or by baeocytes, which arise from intercalary or basal (proximal) cells, sometimes enlarged before baeocytic division, pear-shaped, with thickened, striate sheaths at the base; baeocytes are more or less spherical, to 3 µm in diameter, and arise in large number from mother cells. Variable species, from which were described several, not well-defined varieties (var. *spirorbicola* Hansgirg 1892; var. *nitida* Batters 1896). (Life cycle described in LeCampion & Golubić 1985 and Pantazidou 1991-diss.).
Occurrence: Marine, along calcareous coasts, old mollusc shells or Crustacean shells and calcified macroalgae, mainly in subtidal zone, down to a depth of 60 m (–100 m ?); known from Mediterranean Sea region, Baltic Sea, Nord Sea, Atlantic coasts (both European and American), Sinai, Indonesia, Chile and Pacific coasts (Tuamotu archipelago, Lord Howe and Ellice Islands); probably widely distributed over the whole world in corresponding habitats.

**12. Hyella tenuior** Ercegović 1932 (fig. **637**)

Thallus euendolithic, forming on the surface bluish, blue-green, olive-grey, reddish to dark violet coloured spots, to 12 mm in diameter. Boring pseudofilaments grow more or less parallely and perpendicularly into the substrate, irregularly wavy, to 400 µm long and 8 µm wide, rarely laterally branched; pseudofilaments mainly cylindrical, composed of colourless, hyaline, not striate, clearly delimited mucilaginous sheaths and one row of cells, distinctly distant from one another; pseudobranches always intercalary. Basal cells cylindrical isodiametric or subspherical to spherical, 2.5–6 × 2.5–4 µm, intercalary cells usually elongate, cylindrical, drop-shaped, club-shaped or pear-shaped, 5–25 × 2–5 µm, with blue-green, blue-reddish to reddish and with homogeneous, finely granulated content; terminal cells elongate cylindrical, rarely pear-shaped (to 43 µm long ), widely rounded at the apex. Reproduction by intense cell division in the upper part of pseudofilaments (baeocytes ?); typical baeocyte formation (division of ensheathed cells) not observed.
Occurrence: Marine, boring into limestone rocks and shells mainly in the lower (subtidal) coastal zone (down to 45 m) but also in the upper intertidal zone (mainly in permanently shady localities); wide distributed in the Mediterranean Sea region, Croatia (Dalmatia), France (Marseille) and Greece (Aegean Sea – Attiki Peninsula, Peloponnesos, Cycladic Islands), Egypt, known also from the Atlantic (USA, Jamaica, Belize) and Indian Oceans (Aldabra Atoll).

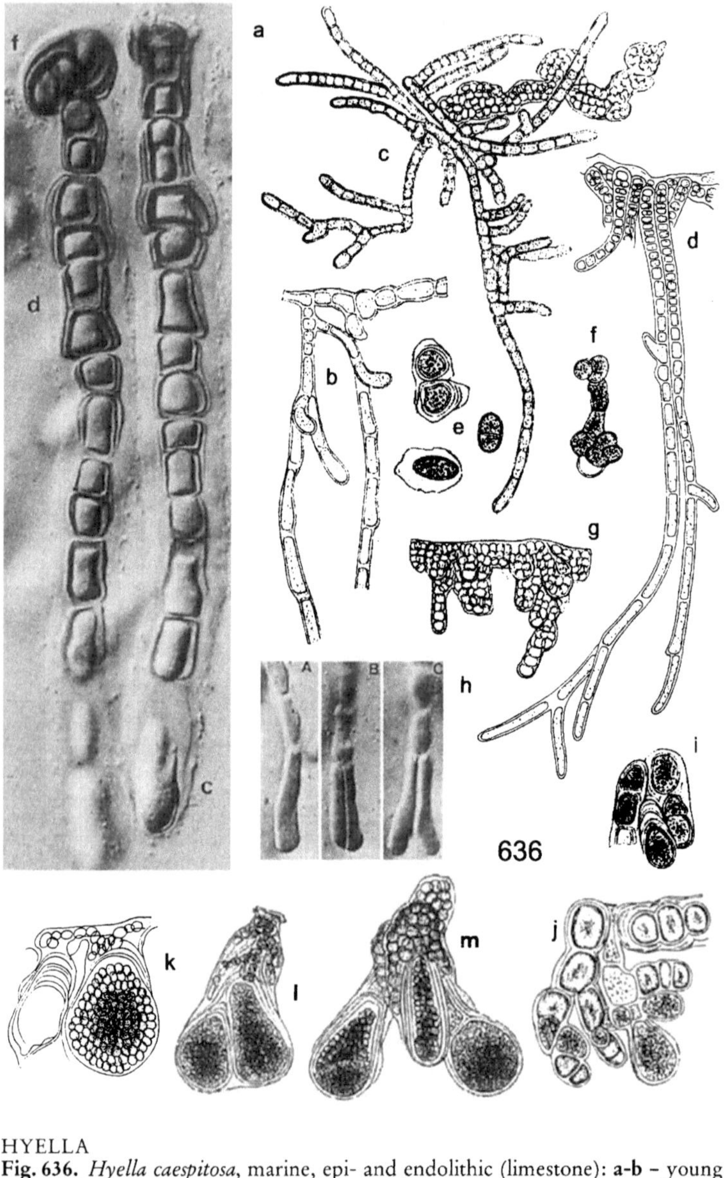


HYELLA

**Fig. 636.** *Hyella caespitosa*, marine, epi- and endolithic (limestone): **a-b** – young pseudofilaments, **c-d** – penetrating pseudofilaments with baeocyte production near the surface, **c-f** – initial stages, **g** – part of the epilithic thallus, **h** –

division of apical cell, **i-m** – cells dividing in baeocytes (however, comp. with *Cyanosaccus*); **a,h** – from LeCampion-Alsumard & Golubić (1985), other after Bornet & Flahault from Kirchner (1900) and Geitler (1942).

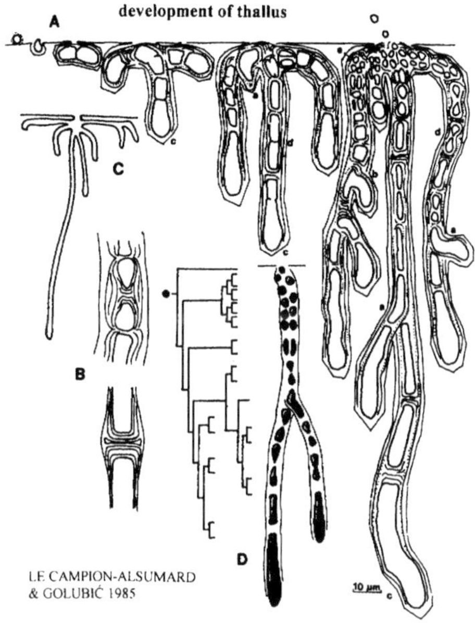


**Fig. 636n.** *Hyella caespitosa* occurs sometimes together with *H. balani*; both species are, however, easy distinguishable: **a** – *H. caespitosa*, b – *H. balani*, from LeCampion-Alsumard & Golubić (1985).

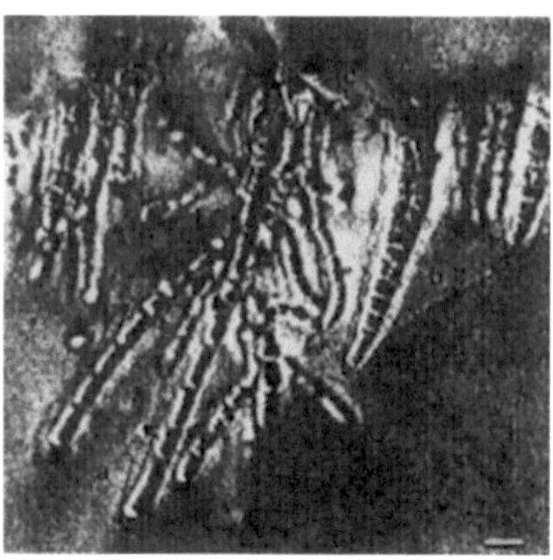

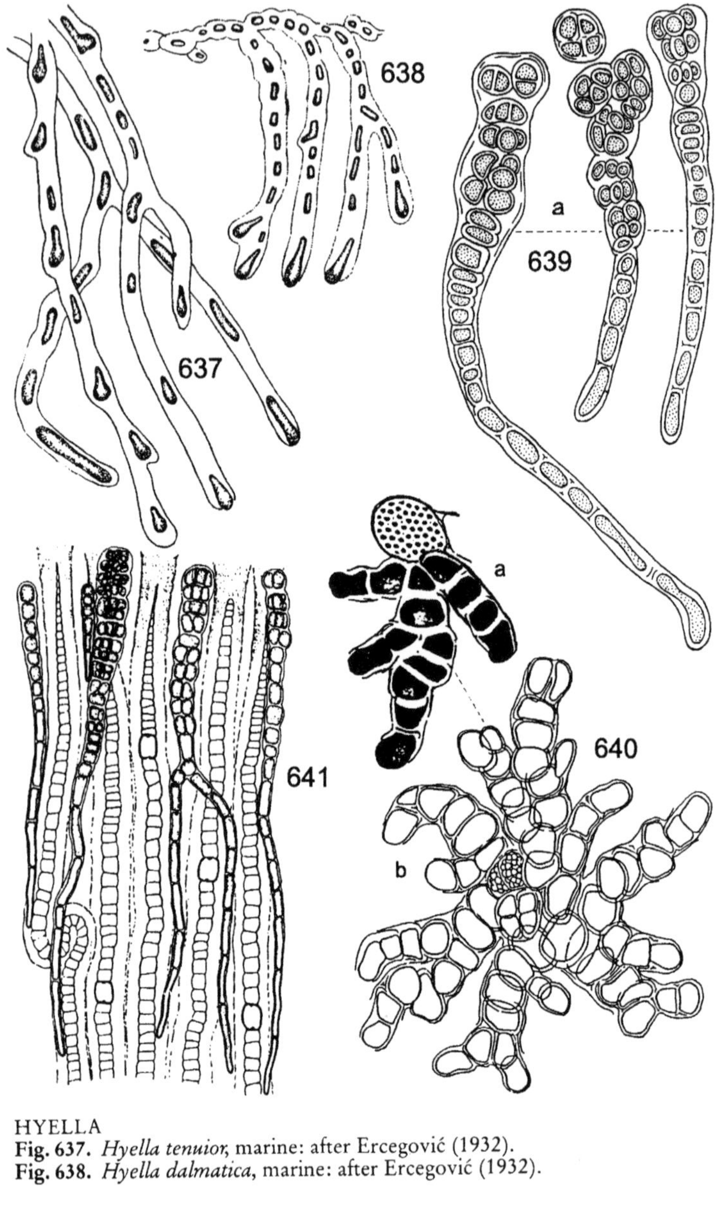


HYELLA

**Fig. 637.** *Hyella tenuior*, marine: after Ercegović (1932).

**Fig. 638.** *Hyella dalmatica*, marine: after Ercegović (1932).

**13. Hyella dalmatica** Ercegović 1932 (fig. **638**)

Thallus epi- and euendolithic, forming yellow-brown or dirty spots on the surface of calcareous rocks. Epilithic pseudofilaments rare, short, creeping, composed of one or a few irregular rows of cells. From these arise boring endolithic pseudofilaments, which are uniseriate, parallely and more or less perpendicularly penetrating the stony substrate, to 300 µm long, usually with lateral branching. Gelatinous sheaths wide, hyaline, colourless, or (in epilithic parts) yellow-brownish, not striate, delimited. Cells in epilithic pseudofilaments spherical or more or less ellipsoidal, in endolithic pseudofilaments more or less cylindrical-rounded or irregular, to 30 µm long and 2–8 µm wide, distinctly distant from one another; apical cells drop- or club-shaped, elongate (to 40 µm long), widely rounded at the apex. Cell content blue-green. Reproduction by baeocytes, which arise from vegetative cells in the upper (epilithic) part of thallus (?).
Occurrence: Marine, on calcareous rocks, in the middle part of lithophytic zone; known from the Mediterranean Sea (Croatia – Dalmatia, France).

**Species outside Europe:**
- *Hyella arbuscula* Al-Thukair 1992: Marine, described from Arabian/Persian Gulf (Saudi Arabia); (fig. **643**).
- *Hyella conferta* Al-Thukair et Golubić 1991: Marine, described from Arabian/Persian Gulf (Saudi Arabia); (fig. **643**).
- *Hyella immanis* Al-Thukair et Golubić 1991: Marine, endolithic, described from Arabian/Persian Gulf (Saudi Arabia), possibly also on limestone coasts of E. Mediterranean Sea (Pantazidou 1991) and in S. Africa (Silva & Pienaar 1997a); (fig. **640**).
- *Hyella inconstans* Al-Thukair et Golubić 1991: Marine, described from Arabian/Persian Gulf (Saudi Arabia), recorded also from coasts of S. Africa; (fig. **643**).
- *Hyella littorinae* Setchell et Gardner in Gardner 1918: Marine, described from California, recorded also from Sea of Japan; (fig. **642**).
- *Hyella racemus* Al-Thukair et al. 1994: Marine, euendolithic, described from Arabian/Persian Gulf (Saudi Arabia) and common in ooids of the Bahamas; (fig. **643**).
- *Hyella reptans* Al-Thukair et Golubić 1991: Marine, described from Arabian/Persian Gulf (Saudi Arabia), known also from S. Africa; (fig. **643**).

**639**

b

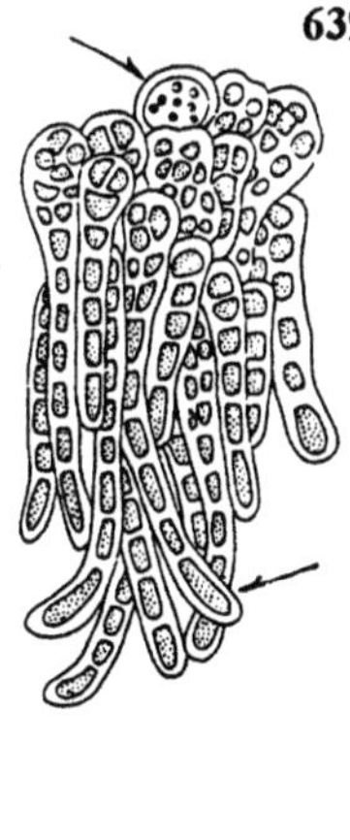

**Fig. 639.** *Hyella* sp., marine, from Japan: **a-b** – after Umezaki (1961, sub *Entophysalis deusta*).
**Fig. 640.** *Hyella immanis*, marine, from Arabian Gulf and Aegean Sea: **a** – after Al-Thukair (1992), **b** – after Al-Thukair & Golubić (1991).
**Fig. 641.** *Hyella seriata*, marine, epiphytic on cyanoprokaryotes, from California: after Hollenberg (1939, sub *Myxohyella seriata*).

642

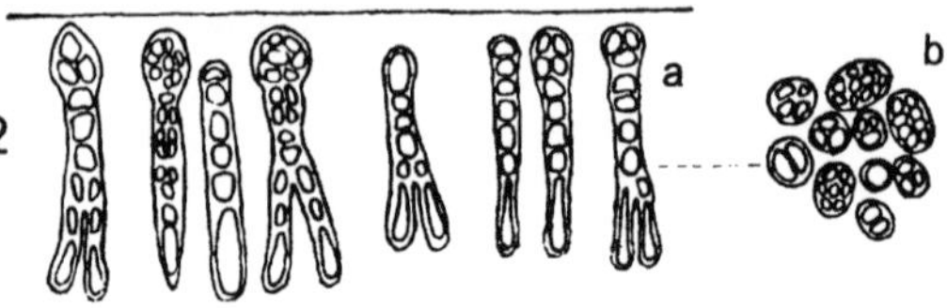


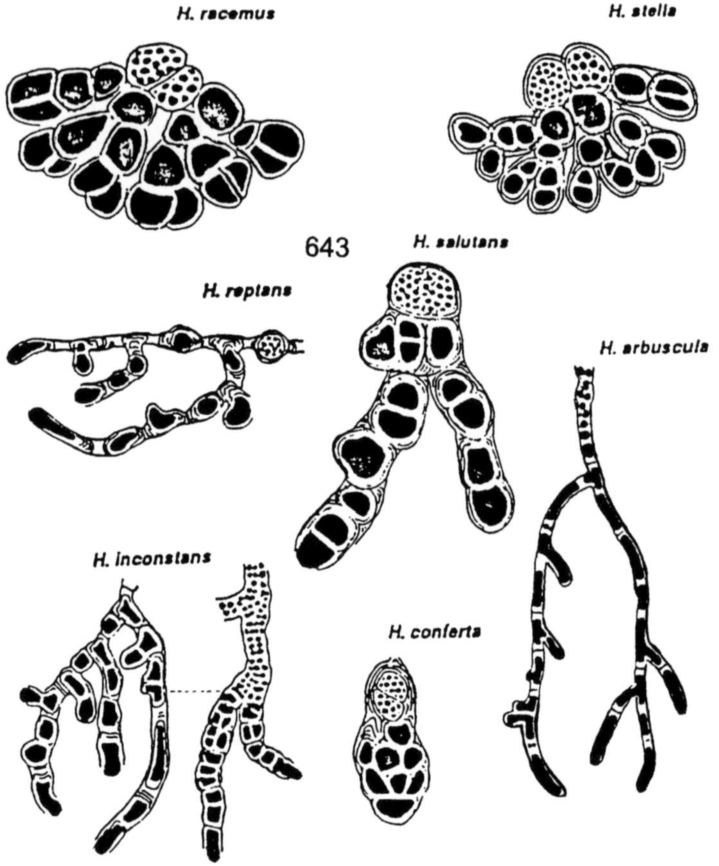


HYELLA
Examples of *Hyella* species from the coast of warm seas:
**Fig. 642.** *Hyella littorinae*, described from California: **a** – side view, **b** – top view; after Setchell & Gardner from Geitler (1932).
**Fig. 643.** Seven *Hyella* species, described from Arabian Gulf by Al-Thukair & Golubić (1991), Al-Thukair (1992) and Al-Thukair & al. (1994).

- *Hyella salutans* Al-Thukair et Golubić 1991: Marine, described from Arabian/ Persian Gulf (Saudi Arabia), known also from S. Africa; (fig. **643**).
- *Hyella seriata* (Hollenberg) Komárek et Anagnostidis 1995 [syn.: *Myxohyella seriata* Hollenberg 1939]: Marine, growing epiphytically on colonies of *Isactis* and *Kyrtuthrix* and penetrating deeply between the filaments; described from USA (coasts of S. California), known also from Sea of Japan; (fig. **641**).
- *Hyella simplex* Chu et Hua 1982: Marine, described from China (South China Sea - Xisha island).
- *Hyella stella* Al-Thukair et Golubić 1991: Marine, described from Arabian/ Persian Gulf (Saudi Arabia); (fig. **643**).
- *Hyella vacans* Gektidis et Golubić 1996: Marine, endolithic, described from Lee Stocking Island, Bahamas.

**Unrevised species:**
- Anagnostidis & Pantazidou (1988) described an euendolithic, but not taxonomically defined *Hyella* sp. boring into calcareous stones submerged in freshwater streams in Greece (Boeotia - Atalanti).
- LeCampion-Alsumard (1991) reported three marine, endolithic *Hyella*-species from Lizzard Island, Great Barrier Reef, Australia, without taxonomic definition.
- *Entophysalis deusta* sensu Umezaki 1961: Marine, reported from Japan; evidently a *Hyella*-species (fig. **639**).
- *Entophysalis deusta* «status hyelloides» sensu Anagnostidis 1968: Marine, reported from Aegean Sea coasts in Greece; belongs to a *Hyella*-species.
- Pantazidou (1991-diss.) described two not defined species from the marine coasts in Greece (*Hyella* sp. 1,2). Both species are closely related to *H. immanis* and *H. salutans* respectively.
- *Hyella intermedia* Varma et Mitra 1966: Freshwater, described from India, not reported since; according to description and drawings more likely *Myxohyella* sp. or *Chlorogloea* sp.
- *Hyella linearis* Setchell et Gardner 1918 [syn.: *Myxohyella linearis* (Setchell et Gardner) Geitler 1925]: Marine, from USA (Oregon).
- *Hyella voluticola* Chodat 1897: Insufficiently described, needs revision.
- *Hyella* sp. in Komárek 1985 (populations «Managua I -II»): Freshwater, from stones in small streams in limestone regions; known from the south of the USA and from Cuba; («*Hyella fontanoides*» Golubić in litt.).
- Chu & Wu (1984) reported numerous *Hyella*-species with provisional names from Xisha islands, China.
- A new *Hyella* species without formal description was reported from the island Mauritius by Silva & Pienaar (1997b).

# Literature


Abdelahad N.: On four *Myxosarcina*-like species (Cyanophyta) living in the Inferniglio cave (Italy). – Arch. Hydrobiol./ Algolog. Stud. **54**: 3–19 (1989).
– Bazzichelli G.: The genus *Gloeocapsa* Kütz. (Cyanophyta) on calcareous rock surfaces in the upper Valley of the river Anine (Latium, Italy). – Crypt. Bot. **2/3**: 155–160 (1991).
Agardh C.A.: Systema algarum. – Lund (1824).
Akiyama M.: Some soil algae from Japan. – Bull. Shimane Univ. (Nat. Sci.) **14**: 92–121 (1965).
Almestrand A.: Schwierigkeiten bei der Bestimmung von *Microcystis*-Arten. – Schweiz. Ztschr. Hydrol. **23**,1: 207–208 (1961).
Albertano P., Di Somma D., Leonardi D., Canini A. & Grilli Caiola M.: Cell structure of planktic cyanobacteria in the Baltic Sea. – Arch. Hydrobiol./ Algolog. Stud. **83**: 29–54 (1996).
– – & Capucci E.: Cyanobacterial picoplankton from the Central Baltic Sea: cell size classification by image-analyzed fluorescence microscopy. – J. Plankton Res. **19**(10): 1405–1416 (1997).
Al-Thukair A.A.: Cited according to manuscript (1992).
– & Golubić S.: New endolithic cyanobacteria from the Arabian Gulf. I. *Hyella immanis* sp. nov. – J. Phycol. **27**: 766–780 (1991a).
– – Five new *Hyella* species from the Arabian Gulf. – Arch. Hydrobiol./ Algolog. Stud. **64**:167–197 (1991b).
– – Characterization of *Hyella caespitosa* var. *arbuscula* var. nov. (Cyanophyta, Cyanobacteria) from shoaling ooid sand grains, Arabian Gulf. – Nova Hedwigia, Beih., **112**: 81–89 (199.).
– – & Rosen G.: New endolithic cyanobacteria from the Bahama bank and the Arabian Gulf: *Hyella racemus* sp. nov. – J. Phycol. **30**: 764–769 (1994).
Alvarez S.B. & Bazán G.I.: Cianofíceas continentales de la provincia de La Pampa (Argentina) I. [Freshwater blue-green continental algae of the province of La Pampa]. – Rev. Fac. Agronomia, **7** (2): 43–62 (1994).
Ålvik G.: Plankton-Algen norwegischer Austernpollen I. Systematik und vorkommende Arten. – Bergen's Mus. Årb. **1934**: 1–47 (1934).
Anagnostidis K.: Untersuchungen über die Cyanophyceen einiger Thermen in Griechenland. – Inst. Syst. Bot. Pflanzengeogr., 322 pp., Thessaloniki (1961).
– Untersuchungen über die Salz- und Süsswasser- Thiobiocönosen (Sulphuretum) Griechenlands. – Wiss. Jahrb. phys.-math. Fak. Univ. Thessaloniki **10**: 406–866 (1968).
– & Economou-Amilli A.: Microorganisms from the volcano Nea Kammeni Island (Santorini). – Second Internat. Sci. Congr. Santorini, «Thera and the Aegean World», London, **1**:703–723 (1978).
– & Roussomoustakaki M.: Epilithic and chasmolithic microflora (Cyanophyta, Bacillariophyta) from marbles of the Parthenon (Acropolis-Athens, Greece). – Nova Hedwigia **38**: 227–287 (1983).
– & Komárek J.: Modern approach to the classification system of cyanophytes, 1 – Introduction. – Arch. Hydrobiol./ Algolog. Stud. **38/39**: 291–302 (1990).
– & Pantazidou A.: *Cyanosaccus aegaeus* n. sp., a new marine endolithic cyanophyte from the Aegean Sea, Hellas (Greece). – Arch. Hydrobiol./ Algolog. Stud. **38/39**:105–114 (1985).
– – Endolithic cyanophytes from the saline thermal springs of Aedipsos, Hellas (Greece). – Arch. Hydrobiol./ Algolog. Stud. **50–53**: 555–559 (1988a).
– – *Hyella kalligrammos* sp. nov., *Hyella maxima* (Geitl.) comb. nov., and other

freshwater morphotypes of the genus *Hyella* Born. et Flah. (Chroococcales, Cyanophyceae). – Arch. Hydrobiol./ Algolog. Stud. **50–53**: 227–247 (1988b).
– – Marine and aerophytic *Cyanosarcina, Stanieria* and *Pseudocapsa* (Chroococcales) species from Hellas (Greece). – Arch. Hydrobiol./ Algolog. Stud. **64**: 141–157 (1991).
– & Roussomoustakaki M.: Cyanophytes from metal burdened substrates. – Arch. Hydrobiol./ Algolog. Stud. **50–53**: 561–564 (1988).
– – *Isocystis halobia* spec. nova, a benthic nostocalean cyanophyte from the heliothermal saltwork mats of Messolongion, Hellas (Greece). – Arch. Hydrobiol./Algolog. Stud. **64**: 299–332 (1991).
– & Zehnder A.: Beitrag zur Kenntnis der Blaualgenvegetation der Thermen von Baden und Leukerbad (Schweiz). – Schweiz. Ztschr. Hydrol. **26**, 1: 170–176 (1964).
Andreoli C., Rascio N., Dalla Vecchia F. & Talarico L.: An ultrastructural research on natural populations of picoplankton from two brackish water environments in Italy. – J. Plankton Res. **11**: 1067–1074 (1989).
Ardré F.: Quelques cyanophycées dermocarpoides de la côte portugaise. – Rev. Génér. Bot. **67** (1960).
– Contribution a l'étude des algues marines du Portugal. I. Flore. – Port. Acta biol.
(B) **10**:137–556 (1969).
Asencio A.D. & Aboal M.: Cyanophytes from *Andragulla abrigo* (Murcia, SE Spain) and their environmental conditions. – Arch. Hydrobiol./ Algolog. Stud. **83**: 55–72(1996)
Azevedo M.T. de P. & Kováčik L.: *Rhabdogloea brasilica* sp. nov. (Chroococcales, Synechococcaceae): morphological and morphometric variability under cross-gradient cultures. – Arch. Hydrobiol./Algolog. Stud. **83**: 83–92 (1996).
– & Sant'Anna C.L.: *Cyanostylon gelatinosus*, a new species (Chroococcaceae, Cyanophyceae) from Sao Paulo State, Brazil. – Arch. Hydrobiol./Algolog. Stud. **75**: 75–78 (1994).
– – *Hormothece geitleriana:* A new edaphic chroococcal Cyanophyceae from Sao Paulo State, Brazil. – Arch. Hydrobiol./Algolog. Stud. **75**: 79–83 (1994).
Bachmann H.: Vergleichende Studien über das Phytoplankton von Seen Schottlands und der Schweiz. – Arch. Hydrobiol. Planktonk. **3**: 1–91 (1907).
– Beiträge zur Algenflora des Süsswassers von Westgrönland. – Mitt. naturf. Ges. Luzern **8**: 1–181 (1921a).
– Hydrobiologische Untersuchungen am Rothsee. – Ztschr. Hydrol. **5**(3–4): 39–81 (1921b).
– Charakterisierung der Planktonvegetation des Vierwald-stättersees mittels Netzfängen und Zentrifugenproben. – Verh. naturf. Ges. Basel **35**(1): 48–167 (1923).
– Notizen zur Morphologie, Systematik und Ökologie einiger *Chamaesiphon*-Arten aus Fließgewässern. – Schweiz. Ztschr. Hydrol. **29**(1): 211–225 (1967a).
– Ökologische Untersuchungen an den Aufwuchsalgen der obersten Donau und ihrer Quellflüsse. I. Voruntersuchungen. – Arch. Hydrobiol./Suppl. **30** (Donauforschung 2): 364–399 (1967b).
– Ökologische Untersuchungen an den Aufwuchsalgen der obersten Donau und ihrer Quellflüsse. II. Die räumliche und zeitliche Verteilung der Algen. – Arch. Hydrobiol./ Suppl. **34** (Donauforschung 3): 24–73 (1968a).
– Ökologische Untersuchungen an den Aufwuchsalgen der obersten Donau und ihrer Quellflüsse. III. Die Algenverteilung und ihre Beziehungen zur Millieuofferte.- Arch. Hydrobiol./Suppl.**34** (Donauforschung 3): 130–149 (1968b).

– Ökologische Untersuchungen an den Aufwuchsalgen der obersten Donau und ihrer Quellflüsse. IV. Systematisch- autökologischer Teil. – Arch. Hydrobiol./ Suppl. **34** (Donauforschung 4): 251–320 (1968c).
Bailey-Watts A.E., Bindloss M.E. & Belcher J.H.: Freshwater primary production by a blue-green alga of bacterial size. – Nature (London) **220**: 1344–1345 (1968).
– & Komárek J.: Towards a formal description of a new species of *Synechococcus* (Cyanobacteria/ Cyanophyceae) from the freshwater picoplankton. – Arch. Hydrobiol./ Algolog. Stud. **61**: 5–19 (1991).
Beck-Mannagetta G.: Algenfunde im Riesengebirge. Ein Beitrag zur Kenntnis der Algenflora des Riesengebirges. – Věstník Král. Čes. Spol. Nauk. Tř. II, Roč. 1926: 1–18 (1929).
– Algenfunde im Riesengebirge. – Lotos (Prag) **77**: 93–100 (1929).
Behre K.: Die Algenbesiedlung der Unterweser unter Berück-sichtigung ihrer Zuflüsse (ohne Kieselalgen). – Veröff. Inst. Meeresforsch. (Bremerhaven) **7**: 71–263 (1961).
Beljakova R.N.: Morskie sinezelenye vodorosli dalnevostočnogo morskogo zapovednika [Cyanophyta marina in reservato marino orientis extremi vigentia.] – Novit. Syst. Plant. non Vasc. **24**: 3–8 (1987).
– Vidovoj sostav morskich sinezelenych vodoroslej ostrovov jugo-zapadnoj časti Tichogo okeana. [Compositio specierum cyanophytorum in insulis partis austrooccidentalis oceani Pacifici vigentium.] – Novit. Syst. Plant. non Vasc. **24**: 8–13 (1987).
– Sinezelenye vodorosli o-va King-D'ord', Antarktika. [Cyanophyta insulae King-Georg, Antarctis.] – Novit. Syst. Plant. non Vasc. **24**: 13–22 (1987).
Bernard C.: Protococcacées et Desmidiées d'eau douce.- Buitenzorg, 230 pp. (1908).
Bethge H.: *Chroococcus planctonicus*, eine neue planktonische Cyanophycee. – Ber. Dtsch. Bot. Ges. **53**: 265–269 (1935).
Biswas K.: Road slimes of Calcutta. – J. Dept. Sci. Univ. Calcutta **7**: 1–10 (1925).
– Flora of the Salt Lakes, Calcutta. – J. Dept. Sci. Univ. Calcutta **8**: 1–47 (1926).
Böcher T.W.: Studies on the sapropelic flora of the lake Flynderso with special reference to the Oscillatoriaceae. – Kgl. Danske vidensk. Selsk., Biol. Medd. **21**: 1–46 (1949).
Börgesen F.: Freshwater algae. – In: Botany of the Faeröes I. (Copenhagen), p. 198–259 (1901).
– & Ostenfeld C.H. (1903): Phytoplankton of lakes in the Faeröes. – In: Botany of the Faeröes II. (Copenhagen), p. 613–624 (1903).
Bornet E. & Flahault C.: Revision des Nostocacées hétérocystées. – Ann. Sci. nat. – Bot., sér.7, **3**:323–381, **4**:343–373, **5**:51–129, **7**:177–262 (1888).
– & Thuret G.: Notes algologiques. I. – Paris (1876).
Bosse van A.W.: Algues de l'Expedition danoise aux fles Kei.. – Papers from Dr. Th. Mortensen's Pacific Expedition 1914–16, 60–63.
Bourrelly P.: Initiation pratique a la systématique des algues d'eau douce. – Bul. Microb. appliquée **2**(4): 47–61 (1954a).
– Excursion Phycologique en Auvergne. – Not. Bot. Itin. Comm., VIII. Congr. Internat. de Botanique, Paris-Nice, 1954 (1954b).
– Cyanophycées de la Côte d'Ivoire. – Schweiz. Ztschr. Hydrol. **23**(1): 209–210 (1961a).
– Quelques algues d'eau douce de la région de Konakry. – Bull. Res. Counc. of Israel, **10D**: 9–14 (1961b).
– Quelgues Algues rares et nouvelles du Lac de Grand-Lieu. – Rev. Algol., N.S. **6**(1): 57–68 (1961c).

- Algues d'eau douce de la République de Côte d'Ivoire. - Bull. de 1. I.F.A.N. **23**(A): 283-374 (1961d).
- Quelques Algues d'eau douce du Canada. - Int. Revue ges. Hydrobiol. **51**(1): 45-126 (1966).
- Les algues d'eau douce III. - 512 pp., N. Boubée & Cie., Paris (1970).
- Note sur les genres *Pleurocapsa* et *Scopulonema*. - In: Desikachary T.V. (ed.), Proc. Symp. Taxonomy and bio-logy of blue-green algae, Univ. Madras, p. 36-40 (1972).
- Algues d'eau douce de la Nouvelle Calédonie recueillés par la Mission F. Starmuehlner en 1965 (Diatomées excluses). 1 Partie: Cyanophycées, Rhodophycées, Xanthophycées, Pheophycées, Chlorophycées (sans Desmidiées). - Rev. Hydrobiol. trop. **17**(1): 13-51 (1984).
- Les algues d'eau douce III. 2. éd. - 606 pp., N. Boubée & Cie., Paris (1985).
- & Couté A.: Algues d'eau douce de l'Ile de la Réunion. - Cryptogamie/ Algologie 7(2): 87-121 (1986).
- & Denizot M.: Quelques algues marines des niveaux élevés de la d'Étretat (Seine-maritime). - Le botaniste, Sér. L: 43-57 (1967).
- & Manguin E.: Contribution a la flore algale d'eau douce des Açores. - In: Lechevalier M.P. (ed.), Peuplement des Iles Atlantides, 447-500 (1947).
- – Algues d'eau douce de la Guadeloupe. - 281pp., Paris (1952).

Boye-Petersen J.: The freshwater Cyanophyceae of Iceland. - Arb. Bot. Kobenhavn (The botany of Iceland, 2) **101**, 7:251-324 (1923).
- Algae collected by Eric Hultén on the Swedish Kamchatka expedition 1920-1922 especially from hot springs. - Kgl. Danske Vidensk. Selsk., Biol. Medd. **20**: 1-222 (1946).

Branco L.H.Z., Sant'Anna C.L., Azevedo M.T.P. & Sormus L.: Cyanophyte flora from Cardoso Island mangroves, Sao Paulo State, Brazil. 1. Chroococcales. - Arch. Hydrobiol./Algolog. Stud. **80**: 99-111 (1996).
- Silva S.M.F. & Sant'Anna C.L.: *Stichosiphon mangle* sp. nova, a new cyanophyte from mangrove environments. - Arch. Hydrobiol./ Algolog. Stud. **72**: 1-7 (1994).

Brand F.: Das Formenkreis von *Gloeocapsa alpina*. - Bot. Cbl. **83**: 224 (1900).

Broady P.A.: Ecological and taxonomic observations on sub-aerial epilithic algae from Princess Elizabeth Land and MacRobertson Land, Antarctica. - Br. Phycol. J. **16**: 257-266 (1981).
- The ecology of chasmolithic algae at coastal locations of Antarctica. - Phycologia **20**: 259-272 (1981).
- Ecology of non-marine algae at Mawson Rock, Antarctica. - Nova Hedwigia **36**: 209-229 (1982).
- Taxonomic and ecological investigations of algae on steam-warmed soil on Mt. Erebus, Ross Island, Antarctica. - Phycologia **23**(3): 257-271 (1989).
- Ecology and taxonomy of the terrestrial algae of the Vestfold hills. - In: Antarctic oasis, p. 165-202, Acad. Press, Australia (1986).
- Survey of algae and other terrestrial biota at Edward VII Peninsula, Marie Byrd Land. - Antarctic Sci. **1**(3): 215-224 (1989).
- & Ingerfeld M.: *Placoma regulare* sp. nov. (*Entophysalidaceae*, Cyanobacteria) from New Zealand streams. - Phycologia **30**(6): 547-555 (1991).

Büdel B.: Blue-green phycobionts in the lichen family Lichinaceae. - Arch. Hydrobiol./ Algolog. Stud. **38/39**: 355-357 (1985).
- Zur Systematik und Biologie der Flechtengattungen *Heppia* und *Peltula* im südlichen Africa. - Bibl. Lichenol. **23**: 1-105 (1987).
- & Henssen A.: *Chroococcidiopsis* (Cyanophyceae), a phycobiont in the lichen family Lichinaceae. - Phycologia **22**(4): 367-375(1983).

- - *Trebouxia aggregata* und *Gloeocapsa sanguinea*, Phycobionten in *Eucopsis granatina* (Lichinaceae). - Pl. Syst. Evol. **158**: 235–241 (1988).
- Lüttge U., Stelzer R., Huber O. & Medina E.: Cyanobacteria of rocks and soils of the Orinoco lowlands and the Guayana uplands, Venezuela. - Bot. Acta **107** (6): 422–431 (1994).
- Mollenhauer D. & Mollenhauer R.: *Synechococcus elongatus* - cryptoendolithic growth within bleached sandstone from creeks in the midland area Spessart (Germany). - Arch. Hydrobiol./ Algolog. Stud. **64**: 357–360 (1991).
- & Rhiel E.: A new cell wall structure in a symbiotic and free-living strain of the blue-green alga genus *Chroococcidiopsis* (Pleurocapsales). - Arch. Microbiol. **143**: 117–121 (1985).
- & Wessels D.C.J.: Rock inhabiting blue-green algae/ cyanobacteria from hot arid regions. - Arch. Hydrobiol./ Algolog. Stud. **64**: 385–398 (1991).

Buell H.F.: A community of blue-green algae in a Minnesota pond. - Bull. Torrey bot. Cl. **65**: 377–396 (1938).

Butcher R.W.: Contributions to our knowledge of the smaller marine algae. - J. Marine Biol. Assoc. U.K. **31**: 175–191 (1952).

Čado I.: Endolithofiti po klifovite i podvodnite steni na Ochridskoto ezero. [Die Endolithophyten an den Kliffen und Unterwassersteinen des Ochridsees.] - Rec. Trav., Fac. Phil. Univ. Skopje, St. Hydrobiol. Ohrid, VI, **3**(19): 1–13 (1958a).
- Sostav i zonacija na mikrofitite po kraj're'nite steni na Ochridskoto ezero. [Composition and zonation of microphytes on littoral rocks of the Ochrid lake.] - Rec. Trav., Fac. Phil. Univ. Skopje, St. Hydrobiol. Ohrid VI, **5** (21): 1–34 (1958b).
- *Aphanothece microscopica* NÄG. a member of the community of water-plants of Ohrid lake. - Rec. Trav., Sec. Sci. Nat. Univ. Skopje, St. Hydrobiol. Ohrid VII, **10**(43): 1–2 (1959).

Cambra J.: Flore et végétation algologiques des eaux épicontinentales de la Réserve Naturelle des vallées d'Ordesa (Pyrénées aragonaises). - Candollea **42**: 475–490 (1987).
- Nolla J. & Sabater S.: Composicion fitoplanctonica en embalses de pequeno volumen del este de la Peninsula Iberica. - Bol. Soc. Brot., Sér. 2, **62**: 5–18 (1989).

Campbell S.E. & Golubić S.: Benthic cyanophytes (cyanobacteria) of Solar Lake (Sinai). - Arch. Hydrobiol./Algolog. Stud. **38/39**: 311–329 (1985).

Campos I.F.P. & Senna P.A.C.: Nostocophyceae (Cyanophyceae) da Lagoa Bonita, Distrito Federal, Brasil. Parte I. - Acta bot. bras. **2**(1–2): 7–30 (1988).

Cantonati M., Rott E. & Pipp E.: Ecology of cyanophytes in mountain springs of the River Sarca catchment (Adamello-Brenta Regional Park, Trentino, Northern Italy). - Arch. Hydrobiol./ Algolog. Stud. **83**: 145–162 (1996).

Castenholz R.W.: Species usage, concept and evolution in the Cyanobacteria (Blue-green algae). - J. Phycol. **28**: 737–745 (1992).
- & Waterbury J.B.: Oxygenic Photosynthetic Bacteria (sect. 19), Group I. Cyanobacteria. - In: Staley J.T. (ed.), Bergey's Manual of Systematic Bacteriology, Vol. **3**, Williams & Wilkins Publ., Baltimore, 1710–1799 (1989).

Cepák V.: Morphology of DNA containing structures (nucleoids) as a prospective character in cyanophyte taxonomy. - J. Phycol. **29**(6): 844–852 (1993).
- Nucleoid morphology in the coccal cyanophyte *Cyanothece halobia* (Chroococcales, Cyanophyta) - a DAPI fluorescence study. - Phycologia **35**(6): 523–527 (1996).

Chang P.T. : Zwei neue *Synechococcus*-Arten aus dem Zürich-see. - Schweiz. Ztschr. Hydrol. **42**(2): 247–254 (1980).

Chodat R.: Études de biologie lacustre. – Bull. Herb. Boiss., sér. 1, **6**: 155–188 (1898).
Chou H.-M. & Huang T.-C.: Ultrastructure of the aerobic, nitrogen-fixing unicellular cyanobacterium *Synechococcus* sp. RF-1. – Arch. Hydrobiol./ Algolog. Stud. **64**: 53–59 (1991).
Christensen T.: Algae. A taxonomic survey.1. – 228 pp., Odense (1980).
Chu H.: Some new Myxophyceae from Szechwan province, China. – Ohio J.Sci. **3**: 96–101 (1952).
– & al.(ed.): Chroococcophyceae. – In: Flora algarum sinicarum aquae dulcis II., Science Press, 161 pp. (1991).
– & Wu B.-T.: Studies on the lime-boring algae of China. – Hydrobiologia **116/117**: 227–228 (1984).
Claus G.: Beiträge zur Kenntnis der Algenflora der Abaligeter Höhle. – Hydrobiologia **19**(2): 192–222 (1962).
– Comments on the Species of the Genus *Tetrapedia* Reinsch (Chroococcaceae, Cyanophyta). – Hydrobiologia **21**(3–4): 266–274 (1963).
Cmiech H.A., Leedale G.F. & Reynolds C.S.: Morphological and ultrastructural variability of planktonic Cyanophyceae in relation to seasonal periodicity. III. *Gomphosphaeria naegeliana*. – Br. phycol. J. **22**: 339–343 (1987).
Cocquyt C., Vyverman W. & Compere P.: A check-list of the algal flora of the East African Great Lakes (Malawi, Tanganyika and Victoria). – Scripta Bot. Belg. **8**: 1–12, 49–55 (1993).
Comas A. & Krienitz L.: Comparative LM- and SEM-studies on *Coelastrum* (Chlorophyta, Chlorellales) under culture conditions. – Arch. Hydrobiol./ Algolog. Stud. **87**: 87–98 (1997).
Compère P.: Algues du Sahara et de la région du lac Tchad. – Bull. Jard. bot. nat. Belg. **37**:109–288 (1967).
– Contribution a l'étude des eaux douces de l'Ennedi. – Bull.de l'I.F.A.N. **32**(1): 18–64 (1970).
– La position systématique de *Phylloderma sacrum* Suringar, Cyanophycée alimentaire du Japon. – Bull. Jard. Bot. Nat. Belg. **40**: 423–427 (1970).
– Algues de la région du lac Tchad. II. Cyanophycées. – Cah. O.R.S.T.O.M., Sér. Hydrobiol. 8 (3–4): 165–198 (1974).
– Algues des déserts d'Iran. – Bull. Jard. Bot. Nat. Belg. **51**(1–2): 3–40 (1981).
Copeland J.J.: Yellowstone thermal Myxophyceae. – Ann. New York Acad. Sci. **36**: 1–232 (1936).
Corpe W.A. & Jensen T.E.: An electron microscopic study of picoplanktonic organisms from a small lake. – Microb. Ecol. **24**: 181–197 (1992).
Croasdale H.: Freshwater algae of Ellesmere Island, N.W.T. – Nat. Mus. Nat. Sci., Publ. Bot. **3**: 131 pp. (1973).
Cronberg G.: *Cyanodictyon tubiforme*, a new chroococcal blue-green alga from lake Börringesjön, Scania, Sweden. – Arch. Hydrobiol./ Algolog. Stud. **50–53**: 191–194 (1988).
– *Cyanothamnos planctícus* gen. et sp. nov., a new colonial cyanophyte from an eutrophic Scanian lake, Sweden. – Arch. Hydrobiol./ Algolog. Stud. **64**: 61–70 (1991).
– & Komárek J.: Planktic cyanoprokaryotes found in south Swedish lakes during the 12th International Symposium of Cyanophyte Research, 1992. – Arch. Hydrobiol./Algolog. Stud. **75**: 323–352 (1994).
– & Weibull C.: *Cyanodictyon imperfectum*, a new chroococcal blue-green alga from Lake Trummen, Sweden. – Arch. Hydrobiol./ Algolog. Stud. **27**: 101–110 (1981).
Crow W.B.: The taxonomy and variation of the genus *Microcystis* in Ceylon. – The New Phytologist **22**(2): 59–68 (1923a).

- Freshwater plankton algae from Ceylon. – The J. Bot. **61**:110–171 (1932b).
Daily W.A.: The Chroococcaceae of Ohio, Kentucky, and Indiana. – Am. Midl. Natur. **27**(3): 636–661 (1942).
- Notes on the algae I,II. – Butler Univ. Bot. Stud. **8**: 118–120 (1946).
Dangeard: Chroococcales, genre *Mantellum* gen. nov. – Le Botaniste **30**: 129–131, Pl.2 (1941).
De Philippis, Margheri M.C., Bastianini A. & Vincenzini M.: Phenotypic diversity among exopolysaccharide producing marine cyanobacteria of the *Cyanothece* group. – Abstr. Symp.«Marine Cyanobacteria and related organisms», Nov.24–28, Inst. Océanogr., Paris, p. 8 (1997).
Desikachary T.V.: Cyanophyta. – I.C.A.R. Monographs on Algae, 686 pp., New Delhi (1959).
De Toni G.: Noterelle di nomenclatura algologica. VIII. Terzo elenco di Missoficee omonime. – Brescia 5 (1936).
Diaz M.M. & Wenzel M.T.: Cyanophyta del Parque Nacional Nahuel Huapi. I. Cuenca del Rio Manso superior (provincia de Rio Negro, Argentina). – Darwiniana **32**(1–4): 279–291 (1993).
Doers M.P. & Parker D.L.: Properties of *Microcystis aeruginosa* and *M. flos-aquae* (Cyanophyta) in culture: taxonomic implications. – J. Phycol. **24**: 502–508 (1988).
Dor I.: Algues des sources thermales de Tibériade. – Bull. Sea Fish. Res. St. Haifa **48**: 1–29 (1967).
- Carl N. & Baldinger I.: Polymorphism and salinity tolerance as a criterion for differentiation of three new species of *Chroococcidiopsis* (Chroococcales). – Arch. Hydrobiol./ Algolog. Stud. **64**: 411–421 (1991).
- & Danin A.: Cyanobacterial desert crusts in the Dead Sea Valley, Israel. – Arch. Hydrobiol./ Algolog. Stud. **83**: 197–206 (1996).
- & Hornoff M.: Salinity-temperature relations and morphotypes of a mixed populations of coccoid cyanobacteria from a hot hypersaline pond in Israel. – Mar. Ecol. **6**: 13–25 (1985a).
-- Studies on *Aphanothece halophytica* Frémy from a solar pond: comparison of two isolates on the basis of cell polymorphism and growth response to salinity, temperature and light conditions. – Bot. Mar. **28**: 389–398 (1985b).
Drouet F.: Notes on the flora of Columbia, Missouri, III. – Rhodora **38**: 191–195 (1936a).
- Myxophyceae of the G. Allan Hancock Expedition of 1934, collected by Wm. R. Taylor. – The Hancock Pacific Expeditions **3**(2): 15–30 (1936b).
- Myxophyceae of the Yale North India Expedition coll. by G.E. Hutchinson. – Trans. Amer. Micr. Soc. **57**(2): (1938)
- Studies in Myxophyceae I. – Bot. Ser. Field Mus. nat. Hist. **20**(6): 125–141 (1942).
- & Daily W.A.: Nomenclatural transfers among coccoid algae. – Lloydia **11**: 77–79 (1948).
-- A synopsis of the coccoid Myxophyceae. – Butler Univ. Bot. Stud. **10**: 220–223 (1952).
-- Revision of the coccoid Myxophyceae. – Butler Univ. Bot. Stud. **12**: 1–218 (1956).
Dural B., Aysel V., Lök A. & Güner H.:Benthic algal flora of the natural and artifical substrata of Hekim Island (Izmir, Turkey). – Arch. Hydrobiol./Algolg. Stud. **85**: 31–48 (1997).
Duvigneaud P. & Symoens J.J.: Observations sur la strate algale des formations herbeuses du sud du Congo belge. – Lejeunia, Rev. de Botanique, **13**: 67–98 (1949).

Economou-Amilli A. & Anagnostidis K.: Periphyton algae and bacteria from springs of Levadia Boeotia, Greece. - Hydrobiologia **80**: 67–89 (1981).

- & Spartinou M.: The diversity of *Cyanodictyon imperfectum* (Chroococcales, Cyanophytes) in Lake Amvrakia, Greece. - Arch. Hydrobiol./ Algolog. Stud. **64**:105–114 (1991).

Ehresmann D.W. & Claus G.: *Wolskyella solitaria*, nov. spec., a new marine blue-green alga. - Nova Hedwigia **16**: 409–416 (1968).

Elenkin A.A.: O novych semejstvach sinezelenych vodoroslej grupy Stereometreae Elenk. (podgrupy Excavatae Elenk.) porjadka Chroococcales Geitler (1925). [Über neue Familien der Cyanophyceen aus der Gruppe der Stereometrae Elenk. (Untergruppe Excavatae Elenk.) der Ordnung der Chroococcales Geitler (1925).] - Acta Inst. bot. Acad. Sci. USSR, Ser. **2** (1–2): 23–34 (1933a).

- O sistematičeskoj podrazdelenii porjadka Chroococcales Geitler (1925). [Über die systematische Einteilung der Ordnung Chroococcales Geitler (1925).] - Acta Inst. bot. Acad. Sci. USSR, Ser. **2** (1): 17–21 (1933b).

- Monographia algarum cyanophycearum aquidulcium et terrestrium in finibus URSS inventarum. [Sinezelenye vodorosli SSSR]. Obščaja časť'. – 684 pp., Izd. AN SSSR, Moskva-Leningrad (1936).

- Monographia algarum cyanophycearum aquidulcium et terrestrium in finibus URSS inventarum. [Sinezelenye vodorosli SSSR]. Pars spec. (1–2). – 1908 pp., Izd. AN SSSR, Moskva-Leningrad (1938).

Emoto Y. & Hirose H.: Studies of thermal flora of Japan (XIV). Bacteria and algae from the Naguro thermal springs. - Bot. Mag. Tokyo **56**: 25–42 (1942a). (In Japanese)

-- Studies of thermal flora of Japan (XV). Thermal bacteria and algae from Nakayama thermal springs, Miyagi-Prefecture. - J. Balneol. Soc. Jap. **2**: 29–39(1942b). (In Japanese)

-- Studies of thermal flora of Japan (XVI). Bacteria and algae of the Onikobe thermal springs. - Bot. Mag. Tokyo **56**: 120–136 (1942c). (In Japanese)

-- Studies of thermal flora of Japan (XIX). Thermal bacteria and algae from thermal springs of Hatimantai und Yakeyama. - Bot. Mag. Tokyo **56**: 332–342 (1942d). (In Japanese)

-- Studies of thermal flora of Japan (XXI). Thermal bacteria and algae from hot springs of Siobara in Totigi-Prefecture (2). - J. Balneol. Soc. Jap. **2**: 86–95 (1942e). (In Japanese)

-- Studies of thermal flora of Japan (XXIV). Thermal flora of hot springs in the peninsula Izu I. Thermal bacteria and algae from hot springs at the eastern coast. - J. Balneol. Soc. Jap. **2**: 1–14 (1942f). (In Japanese)

-- Studies of thermal flora of Japan (XXVIII). Bacteria and algae of western in Nagano Prefecture. - Ecol.Rev. 11: 107–112 (1952a). (In Japanese)

-- Studies of thermal flora of Japan (XXIX). Thermal flora of hot springs in Nagano Prefecture. 3. Bacteria and algae of the Yamanouchi (Hirao) thermal springs. - J. Balneol. Soc. Jap. **5**: 17–25 (1952b). (In Japanese)

-- Studies of thermal flora of Japan (XXX). Thermal flora of the Ikao thermal springs. - J. Balneol. Soc. Jap. **5**: 26–34 (1952b). (In Japanese)

- & Yoneda Y.: Studies on the thermal flora of Japan VI. Bacteria and algae of the Ibusuki thermal springs. - Ecol. Rev. **6**: 257–274 (1940).

-- Bacteria and algae of the thermal springs in Shimane prefecture (II). - J. Jap. Bot. **17**(12): 704–720 (1941).

Ercegović A.: Litofitska vegetacija vapnenaca i dolomita u Hrvatskoj. [La végétation des lithophytes sur les calcaires et les dolomites en Croatie]. - Acta Bot. Univ.Zagreb **1**: 64–114 (1925).

- Trois nouveaux genres des Cyanophycées lithophytes. – Acta Bot. Univ. Zagreb **2**: 78–84 (1925).
- Sur quelques nouveaux types des Cyanophycées lithophytes de la côte adriatique. – Arch. Protistenk. **66**: 164–174 (1929a).
- Sur la tolérance des cyanophycées vis-a-vis des variations brusques de la salinité de l'eau de mer. – Acta Bot. Univ. Zagreb **5**: 48–56 (1929b).
- Sur quelques genres peu connues des Cyanophycées lithophytes de la côte Yugoslave de l'Adriatique. – Arch. Protistenk. **71**: 361–376 (1930).
- *Podocapsa* et *Brachynema*, deux genres nouveaux Chamaesiphonales de la côte adriatique de Dalmatie. – Acta bot. Inst. bot. Zagreb **6**: 33–37 (1931).
- Études écologiques et sociologiques des Cyanophycées lithophytes de la côte Yougoslave de l'Adriatique. – Bull.internat. Acad. Yougosl. Sci. Arts, Cl. mat.-nat., **26**: 33–56 (1932).
- Ekološke i sociološke studije o litofitskim cijanoficejama na jugoslav. obale Jadrana. – Knjiga «Rada» Jug. Akad. znanosti i umjetnosti, p.136–165 (1932).

Fallon R.D. & Brock T.D.: Overwintering of *Microcystis* in Lake Mendota. – Fresh. Biol. **11**: 217–226 (1981).

Fedorov V.G.: O nekotorych sinezelenych vodorosljach iz Zapadnoj Sibiri. [Nonnullae Cyanophyceae e Sibiria occidentali]. – Nov. Sist. Nizš. Rast. **1967**: 11–15 (1967).
- Novye sinezelenye vodorosli iz Zapadnoj Sibiri. [Cyanophyta nova e Sibiria occidentali]. – Nov. Sist. Nizš. Rast. **6**: 15–22 (1969).

Feldmann J.: Les algues marines de la côte des Alberes. I. Cyanophycées. – Rev. Algol. N.S. **9**:145–172 (1937).
- Deux nouvelles Cyanophycées marines de Roscoff. – Bull. Soc. Bot. France **100**(7–9): 292–295 (1953).
- Une nouvelle Cyanophycée endophyte (*Myxohyella Dilseae*, nov. sp). – Bull. Soc. bot. France, 1956, **103**(9–10): 573–576 (1956).
- Les Cyanophycées marines de la Guadeloupe. – Rev. Algol. **IV**(I): 25–40 (1958).
- Cyanophycées. – In: Additions a l'Inventaire de la Flore Marine de Roscoff, Edit. Stat. biolog. Roscoff, p. 20–23 (1964).
- & Feldmann G.: Observations sur les genres *Dermocarpa* et *Dermocarpella* (Cyanophyceae). – Österr. bot. Ztschr. **100**(4/5): 505–514 (1953).

Fjerdingstad E.: The Algal Flora of some «Tintenstriche»in the «Alpes-Maritimes»(France). – Hydrologie **27**(1): 167–171 (1965).
- The microphyte communities of two stagnant fresh-water ditches rich in $H_2S$. – Dansk Botanisk Arkiv **14**: 34–39.

Fott B.: Mikroflora der Orava-Moore. – Preslia **24**: 189–209 (1952).
- Sinice a řasy. – 482 pp., Nakl. ČSAV, Praha (1959).
- Algenkunde. – 2nd ed., 581 pp., VEB G. Fischer, Jena (1971).
- On the occurrence and the taxonomy of the genus *Microcrocis* P. RICHTER (Cyanophyceae). – Preslia (Praha) **44**: 97–99 (1972).

Frank T.H. & Landmann A.G.: Morphology and taxonomy of *Microcrocis* sp. (Chroococcales, Cyanophyta) from Lake Vechten, The Netherlands, with a note on its observed movements. – Arch. Hydrobiol./ Algolog. Stud. **46–49**:397–410 (1988).

Frémy P.: Les Cyanophycées des Côtes d'Europe. – Mém. Soc. Nat. Sci. Nat. Math. Cherbourg **41**: 1–236 (1929–1933).
- Contribution a la Flore algologique du Congo belge. – Bull. Jardin Botanique, **9**(3):109–138 (1930a).
- Les Myxophycées de l'Afrique équatoriale française. – Arch. Bot. **2**: 1–508 (1930b).

- Trois Cyanophycées nouvelles de l'Afrique du Nord. – Bull. hist. Nat. Afr. Nord **26**: 89–101 (1935).
- Algues vivant dans de la Craie. – Bull. Soc. Linn. Normandie, Ser. 9, **3**: 19–22 (1942).
- Cyanophyceae. – Explor. Parc Nat. Albert, Mission H. Dumas, Fasc. **19**: 17–51 (1949).

Friedmann E.I.: *Chroococcidiopsis kashayii* sp. n. and the genus *Chroococcidiopsis*. – Österr. Bot. Ztschr. **108**: 354–367 (1961).
- Contributions to the knowledge of *Cyanophanon mirabile* Geitler. – Brit. Phycol. Bull. **2**(5): 369–371 (1964).
- Light and scanning electron microscopy of the endolithic desert algal habitat. – Phycologia **10**: 411–428 (1971).
- Endolithic microbial life in hot and cold deserts. – Origins of Life **10**: 223–235 (1980).
- Hua M. & Ocampo-Friedmann R.: Cryptoendolithic lichen and cyanobacterial communities of the Ross Desert, Antarctica. – Polarforschung **58**: 251–260 (1988).
- Lipkin Y. & Ocampo-Paus R.: Desert algae of the Negev (Israel). – Phycologia **6**(4):185–196 (1967).
- & Ocampo R.: Endolithic blue-green algae in the Dry Valleys: Primary producers in the Antarctic desert ecosystem. – Science **193**: 1247–1249 (1976).
- – Endolithic microorganisms in the Antarctic cold desert. – Science **215**: 1045–1053 (1982).
- – Blue-green algae in arid cryptoendolithic habitats. – Arch. Hydrobiol./ Algolog. Stud. **38/39**: 349–350 (1985).
- & Ocampo-Friedmann R.: Endolithic microorganisms in extreme dry environments: Analysis of a lithobiontic microbial habitat. – In: Klug M.J. & Reddy C.A. (eds.), Current perspectives in microbial ecology, p. 177–185. American Society for Microbiology, Washington, D.C.(1984)

Fritsch F.E.: Freshwater Algae, National Antarctic Expedition. – Nat. Hist. **6**: 1–66 (1912a).
- The freshwater algae of the South Orkneys. – Rep. Sci. Results Scottish Nat. Antarctic Exped. **3**: 95–134 (1912b).
- The encrusting algal communities of certain fast flowing streams. – New Phytol. **28**: 165–196 (1929).
- Structure and reproduction of algae. I. – Cambridge, 791 pp. (1935).

Garcia-Pichel F., Nuebel U., Muyzer G. & Prufert-Bebout L.: Combined molecular and traditional approaches to cyanobacterial phylogeny: case studies of «*Microcoleus chthonoplastes*» and «*Aphanothece halophytica*». – Abstr. Symp. «Marine Cyanobacteria and related organisms», Nov. 24–28, Inst. Océanogr., Paris, p. 9 (1997).

Gardner N.L.: Cytological studies in cyanophyceae. – Univ. Calif. Publ. Botany **2**(12): 237–296 (1906).
- New Pacific coast marine algae III. – Univ. Calif. Publ. Bot. **6**: 455–486 (1918).
- New Myxophyceae from Porto Rico. – Mem. New York Bot. Gard. **7**: 1–144 (1927a).
- On a collection of Myxophyceae from Fukien province, China. – Univ. Calif. Publ. Bot. **14**: 1–20 (1927b).

Geitler L.: Cyanophyceae. – In: Pascher's Süssw.-Fl. **12**:1–450, G. Fischer-Verl., Jena (1925a).
- Synoptische Darstellung der Cyanophyceen in morphologischer und systematischer Hinsicht. – Beih. bot. Cbl. **2**, 41: 163–184 (1925b).

- Über neue oder wenig bekannte interessante Cyanophyceen aus der Gruppe der Chamaesiphonaceae. - Arch. Protistenk. **51**: 321–360 (1925c).
- Neue Blaualgen aus Lunz. - Arch. Protistenk. **60**: 440–448 (1927).
- Über die Tiefenflora an Felsen im Lunzer Untersee. - Arch. Protistenkd. **62**: 96–104 (1928).
- Cyanophyceae. - In: Rabenhorst's Krypt.-Fl. **14**: 1–1196, Akad. Verlagsges., Leipzig (1932).
- Diagnosen der neuen Blaualgen von den Sunda-Inseln. - Arch. Hydrobiol./ Suppl. 12 (Tropische Binnengew.) **4**:622–634 (1933).
- Kleine Mitteilungen über neue oder wenig bekannte Blaualgen. - Österr. Bot. Ztschr. **84**: 287–291 (1935).
- Schizophyta (Klasse Schizophyceae). - In: Engler & Prantl, Nat. Pfl.-Fam., ed.2, 1b: 232pp. (1942).
- Eine ökologisch merkwürdige Cyanophycee, *Synechococcus diatomicola* n. sp. - Österr. Bot. Ztschr. **100**: 292–301 (1953).
- Lebensweise, Fortpflanzung und Wachstumsvorgänge bei *Cyanophanon mirabile*, n. gen., n. sp., einem merkwürdigen Blaualgen-ähnlichen Organismus. - Öster. Bot. Ztschr. **102**(2–3): 235–272 (1955).
- Zur morphologischen Variation und Ökologie auf kleinstem Raum von *Cyanophanon mirabile*. - Österr. Bot. Ztschr. **104**(1–2): 100–114 (1957).
- Schizophyceen. - In: Handb. d. Pflanzenanatomie 6, 1, Spez. Teil., 131 pp., 2. Aufl. - Gebr. Borntraeger, Berlin-Nikolassee (1960).
- Aktive Bewegung der Cyanophycee *Synechococcus* und die Koloniebildung in einer Ebene bei *Chamaesiphon* und *Salpingoeca*. - Österr. Bot. Ztschr. **109**(3): 349–363 (1962).
- *Cyanophanon minus* n.sp. - Österr. Bot. Ztschr. **113**(3/4): 459–464 (1966).
- Entwicklungsgeschichtliche und systematische Untersuchungen an einigen Cyanophyceen. - Nova Hedwigia **13** (3/4): 403–421 (1967).
- Beiträge zur epiphytischen Algenflora des Neusiedler Sees. - Österr. Bot. Ztschr. **118**: 17–29 (1970).
- Zur vergleichenden Morphologie der Cyanophycee *Clastidium setigerum* und verwandter Gattungen. - Pl. Syst. Evol. **124**: 161–171 (1975).
- Beeinflussung der Blattzellen von *Fontinalis antipyretica* durch den Bewuchs von *Cocconeis placentula*. - Plant Syst. Evol. **128**: 37–45 (1977).
- *Chroococcopsis epiphytica* n. sp., eine neue Cyanophycee aus dem Gebiet des Neusiedler Sees. - Pl. Syst. Evol. **136**: 53–62 (1980).
- & Ruttner F.: Die Cyanophyceen der Deutschen limnologischen Sunda-Expedition, ihre Morphologie, Systematik und Ökologie. - Arch. Hydrobiol./ Suppl.14 (Tropische Binnengew.) **6**: 308–369, 371–483 (1935).

Ginsburg-Ardré F.: *Dermocarpa, Xenococcus, Dermocarpella* (Cyanophycées): Nouvelles observations. - Österr. Bot. Ztschr. **113**: 362–367 (1966).

Godward M.: An ecological and taxonomic investigation of the littoral algal flora of lake Windermere. - J.Ecol. **25**: 496–568 (1937).

Gold-Morgan M., Montejano G. & Komárek J.: Freshwater epiphytic cyanoprokaryotes from central Mexico 2. Heterogeneity of the genus *Xenococcus*. - Arch. Protistenk. **144**:383–405 (1994).

- - - Freshwater epiphytic Chamaesiphonaceae from Central Mexico. - Arch. Hydrobiol./ Algolog. Stud. **83**: 2.7–271 (1996).

Golubić S.: Vegetacija cijanofita u lukama sjevernog Jadrana. - Thalassia Jugoslavica **2**(2):5–36 (1960).
- Zur Revision der Gattung *Gloeocapsa* Kützing (Cyanophyta).- Schweiz. Ztschr. Hydrol. **27**: 218–232 (1965).
- Algenvegetation der Felsen. - Die Binnengewässer **23**: 183 pp., Stuttgart (1967a).

– Die litorale Algenvegetation des Titisees. – Arch. Hydrobiol./Suppl. **33**(2): 172–205 (1967b).
– Zwei wichtige Merkmale zur Abgrenzung der Blaualgengattungen. – Schweiz. Ztschr. Hydrol. **29**:176–184 (1967).
– Halophily and halotolerance in cyanophytes. – Origins of Life **10**: 169–183 (1980).
– Al-Thukair A.A. & Gektidis M.: New euendolithic cyanobacteria from the Arabian Gulf and the Bahama Bank: *Solentia sanguinea* sp. nova. – Arch. Hydrobiol./Algolog. Stud. **83**: 291–301(1996).
– & Campbell S.: Analogous microbial forms in recent subaerial habitats and in precambrian cherts: *Gloeothece coerulea* Geitler and *Eosynechococcus moorei* Hofmann. – Precambr. Res. **8**: 201–217 (1979).
– Friedmann E.I. & Schneider J.: The lithobiontic ecological niche, with special reference to microorganisms. – J. Sed. Petrol. **51**: 475–478 (1981).
– Perkins R.D. & Lukas K.J.: Boring microorganisms and microborings in carbonate substrates. – In: Frey R.W.(ed.), The study of trace fossils: 229–269, Springer-Verlag, New York – Heidelberg – Berlin (1975).
– Yun Z. & Campbell S.E.: Early evolution of morphological complexity in Procaryote (Cyanophyta, Cyanobacteria). – In: Mlíkovský J. & Novák V.J.P. (eds.), Evolution and morphogenesis, p. 355–368, Academia, Praha (1985).
Gonzalez Guerrero P.: Cianofíceas y algas continentales de Cádiz. – An. Jard. bot. Madrid **6**: 237–335 (1946).
Gonzalves E.A. & Kamat N.D.: The Myxophyceae of the Karnatak I. – J. Univ. Bombay **3**: 22–36 (1958).
Grilli Caiola M., Pellegrini S., Gerola M. & Ribaldone A.: *Bdellovibrio*-like bacteria in *Microcystis aeruginosa*. – Algolog. Stud. **64**: 369–376 (1991).
Gromov B.V. & Mamkaeva K.A.: Osobennosti kletočnogo cikla *Chamaesiphon confervicola* – cianobakterii obrazujuščej ekzospory. [Characteristic of the cellular cycle of *Chamaesiphon confervicola*, a cyanobacterium producing exospores.] – Mikrobiologija (Moskva) **49**(4): 551–554 (1980).
Gruia L.: Alge noi pentru flora R.P.R., IV. [Algues nouvelles pour la flore de la R.P.R. IV.]. – Comun. Acad. Rep. pop. Romine **13**(7): 619–628 (1963).
Guarrera S.A.: Cyanophyta, Chroococcales. – In: Guarrera S.A., Gamundi de Amos I. & Rabinovich de Halperin D.(eds.), Flora criptog. de Tierra del Fuego **1**(1): 7–62 (1986).
– Cabrera S.M., Lopez F. & Tell G.: Fitoplancton de las aguas superficiales de la provincia de Buenos Aires. I. Area de la Pampa deprimida. – Rev. Mus. La Plata (Nueva Serie), Secc. Bot., **10**(2-3): 200-206 (1986).
– Casco M.A., Echenique R.O. & Labollita H.A.: Las algas del sistema del Rio Limay (R. Argentina) I. Cyanophyta Chroococcales y Chamaesiphonales. – Rev. Mus. La Plata (N.S.) Bor. XIV **96**:163–182 (1987).
– & Ferrario M.E.: Algeous taxones novedosos para la ficoflora Argentina. – Physis, Sec. B, **38**(94): 9–19 (1979).
Guerrero G.P.: Cianofíceas y algas continentales de Cádiz. – An. Jard. bot. Madrid **6**(1): 237–335 (1946).
Hällfors G. & Munsterhjelm R.: Some epiphytic Chamaesiphonales from fresh and brackish water in southern Finland. – Ann. bot. fenn. **19**: 147–176 (1982).
Halperin D.R : «Cyanophyta» nuevas para la Argentina. – Darwiniana **13**(1): 115–131.
– & Wenzel M.T.: Cianofíceas marinas de la provincia Rio Negro, Argentina I. – Darwiniana **28**(1–4): 87–103 (1987).
– Zulpa de Caire G. & Zaccaro de Mule M.C.: Contenido proteico de *Aphano-*

*thece stagnina* (Sprengel) A. Braun (Cyanophyta, Chroococcaceae). – Physis **33**(87): 159–164 (1974).

Hansgirg A.: Prodromus der Algenflora von Böhmen. 2. – Arch. Naturwiss. – Landesdurchforsch. Böhmen **8**(4): 268 pp. (1892).

Hauck F.: Die Meeresalgen Deutschlands und Österreichs. – In: Rabenhorst's Krypt.-Fl. **2**:487–519, Kummer, Leipzig (1885).

Healey F.P.: Interacting effects of light and nutrient limitation on the growth rate of *Synechococcus linearis* (Cyanophyceae). – J.Phycol. **21**: 134–146 (1985).

Hegewald E., Jeeji-Bai N. & Hesse M.: Taxonomische und floristische Studien an Planktonalgen aus ungarischen Gewässern. – Arch. Hydrobiol./ Algolog. Stud. **13**: 392–432 (1975).

Henssen A. & Büdel B.: *Peccania cerebriformis* und *Psorotichia columnaris*, zwei neue Lichinaceen von Lanzarote. – Int. J. Myc. Lich. **1**(3): 261–271 (1984a).

– – *Phyllisciella*, a new genus of the Lichinaceae. – Nova Hedwigia **79**:381–398 (1984b).

Hernandez-Mariné M., Turon X. & Catalan J.: A marine *Synechocystis* (Chroococcales, Cyanophyta) epizoic on didemnid Ascidians from the Mediterranean Sea. – Phycologia **29**(3): 275–284 (1990).

Heynig H.: Zur Kenntnis des Planktons mitteldeutscher Gewässer VI. – Arch. Protistenk. **112**: 85–98 (1970).

– Einige interessante Phytoplankter aus Gewässern des Bezirks Halle (DDR). – Arch. Protistenk. **122**: 1–8 (1979).

Hickel B.: *Cyanodictyon reticulatum* (Lemm.) Geitler (Cyanophyta) a rare planktonic blue-green alga refound in eutrophic lakes. – Arch. Hydrobiol./ Algolog. Stud. **27**:111–118 (1981).

– *Cyanonephron styloides* gen. et sp. nov., a new chroococcal blue-green alga (Cyanophyta) from a brackish lake. – Arch. Hydrobiol./ Algolog. Stud. **38/39**: 99–104 (1985).

– Two new chroococcal cyanophytes from a brackish environment (Schlei-Fjord), Germany. – Arch. Hydrobiol./Algolog. Stud. **64**: 97–104 (1991).

Hindák F.: Mass development of the chroococcal blue-green alga *Synechocystis salina* Wisl. in ionide and bromide rich waters from crude oil-well. – Biológia (Bratislava) **23**(10): 841–844 (1968).

– Einige neue und interessante Planktonblaualgen aus der Westslowakei. – Arch. Hydrobiol./ Algolog. Stud. **15**: 330–353 (1975).

– Coccal blue-green algae from the thermal springs at Piešťany and Sklené Teplice Spa in Slovakia. – Arch. Hydrobiol./Algolog. Stud. **21**: 359–376 (1978).

– On some planktonic coccoid blue-green algae characteristic by Fe-precipitates. – Arch. Hydrobiol./ Algolog. Stud. **32**:241–258 (1982).

– Niektoré zaujímavé sinice (Cyanophyta) z planktóna našich vôd. [Several interesting cyanophytes from the plankton of our water bodies.] -Inf. Bull. MLVH SSR, T. Štrba (1983a).

– Morphological variability and taxonomy of some coccoid blue-green algae (Cyanophyta). – Schweiz. Ztschr. Hydrol. **45**(1): 311–320 (1983b).

– On the taxonomy of the cyanophycean genus *Rhabdogloea* Schröder = *Dactylococcopsis* Hansg. sensu auct. post. – Arch. Hydrobiol./Algolog. Stud. **35**: 121–133 (1984).

– The cyanophycean genus *Lemmermanniella* Geitler 1942. – Arch. Hydrobiol./ Algolog. Stud. **40**: 393–401 (1985).

– Contribution to the taxonomy of some cyanophyte genera. – Preslia **60**: 289–308 (1988a).

– *Tetrarcus* Skuja and *Merismoarcus* gen nov. – members of Cyanophyta/Cyanobacteria or bacteria? – Biológia **43**(9): 745–754 (1988b).

- Several interesting planktic cyanophytes. – Arch. Hydrobiol./ Algolog. Stud. **66**: 1–15 (1992a).
- On the taxonomy of the genus *Merismopedia* and related genera (Cyanophyta). – Arch. Hydrobiol./Algolog. Stud. **67**: 3–19 (1992b).
- Some neglected planktic colonial cyanophytes (Chroococcales, Cyanophyceae). – In: Salánki J.& Bíró P. ed., Limnological Bases of Lake Management, Tihany, p. 194–198 (1993).
- To the taxonomy of the chroococcal genus *Pannus* Hickel 1991 (Cyanophyta/ Cyanobacteria). – Arch. Hydrobiol./Algolog. Stud. **69**: 1–10 (1993).
- Cyanophytes colonizing mucilage of chroococcal water blooms. – Nova Hedwigia **112**: 69–82 (1996).
- Two new species of the genus *Radiocystis* (Chroococcales, Cyanophyta/Cyanobacteria). – Arch. Hydrobiol./Algolog. Stud. **83**: 367–375 (1996).
- & Moustaka M.T.: Planktic cyanophytes of Lake Volvi,Greece. – Arch. Hydrobiol./ Algolog. Stud. **50–53**: 497–528 (1988).

Hirose H. & Hirano M.: Class Cyanophyceae. – In: Hirose H. & Yamagishi T. (eds.), Illustrations of the Japanese fresh-water algae, 2nd ed., p. 1–151, Tokyo, Uchidarokakuho Publ. Co. (1981).

Hof T. & Frémy P.: On Myxophyceae living in Strong brines. – Rec. Trav. Bot. Néerl. **30**: 140–162 (1933).

Hofbauer B. & Jüttner F.: Occurrence of propylthio compounds in the aquatic ecosystem (Lake Neusiedl, Austria) as a chemical marker for *Microcystis flos-aquae*. – FEMS Microbiol. Ecol. **53**: 113–122 (1988).

Hoffmann L.: Cyanophycées aériennes et subaériennes du Grand-Duché de Luxembourg. – Bull. Jard. Bot. Nat. Belg. **56**: 77–127 (1986).

Hollerbach M.M.: De statibus *Gloeocapsae magmatis* (Bréb.) Kütz.-Not. Syst. Crypt. Horti Bot. Petropol. **3**:1–8 (1924).
- Nouvelles especes de Cyanophyceae decouvertes dans le sols. – Acta Inst. Bot. Acad. Sci. URSS, Ser. 2, **2**: 49–50 (1935).
- Kosinskaja E.K. & Poljanskij V.I.: Sinezelenye vodorosli. [Blue-green algae]. – In: Opredelitel' presnovodnych vodoroslej SSSR **2**: 1–625, Izd. «Sovetskaja nauka», Moskva (1953).

Hollenberg G.J.: Some new Myxophyceae from Southern California. – Bull. Torrey Bot. Club **66**: 489–494 (1939).

Holsinger E.C.T.: The plankton algae of three Ceylon Lakes. – Taxonom. Notes, 8–24 pp.

Hortobágyi T.: Qualitative Untersuchungen des Phytoplanktons des toten Armes «Nagyfa» der Tisza. – Folia cryptog. 1937, **3**(2): 151–208 (1939a).
- Algen der Cinkotaer stehenden Wasser I. – Borbásia **1**,9:136–139 (1939b).
- Relations phylogénétiques entre Cyanophyta: Coelosphaeriaceae, Gomphosphaeriaceae et Woronichiniaceae. – Acta bot. Acad. Sci. hung. **3**, 1–2: 9–17 (1957).
- Das Phytoplankton des Szelider Sees (Exc. Bacillariophyceae). – Die Binnengewässer Ungarns, Band I. p. 213–300 (1959a).
- Algák a Balatonból. – Annal. Biol. Tihany **26**: 329–342 (1959b).
- Adatok a Bükki algavegetáció ismeretéhez. – Bot. Közlem. **48**(1–2): 22–24 (1959c).
- Die Algen der Fischteiche von Buzsák V. – Nova Hedwigia, **6**(3/4): 353–390 (1963).
- Egy új Kékalga a Dunából: *Merismopedia danubiana* Hortob. n. sp. – Bot. Közlem. **61**(1): 7–23. (1974)

Hua M. S., Friedmann E.I., Ocampo-Friedmann R. & Campbell S.B.: Heteropolarity in unicellular cyanobacteria: structure and development of *Cyanocystis violacea*. – Pl. Syst. Evol. **164**: 17–26 (1989).

Huber J. & Jadin F.: Sur une algue perforante d'eau douce. – C.R. Acad. Sci. Paris **115**: 262–264 (1892).
Huber-Pestalozzi G.: Das Plankton natürlicher und künstlicher Seebecken Südafrikas. – Verh. Internat. Verein. theoret. angew. Limnolog. **4**: 343–390 (1929).
– Das Phytoplankton des Süsswassers. Systematik und Biologie. 1. – Die Binnengewässer 16: 1–342, E. Schweizerb. Verlagsbuchh., Stuttgart (1938).
Iltis A.: Algues des eaux natronées du Kanem (Tchad). Premiere partie. – Cah. O.R.S.T.O.M., ser. Hydrobiol., **VI**(3–4): 173–246 (1972).
Islam A.K.M.N.: Some new and rare freshwater algae from Iraq. – Internat. Rev. ges. Hydrobiol. **70**: 755–766 (1985).
– & Nahar L.: Preliminary studies on the phytoplankton of polluted waters. II. Blue green algae.- Sci. Res. **IV** (2&3):141–149 (1967).
– & Uddin M.A.: Blue-green algae from Dacca, Bangladesh. I. Chroococcaceae and Pleurocapsaceae. – J. Asiatic Soc. Bangladesh (Sc) **2**(2): 75–81 (1977).
Iyengar M.O.P. & Desikachary T.V.: *Johannesbaptistia pellucida* (Dickie) Taylor & Drouet from Madras. – J. Indian bot. Soc. **25**: 117–121 (1946).
Jaag O.: Untersuchungen über die Vegetation und Biologie der Algen des nackten Gesteins in den Alpen, im Jura und schweizerischen Mittelland. – Beih. Kryptogamenfl. Schweiz 9, 3: 560pp. (1945).
Jao C.-C.: Studies on the fresh-water algae of China. XII. The attached algal communities of the Kialing river. – Sinensia **15** (1–6): 61–74 (1944).
– Studies on the fresh-water algae of China. XIII. New Myxophyceae from Kwangsi. – Sinensia **15** (1–6):75–90 (1944).
Jensen T.E.: Cell inclusions in the Cyanobacteria. – Arch. Hydrobiol./ Algolog. Stud. **38/39**: 33–73 (1985).
Kaas H.: Algal studies of the Danish Wadden Sea. III. Blue-green algae in tidal flat sediments (sand flats and lower salt marsh) at Rejsby; taxonomy and ecology. – Opera Bot. **79**: 38–61 (1985).
– Algal studies of the Danish Wadden Sea. IV. Blue-green algae in higher salt marsh areas; taxonomy and ecology. – Cited from the manuscript (1985).
Kann E.: Die eulitorale Algenzone im Traunsee (Oberösterreich). – Arch. f. Hydrobiol. **55**(2): 129–192 (1959).
– Der Algenaufwuchs in einigen Bächen Österreichs. – Verh. Internat. Verein. Limnol. **16**: 646–654 (1966a).
– Einige Bemerkungen zur Gattung *Chamaesiphon*. – Verh. Internat. Verein. Limnol. **16**(3): 1569–1573 (1966b).
– Zur Systematik und Ökologie der Gattung *Chamaesiphon* (Cyanophyceae) 1. Systematik. – Arch. Hydrobiol./ Algolog. Stud. **7**: 117–171 (1972).
– Systematik und Ökologie der Algen österreichischer Bergbäche. – Arch. Hydrobiol./Suppl. **53** (Monogr. Beitr.): 405–643 (1978).
Karim A.G.A.: Studies on the freshwater algae of the Sudan I. On the ecology of the algae of Wadi Gabol – Jebel Marra. – Hydrobiologia **32**(1–2): 33–46 (1968).
Karlsson B., Vaara T., Lounatmaa K. & Gyllenberg H.: Three dimensional structure of the regularly constructed surface layer from *Synechocystis* sp. CL-II. – J. Bacteriol. **156**(3): 1338–1343 (1983).
Kato T. & Watanabe M.: Allozyme divergence of *Microcystis* strains from Lake Kasumigaura. – Proc. Internat. Phycol. Forum, Univ. of Tsukuba, 73pp. (1993).
– Watanabe M.F. & Watanabe M.: Allozyme divergence in *Microcystis* (Cyanophyceae) and its taxonomic inference. – Arch. Hydrobiol./ Algolog. Stud. **64**: 129–140 (1991).
Kirchner O.: Beiträge zur Algenflora von Württemberg. – Württemb. naturw.

Jahreshefte **36** (1880). – Schizophyceae. – In: Engler A. & Prantl K. (eds.), Die natürlichen Pflanzenfamilien I, **1a**: 45–92 (1900).
Kiselev I.A.: Fitoplankton severo-vostočnogo Kaspija s ego zalivami Komsomolec (Mertvyj Kultuk) i Kajdak po materialam ekspedicij AN SSSR v 1934 i 1935 gg. [Phytoplankton of NE Caspian Sea and bays Mertvyj Kultuk and Kajdak; samples collected of the expeditions of the AN USSR in 1934 and 1935]. – Trudy Kompleks. Izuč. Kasp. Morja **3**, Izd. AN USSR (1940).
– K morfologii, ekologii, sistematike i geografičeskomu rąsprostraneniju sinezelenoj vodorosli *Cyanothrix gardneri* (Frémy). – Bot. Ž. SSSR **32**(3): 111–118 (1947).
Kiseleva E.I.: Materialy k izučeniu mikroflory risovych polej okresnosti goroda Samarkanda. [Materials to the investigations of microflora of paddy fields in the vicinity of Samarkand]. – Ž. Russk. Bot. Obšč. **16**(4): 353–378 (1931).
Kiss I.: Snow-blooms in the Caucasus. – Bot. Közlem. **48**(1–2): 25–28 (1959).
Kogan S.I.: Vodorosli vodojemov Turkmenskoj SSR. – Izdat. «Ilym», Aščhabad, 48 pp.(1973).
Kol E.: Über die Kryovegetation der Hohen Tatra. I.- Folia Cryptogam. (Szeged) **1**,6: 613–622 (1928).
– Blauer Schnee im Gebiet der Kleinen-Küküllö. – Annal. Histor.-Nat. Musei National. Hung. **6**: 93–96 (1955).
– Kryobiologie. – In: Elster H.J. & Ohle W. (eds.), Die Binnengewässer **24**: 216 pp., Schweizerbart'sche Verlagsbuchh., Stuttgart (1968).
Komárek J.: Neskol'ko interesnych sinezelenych vodoroslej černomorskogo poberežja v okrestnostjach Burgasskoj buchty (Bulgaria). [Some interesting blue-green algae from Bulgarian coast of Black Sea near Burgas.]. – Acta Univ. Carol.- Biologica **2**(1): 91–123 (1956).
– Das *«Microcystis»* – Problem. – Taxon (Utrecht) **6**:145–149 (1957).
– Die taxonomische Revision der planktischen Blaualgen der Tschechoslowakei. – In: Komárek J. & Ettl H., Algologische Studien, p.10–206, Naklad. ČSAV, Praha (1958).
– *Dactylococcopsis* Hansgirg ex R. et F. Chodat 1925, nomen conservandum propositum. – Taxon (Utrecht) **18**: 339–345 (1969a).
– On the validity of the genus *Dactylococcopsis* (*Cyanophyceae*). – Österr. Bot. Ztschr. **117**: 248–257(1969b).
– Das Studium der Blaualgenvariabilität in Kulturen: Möglichkeiten für die Taxonomie. (Abstr.) – In: Zehnder A., Schweiz. Ztschr. Hydrol. **32**: 483–484 (1970a).
– Generic identity of the *«Anacystis nidulans»* strain Kratz-Allen/Bloom. 625 with *Synechococcus* Näg. 1849. – Arch. Protistenk. **112**: 343–264 (1970b).
– Reproduction process and taxonomy of unicellular endosporine blue-green algae. – In: Desikachary T.V.(ed.), Proc. Symp. taxonomy and biology of blue-green algae, Univ. Madras, p. 41–47 (1972).
– Blaualgen aus dem Naturschutzgebiet Řežabinec bei Ražice. – Nova Hedwigia **26**: 601–643 (1975a).
– *Geitleribactron* – eine neue, *Chamaesiphon*- ähnliche Blaualgengattung. – Plant Syst. Evol. **123**:263–281 (1975b).
– Taxonomic review of the genera *Synechocystis* Sauv. 1892, *Synechococcus* Näg. 1849, and *Cyanothece* gen. nov. (Cyanophyceae). – Arch. Protistenk. **118**: 119–179 (1976).
– *Rhabdogloea*, the correct name of cyanophycean *Dactylococcopsis* sensu auctt., non Hansgirg (1888). – Taxon (Utrecht) **32**(3): 464–466 (1983).
– Sobre las cianofíceas de Cuba: (3) Especies planctónicas que forman florecimientos de las aguas. – Acta Bot. Cubana **19**: 1–33 (1984).
– Do all cyanophytes have a cosmopolitan distribution? Survey of the freshwater

cyanophyte flora of Cuba. – Arch. Hydrobiol./ Algolog. Stud. **38/39**: 359–386 (1985).
– Studies on the Cyanophytes of Cuba 7–9. – Folia geobot. phytotax. (Praha) **24**: 131–206 (1989).
– A review of water-bloom forming *Microcystis*-species with regard to populations from Japan. – Arch. Hydrobiol./Algolog. Stud. **64**: 115–127 (1991).
– Validation of the genera *Gloeocapsopsis* and *Asterocapsa (Cyanoprokaryota)* with regard to species from Japan, Mexico and Himalayas. – Bull. Nat. Sci. Mus. Tokyo, Ser. B, **19**(1):19–37 (1993).
– Current trends and species delimitation in the cyanoprocaryote taxonomy. (Introduction lecture of the 12th IAC symposium, Sjöarp, August 1992). – Arch. Hydrobiol./Algolog. Stud. **75**:11–29 (1994).
– Studies on the Cyanophytes (Cyanoprokaryotes) of Cuba 10. New and little known Chroococcalean species. – Folia geobot. phytotax **30**: 81–90 (1995).
– Towards a combined approach for the taxonomy and species delimitation of picoplanktic cyanoprokaryotes. – Arch. Hydrobiol./Algolog. Stud. **83**: 377–401 (1996).
– & Albertano P.: Cell structure of a planktic cyanoprokaryote, *Tychonema bourrellyi*. – Arch. Hydrobiol./ Algolog. Stud. **75**: 157–166 (1994).
– & Anagnostidis K.: Modern approach to the classification system of cyanophytes, 2 -Chroococcales. – Arch. Hydrobiol./ Algolog. Stud.**43**: 157–226 (1986).
– – Modern approach to the classification system of cyanophytes, 4 – Nostocales. – Arch. Hydrobiol./ Algolog. Stud. **56**: 247–345 (1989).
– – Nomenclatural novelties in chroococcalean cyanoprokaryotes. – Preslia, Praha, **67**: 15–23 (1995).
– & Čáslavská J.: Thylakoidal patterns in oscillatorialean genera. – Arch. Hydrobiol./ Algolog. Stud., **64**: 267–270 (1991).
– & Cronberg G.: Some chroococcalean and oscillatorialean cyanoprokaryotes from central and south African water reservoirs. – Nova Hedwigia, in press (1998).
– & Hindák F.: Taxonomy of the new isolated strains of *Chroococcidiopsis* (Cyanophyceae). – Arch. Hydrobiol./Algolog. Stud. **13**: 311–329 (1975).
– – Taxonomic review of natural populations of the cyanophytes from the *Gomphosphaeria*- complex. – Arch. Hydrobiol./Algolog. Stud. **50–53**: 203–225 (1988).
– – The genus *Eucapsis* (*Cyanophyta/Cyanobacteria*) in Czechoslovakia. – Acta Hydrobiol. (Kraków) **31**: 25–34 (1989).
– & Ludvík J.: *Synechococcus diatomicola* isolated from the lake Gerzensee (Switzerland). – Schweiz. Ztschr. Hydrol. (Basel) **45**: 274–278 (1983).
– – – The cell structure of two Chamaesiphonaceae, *Cyanophanon minus* and *Stichosiphon sansibaricus*. – Arch. Hydrobiol./ Algolog. Stud.**38/39**: 73–90 (1985).
– & Kalina T.: Bemerkungen zur Coenologie und Systematik des Gattung *Homoeothrix* (Thur. ex Born. et Flah.) Kirchn. – Österr. Bot. Ztschr. **112**: 424–441 (1965).
– & Kling H.: Variation in six planktic cyanophyte genera in Lake Victoria (East Africa). – Arch.Hydrobiol./ Algolog. Stud. **61**: 24–45 (1991).
– & Komárková-Legnerová J.: Variability of some planktic gomphosphaerioid cyanoprokaryotes in northern lakes. – Nordic J. Bot. **12**: 513–524 (1992).
– – *Radiocystis fernandoi*, a new planktic cyanoprokaryotic species from tropical freshwater reservoirs. – Preslia (Praha) **65**: 355–357 (1993).
– Kopecký J. & Cepák V.: The background for taxonomy of the simplest

cyanoprokaryotic genera *Cyanobium, Cyanobacterium* and *Synechococcus.* – Cryptogamie/Algologie, in press (1998).
– & Ludvík J.: Cell structure and reproduction process in the blue-green alga *Chamaesiphon.* – Plant. Syst. Evol. **139**: 267–277 (1982).
– – & Pokorný V.: Cell structure and endospore formation in the blue-green alga *Chroococcidiopsis.* – Arch. Hydrobiol./Algolog. Stud. **12**: 205–223 (1975).
– & Montejano G.: Taxonomic evaluation of several *Chlorogloea* species (Cyanoprocaryota) from inland biotopes. – Arch. Hydrobiol./ Algolog. Stud. **74**: 1–26 (1994).
– & Novelo E.: Little known tropical *Chroococcus* species (Cyanoprokaryotes). – Preslia (Praha) **66**: 1–21 (1994).
Komarenko I.E. & Vasil'eva I.I.: K issledovaniu vodoroslej (mikroflory) bassejna r. Olenok. [Remarks to algae (microflora) of the basin of the river Olenok.] – Sb.«Ljubite i ochranjajte prirodu Jakutii», Jakutsk, p.103–110 (1975).
Komárková-Legnerová J.: *Epigloeosphaera,* a new cyanophyte genus from Nordic lakes. – Arch. Hydrobiol./ Algolog. Stud. **62**: 7–12 (1991).
– & Cronberg G.: *Lemmermanniella pallida* (Lemm.) Geitl. from South-Swedish Lakes. – Arch. Hydrobiol./ Algolog. Stud. **40**: 403–413 (1985).
– – New and recombined filamentous Cyanophytes from lakes in South Scania, Sweden. – Arch. Hydrobiol./ Algolog. Stud. **67**: 21–31 (1992).
– – Planktic blue-green algae from lakes in South Scania, Sweden. Part I. Chroococcales. – Arch. Hydrobiol./ Algolog. Stud. **72**: 13–51 (1994).
– & Tavera R.: Cyanoprokaryota (Cyanobacteria) in the phytoplankton of Lake Catemaco (Veracruz, Mexico). – Arch. Hydrobiol./Algolog. Stud. **83**: 403–422 (1996).
Kondrateva N.V.: Sin'o-zeleni vodorosti dejakich obrobljuvanich gruntiv z okolic Kieva. [Blue-green algae of some cultivated soils in the environments of Kiev.] – Ukr. bot. Ž. **16**(1): 74–85 (1959).
– Sin'o-zeleni vodorosti – Cyanophyta [Blue-green algae – Cyanophyta]. – In: Vizn. prisnov. vodorost. Ukr. RSR 1, **2**: 524 pp., Naukova Dumka, Kiev (1968a).
– Voprosy morfologii i sistematiki *Microcystis aeruginosa* Kuetz. emend. Elenk. i blizkich k nemu vidov. [Problem of morphology and systematics of *Microcystis aeruginosa* Kuetz. emend. Elenk. and related species.] – In: «Cvetenie vody», p. 13–42, Kiev (1968b).
– Kovalenko O.V. & Prichod'kova L.P.: Sin'ozeleni vodorosti – Cyanophyta. [Blue-green algae – Cyanophyta]. – Vizn. prisnov. vodorost. Ukr. RSR 1, **1**: 1–388, Kiev, Naukova Dumka, (1984).
Kosinskaja E.K.: Opredelitel' morskich sinezelenych vodoroslej. [Determination key for marine blue-green algae.] – 278 pp., Izd. AN SSSR, Moskva-Leningrad (1948).
Kováčik L.: Type of reproduction of *Aphanocapsa* in comparison with the genera *Chroococcus, Microcystis* and *Merismopedia.* – Schweiz. Ztschr. Hydrol. **45**(1): 279–282 (1983).
– Biológia a reprodukčný proces kokálnych siníc. [Biology and reproduction process in coccal cyanophytes.] – Diss., 74 pp., Inst. of Bot. ČSAV, Třeboň (1984).
– Cell division in simple coccal cyanophytes. – Arch. Hydrobiol./ Algolog. Stud. **50–53**: 149–190 (1988).
Krieger W.: Süßwasseralgen aus Griechenland. – Ber. Dtsch. Bot. Ges. **61**(5): 250–270 (1944).
Kufferath H.: Algues et Protistes muscicoles et terrestres récoltées sur la montagne de Barba (Costa-Rica). – Ann. Crypt. exot. **2**(1): 50–52 (1929).

– Fleur d'eau rouge permanente a Myxophycées dans un étang a Boirs-sur-Geer. – Inst. Royal Sci. Nat. Belgique – Bull. **26**(2): 1–22 (1950).
Kützing T.F.: Tabulae Phycologicae oder Abbildungen der Tange I. – 54 pp. (1845–1849).
– Species algarum. – 922 pp., F.A.Brockhaus, Leipzig (1949).
Laamanen M.: Cyanoprokaryotes in the Baltic Sea ice and winter plankton. – Algolog. Stud. **83**: 423–433 (1996).
Lafargue F. & Duclaux G.: Premier exemple, en atlantique tropical, d'une association symbiotique entre une ascidie didemnide et une cyanophycée chroococcale: *Trididemnum cyanophorum*, nov. sp. et *Synechocystis trididemni* nov. sp. – Ann. Inst. Oceanogr. (Paris) **55**: 163–184 (1979).
Lagerheim G.: Bidrag till Sveriges algflora. – Öfvers. Kgl. Svensk. Vetensk.-Ak. Förh. (Stockholm) 40:37–78 (1883).
Le Campion-Alsumard T.: Contribution a l'étude des cyanophycées lithophytes des étages supralittoral et médiolittoral (Region de Marseille). – Tethys **1**: 119–172 (1969).
– Cyanophycées marines endolithes colonisant les surfaces rocheuses denudées (étages suppralittoral) et médiolittoral de la région de Marseille. – Schweiz. Ztschr. Hydrol. **32**: 552–558 (1970).
– Quelques remarques sur la culture et le cycle de dévelopment de la cyanophycée lithophyte: *Hyella caespitosa* Bornet et Flahault. – Tethys **4**: 391–396 (1972).
– Les Cyanophycées endolithes marines: systématique, ultrastructure, ecologie et biodestruction. – Thése état, Univ. Aix-Marseille, 198pp. (1979).
– Three *Hyella* taxa (endolithic cyanophytes) from tropical environments (Lizard Island, Great Barrier Reef) (1991).
– & Golubić S. (1985a): *Hyella caespitosa* Bornet et Flahault and *Hyella balani* Lehman (Pleurocapsales, Cyanophyta). – Arch. Hydrobiol./ Algolog. Stud. **38/39**: 115–118 (1985a).
– – Ecological and taxonomic relationships between endolithic cyanophytes *Hormathonema* and *Solentia*. Abstract. – Arch. Hydrobiol./Algolog. Stud. **38/39**: 119–148 (1985b).
– – & Pantazidou A.: On the euendolithic genus *Solentia* Ercegović (Cyanophyta/ Cyanobacteria). – Algolog. Stud. **83**:107–127 (1996).
Lederer F.: A new species of *Cyanodictyon* (Cyanoprokaryota, Chroococcales) from peat-bogs in the Šumava Mts., Czech Republic. – Preslia **67**: 117–121 (1995a).
– Several little known Cyanobacteria/Cyanoprocaryota from peat-bogs in the Šumava Mountains, Czech Republic. – Arch. Hydrobiol./ Algolog. Stud. **79**: 57–65 (1995b).
Lemmermann E.: Der grosse Waterneversdorfer Binnensee. – Forschungsb. Biol. Stat. Plön **6**: 167–204 (1898).
– Algen I. – Krypt.-Fl. Mark Brandenburg 3: 1–256 (1907–1910).
Lewin R. & Robinson X.: Greening of polar bears. – Nature 1234: 278–280 (1979).
Ley S.-H.: New Myxophyceae from Northern Kwangtung. – Bot. Bull. Ac. Sin. **1**: 77–79 (1947a).
– Heleoplanktonic algae of North Kwangtung. – Bot. Bull. Ac. Sin. **1**: 270–282 (1947b).
– The subaerial algae from the Paracel Islands in the South China Sea. – Bot. Bull. Ac. Sin. **2**: 235–242 (1948).
Li Y.-Y.: New cyanophytes from Xizang (Tibet). – Acta Phytotax. Sinica **22** (2): 167–174 (1984).

Liebetanz B.: Hydrobiologische Studien an Kujawischen Brackwässern. – Bull. Acad. Polon. Sci. Lett., Sér. B: 114 pp. (1925).
Lindstedt A.: Die Flora der marinen Cyanophyceen der schwedischen Westküste. – 121 pp., Lund (1943).
Ling H.U., Thomas D.P. & Tyler P.A.: Microalgae. – In: Zeidler & Ponder (eds.), Natural History of Dalhousie Springs, p. 47–52 (1989).
López-Rodríguez M. del C. & Pérez-Cirera J.L.: Cianófitos en medios alterados porla contaminación industrial en el noreoste de la Península Ibérica. – Anal. Jardín Botanico de Madrid **54**: 43–49 (1996).
Lukas K.J. & Golubić S.: New endolithic cyanophytes from the North Atlantic ocean: I. *Cyanosaccus piriformis* gen. et sp. nov. – J. Phycol. **17**: 224–229 (1981).
– – New endolithic cyanophytes from the North Atlantic ocean: II. *Hyella gigas* Lukas et Golubić sp. nov. from the Florida continental margin. – J. Phycol. **19**: 129–136 (1983).
– & Hoffman E.J.: New endolithic cyanophytes from the North Atlantic Ocean. III. *Hyella pyxis* Lukas & Hoffman sp. nov. – J. Phycol. **20**: 515–520 (1984).
Lukavský J.: First record of cryoseston in the Bohemian Forest Mts. (Šumava). – Arch.Hydrobiol./ Algolog. Stud. **69**:83–89 (1993).
– & Cepák V.: DAPI fluorescent staining of DNA material in cyanelles of the rhizopod *Paulinella chromatophora* Lauterb. – Arch. Protistenkd. **142**: 207–212 (1992).
Lund J.W.G. (1942): Contributions to our knowledge of British Algae – VIII. – The J. Bot., April, p. 57–73 (1942).
– Algological Notes I-III. – The Naturalist, April – June, p.45–49 (1950a).
– Algological Notes IV-VI. – The Naturalist, October-December, p. 143–148 (1950b).
Lundberg F.: Beiträge zur Kenntnis der Algenflora von Schweden. I. – Bot. Notiser 1931: 269–298 (1931).
Margalef R.: Contribution al conocimento hidrobiológico del País Vasco-Navarro. – Consejo Superior de Invest. Cient. Estacion de Estudios pirenaicos; Estudio de la Fauna y Flora Vasco-Navarras, Zaragoza, 7–44 pp. (1946).
– Materiales pro la hidrobiologia de Ibiza. – P. Inst. Biol. Apl. **8**: 5–70 (1951).
– Algas de agua dulce del noroeste de España. – P. Inst. Biol. Apl. **22**: 43–152 (1956).
Meneghini J.: Conspectus Algarum Euganeorum. – Comm. Med. **4**: 231–335 (1837).
Messikommer E.: Hydrobiologische Studie an der Moorreservation der Schweizerischen Naturforschenden Gesellschaft in Robenhausen-Wetzikon. – Viertelj. naturf. Ges. Zürich **88**, suppl. 2: 70 pp. (1943).
Meyer B.: A new species of *Cyanodictyon* (Cyanophyceae, Chroococcales) planktic in eutrophic lakes. – Arch. Hydrobiol./ Algolog. Stud. **75**: 183–188 (1994).
Micka J. & Holleová J.: Biologické problémy při úpravě vody z nádrže Janov [Biological problems during water treatment in the Janov reservoir]. – Aktuální otázky vodárenské biologie 45–48 (1997).
Mollenhauer D.: Beiträge zur Kenntnis der Gattung *Nostoc* I. – Abh. Senckenb. Naturf. Ges. **524**:1–80 (1970).
Montejano G., Gold M. & Komárek J.: Freshwater epiphytic cyanoprokaryotes from central Mexico I. *Cyanocystis* and *Xenococcus* – Arch. Protistenk. **143**: 237–247 (1993).
– – – Freshwater epiphytic cyanoprokaryotes from central Mexico III. The genus *Stichosiphon*. – Arch. Protistenk. **148**: 3–16 (1997).
Montoya H.T.: Algas de la Laguna de Villa (Lima): Cyanophyta, Chlorophyta y

Chrysophyta (Xanthophyceae). – Bol. de Lima **31**(6): 75–89, **32**(6): 49–62 (1983).
– Barberena C. & Quesquen R.: Algas de las salinas de Huacho, Perú, con enfasis en *Aphanothece halophytica* Frémy. – Bolet. Inst. Mar del Peru, Memor. 2. Congr. COLACMAR I: 67–75 (1988).
– & Golubić S.: Morphological variability in natural populations of mat-forming cyanobacteria in the salinas of Huacho, Lima, Peru. – Arch. Hydrobiol./ Algolog. Stud. **64**: 423–441 (1991).
Moore G.T. & Carter N.: Algae from lakes in the north-eastern part of North Dakota. – Ann. Missouri Bot. Gard. **10**: 393–422 (1923).
Nägeli C.: Gattungen einzelliger Algen. – Neue Denkschr. Allg. Schweiz. Natur. Ges. **10**: 139 pp., Zürich (1849).
Naumann E.: Notizen zur Biologie der Süsswasseralgen II. Über *Paracapsa siderophila* n. g. et n. sp., als Ursache einer auffälligen limnischen Eiseninkrustation. – Ark. Bot. **18**(21) (1924).
Necchi Júnior O.. & Sant'Anna C.L.: Taxonomic study of some Chamaesiphonales (Cyanophyceae) from the State of Sao Paulo, Southeastern Brazil. – Revista brasil. Bot. **9**: 201–206 (1986).
Negoro K.: Über die Algenvegetation der Thermen von Kusatu, Gunma Präfektur, Japan. – Cit. from Inst. Bot., Univ. Tokyo f. Literatur u. Wissenschaft, p. 302–312 (1943). (Japanese with German summary.)
– Untersuchungen über die Vegetation der mineralogen azidotrophen Gewässer Japans. – Sci. Rep. Tokyo Bunrika Daigaku, sect. B., **6** (101): 231–274 (1944).
– The plankton of lake Biwa. – The Shiga Prefect. Inst. of Public Health and Environm. Science, 156 pp. (1982).
Niell F.X. & Anadon R.: Seasonal data on morphology and ecology of *Merismopedia*-like marine algae. Taxonomical implications of the observed changes. – Bot. marina **21**: 39–47 (1978).
Noguerol-Seoane A.N. & Rifón Lastra A.B.: Aportación al conocimiento de la ficoflora epilíthica en monumentos del noroeste de España. Estudio del monasterio de Samos (Lugo). – Anales Jard. Bot. Madrid **54**: 37–42 (1996).
– – Epilithic phycoflora on monuments, a survey of San Esteban de Ribas de Sil Monastery (Ourense, NW Spain). – Cryptogamie, Algol. **18**(4): 351–361 (1997).
Norris R.E.: Algal consortisms in marine plankton.- Proc. Seminar on Sea, Salt and Plants, p. 178–189 (1967).
Nováček F.: Nový druh rodu *Gloeocapsa, Gloeocapsa dvořákii* nov. (spec. nova). [New species of the g. *Gloeocapsa, G.dvořákii* spec. nova.] – Zpr. Komise Přírodov. Výzk. Moravy a Slezska, Odd. bot., **7**: 1–11 (1929).
– Příspěvek k poznání životního cyklu *Gloeocapsa sanguinea* AG. [Remarks to the life cycle of *Gloeocapsa sanguinea* AG.] – Sborn. Kl. přírodověd. v Brně **13**: 1–13 (1930).
– *Entophysalis atroviolacea* Nčk., nový druh rodu *Entophysalis* – ad interim. (De nova specie generis *Entophysalis – Entophysalis atroviolacea* Nčk. notula.) – Acta Soc. Sci. nat. Moravicae 7,3: 1–12 (1932).
– Additamentum ad oecologiam morphologiamque Cyanophycearum ad rupes serpentinicas prope Mohelno Moraviae occidentalis epilithice habitantium. I.Chroococcales. – In: Mohelno, Arch. Sv. Ochr. Přír., Brno, **3a**: 1–178 pp. (1934).
Nygaard G.: Plankton from two lakes of the Malayan region. – Vidensk. Medd. Dansk Naturh. Foren. **82**: 197–240 (1926).
– Contributions to our knowledge of the freshwater Algae of Africa, 9. Freshwater Algae and phytoplankton from the Transvaal. – Trans. Roy. Soc. South Africa **20**(2): 101–148 (1932).

– Hydrobiological studies on some Danish ponds and lakes. II. – Det Kong. Danske Vidensk. Selskab, Biolog. Skrifter 7(1): 1–294 (1949).
– New or interesting plankton algae. – Kgl. danske vidensk. Selsk., Biol. Skrifter **21**(1): 1–107 (1977).
– Freshwater phytoplankton from the Narssaq Area, South Greenland. – Bot. Tidsskrift **73**(3–4):191–201 (1984).
Ohtani S. : Freshwater algae in the Okayama Prefectural Nature Conservation Center. – Bull. Okayama Pref. Nature Conservation Center Suppl. (**1**): 103–150 (1996).
– & Kanda H.: Epiphytic algae on the moss community of *Grimmia lawiana* around Syowa Station, Antarctica. – Proc. NIPR Symp., Polar Biol., **1**: 255–264 (1987).
Padmaja T.D.: Studies on coccoid blue-green algae – II.- In: Desikachary T.V.(ed.) Proc. Symp. Taxonomy and biology of blue-green algae, Univ. Madras, p. 75–127 (1972).
– & Desikachary T.V.: Studies on coccoid blue-green algae. I. *Synechococcus elongatus* and *Anacystis nidulans*. – Phykos (Delhi) 7(1 & 2): 62–89 (1968).
Palik P.: Beiträge zur Kenntnis der Lithophyten Algenvegetation des Bükkgebirges. – Index Horti Bot. Univ. Budapest **3**: 147–149 (1938).
– Beitrag zur Kenntnis der Algenvegetation der Bálintquelle des Bades Félixfürdö. – Hung. Acta Biol. **1/2**: 40–55 (1949).
– A new blue-green alga from the cave Baradla near Aggtelék. – Annal. Univ. Scient. Budapest. de Rolando Eötvös nominatae, sec. Biolog., **3**: 275–285 (1960a).
– Contributii la flora algelor din turbaria Mohos (Tusnad). – Stud. si Cercetári de Biologie, ser. Biol (1960b).
Palinska K.A., Liesack W., Rhiel E. & Krumbein W.E.: Phenotype variability of identical genotypes: the need for a combined approach in cyanobacterial taxonomy demonstrated on *Merismopedia*-like isolates. – Arch. Microbiol **166**: 224–233 (1996).
Pantazidou A.: [Studies on euendolithic cyanophytes from marine coastal carbonate substrates of Greece]. – Doctor Thesis, Univ. Athens, 298 pp. (1991).
Parker D.L.: Improved procedures for the cloning and purification of *Microcystis* cultures (Cyanophyta). – J. Phycol. **18**: 471–477 (1982).
Pascher A.: Über Symbiosen von Spaltpilzen und Flagellaten mit Blaualgen. – Ber. Dtsch. Bot. Ges. **32**(5): 339–352 (1914).
– Über die Teilungsvorgänge bei einer neuen Blaualge,
*Endonema*. – Jahresb. wiss. Bot. **70**: 329–347 (1929).
Pechar L. & Kalina T.: Ecological, morphological and ultrastructural differentiation between two types of *Aphanizomenon flos-aquae*: a study of natural populations. Mscr.
Pellegrini S., Grilli Caiola M. & Ribaldone A.: Developmental cycle of *Microcystis aeruginosa* in culture. – Ann. Microbiol. Enzymol. **38**: 239–247 (1988).
Pennisi E.: Static evolution. – Science News **145**(11):168–169 (1994).
Pevalek I.: Prilog poznavanju alga Kamnišnih Alpa. [Remarks to the knowledge of algae of Kamnišni Alps Mts.] – Glasn. Hrv. prip. družstv. 1925: 1–5 (1925).
Pinevich A.V., Matthijs H.C.P., Averina S.G. & Gavrilova O.V.: Picocyanophyte (cyanobacterium) from the boreal inland water accumulates phycoerythrin as a major biliprotein. – Arch Hydrobiol./Algolog. Stud. **87**: 99–108 (1997).
Pitschmann H.: Beiträge zur Thermalalgenflora der Insel India. – Ber. nat.-med. Ver. Innsbruck **57**:185–193 (1969).
Pollingher U.: A long-term study of *Chroococcus* population in Lake Kinneret (Israel). – Arch. Hydrobiol./ Algolog. Stud. **64**: 377–383 (1991).
Potts M.: Blue-green algae (Cyanophyta) in marine coastal environments of the

Sinai Peninsula; distribution, zonation, stratification and taxonomic diversity. – Phycologia **19**: 60–73 (1980).
Preisig H.R.: Vergleichende Studien an Algenpopulationen in Kiesgrubenweihern. – Diss., 271pp., Univ. Zürich (1979).
Prescott G.W.: Algae of the Western Great Lakes Area. – Cranbrook Inst. Sci., Bull. 31: 946 pp., Bloomfield Hills, Michigan. (1951)
– & Vinyard W.C.: Ecology of Alaskan Freshwater algae. V. Limnology and flora of Malikpuk lake. – Trans. Amer. Micros. Soc. **84**(4): 427–478 (1965).
Pringsheim E.G.: Cyanophyceen – Studien. – Arch. Mikrobiol. **63**: 331–355 (1968a).
– Kleine Mitteilungen über Flagellaten und Algen. XVI. *Lauterbornia (Anacystis) nidulans* (Richter) nov. gen., nov. comb. Cyanophyceae. – Arch. Mikrobiol. **63**: 1–6 (1968b).
Proškina-Lavrenko A.I.: Novye vidy vodoroslej iz solenich vodoemov SSSR. 2. [New algal species from salt waters of USSR. 2]. – Bot. Mater. Otd. Spor. Rast. Bot. Inst. AN SSSR 7:69–74 (1951).
– & Makarova I.V.: Vodorosli planktona Kaspijskogo morja. [Planktic algae of the Caspian sea.] – 291 pp., Izd. «Nauka», Leningrad (1968).
Ramírez J.J.: Autecology of *Cyanocatena bicudoi* sp. nova, a new Cyanophycea from Parque Norte lagoon, Colombia. – Arch Hydrobiol./Algolog.Stud. **80**: 21–34 (1996).
Rao C.S.: The Myxophyceae of the Delhi Province, India. I. – Proc. Indian Acad. Sci. B, **11**(3): 125–131 (1940).
Rayss T.: Materiaux pour la flore algologique de la Palestine, I. Les Cyanophycées. – Palest. J. Bot. (Jerusalem) ser. **3**: 94–113 (1944).
Reynolds C.S., Jaworski G.H.M., Cmiech H.A.& Leedale G.F.: On the annual cycle of the blue-green alga *Microcystis aeruginosa* Kütz. Emend. Elenkin. – Phil. Trans. R. Soc. London B., **293**:419–477 (1981).
Rich F.: Scientific results of the Cambridge Expedition to the East African Lakes 1930-1-7. The Algae. – Linn. Soc. J. – Zool. **38**: 249–275 (1933).
Richter P.: Neue Algen der Phykotheka universalis, Fasc. XIII. – Hedwigia **43**: 22–26 (1895).
Rino J.A.: Contribuiçao para o conhecimento das algas de água doce de Moçambique – I. – Rev. Cienc. biol. **2**, ser.A: 51–102 (1969).
Rippka R. & Cohen-Bazire G.: The Cyanobacteriales: a legitimate order based on the type strain *Cyanobacterium stanieri*? – Ann. Microbiol. (Inst. Pasteur) **134B**: 21–36 (1983).
– Deruelles J., Waterbury J.B., Herdmann M. & Stanier R.Y.: Generic assignments, strain histories and properties of pure cultures of cyanobacteria. – J. gen. Microbiol. **111**: 1–61 (1979).
– & Herdman M.: Division patterns and cellular differentiation in Cyanobacteria. – Ann. Microbiol. (Inst. Pasteur) **136A**: 33–39 (1985).
– Waterbury J. & Cohen-Bazire G.: A cyanobacterium which lacks thylakoids. – Arch. Microbiol. **100**: 419–436 (1974).
Roll J.: Algues nouvelles, trouvées dans le plancton de la riviere Dniepre. – Ann. Protistolog. **1**(4):1–5 (1928).
Romo S., Bécares E. & Compere P.: Algae from Bioko Island (Aquatorial Guinea, West Africa). – Arch Hydrobiol./ Algolog. Stud. **76**: 79–95 (1995).
Rosenvinge L.K.: Gronlands Havalger. Meddelelser om Gronland III. – Kjobenhavn (1893).
Rott E.: A contribution to the phytoplankton species composition of Parakrama Samudra, an ancient man-made lake in Srí Lanka. – In: Schiemer F. (ed.), Limnology of Parakrama Samudra – Srí Lanka. A case study of an ancient man-

made lake in the tropics, 236 pp. Dr. W.Junk Publishers, The Hague-Boston-London (1983).
Roussomoustakaki M. & Anagnostidis K.: *Cyanothece halobia*, a new planktic chroococcalean cyanophyte from Hellenic heliothermal saltworks. – Arch. Hydrobiol./Algolog. Stud. **64**:71–95 (1991).
Sanchez-Castillo M.P.: Algas de las lagunas de alta montana de Sierra Nevada (Granada, Espana). – Acta Botan. Malacitana **13**: 21–52 (1988).
Sant'Anna C.L., Azevedo M.T. P., Branco L.H.Z., Braga M.R.A., Corderiro-Marino M. & Guimaraes S.M.P.B.: Cianofíceas marinhas bentônicas das Praias de Peruíbe e dos Sonhos, Município de Itanhaém, SP, Brasil, III. – Rev. Brasil. Biol. **55**(3): 389–407 (1995).
– Bicudo R.M.T. & Pereira H.A.S.L.: Nostocophyceae (Cyanophyceae) do Parque Estadual da Ilha do Cardoso, Estado de Sao Paulo, Brasil. – Rickia **10**: 1–27 (1983).
– Pereira H.A.S.L. & Bicudo R.M.T.:Contribuiçao ao conhecimento das Cyanophyceae do Parque Estadual das Fontes do Ipiranga, Sao Paulo, Brasil. – Rev. Brasil. Biol. **38**(2): 321–337 (1978).
– Silva S.M.F. & Branco L.H.Z.: Cyanophyceae da Gruta-que-chora, Município de Ubatuba, Estado de Sao Paulo, Brasil. – Hoehnea **18**(2): 75–97 (1991).
– Xavier M.B. & Sormus L.: Estudo qualitativo do fitopláncton da Represa de Serraria, Estado de Sao Paulo, Brasil. – Rev. Brasil. Biol. **48**(1): 83–102 (1988).
Sarma T.A.: Life cycle of *Stichosiphon indica* in cultures. – New. Phytol. **88**: 107–110 (1981).
Sauvageau C.: Sur les algues d'eau douce recoltées en Algérie pendant le session de la societé botanique en 1892. – Bull. Soc. Bot. Fr. **34**: 104–128 (1892).
Schanz F.: Zur Ökologie der Algen in Quellbächen des Schweizerischen Nationalparks (von Buffalora bis Il Fuorn). – Ergebn. wiss. Untersuch. Schweiz. Nationalp. **16**(81): 47–138 (1983).
Schiller J.: Über Cyanophyceen aus kleinen künstlichen Wasserbecken und aus dem Ruster Kanal des Neusiedler Sees. – Österr. Akad. Wissensch., Math.-naturw. Kl., Abt. I., **163** (3):109–139 (1954).
– Die Mikroflora der roten Tümpel auf den Koralleninseln «Los Aves» in Karibischen Meer. – Ergebn. Dtsch. limnol. Venezuela-Exped. 1952, **1**: 197–216 (1956).
Schmidle W.: Schizophyceae, Conjugatae, Chlorophyceae.- In: Engler (ed.), Beiträge zur Flora von Afrika – XXII, Bot. Jahrb. **30**: 240–254 (1902).
Scholtissek B., Jürgens U.J. & Weckesser J.: Purification and electrophoretic pattern of the lipopolysaccharide of *Microcystis* sp. PCC 7806. – Arch. Hydrobiol./ Algolog. Stud. **64**: 361–368 (1991).
Schulz-Baldes M. & Lewin R.A.: Fine structure of *Synechocystis didemni* (Cyanophyta/ Chroococcales) – Phycologia **15**(1): 1–6 (1976).
Schwabe G.H.: Umraumfremde Quellen. -Mitt. Dtsch. Ges. Nat.- Völkerk. Ostasiens (Shanghai), Suppl. **21**:1–300 (1944).
Schwabe G.H.: Grundprobleme der Cyanophytentaxonomie. – Gewässer und Abwässer, H. **36**: 7–39 (1964).
– & Simonsen R.: Cyanophyceen und Diatomeen aus der Krateroase Wau en-Namus (Fezzan, Zentrale Sahara). – Internat. Revue ges. Hydrobiol. **46**: 255–268 (1961).
Seckt H.:Estudios hidrobiológicos en la Argentina. Schizophyceae. – Bol. Acad. Nac. Cienc. Córdoba **25**: 383–429 (1922).
Senna P.A.C.: Estudo das Nostocophyceae (Cyanophyceae) do Distrito Federal: Lagoas Joaquim Medeiros e dos Carás, 1. – Rev. brasil. Biol. **52**(2): 259–274 (1992).

- Cyanophyceae de la Région Est du District Fédéral, Brésil, 1. - Bull. Jard. Bot. Nat. Belg./Bull. Nat. Plant. Belg. **63**: 81–100 (1994).
- & Ferreira L.V.: Nostocophyceae (Cyanophyceae) da Fazenda Água Limpa, Distrito Federal, Brasil. Parte I - Famílias Chroococcaceae e Oscillatoriaceae. - Rev. brasil. Bot. **9**: 91–108 (1986).
- Peres A.C. & Komárek J.: *Coelomoron tropicale*, a new cyanoprokaryotic species from the tropical zone of Sao Paulo State, Brazil. - Nova Hedwigia, in press (1997).

Setchell W.A.: American Samoa: Part I · Vegetation of Tutuila Island. Part II. Ethnobotany of the Samoas. Part III. Vegetation of Rose Atoll. - p. 258–259 (1924).
- & Gardner N.L.: Algae of Northwestern America. - Univ. Calif. Publ., Botany (Berkeley) **1**: 165–418 (1905).
- - New Pacific Coast Algae. - Univ. Calif. Publ., Botany (Berkeley) (1918).
- - The marine algae of the Pacific coast of N. America, Part I, Myxophyceae. - Univ. Calif. Publ., Botany (Berkeley) **8**: 1–138 (1919).
- - New marine algae from the Gulf of California. - Calif. Acad. Sci., Ser. 4, **12**(29): 695–949 (1924).

Siemińska J.: Algae from the Toporowy Staw-Wyzni Lake in the Tatra Mts. - Acta Hydrobiol. (Kraków) **9**(1–2): 169–185 (1967).

Širšov P.P.: Pro dvi cikavi vodorosti z r. Kodymi (dopliv r. Bugu) [Über zwei interessante Algen aus dem Fluss Kodyma (Nebenfluss des Süd-Bug)]. - Visnik Dnipropetr. Gidrobiol. St. **1**: 193–196 (1929).

Škoda B.: Contributions to the biochemical taxonomy of the genus *Chlorella* BEIJERINCK s. l. - pigment composition. II. Biochemotaxonomical differences in pigmentcomposition of the strains growing under nitrogen deficient nutritional conditions. - Arch. Hydrobiol./ Algolog. Stud. **87**:109–136 (1997).

Skuja H.: Vorarbeiten zu einer Algenflora von Lettland. II. - Acta Horti Bot. Univ. Latv. **1**: 23–52 (1926).
- Süsswasseralgen von den westestnischen Inseln Saaremaa und Hiiumaa. - Acta Horti Bot. Univ Latv. **4**:1–76 (1930).
- Algen. - In: Bosshard W. (ed.), Bot. Ergebn. Deutsch. Zentral-Asien Exped., Feddes Repet. **31**:(1933).
- Algae. - In: Handel-Mazzetti H. (ed.), Botanische Ergebnisse der Expedition der Akademie der Wissenschaften in Wien nach Südwest-China 1914/1918., **1**: 1–106, J. Springer, Wien (1937).
- Beitrag zur Algenflora Lettlands II. - Acta Horti Bot. Univ. Latv. **11/12**: 41–169 (1939).
- Taxonomie des Phytoplanktons einiger Seen in Uppland, Schweden. - Symb. Bot. Upsal. **9** (3): 1–399 (1948).
- Zur Süßwaseralgen-Flora Burmas. - Nova Acta Reg. Soc. Sci. Upsal., Ser. **4,14** (5): 1–188 (1949).
- Taxonomische und biologische Studien über das Phytoplankton schwedischer Binnengewässer. - Nova Acta Reg. Soc. Sci. Upsal., Ser 4, **16** (3): 1–404 (1956).
- Grundzüge der Algenflora und Algenvegetation der Fjeldgegenden um Abisko in Schwedisch-Lappland. - Nova Acta Reg. Soc. Sci. Upsal., Ser. 4, **18** (3): 1–465 (1964).
- Zur Kenntnis der Algen Neuseeländischer Torfmoore. - Nova Acta Soc. Sci. Upsal., Ser. 5C, **2**: 1–159 (1976).

Škorbatov L.A.: Myxophycearum in provincia Charkoviensi (Ukrainae) inventarum novae species et varietates. - Not. Syst. Inst. Cryptog. Horti Bot. Petropol. **2**,6: 87–89 (1923).

Šmarda J.: S-layer of chroococcal cell walls. – Arch. Hydrobiol./ Algolog. Stud. 64: 41–51 (1991).
– Šmajs D. & Komrska J.: Advances in cyanobacterial S-layers. – Arch. Hydrobiol./ Algolog. Stud. **83**: 501–510.
– Čáslavská J. & Komárek J.: Cell wall structure of *Synechococcus aquatilis* (Cyanophyceae). – Arch. Hydrobiol./ Algolog. Stud. **23**: 154–165 (1979).
Smith G.M.: Phytoplankton of the Inland Lakes of Wisconsin. I. – Wisc. Geol. Nat. Hist. Surv. **57**, Ser. sci., 12: 1–243 (1920).
– The plankton algae of the Okoboji region. – Trans. Amer. Microsc. Soc. **45**: 156–233 (1925).
– The freshwater algae of the United States. – McGraw-Hill Book Co., 716 pp., New York (1933).
– Fresh water algae of the United States of America, 2nd ed. – 719 pp., McGraw-Hill, New York (1950).
Srinivasan K.S.: Note of *Johannesbaptistia* J. de Toni from Chilka Lake. – Phykos **1**(2): 101–104 (1962).
Stal L.J.(coord.): Development and Fate of Cyanobacterial Blooms in the Baltic. (Poster) – Europ. Comm. Environm. RTD Progr. DG XII.
– & Caumette P.: Microbial Mats. Structure, Development and Environmental Significance. – Springer-Verlag, (1994).
Stam W.T.: A taxonomic study of number of blue-green algal strains (Cyanophyceae) based on morphology, growth and deoxyribonucleic acid homologies. – Groningen, 116 pp. (1978).
– Relationships between a number of filamentous blue-green algal strains (Cyanophyceae) revealed by DNA-DNA hybridization. – Arch. Hydrobiol./ Algolog. Stud. **25**: 351–374 (1980).
– & Holleman H.C.: The influence of different salinities on growth and morphological variability of a number of *Phormidium* strains (Cyanophyceae) in culture. -Acta Bot. néerl. **24**:379–390 (1975).
– – Cultures of *Phormidium, Plectonema, Lyngbya* and *Synechococcus* (Cyanophyceae) under different conditions: their growth and morphological variability. – Acta Bot. néerl. **28**(1): 45–66 (1979).
Stanier R.Y. & Cohen-Bazire G.: Phototrophic prokaryotes: the cyanobacteria.1 – Ann. Rev. Microbiol. **31**:225–274 (1977).
– Kunisawa R., Mandel M. & Cohen-Bazire G.:Purification and properties of unicellular blue-green algae (order Chroococcales). – Bact. Rev. **35**: 171–205 (1971).
– & al.: Introduction to the phototrophic prokaryotes. – In: Starr M.P. & al. (eds.), The prokaryotes, Vol. **1**: 197–211, Springer Verlag, Berlin-Heidelberg-New York (1981).
Starmach K.: Über polnische *Chamaesiphon*-Arten. – Acta Soc. Bot. Poloniae **6**: 30–45 (1929).
– Dwa nowe gatunki sinic z Tatr. [Über zwei neue Cyanophyceen-Arten aus dem Tatra-Gebirge.] – Acta Soc. Bot. Polon. **11** (3): 287–297 (1934).
– Zapiski Algologiczne. I-II. [Algologische Notizen. I-II]. – Acta Soc. Bot. Polon. **13**(1):23–37 (1936).
– Cyanophyta – sinice. – Flora slodkow. Polski **2**: 753 pp., PAN, Państw. Wyd. Nauk., Warszawa (1966).
– *Hildenbrandia rivularis* (Liebm.) J.Ag., *Chamaesiphon fuscoviolaceus* n. sp. i glony towarzyszące w potoku Lubogoszcz w Beskidzie Wyspowym. [ *Hildenbrandia rivularis* (Liebm.) J.Ag., *Chamaesiphon fuscoviolaceus* n. sp., and accompanying algae in the stream Lubogoszcz in the Beskid Wyspowy (Polish Western Carpatians).] – Fragm. flor. geobot. (Kraków) **15**(4): 487–501 (1969).

– Glony osiadle w Wielkim Stawie w Dolinie Pi∞ciu Stawów Polskich w Tatrach. [Benthic Algae of the Great Lake in the Valley of the Five Polish Lakes in the Tatra Mountains.] – Fragm. Florist. Geobot. **19**, 4:481–511 (1973).
– Glony w Wąwozie Szopczanskim w Pieninach. [Algae in the Szopczanski Gorge in the Pieniny Mts.] – Fragm. Florist. Geobot. **21**, 4: 537–549 (1975).
– Algae of the catchment area of Kryniczanka stream. – Stud. Osr. Dokum. Fizjog. PAN (Kraków) **27**: 373–400 (1989).
Ström K.M.: Algological Notes V – IX. – Nyt Magaz. Naturvidensk. **61**: 127–138 (1923a).
– Snow Algae (Cryoplankton) from the Sarek Mountains. – Naturwiss. Untersuch. des Sarekgeb. in Schwedisch-Lappland **3**(5): 522–524 (1923b).
Takamura N., Watanabe M.M., Nozaki H. & Kohata K.: Taxonomy of picophytoplankton in freshwater. – Abstr. XVth Internat. Bot. Congr., Yokohama, no. 2390, p. 325 (1993).
Tandeau de Marsac N.: Occurrence and nature of chromatic adaptation in Cyanobacteria. – J. Bacteriol. **130**(1): 82–91 (1977).
Tarnavschi I.T., Jitariu G., Radulescu D. & Mitroiu N.: [Contribution a l'étude de la flore et de la végétation algologique turficole du bassin de Dorna (région de Suceava). – Bull. Stiint., sect. Biol. Aquic., **8**(2): 273–327 (1956).
Tarnogradskij D.: Microflora and Microfauna of peat bogs in Caucasus. – Works North Caucas. Hydrobiol. Station **6** (3):1–59 (1959).
Tavera R. & Komárek J.: Cyanoprokaryotes in the volcanic lake of Alchichica, Puebla State, Mexico. – Arch. Hydrobiol./Algolog. Stud. **83**: 511–538 (1996).
Taylor W.R.: The marine algae of Florida with special reference to the Dry Tortugas. – Pap. Tortugas Lab., Carnegie Inst. Wash. **25**: 1–219 (1928).
Teiling E.: Aeruginosa oder flos-aquae. Eine kleine *Microcystis*-Studie. – Svensk Bot. Tidskr. **35**(4): 337–349 (1941).
– Schwedische Planktonalgen 3. Neue oder wenig bekannte Formen. – Bot. Notiser **1942**:64–68 (1942a).
– Schwedische Planktonalgen 4. Phytoplankton aus Roslagen. – Bot. Notiser **1942**: 207–217 (1942b).
– Vara viktigare fytoplankter. – Medel. Biol. Förening. 1944: 15–58 (1944).
– Zur Phytoplanktonflora Schwedens. – Bot. Notiser **1**:61–88 (1946).
Tell G.: *Cyanophyta* epifitas de las Chascomús, El Burro, Yalca y Vitel (Prov. de Buenos Aires, Argentina). – Darwiniana **18**: 129-152 (1973).
– & Velez C.G.: Nuevos aportles al conocimiento de algas de agua dulce de Tierra del Fuego (Argentina). – Physis, Sec. B. **41**(100): 47–54 (1982).
Thompson R.: A preliminary survey of the fresh-water algae of Eastern Kansas. – Bull. Univ. Kansas **39**,11: 5–83 (1938).
Thuret G.: Essai de classification des Nostochinées. – Ann. Sci. Nat. Bot., Sér. 6, **1**: 372–382 (1875).
Tiffany L.H.: The plankton algae of the west end of Lake Erie. – Contr. Ohio State Univ. **6**: 112 pp. (1934).
Tikkanen T.: Kasviplanktonopas. – 278 pp., Helsinki (1986).
Tilden J.: Minnesota algae. The Myxophyceae of North America and adjacent regions including Central America, Greenland, Bermuda, The West Indies and Hawaii. – Bot. Ser. **8**: 1–328, Minneapolis (1910).
Tiwari G.L.: A study of the blue-green algae from paddy field soils of India. – Hydrobiologia **39**: 335–350 (1972).
– A study of blue-green algae of paddy field soils of India – II. Taxonomic consideration on non-heterocystous blue-green algae. – Nova Hedwigia **26**: 765–798 (1975).
Tripathy P., Roy A. & Adhikary S.P.: Survey of epilithic blue-green algae

(cyanobacteria) from temples of India and Nepal. – Arch. Hydrobiol./Algolog. Stud. **87**: 43–57 (1997).
Tsangridis A.: On Cyanophyta and Chlorophyta from rice-fields Kalochorion – Thessaloniki. Taxonomic, floristic, phytogeographical, ecological research. – Diss. thesis, 221 pp., Univ. Athens (1982).
Uherkovich G.: Das Leben der Tisza. XII. Weitere synoptische Beobachtungen über die Algenvegetation der Tisza (Theiss) zwischen Tiszabecs und Tiszacsege und ihrer Nebenflüsse. – Acta Univ. Szeged. **7**(1–2): 103–119 (1961).
– Beiträge zur Kenntnis der Algenvegetation der Natron bzw. Soda Szik. Gewässer Ungarns. I. Über die Algen des Feher-Teiches Kunfehertó. – Acta Bot. Hung. **9**: 263–279 (1965).
– Data of the periphyton of the Balaton. – Acta Biol. Szeged. **25**(3–4): 29–34 (1979).
– & Rai H.: Algen aus dem Rio Negro und seinen Neben-flüssen. – Amazoniana **6**(4): 611–638 (1979).
Umezaki I.: Marine Cyanophyceae from Hokkaido. – Acta phytotax. geobot. **16**:84–90 (1956).
– The marine blue-green algae of Japan. – Mem. Coll. Agric. Kyoto Univ. 83, Fish. Ser., **8**: 1–149 (1961).
– & Modelo R.B.: The marine blue-green algae in the Visayas of the Philippines. – J. Jap. Bot. **62**(4): 104–117 (1987).
– & Watanabe M.: Enumeration of the Cyanophyta (blue-green algae) of Japan 1. Chroococcales and Oscillatoriales. – Jpn. J. Phycol. (Sorui) **42**: 175–219 (1994).
Vaara T.: The outermost surface structure in chroococcal cyanobacteria. – Can. J. Microbiol. **28**: 929–941 (1982).
van den Hoek C.: Nomenclatural typification of some unicellular and colonial algae. – Nova Hedwigia **6**: 277–296 (1963).
van Goor A.C.I.: Die Euglenineae des holländischen Brackwassers etc. – Rec. Trav. Bot. Néerl. **22** (1924).
Varma A.K.: Morphology and life-history of *Scopulonema concharum* (Hansg.) Geitl. – Nova Hedwigia **15**: 537–541 (1968).
Vasser S.P., Lenova L.I., Stupina V.V., Carenko P.M. & Navrockaja I.L.: Biopobreždenie promyšlennych materialov bakterijami, vodorosljami i lišajnikami. – Akad. Nauk Ukr. SSR, Institut Botaniky, 1–40 (1988).
Vasishta P.C.: More Cyanophyceae of Hoshiapur: II. – J.Bombay Nat. Hist. Soc. **60**: 671–678 (1963).
– Thermal Cyanophyceae of India I. – Phykos (Delhi) 1968, **7**: 198–241 (1968).
Vela A. & Hernández-Mariné M.C.: Algas edáficas del Cabo Norfeu (Gerona). – Act. VIth Simp Nac. Bot. Cript., p.185–198 (1987).
Vialov O.S. & Zdobikova N.V.: Sweet-water algae of Antarctica. – Acta Soc. Bot. Poloniae **30**(3–4): 765 (1961).
Villeret S.: Contributions a la flore algologique d'eau douce des Iles d'Ouessant (Bretagne). – Revue génér. Bot. **60**:1–12 (1953).
– Quelques especes d'Algues d'eau douce rares ou nouvelles. – Bull. Soc. Bot. Fr. 1953, **100**(1–3): 71–74 (1953).
Virieux M.J.: Quelques Algues et quelques Péridiniens de Franche – Comté. – Bull. Soc. d'Hist. Nat. Doub. **27**: 1–12 (1912).
– Recherches sur le plancton des lacs du Jura central. – Ann. Biol. Lacustre **8**: 5–192 (1916).
Vodeničarov D. G.: *Gloeothece heufleri* Grun. v Bolgarii. [*Gloeothece heufleri* Grun. in Bulgaria.] – Bot. Mater. Otd. Spor. Rast. **14**: 111–114 (1961).
– [Hydrobotanische Untersuchungen über das Urdino-Kar (N-W Rila-Gebirge)]. – Ann. Univ. Sofia **54–55**: 141–211 (1962).

Voronichin N.N.: Materialy dlja flory presnovodnych vodoroslej Kavkaza. I. Schizophyceae. [Materials to the flora of freshwater algae from Caucasus Mts I. Schizophyceae.] – Trudy Petrogr. Obšč. Jestestvcisp. T. **53**(3): 211–263 (1924).
– Materialy k izučeniju algologičeskoj rastitelnosti ozer Kulundinskoj stepi. [Materials to the algal flora of lakes from Kulundinskaja step.] – Izv. Glavn. Bot. Sada **1929**: 12–40 (1929a).
– Opyt sravnitelnogo izučenija mikroflory ozera Don-ty (obl. Kami) i ego otloženij. [Comparative research of the microflora of the lake Don-ty (region of Kama) and its sediments.] – Izv. Sapropel. Komit. **5**: 117–186 (1929b).
Walsby A.E.: Structure and function of gas vacuoles. – Bact. Rev. **36**: 1–32 (1972).
– The gas vesicles of aquatic prokaryotes. – In: Stanier R.Y., Rogers H.J. & Ward J.B. (eds.), Symp. Soc. gen. Microbiol. **28**: 327–357, Univ. Press, Cambridge (1978).
– Cyanobacteria: planktic gas vacuolate forms. – In: Starr M.P. et al. (eds.), The prokaryotes, Vol. **1**: 224–235, Springer Verlag, Berlin – Heidelberg – New York (1981).
– Van Rijn J. & Cohen Y.: The biology of a new gas vacuolate cyanobacterium *Dactylococcopsis salina*, new species in Solar Lake, Sinai, Egypt. – Proc. R. Soc. Lond., B. Biol. Sci., **217** (1209): 417–448 (1983).
Wang C.-Z.: *Cyanoarbor*, a new genus of the Entophysalidaceae (Cyanophyta). – Acta Phytotax. Sinica **27**(2): 129–131 (1989).
Watanabe M., Haga M. & Ioriya T.: Algae in the mountain streams of the Satsunai-gawa river 2.Notes on blue-green algae. – Jap. J. Phycol. **32**: 60–64 (1984).
– & Komárek J.: Blue-green algae from Kathmandu. – In: Watanabe M. & Malla B.S. (eds.), Cryptogams of the Himalayas, I: The Kathmandu valley: 1–13, Dep. Bot. Nat. Sci. Mus., Tsukuba (1988).
– – Several cyanoprokaryotes from Sagarmatha National Park, Nepal Himalayas. – Bull. Nat. Sci. Mus., Tokyo, Ser. B., **20**(1): 1–31 (1994).
Watanabe Y., Watanabe M.F. & Watanabe M.: The distribution and relative abundance of bloom forming *Microcystis* species in several eutrophic waters. – Jap. J. Limnol. **47**: 87–93 (1986).
Watanabe M.M., Watanabe M. & Yamagishi T.: Freshwater algae from Papua New Guinea (3). Blue-green algae from Mt. Wilhelm. – In: Kurokawa S. (ed.), Studies on Cryptogams of Papua New Guinea, p. 67–85, Acad. Sci. Book Inc. Tokyo (1979).
Waterbury J.B.: Order Pleurocapsales Geitler 1925, emend. Waterbury and Stanier 1978. – In: Staley J.T., Bryant M.P., Pfennig N. & Holt J.G. (eds.), Bergey's Manual of Systematic Bacteriology **3**: 1746–1770 (1989).
– & al.: Wide spread occurrence of a unicellular, marine, planktonic cyanobacterium. – Nature **277** (5694): 293–294 (1979).
– & Rippka R.: Subsection I. Order Chroococcales Wettstein 1924, emend. Rippka et al., 1979. – In: Staley J.T., Bryant M.P., Pfennig N. & Holt J.G. (eds.), Bergey's Manual of Systematic Bacteriology **3**: 1728–1746 (1989).
– & Stanier R.Y.: Two unicellular cyanobacteria which repro- duce by budding. – Arch. Microbiol. **115**:249–257 (1977).
– – Patterns of growth and development in pleurocapsalean Cyanobacteria. – Microb. Rev. **42**(1): 2–44 (1978).
Wawrik F.: Planktonorganismen aus Waldviertler Fischteichen II. – Arch. Protistenk. **106**: 369–377 (1963).
– Eisschlub- und Eisbruchvegetationen in den Teichen des nördlichen Waldviertels 1977/1978. [Planktonvegetation in the Ponds of the Northern Waldviertel

during Periods of Closing and Breaking of Ice 1997/1978]. – Arch. Protistenk. **122**: 247–266 (1979).
– Phytoplankton – Erstbesiedlung eines Teiches in Niederösterreich. – Phyton (Austria) **27**(2): 195–204 (1987).
Weber B., Wessels D.C. & Büdel B.: Biology and ecology of cryptoendolithic cyanobacteria of a sandstone outcrop in the Northern Province, South Africa. – Arch. Hydrobiol./Algolog. Stud. **83**: 565–579 (1996).
Weber van Bosse A.: Étude sur les algues parasites des paresseux. – Nat. Verh. Holland Maatsch. Wetensch. Haarlem, pl. 1–2 (1887).
– Algues de l'Expédition danoise aux iles Kei. – Pap. Th. Mortensen's Pacif. Exped. 1914–16, p. 61–64 (1925).
Welsh H.: A contribution to our knowledge of the blue-green algae of South West Africa and Bechuanaland. – Nova Hedwigia **9**(1–4): 131–162 (1965).
Wenzel M.T. & De Halperin D.R.: *Johannesbaptistia pellucida* (Cyanophyta, Entophysalidaceae): genero y especie nuevos para Argentina. – Darwiniana **28** (1–4): 303–306 (1987).
Werner V.R.: Cianofíceas planctônicas da Lagoa de Tramandaí e da Lagoa do Armazém, Rio Grande so Sul, Brasil. – Iheringia. Sér. Bot., **37**: 33–70 (1988).
– & Rosa Z.M.: Cyanophyceae da Estacao ecológica do taim, Rio Grande do Sul, Brasil. – Rev. brasil. Biol. **52**(3): 481–502 (1992).
Wessels D.C.J. & Büdel B.: A rockpool lichen community in Northern Transvaal, South Africa: composition and distribution patterns. – Lichenologist **21**: 259–277 (1989).
West G.S.: Fresh water algae. – Annals South African Mus. **9**(II): 81(1912).
West W. & Anandale N.: Descriptions of three new species associated with Indian freshwater Polyzoa. – J. and Proc. Asiatic Soc., Bengal, **7**: 83–84 (1911).
– & West G.S.: On the periodicity of the phytoplankton of some British Lakes. – Linn. Soc. J. – Bot. **40**: 395–432 (1912).
Whelden R.M.: Some observations on freshwater algae of Florida. – J. Elisha Mitchell Sci. Soc. **57**(2): 261–272 (1941).
– Algae. – In: Polunin N. (ed.), Botany of Canadian eastern Arctic, II. Thallophyta and Bryophyta. – Bull. Nat. Mus. Can. No. **97**: 13–127 (1947).
Whitton B.A.: The biology of Rivulariaceae. – In: Fay P. & Van Baalen C. (eds.), The Cyanobacteria, p. 513–534, Elsevier, Amsterdam-New York-Oxford (1987).
Wille N.: Algologische Notizen I-VI. – Nyt Magaz. Naturvidensk. **38**(1): 1–27 (1900).
– Algologische Notizen XXIX. Studien in Agardh's Herbarium 8–15. – Nyt Magaz. Naturvidensk. **56**: 33–60 (1918).
Wilmotte A. & Golubić S.: Morphological and genetic criteria in the taxonomy of Cyanophyta/ Cyanobacteria. – Arch. Hydrobiol./Algolog. Stud. **64**: 1–24 (1991).
– & Stam W.T.: Genetic relationships among cyanobacterial strains originally designated as «*Anacystis nidulans*» and some other *Synechococcus* strains. – J. gen. Microbiol. **130**:2737–2740 (1984).
– – & Demoulin V.: Taxonomic study of marine oscillatoriacean strains (Cyanophyceae, Cyanobacteria) with narrow trichomes. III. DNA-DNA hybridization studies and taxonomic conclusions. – Arch. Hydrobiol./ Algolog. Stud. **87**: 11–28 (1997).
Wojciechowski I.: Die Plankton-Flora der Seen in der Umgebung von Sosnowica (Ostpolen). – Ann. Univ. M. Curie-Sklodowska Lublin, Polonia, **26**(20): 233–263 (1971).
Wujek D.E. & Gretz M.R.: Ultrastructure of the blue-green alga *Chamaesiphon*. – Cytobios **40**: 87–93 (1984).

Yacubson S.: Catálogo e iconografia de las Cyanophyta de Venezuela. - Bol. Centro Investig. biolog. **5**: 79 pp. (1972).
- El fitoplancton de la laguna de San Javier del Valle (Estado Mérida), Venezuela. - Rev. Algol., N.S. **11**(1-2): 91-131 (1974).
Yamagishi T.: Plankton algae in Taiwan (Formosa). - Uchida Rokakuho Publ. Co., Tokyo 253pp. (1992).
- & Akiyama M. (eds.): Photomicrographs of the Freshwater Algae Vol. **18**: 1-100, Tsukuba (1997).
- Ooshima K. & Watanabe M.: The Chlorophyceae and Cyanophyceae collected from Fuji Goko. - Gener. Educ. Rev. Coll. Agric. Veter. Med., Nihon Univ., **18**: 19-35 (1982).
Yoneda K.: The Cyanophycean genus *Synechococcus* of Japan. - Acta Phytotax. Geobot. **13**(1): 89-105 (1943).
Yopp J.H., Tindall D.R., Miller D. M. & Schmid W.E.: Isolation, purification and evidence of the obligate halophilic nature of the blue-green alga, *Aphanothece halophytica* Frémy (Chroococcales). - Phycologia **17**: 172-178 (1978).
Zalessky M.M.: Sur les nouvelles algues découvertes dans le sapropélogéne du lac Beloe. - Rev. Génér. Bot. **38**:31-42 (1926).
Zehnder A.: Zur Systematik der *Chroococcaceae*. - Schweiz. Ztschr. Hydrol. **22**(1): 365-379 (1960).
Zhuang S., Schubert H. & Schiewer U.: Influence of irradiance and temperature on the Cyanobacterium *Aphanothece stagnina* Sprengel isolated from the Darss-Zingst Estuary (Southern Baltic) under continuous turbidostat culture. - Arch. Hydrobiol./ Algolog. Stud. **70**: 1-13 (1993).

# Index

Normal print - valid names. *Italics* - synonyms and names of taxa which need revision or confirmation, uncertain species. * - taxa known only from outside Europe. **Bold print of numbers** - dsecriptions. *Numbers in italics* - descriptions to illustrations.

## Scientific names of other than chroococcal cyanoprokaryotic organisms

If you have any concerns about our products,
you can contact us on
**ProductSafety@springernature.com**

In case Publisher is established outside the EU,
the EU authorized representative is:
**Springer Nature Customer Service Center GmbH**
**Europaplatz 3, 69115 Heidelberg, Germany**

Printed by Libri Plureos GmbH
in Hamburg, Germany